DISCARDED

AIChE Symposium Series No. 320
Volume 94, 1998

Third International Conference on

FOUNDATIONS of COMPUTER-AIDED PROCESS OPERATIONS

Proceedings of the Conference
held at Snowbird, Utah
July 5-10, 1998

Editors

Joseph F. Pekny
Purdue University

Gary E. Blau
Dow AgroSciences

Production Editor, CACHE Publications

Brice Carnahan
University of Michigan

CACHE

American Institute of Chemical Engineers

1998

© 1998
American Institute of Chemical Engineers (AIChE)
and
Computer Aids for Chemical Engineering Education (CACHE)

Neither AIChE nor CACHE shall be responsible for statements or opinions advanced in their papers or printed in their publications.

Library of Congress Cataloging-in-Publication Data

Foundations of computer-aided process operations / Joseph F. Pekny, Gary E. Blau, and Brice Carnahan, editors
       p. cm. -- (AIChE symposium series : no. 320)
  Includes bibliographical references and indexes.
  ISBN 0-8169-0776-5
       1. Chemical process operations – Data processing – Congresses.
  I. Pekny, Joseph F. II. Blau, Gary E. III. Carnahan, Brice.
  IV. Series.
  TP155.75.F68   1998   98-52902
  660'.2815--dc21   CIP

    All rights reserved whether the whole or part of the material is concerned, specifically those of translation, reprinting, re-use of illustrations, broadcasting, electronic networks, reproduction by photocopying machine or similar means, and storage of data in banks.
    Authorization to photocopy items for internal use, or the internal or personal use of specific clients, is granted by AIChE for libraries and other users registered with the Copyright Clearance Center Inc., provided that $3.50 per copy is paid directly to CCC, 222 Rosewood Dr., Danvers, MA 01923. This consent does not extend to copying for general distribution, for advertising, for promotional purposes, for inclusion in a publication, or for resale.
    Articles published before 1978 are subject to the same copyright conditions. AIChE Symposium Series fee code: 0065-8812/1998. $15.00 per article.

PREFACE

FOCAPO III

This volume contains the proceedings of the Third International Conference on the Foundations of Computer Aided Process Operations, FOCAPO III, held in Snowbird, Utah (USA) from July 5th to 10th 1998. The first conference on this theme (FOCAPO I) was held in Park City, Utah in 1987 and the second (FOCAPO II) was held in Crested Butte, Colorado in 1993. Two other closely related series of conferences are regularly held on Chemical Process Control (CPC) and on the Foundations of Computer Aided Process Design (FOCAPD). All three conference series are sponsored jointly by the Computing and Systems Technology (CAST) Division of the American Institute of Chemical Engineers and the non-profit CACHE Corporation. In addition, FOCAPO III was also sponsored by APICS: The Educational Society for Resource Management.

At the time of FOCAPO III, computer-integrated process operations are becoming a centerpiece of corporate business. The installation of Enterprise Resource Planning (ERP) systems at many companies promises to make technologies that were just beginning to be discussed at FOCAPO I and in their infancy at FOCAPO II an indispensable part of process management. This continuing ERP investment has given rise to multibillion-dollar service companies to make process integration a reality, a generalization of the notion of process operations relative to that considered at FOCAPO I and II, and the realization that corporate survival is tied to the effective use of information. Despite this mainstreaming of the topics covered by FOCAPO III, the conference clearly underscored the need for further research and development to address fundamental challenges to exploiting further computer integration.

As with past such conferences, the objectives of FOCAPO III included bringing together a group of academic and industrial participants to survey current problems facing process operations, the state of relevant technology, and needed advances to combat ever-increasing complexity. To cover these objectives, sessions were conducted to address computers in the 21st century, plant wide optimization, pilot plants/market-development plants, emerging and high growth processes, planning and scheduling, environmental issues, product integrity/quality, and next-generation enabling technology and deployment. The keynote session provided an international perspective on computers in process operations with a speaker each from America, Asia, and Europe.

FOCAPO III brought together 157 attendees from 21 countries, of whom 67 were from academic institutions and 90 were from operating companies, vendors, and service companies. Attendees represented all populated continents, with 107 from North America, 33 from Europe, 13 from Asia, 2 from South America, 1 from Australia, and 1 from Africa. Several of the attendees were actively engaged in operations management and indicated that the conference was of interest because of the increased corporate visibility of process operations. Nearly every attendee was a speaker, session chair, participant in the contributed paper session, or commentator.

Vigorous programmed and informal discussions took place concerning challenges and needs for process oriented industries such as food and beverage, chemical, pharmaceutical, metal and mining, and energy. These discussions dealt with issues such as process optimization, information system needs, software tool development, supply chain management, modeling, exploiting data rich environments, and the barriers to deploying advanced technology. Whereas past FOCAPO conferences emphasized plant operations, in addition, FOCAPO III discussions emphasized simultaneous consideration of financial, market, and manufacturing factors in developing successful business practices. The intense global competition present in industry accentuated the importance of bringing new ideas to practice as rapidly as possible.

Interestingly, many attendees kept notes during sessions using laptop computers that were more powerful than the mainframe computers of the previous decade and that weighed less than the combined proceedings of FOCAPO I and II. Registration for and organization of the conference was accomplished largely using the World Wide Web, a medium that did not widely exist during

FOCAPO II. An obvious but compelling conclusion of the conference was that the technological advances responsible for such observations were dramatically changing the process industries and the influence of these continuing advances was accelerating.

The editors wish to acknowledge those who participated in FOCAPO III:

- The conference participants, especially the session chairs, speakers, authors, and commentators

- Financial contributions from:

    The National Science Foundation
    Dow Chemical
    Eastman Chemical

- Jeff Siirola (Eastman Chemical) in his role as CACHE conferences chairman and for all his advice

- David Himmelblau (University of Texas) and the staff of the CACHE office

- Robin Craven (Alliance, LLC) for handling conference planning and logistics

- George Applequist, Brenda Wisnewski, and Janet Jones (all at Purdue University) for their help with the conference web site

- Brice Carnahan (University of Michigan) who developed the proceedings format guidelines and was responsible for its publication

- Edvins Daiga and Scott Couzens (University of Michigan students) who handled much of the editing and electronic assembly of the proceedings

- Ignacio Grossmann (Carnegie Mellon University) and Venkat Venkatasubramanian (Purdue University) for their work on the Contributed Session which featured over 65 papers

- J. D. Seader (University of Utah) who helped arrange for preliminary copies of papers for conference participants

- David Bowman (APICS) and Edward Schuster (Welchs) for arranging APICS cosponsorship

- The Technical Planning Committee:

Kun Soo Chang	Pohang University	Korea
James Downs	Eastman Chemical	USA
Ignacio Grossmann	Carnegie Mellon University	USA
Iori Hashimoto	Kyoto University	Japan
Simon Jones	BASF	Germany
Hank Kohlbrand	Dow Chemical	USA
Lowell Koppel	Aspen Technology	USA
Mark Kramer	Gensym Corp	USA
Philip Law	Swiss Federal Institute of Technology (ETH)	Switzerland
Kristian Lien	Norwegian University	Norway
Wolfgang Marquardt	Aachen Technical University	Germany
Bernard McGarvey	Eli Lilly & Co	USA
Gregory McRae	Massachusetts Institute of Technology	USA
Manfred Morari	Swiss Federal Institute of Technology (ETH)	Switzerland
John Perkins	Imperial College	England
Malcolm Preston	ICI PLC	England
Luis Puigjaner	Universidad Politecnica de Catalunya	Spain
Moe K. Sood	Mobil	USA
George Stephanopoulos	Massachusetts Institute of Technology	USA

Whereas FOCAPO I served as an experiment and FOCAPO II sought to define a discipline of process operations, the broad participation in FOCAPO III along with contemporary corporate activity indicates that continuing rapid advances in computational infrastructure have brought process operations technologies to the forefront of corporate management. Future FOCAPO conferences will be conducted in an even more integrated world, where business without computers will be impossible and the efficiency of process operations will closely control economic success. Future speakers will have much to talk about and we look forward to listening.

Additional information about the conference may be found at the World Wide Web location:

http://che.www.ecn.purdue.edu/FOCAPO98.

Chairmen

Joseph F. Pekny
Purdue University

Gary E. Blau
Dow AgroSciences

TABLE OF CONTENTS

INVITED PAPERS

KEYNOTE SESSION

To Innovate Chemical Plant Operation by Applying Advanced
Technology and Management.. 1
    Yukikazu Natori and I. Bhieng Tjoa

Towards Intentional Dynamics in Supply Chain Conscious Process Operations 5
    Ton Backx, Okko Bosgra and Wolfgang Marquardt

PLANT-WIDE OPTIMIZATION

Session Summary: Plant-wide Optimization... 6
    Lorenz T. Biegler

Plant-wide Optimization: Keeping Chemical Engineering in Mind... 8
    Herman De Meyer

Plant-wide Optimization: Opportunities and Challenges... 15
    John D. Perkins

PRODUCT INTEGRITY/QUALITY

The Challenges of Product Consistency and Integrity in Slow and Fast Processes................... 27
    Hugo Patiño

Multivariate Statistical Treatment of Historical Data for
Productivity and Quality Improvements.. 31
    John F. MacGregor and Theodora Kourti

Refinery Quality Control by Overall Integration 42
    (Frank) X. X. Zhu and Gavin Towler

EMERGING AND HIGH-GROWTH PROCESSES

Improved Technology Management through Physico-chemical Process Modeling 44
    Erik W. Egan

PLANNING AND SCHEDULING

Session Summary: Planning and Scheduling 60
    Stephen P. Lombardo

Synthesizing Enterprise-wide Optimization with Global Information
Technologies: Harmony or Discord? 62
    Conor M. McDonald

Single- and Multisite Planning and Scheduling: Current Status and Future Challenges 75
    Nilay Shah

Towards the Convergence of Theory and Practice: A Technology Guide for
Scheduling/Planning Methodology 91
    J. F. Pekny and G.V. Reklaitis

ENVIRONMENTAL ISSUES

From Waste Treatment to Pollution Prevention and Beyond:
Opportunities for the Next 20 Years...112
    Henry T. Kohlbrand

NEXT GENERATION ENABLING TECHNOLOGY

Computation and Competitiveness: Managing Technology in the Information Age 126
    Michael P. Ramage

Culture Shift: Positioning Technical Computing to Enable Sustained
Profitability in the Specialties Business 127
    Gary E. Blau and Kay E. Kuenker

CONTRIBUTED PAPERS

    Session Summary: Contributed Papers 136
        Ignacio E. Grossmann and Venkat Venkatasubramanian

PLANNING

    Issues in Next-Generation Supply Chain Management..142
        Bala Ramachandran

    A New Approach to Batch Process Planning...146
        Yang Gul Lee and Michael F. Malone

    Auction-driven Coordination for Plantwide Optimization 152
        Rinaldo A. Jose and Lyle H. Ungar

    Reliability and Maintenance Considerations in Process Operations under Uncertainty...........158
        Efstratios N. Pistikopoulos and Constantinos G. Vassiliadis

    Planning Under Uncertainty: A Parametric Optimization Approach 164
        Efstratios N. Pistikopoulos and Vivek Dua

    Pilot Plant Operations in Pharmaceutical Research and Development
    in the Next Century..171
        Jonathan M. Vinson, Owen D. Keck, Prabir K. Basu,
        R. Brian Houston, Linas Mockus and Alan R. Noren

    An Industrial Perspective of Supply-Chain Optimization and Simulation............................178
        Oscar Rosen

    Total Site Optimization of a Petrochemical Complex...185
        Metin Turkay, Tatsuyuki Asakura, Kaoru Fujita,
        Chi Wai Hui, Yukikazu Natori, Yoshihisa Masaiwa,
        Haruyoshi Oonishi and I. Bhieng Tjoa

    Operational Planning for Chemical Processes using Geographic
    Information and Environmental Modeling 190
        Derya B. Özyurt and Matthew J. Reallf

    Bringing New Specialty Chemicals to Market - A Nightmare
    in Planning and Scheduling...197
        Karl Schnelle and Gary Blau

Improving Batch Manufacturing Process Operations using
Mathematical Programming Based Models 204
    Paul R. Bunch, Doug L. Watson and Joseph F. Pekny

Disjunctive Multiperiod Optimization Models for Process Networks 210
    Susara A. van den Heever and Ignacio E. Grossmann

A New Conceptual Approach for Enterprise Resource Management Systems 217
    M. Badell and L. Puigjaner

SCHEDULING

Scheduling of a Multi-Product Polymer Batch Plant 224
    Christian Schulz, Sebastian Engell and Rüdiger Rudolf

Maximum Dispersion Algorithm for Multi-Site Delivery Scheduling 231
    S. D. Mokashi and A. C. Kokossis

An Optimal Control Approach for Scheduling Mixed Batch/Continuous
Process Plants with Variable Cycle Time 237
    H. P. Nott and P. L. Lee

Integrated Scheduling in Steel Plants 243
    Vipul Jain and Ignacio E. Grossmann

Large Scale Multi-facility Planning using Mathematical Programming Methods 249
    Paul R. Bunch, Rex L. Rowe and Michael G. Zentner

Machine Learning Approaches for Enhancing the Optimization of
Batch Production Scheduling and Process Operation Problems 254
    Matthew J. Realff and Kenneth J. Kirschner

Using Dynamic Modeling to Improve Scheduling Capability 261
    Bernard McGarvey and Bruce Bickle

Minimizing Production Cycle Time at a Formulations and Packaging Plant 267
    Matt Bassett

Integration of Production Scheduling and Activity Based Costing Models 273
    Vangelis Lionis, Guillermo E. Rotstein, Ebrahim Mohamed and Robert A. Weiss

A Task-resource Based Framework for Process Operations Modeling 279
    Gabriela S. Mannarino, Horacio P. Leone and Gabriela P. Henning

Batch Production Scheduling with Flexible Recipes: The Single Product Case 286
    Moisès Graells, Espen Løberg, Antonio Delgado, Enrique Font and Luis Puigjaner

Solving Trim-Loss Problems with Variable Raw Paper and Trim-Loss Widths 291
    Iiro Harjunkoski and Tapio Westerlund

CONTROL/DYNAMICS

CLIFFTENT: Determining Full Financial Benefit from Improved Dynamic Performance 297
    Pierre R. Latour

Model Predictive Control and Identification: A New Approach to
Closed-loop Identification and Adaptive Control 303
    Alexander Schwarm, S. Alper Eker and Michael Nikolaou

Operational Implications of Optimality 308
    John Bagterp Jørgensen and Sten Bay Jørgensen

Dynamic Simulation and Control Strategy Evaluation for MTBE Reactive Distillation 315
    Douglas A. Bartlett and Oliver M. Wahnschafft

Real-Time Optimization of a FCC Recovery Section 322
    John Brydges, Andrew Hrymak and Thomas Marlin

IDENTIFICATION/DIAGNOSIS

One Step Collective Gross Error Identification and Compensation in
Linear Dynamic and Steady-State Data Reconciliation 328
    Miguel Bagajewicz and Qiyou Jiang

Improved Statistical Process Control Using Wavelets 332
    Sermin Top and Bhavik R. Bakshi

Process Operation Improvement Based on Multivariate Statistical Analysis 338
    Dae-Hee Hwang, Chonghun Han and Tae-Jin Ahn

Optimal Alarm Logic Design for Process Networks 345
    Chuei-Tin Chang and Chii-Shang Tsai

Adaptive Random Search and Short-cut Techniques for
Process Model Identification and Monitoring 351
    Gheorghe Maria

Process Monitoring using the Clustering Method and
Functional-Link-Associative Neural Network..360
    Kyung Joo Mo, Dongil Shin, Sooyoung Eo, En Sup Yoon and Kun Soo Chang

Applying a Procedural and Reactive Approach to
Abnormal Situations in Refinery Control..366
    David J. Musliner and Kurt D. Krebsbach

Performance Assessment and Diagnosis of Refinery Control Loops......................................373
    N.F. Thornhill, M. Oettinger and P. Fedenczuk

Fault Isolation in Industrial Processes using Fisher's Discriminant Analysis 380
    Evan L. Russell and Richard D. Braatz

Enhancement of Gross Error Detection when Data are Serially Correlated 386
    Ruth Kongsjahju and Derrick Rollins

Faster Development of Fermentation Processes - Early Stage Process Diagnosis 391
    João A. Lopes and José C. Menezes

DESIGN

Operational Design and its Benefits in Real-Time Use 397
    Yuji Naka, Rafael Batres and Tetsuo Fuchino

Use of Dynamic Event-Driven Simulation for the Chemical Industry 404
    Heinz Ensen, Joerg Krames, Dirk Wollenweber and Christine Edinger

Using Homotopy Continuation and Gröbner Bases for Finding
All Solutions of Steady State Process Design Problems..410
    D. A. Harney and N. L. Book

Design and Optimization of Chromatographic Processes in Production Scale 417
    Jochen Strube, Henner Schmidt-Traub, Michael Schulte and Reinhard Ditz

An Advanced Process Analysis System for Pollution Prevention...421
    Xueyu Chen, Kedar Telang, Ralph W. Pike, Jack R. Hopper,
    Jamal Saleh, Carl L. Yaws and Thomas A. Hertwig

The Cost of Crossing Reaction Equilibrium in a System that is Overall Adiabatic 428
  Willie Nicol, Diane Hildebrandt and David Glasser

Flexibility Analysis of Natural Gas Plants 433
  Ana M. Eliceche, Laura Fernandez and Maria P. Sanchez

On Combining Operating Features in the Design of a Batch Process.................................439
  Girish Joglekar, Steven M. Clark, Vikram G. Kalthod and Guy Maineult

Nonlinear Parameter Estimation Using Interval Analysis..445
  Chao-Yang Gau and Mark A. Stadtherr

A Systematic Procedure for Optimal Operation of Heat Exchanger Networks 451
  Bjørn Glemmestad and Truls Gundersen

On the Efficiency of Material Flow Simulation for Process and
Multipurpose Plant Optimization...458
  Philippe Solot, Linus Willmann and Tibor Dudás

Process Optimization and Process Robustness using Experimental Design: STAVEX..........464
  Y. L. Grize and W. Seewald

A New Concept to Improve Operation and Performance of an
Industrial Size Tube Reactor by Means of CFD-Modeling..469
  H. J. Warnecke, M. Schäfer, J. Prüß and M. Weidenbach

INFORMATION SYSTEMS

Information Models for Batch and Real-time Chemical Process Data..................................474
  Neil L. Book and Arvind Sharma

Application Programming Interfaces for EXPRESS Information Models 480
  Milind Madhav Khandekar and Neil L. Book

A Process Plant Knowledge Repository for Multiple Applications 487
  Michael J. Elsass, David C. Miller, James F. Davis and John R. Josephson

An Initiative for Integrated Computer-aided Process Engineering..494
  Andreas A. Linninger, Helmut Krendl and Helmut Pinger

A Process Engineering Information Management System using
World Wide Web Technology 501
    R. Andrews and J. W. Ponton

From Engineering Analyses to Operator Training 507
    Peter Stanley

OLE Automation: Bringing the Power of Process Simulation to
Plant Operations Personnel 513
    James McGill

An Open Software Architecture for Process Modeling and Model-Based Applications 518
    Anthony Kakhu, Benjamin Keeping, Yuzhao Lu and Constantinos Pantelides

AUTHOR INDEX 525

SUBJECT INDEX 528

TO INNOVATE CHEMICAL PLANT OPERATION BY APPLYING ADVANCED TECHNOLOGY AND MANAGEMENT

Yukikazu Natori and I. Bhieng Tjoa
Development and Engineering Research Center
Mitsubishi Chemical Corporation

Abstract

Over the past ten years, we have seen fast pace of computer supported applications for improving operation of chemical plants. The applications range from data collections, DCS, advanced process control, real-time optimization, scheduling and planning to safety and operator training simulation. At Mitsubishi Chemical Corporation (MCC), we have implemented these technologies for improving our operations. Through these implementations, we have gained insight on the capability, limitation and new applications for these technologies. Here we will share some of our experiences on industrial applications, and discuss the needs for further research and collaborations between academics and industry in order to bring the next generation of relevant technology.

Keywords

Mathematical programming, Modeling, Production scheduling and planning, Real time optimization.

Introduction

Applying advanced technology for improving plant operations is very challenging for many practical reasons due to the operational nature of the activities. Various computer aided technologies have been developed and applied to activities ranging from plant operation to production scheduling, planning and distribution. The industrial trend in applying these technologies for process operations is moving toward a more rigorous problem representation for better accuracy and therefore delivering a more reliable solution.

At MCC, we began applying advanced computing technologies for chemical plant operation in the mid 80's. At our Mizushima plant site, the first step toward modernization was taken by introducing host computers for running TICS (Total Information Control System), DCS and process computers. TICS is used for managing data related to production, accounting and maintenance activities. DCS and process computers were installed as part of infrastructure modernization which allows us to collect plant data and make improvement on implementations of basic control system such as PIDs and cascade controllers. In the late 80's, steady state simulation systems were widely used by our engineers along with other engineering tools such as Computer Aided Engineering (CAE) to perform rigorous analysis on process study and improvements. In the earlier 90's, we applied other advanced technologies such as pinch analysis, dynamic simulation, advanced process control, integrated database and a new maintenance management system for plant operations and process improvements.

Recognizing the benefits of applying the above advanced technologies and the need to support the activities, Production Technology and Engineering Center (PTEC) was established in 1992 with the primary objectives to support and improve the existing plant operations. Since then, we expanded our activities to include technologies for intelligent laboratory system, rigorous physical property, micro kinetics, process synthesis, process optimization, concurrent development and engineering work. Some areas of these applications clearly are beyond the boundary of process operations and maintenance toward process research and development activities.

In current business environment, we no longer can afford to separate plant operation work from research and process development because the keys to business competitiveness in a global business environment are determined by the speed of implementing new technology and the knowledge of the technology itself. In recognizing the need for faster business cycle of our activities, our R&D Center and PTEC were merged into Development and Engineering Research Center (DERC) in 1996. We also established DERC branches outside Japan in order to accelerate our technology development further by enabling us to collaborate more effectively with academics as well as our industrial collaborators around the world. The first two offices are located in the West and East Coast of the United States.

There are a lot of practical challenges in technical as well as non technical areas which need to be solved in order to achieve sustainable benefits from applying advanced technology. Here, we will share our experiences on applications of computer aided technologies for process operations in our center and identify important areas where further improvement can be made.

Activities at DERC

In the early stage of computer aided technology, MCC traditionally developed applications in-house. In the late 80's, we started doing more benchmarking on technologies that were available commercially such as advanced process control and process simulation packages. This exercise is very useful because it gives us a better indication on the current state of a commercially available technology and the direction on how it will evolve. This information is one of many factors which are used by the management for making decision to switch from in-house development to a commercially package. When in-house development of a technology has reached a stage where it is no longer economical or holds a strategic competitive advantage, we may replace it with the commercial one. This move allows us to refocus our limited resources on emerging technology which may give us a more competitive advantage.

In the early 90's we focused our effort on improving our technology development in each engineering activity such as process control, simulation and optimization. We also realized that many technologies had reached maturity to be applied for industrial applications, and the commercial software was available. Therefore we started to introduce many of these technologies for our applications. To illustrate our activities, we will discuss some of the applications that we implemented at our Mizushima plant site.

We implemented a real-time optimization application as an operator support system at our power plant in the early 90's. The purpose of this application was to help the operator to run the power plant operation by providing advise on moving operating condition that resulting in saving fuel consumption. In order to ease the level of complexity for the operator, the model which was built using SPEEDUP (from AspenTech) was linked to a user friendly interface which was built using G2 from Gensym Corporation. This was one of the early real-time application using SPEEDUP (Fujita et.al., 1994), and it was a unique application by combining two software, each with its own strength, to create a practical application. Further improvement to power plant operation was continued by focusing on improving steam control and monitoring systems. During this period, we also implemented [DMC]™ model predicted process control and [DMO]™ on-line closed-loop real-time optimization (CLRTO) systems for our ethylene plant (Emoto et.al., 1994). By collaborating closely with our industrial partner, we implemented for the first time the largest composite model predicted control for feed maximizer to the ethylene furnaces. Similarly, the CLRTO was the largest application reported at that time. In our plant site, the ethylene plant is considered as part of power plant because it produces large amount of high pressure steam. Therefore we recently also implemented on-line CLRTO to our power plant by replacing the SPEEDUP application with [DMO] ™ (Emoto et.al., 1998). With this application, we can coordinate the CLRTO of both plants in order to gain more benefits.

Through implementation of these technologies we have seen major improvements to our plant operation. Since the degrees of freedom in the plant are not so much, we may have reached near the maximum potential benefits without expanding the scope of application to higher level of integration. In the area of business operation, further benefits can be realized by exploiting the interactions of plant operation to production scheduling, planning and distribution. In the area of process development, further benefits can be achieved by improving our understanding at the chemistry level through application of micro-kinetics study of catalyst activity and through process synthesis (O'Young and Natori, 1996).

Unlike the implementation of applications that was discussed earlier where commercial technology is quite advance and the availability of process model library is quite common, the commercial software in the area of business operation has not reached the same level of sophistication. Rigorous modeling of production scheduling using mathematical programming techniques often requires special expertise. The formulation of logical operational constraints is often non-trivial if one wants to solve the problem efficiently due to combinatorial nature of the problem. Since there are large potential benefits for applying these technologies, we currently focus our effort on supply chain applications.

Our effort in developing technology for business optimization has given us tangible benefits. Continuing with our previous examples, we have implemented an electric purchasing scheduling system based on a multi-period mixed integer linear programming utility system model for the power plant operation (Hui and Natori, 1996). By coordinating the CLRTO system with the scheduling system, the power plant can be operated at a better operating condition by taking advantage the interaction between the yearly electric buying contract with the current operation target. We also implemented a feedstock scheduling system for the ethylene plant, the system was developed in-house based on a mixed integer nonlinear formulation which covers from vessel scheduling to product inventory management (Tjoa *et.al.*, 1997). With the system, we can prepare daily optimal feedstock composition for the ethylene plant operation by considering the current and future availability of feedstock and production targets. Therefore we can provide the best condition for the current CLRTO operation by exploiting the interaction with the monthly production operation target.

Recently we implemented a total site planning system which covers the entire processes at our Mizushima plant site (Turkay *et.al.*, 1998). The system was developed in-house based on a mixed integer linear programming formulation. Since the model is quite detail representation of the processes, it can be used for many different applications such as monthly utility and production planning, yearly electric purchasing planning, man power allocation for plant maintenance scheduling, etc.

Through these implementations, we gained not only improvements for our plant operation but also technology and expertise that can be applied to our research and development activities. This is possible because we recognize that modeling and optimization technology as one of our core technologies, and we are building a strong group in this area. Here we will share our practical experiences that contribute to our success.

Practical Issues in Process Operation Applications

Understanding the fundamental nature of a process operation is one of the most important factors for a successful application of advanced technology. The decision to introduce a new technology for process operations in industrial practices is rarely done solely based on the technical merit. There are practical issues related to the nature of process operations practices that need to be considered. There is management as well as technical issues. In the technical area, we need to consider functionality, integration with legacy applications, easiness of use, reliability, training, technical supports, etc.

A computer aided application rarely can satisfy the end user's needs during its lifetime due to either the limitations on the original design or the new requirements. During its lifetime, we often make evaluations whether it's necessary to update the application by either upgrading the existing system or introducing a completely new system. A new application often has new requirements: hardware, software as well as personnel training for using the new system effectively. Hardware and software requirements are usually quite straight forward, but the integration with the existing environment is not. In addition, a personnel training for acquiring the new skill is also not a trivial task. Therefore, a careful evaluation is needed to ensure a successful implementation of a new application.

We also need to consider issues such as easiness of use, reliability and technical support to insure sustainable deployment of the application. Easiness of use is important to make the end user feels comfortable and motivated to use the new system. However, reliability is the most important factor whether a new application will be accepted as part of operational tools in industrial practices. In a process operation, the end user is rarely an expert in technology that is used to develop the application. However, the user is very knowledgeable about the problem itself, and his or her objective is often related to his or her ability to perform the job well. Therefore the user would not like to deal with uncertainty and interruptions due to the system reliability. A failure to satisfy this criterion often causes discontinuing the use of the application, and a second opportunity for introducing a new application to the same user will be much more difficult. A technical support after the commissioning is also crucial, it helps the end user in making the best use of the application further.

Although understanding the technical aspect for operational applications is a necessary condition for a successful implementation, it is not a sufficient condition. Other factors such as company commitment, organization structure, management style and the right combination of human resources also play critical roles.

Conclusion

Innovating chemical plant operation by applying an appropriate advanced technology is crucial for an operating company to remain competitive. Since the speed of innovation cycle is critical for maintaining the competitive advantage, it is important for us to consider issues that effects our ability to recognize, understand and implement an advanced technology for a suitable application. Therefore company commitment and leadership play very important role in differentiating the actual success derived from the implementation of an advanced technology.

References

Emoto, G., Y. Ota, H. Matsuo, M. Ogawa, D. B. Raven, R. F. Preston and J. S. Ayala (1994), Integrated Advanced Control and Closed-Loop Real-Time Optimization of an Olefins Plant. *In Proc. Of IFAC Symposium ADCHEM 94,* Kyoto, Japan.

Emoto, G., A. Tsuda, T. Takeshita, M. T. Monical, S. Nakagawa and K. Fujita (1998), Integrated Large-Scale Multivariable Control and Real-Time Optimization of a Power Plant. To be presented at *IEEE Conf. On Control Applications,* Trieste, Italy.

Fujita, K., Y. Natori and T. Takeshita (1994), Steam Balance Optimization in Chemical Plant. *In Proc. Of IFAC Symposium ADCHEM 94,* Kyoto, Japan.

Hui, C. W. and Y. Natori (1996), An industrial application using mixed-integer programming technique: A multi-period utility system model. *Computers Chem. Engng.* **20,** Suppl., S1577-S1582.

O'Young, D.L., and Y. Natori (1996), Process Synthesis: technology, environment and applications. *Computers Chem. Engng.* **20**, Suppl., S381-S387.

Tjoa, I.B., Y. Ota, H. Matsuo and Y. Natori (1997), Ethylene plant scheduling system based on a MINLP formulation. *Computers Chem. Engng.* **21**, Suppl., S1073-S1077.

Turkay, M., T. Asakura, K. Fujita, D. Hui, Y. Natori, Y. Masaiwa, H. Oonishi and I. B. Tjoa (1998), Total Site Optimization of a Petrochemical Complex. To be presented at *FOCAPO 98*, Snowbird, Utah.

TOWARDS INTENTIONAL DYNAMICS IN SUPPLY CHAIN CONSCIOUS PROCESS OPERATIONS

Ton Backx
Eindhoven University of Technology, Netherlands
and AspenTech Europe, Netherlands

Okko Bosgra
Delft University of Technology, Netherlands

Wolfgang Marquardt
RWTH Aachen University of Technology, Germany

Abstract

Chemical and refinery process operations have to deal with an increasingly transient and competitive marketplace. The traditional strategy of operating a plant in isolation from its environment within and outside the supply chain with exogenous influences interpreted as disturbances is not any more appropriate. Rather, manufacturing must quickly adapt to the transient environment to exploit economical potentials to the degree possible. At least, the traditional plant focussed operational strategy must account for the dynamics in the disturbances by employing predictions of their future time-varying behavior in real-time optimization to result in an intentionally dynamic operational strategy. Further, plant performance can be significantly improved, if knowledge on the status and on the future policies and goals of some partners in the supply chain can be employed for plant optimization. Such a cooperative mode of interaction in the supply chain will lead from plant focussed to supply chain conscious, often intentionally dynamic plant operation. Advanced concepts are identified and discussed, a number of relevant research issues are suggested.

Editor's Note: The full text of this paper is available at the World Wide Web location http://che.www.ecn.purdue.edu/FOCAPO98

Keywords

Process operations, Supply chain dynamics, Dynamics in operations, Planning, Scheduling, Real-time optimization, Decomposition, Decentralization, Decision structure, Control structure, Information structure.

SESSION SUMMARY: PLANT-WIDE OPTIMIZATION

Lorenz T. Biegler
Chemical Engineering Department
Carnegie Mellon University
Pittsburgh, PA 15213 USA

Summary

Over the past decade, real time optimization (RTO) has become an essential feature in the petrochemical industry. Examples abound in the successful implementation of on-line optimization for refineries, ethylene plants, chemical plants and power plants. Features of real time optimization include the formulation of large scale process models, application of large nonlinear equation and optimization solvers and, last but not least, an appropriate interaction with the process control system. An important consequence of these features is the closed loop nature of on-line optimization that links the data acquisition, the plant model, the optimizer that determines new setpoints and the control system that enforces them.

As a result the ingredients of a successful RTO system can be classified as:

- Nonlinear programming (NLP) solvers and large-scale algorithm development
- Process model formulation and reduction, which requires a strong chemical engineering component
- Closed loop interactions of the optimization and control system, which provides the systems component
- Implementation, interpretation and validation on the actual process

In most applications, the nonlinear programming algorithm of choice is Successive Quadratic Programming (SQP). In fact, the evolution of RTO is preceded directly by the development of SQP algorithms in the applied math and the process engineering communities. The first appearance of SQP can be traced back to Wilson and Beale in the early sixties, but numerical difficulties hampered widespread application. By 1977, the application of quasi-Newton methods and analysis of exact penalty functions led to the first efficient SQP algorithms by Han and Powell. From this starting point, the next decade saw algorithmic developments by Conn, Fletcher, Gill, Murray, Nocedal and many others, which led to advanced features including convergence properties for a variety of *merit functions*, applications of *trust region* and *line search* globalizations for constrained optimization, and efficient *factorization* and *decomposition* for large-scale problems. An excellent survey of these developments is provided by Sargent (1997). Several other papers in this volume point to further details of these features. Applications of SQP in process engineering begin in 1980 and include contributions from Westerberg and co-workers, Sargent, Biegler, Lucia, Macchietto, Stadtherr, Morton, Kalitventzeff and others. Many of these are summarized in Biegler et al. (1997).

Spurred by the development of efficient NLP algorithms, the evolution of RTO can be seen through numerous commercial applications. In a very interesting survey by Ayala (1997), the following applications milestones show the growth of RTO through large-scale SQP:

1980s	In house developments at DSM, ICI, Shell ($\leq$ 20,000 eqns.)
1986	Shell "Opera" package ethylene plants
1988	First DMO application Sunoco Hydrocracker
1991	Lyondell integrated refinery
1994	Mobil and Mitsubishi Chemical applications (over 200,000 eqns.)
1996	Aspen/DMC/Setpoint mergers

As a result of this development, the DMO and RTOPT products from Aspen Technologies, Inc. command more than 80% of the RTO market. These applications can be classified further as follows:

Area	Operating	Under Development
Chemical	7	3
Ethylene	13	5
Refining	11	13
Total	31	21

The paper in this volume by Prof. John Perkins reviews previous work in RTO and points to future directions by focusing on interactions of design, control and the on-line optimizer. Also discussed are the roles of process dynamics and uncertainty, and the integration of both aspects through back-off calculations and dynamic optimization. As noted in his paper, the survey of Marlin and Hrymak (1997) also serves as a suitable companion to this work.

Other presentations at the session, that are not included in the proceedings but worthy of note, include the one by Dr. Herman deMeyer who emphasized the role of modeling and the importance of capturing the essence of the problem. In his presentation "Plantwide Optimization: Keeping Chemical Engineering in Mind," he noted the importance of complexity in plantwide modeling, the availability of process information and the use of top-down approaches. Illustrated with several process examples, the presentation stressed the importance of setting the models and problem scope properly *before* embarking on the optimization exercise.

Finally, Prof. Thomas Marlin provided an excellent discussion on the claims and validation of RTO systems. These involve much more than solving large nonlinear problems and getting an improved objective function from the computer program; they require the demonstration of that solution on the actual process. To raise the level of awareness among process engineers (and copying the trends of popular management theories) he introduced the concept of Total Profitability Management (TPM) which demands that process managers know the value of what they are producing and what it is costing them, on a continuous basis.

References

Ayala, J., "Realtime Optimization," Proc. Aspenworld, Boston, MA (1997)

Biegler, L. T., I. E. Grossmann and A. W. Westerberg, Systematic Methods of Chemical Process Design, Prentice-Hall, Upper Saddle River, NJ (1997)

Marlin, T. E., and A. N. Hrymak (1997), "Real-time operations optimization of continuous processes," In J.C. Kantor, C. E. Garcia and B. Carnahan, *Chemical Process Control V*, AIChE Symposium Series No. 316, Vol. 93, pp. 156-164.

Sargent, R. W. H., "The Development of Sequential Quadratic Programming," in Biegler, L. T., T. Coleman, A. R. Conn and F. Santosa (eds.), Large Scale Optimization with Applications, IMA Volumes in Mathematics and Applications, *Springer Verlag, New York (1997) Volume 93: Part II: Optimal Design and Control*

PLANT-WIDE OPTIMIZATION: KEEPING CHEMICAL ENGINEERING IN MIND

Herman De Meyer
Bayer
Antwerpen N.V.

Keywords

Dynamic simulation, Plant-wide optimization, Model development.

Introduction

For Bayer, a company which operates on the global marketplace for more than a century and is established in more than 150 countries with 350 companies, innovation and technology leadership has been a key to its competitive advantage.

Market pressures have recently forced the company to refocus on core-business. For the realisation of its ambitious economic goals and the continued commitment to its environmental responsibility the company will continue to rely heavily on the corporate technology group with its rich history of innovation. As a result, the model-based analysis of existing entire plants including the rethinking of their operation as well as the computer aided design of new chemical plants world-wide received renewed impulses.

The technological developments towards tighter integrated chemical plants, with highly interdependent subsystems and recycles require development of model based automation as well as operating and maintenance procedures. Individual subsystems of the plant can only be viewed as nodes in a functional web and the plant must be looked at in its entirety. In order to obtain an abstract, yet quantitative, view of the interactions involved in such plants, physical modelling, simulation and optimization are the techniques of choice that are being used. Process models, sufficiently detailed to reliably represent actual plant behaviour, allow analysis, experimentation and the exploration of the operability limits of the plant, sometimes even before it is build.

Although simulation and optimization have been familiar tools in chemical industry for a long time, it is only recently that cheap computer power has allowed the mathematical techniques to mature and software to be developed to match the scale and complexity needed for full chemical plant simulation. Therefore, the heuristic and ad hoc approach to redefinition of plant operating procedures and optimization, based on plant operation experience, is being complemented and even replaced by a more strategic methodology based on simulation.

At the same time, the recognition that chemical reactors are as a rule at the heart of the operations and can only be controlled well if their mechanisms are understood, is at the basis of most projects. The optimal plant can only be build around reactors whose behaviour is understood.

Simulation and optimization of a complete chemical processing plant is, however, a far larger task than just combining unit operations because all interactions have to be taken into account.

The Process of Plant-wide Modeling and Optimization

In this paper I wish to focus on the process of modelling and optimization of a complete chemical plant and discuss the issues and difficulties involved from an industrial point of view. Rather than presenting the (polished) results of successful projects of this type, of which we have realised quite a few, it may be more instructive to spend some time analysing some exemplary aspects of the work in order to define expectations and engineering challenges for the future.

In assigning priorities to projects of plant analysis and optimization, Business Units tend to place emphasis on processes that are at the core of their operations but for which the underlying systematics on the level of the physics, the logistics and the chemical reactions involved may be less well understood. The logic that is the basis for taking major technological decisions for such processes may sometimes be entirely heuristic in nature. In situations were this is not satisfactory a project for in-depth analysis of the process then needs to be set up.

Budgeting for such plantwide projects is based on the expectation of the overall efficiency of the modelling, validation, analysis, scale-up and optimization steps and entrails a timeframe for the completion of the task, as a rule with considerable uncertainty margins. Judging the complexity and the difficulties involved is to a large extend based on successful unit operation modelling projects. It is perhaps indicative of the importance of plantwide modelling and simulation that most of the uncertainty and most of the unexpected difficulties and opportunities lie hidden in the interactions of the subsystems, the part one misses out on in unit operation simulation.

As the project progresses and time becomes more precious, the difficulties of scale become more and more apparent. Rescheduling specialised lab-space (as a rule these labs have limited timeslots available) to gather extra data on kinetics or physical properties of species that - after all- are of importance and remodelling batch reactors to interpret the lab results are just a few examples of decisions that are to be taken because large systems are less transparent and blur our feeling for details more easily than do small simulations.

A system we would call "large" to be simulated would typically include half a dozen reactors and contain about as much reactive unit operations (absorbers, extractors,...). A few dozen recycle streams to the reactive operations and 40 - 100 significant components, at least in the reactive mixtures, would be typical. With 150 unit operations, 20 of which are critical in terms of chemistry, the product quality or dynamics, the modelling, simulation and optimization may even require special techniques and software to be able to cope with the numbers.

Handling the incompleteness of the available information at every stage is apparently the main cause for the adoption of an iterative methodology. It is our experience that as the plant model grows to completion, gaining additional information in discrete steps forces very involved upgrading (e.g. when extra component families need to be introduced) and require much extra effort and may even become a source of frustration.

For this is also the point in time when the scope and magnitude of potential benefits start to be apparent and the quantification has to be given a secure basis.

This is crucial to many a project because - although successful as a whole - it is often turned into a battle to keep it affordable. The difficulties involved in this phase involve many different aspects such as mathematical tricks to assure renewed convergence or unique minima but also the review of earlier assumptions and setting up the necessary additional information for renewed validation on a very detailed level.

The mounting pressure to produce results often makes it a challenge to stay focused on the chemical engineering aspects!

For many of these tasks, which one might expect to be not only typical for what Bayer is doing, commercial packages offer neither procedural nor systematic fundamental support.

Complexity in Plant-wide Modeling and Optimization

As in any project of this type, the first steps in the series of activities to be set up, are classics such as: gather physical property data on the chemicals involved, get kinetics on the reactions and construction details of the fixed items of equipment.

If one were to present the results of control or optimization projects, the above statement would suffice to state that a good job was done to collect all those essentials.

Since we want to analyse the process of getting to this point, the least one could say is that it would be considered luck and a luxury if the above statement captures all that is involved.

Some of the complexities that I want to illustrate, being incomplete by definition, and that are hard to cope with when numbers become large have to do with:

- Timely availability of additional information, leading to remodelling, shifting of focus and renewed validation.
- Numerics that in our opinion do not support a top-down approach very well. Starting with rather crude, often lumped, models and gradually refining as insight is gained, might seem a logical way of doing things but are hard to realise.
- Incompleteness of information, forcing "quantised leaps" and retracing steps.

Cumbersome and, in general, hard to avoid in any project, the work involved when large plant-wide simulations are undertaken all too often demands thinking in terms of mathematics rather than in terms of engineering and hampers the speed of response to legitimate questions about the process.

Let me be concrete and consider examples of what we are concerned with as a rule, without going into the details of the chemistry.

Timely Availability of Information

Consider part of a larger process as illustrated by its flowsheet in Fig. 1.

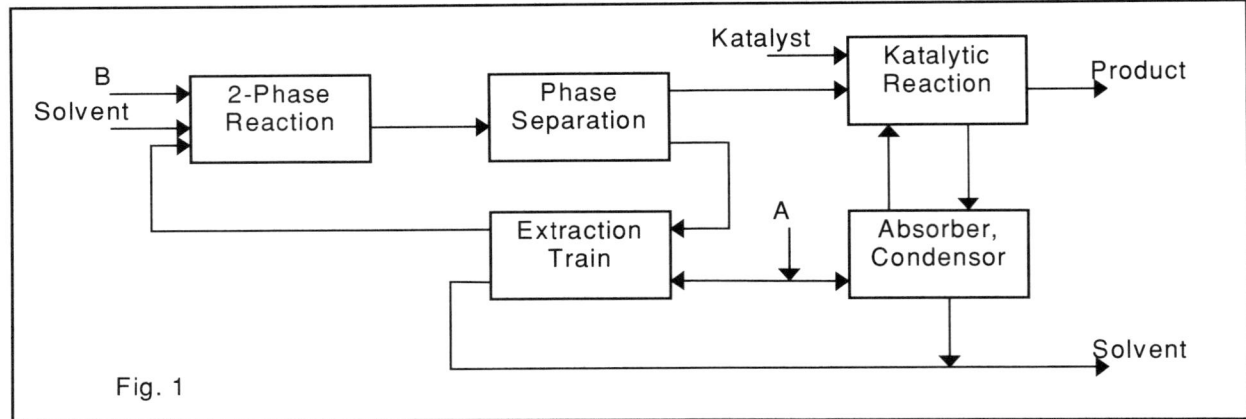

Figure 1. Optimization of a two-phase chemical reactor.

The chemical reaction of the 2-phase reactor was considered known at the start of the optimization project. Lab measurements had shown clear second and first order dependence on the reactants, both of which posed no problem to track analytically during batch experiments. Consequently the following reaction was to be used with confidence in the simulation:

$$2A + B \rightarrow F + E \quad (1)$$
$$F' = k\, C_A^2\, C_B \quad (2)$$

where F is the desired compound and component E is also used as a solvent.

Since the component B was never detected in the reactor-exit, even when A/B ratio's sank below the stoichiometric limit, a reaction with undefined but fast kinetics was postulated so as not to influence the rate-determining step:

$$A + B \rightarrow C \quad (3)$$

Only after the kinetics of the second, catalytic reactor, unknown at the start of the project, was elucidated (it involves some 60 components and 45 reactions) one could take a fresh look at the modelled process as a whole and, in this case, compare it to the real process for complete validation.

As it turned out, neither the balance of component F (which is the main reacting partner in the catalytic reactor), the solvent balance nor the heat balance of the catalytic reactor could be brought to match the validation data from the real plant.

As a consequence, all degrees of freedom left by the kinetic experiments were exhausted using the batch reactor model, that had allowed to interpret the lab-results - a toy containing some 50000 equations, to find ways to account for the differences observed.

As it became clear this would be not be successful, the chemistry of the "simple" 2-phase reactor was questioned. Extending the reaction invariant sufficiently to mathematically generate the minimum number of new degrees of freedom to allow the older lab-results to stand on the one hand and to compensate for the observed differences on the other hand, required the existence of at least 3 extra components (or families of them).

In view of the type of time restrictions mentioned above, using fictitious components the simulations could go ahead for the rest of the plant.

Going back to the lab, generating new analytics and the creation of a dynamic model for the batch reactor to interpret the results of the experiments and test the validity of proposed mechanisms, led to the following chemistry for the 2-phase reactor, after 6 months work and countless optimization runs to determine the kinetic parameters:

$$\begin{aligned}
&A + B \rightarrow C \\
&C \leftrightarrow D + E \quad \text{at the interface} \\
&A + D \rightarrow F \\
&C + D \rightarrow G + E \quad \text{known impurity} \\
&C + D \leftrightarrow H \\
&H + D \leftrightarrow I \quad \text{extra component family 1} \quad (4)\\
&I + D \leftrightarrow J \\
&A + H \leftrightarrow K + E \quad \text{extra component 2} \\
&I \rightarrow L + E \quad \text{extra component 3} \\
&J \rightarrow M\!\downarrow + E \quad \text{solid formation}
\end{aligned}$$

It will be no surprise that the richness of these results forced us to build a completely new model of the reactor. Not only three new components, this time by name, were indeed found, but especially the role of the interface (mixing) and the possibility of solid formation, which was previously attributed to impurities in the feed, became apparent.

This last information proved very valuable in redesigning a reactor for a new plant where, due to a switch-over to a new supplier of a the product B, conditions proved just right for solid formation which never happened in any production site before!

The appearance of an important impurity, "early" in the reactor chain, lead to a complete revision of operational limits for the reactors further downstream.

The optimization of the plant without this information lead to completely different results, which were interpreted, afterwards, as a strategy to maximise the destruction of impurities rather than minimising its formation.

The point is that no tools, except exhaustive simulation, could be brought in to shorten the whole cycle. This makes the modelling very expensive, irrespective of the result.

One way to start tackling this problem is to be able to account in a systematic way for the uncertainties that are inherent to the information used for modelling. Inverting the arguments would have quickly and systematically showed in this case that the sought parameter-combination in the original model was not within the bounds set by the lab-results. Although we do have code to look at such questions now (macro, MIT), handling large problems is still not "state of the art".

Moreover, the innovation cycle of commercial simulation and optimization packages is quite high, so add-ons quickly create large amounts of maintenance work just to keep up. In order to be easy to use, a high degree of integration has to be maintained.

Top-down Modeling Approach and Numerics

At this point it is interesting to explore the possibility of a more top-down approach to modelling, in which elementary things such as closed mass-, component- and energy balances are guaranteed first. It is a sufficiently detailed approach, even for optimization, when looking at heat-integration and many planning activities.

This allows to set up rather crude models at first and to postpone the introduction of more complicated (physical, chemical) issues until they are apparently needed. One should, however, not forget that e.g. "forcing" the balances just shifts the imponderability from one representation to another and it may even be harder to extract hints from it to combine with engineering intuition as to what phenomena could open interesting new ways to explore. It remains a fact that the initial effort to set up the models and optimization is a far simpler one. Although it is common practise to work in this way, most commercial packages, in our opinion, do not support this approach very well. Especially with plant-wide simulations, the point at which one should start looking at the whole plant, and abandon (postpone) the introduction of further details on the level of the unit operations is a matter of experience and feeling.

To illustrate a number of these issues, I will introduce the flowsheet of a unit in Fig. 2 which is part of a design for a very highly integrated plant that was scaled-up and optimised from lab-experiments.

The following reactions are important in this section:

$$\begin{aligned} A &\rightarrow B \\ B &\rightarrow C + D \\ D + E &\leftrightarrow F + G \\ C + G &\leftrightarrow H \\ H &\rightarrow I + J\downarrow \end{aligned} \quad (5)$$

The desired chemical is C. The components G and E are volatile. Component E is in excess and a solvent is present as well. Component H may deteriorate, as later was discovered, and build the solid J.

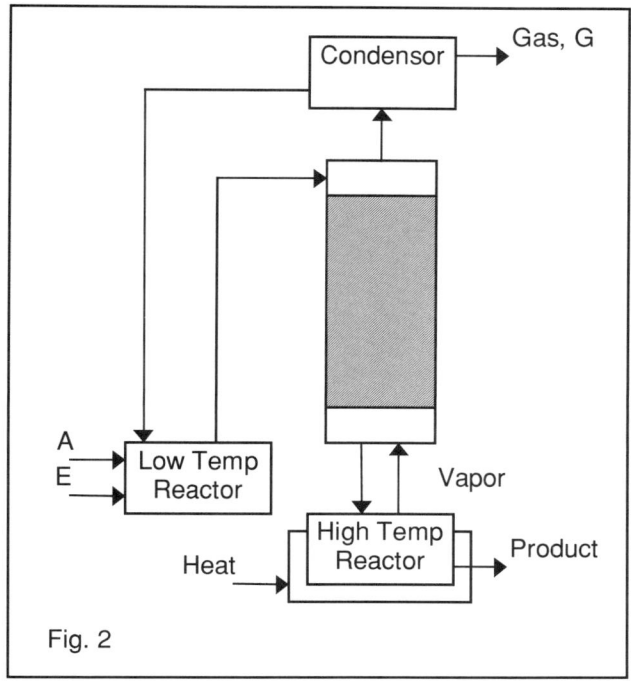

Figure 2. Part of a design for a highly integrated plant.

The design of the plant is "modern" in the sense that multi-role chemicals are used and completely recycled. These compounds perform different functions at different stages of the process: they act as carriers for active chemical groups to combine to the target products, they capture acidic species when they are no longer wanted and are solvents for the main products of the reaction. Simulation/optimization is "a must" to refine the design.

The low temperature reactor essentially produces the product B; the product A has to be stabilised by an excess of compound E, which is recycled as a liquid from the condensation. The further reaction of A to the desired compound C is, however, to be performed in the absence of E for this component attacks one of the secondary products of the reaction, which in turn may destroy the wanted component and reduce the yield of the reaction considerably.

The flowsheet as illustrated in Fig. 2 is but one of the promising options retained in a design study for a plant to produce an intermediate for one of our pharmaceuticals.

The logic is to use the energy required to bring the reaction mixture to high temperature to evaporate the undesired component E and recycle it to the low temperature reactor.

The carrier of the heat would be the solvent, condensing back into the second reactor. The mixture, with its ionic species in a non-aqueous solvent, is highly non-ideal but the compound E is very volatile. Stripping E should be easy and quick and there is, at first sight, no need to model the packed stripping column as a reacting vessel with almost no residence time in the liquid phase.

This top-down approach allows simple models to be used, and optimiser runs in which the column is participating in the flowsheet require no extra attention versus the ones in which it is not.

Using this design, the yield was maximised and the cost of the product minimised (compound E is expensive). The reactor was to be heated with steam (which fitted nicely in the heat integration scheme) and standard apparatus could be used for it. This was desirable in terms of maintenance of the plant, where in all 8 reactors of the same design were needed.

As already indicated above, it was only at a late stage in the project that detailed lab-measurements indicated that it was not the restricted stirrer capability of the lab-reactors that lead to crust formation, but the chemistry of the mixture as shown in the last reaction of the reaction set (5).

The effect on the design decisions was immediately apparent: a restriction on the wall temperature of the high temperature reactor must be imposed.

The cascade of consequences that follow from this information are unimportant in detail, but illustrate my point:

- It was no longer possible to use low-pressure steam to heat the reactor.
- An alternative heat-carrier had to be selected.
- Heat integration schemes had to be totally reviewed.
- The same hardware could no longer be used for this reactor, internal heating and stirrer arrangement needed redesign.
- Control-loops would be slower
- Modelling the reactor was to be redone (heat exchanger, control, …).
- Heat recuperation for the whole plant was to be remodelled, the integration had to account for this particular reactor in a completely different way.

The most aggravating consequence, however, is that in this example restriction of heat input to the stripping column - and rate of change of heat input in case of disturbances - cause this column to exhibit hysteresis. This results in the existence of two steady states: one in which a low concentration of E exist in the reactor (the desired one) but a state in which the concentration of E is very high also exists.

In the latter case, the reaction in the stripper cannot be neglected and actually "eats away" enough of component C to allow the reactor to recover slowly, although the disturbance lasts a long time.

It is clear that achieving convergence of the plant flowsheet, the addition of enough detail by intensive remodelling of the column (including a carefully controlled preheater), the reactor and the heat recuperation equipment as well as the introduction of additional components and the renewed validation entrails a cost.

Although it is a textbook example in this case, it took 30% of the projects planned manpower, to get to the point were two steady states were recognised and a clean initialisation of the whole flowsheet (in Aspentech's Speedup) was again established with the new models. Of this time, only a tiny fraction was needed to do the creative work, that we consider chemical engineering to be, and come up with alternative hardware in this case.

It is frustrating that the interesting questions, such as: "what if we did change the solvent?" are as yet out of reach because of these reasons (but not only because of them).

The point is that more support for a top-down approach is needed and, for instance through using more formal methods to analyse the flowsheet changes, detect ambiguities at the point in time where they are created, rather than by analysing the numerical difficulties of simulators or optimizers.

Incompleteness of Information

The ability to cope with incompleteness of information, especially during the early phases of a project when even physical properties of some components and mixtures are uncertain, is a typical human one. Software of the actual generation is lost as completely as the data are incomplete. As estimation of the most critical data such as kinetics of reacting species and physical properties of the mixtures on a quantum-mechanical basis is still in its infancy and support for the engineer to help him in this area will, for some time to come, only be found in a more structural, organisational approach of the whole simulation.

To illustrate the dramatic effects and the "quantised" nature the incompleteness of information can have I will consider the flowsheet of a reactor-train as in Fig. 3.

The chemical system of this polymeriser is rather complicated, and in order to optimize the ratio of the M_i components in the product, the feed strategy for A and F was calculated as a function of the B-rate. The stopper, O, concentration strategy was used to maximise the yield of certain products N_{ik} depending on the desired product type.

As it turned out, this strategy was the one in use, and raised some questions as to what all the optimization effort was about.

Just to get a feeling for the relative intractability of the reacting system, the following lines show which reactions take place. There is a marked difference in rate between certain reaction groups. The choice of temperature in the reactors, therefore, is a degree of

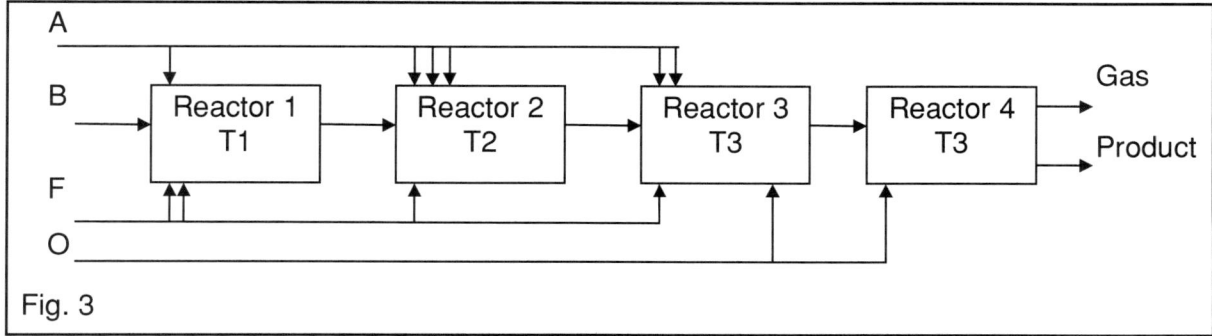

Figure 3. Reactor train.

freedom used to make different reactions dominant as one goes along the reactorcascade.

Our concern is the role of the concentration of A along the chain of reactions.

$$
\begin{aligned}
A + B &\rightarrow C \\
A + C &\rightarrow D + E \quad \text{(E volatile)} \\
D + F &\rightarrow G \\
A + G &\rightarrow H_1 \\
&\rightarrow H_2 \\
A + H_1 &\rightarrow I_{10} \\
&\rightarrow I_{20} \\
A + H_2 &\rightarrow I_{30} \\
&\rightarrow I_{40} \\
H_i + I_k &\rightarrow I_{ik} \\
I_{jk} &\rightarrow K_i + L_{ik} \\
K_i + A &\rightarrow M_i \\
K_i + M_k &\rightarrow M_{ik} \\
O + F &\rightarrow P
\end{aligned}
\qquad (6)
$$

For the determination of the reaction rates, the parameter estimation was turned into an restricted optimization problem. The existence of isomerization reactions of the type

$$M_i \leftrightarrow M_k \qquad (7)$$

was taken into account, but without a fixed kinetic formula. This forced a search for the best-suited expression along with the determination of its parameters.

As is often the case, the optimization suffered from multiple local minima. As more experimental data were incrementally added, the system converged on one of the minima as the global minimum and it turned out that isomerization was to be neglected.

It turned out later that the system, which was a "classical" commercial SQP, when it was not forced to take isomerizaton into account drowned the isomerization effects.

When not only using the variances of the measurements, but by enforcing the expected uncertainties in the kinetic parameters directly by bounding quite a different set of parameters was found and isomerization was contributing.

The effects on the operation of the reactor sequence is dramatic:

the temperature in the first reactor was increased by 20°C (from its year-long untouchable value), the temperature in the second reactor was increased by an astounding 50°C while the last reactors temperature was dropped by 20°C.

The feed strategy was changed also and its rate-dependency retained its importance.

The new operating strategy, including new controllers, remind us of the difficulty of judging what to neglect rather than of what to include in a model, and once more of the need to find new ways to avoid to have to do the same work a number of times.

Including uncertainty on a systematic basis as a way of using more information than we actually have, or to avoid missing out on information we are not very confident of might be one of the avenues to explore further.

Conclusion

By referring to a few case studies, I hope to have given some insight in the processes, the restrictions and the ad-hoc nature of the way we overcome such difficulties.

With the emphasis increasingly placed on shortening review and development cycles, the focus of chemical engineering is shifting from plant analysis to supporting plant design.

A lot of research effort will have to be concentrated on solution methodologies that support such projects better when numbers become large.

All too often, the chemical engineering aspects that should come in at the different levels of detail involved: from generating novel alternatives for entire subsystems and unit operations (e.g. combining reaction and extraction operations) down to the level of the physico-chemical basis (such as changing solvents) are drowned. Bringing in people external to the project to "scrutinise" the partial results is the only way to avoid this conflict.

Success and affordability will have to mean to let us keep chemical engineering in mind!

PLANT-WIDE OPTIMIZATION: OPPORTUNITIES AND CHALLENGES

John D. Perkins
Centre for Process Systems Engineering
Imperial College
London, SW7 2BY, UK

Abstract

This paper discusses the use of detailed on-line mathematical models of process physics and chemistry, in conjunction with techniques for data analysis and for mathematical optimization, to improve the economic performance of plant in the process industries. First, the current approach to real-time optimization (RTO) of large scale continuous processes based on steady state models is briefly reviewed. (A more comprehensive recent review is available in Marlin and Hrymak (1997)). Next, the issues involved in the design of RTO systems in this class are discussed, particularly with respect to the structural decisions involved, for example the choice of measurements to be used to monitor plant performance and update the optimization model, and the level of model complexity to be used in the RTO system. The potential opportunities opened up by an increased capability to handle dynamic data, models and optimization are considered next. The paper concludes with a discussion of future needs in this area.

Keywords

Real-time optimization, Economics, Steady-state models, Dynamic systems, System design.

Introduction

There is a long tradition in the process industries of using fundamental knowledge captured in the form of mathematical models to aid plant operations (Perkins and Barton (1987)). As confidence in our ability to develop adequate representations of plant behaviour has grown, combined with the ever increasing availability of low cost high powered computing, there has been a trend to make process models available on-line, and to use them to inform real-time decisions about plant operations.

In this paper, one particular class of on-line modelling applications is discussed, where a process model is used as the basis for a closed-loop optimization of plant performance. The arrangement is illustrated in Figure 1. Information on the status of the plant is gathered, and used to assess the consistency of current conditions with the assumptions built into the mathematical representation. Where appropriate, adjustments are made to the model to reflect the latest situation. Armed with an up-to-date mathematical representation, and information on current operational requirements, an optimization problem is solved to determine the best operating strategy for the plant. If appropriate, this strategy is implemented. The plant and operational requirements are monitored to detect changes which might necessitate a redetermination of the optimal operating strategy, and the cycle is repeated at appropriate intervals.

In the past ten years, there has been a significant growth in the application of on-line optimization systems in the process industries. Large-scale applications are becoming commonplace, and several vendors offer products and services in the area of model-based real-time optimization. The technology underlying these developments has matured to a point where its application, while still requiring significant engineering expertise, is reasonably routine. The components of the current approach will be reviewed next. Another recent and more comprehensive review of the issues involved in the current approach to RTO may be found in Marlin and Hrymak (1997).

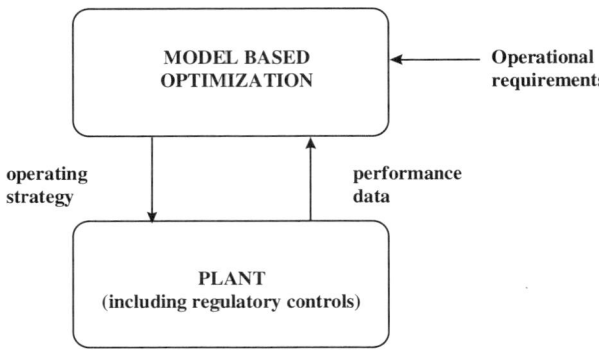

Figure 1. Real-time optimization.

Current Approaches to Real-time Optimization

Current industrial applications of model-based real-time optimization (RTO) address complex plants. Marlin and Hrymak (1997) list the following features of plant which favour the application of RTO:

- adjustable optimization variables exist after higher priority safety, quality and production rate objectives have been achieved;
- profit changes significantly as values of the optimization variables are changed;
- disturbances occur frequently enough for real-time adjustments to be required;
- determining the proper values for the optimization variables is too complex to be achieved by selecting from several standard operating procedures.

There is a fair amount of published evidence supporting the conclusion that a successful RTO application delivers about 3% of the value added by the plant in economic benefits (Cutler and Perry (1983); Lauks et al (1992)). It should be borne in mind that published applications to date cover a very narrow spectrum of the full range of manufacturing plants employed in the process industries, viz. large scale continuous plants in the petroleum and petrochemical sectors. The benefits figure quoted above is at best only typical of plants for that spectrum.

It appears that the cost of engineering an RTO system is most easily justified with current technology when the cash flows associated with large scale operation are available. Since these operations are typically continuous, the focus of RTO applications has usually been on the optimization of **steady-state** conditions. This has the benefit of greatly simplifying the modelling task associated with RTO, but raises other issues associated with model validation.

The components of the current generation of RTO systems based on steady-state models are enumerated below, where a typical optimization cycle is defined (based on van Wijk and Pope (1992)).

1. Check that plant is steady.
2. Assemble plant measurement data for input to the optimization system.
3. Check input data and ignore or adjust bad values.
4. Match the optimization model to current plant operating conditions. Model parameters (e.g. heat transfer coefficients, pressure drops) are calculated based on current measurement values.
5. Check the current status of all control loops to determine the degrees of freedom available for optimization.
6. Run the optimization case to determine a new set of operating targets.
7. Review new operating targets prior to implementation.
8. After review, all validated targets are automatically implemented.

The cycle is repeated after a suitable interval, chosen to be sufficiently long to give a reasonable opportunity for the plant to settle to the new steady state determined by the optimizer.

In the procedure defined above, steps 1 to 3 are concerned with ensuring that a valid set of plant operating data is assembled to enable matching of the optimization model to current conditions. Thus, since the model is typically based on the assumption of steady-state, the plant status is assessed to check the validity of this assumption. The techniques employed to do this are straightforward, being based on the analysis of trends in key measurements. However, as with many aspects of RTO implementation, there is an art in the design of these tests. If the requirements are made too stringent, then many opportunities to apply the optimizer will be missed. On the other hand, the use of data sets representing operation inconsistent with the assumption of steady-state built into the model has obvious dangers.

The identification and removal of measurement values containing gross errors (step 3) is an important prelude to fitting the model to the current plant condition. A combination of checks based on knowledge of the status of plant elements (e.g. equipment or instruments off-line or known to be malfunctioning), and more formal statistical tests (Crowe (1994)) based on assumed distributions of measurements and/or redundancy in the data set, is employed to exclude unreliable information.

Parameter calculation at step 4 may be implemented using statistically-based parameter estimation techniques. It is also feasible to combine gross error detection with simultaneous parameter estimation (Tjoa and Biegler (1991)). Alternatively, local calculations for process units where sufficient measurements are available may be employed. For example, a heat transfer coefficient for a heat exchanger may be determined from local measurements of flow rates and inlet and outlet temperatures. Of course, a local approach such as this will generally be suboptimal in its use of available information, especially in highly integrated processes with strong interactions between subsystems.

Having established a mathematical representation of the current plant status which is consistent with available information, an optimization problem is set up and solved (steps 5 and 6). The problem may be represented as:

$$\max_{\underline{x}} \; J(\underline{x}, \underline{p}) \quad (1)$$

subject to

$$\underline{f}(\underline{x}, \underline{p}) = 0 \quad (2)$$

$$\underline{g}(\underline{x}, \underline{p}) \geq 0 \quad (3)$$

In this formulation, $\underline{p}$ represents the model parameters estimated from plant data, and $\underline{x}$ includes all remaining model variables. Equation (2) is the collection of equality constraints representing the mathematical model of the process. Typically, the dimensions of $\underline{x}$ and of $\underline{f}$ are large (of order 10^3 to 10^5 in current applications). The difference between these two dimensions corresponds to the number of optimization degrees of freedom. This number is typically in the range 10 to 100. The inequality constraints (equation (3)) play a significant role in the problem formulation, since it is quite common for the solution to the optimization problem to be at least partly constrained. Since, therefore, the position of the optimum is determined by a subset of these inequalities which are active at the solution, it is important that the model contain as accurate a representation of these constraints as possible.

Advances in technology to handle large scale nonlinear programming problems have enabled the direct solution of the problem above, based on an "open" or equation-based representation of the plant model (equation (2)). While this is probably now the most commonly used approach in industrial applications, implementations based on sequential-modular modelling technology, or on hybrid approaches have also been successful (Lauks et al (1992)).

To cope with the complexity and sheer size of the models employed in RTO applications, ideas from process modelling have been used. For example, the models are typically assembled from standard modules representing unit operations. Use of the natural structure of the flowsheet to impose a structure on the model greatly facilitates model building, and more importantly model maintenance, which is a key issue in industrial applications.

Having solved the optimization problem, the results are reviewed to decide whether they should be implemented on the plant. Again, a variety of approaches is available ranging from allowing the plant operators to review the recommendations from the RTO systems (van Wijk and Pope (1992)) to formal statistical tests (Miletic and Marlin (1996)). The basis for the latter approach is the observation that the RTO system could determine that a change in operation is necessary purely because of the effect of noise in plant measurements, rather than because of a change in process conditions. Miletic and Marlin have devised a significance test to distinguish between the common cause variability due to sensor noise and high frequency disturbances, and the variability due to lower frequency non-stationary disturbances. The application of this test to a simulated case study showed a significant reduction in induced variability due to the RTO system, and a corresponding increase in the profit achieved.

Structural Decisions in RTO Systems

In implementing a RTO system using the approach discussed above, a number of structural decisions are necessary. These decisions are mainly associated with the linkages between components of the overall RTO system, see Figure 2. In this section, work aimed at providing help with these decisions to implementers of RTO systems is reviewed.

The key decisions implied by Figure 2 are as follows:

- the structure of the regulatory control system in terms of set points, measured and manipulated variables;
- the variables selected to transmit the results of the RTO from the optimizer to the plant ("Set Points");
- the model parameters to be estimated from plant data in real time;
- the plant measurements used to estimate model parameters;
- the level of model complexity to be used in the RTO system.

In making choices in each of these areas from the range of feasible possibilities, the aim should be to achieve the best possible performance of the closed-loop system, by mitigating the effects of the inevitable errors introduced through measurement noise and plant/model mismatch on optimized plant performance.

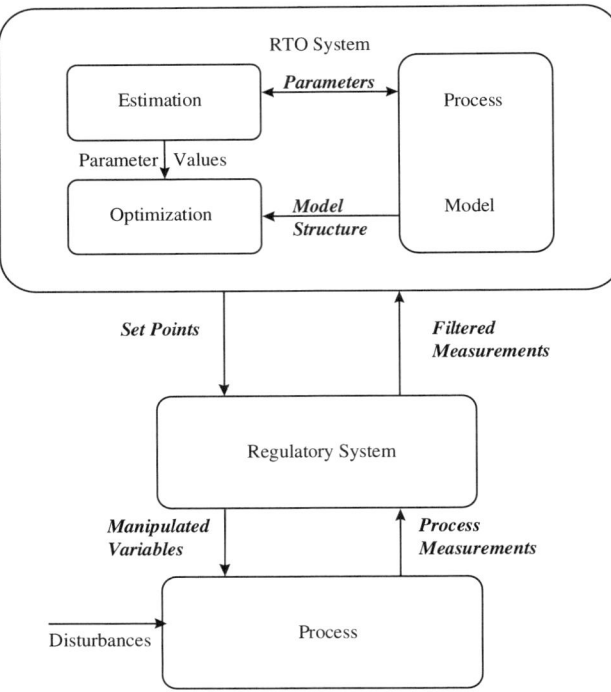

Figure 2. Structural decisions for an on-line optimizer.

In principle, the design of a RTO system is a significant combinatorial problem. De Hennin (1994) has shown that the number of possible structures for a given regulatory control system and optimization model is given by

$$1 + \sum_{n=1}^{N} \left[\frac{N!}{n!(N-n)!} \sum_{m=n}^{M} \frac{M!}{m!(M-m)!} \right]$$

where N is the total number of model parameters which may be selected for estimation in real time, and M is the total number of candidate measurements that may be used for parameter estimation. Table 1 shows the number of candidate structures for some typical cases.

When the decisions associated with the structure of the regulatory layer are added, together with the determination of the appropriate model complexity, the number of possible options becomes very large indeed. Several approaches have been proposed to handle the decisions involved in implementation in a logical way.

Krishnan et al (1992) consider the configuration of the parameter estimation subsystem based on a fixed regulatory control structure, and an optimization model which is sufficiently accurate to justify the assumption that all significant plant/model mismatch may be accounted for by errors in the model parameters. Their methodology is illustrated in Figure 3. First, a subset of the parameters in the model is selected on the basis that there is anticipated to be significant variability or uncertainty in their values, and that changes in their values cause significant variations in the optimal operating conditions determined from the model. Having decided which parameters should be estimated on-line, a series of techniques is proposed to identify the best set of measurements to perform the task. A minimal requirement is that the parameters to be estimated should be **observable** from the measurements, (Stanley and Mah (1981)). Structural properties of the model may be employed to test the observability of a given structure and to eliminate candidate measurements which are unlikely to be of value. On the assumption that a statistically-based parameter estimation technique is to be employed, the potential contribution of candidate measurements to reducing the confidence region of the derived parameter values is estimated using an approximate decomposition based on singular value analysis. The principle underlying this approach is that whilst increasing the number of measurements may improve the quality of the parameter estimates, each extra measurement has costs associated with installation and maintenance and may actually cause estimates to deteriorate where there is plant-model mismatch. Thus, a minimum set of measurements giving adequate parameter estimates is sought.

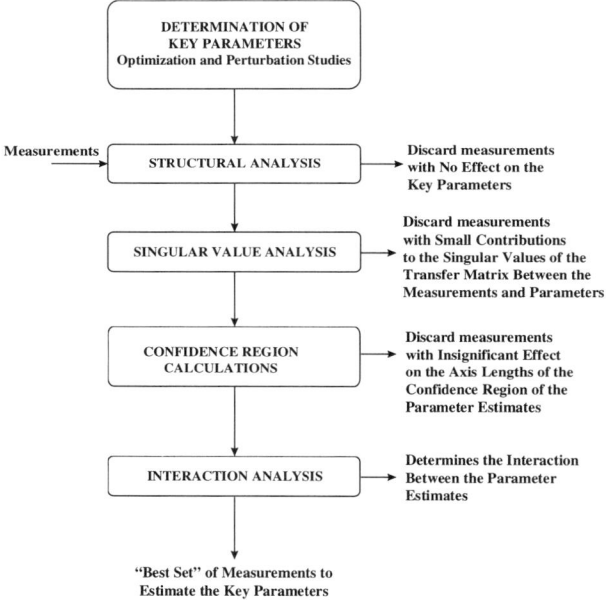

Figure 3. Methodology for selecting a robust parameter estimation scheme.

Once a candidate measurement set has been identified, further statistical tests are employed to verify the choice. Application to an industrial case study illustrates the potential of the method (Krishnan et al (1993)).

Table 1. Number of Possible RTO Structures as a Function of Number of Candidate Measurements (M) and Parameters that Might be Estimated (N).

N	M=2	3	4	5	6	7	8	9	10
2	8	19	42	89	184	375	758	1525	3060
3		35	84	188	403	841	1726	3506	7077
4			148	351	785	1689	3542	7303	14891
5				607	1423	3175	6844	14402	29804

A complementary approach to the selection of appropriate model parameters for a given model, which may also be used to validate entire RTO structures for a given regulatory control system, is based on the concept of **adequacy** (Forbes et al (1994)). In order for a given structure to be adequate, it should be possible to find a feasible set of parameter values in the model such that necessary conditions for optimality are satisfied by the model at the optimum plant conditions. The concept is illustrated in Figure 4, where set points corresponding to optimal plant conditions, r^* (a subset of $\underline{x}$ in equations (1) to (3)) result when a feasible parameter set $\underline{p}^*$ is fed to the model-based optimizer.

predict accurate objective function gradient values (Durbeck (1965)), as well as being able to predict the location of active constraints.

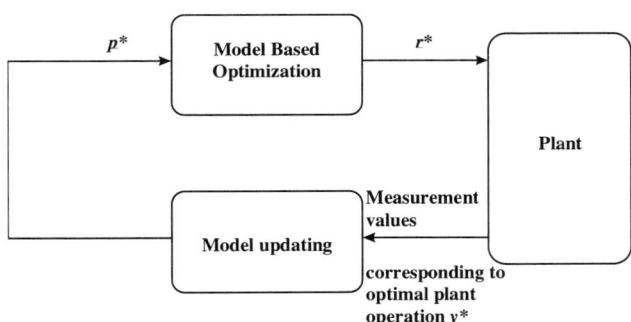

Figure 5. Augmented model adequacy.

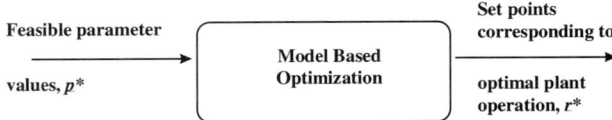

Figure 4. Point-wise model adequacy.

The concept of adequacy may be extended to consider also measurement choices, and parameter updating strategies, see Figure 5.

Here, the parameter values $\underline{p}^*$ must not only result in the optimal plant conditions when fed to the model-based optimization, but also should result when the values of the chosen measurements corresponding to optimal plant conditions are fed to the parameter estimation scheme. Screening tests may be devised based on these concepts, which emphasize the importance of selecting models (with embedded estimated parameters) which are able to

It is possible to avoid the need to devise a model satisfying these stringent requirements by implementing the parameter estimation and optimization tasks in a more integrated way, rather than using the sequential approach favoured in most applications (Haimes and Wismer (1972), Roberts (1979), Roberts and Williams (1981)). Roberts and Williams (1981) propose a modified two-step approach in order to remove the effect of plant-model mismatch. By using the equality constraints (equation (2)), a large number of the system variables, $\underline{x}$ may be eliminated so that the optimization problem may be formulated in the reduced space of set points, $\underline{r}$, as follows.

$$\min_{r} \Phi(\underline{r}, \underline{p}) \quad (4)$$

$$\underline{h}(\underline{r}, \underline{p}) \geq 0 \quad (5)$$

For a given set of plant measurements, $\underline{y}$, a risk function for the parameter estimation problem, R, may be formulated, and the parameter estimation problem represented as:

$$\min_{\underline{p}} R(\underline{r}, \underline{p}, \underline{y}) \quad (6)$$

Co-ordination between the parameter estimation and optimization steps is achieved by goal modification of the optimization problem. The objective is reformulated as

$$\min_{\underline{r}} \left[\Phi(\underline{r}, \underline{p}) - \underline{\lambda}^T \underline{r} \right] \quad (7)$$

where

$$\underline{\lambda} = \frac{\partial^2 R}{\partial r \partial p} \left(\frac{\partial^2 R}{\partial p^2} \right)^{-1} \frac{\partial \Phi}{\partial p} \quad (8)$$

The computational requirements for this modified approach are rather higher than those of the standard method. In addition, derivatives of the actual plant measurements with respect to the set point variables are required, which can only readily be generated using finite difference methods implying plant perturbations at each optimization step. There is some evidence that the performance of the algorithm is sensitive to errors in this information (De Hennin (1994)), but that where accurate derivatives are supplied, the algorithm outperforms the standard two-step approach where there is plant-model mismatch. Despite the robustness advantages of this modified approach, it does not seem to be widely employed in industrial applications in the process industries. It appears that the disadvantage of having to generate sensitivity information by imposing perturbations on the plant outweighs the advantage of obviating the extra engineering effort needed to try to formulate an adequate model.

Rather than seek to guarantee the potential of the RTO system to deliver the true plant optimum, an alternative approach is to recognize that the effect of the errors inherent in the instrumentation system on the plant and in the model used for the optimization is to cause the "optimum" determined by the RTO system to be erroneous, and to seek to design the system so as to make the **economic** effect of this error as small as practicable (De Hennin et al (1994); Forbes and Marlin (1994)).

The key to the implementation of this approach is to identify the various error sources in the RTO system, and to estimate their effect on closed loop performance. As well as measurement errors, and the variability in process parameters (which may or may not be estimated in real time), there is the effect of plant/model mismatch to consider. Given a reference model of the process to be optimized, together with mathematical or statistical representations of the uncertainties and variabilities identified above, any candidate structure (choice of measurements, estimated parameters, set point variables and optimization model) may in principle be evaluated using Monte Carlo simulation techniques, (Loeblein (1997)). However, the computational requirements for any one structure are large, making the evaluation of a large number of structures during system design using this approach infeasible.

As an alternative, an approximate analytical technique has been developed permitting evaluation of a large number of potential RTO structures (Loeblein and Perkins (1996)). The approach uses a local approximation of the plant (or reference model) and optimization models based on Taylor series expansion for a fixed set of active constraints, and Gaussian distributions to represent measurement noise and parametric uncertainty. Deterministic drift in parameter values is also accommodated. The result is an analytical prediction of the **average deviation from optimum**, that is the average over all the statistical variations of the difference between the actual performance of the plant when the results of the optimization are applied, and the process optimum in the absence of uncertainties.

The basic principle is illustrated in Figure 6 (taken from Loeblein and Perkins (1998a)) for a fully constrained two degree of freedom system. The set points calculated by the RTO system lie in a corner of the feasible region defined by the process constraints (equation (3)). Because of variability and uncertainty in the input data a region of calculated set points from the optimizer (here generated by Monte Carlo simulation) results. To calculate the

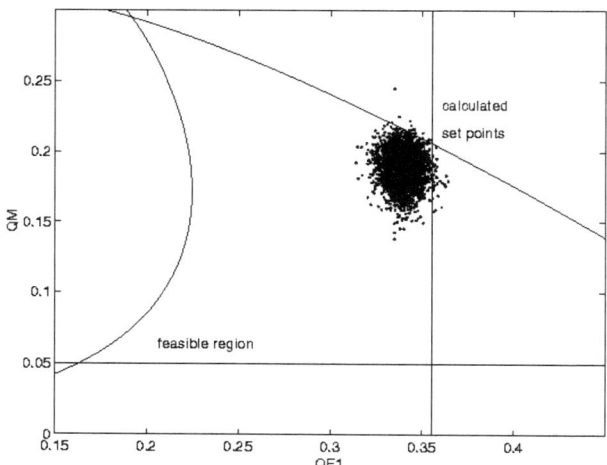

Figure 6. Back-off from active constraints and variation of the predicted set points in the feasible region.

average deviation from optimum, the average economic performance of the plant over this region is compared with the ideal performance that would be achieved by operating the plant in the appropriate corner of the feasible region.

Note that in order to guarantee the feasibility of the closed-loop system with high probability, a **back-off** from the active constraints has been introduced. The appropriate size of the back-off is determined as a part of the procedure.

To illustrate some of the issues in the design of RTO systems, and the application of the above approach, a simple case study involving the on-line optimization of two reactors in series will be discussed next. The plant is illustrated in Figure 7. The two reactors are designed to make product B from raw material A by the exothermic, irreversible reaction $A \rightarrow B$. An undesired by-product C is also produced from the side reaction $B \rightarrow C$. Both reactors are externally cooled, and the distribution of the raw material feed may be varied. A detailed reference model for this example is given in Loeblein and Perkins (1998a).

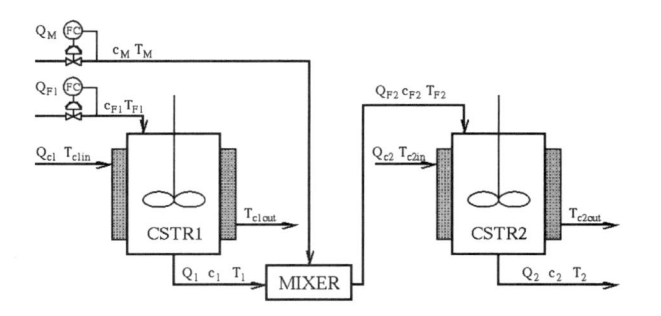

Figure 7. A two reactor case study.

In the reference model, seven parameters are assumed to have significant uncertainty: the temperatures of the two process feeds, the flowrates of coolant to the two reactors, the two heat transfer coefficients and the pre-exponential factor for the rate constant of the side reaction. Eight candidate measurements, each available at a given annualised cost are identified: six temperatures (of the process feeds, and the process and utility outlets from the two reactors) and the two coolant flowrates. For a given optimization model, the total number of RTO structures with 8 candidate measurements and 7 candidate parameters which could be estimated on-line is 22,564.

Each of these structures was evaluated using the approximate method outlined above for three candidate optimization models: the rigorous, reference model for the system; and two approximate models in which the side reaction was neglected, and the heat transfer between the jackets and the reactor contents was only approximately modelled. In approximate model 1, the cooling medium was assumed to have a uniform temperature throughout, whereas in approximate model 2 a linear approximation to the logarithmic mean driving force used in the rigorous model was employed.

The results of the evaluation show that, while the use of approximate model 1 causes a significant loss of performance compared to the reference model, model 2 gives performance which is comparable. The predicted performance of the optimal structures for the three models is shown in Table 2, and compared with the results from Monte Carlo simulation of the nonlinear closed-loop system to provide some validation of the approximate method.

Table 2. Predicted and Achieved RTO Results (compared to nominal optimal profit of 82.4)

Model	Average Profit	
	Predicted	Simulated
Reference	77.5	77.6 ± 2.0
Model 1	73.6	73.8 ± 3.7
Model 2	77.0	77.0 ± 2.0

The selected optimal structure, both for the reference model and model 2, uses all temperature measurements except the coolant outlet from the first reactor, and the coolant flowrate measurement in the second reactor, to estimate the coolant flow and heat transfer coefficient in the second reactor, and the two process feed temperatures.

While being a very simple case study in comparison to industrial practice, a number of features of RTO are illustrated. First, there is an inevitable loss of performance associated with the effects of uncertainties and measurement noise, when using the standard two step approach. Second, the effects of these variabilities on optimizer performance may be significantly affected by the decisions involved in the design and implementation of such systems. Third, it may well not be optimal from an economic viewpoint to install instruments for all available measurements on the plant; a carefully selected subset is more likely to give a better return. Finally, to consider all possible options in the design of such systems represents a significant combinatorial problem. Even a system of modest size generates of order 10^5 alternatives.

So far, we have not considered the structural decisions involved in the configuration of the regulatory control system. Regulatory control structure selection is an area of active research in itself. Given the usual 'bottom-up' approach to RTO implementation, where a RTO system is built on top of a regulatory control system which has already been designed and tuned to give good performance, it is not appropriate in this contribution to give a comprehensive review of the area. Instead, we focus on work which has explored the interrelationships between optimization and regulatory control performance.

The observation that the impact of seeking optimal steady state performance from plant is likely to drive operations close to or ideally on to process constraints dates back at least to the 1970s (Maarleveld and Rijnsdorp (1970)). In order to guarantee feasibility of the plant during dynamic excursions caused by disturbances, a back-off from the active constraints at the process optimum can be introduced. Of course, this back-off has an associated economic penalty, compared to the idealised performance that could be achieved in the absence of disturbances, and the magnitude of this penalty depends on the quality of the regulatory control system, (Narraway et al (1991)). It is important to recognise that this back-off is not the same as that illustrated in Figure 6, which was introduced to cope with the effects of uncertainty and variability on **optimizer** performance. Here, we are seeking to accommodate dynamic excursions of the plant within the region of feasible plant performance. In a recent study, Loeblein and Perkins (1998b) propose a method to evaluate the economic performance of model predictive control (MPC) in the presence of constraints using the concept of dynamic back-off and combine it with the economic analysis of RTO system performance outlined above to give a prediction of the economic performance of the combined system. The use of MPC technology is particularly appropriate where the tracking of process constraints is a key objective of the control system (Cutler and Perry (1983)).

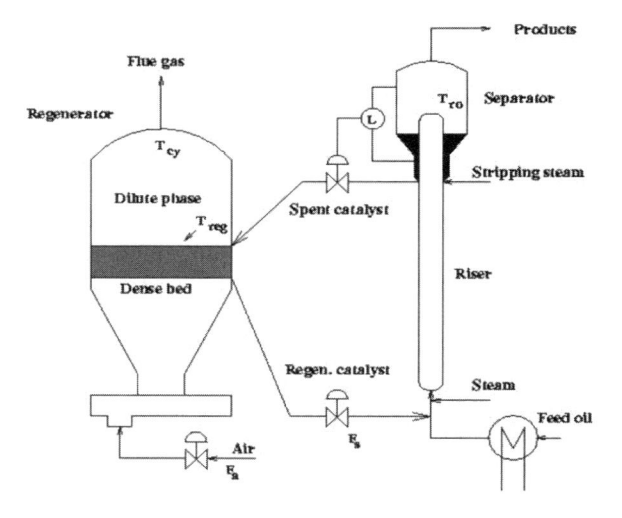

Figure 8. A fluidised catalytic cracking reactor.

The methodology has been applied to a case study based on a fluidized catalytic cracking (FCC) plant, see Figure 8. The regulatory control of FCC plants has been the subject of numerous studies, the most recent and comprehensive being those of Hovd and Skogestad (1993) and of Arbel et al (1996). Loeblein and Perkins (1998b) consider the integration of two promising regulatory control structures: the 'conventional' and 'riser-regenerator' structures, implemented using multivariable MPC technology, with a RTO system whose structure is chosen to have the best overall predicted economic performance.

Nominal optimization of the process with respect to two degrees of freedom, the catalyst circulation rate and the air flow to the regenerator, results in an optimal operating point constrained only by the metallurgical limit on the cyclone temperature. Thus, the optimum is only partially constrained. The back-offs required from this constraint to accommodate dynamic disturbances, and the uncertainty and measurement error at the optimization level, are shown in Table 3.

Table 3. Predicted Performance of Components RTO System for Different Regulatory Control Structures.

Regulatory Control Structure	Dynamic Back-off (K)	Total Back-off (K)	Average deviation from optimum ($/day)
Conventional	1.7	3.3	−1,700
Riser-regenerator	2.2	2.3	−1,200

It can be seen that the 'conventional' regulatory control structure is predicted to give the better economic performance in isolation, as measured by the dynamic back-off. However, use of the 'riser-regenerator' structure permits a more effective integrated RTO system to be implemented, the difference in performance corresponding to a 1.4% difference in the value added by the plant. This example illustrates that a 'bottom up' approach to the implementation of RTO systems may not lead to the best overall performance, since regulatory control structures chosen to given excellent economic performance in the face of high frequency disturbances may not deliver optimal performance when combined with a RTO system.

The potential importance of integrating the design of the regulatory control system with that of the RTO system is also emphasized by Ydstie and Coffey (1998). They show that some *decentralised* regulatory control structure choices may lead to suboptimal performance of the RTO system due to the presence of uncontrolled dynamics. A procedure, based on the application of concepts from passivity theory to process systems (Ydstie and Alonso (1997)), is proposed to devise structures that are able to deliver optimal RTO performance in the face of disturbances.

RTO of Dynamic Systems

The presence of periodic changes in operational requirements, as well as of disturbances emanating from the environment of the plant, or from time dependent features of the plant itself, cause the operation of any process to be dynamic. In addition, the operation of some equipment is inherently dynamic (e.g. in batch and semicontinuous plants), and so it is natural to explore the practicality of extending the approach to RTO discussed above to handle the dynamics of process systems.

There are several potential opportunities. First, that dynamic data generated during transient operation of a continuous system designed to be operated at steady-state might be used to monitor process performance and in particular to update model parameters has been recognised by a number of authors (e.g. Lin et al (1989), (1990)). Second, the possibility of not only optimizing the steady state performance of continuous processes, but also the dynamic transitions to new steady states either in response to new steady-state optimization results or to changes in operational requirements e.g. grade changes on continuous multiproduct plants, is being investigated (e.g. Verwys (1997)). Third, the implementation of dynamic analogues to the steady-state RTO approach is a feasible option, (Terwiesch et al (1994)).

The use of dynamic models adds significant extra complexity to virtually all aspects of the design and implementation of an RTO system. At the most fundamental level, the relationship between establishing the current state of the system and the taking of control action to improve that state is not straightforward. The most intuitively appealing and simplest approach is based on the concept of 'certainty equivalence' where data from past plant measurements are used to estimate the current plant state, and this state is used as the initial condition for the deterministic optimization of the future trajectory of the plant. In general, it is known that this approach is suboptimal since on the one hand it neglects the possibility of using control action to improve knowledge of the process condition ('probing'), and on the other, the assumption that the current state estimate has no uncertainty associated with it may be overly optimistic, and more 'caution' may be required in the exploitation of this knowledge to decide current and future actions. Some of the difficulties that may be encountered using certainty equivalence approaches are discussed by Mayne (1997).

The dynamic modelling of process systems is a less mature activity in industry than its steady state counterpart. However, the advent of process engineering tools for dynamic modelling and simulation over the past two decades has encouraged some industrial users to build high fidelity dynamic models of their processes, and to use these models to optimize dynamic operations (see Ishikawa et al (1997) for a recent example).

Algorithmic issues associated with the implementation of the estimation and optimization tasks in the RTO system are also more challenging, due to the need to analyse and optimize trajectories of model variables over time rather than just point values. At the simplest level, the size of problem, in terms of number of variables, is greatly increased compared to the steady-state case. The dynamic analogue of equation (2) is usually a system of differential-algebraic equations:

$$\underline{F}(\underline{x}, \underline{\dot{x}}, \underline{p}) = \underline{0} \qquad (9)$$

Leaving aside the issue of whether more process variables are required to capture plant dynamics than steady-state behaviour, the solution to equation (2) is a trajectory in time for the $\underline{x}$-variables (some of whose time variation in the period of interest will be specified to take up degrees of freedom) rather than a single value for $\underline{x}$ corresponding to steady-state conditions. The total number of variables involved in computational schemes for models of the form of equation (9) will be of order the product of the dimension of $\underline{x}$, and the number of discretisation points chosen to represent the time variation, clearly a much larger number than for the corresponding steady-state model.

The estimation problem for such systems involves using process measurements not only to estimate $\underline{p}$, but also the current value of $\underline{x}$, consistent with historical data from the process. Again, a point calculation is replaced by the analysis of trajectories. The classical approach to estimation of states and parameters is a recursive one based on variations of the Kalman filter (Jazwinski (1970)). The use of these approaches on nonlinear systems for the simultaneous estimation of state and parameters is known to be unreliable, and more recently methods based on analogues to receding horizon controllers have begun to emerge as more convincing alternatives (e.g. Robertson et al (1996)).

These approaches rely on an ability to solve dynamic optimization problems reliably, as indeed do the optimization tasks in the RTO system themselves. In this area, there have also been significant advances in algorithms in the past decade. The state of the art in the late 1980s was reviewed by Biegler (1990). Steady development of algorithms, and of software, often based on careful tailoring of components to address the special features of the large scale problem involved (Biegler (1992)), since that time has encouraged early industrial adopters to explore the utility of these methods on industrial case studies (e.g. Verwys (1997); Ishikawa et al (1997)) often resulting in significant economic benefits.

The most appropriate approach to account for uncertainty in a dynamic RTO system is an open question. This is a complex issue involving both theoretical and computational considerations. A comprehensive review of most approaches that have been attempted is provided by Terwiesch et al (1994). The main sources of uncertainty are the noise in process measurements used to construct state and parameter estimates, incomplete knowledge of future disturbances which may affect the process over the time horizon of interest, and mismatch between plant and model behaviour. For the 'parametric' uncertainty associated with the first two sources, a variety of modelling approaches and solution strategies are available (Terwiesch et al (1994)).

To help with the analysis of the effects of plant-model mismatch, it should be possible to extend the concepts of model adequacy to the dynamic case, to provide some guidance on important features of the plant which should be captured in a dynamic optimization model, but to my knowledge this has not yet been done. Alternatively, the careful integration of parameter estimation and optimization offers some hope of eliminating the effects of plant-model mismatch (Becerra and Roberts (1996)). As in the steady-state case, the choice of the most appropriate structure for the RTO system is an issue (Loeblein et al (1997)).

Given the number of open questions in the application of RTO to dynamic systems, it should be anticipated that significant advances in theory and practice will be seen.

Challenges and Opportunities

RTO in general is a computationally intensive activity, where the model complexity implemented is a function of computational power available as well as the application need. While in a sense the current generation of industrial RTO systems represents a mature technology, there will always be room for the adoption of better algorithms to handle the component tasks, either to improve efficiency or to broaden the scope of the RTO system.

In terms of efficiency improvements, interior point methods of various kinds are being investigated for the kinds of large scale optimization problems addressed in RTO applications (Albuquerque et al (1997). The incorporation of integer decisions (Floudas (1995)) is a development which would enhance the capabilities of RTO systems. As well as the fact that many decisions in process operations can only take discrete values, the integration of logic with mathematical optimization suggests some intriguing possibilities in the context of RTO, (Bemporad and Morari (1998)). More effective algorithmic approaches to optimization under uncertainty (Pistikopoulos (1995)) could potentially improve the flexibility and robustness of RTO schemes, particularly when using dynamic models. Finally, integration of the various tasks involved in RTO from an algorithmic point of view (parameter estimation with gross error detection, parameter estimation with optimization) has so far been investigated in an *ad hoc* way. Systematic analysis of the most appropriate decomposition of the overall RTO problem may lead to new insights.

In addition to the algorithmic opportunities identified above, there are a number of systems integration issues deserving of further research and development. There is a need for more effective techniques and tools to provide guidance and support to designers of RTO systems. The selection of the appropriate level of model complexity to employ, and of the instruments to deploy to monitor current plant performance are two examples of difficult implementation issues where more help could usefully be available.

The current generation of RTO systems are bespoke applications, each one requiring significant engineering effort to tailor the implementation to the particular plant under consideration. Utilisation of ideas from process modelling, most notably the development of a library of standard unit operations models with underpinning thermodynamic methods, has reduced the effort involved in model development to some extent. However, there is a need to find ways of making the application of these technologies more routine. Discussion with industrial users of RTO suggest that there is a lower limit, imposed by the cost of engineering RTO systems, to the scale of application that can be contemplated. Thus, even in the more commodity-oriented process sectors, development of more routinely applicable technologies will open up new applications. In sectors where responsiveness and flexibility in manufacturing systems are key requirements, there is a much more obvious need to make use of RTO technology routine. Here, perhaps the RTO system for a particular flexible manufacturing unit may usefully be thought of as part of the control and automation system whose structure is largely fixed, only a small amount of data input being required to set up the unit for the next manufacturing run.

Finally, developments allowing greater exploitation of the knowledge encapsulated in the RTO model by operating personnel would be valuable. Within the context of RTO itself, information on the motivation for a proposed change in operating strategy is a common request from current users, as well as assessment of the confidence that can be placed in the proposal. More generally, there are a variety of ways in which on-line models can be used to assist operations personnel. More flexible implementations of RTO systems permitting use of the model for a variety of purposes, both in response to inquiries from operations personnel and automatically, would greatly enhance the value of the on-line system.

Conclusions

There have been significant advances in the technology underlying RTO and in its application in the process industries in the past decade. For large, continuous processes, a mature technology has emerged based on steady-state modelling.

There remain significant research and development challenges in the field to facilitate the broader application of RTO in the process industries. As well as algorithmic developments, difficult issues in systems design and implementation will need to be addressed if the scope of application of RTO is to broaden from its current niche to encompass sectors where rapid product innovation demands flexible and responsive manufacturing systems. Thus, there are many interesting new opportunities for the academic community, as well as opportunities for technology suppliers and operating companies in the area of real-time optimization of process systems.

Acknowledgements

My thanks to two of my current students, Andrew Cobden and Costas Vafiadis, for help in assembling this paper. The whole job of preparing the manuscript has been masterminded by Mrs Jo Lines, in her usual cool, calm and competent way, (despite unreasonable pressure generated by me). My own interest in this subject has been nurtured by very fruitful collaborations with Christian Loeblein, Stephen de Hennin, Geoff Barton and Shiela Krishnan in particular. Dominic Bonvin and Srini Srinivasan at EPFL Lausanne have helped introduce me to the complexities of on-line batch optimization. Financial support for research in the area from BP plc, Electrolytic Zinc of Tasmania, the European Commission and the Engineering and Physical Sciences Research Council in the UK has also been enormously helpful.

References

Albuquerque, J. S., V. Gopal, G. H. Staus, L. T. Biegler, and B. E. Ydstie (1997). Interior point SQP strategies for structured process optimization problems. *Comp. &. Chem. Eng.*, **21 S**, S853-859.

Arbel, A., I. H. Rinard, and R. Shinnar (1996). Dynamics and control of fluidized catalytic crackers. Part 3. Designing the control: choice of manipulated and measured variables for partial control. *Ind. Eng. Chem. Res.*, **35**, 2215-2233.

Becerra, V. M., and P. D. Roberts (1996). Dynamic integrated system optimization and parameter estimation for discrete time optimal control of nonlinear systems. *Int. J. Cont.*, **62**, 257-281.

Bemporad, A. and M. Morari (1998). Control of systems integrating logic, dynamics and constraints. *Technical Report AUT-98-04*, Institute für Automatik, ETH Zurich.

Biegler, L. T. (1990). Strategies for simultaneous solution and optimization of differential-algebraic systems. In J. J. Siirola, I. E. Grossmann and G. Stephanopoulos. *Foundations of Computer-aided Process Design*, (CACHE/Elsevier), pp. 155-180.

Biegler, L. T. (1992). Tailoring optimization algorithms to process applications. *Comp. & Chem. Eng.*, **16 S**, S81-95.

Crowe, C. M. (1994). Data reconciliation - progress and challenges. *Proceedings of PSE-94*, Kyougju, Korea. pp. 111-121.

Cutler, C. R., and R. T. Perry (1983). Real-time optimization with multivariable control is required to maximize profits. *Comp. & Chem. Eng.*, **7**, 663-667.

De Hennin, S. R., J. D. Perkins, and G. W. Barton (1994). Structural decisions in on-line optimization. *Proceedings of PSE-94*, Kyougju, Korea. 297-302.

De Hennin, S. R. (1994). *Structural decisions in on-line process optimization*. PhD Thesis, University of London.

Durbeck, R. C. (1965). Principles for simplification of optimizing control models. PhD Thesis, Case Western Reserve University.

Floudas, C. A. (1995). *Nonlinear and Mixed-Integer Optimization*. (O.U.P.).

Forbes, F., T. Marlin, and J. MacGregor (1994). Model selection criteria for economics-based optimizing control. *Comp. & Chem. Eng.*, **18**, 497-510.

Forbes, F., and T. E. Marlin (1994). Design criteria for model-based real-time optimization systems. *Proceedings of PSE-94*, Kyougju, Korea. 133-140.

Haimes, Y. Y., and D. A. Wismer (1972). A computational approach to the combined problem of optimization and parameter estimation. *Automatica*, **8**, 337-347.

Hovd, M., and S. Skogestad (1993). Procedure for regulatory control structure selection with application to the FCC process. *AIChE J.*, **39**, 1938-1953.

Ishikawa, T., Y. Natori, L. Liberis, and C. C. Pantelides (1997). Modelling and optimization of an industrial batch process for the production of dioctyl phthalate. *Comp. & Chem. Eng.*, **21 S**, S1239-1244.

Jazwinski, A. H. (1970). *Stochastic Processes and Filtering Theory*. (Academic Press, New York).

Krishnan, S., G. W. Barton, and J. D. Perkins (1992). Robust parameter estimation in on-line optimization - Part I Methodology and simulation case study. *Comp. & Chem. Eng.*, **16**, 545-562.

Krishnan, S., G. W. Barton, and J. D. Perkins (1993). Robust parameter estimation in on-line optimization - Part II Application to an industrial process. *Comp. & Chem. Eng.*, **17**, 663-669.

Lauks, V. E., R. J. Vasbinder, P. J. Vallenburg, and C. van Leuwen (1992). On-line optimization of an ethylene plant. *Comp. & Chem. Eng.*, **16 S**, S213-220.

Lin, J., C. Han, P. D. Roberts, and B. Wan (1989). New approach to stochastic optimizing control of steady state systems using dynamic information. *Int. J. Control*, **50**, 2205-2235.

Lin, J., M. Wong, and P. D. Roberts (1990). Improvements in the formulation and solution approach for stochastic optimizing control of steady state industrial processes. *Int. J. Control*, **52**, 517-548.

Loeblein, C., and J. D. Perkins (1996). Economic analysis of different structures of on-line process optimization systems. *Comp. & Chem. Eng.*, **20 S**, S551-556.

Loeblein, C., J. D. Perkins, B. Srinivasan, and D. Bonvin (1997). Performance analysis of on-line batch optimization systems. *Comp. & Chem. Eng.*, **21 S**, S867-872.

Loeblein, C. (1997). Analysis and structural design of on-line process optimization systems. PhD Thesis, University of London.

Loeblein, C., and J. D. Perkins (1998a). Economic analysis of different structures of on-line process optimization systems. *Comp. & Chem. Eng.*, **22**, 1257-1269.

Loeblein, C., and J. D. Perkins (1998b). Analysis and structural design of integrated on-line optimization and regulatory control systems. IRC Report Series, B98-10, Imperial College, London.

Maarleveld, A., and J. E. Rijnsdorp (1970). Constraint control on distillation columns. *Automatica*, **6**, 51-58.

Marlin, T. E., and A. N. Hrymak (1997). Real-time operations optimization of continuous processes. In J.C. Kantor, C. E. Garcia and B. Carnahan, *Chemical Process Control*, (CACHE/AIChE), pp. 156-164.

Mayne, D. Q. (1997). Nonlinear model predictive control: an assessment. In J. C. Kantor, C. E. Garcia and B. Carnahan. *Chemical Process Control*, (CACHE/AIChE), pp. 217-231.

Miletic, I., and T. Marlin (1996). Results analysis for real-time optimization: deciding when to change the plant operation. *Comp. & Chem. Eng.*, **20 S**, S1071-1082.

Narraway, L. T., J. D. Perkins, and G. W. Barton (1991). Interaction between process design and process control: economic analysis of process dynamics. *J. Proc. Cont.*, **1**, 243-250.

Perkins, J. D., and G. W. Barton (1987). Modelling and simulation in process operation. In G. V. Reklaitis and H. D. Spriggs, *Computer Aided Process Operations*, (CACHE/Elsevier), pp. 287-316.

Pistikopoulos, E. N. (1995). Uncertainty in process design and operations. *Comp. & Chem. Eng.*, **19 S**, S553-563.

Roberts, P. D. (1979). An algorithm for steady-state optimization and parameter estimation. *Int. J. Systems. Sci.*, **10**, 719-734.

Roberts, P. D., and T. W. C. Williams (1981). On an algorithm for combined system optimization and parameter estimation. *Automatica*, **18**, 199-209.

Robertson, D. G., J. H. Lee, and J. B. Rawlings (1996). A moving-horizon based approach for least squares state estimation. *AIChE J*, **42**, 2209-2224.

Stanley, G. M., and R. S. H. Mah (1981). Observability and redundancy in process data estimation. *Chem. Eng., Sci.*, **36**, 259-272.

Terwiesch, P., M. Agarwal, and D. W. T. Rippin (1994). Batch unit optimization with imperfect modelling: a survey. *J. Proc. Cont.*, **4**, 238-258.

Tjoa, I., and L. Biegler (1991). Simultaneous strategies for data reconciliation and gross error detection of nonlinear systems. *Comp. & Chem. Eng.*, **15**, 679-690.

van Wijk, R. A., and M. R. Pope (1992). Advanced process control and on-line optimization in Shell refineries. *Comp. & Chem. Eng.*, **16 S**, S69-80.

Verwys, J. W. (1997). *Personal Communication*.

Ydstie, B. E., and A. A. Alonso (1997). Process systems and passivity via the Clausius-Planck inequality *Systems and Control Letters*. **30**, 253-264.

Ydstie, B. E., and D. P. Coffey (1998). Distributed control and real-time optimization of a chemical process. *Presented at DYCOPS-5*, Corfu, Greece.

THE CHALLENGES OF PRODUCT CONSISTENCY AND INTEGRITY IN SLOW AND FAST PROCESSES

Hugo Patiño
Coors Brewing Co.
Golden, CO 80401 U.S.A.

Abstract

The malting, brewing and packaging processes contribute to the quality of the final product that our consumers experience. The malting process is highly sensitive on the characteristics of the barley used, and each annual harvest brings with it differences that the maltster needs to compensate for prior to, during and after the malting process. The brewing process follows, where the characteristics of the barley malt, hops and water interact with yeast and significantly influence the process results during ingredient the extraction, fermentation and aging process as well as the product's chemistry and sensory performance. While malting and brewing represent relatively slow processes that last over a month each, the final packaging operation represents yet a different challenge, where filler speeds are in the order of 1200 to 1800 packages per minute. Consistency from package to package is critically important from a consumer quality and regulatory standpoint. This paper illustrates some of the challenges that these type of processes present for the control of product and package consistency.

Keywords:

Pearson correlation, Predictive measures, Process control.

Introduction

Consistency in consumer-perceived quality is a foundation for the long-term viability of any company. Achieving this consistency requires one to know the key attributes that drive consumer-perceived quality and the processes that influence these attributes.

Within these processes, there are some control variables that represent attributes in the process that are measurable and that have been found through years of research to best correlate or influence the final product attributes. Some of these product attributes can be influenced in different ways at different points in the process, making the control strategies non-unique and complex.

Similarly, several product attributes are correlated in different ways making decision making on process changes more complex as driving one attribute up will drive another one in a different direction.

It is the goal of this paper to illustrate some of these complexities in the brewing process and to look at some of the tools that brewers have used to address these challenges when the speed of the processes involved is so drastically different.

An Overview of the Brewing Process

In the context of this paper, the brewing process begins in the barley fields. The quality of the barley used greatly influences the final product. Examples of barley attributes that influence beer quality are the levels of protein, extract and glucans. Barley is harvested once a year in countries like the U.S. and Canada, and sometimes twice a year in other countries like Mexico. The barley goes through a period of dormancy that requires the kernels to rest for a period of several months before this ingredient is ready for use in the malting process.

The malting process makes the contents of the barley kernel accessible for the brewing process. In this process, barley is germinated under controlled time, temperature and humidity conditions until the appropriate degree of "modification" of this kernel has been achieved through various enzyme systems. The malting process is adjusted to compensate for differences in barley performance from year to year and in fact from growing area to growing area within a year. The goal is to drive consistency in the chemistry of the malted barley as it enters the brewing process. The malting process takes approximately a week, but the malted barley requires a period of storage of approximately one month before it is ready for use in the mashing process.

Extraction of the contents of the malted barley begins in the mashing operation where typically for American beers the use of so-called adjuncts (either corn, rice in raw, flour or syrup forms) provide a second source of starch. The starch is converted into sugars through the enzymes in the malted barley. Similarly, proteins present in the malted barley are solubilized to provide the nutrients that the yeast will need in fermentation. Hops provide the bitter flavor and a unique aroma characteristic to the final product. The quantity and type of malt, adjuncts and hops as well as how they are used in the process can greatly influence the final product attributes. This extraction process takes approximately one-half of a day to complete.

The unfermented product is now ready for fermentation by yeast. The fermentation process takes approximately one week. The strain of yeast used and the temperature, pressure and oxygen levels used during the fermentation process represent another set of variables that will greatly influence the final product attributes. This fermentation process is followed by an aging or flavor maturation process at cold temperatures that takes typically between one to six weeks.

The product is now ready for filtration, adjustment of alcohol level and carbonation. The fresh beer flavor that we have in a batch of beer if therefore the result of a large number of inputs that took place over several months. Being able to anticipate trends or deviations the earliest in the process is a key to providing the greatest consistency to the final product. A wide array of chemistry measurements are tracked and used for preventive and corrective action. On the other hand, the interaction of these chemical compounds and hundreds of others that go unquantified determine the final product's sensory characteristics. Several sensory tools are used, therefore, to track and provide the basis for corrective actions as well. The merging of sensory and chemistry attributes and translating these back into process measures is a great area of learning.

As we go from the finished beer tank into the packaging operation, the pace drastically changes as large batches of potentially hundred of thousands of liters go into individual containers where packages are filled at rates around 1500 packages per minute. For each of these packages a tight level of control is expected starting with regulatory requirements such as fill levels, product quality attributes such as oxygen and microbiological levels as these influence product stability as well as package functionality characteristics, such as detection of micro-leakage, seaming, or ease of opening features such as the torque required to open a bottle crown.

Some Challenges of "slow" Processes

Typically a large number of brews is made out of one malt blend. Similarly, several brews go into one fermenter. Further blending of batches may take place between fermentation and aging. Finally, several aging tanks may be combined into one finished beer tank. Thus, data analysis needs to account for the timing and the blending that takes place across the process. Malt blend uniformity within a batch and grain sampling issues are sources of variability in the data. Stratification in the subsequent steps in the process is typically not a major source of variability. Some changes may have a cumulative effect, such as wort nutritional deficiencies for yeast.

In addition to meeting regulatory requirements (e.g. caloric and alcohol content), achieving desired consumer-perceivable sensory expectations is a key area of focus. Consumer language needs to be developed. Internal panel descriptors are also needed. Panelists are screened and calibrated. Approximately a dozen consumer attributes are measured including overall likeability. Similarly, over thirty attributes are measured by internal panelists covering aroma, flavor, after-flavor and textural characteristics. Consumer work, however is done on a limited basis. Correlating the consumer data with the internal sensory attributes helps us to determine consumer relevance for the internally measured attributes. Similarly, correlation between consumer and internal sensory attributes with process and product physical or chemical measurement help drive process and design changes.

There are several ways to illustrate these challenges. We will illustrate two sensory different parameters that are impacted by unique process issues.

Malting Process Example

Darker beers typically require of more than one malted barley type, e.g. pale and crystal or caramel malts. The second of these is typically used in small amounts and provides primarily color and caramel-type flavor. The enzyme potential of this malt is considerably lower than that of pale malt. The relative balance of the two malts is important as this balance influences the relative amount of sugar to alcohol in the final product. The brewing process on the other hand can also influence these attributes.

Over thirty ingredient, process and product attributes which may have an impact in this particular case are measured. More than one combination of these can influence the final product. As illustrated in Table 1 in this example, the increase in alcohol to real extract ratio (i.e. residual sugar) was linked to a decrease in the caramel malt flavor of the product and a decrease in the malt sweet aroma and flavor as well.

The increase in alcohol/real extract ratio in turn was linked to the pounds of caramel malt used. Learning from and anticipating these slight sensory changes will assist in further improving sensory consistency of the final product.

Fermentation and Aging Processes Example

In the case of this product, only one malt type is involved. The flavor of pale American lager beers is particularly sensitive to slight ingredient or process variations. For these products, a very tight control of the degree of fermentability (i.e. the alcohol to real extract ratio) is important. As discussed in the previous example, that ratio can be influenced by the malting and mashing process used. It can also be influenced by fermentation conditions of time and temperature. The list of ingredient, process and product attributes is again quite long in this example. Table 2 illustrates the relationship found between the conditions at the end of the aging process vs some product characteristics. Specifically, a higher alcohol level was found to be negatively correlated with the product's body while positively correlated with fruity aroma and flavor, as well as higher astringency.

The complicating factor here is the lag time involved between the time when the malt is made, the brew date and the weeks of fermentation and aging vs the time when the final product is evaluated. Predictive tools will make process changes more proactive.

Examples of Challenges of Fast Processes

The packaging operation represents a very different set of challenges. Beer chemistry, taste, package integrity and other important parameters can be impacted. The processes run, however, at much higher speeds. We will illustrate two examples in the packaging operation that look at two different aspects of final product quality.

Filler Control

A typical filler consists of over 100 individual filling valves. During filling, several stages occur to purge out oxygen while counter-pressure is applied for a smooth filling with a control level of foaming. Each valve is in a way unique because it has its own mechanical elements. In essence, we can consider that one filling line has over 100 individual fillers supplying the final product. Some of the key expectations include the amount of product in the final package ("fill level") as well as oxygen level in the final product which influences product freshness. There are a variety of control variables that are correlated with each other. For instance, a "bubble breaker" device is positioned above a transfer conveyor after filling and prior to seaming. Through the use of a jet of carbon dioxide or steam, foaming is created which helps purge out oxygen present on the surface of the filled can. A second device called the "after gasser" is placed in the seamer. This device helps drive out the oxygen-rich foam out of the top of the can just prior to seaming. Optimal settings need to be determined to produce the lowest oxygen level with the minimum amount of beer loss while delivering the right fill level. These settings can also be influenced by beer chemistry characteristics, such as density, viscosity and carbonation levels.

Individual control of each filler valve is enabled through the use of a device known as a "filler valve monitor" that provides a measure for the approximate amount of product coming out of a given valve. Oxygen level, on the other hand, can only be measured by destructive testing and the measurement takes several minutes. Clearly, a large volume of product is produced between consecutive measurements.

Consistent fill level is a necessary condition for consistent oxygen level control. The filler valve monitor is mounted over the transfer chain between the filler and the closer. It operates on the basis of gamma rays and looks at the beer level of each can. This instrument has been effectively used to identify valves that produce cans with major under and overfills. The information is displayed in histograms and reports are also generated which identify the valves that are producing the lowest and highest fills. Better understanding of the measurement error structure is needed to enable predictions of average fill and making optimal decisions on when a valve needs to be repaired.

Anticipating performance through better understanding how filler controls operate and tighter control of individual filler valve performance will help improve package to package consistency.

Package Seam Consistency

Very slight changes in packaging materials and filler/closer conditions can influence the consistency of the seam in the final package. This consistency of seaming for bottles is now measurable through a device which inspects the integrity of bottles after they are filled and sealed. It is a non-contact non-destructive inspection systems which inspects 100% of product and can operate at speeds up to 2000 bottles per minute. It operates based on acoustic technology. A transmitter emits a sonic impulse upon each bottle which causes the crown to vibrate. The reciever listens to the tone generated by the crown. The pressure inside the bottle along with the composition of the crown determines the acoustic level or tone. The system detects several types of defects, such as defective crowns, missing crowns, double crowns, missing sealant in the crowns, cracked bottles, chipped bottles and defects in bottle molds. It can also detect differences in tightness between different stations in the closer.

While the instrument is a very useful screening tool to determine go vs no-go decisions on every package, data analysis can also reveal some more subtle process issues. These include differences between the closing stations withing the closer (which in turn impact the ease of opening for twist-off crowns), variability in the composition of the crown materials, and others. Real-time intelligent data analysis can increase the amount of actionable information that can be used for improved process control. This represents an area of opportunity for further development.

Table 1. Malting Example Pearson Correlation Coefficients
(* Correlation significant @ 0.05 level, 1-tailed)
(**Correlation significant @ 0.01 level, 1-tailed).

	Alcohol / Real Extract	Caramel Flavor	Malt Aroma	Malt Flavor
Alcohol / Real Extract	1.00	-0.42**	-0.23**	-0.21*
Caramel Flavor	-0.42**	1.00	0.17	0.08
Malt Aroma	-0.23**	0.17	1.00	0.65**
Malt Flavor	-0.21*	0.08	0.65**	1.00

Table 2. Fermentation Example Pearson Correlation Coefficients
(* Correlation significant @ 0.05 level, 1-tailed)
(**Correlation significant @ 0.01 level, 1-tailed).

	Alcohol / Real Extract	Fruity Flavor	Astrin-gency	Body
Alcohol / Real Extract	1.00	0.33**	-0.13**	-0.24*
Fruity Flavor	0.33**	1.00	-0.27**	-0.44**
Astrin-gency	-0.13**	-0.27**	1.00	0.46**
Body	-0.24*	-0.44**	0.46**	1.00

MULTIVARIATE STATISTICAL TREATMENT OF HISTORICAL DATA FOR PRODUCTIVITY AND QUALITY IMPROVEMENTS

John F. MacGregor and Theodora Kourti
McMaster Advanced Control Consortium
Department of Chemical Engineering, McMaster University
Hamilton, Ontario, Canada L8S 4L7

Abstract

Historical data collected routinely on most processes provide a potentially valuable source of information for improving process operability and product quality. This paper discusses recent approaches to extracting and utilizing information from these data using multivariate statistical latent variable methods such as Principal Component Analysis (PCA) and Projection to Latent Structures (PLS). The problems considered include: (i) the exploration and analysis of historical data for process troubleshooting; (ii) process monitoring and fault diagnosis; (iii) the development of soft sensors; (iv) finding process conditions capable of yielding a product with specified properties; (v) establishing multivariate product specification regions; (vi) start up and grade transitions; (vii) extraction of information from multivariate sensors and (viii) process optimization. Several of the approaches are illustrated with industrial examples.

Keywords

Multivariate projection methods, Historical data analysis, Process monitoring, Fault detection, New product design, Soft sensors, Image analysis.

Introduction

Given the relatively mature state of the chemical process industries, competition is increasingly being based on market differentiation and improved quality of a company's products. Therefore, great emphasis is being placed on improving the operations of existing processes in order to reduce variability and improve quality. This involves developing better methods for the analysis of historical operating policies, process troubleshooting, process monitoring, fault detection and isolation, and process and product optimization.

Traditional approaches taken by the process systems engineering community to tackle these problems have usually involved the building of fundamental models and then optimizing these models or using them for process monitoring and fault detection. Although such approaches have many advantages, they also have significant limitations. The time and effort necessary to build such models or customize existing general models to a specific plant is often not available in industry. Furthermore, the theoretical equations only provide the structure of a model for that part of the system that we are most familiar with, namely, the mass, energy and momentum balances. The large number of parameters in these models must still be either fixed using literature information or estimated using past plant data. The success of this model fitting step is often crucial in an application such as process optimization since it must always be remembered that we are optimizing the model and not the process. For applications such as process monitoring or soft sensor/inferential modelling there often exist measurements on the product quality variables, and on the mechanical and electrical parts of the process (e.g., agitator power, pump speeds, etc.) which can be very important to the application but are rarely included as variables in our theoretical models. Empirical models based on plant data must then be used to supplement the fundamental models. In this paper we confine ourselves exclusively to the use of empirical models based on historical data, and examine their potential and their limitations for improving operations for both batch and continuous processes.

With process computers routinely collecting data from on-line sensors on hundreds to thousands of variables every few seconds, large databases are accumulated in industry. The exploitation of these data is a critical component in the successful operation of any industrial process over the long term. Establishing methods whereby we can learn from our past experiences to eliminate problems in our process or in our operating procedures, monitor our current operations in real time, and find improved conditions for both our present and new products is of paramount importance in competitive environment today. Up to now, rarely something is done with them, for the reasons that follow. The sheer size of the data sets is overwhelming. The data are highly correlated (many variables being collinear) and non-causal in nature. The information contained in any one variable is often very small due to the low signal / noise ratios. There are often missing measurements on many variables. In order to utilize these databases, an empirical modelling method must be able to deal effectively with all these difficulties. For this reason we have focused our efforts on models developed by using latent variable methods such as Principal Component Analysis (PCA), and Projection to Latent Structures (PLS). These methods address all of the above problems in a very straightforward manner, and provide analysis tools that are easy to present and interpret.

An overview of latent variable methods is given in section 2. Their application to the analysis and exploration of historical databases and process troubleshooting, for the purpose of uncovering understanding past problems and learning how they impact the process, is outlined in section 3. Process monitoring, fault detection and diagnosis which enable us to utilize all the process data as well as the laboratory quality data is discussed in section 4. The difficulties and issues involved in building inferential models or "soft sensors", for inferring productivity and quality variables that are not easily measured on-line are discussed in section 5. Section 6 considers the problem of product design or finding a new window of operating conditions which will yield a desired new product grade in an existing plant or the same product in a different plant. Other areas in which these data-based approaches offer great potential, such as in establishing multivariate product specifications, finding improved start-up and grade transition policies, extracting information from on-line imaging systems, and data-based optimization are briefly presented in section 7. We then summarize with some general conclusions and future directions.

Latent Variable Methods

The main characteristic of process databases is that they consist of measurements on a large number of variables (hundreds) but under plant operating procedures these variables are highly correlated, and the effective dimension of the space in which they move is very small (usually less than 10 and often only 2 to 4). This is a result of the fact that there are usually only a few process disturbances or independent process changes that routinely occur, and the hundreds of measurements are only different reflections of these few underlying events. Consider the historical process data to consist of a (n×k) matrix of process variable measurements X and a corresponding (n×m) matrix of productivity and quality data Y. The latter are often available only at less frequent periods from a quality control lab. Latent variable models assume that these data spaces are effectively of very low dimension (i.e., non full rank) and are observed with error. For linear spaces the models take the form:

$$X = T P^T + E \qquad (2.1)$$

$$Y = T Q^T + F \qquad (2.2)$$

where **E** and **F** are error terms and **T** is a (n×A) matrix of latent variable scores, and **P** and **Q** (k×A) and (m×A) are loading matrices which show how the latent variables are related to the original **X** and **Y** variables. The dimension, A, of the latent variable space if often quite small and determined by cross-validation or some other procedure (Jackson, 1991; Wold, 1978). These models reduce the dimension of the problem through a projection of the high dimensional **X** and **Y** spaces onto the low dimensional latent variable space **T** which contains most of the important information. By working in this low dimensional space of the latent variables (t_1, t_2, ... t_A), the problems of process analysis, monitoring and optimization are greatly simplified. There are several latent variable methods. Principal Component Analysis (PCA) models only a single space (**X** or **Y**) by finding the latent variables which explain the maximum variance. Projection to Latent Structures or Partial Least Squares (PLS) maximizes the covariance of **X** and **Y** (i.e., variance of **X** and **Y** explained, plus correlation between **X** and **Y**). Reduced Rank Regression (RRR) maximizes the variance of **Y** and the correlation between **X** and **Y**. Canonical Variate Analysis (CVA) or Canonical Correlation Regression (CCR) maximizes only the correlation between **X** and **Y**. Discussion of these latent variable models is covered in references such as Martens and Naes (1989), Wold et al. (1984), Höskuldsson (1988, 1996), Burnham et al. (1996). Which method to use will depend upon the objectives of the problem. However, all of them lead to a great reduction in the dimension of the problem. This latter point is crucial in most of the applications discussed in the following sections, and also for the problem of treating missing data (Nelson et al., 1996).

In the case of continuous processes one usually is dealing with matrices of n observations on k process variables (**X**) and m output variables (**Y**). In batch processes one will also encounter three dimensional data arrays **X** (n×k×L) where k process variables are measured at L time intervals for each of n batches. Multi-way extensions of PCA and PLS, which are equivalent to unfolding the arrays into matrices in one of several ways, are also available (Wold et al., 1987; Nomikos and MacGregor, 1994, 1995). Finally data for both continuous and batch processes can often be partitioned into meaningful blocks corresponding to sections of the process or type of variable. Multi-block latent variable extensions (Westerhuis et al., 1998; Kourti et al, 1995) often offer an easier interpretation and diagnosis of the data when the process is large.

Although the latent variable model in (2.1), (2.2) is linear, nonlinear versions are available (Wold et al., 1992) and have been used. In situations such as process monitoring where the process is being operated to achieve a certain consistent product quality we have found linear models to almost always be adequate. This applies to both continuous and batch operations. Even when one is considering a much wider range of variation where moderate nonlinearities are present, simple transformations such as the logarithm of some of the variables often allow one to use the linear model. Therefore, we will focus on linear models in the remainder of this paper although in most cases extensions to nonlinear models is straightforward.

Exploration and Analysis of Process Databases

In this section we consider the use of multivariate latent variable methods to explore databases in order to identify periods where unusual operating behaviour was present and to try to diagnose possible causes for these problems. By examining the behaviour of the process data in the projection spaces defined by the small number of latent variables (t_1, t_2, ..., t_A), regions of stable operation, sudden changes or, slow process drifts may be readily observed. An interpretation of the process movements in this reduced space can be found by examining the loading vectors (p_1, p_2, ..., p_A), or (w_1, w_2, ..., w_A) in the case of PLS, or the contribution plots. Although the methods will not be able to unequivocally identify the cause of any problem (because of the highly correlated and non causal nature of the data), they can almost always identify unusual operating periods, and can usually isolate the region of the plant and the group of process variables that are related to the problem. In this way they can serve as a powerful tool for focusing the attention of the operations engineers to a much smaller space and thereby allow them to better use their engineering knowledge to diagnose the cause of any problems, and to use this information to improve the process.

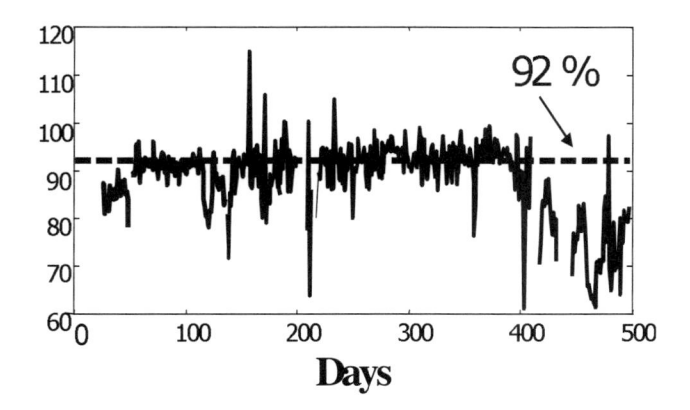

Figure 1. Recovery history of product A.

To illustrate the use of latent variable methods to explore and analyze historical data we consider a troubleshooting problem on a continuous recovery process (Kourti et al., 1996). More examples and references from other types of processes can be found in Kourti and MacGregor (1995). The feed stream consisting of three major components A, B and C is passed through a series of 12 separators. There are three products from the process, the most valuable being Product #1, a stream with high purity in component A. The principal operating objectives for the plant are to maintain the concentration of A in product #1 at a specified level (more or equal to 99.5 %) while achieving a certain minimum recovery (at least 92 %) of A in this stream. However, for the last three months of operation, the recovery dropped significantly below 92 % (Fig. 1). The company supplied daily averages on 447 process and product variables for a period of 498 days. The data set had several missing data.

Multivariate projection methods were utilized for the analysis. PLS models were built between the process variables and purity and recovery. All the historical data were utilized for the initial model. The set of 442 process variables was projected to 7 principal components that could explain 93 % in purity variation and 93 % in recovery. Projection of all the process variables on the first two principal components (Fig. 2) indicated that for the last three months (points 400 – 490), where the recovery was low, the process behavior had changed; the location of the projection of points 400 - 490 was different than the location of the first 400 points. Contribution plots (MacGregor et al., 1994) were used to identify the process variables that numerically contributed to this different behavior. Figure 3 shows the contributions in the move along the direction of t_1, between days 389 to 464. Notice that 4 variables have the highest contribution to this move.

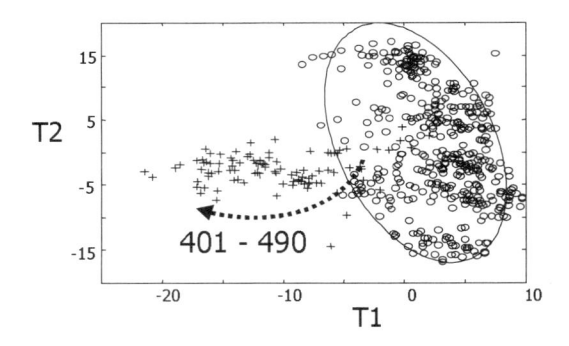

Figure 2. Projection of process history on latent variable space.

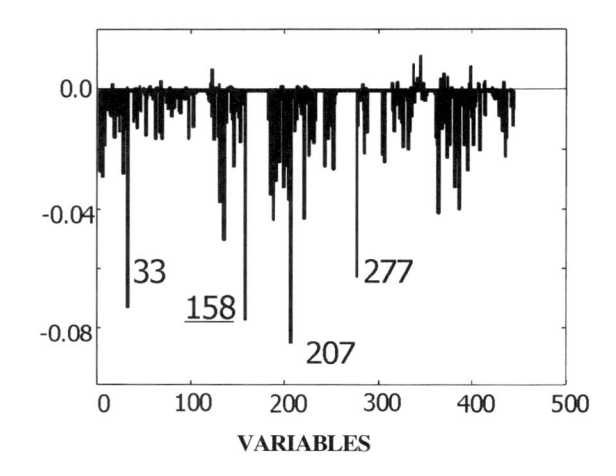

Figure 3. Contribution plots, days 469 – 384.

Three of these variables (33, 207, 277) are process conditions that can be controlled to improve the process performance; the forth variable (158) is correlated to them, mainly a result of these process conditions. From the 442 variables, the projection methods isolated 3 process conditions that were not properly controlled as responsible for the change in recovery. It is worth mentioning here that company engineers had spent considerable amount of time to determine what process conditions were leading to low recovery, before applying multivariate analysis. They had independently determined a set of variables. The multivariate analysis pointed to those same variables, and in a very small fraction of the time spent by the engineers.

Process Monitoring

Although the analysis of historical databases is an important first step towards process improvement, establishing multivariate control charts to detect special events as they occur, and to diagnose possible causes for them while the information is fresh is an essential part of statistical process control (SPC).

Traditionally, univariate SPC charts (Shewhart , CUSUM, EWMA) have been used in industry to separately monitor either a few process variables, or key measurements on the final product which in some way define the quality of that product. The difficulty with this approach is that these quality variables are not independent of one another, nor does any one of them adequately define product quality by itself. Product quality is only defined by the correct simultaneous values of all the measured properties, that is, it is a multivariate property (Kourti and MacGregor, 1995). Natural extensions of the Shewhart chart to situations where one observes a vector of several variables at each time period are the multivariate χ^2 and T^2 charts, employed only by a few companies (Kourti and MacGregor, 1995).

With the traditional approach of statistical quality control (SQC) where only product quality data (Y) are monitored, all of the data on the process variables (X) are being ignored. For a true Statistical Process Control, one must look at all of these process data as well. There are often hundreds of process variables, and they are measured much more frequently (and usually more accurately) than the product quality data. Furthermore, any special events which occur will also have their fingerprints in these process data. Sometimes product quality is only determined by the performance of the product later, in another process (e.g. catalyst conditioning- performance of catalyst is assessed later in polymer production). It would be useful to know if the product is good before using it; monitoring the process would help in the early detection of poor quality product.

There are several other reasons why monitoring the process is advantageous. Sometimes, only a few properties of the product are measured, but these are not sufficient to define entirely the product quality. For example, if only rheological properties of a polymer are measured, any variation in end-use application that arise due to variation in chemical structure (branching, composition, end-group concentration) will not be captured by following only product properties. In these cases the process data may contain more information about events with special causes that may affect the product quality (product performance). Finally, even if product quality measurements are frequently available, monitoring the process may help in diagnosing assignable causes for an event. When monitoring product quality, even if it is determined which quality variable caused the multivariate chart to go out of limits, it may still be difficult to determine what went wrong in the process. Several combinations of process conditions may cause the same product property to change. Monitoring the process would help identify one combination of process variables and therefore determine the underlying cause more easily.

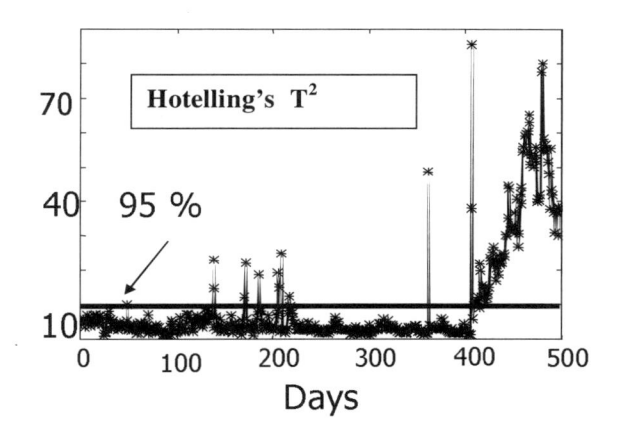

Figure 4. Hotelling's T^2.

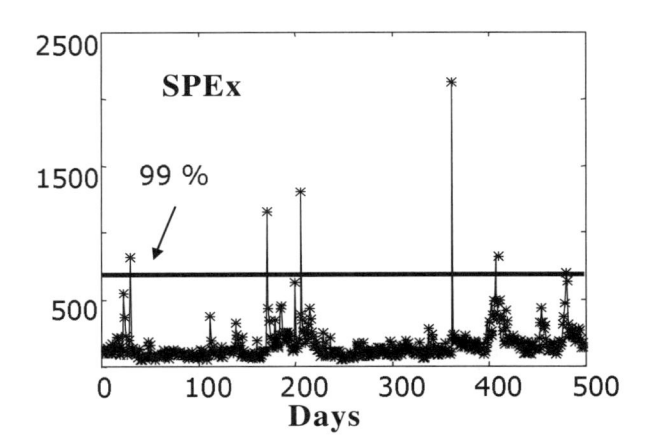

Figure 5. SPEx for the recovery process.

Projection methods can be used to develop process monitoring charts that utilize process information. The philosophy applied in developing multivariate SPC procedures based on projection methods, is the same as that used for the univariate or multivariate Shewhart charts. An appropriate reference set is chosen which defines the normal operating conditions for a particular process. In other words a PCA or PLS model must be built based on data collected from various periods of plant operation when performance was good. Any periods containing variations arising from special events that one would like to detect in the future are omitted at this stage. The choice of the reference set is critical to the successful application of the procedure as discussed in Kresta et al. [1991].

Two complementary multivariate control charts are required for process monitoring using projection methods (Kourti and MacGregor, 1995, 1996). The first one is a T^2 chart on the first A latent variables t_i (t_i could be either the scores from a PCA model on the process (X), or the scores from a PLS model between the process (X) and product quality (Y)).

$$T^2 = \sum_{i=1}^{A} \frac{t^2_i}{s^2_{t_i}} \quad (4.1)$$

where s_{ti} is the estimated variance of the corresponding latent variable t_i. This chart will check if a new observation vector of measurements on k process variables projects on the hyperplane within the limits determined by the reference data. The second chart is a SPE_X chart

$$SPE_X = \sum_{i=1}^{k} (x_{new,i} - \hat{x}_{new,i})^2 \quad (4.2)$$

where $\hat{x}_{new}$ is computed from the reference PLS or PCA model. This latter plot will detect the occurrence of any new events which cause the process to move away from the hyperplane defined by the refence model. It should be emphasized here that the models built for process monitoring model only common cause variation and not causal variation. The main concepts behind the development and use of these multivariate SPC charts for monitoring continuous processes were laid out by Kresta et al. (1991), Wise et al. (1991), Wise and Ricker (1991), and MacGregor et al. (1991a, b). Several illustrations of the methods were also presented in those papers, along with the algorithms and details on estimating control limits.

Figures 4 and 5 give the Hotelling's T^2 and SPEx charts, respectively, for the recovery process data. The 95 % and 99 % confidence limits were determined based on good operation when the recovery is around 92 %. Had the charts been on-line the deviation from normal could have been detected based on process data only, immediately when it occurred around observation 400.

When the data are serially autocorrelated the X and Y matrices can be augmented with time-lagged values, in order to account for the dynamics of the process and the disturbances. Multivariate time series analysis is discussed for PCA by Jollife (1986), Jackson (1991) and for PLS by Wold et al. (1984), MacGregor et al. (1991c). Dead times between variables, are being accounted for by time shifting. An industrial example where plant data had been both time shifted to account for dead times between X and Y, and lagged to account for autocorrelations in Y, is described in Dayal et al. (1994). Multivariate Process Monitoring Charts based on projection methods can also be constructed for Batch processes. Examples of such applications in industrial processes can be found in Nomikos and MacGregor (1995a, 1995b) and Kourti et al. (1995, 1996).

Soft Sensors / Inferential Models

In many monitoring and control situations we are often lacking on-line sensors capable of measuring many of the responses of interest. This is particularly true in the chemical industry where product quality is a major concern. We might wish to control the molecular weight distribution of a polymer, or even the average molecular weight as indicated by a melt flow index. However, the measurement equipment for such quality variables is very expensive, is difficult to put on-line, and costly to maintain. As a result we often try to develop soft sensors or inferential models which use other readily available on-line measurements such as temperatures, and can be used to infer the properties of interest in an on-line manner. These soft sensors can either replace the hardware sensor or be used in parallel with it to provide redundancy and verify whether the hardware sensor is drifting.

These inferential models are usually built by fitting either empirical or theoretically-based models to plant data. Empirical models are usually obtained via multiple linear regression, neural networks, or latent variable regression (eg. PLS). Usually, some of the parameters have to be updated on a regular basis or periodically to ensure that the model tracks the lab measurements. Even though the process is dynamic in nature, the inferential models can often be steady-state in form. This is the case whenever some of the independent variables used as regressors are measured in the same vicinity as the soft sensor location, and therefore already contain information on the dynamic behaviour of the system. When the regressor variables are all located well upstream of the soft sensor location then a dynamic model involving lagged regressor variables is usually necessary.

In spite of its apparent simplicity, this problem of building empirical soft sensor models is fraught with difficulty and in most cases very poorly done. The key to building good soft sensors is to use the correct type of data for building them. In general one should collect the data in a manner that closely resembles the situation under which the eventual model will be used. For example, if the soft sensor is to be used for feedback control of the inferred property, the model must be built based on data collected while operating under a similar feedback control structure. One does not want to build a causal model using data where a designed experiment or independent variations have been made in the major manipulated variables. Models built using these two different sets of data will be very different because the correlation structure among the variables in the two data sets will be very different and any empirical model is only modelling the correlation structure existing in the particular data set used. These issues are discussed and illustrated in MacGregor et al. (1991a) and Kresta et al. (1994). We further illustrate the problem here with an example which is a great simplification of an actual industrial study.

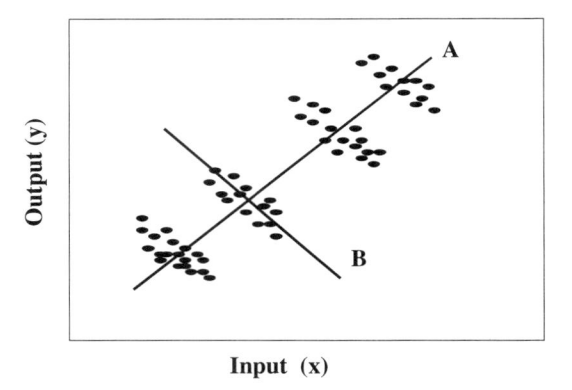

*Figure 6. Input / output relations for four different grades. **A** is the regression line for 4 grades. **B** is a regression line within a grade.*

Consider the case of a single quality variable y and a single regressor variable x which is also used as a manipulated variable by operators or by a feedback controller to keep y near a target. The data is generated according to the following scheme.

$$y_{k+1} - y_{sp} = g(x_k - x_{ss}) + D_k \quad (5.1)$$

$$x_k = x_{ss} + K_p(y_k - y_{sp}) + d_k \quad (5.2)$$

D_k represents the effect of process disturbances and is taken to be normally distributed with mean zero and variance 4.0 ie. N(0, 4.0). d_k represents variations in implementing the feedback (eg. operator variation) and is taken as N(0, 0.5). g and K_p are the process and controller gains, respectively. It is assumed that four different grades of product are made (e.g., polymer with different melt flow indices) corresponding to different set-points y_{sp} and corresponding steady-state values x_{ss}. The data collected in this manner is shown in Fig. 6. Also shown are two regression lines labelled A and B. The model A was fit to all the data collected from the four product grades. This model gives a good representation of the true causal effect of x (due to changing x_{ss}) on y, and might be very useful for suggesting new values of x_{ss} that would give a desired product quality y_{sp}. However, this model would be very poor as a soft sensor for y at any particular grade, the precise situation in which would want a soft sensor for control. The model given by line B on the other hand is what we would need as a soft sensor for use in future control of y. This is a non-causal model obtained using only the data at one grade or the mean adjusted pooled data from all grades (i.e., collected under conditions similar to the intended end use of the model).

In general there are many regressor variables and the situation is not as obvious as this simple example, but similar effects will be present depending upon how the data is collected. In cases where a very detailed fundamental dynamic model is available, rather than using that complex model directly as an inferential model, empirical approximating models have often been fitted to simulated data, and these simple empirical models used as soft sensors. However, these often fail because the simulated data was generated in a causal manner by introducing independent variations in all manipulated and disturbance variables. This would give a causal model similar to model A in Fig. 6 , and would be unsuitable for use as a soft sensor in a feedback environment. Instead, one should simulate the data from the fundamental model with all controllers in place and with disturbance similar to those existing in the actual plant. In general, one can often achieve good inferential models by first obtaining a model with data collected in a manner approximating the end use situation, and then refitting it using data collect with the soft sensor in place (Kresta et al., 1994).

Most of these problems disappear if theoretically-based models are used directly in the soft sensor. These models have a valid structure regardless of the type of data used to estimate its parameters, and regardless of how it will be used as a soft sensor. An industrial application for melt index and density soft sensors based on fundamental models for a linear low density polyethylene process in presented by McAuley and MacGregor (1993). When using empirical models for soft sensors, latent variable models such as PLS which model both the X and Y space offer some important advantages over standard regression models or neural networks ; in particular, their ability to easily handle missing data and to easily detect whether or not any new observation vector of regressor variables is an outlier.

Product Design

Here, we move away from the process monitoring and control topics of the previous sections, and look more at problems of optimizing a process/product through the use of historical data collected on different product grades. Two related problems are considered. The first problem involves finding a window of process operating conditions for a given plant within which one can produce a product having a specified set of quality characteristics. The second problem involves finding a window of operating conditions in a second plant (perhaps quite different in design from the first plant), that will enable one to manufacture a product with identical quality characteristics to one that is already manufactured in a first plant. If one has a good theoretical model of the process, then these problems of optimizing plant operating conditions to make a new product or to transfer or scale-up production to another plant are easily handled as a constrained optimization problem. If theoretical models are not available, then an alternative is to use Response Surface Methodology (Box, Hunter and Hunter, 1978) which involves statistically designed experiments run on the process to move the process to the desired conditions. However, even before one performs experiments, there exists information within the historical database on past operating conditions for a range of existing product grades. In this section, multivariate statistical methods using these existing data are used to solve the problems. Details on the approaches and on both simulated and industrial examples are given in Jaeckle and MacGregor (1998) and Jaeckle (1998). A brief overview of the method and its power and limitations is presented here.

Consider the problem of finding process conditions to manufacture a new product grade with a desired set of quality characteristics y_{des}. It is assumed that the plant has already manufactured a number of other product grades. r independent quality characteristics of these grades **Y** and m process conditions **X**, corresponding to these grades are available in the database. Using a Principal Component Regression approach (see Jaeckle and MacGregor (1998) for alternative approaches) models can be built for the existing process data. A PCA (or SVD) model for X based on A significant components gives:

$$\hat{X} = U_A \Sigma_A V_A^T \qquad (6.1)$$

and a PCR model relating Y to X is given by

$$\hat{Y} = U_A \Sigma_A B \qquad (6.2)$$

where $B = \Sigma_A^{-1} U_A Y$. Using these models the problem is now to find a window of process operating conditions x_{pred} which both yield the desired product quality y_{des} and are consistent with the past operating policies and constraints of the plant (i.e., consistent with the $\hat{X}$ model in the equation (6.1)). The solution to this involves inverting model (2) for a new value of singular variables (scores) u_{new} as

$$u_{new}^T = y_{des}^T (B^T \Sigma_A^2 B)^{-1} B^T \Sigma_A \qquad (6.3)$$

and then finding the process conditions x_{new} corresponding to u_{new} but which satisfy the model (6.1) for X, i.e. :

$$x_{new}^T = u_{new}^T \Sigma_A V_A \qquad (6.4)$$

In the case where the dimension of the quality space (r) is less than A (A is number of independent dimensions in X) then x_{new} represents only one possible solution. The complete solution is given by (see Jaeckle and MacGregor (1998) for details)

$$\mathbf{x}_{des}^T = \mathbf{x}_{new}^T + \mathbf{x}_{null}^T \qquad (6.5)$$

where $\mathbf{x}_{null}$ is any vector lying within the (A-r) dimensional space of **X** which has no effect on $\hat{\mathbf{Y}}$. The complete solution (5) for $\mathbf{x}_{des}$ therefore defines a line or plane in the space of the m process operating variables, and any point in that space should, according to the models (6.1) and (6.2), result in the desired product $\mathbf{y}_{des}$, and still satisfy all past operating policies of the plant.

The key elements in this empirical model approach is the use of latent variable models that both reduce the space of **X** and **Y** to a lower dimensional orthogonal set of latent variables (**U**), and provide a model for **X** as well as **Y**. The model (6.1) for **X** is essential in providing solutions (6.4) and (6.5) which are consistent with past operating policies. In this sense, PCR and PLS are acceptable approaches, while MLR, Neural Networks, CCR and RRR are not.

This methodology has been demonstrated via simulation using both linear and nonlinear approaches to the design of new grades of low density polyethylene (Jaeckle and MacGregor, 1998). It has also been applied to two industrial batch polymerization processes and to a continuous industrial manufacturing process in Jaeckle (1998). The major limitations of this approach is that one is restricted to finding solutions within the space and bounds of the process space **X** defined by previously produced grades. There may indeed be equivalent or better conditions in other regions where the process has never been operated before, and hence where no data exists. Fundamental models would be needed if one hopes to find such novel conditions.

Other Areas and Future Directions

In this section, we introduce a few additional areas where the use of multivariate statistical models and historical databases offer significant potential for improving process operations. Due to lack of space in the paper, we only outline the problems and the potential benefits from their solution. These are current areas of research and references are provided where appropriate.

Establishing Multivariate Specification Regions

In the area of quality control, it is very important for a manufacturer of any materials to set up process control (SPC) schemes, establish control limits, and evaluate process capability. However, for any customer that is going to process these materials into higher value added products, the main concern is that these incoming materials consistently meet a well defined set of specifications which will ensure that they will cause no quality problems in his final product. In spite of the great emphasis in recent years on SPC to reduce process variability, and on the use of process capability indices (C_{pk}) which measure supplier variability relative to specification limits, almost no attention has been given to methods for quantitatively establishing meaningful specification limits. In the introduction of the SPC Reference prepared by the big three automotive companies (SPC, 1991), this is recognised in the following quote: " ... capability indices include product specifications in the formula. If the specification is inappropriate, or not based on customer requirements, much time and effort may be wasted in trying to force the process to conform."

In general, specification limits are usually established in a very subjective manner, and are invariably set on a univariate basis, that is, as upper and lower limits on each quality variable separately. However, for an incoming material or product, it is almost irrelevant whether an individual quality property meets a specification. Quality is a multivariate property and an incoming material is of high quality only if it has the correct combination of <u>all</u> properties simultaneously. Therefore, procedures are required to properly establish meaningful multivariate specification regions. Figure 7 illustrates the issue in the case of only two quality variables. If these variables are correlated, then a reasonable multivariable region that would lead to a high quality customer product might be given by any product property combination lying within the ellipse. If univariate specification limits are set on each variable separately, they would define one of the two rectangles in Fig. 7. The larger rectangle would include all good material but would also include a large amount of poor material. Therefore, the individual specification limits are usually moved in to the smaller rectangle to include much less poor material, but at the expense of excluding a very substantial amount of perfectly good material. With a large number of quality variables, this effect is magnified.

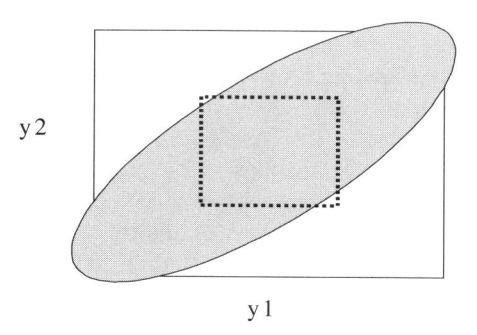

Figure 7. Specifications for product quality properties y1 and y2. The ellipse gives the property combination that would lead to high quality customer product. Rectangles define univariate specification limits.

With the availability of process and product databases at both the producer and customer plants, then as long as one can adequately track the lots of material through the customer plant to his final product, one is in a position to quantitatively define proper multivariate specification regions. Again due to the large number of process and quality variables, the high degree of correlation among them, and the presence of errors and missing data, multivariate statistical projection methods provide an ideal way of approaching this problem. DeSmet (1993) provides some preliminary research on this topic and Duchesne (1998) discusses some on-going research. The importance of this problem cannot be understated. Companies which can establish proper specification regions can potentially make large gains in market share. Furthermore, these multivariate specification regions provide the ultimate objective function for multivariate control systems (DeSmet, 1993).

Start-up/Grade Transition Problems

Many continuous manufacturing operations, particularly multi-product plants, go through a large number of start-ups or grade transitions. There are two problems which arise. The first is a steady-state issue, namely, how can we tell when we have reached an operating region where we are producing a product grade with quality which is consistent with the historical high quality material made for that grade? This would appear to be a trivial problem in that one can simply measure all the product quality variables and see if one has attained this region, or simply reset all the process variable set points to the same values and one should get the same product. However, companies rarely measure the complete set quality variables that define what their customer sees. Furthermore, process disturbances are present, and in multi-product campaigns, some process variable settings or process changes are never reset in the same way when one returns to any given grade.

The second problem is that of optimizing the grade transition/start-up trajectories for the manipulated variables. Ideally these optimal transition policies can be obtained using fundamental models and constrained optimization. However, if theoretical models are not available, but a good database on prior transitions and steady-states is available then multivariate PCA/PLS methods discussed previously can be used to analyze and improve these transition problems (Duchesne, 1998).

Multivariate Sensor and Image Analysis for On-Line Monitoring

In analytical labs, biological labs, and medical facilities, the use of multiple instruments, each providing a multivariate output (e.g., NIR), and multi-spectral imaging (e.g., color images, MRI) have become commonplace. It is only a matter of time before similar banks of multivariate sensors and colour imaging cameras are used on-line to monitor and control industrial processes. In particular, colour video cameras can easily be installed to monitor the state of combustion in furnaces, and to detect defects and quality problems in sheet and film forming processes.

One of the major limitations has been knowing how to handle the huge amount of highly correlated data collected from these sensors, and how to efficiently extract the subtle information contained in the data. Recent work on multivariate image analysis methods using multi-way PCA (Geladi and Grahn, 1996) and their extension to on-line monitoring (Bharati and MacGregor, 1998) provide a breakthrough in this area, and could lead to major use of these imaging sensors for monitoring and controlling industrial processes.

Data Driven Process Optimization

Most process optimization is carried out using fundamental models and constrained optimization algorithms. More recently with the availability of empirical modelling tools such as Neural Networks regression, there has been a lot of interest in exploiting historical databases for this purpose. However, for process optimization causal information must be extracted from the data, so that a change in the operating variables can be made that will lead to a better quality product, or higher productivity and profit. The problem is that databases obtained from routine operation contain mostly non-causal information. Some of the problems with models built from such highly correlated process data containing unmeasured disturbances and feedback controllers are outlined in MacGregor et al (1991a). Similar problems in the analysis of undesigned agricultural data led Sir R.A. Fisher to state that all one can do with such happenstance data " is a post-mortem to see what they died of" (Box, 1970 personal communication).

Therefore Fisher proposed using experimental designs a means of obtaining causal information, and Box proposed Response Surface Methods (RSM) and Evolutionary Operation as means of using sequential designs to find directions of steepest ascent and local optima (Davies, 1961).

In spite of this several authors have proposed approaches to optimization and control based on interpolating historical bases (DeSmet, 1993; Saraiva and Stephanopoulos, 1994; Saraiva, 1995 and Chen and McAvoy, 1996). However, in all these cases their success was based on making strong assumptions which allowed the database to be reorganized and causal information to be extracted. Jaeckle (1998, thesis) investigated an approach referred to as "similarity optimization" which combined multivariate statistical methods for reconstructing unmeasured disturbances and nearest neighbour methods for finding similar conditions with better performance. However, it too was shown to fail for many of the same reasons.

In general it was concluded that one can only optimize the process if there exist manipulated variables that change independently of the disturbances and if disturbances are piecewise constant, a situation that would be rare in historical process operations.

Conclusions

In this paper we have illustrated both the tremendous potential for using historical data bases to solve a number of important problems related to process operations and product quality and the often severe limitations that exist in interpreting such data with empirical models. The use of latent variable models is shown to be particularly appealing, because they greatly reduce the dimension of the problem and allow for easy graphical representation and interpretation, they allow for easy treatment of missing data and they provide a model for the X space as well as the Y space. This latter feature is crucial in many of the problems such as process monitoring and product design and is often of great value in soft sensor applications and process analysis.

The use of these latent variable approaches for integrating historical data bases has started to find wide acceptance in industry particularly for the problems of process analysis, monitoring and soft sensors. However, there are very many other potential problems that can be solved by extracting information from historical data bases. In this paper we have focused on a few of these and discussed the potential benefits to be gained and some approaches to their solution.

References

Bharati, M.H., and MacGregor, J.F. (1998). Multivariate Image Analysis for Real Time Process Monitoring and Control. Submitted to *AIChE Journal*.

Bharati, M. (1997). *Multivariate Image Analysis for Real Time ProcessMonitoring*. M. Eng., McMaster University, Hamilton, Ont. Canada.

Box, G.E.P. (1970). Personal communication.

Box, G.E.P., W.E. Hunter and J.S. Hunter (1978). *Statistics for Experimenters*. John Wiley and Sons.

Burnham, A.J., Viveros, R., MacGregor, J.F. (1996). Frameworks for Latent Variable Multivariate Regression. *Journal of Chemometrics*, **10**, 31-45

Chen, G., and T.J. McAvoy (1996). Process Control Utilizing Data Based Multivariate Statistical models. *Canadian J. Chem. Eng.* **74**, 1010-1024.

Davies, O.L (1961, Editor) *Statistical Methods in Research and Productionwith Specific Reference to Chemical Industry*. Hafner Publishing Company, New York.

Dayal, B., J.F. MacGregor, P.A. Taylor, R. Kildaw, and S. Marcikic. (1994). Application of Feedforward Neural Networks and Partial Least Squares Regression for Modelling Kappa Number in a Continuous Kamyr Digester. *Pulp and Paper Canada*, **95 (1)**, pp. 26-32.

DeSmet J. (1993). Development of Multivariate Specification Limits usingPartial Least Squares Regression. M. Eng., McMaster University, Hamilton, Ont.Canada.

Duchsne, C. (1998). *Product Quality Improvement Using Databases*. Thesis Proposal. Chem. Eng. McMaster University.

Geladi, P. and Grahn, H. *Multivariate Image Analysis*, John Wiley and Sons, Chichester, UK., 1996

Höskuldsson, A. (1988). PLS regression methods. *J. Chemometrics*, **2**, 211-228.

Höskuldsson, A. (1996). *Prediction Methods in Science and Technology: Volume 1. Basic Theory*. Thor Publishing, Denmark.

Jackson, J.E. (1991). *A User's Guide to Principal Components*, John Wiley and Sons, Inc., New York.

Jaeckle, J.M., and MacGregor, J.F. (1997). Product Design Through Multivariate Statistical Analysis of Process Data. Accepted *AIChE Journal*.

Jaeckle, C. (1998). *Product and Process Improvement Using Latent Variable Methods*. Ph.D. Thesis, McMaster University, Hamilton, Ontario, Canada.

Jollife, I.T. (1986). *Principal Component Analysis*, Springer Series in Statistics, Springer Verlag, New York.

Kourti, T., and MacGregor, J.F. (1995). Process Analysis, Monitoring and Diagnosis Using Multivariate Projection Methods - A Tutorial. *Chemometrics and Intelligent laboratory Systems*, **28**, 3-21.

Kourti, T., Nomikos, P. and MacGregor, J.F. (1995). Analysis, Monitoring and Fault Diagnosis of Batch Processes Using Multiblock and Multiway PLS. J*ournal of Process Control*, **5**, pp. 277-284.

Kourti, T., Lee, J. and MacGregor, J.F. (1996). Experiences with Industrial Applications of Projection Methods for Multivariate Statistical Process Control. *Computers in Chemical Engineering*, **20 Suppl**. S745-S750.

Kourti, T., and MacGregor, J.F. (1996). Recent Developments in Multivariate SPC Methods for Monitoring and Diagnosing Process and Product Performance. *Journal Of Quality Technology*, **28 (4)**, 409-428.

Kresta, J., MacGregor, J.F and Marlin, T.E. (1991). Multivariate Statistical Monitoring of Process Operating Performance. *Can. J. Chem. Eng.* **69**, pp. 35-47.

Kresta, J.V., Marlin, T.E., MacGregor, J.F. (1994). Development of Inferential Process Models using PLS. *Computers Chem. Engng.*, **18**, 597-611.

H. Martens and T. Naes (1989). *Multivariate Calibration*, Wiley, New York.

MacGregor, J.F., T.E. Marlin, J. Kresta and B. Skagerberg (1991a). Multivariate statistical methods in process analysis and control. In Y. Arkun and H. Ray, (Eds.), AIChE Symposium Proceedings of the Fourth International Conference on Chemical Process Control, Padre Island, TX, February 17-22; *AIChE Publication*, **No. P-67**, pp. 17-22, New York.

MacGregor, J.F., B. Skagerberg and C. Kiparissides (1991b). Multivariate statistical process control and property inference applied to low density polyethylene reactors. *IFAC Symp. ADCHEM'91, Toulouse*, France, Oct. 1991., pp. 131-135, Pergamon Press.

MacGregor, J.F., Kourti, T. and Kresta, J.V. (1991c). "Multivariate Identification: A Study of Several

Methods ", IFAC Intern. Symp. on Advanced Control of Chemical Processes Proceedings, K. Najim and J.P. Babary, eds., Toulouse, France, pp. 369-375

MacGregor, J.F., Jaeckle, C., Kiparissides, C. and Koutoudi, M. (1994). "Process Monitoring and Diagnosis by Multi-Block PLS Methods". AIChE Journal, 40 (5), pp. 826-838.

McAuley, K. and MacGregor, J.F. (1993).On-Line Inference of Polymer Properties in an Industrial Polyethylene Reactor. *AIChE*, **37** (825-835).

Nelson, P.R.C., Taylor, P.A., MacGregor, J.F. (1996). Missing Data Methods in PCA and PLS: Score Calculations with Incomplete Observations. *Chemometrics and Intelligent Laboratory Systems*, **35**, 45-65.

Nomikos, P. and MacGregor, J.F. (1994). Monitoring of Batch Processes using Multi-way Principal Component Analysis. *AIChE Journal,* **40 (8)** pp. 1361-1375.

Nomikos, P. and MacGregor, J.F. (1995a). Multivariate SPC charts for Monitoring Batch Processes. *Technometrics*, **37(1)** pp.41-59.

Nomikos, P., and J.F. MacGregor. (1995b). Multiway Partial Least squares in Monitoring Batch Processes. *Chemometrics and Intelligent Laboratory Systems*, **30** 97-108. (presented at. First International Chemometrics Internet Conference. September 1994)

Saraiva, P.M. (1995) . Inductive and Analogical Learning: Data Driven Improvement of Process Operations. pp. 377-435 Vol. 22. *Advances in Chemical Engineering*, Academic Press.

Saraiva, P.M. and Stephanopoulos, G. (1994). Data Driven Learning Frameworks for Continuous Process Improvements. PSE Korea..

Skagerberg, B., MacGregor, J.F., and Kiparissides, C. (1992). Multivariate Data Analysis Applied to Low Density Polyethylene Reactors. *Chemometrics and Intelligent Laboratory Systems*, **14**, pp. 341-356.

SPC(1991). *Fundamental Statistical Process Control.* Reference Manual. Automotine Industry Action Group., prepared by staff from Chrysler, Ford and General Motors.

Westerhuis, J., Kourti, T., and MacGregor, J.F. (1998). Analysis of Multiblock and Hierarchical PCA and PLS models. *Accepted in Journal of Chemometrics.*

Wise, B.M., D.J. Veltkamp, N.L. Ricker, B.R. Kowalski, S. Barnes and V. Arakali (1991). Application of multivariate statistical process control (MSPC) to the West Valley slurry-red ceramic melter process. Waste Management '91 Proc., Tucson, Arizona.

Wise, B.M. and N.L. Ricker (1991). Recent advances in multivariate statistical process control improving robustness and sensitivity. IFAC Symp. ADCHEM'91, Toulouse, France, Oct. 1991., pp. 125-130, Pergamon Press.

Wold, S. (1978). Cross-validatory estimation of the number of components in factor and principal components model. *Technometrics*, **20** (4), 397-405.

Wold, S., C. Albano, W.J. Dunn III, U. Edlund, K. Esbensen, P. Geladi, S. Hellberg, E. Johansson, W. Lindberg and M. Sj¬str¬m (1984). Multivariate Data Analysis in Chemistry. In: *Chemometrics. Mathematics and Statistics in Chemistry (B. Kowalski, Ed.), NATO - ASI, D. Reidel Publishing Company,* pp. 17-95.

Wold, S., P. Geladi, K. Esbensen, and J. Ohman (1987). Multiway principal components and PLS analysis. *Journal of Chemometrics,* **1**, 41-56.B

Wold, S. (1992). Nonlinear partial least squares modelling. II Spline inner relation. *Chemometrics and Intelligent Laboratory Systems,* **14**, 71-84.

REFINERY QUALITY CONTROL BY OVERALL INTEGRATION

(Frank) X. X. Zhu and Gavin Towler
Department of Process Integration, UMIST
Manchester, PO Box 88, M60 1QD, England

Introduction

Due to environmental concern and new regulations, producing better quality and clean products has become essential for the petroleum refining industry. For example, future environmental regulations require refiners to produce gasoline with less than 200 ppm sulphur (in year 2000) and 100ppm sulphur (in year 2005). Oil refining has already been operating with low margins due to relatively low energy prices. The requirement of better quality refined products calls for significant capital investment, which makes the current situation for oil refining industry even tougher. The question is how the refining industry can satisfy the higher quality specifications while remaining competitive. The most promising way to meet the challenge is to consider overall integration by integrating process change options, existing processes, and supporting systems (hydrogen system, utility system, heat recovery system etc.).

Integration of New Processes

To comply with the requirement of producing products of better quality, refiners have looked for new hydrotreating options. Possible options described by Upson and Schnaith (1997) include:

- Hydrotreating gasoline and diesel
- Recracking the heavy-gasoline fraction
- Using liquid-liquid extraction to remove sulphur compounds
- Using sulphur adsorption to remove sulphur

Hydrotreating the heavy fraction results in a major decrease in a sulphur content but a loss of octane and it is a costly option. To combine sulphur removal from the light-gasoline fraction with hydrotreating of the heavy gasoline fraction, the extractive Merox process reduces the amount of heavy fraction that is treated and thus reduces the octane loss. Liquid-liquid extraction is an alternative separation technique to segregate a sulphur-rich stream. The use of such an extraction process results in a smaller hydrotreater and less olefin and octane loss to achieve the same degree of desulfurisation. An emerging technique for desulfurisation is the adsorption of sulphur compounds from gasoline stream. In the adsorption process, sulphur would be totally removed from gasoline with little loss in gasoline yield and no loss in octane.

Integration of Hydrogen System

The process options for deeper hydrotreating requires more hydrogen consumption. We often find that the supply of hydrogen to the refinery as a whole is a limiting factor. This may lead us to consider adding hydrogen capacity, which can be a significant capital investment.

The supply of hydrogen system is a system-engineering problem, similar to the supply of heat. The refinery has a network of processes that consume and produce hydrogen, similar to the network of heat exchangers in an individual process. The problem of hydrogen management is therefore to optimise the hydrogen network that will allow us to achieve the best overall operation of the refinery. A two-stage approach is developed for analysis and optimisation of refinery hydrogen distribution networks (Towler and Alves, 1997). In the first stage, an economic analysis of the refining margins through use of hydrogen cost composite curves is used to set targets for hydrogen recovery and hydrogen plant production. These targets are then used to guide the optimisation of the hydrogen distribution system. This method has been successfully applied in several industrial cases.

Overall Refinery Optimisation

To improve economics further while using hydrotreating to satisfy the requirement of refinery product quality, interactions between process options, the hydrogen distribution system, utility system and heat recovery systems are exploited. The objective is to shift limiting factors from more expensive places to cheaper places in order to reduce capital investment for making process changes.

Within the manufacturing industry, the economics of petroleum refining are considered to extreme complex. In the oil refining industry, many products are produced simultaneously from one or several feed stocks. The feed and product streams to and from distillation, treating and upgrading units, followed by product blending, resemble a spider web in which the many threads together keep the web in balance. The ability to process different crudes, some of which require segregation, multiplies the number of processing possibilities and thereby the complexity of the problem.

To create order in this chaos of possibilities and limitations, a level-by-level optimisation approach has been proposed (Zhu et al, 1997). A master MILP model is the heart of the overall approach, which captures key variables, and important aspects that are shared by other systems. This master model determines the overall trade off by integrating existing processes, process change options, hydrogen distribution system and the utility system. It also indicates the marginal values of additional processing capacity. This master model communicates with other lower level NLP models through a feedback mechanism to make a final solution compatible with an exiting plant capacity limitations. As a result of applying this level-by-level optimisation approach, limiting factors can be removed with no or little capital investment while product quality can be satisfied.

Conclusions

The trend of improving refinery product quality is to look for hydrotreating options, which requires significant capital investment. This is a significant challenge. To meet this challenge, the refinery industry needs to exploit opportunities in considering overall integration of a refinery in terms of selection of raw material and products, and optimisation of production, process change options, hydrogen distribution system and utility system etc. By doing this, the refinery industry can improve its competitive edge while continue to provide cheaper and cleaner products with better economic margins.

References

Towler, G.P. and Alves, J.J., Analysis and design of refinery hydrogen systems. AIChE Conference, March 12, 1997.

Upson, L.L. and Schnaith, M.W., Low-sulphur specifications causes refinery to look at hydrotreating options. Oil and Gas, Dec. 8, 1997.

Zhu, X.X., Towler, G.P., N. Zhang and J. Zhang, Overall refinery optimisation and debottlenecking, Consortium Meeting of Department of Process Integration, UMIST, 1998.

IMPROVED TECHNOLOGY MANAGEMENT THROUGH PHYSICO-CHEMICAL PROCESS MODELING

Dr. Erik W. Egan
Reaction Design
11436 Sorrento Valley Road, San Diego, CA 92121

Abstract

Today we make more decisions than ever which challenge us to locate or generate the right knowledge, at the right time, organized in a form suitable for managing technical progress. Physico-chemical modeling, when effectively implemented, improves the knowledge content of decisions and limits the risk of technology management error. Effective implementation requires four critical elements. First, a hybrid approach must be created which integrates modeling capability with empirical methods to leverage the synergy between the two for maximum value. Second, the hybrid capability must be accessible directly by technology managers, developers and practitioners to facilitate sharing of "common" knowledge. Third, the uncertainty associated with assumptions and inputs to the empirical and modeling components must be quantified to improve a user's ability to direct resources to the most "uncertain" topics and to integrate risk assessment into their decisions. Fourth, and most important, this hybrid approach must be built into the technology plan through all of its phases, from evaluation to production. A prototype structure will be described and examples of partial use of the approach in the microelectronics industry will be described. Observations of how full implementation would improve the return on investment will also be noted.

Keywords

Chemical process, Chemical reaction, Chemical kinetics, Modeling, Simulation, Software, Microelectronics, Uncertainty, Technology management, Decision-making.

Introduction

For their 75th anniversary edition in 1997, the *Harvard Business Review* asked five prominent thinkers and futurists to forecast the business challenges heading into the next century. In response, Peter Drucker (1997) states that "the only [quantitative] advantage of the developed countries is in the supply of knowledge workers." He goes on that "knowledge is different from all other kinds of resources. It constantly makes itself obsolete ... and the knowledge that matters is subject to rapid and abrupt shifts. The productivity of knowledge and knowledge workers ... is likely to become the decisive factor."

In the face of such rapid change in a knowledge-based economy, Peter Senge (1997) concludes, "successful organizations are building competitive advantage through continually creating and sharing new knowledge." Based on extensive research at the MIT Organizational Learning Center, "companies are finding that enduring institutional learning arises only from three interrelated activities: research, capacity building and practice." The combination of "generalizeable theory and method" and extension of people's knowledge and skills, plus the ability to integrate these with practical, every-day processes, is critical to remedying "the deep systemic problems that afflict our institutions."

Drucker also makes the distinction that "knowledge makes resources mobile. Knowledge workers, unlike manufacturing workers, own the means of production: they carry their knowledge in their heads and therefore can take it with them." Thus, not only must organizations accomplish Senge's integration of individual learning with organizational methodologies to succeed, but they must develop a memory and knowledge-base that is immune to the loss of key knowledge workers.

Given the importance of knowledge to organizations and their competitiveness, it is worthwhile to examine what knowledge is, how we create it, how we can use it effectively, and how we can keep it. Knowledge is clearly not simply a pool of *data* or a collection of *information*, though both are critical components of it. *Data*, of which most organizations have reams, only becomes useful *information* when it is provided a context, or a question is asked of it. *Knowledge* is ultimately gained only when collections of information are assembled into a structure, or interpreted in terms of a *model*.

The creation of knowledge involves a vital interaction of data and information with descriptions of how the world works. We start by collecting data and asking questions of it to establish a set of information about a system. One or more models are then formulated to explain these observations and to make conclusions about how the system is operating, what the controlling factors are, and how the system might behave in the future or under alternative scenarios. Such models may range from mental versions to rigorously derived mathematical descriptions based on established physical, chemical, biological, economical or sociological principles. The more powerful formal models often capture enough detail to accurately predict system behavior before it happens.

Today, many individuals and organizations are discovering that they no longer have sufficiently sophisticated models to deal with faster changing times. Paul Saffo (1997), a director of the Institute for the Future in California, cogently observes that "all our innovations have left us afloat in a growing sea of information, which we must navigate with tools which are far from being up to the task. We don't even fully appreciate our predicament, wrongly labeling it 'information overload' when it is not a consequence of the amount of information confronting us but rather of the gap between the volume of information and the effectiveness of the sense-making tools that technology has built for us." In short, we cannot effectively use all accessible information because we lack the models to do so.

This growing lack of good sense-making tools for creating knowledge should be a great concern for organizations of all sizes and types. With such models, we can proceed to make new observations about how the world works and use them as a basis for making decisions that further our particular technical, sociological, political or business cause. Even when models are not able to make highly precise or decision-worthy predictions, they always are useful for focusing further data and information gathering efforts, or as guides for experimental investigation. Once we have sufficient confidence in a mental or formal model, we commonly use them to make decisions. Businesses routinely rely on such models as tools of the trade in making their most important financial and marketing decisions. An important attribute of these models is their permanence within the organization. A finance or marketing manager or executive may leave the organization, but the knowledge base remains intact with the model they leave behind. Codifying knowledge in a tangible, formal form also has the advantage of providing as excellent training tools for all members of an organization.

Thus, models are fundamental to forming the knowledge base that Drucker contends we must have to compete in the next century. When integrated with technology and business processes, they can be quickly adapted to address "rapid and abrupt shifts" in knowledge requirements when they occur, and they provide an excellent medium for sharing information and providing Senge's link from theory and method to practice. When employed effectively, they improve the productivity of the knowledge worker resource base. When used consistently, models act as Saffo's "sense-making tools" required to sift through and evaluate the tremendous volume of data and information available to us today. And finally, models provide an organization with a dynamic core knowledge base largely immune to organizational changes.

The Microelectronics Industry as Microcosm

In today's knowledge-dominated high-technology industries, knowledge requirements and decision-making demands are more frequently exceeding the individual's *mental modeling* capabilities. The microelectronics industry represents a highly visible example where technology, product and market evolution are changing at a tremendous rate and reaching new levels of complexity. Like industries, microelectronics companies have worldwide competitors filling every role from basic technology research to foundry-based manufacturing. Likewise, they have markets growing on a seemingly boundless number of fronts requiring products for computing, networking, communications, industrial process monitoring and control, automotive systems, aerospace systems, satellite and space systems, defense applications and entertainment, to name several. These technologies dominate most of our lives, and yet they are only beginning to reach the total available market including China, Southeast Asia, and South America.

Added to this product-mix/market complexity is an unprecedented simultaneous confluence of technology challenges (see Hutcheson and Hutcheson, 1997). The well-known historical drive for smaller device dimensions is now being pursued in combination with a move to much larger wafers (from 200 mm to 300 mm), hyper-integration of dissimilar technologies onto the same chip, and the introduction of radically different materials systems. Furthermore, as the semiconductor industry consortium SEMATECH forecast in 1995, the historical 25-30% reduction in cost per function per year, a hallmark of the industry, is in jeopardy (Singer, 1996). To compensate for lower expected cost reduction rates due to yield improvement and larger wafer size, overall equipment effectiveness must improve from the historical 7-10% cost reduction per year rate to 9-15% per year. The net result is that not only must many new processing challenges be taken on at once, but the rate of improvement must also increase dramatically.

Recent changes in on-chip interconnect technology highlight this escalation in compounding factors. In the late 1980's, the industry began a major shift from dry-etch-based planarization techniques to a wet, chemical-mechanical polishing (CMP) process, and is now pushing CMP application to features below the current 0.25 µm minimum size (DeJule, 1997). While this technology was still being learned and fine-tuned experimentally, IBM and Motorola announced last year a shift from aluminum to copper as the metallic conductor of choice for high-performance parts (Singer, 1997; Edelstein et al., 1997; Venkatesan et al., 1997). This shift will soon be followed by a change to so-called low-k dielectrics from today's silicon dioxide (glass) (Bindra, 1998). In this same timeframe, leading-edge production has gone from 0.35 µm minimum feature sizes to 0.25 µm with 0.18 µm in final development today, and 0.15 µm targeted for 2001 (Semiconductor Industry Association, 1997). In stark counterpoint to the drive toward smaller dimensions is a concurrent focus on extending process capability to work on larger, 300 mm wafers (I300I 1998; Seligson, 1998). The International 300 mm Initiative, or I300I, member companies have identified CMP and deposition of copper as technology areas in jeopardy of not having suitable equipment and processes available in time for 300 mm production. Film deposition and plasma processing, including feature etch for interconnect structures, have "marginal" timing for meeting manufacturing needs. Model development to accompany equipment and process work is conspicuously absent from the I300I plan, though some companies have shown the foresight to use them effectively in 300 mm equipment design (Tandon, 1998).

The combination of market and technology demands has also resulted in a reduction in the time between technology generations and an increase in the capital cost of building a new production plant, or "fab". Evidence for the former can be found in successive generations of the Semiconductor Industry Association (SIA) National Technology Roadmap for Semiconductors where the 1994 version targeted the 0.18 µm technology node for 2001 (SIA, 1994), but is now expecting it by 1999 according to the 1997 roadmap (SIA, 1997). Meanwhile, a new 200 mm wafer fab costs over $1.5B today and fabs currently in the planning stages will cost over $2B (Hutcheson and Hutcheson, 1997).

Hence, company decision makers are dealing with more difficult technical frontiers, shorter technology cycles and higher capital costs all at the same time. Despite also having access to more data and information, however, there remain significant knowledge deficits because most companies do not possess or utilize adequate models or sense-making tools.

Materials and Process Modeling Deficiency

The gap is particularly severe in both the materials science and materials processing areas. The rigorous models available for the wide range of processes and materials systems involved in semiconductor manufacturing do not nearly meet the growing needs of the process development and engineering communities. Those that exist are largely limited to use by a small group of modeling experts at a small minority of the semiconductor processing companies and their equipment suppliers. Since these experts rarely have a method for deploying models to their development engineers in a re-usable form, they must continuously support the needs of many groups, each with a different area of emphasis and degree of sophistication. It is still rarer for such models to be used at the capital-intensive production level.

In those infrequent cases where rigorous models are adequate on a process development basis, they are not linked to the device and circuit models used to assess actual integrated circuit performance. Nor are they linked to the factory models used to assess manufacturability, the financial models used to analyze equipment cost of ownership, or the yield models which are used to estimate production costs, product pricing and profit potential.

As a result, semiconductor process technology development rests almost entirely on the talent and frequently heroic efforts of a few key process and device engineers. Their experience and mental models are applied in conjunction with process experimentation to arrive at solutions which meet the minimum process specifications. They have no way of knowing whether the process is globally optimized or how robust it is, except by prior experience. Hence, the slow and expensive experimental approach is repeated many times every technology generation starting from the last best-known conditions. As a bi-product of this approach, knowledge is accumulated in the process engineers' heads, goes home with them every night, and in some cases leaves for good to be used by the competition.

In the absence of a common knowledge reference point, marketing people, product designers and even some technology development managers can fail to have a good grasp of process capabilities and development requirements. Commitments are often made to build parts that require massive process improvement, forcing process engineers to perform heroically once again. Since they, in turn, cannot communicate in objective terms the limits of process capabilities, a cycle of misinformation results and the problem persists indefinitely.

The changes in on-chip interconnect technology alluded to earlier provide a good example of the impact the confluence of complexity, speed, cost and lack of coherent models has on development efforts. So little is known about the fundamental physics and chemistry governing the CMP process that it is essentially an art-form practiced adequately on the aluminum/oxide material system by a few talented process engineers. Shifting production CMP to softer copper metal will be a painful task made perhaps impossibly difficult when non-rigid, elastic, polymers are added for low-k applications. The lack of industry foresight to develop models for CMP at the same time early experimental work was being conducted has resulted in a major deficiency. Thus, process engineers, whose mental models are overextended today, seek out a model-based understanding which they need *immediately*. Both the process engineer and the modeling expert know that such a request is impractical. Starting a modeling effort so late in the technology development lifetime generally yields model-based knowledge that confirm known experimental results, but rarely lead. In the end, modeling gains the reputation of "solving yesterday's problems tomorrow."

As a result, microelectronics company decision makers must make more choices, more frequently with less relevant knowledge than ever before, with their key technology resources walking in and out of their facilities every day, in some cases never to return. Based on the views of Drucker, Senge and Saffo, this does not promise to be a competitive practice for the future.

Foundations and Emerging, High Growth Process Modeling Opportunities

The stated purpose of this conference is to focus on the *foundations* of computing and process operations. In the economy of the next century, the models that provide the framework for supporting knowledge creation will be the foundation of an organization's competitiveness. Physico-chemical models that yield a deep understanding of the technologies supporting chemical and materials processing businesses are not currently adequate to sustain such a competitive knowledge base. Fortunately, many necessary components are emerging from the proof-of-concept stage to be actively used at the industrial level.

Several prevailing conditions will make it possible to supply development and manufacturing operations with the sense-making tools necessary for improved decision-making. First is the well-known, well-documented rate of growth in computing power available on an affordable basis (Hutcheson and Hutcheson, 1997). Raw computing speeds and memory densities are doubling about every 18 months for the same price. This means that computing costs are actually halving every 18 months when measured on a per-function basis. Second, the greatly improved speed and breadth of access to on-line data via company intranets, the internet and the world-wide-web make it possible to share data and information virtually anywhere, anytime on demand. Third, methods are starting to emerge for organizing, processing and assessing the quality of data and information for more comprehensive knowledge generation and evaluation. Fourth, distributed multi-processor architectures and associated distributed software application methods are providing the basic tools for creating fast, accessible software solutions. Finally, computational methods and algorithms are evolving to take advantage of the improved performance to yield more physically and chemically rigorous descriptions of fundamental processes.

Thus, the need to build a model-based knowledge-generation environment and the precursors for producing it are converging simultaneously. To take advantage of the opportunity, however, a number of major elements must yet be established to create the desired comprehensive software solution. First, advanced data analysis and data mining methods must be established to extract decision-worthy knowledge from the sea of available, but disorganized, data and information. The required effort to establish application-specific knowledge bases is probably best suited to a combination of information management specialists, database designers, programmers, and, perhaps most importantly, librarians who can bring to bear successful, tested methods and models for organizing and retrieving information.

Four other significant elements are ideally suited to development by chemical and mechanical engineers, material scientists, computational chemists, database and applications programmers. The first of these is a reactive process modeling (RPM) system which integrates rigorous modeling capability with empirical methods. Since synergy between experiment and modeling always yield maximum value, the design should accommodate data and information from outside sources and be able to generate model-based results internally. The hybrid system will therefore require a collection of tools for extracting data and information from databases dedicated to specific application areas, for analyzing and interpreting experimental data, for generating physical and chemical parameters from first principles, and for modeling target physico-chemical processes.

Once developed, this hybrid capability must satisfy the second requirement that it be accessible directly by technology managers, developers and practitioners. Knowledge leads to generation of new data and information, and forms the basis for decisions. Therefore, those responsible for creating the knowledge must be able to communicate it to the planners and decision-makers in a common language. To accomplish this, an intranet-accessible user environment must be implemented. The knowledge base must be dynamically updateable with built-in validation and verification capability, must provide interfaces to facilitate delivery of knowledge for decisions on technical as well as business issues, and must contain the hooks necessary for linking to enterprise software tools used for financial and market analysis.

The third component is critical to making the knowledge-base a decision-worthy sense-making tool. All RPM capabilities must be coupled to an uncertainty evaluation and decision support functionality that will provide confidence levels on acquired knowledge and help identify problem areas where more work is required to refine conclusions. This capability would provide technology development managers and engineers at all levels with tools for guiding resource deployment, conducting risk assessment and developing decision methodologies consistent with the demands of the 21st century knowledge-based economy. When the capability is inevitably linked to other technical and non-technical modeling tools as noted above, integrating decision-making across strategic planning, contingency planning, competitive analysis, market analysis, and operations planning and scheduling becomes a realistic goal.

The final and most difficult element to implement is the integration of the hybrid model-based approach into traditional technology planning and development strategies. A major shift in organizational philosophy will have to occur for rigorous model development to be included as part of standard business operations. Without such a change, the resources to establish the industrial capability and make it part of industry best practice will not be made available. To reach its true value potential, model-based knowledge must cease to be viewed as a magic bullet employed as a last resort and become visible as an active part of the technology and business planning, development and implementation process.

Reactive Process Modeling (RPM) Framework Design

Neglecting for the moment the significant missionary work required for its adoption, the clear emerging opportunity is to systematically establish an RPM framework. Fig. 1 shows a hypothetical RPM design for general chemical and material processing applications. The key architectural levels are the *model manager*, the application-specific *database*, *uncertainty analysis*, and a collection of physico-chemical and data analysis tools. In principle, any chemically reactive process can be modeled from the most fundamental levels of the chemical mechanism all the way to the most production-oriented implementation of a model for use in closed-loop process control. The green boxes in Fig. 1 indicate tool capability contained in the system, while yellow ovals reflect intermediate outputs within a comprehensive model.

RPM Model Manager

The *model manager* operates at the outer-most level. By customizing its interface to the outside world, any amount of the underlying functionality and knowledge-base can be made available as needed for expert modeling, process development and optimization, process controller development and implementation, or management of technology development. By implementing modern interface building tools within the framework itself, a user that develops new models or analysis tools can also construct the necessary interface components to satisfy the needs of different potential users.

Once such easy access is established, model results that are generated and validated within the system are then automatically deployed to all users for immediate use. If experimental data collected from actual production tools were automatically fed into the system, analyzed, filtered and added to the database, it too would become instantly accessible to all users for further analysis and knowledge generation. Operationally, the *model manager* can enter the modeling structure at any point to create or extract the desired data, information or knowledge. To support the user in such analysis, data processing and graphics capability are implemented at this level for rapid visual and quantitative evaluation and comparison.

RPM Database

The *database* is built on commercially available software to establish and maintain standard methods, and can be expanded as necessary to handle virtually any data type. Experimental information ranging from fundamental physico-chemical data to process-scale results can be included to span the chemical mechanism generation to process model validation needs. Detailed and reduced mechanisms developed for specific operating conditions or reactor geometries can be stored and retrieved as necessary for technology application. Similarly, any level of model, its inputs and its results can be directly accessed. Since validation data and model data coexist within the database, each automatically updated as progress is made, comparison becomes as transparent as choosing which sets to view graphically. Finally, space is allocated to document the source, history and accuracy of experimental data and model results.

RPM Uncertainty Analysis

Accuracy of such data and information is the purview of the *uncertainty analysis* capability. Developers of chemical reactors and processes depend heavily on the quality of the thermodynamic, transport and kinetic data that are used in models that support design calculations. Performance of designs may also be sensitive to the

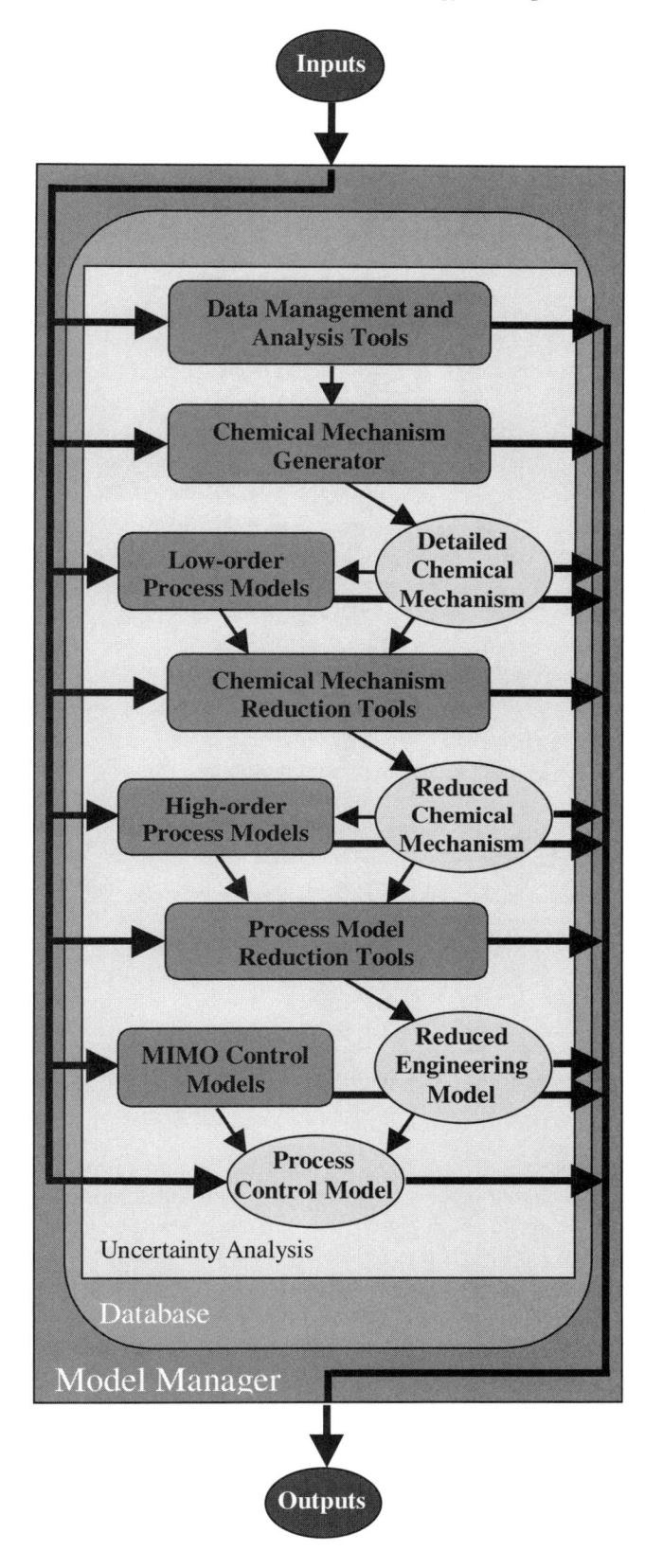

Figure 1. Schematic of hypothetical reactive process modeling (RPM) framework for general chemical processing application.

accuracy with which operating parameters such as temperature, pressure and flow rate can be controlled. In many circumstances, there will also be important dependencies on geometric factors such as design dimensions or manufacturing tolerances. These data are always subject to varying degrees of uncertainty. The basic problem is identifying which of the many sources of uncertainty contribute to uncertainty in performance of the overall processing system. As mathematical models of the underlying processes become more complex, traditional methods of taking uncertainty into account, such as Monte Carlo, become computationally intractable.

McRae and coworkers addressed this problem by developing a new uncertainty analysis method called the Deterministically Equivalent Modeling Method (DEMM; Tatang, 1995; Tatang *et al.*, 1997). DEMM entails the application of polynomial approximation and the method of weighted residuals to systems of equations with stochastic variables. The simplicity of the representation means the uncertainty associated with any number of input variables and output quantities of a model can be described to a high degree of statistical accuracy using just a few parameters each. Hence, all model parameters can have uncertainty data associated with them which can be easily stored in the RPM database for use in all phases of modeling and analysis. Used effectively, the new procedure is useful in identifying suitable experimental efforts, in developing operational control strategies, and in guiding technology planning and execution decisions.

RPM Data Analysis and Modeling Tools

The data analysis and modeling tools themselves provide a hierarchy of capability to accomplish a wide range of integrated tasks. At the purely empirical level, data regression, process-centering methods using experimental design techniques, and functional analysis are all included either by direct implementation or through linkage to other software. Semi-empirical models form the next level of sophistication and include, for example, tools for using Langmuir-Hinshelwood models to extract effective deposition rate expressions from reactor-scale data.

At the most rigorous level of physico-chemical modeling, the tools are dedicated to developing detailed bulk and surface reaction mechanisms, as shown in Fig. 2. The central components are a reaction network generator capable of identifying and evaluating available reaction pathways and low-order process models for validating evolving mechanisms. Experimental data analysis and a collection of computational chemistry tools are then used to deduce thermochemical quantities and transport parameters of reactant and product species. Transition-state algorithms are then applied to compute relevant rate constants. This capability is designed for the expert modeler to establish a sound chemical basis for further advanced process modeling. The mechanism is finally evaluated against appropriate validation data under the guidance of uncertainty analysis.

Once detailed mechanisms are generated, the expert user must then generate one or more reduced versions unique to the specific operating conditions and reactor geometries of interest. This process utilizes the same low-order process models, for example a one-dimensional model of a rotating disk system or a simple boundary-layer model for deposition in a channel flow, to highlight the minimum set of reactions needed to provide a given solution accuracy. A combination of sensitivity-based, rule-based and uncertainty-driven methods provide a range of options which can be applied as needed.

The reduced mechanisms are ideal for combination with detailed, multi-dimensional (high-order) models of a reactor. Typically, these models are built and run using readily available computational fluid dynamics (CFD) codes, though some restrictions may apply for achieving accurate solutions of the process chemistry part of the problem. These tools are suited to handling shear flows, recirculation regions, separating and re-entrant flows, and turbulence. Such models can account for detailed thermal solutions, which may include radiation and heat of reaction effects, and flow and thermal solutions can be coupled to handle buoyancy effects. In some cases the commercial tools are not sufficient for reaching robust solutions of, for example, fully coupled flow-thermal-reaction problems. Special use codes, such as Kiva (Amsden, 1993) for modeling engine combustion or MPSalsa (Salinger et al., 1997) for modeling very stiff deposition problems, are readily incorporated into the system as needed. In addition, the CFD codes do not handle all of the physics of interest to the broad chemical processing community. For example, plasma processes cannot currently be done in CFD codes, though there are specialty tools such as PROTEUS (Meeks and Ting, 1996).

Since the CFD-like models are typically compute intensive and large memory users, they are not ideal for wide-spread, continual re-use in process development and engineering applications. The RPM system therefore includes a set of tools for creating so-called "reduced engineering models." Jensen and coworkers (Banerjee et al., 1998) have demonstrated that multi-dimensional flow and thermal models can be recast in terms of a greatly reduced basis set using orthogonal decomposition and reconstruction techniques. Similar methods have been used by Kevrikides and coworkers (Deane 1991) to describe oscillatory and near-chaotic flows. The results of using such methods are models which remain highly accurate (<5% error relative to their more rigorous "parent") over a large range of process space, but run two or more orders of magnitude faster and require far less memory. These are ideal for wide deployment and use in development, engineering and optimization applications.

These reduced engineering models have also shown promise for extended use in model-based process control schemes. Whether used directly with multiple-input-multiple-output control algorithms, or to generate parameterized models for integration with the algorithms, or to "educate" such algorithms, the models will be useful in developing robust model-based process control for reacting systems. The RPM scheme is intended to accommodate developments in this highly active research area by integrating new methods as they become available. Naturally, the implementation of such control schemes in production would have great financial impact for organizations that work to implement them.

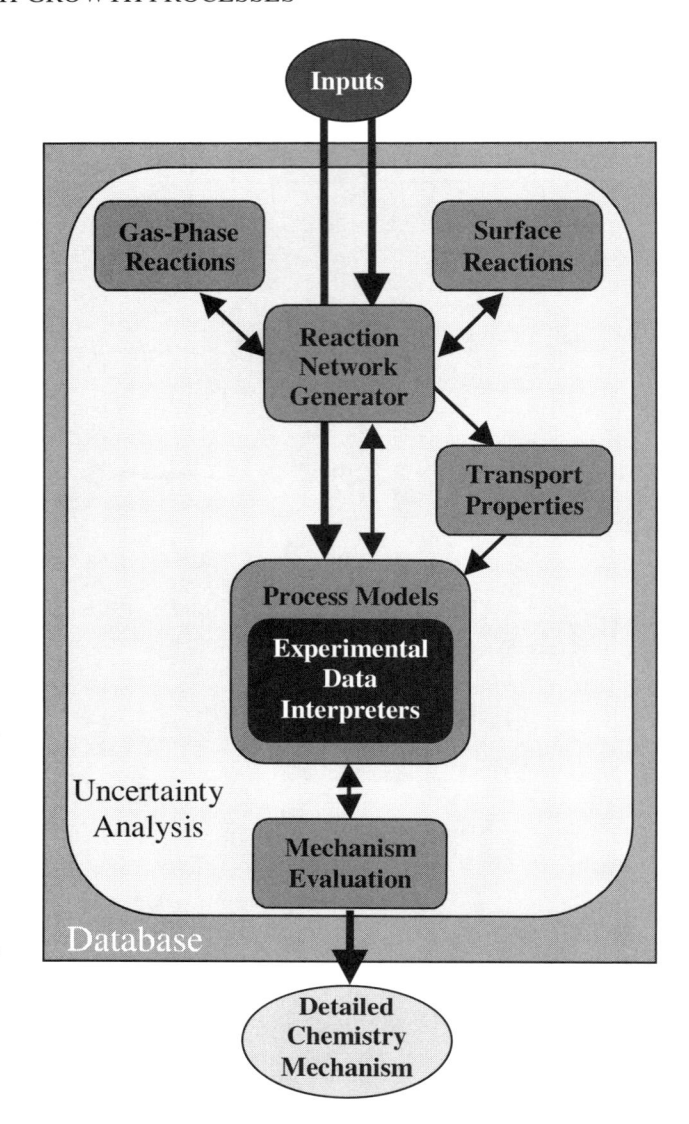

Figure 2. Schematic of proposed chemical mechanism development tool suite.

Extracting Value from RPM

Many organizations will not see the need to spend resources at all levels of the RPM approach, and all will want to introduce the desired capability into their existing methodologies a piece at a time. An important feature of the RPM scheme, therefore, is that models and results can be extracted at any stage of implementation and encapsulated for use by non-experts. For example, significant chemical mechanism information is readily available in the literature today and can be incorporated into the database quite easily. When combined with the simple low-order process models, which are also available today in packages such as The CHEMKIN Collection (Kee et al., 1998), quite useful and powerful process analysis and equipment design support models can be packaged and provided directly to non-expert users.

Frenklach and coworkers have implemented a fine example of this approach under funding from the Gas Research Institute (GRI). Their Java-based prototype is designed for studying natural gas combustion systems using a linkable set of models for perfectly-stirred, plug flow and jet flame reactors, plus mixer and splitter modules for specifying gas stream routings. The system is built on CHEMKIN-III and uses the very detailed GRI-Mech reaction mechanism for natural gas combustion, the most advanced such mechanism yet developed. This modeling environment, the GRI-Mech Calculator 3.0, is shown in Fig. 3 as a screen capture. It is ideal for use by combustor designers and plant engineers to create optimal heating solutions while minimizing unwanted byproducts (e.g., NO_x). By generalizing such capabilities to other chemical systems and models, the more complete RPM design begins to take shape.

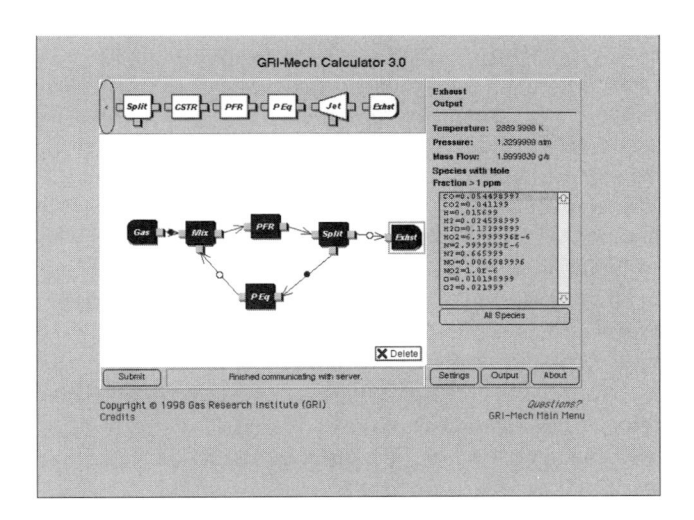

Figure 3. Screen capture of GRI-Mech Calculator 3.0 combustion designers workbench (courtesy of Prof. Michael Frenklach and GRI).

Industry-specific Needs and Customization

The RPM design as presented is generally applicable in all chemical or materials processing organizations. Inevitably, however, industry-based classes of application will require unique capabilities to be built into the system. Also, different organizations will have the opportunity to couple the RPM scheme to other, non-reactive processing models to extend their knowledge-creating and decision-making power to other key technology and business areas.

By way of example, the microelectronics industry has five notable requirements in addition to those encompassed in the "generic" RPM design:

- plasma chemistry and models,
  so-called feature-scale models for etching and deposition,
  - optical properties models,
- advanced surface thermal radiation models, and
- mechanical erosion models in the presence of chemical reaction for studying CMP.

The RPM design must be augmented at all levels to accommodate plasma process modeling. The database must be extended to handle ionization and excited state species, reaction data in the form of collision cross-sections, and diagnostic data from various methods including optical emission spectroscopy. The chemical mechanism generator must be capable of computing electron collision cross-sections and extracting, for example, Arrhenius-type expressions based on specified energy distribution functions. The low-order modeling tools must be adapted to handle solution of the electron density, ionization and excited state chemistry, and ion-surface chemical processes. The AURORA application in The CHEMKIN Collection provides an example of such capability. Mechanism reduction remains largely untouched, although some modification of the rules may be required for rules-based methods. As noted previously, special multi-dimensional plasma modeling codes must be implemented to complement the normally adequate CFD-type capabilities. Subsequent model reduction almost certainly requires a unique approach to take advantage of the special characteristics of plasmas. Once reduced engineering models are generated, however, the process control implementation strategy can be expected to remain essentially the same.

A special attribute of microelectronics and related "patterned surface" material processing is the need to model both etching of materials in lithographically defined areas and the subsequent deposition of other materials into the etched features. For state-of-the-art on-chip interconnect formation, silicon dioxide (SiO_2) dielectric is removed in patterns characterized as lines, trenches and vias. These are then filled with aluminum using either physical vapor deposition (PVD) or chemical vapor deposition (CVD) techniques, though recent advances have substituted copper for aluminum and added electroplating to the metal deposition methods used. Several university groups have developed two-dimensional models for tracking the movement of surfaces under etching and deposition conditions, with the only chemically rigorous implementation being the EVOLVE code (Cale and Mahadev, 1996). Feature-scale models are used stand-alone or in conjunction with low-order and high-order reactor-scale process models to establish a physico-chemical knowledge base for these critical etch and deposition steps.

Since semiconductor devices are built up using sequential layers of various materials with different patterns at each level, wafers in process have complex surface structures producing complex optical properties. Optical properties refer to the emissivity, transmissivity and reflectivity of a material under incident radiation (light) of varying wavelength, intensity and angle of incidence. These properties also depend on the material's temperature. Since many semiconductor processing steps involve high temperatures, absorption and emission of radiation becomes a key component of the heat transfer process. In the case of rapid thermal processing (RTP), the entire operation of the reactor is based on radiative coupling among lamps (e.g., tungsten-halogen) used to heat the system, wafer surfaces and chamber wall materials (e.g., stainless steel and quartz). In such radiation dominated systems, models to predict the optical properties of a wafer with multiple layers of materials and patterns become very important. Hebb and Jensen (1996) have developed a code to handle just such problems involving unpatterned, blanket layer systems. Those capabilities and extensions of it to patterned systems can be added to the RPM system to handle the changing heat transfer conditions in systems where optical properties are evolving due to deposition of new material layers.

As noted above, RTP is a radiation heat transfer dominated process. Hence, fast, highly accurate models of surface-to-surface exchange between specular (polished) and diffuse (rough) surfaces of spectral (range of wavelengths) radiation is a necessary capability. Once again, Jensen and coworkers (Merchant et al., 1996) have led the way with a pair of coupled codes. One is a finite element model for exchange between opaque materials (e.g., in the infrared range), and the other is a Monte Carlo code for modeling exchange with highly transparent materials (e.g., quartz in the visible range). These codes are able to predict temperatures at steady state to within ±4°C at 1000°C-plus operating conditions, and to within approximately ±10°C during transient operation.

The last process unique to microelectronics, CMP, has been highlighted previously as a technology that has missed the opportunity to develop sophisticated models in conjunction with the evolution of process understanding. Nevertheless, when adequate models are developed, they will be welcome additions to the RPM scheme.

Finally, each application area has a set of other models it uses for technology, business, financial, marketing, planning and management. In many cases, the integration of these models to provide a complete picture provides the organization a critical competitive advantage. Shunk (1991) describes how just such value has been successfully derived from integrating factory design and development models with production scheduling, financial and product planning models. Linking the RPM capability to solid state process and device TCAD (technology computer aided design) models, electrical characterization tools, circuit models and design tools, as shown in Fig. 4, could provide the great advantage of full technology system understanding. Further incorporating yield, factory scheduling, production planning, and market and sales forecasting models would provide the technology-to-profits link that is treated using much less rigorous, hence much less knowledgeable and reliable, methods today.

Partial RPM Implementations in Microelectronics

Among the Jensen group at MIT, the Sandia group built and led by Kee until 1996, and a few industrial groups in the U.S., a number of examples have been generated that demonstrate how elements of the RPM system can be successfully used to address microelectronics industry problems. The four cases discussed here have been chosen because they highlight different aspects of the system, or different levels and types of complexity encountered in typical semiconductor manufacturing problems.

PVD Modeling at Motorola

Physical vapor deposition (PVD) is among the processes most amenable to modeling for several reasons. First, the chemistry of the system is quite simple. As shown in Fig. 5 the geometry consists of a cylindrical stainless steel chamber in which an argon plasma is created in the near vicinity of a target of source material. The ions formed in the plasma sputter the source material, for example titanium (Ti), which then traverse the chamber to deposit on reactor walls or the wafer. The only chemistry occurring under these conditions are sticking events on surfaces, describable entirely in terms of sticking factors. The chemistry is more complex only under so-called reactive sputtering conditions. In this case, the chamber also contains, for example, nitrogen which is free to react with the sputtered Ti atoms to form TiN films on the deposition surfaces. Reactive sputtering can be taken one step further by making the target an alloy, of silicon (Si) and metal for example, with a chosen stoichiometry. Again using nitrogen gas, a ternary TiSiN alloy can be deposited. Though simple, correctly representing the chemistry is important to accurately describe the uniformity of deposits, and in the case of alloy materials, the stoichiometry of the films.

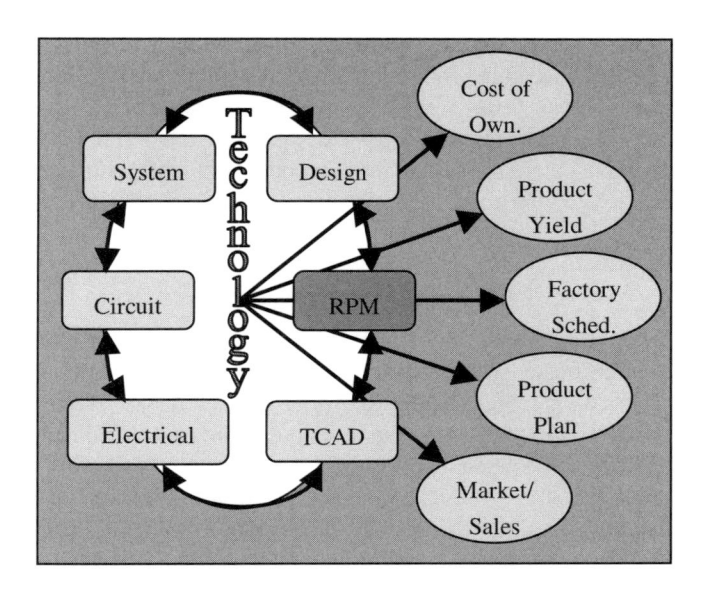

Figure 4. Extended integration of RPM scheme to include related technical, business and operational models.

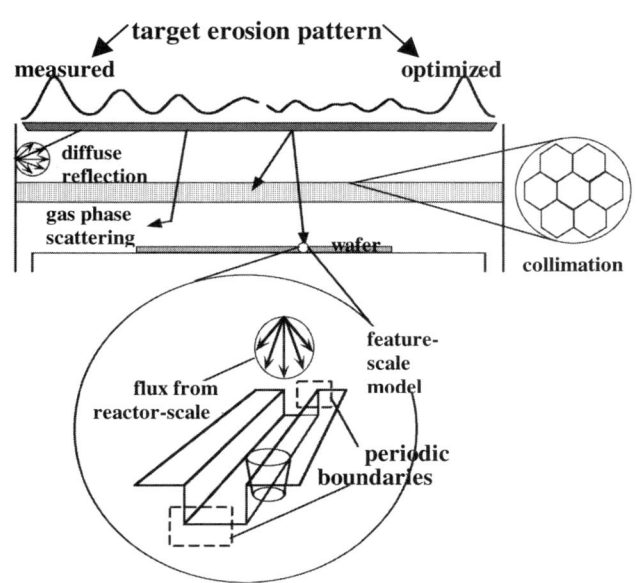

Figure 5. Reactor-scale and feature-scale domains captured in the Motorola PVD model (from Coronell et al., 1998).

Coronell et al. (1998) developed an integrated reactor-scale and feature-scale model for predicting cross-wafer deposition uniformity and within feature coverage under such PVD conditions. Since the PVD process is operated at very low pressures, mean free paths are long and metal atoms undergo very few collisions. It is, therefore, appropriate to model the transport of atoms from the target to other surfaces using Monte Carlo techniques. The Motorola model tracks statistically large numbers of atom trajectories and includes gas phase and surface collision models. Surface interactions include reflections, ranging from fully diffuse to fully specular, and deposition events using a reactive sticking probability formulation. Since collimators are frequently employed to remove species with trajectories not sufficiently normal to the wafer surface, a highly efficient model was implemented to account for collimation effects.

The plasma which produces the actual sputtering of the target is not accounted for in the model. Instead, erosion profiles of used targets are measured and introduced directly as empirical input. The model then calculates what the sputter profile must be to arrive at the final, used profile starting from a new, flat target. Since reactor manufacturers, and to some extent process engineers using the tool, can alter the plasma density profile to change the sputter profile of the target, the Motorola model was also provided with an optimization capability to work backward from an ideal deposition result to the required erosion profile. Example measured and optimized erosion profiles are indicated in Fig. 5.

The feature-scale component of the model is conceptually identical to that for the reactor-scale, but the gas-phase collisions are neglected since the species' mean free paths are much larger than the feature dimensions. The flux at a given feature position is provided as input from the reactor-scale model and surface interactions are tracked until either the atom sticks (deposits) or exits the feature altogether after one or more reflections. Sticking events are counted to insure that statistically significant numbers of atoms are deposited over all lengths of the feature.

The overall model runs on PC-class machines and is ideal for use in studying conventional Ti sputter deposition where effects of collimation, dimensions of the reactor chamber, or the erosion profile of the target are important. As indicated in Table 1, the model shows reasonable agreement with experimental data, though the sidewall profile is flatter than that seen experimentally. Note that both experiment and model show that the bottom coverage is low and the bottom sidewall coverage very low. One of the primary challenges for PVD engineers is devising ways of improving the coverage in these two areas.

In terms of the overall RPM implementation, the Motorola tool would be a "high-order process model." The chemistry requirements are minimal, as noted, though extensions to TiN or TiSiN would require more reaction data to be provided and probably computational chemistry work to evaluate gas-phase and surface reaction probabilities. Since surface processes such as diffusion and re-emission can be important factors, modeling of the surface physics and chemistry fundamentals would also be warranted.

This model has seen extensive use within Motorola, both by modeling experts there and by their process engineering customers. New targets, materials systems and operating conditions are nearly always tested using the model prior to any experimental wafers being run. Because the model is simple to use and extremely fast, it is common for engineers to use the model to guide optimization efforts and test design-of-experiment conditions before each run. Note that in spite of the model being considered "high-order" in terms of spatial detail, its wide application demonstrates that non-expert access is often possible without the model reduction step indicated in Fig. 1.

Table 1. Experiment and Model Results for Ti Sputter Deposition (from Coronell et al., 1998).

Surface Position	Experiment	Simulation
Sidewall - Top	0.60	0.50
Sidewall - Middle	0.31	0.27
Sidewall - Bottom	0.09	0.11
Bottom - Middle	0.35	0.37

Beyond the barrier layer applications described above, process engineers and their model suppliers currently have a particular interest in studying ionized metal plasma (IMP) PVD, copper deposition, and advanced barrier layer materials. In the IMP process, the plasma is modified so that it not only sputters atoms from the target, but also subsequently ionizes a fraction of them on their way to the wafer. Since ions are accelerated normal to all surfaces by the electrically polarized plasma sheath, IMP is "self-collimating" and so yields better bottom coverage than conventional PVD. Since the ions also have significant kinetic energies, they have a sputtering effect on wafer surfaces which are close to normal to the incoming ions. Hence, feature bottoms are preferentially sputtered, with most of the removed material re-depositing on the sidewalls, once again putting material where it is most needed for improved coverage. Since the details of this process are not well understood, models are actively being developed to drive its effective development and use, and to aid decisions regarding how far it can be pushed before it fails to meet device specifications.

The accuracy of PVD models in general, and IMP models in particular, are highly dependent on surface physico-chemical processes. Nucleation, film structure evolution and re-sputtering are microscopic phenomena technologically important by themselves, and they contribute to the ultimate mesoscopic and macroscopic film character process engineers struggle to optimize. To understand the fundamentals of these controlling phenomena better, the Motorola group has teamed with a surface physics group at Los Alamos National Labs (LANL). An initial publication from this work is expected later this year.

CVD Reaction Mechanism Generation and Process Modeling at MIT

Though detailed descriptions of the chemistry are not a significant barrier to accurate PVD modeling, they are the single largest barrier to wider application of reactive process modeling tools in the microelectronics industry. Though growing in number, the cases for which quantitative chemical mechanisms are known still comprise only a fraction of the important systems in use today. Furthermore, with the aforementioned introduction of new material systems and associated chemistries, the need to provide reaction models continually increases. It is widely accepted that fundamental experimental investigation of all of these systems is impractical. As a result, computational chemistry methods, such as those being employed by the Motorola/LANL team, are growing rapidly in both capability and visibility.

The current state-of-the-art for mechanism building as applied to chemical vapor deposition (CVD) processes is best represented by the work of Jensen and coworkers at MIT. They have studied the available capabilities, developed methodologies for establishing mechanisms and applied them to several chemical systems, including germanium deposition from germane (Simka et al., 1996), silicon deposition from dichlorosilane and in-situ boron doped polysilicon from silane and diborane (Jensen et al., 1996), and aluminum deposition from dimethyl-aluminum hydride (Willis and Jensen, 1998).

The dichlorosilane (DCS) case is a good example of a chemistry which has been in wide use for many years, but has recently been less successful in meeting either more stringent demands on old applications (e.g., moving to 300 mm wafers) or targets for new processes. Though there had been empirically derived mechanisms (Oh and Takoudis, 1991; Hierlemann et al., 1995), these were not sufficient to describe process behavior under all conditions of interest.

Futerko and Jensen's treatment of the DCS system takes advantage of both computational and experimental data to establish a more detailed and robust mechanism. The gas-phase portion is generated using a combination of available experimental values and application of computational chemistry methods. For the latter, *ab initio* calculations are done to establish thermodynamic properties of proposed reactants and products. Once these heats and entropies of formation have been established, transition state theory is used to compute Arrhenius form rate constants in the high-pressure limit. Further computations using RRKM and QRRK methods yield necessary pressure dependence of reactions and molecular stabilization rates. The computed rate constants compare favorably to the few available experimental values. Jensen and coworkers have observed that though the thermochemical calculations are generally different from experimental data, often by many kilocalories, the errors are systematic and tend to cancel one another when used to compute rate constants. Hence, the computed rate constants have proven to be quite reliable.

A key conclusion resulting from applying these methods to DCS decomposition is the determination of what decomposition products result. Researchers had speculated that decomposition to either $SiCl_2$ or $SiHCl$ would be the important pathway for ultimate silicon deposition. The MIT approach indicates that the preferred initial path is:

$$SiH_2Cl_2 \rightarrow HSiCl + HCl \quad (1)$$

as is demonstrated in Fig. 6a, which shows the concentration profiles for stagnation flow of 4% DCS in H_2 on a 1000°C non-reacting surface. The production of $SiCl_2$ is two orders of magnitude lower than of $SiHCl$.

To complete the analysis of the deposition system, the all-important surface chemistry must be included. Since the number of possible surface configurations is so large and usually unknown, universally reliable computational surface chemistry methods are not yet, and may never be, available. Nevertheless, combinations of *ab initio*, molecular dynamics and Monte Carlo methods have been used to successfully describe surface processes and to extract rates. When these are used carefully in conjunction with available experimental values, quite accurate representations of the elementary surface chemistry can be obtained.

For the DCS problem, Futerko and Jensen used a competitive site balance model assuming that highly reactive radicals will stick to the surface with unit probability (the so-called collision limit). They added experimental values for surface reactions involving H and Cl and temperature programmed desorption (TPD) data for reaction products leaving the surface. Finally, the one remaining unknown parameter, the sticking coefficient of DCS itself, was fit to available deposition data. As noted above, it would also be possible to obtain this rate using numerical methods given sufficient computational resources.

Futerko and Jensen demonstrated that the complete mechanism provides agreement with data to within a few percent under most conditions, and this result was reproduced by this author using the CHEMKIN application known as SPIN (Kee et al., 1998) under the conditions shown in Fig 6. This code fits into the "low-order process model" category of the RPM structure. The portability of the mechanism, both between users and among modeling software packages, is an important feature which highlights the inherent value of high-quality information.

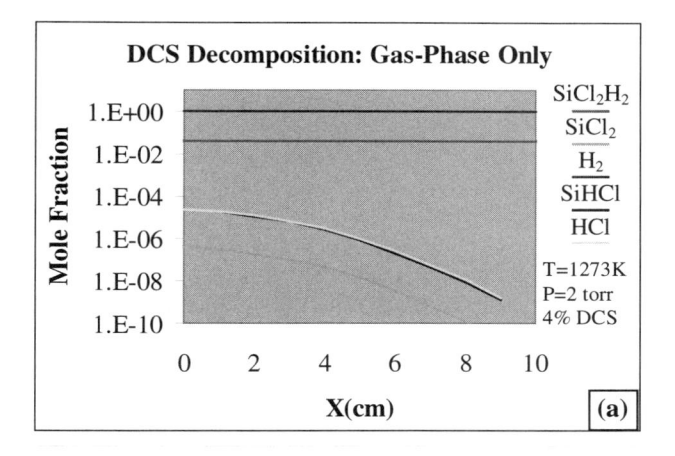

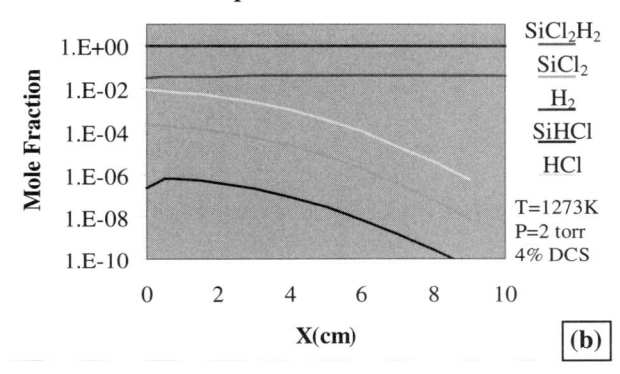

Figure 6. Concentration profiles of dichlorosilane (DCS) under stagnation flow conditions using the Futerko and Jensen mechanism in SPIN (surface is at x=0).

In addition to accurately predicting deposition rate, the models also provide key information about which reactions control the deposition process. The mechanism clearly shows that net deposition occurs via DCS dissociative adsorption to $SiCl_2$, followed by further surface decomposition to incorporate Si and release HCl. Since $SiCl_2$ can also desorb, the surface acts to catalyze the DCS to $SiCl_2$ dissociation which is quite slow in the gas phase. This process is evident in the large amount of $SiCl_2$ observed in Fig. 6b showing the concentration profiles for the same conditions as in Fig. 6a, but on a reacting surface. What is not obvious is that the HCl and $SiCl_2$ produced by these paths essentially reach an equilibrium maintained by a set of reversible deposition-etch reactions:

$$SiCl_2 + S\cdot \leftrightarrow SiCl_2^* \quad (2)$$

$$SiCl_2^* + H^* \leftrightarrow HCl + Cl^* + S\cdot + Si(b) \quad (3)$$

$$H* + Cl* \leftrightarrow HCl + 2S\cdot \qquad (4)$$

where $S\cdot$ represents an open surface site, * implies a surface adsorbed species, and $Si(b)$ is deposited material. Though it had long been concluded that overall deposition from DCS was actually a balance between deposition and etching processes, this mechanism described in detail how the process occurs. Given this knowledge, a process engineer could explore methods to improve deposition rate, uniformity or conformality over features by shifting the deposition-etch balance one way or the other using, for example, additives to scavenge Cl* from the surface to reduce the etching process. Alternatively, adding HCl to the feed gas enhances etching to lower the net deposition rate, a method discovered independently by process engineers through trial and error.

Plasma Etch Modeling at Sandia and Reaction Design

As complex and time intensive as CVD chemical mechanism development is, it is easy compared to establishing reasonable plasma chemistry descriptions. Many factors contribute to the added difficulty. First, in the gas-phase the chemistry is considerably more complex. It is dominated by species interactions with high-energy electrons which produce a "zoo" of ionic and excited state species, many of which are critical process intermediates. To represent the system in this case, two temperatures must be used for computing kinetic rates, one the familiar thermal neutral temperature, the other associated with electron energies. Electron-related reactions require electron collision cross-section data that must be generated either experimentally or computationally as a function of electron energy. Furthermore, since the number of radical, ionic and excited state species created in a plasma is much larger than for CVD systems, identifying key reactions and species becomes significantly more difficult.

Second, the already difficult surface chemistry problem becomes tremendously harder. Mechanisms must capture the chemistries of many more highly surface reactive radicals, and then add in the effects of energetic ions impacting the surface. Ions can sputter material from surfaces, as they do from a PVD target, or merely transfer energy to surface species to enhance surface reaction rates. In multi-dimensional models, even the angle of ion impact must be accounted for to accurately describe the surface kinetics. Since most plasma processes operate at low pressures, surface-to-surface transport is quite rapid. Hence, plasma-surface chemistries for all surfaces, wafer and walls alike, must be accounted for. Finally, the compositions of non-wafer surfaces are not well-known or they change substantially over time so that characterization of the deposition and etch mechanisms becomes extremely difficult.

Useful plasma mechanisms have nevertheless been successfully developed, most notably by the Sandia group led by Meeks and Ho, by Sawin and colleagues at MIT, and by Graves and colleagues at U.C. Berkeley. In their most recent work on aluminum etching using the $BCl_3/Cl_2/Ar$ plasma system, Meeks et al. (1998) started with previously established chemistry for the Ar and Cl_2/Cl systems. In the gas-phase, they added BCl_3-related kinetics using some known values from experiment and computation, plus several estimates derived from knowledge of similar systems and available thermochemistry data. They completed their gas-phase description by incorporating experimental values for ionic species including Cl^+, BCl^+, Ar^+, and Cl^-.

Owing to the complexity of the system, development of plasma surface mechanisms involves significant educated-guess work combined with expert interpretation of macroscopic experimental data. In the case of the $BCl_3/Cl_2/Ar$ system, spontaneous etching of Al by Cl and Cl2 was assumed, and it was proposed that a 40% enhancement to etching would result from energetic ion-surface interactions. Finally, the Meeks-led team analyzed the mechanism in the presence of available data to understand the surface competition processes and to identify the rate limiting steps. This knowledge led, in turn, to refined values for the key kinetic parameters.

All of this analysis was conducted within a zero-dimensional plasma modeling environment using the CHEMKIN-based application AURORA. This model was further used to predict the characteristics of an actual commercial plasma reactor running the $BCl_3/Cl_2/Ar$ chemistry, the results of which are shown in Fig. 7. The mechanism was then implemented in a two-dimensional model of the same commercial reactor using the Sandia code PROTEUS (Meeks and Ting, 1996). Etch-rate was predicted to within 10% of experimental data from the reactor, and etch uniformity was predicted to within a few percent. Owing to the proprietary nature of these results, it is not possible to present them here.

Once again, the portability of the mechanism to models of other reactors and other software packages represents a significant value in creating application-specific knowledge. Process engineers are able to use either the very fast AURORA-type model to locate and optimize operating points on the reactor they are using, or they can exercise multi-dimensional models to address wafer-scale problems in meeting process targets. When these models are linked with feature-scale models within the RPM structure, these same engineers are further able to connect the variables they can control to the technology outputs they seek: trench and via structures with very specific shapes. The inherently rich information contained in the physico-chemically models provides not only the simple answer to the "what process conditions should I use" question, but provides the "why it works" answer as well.

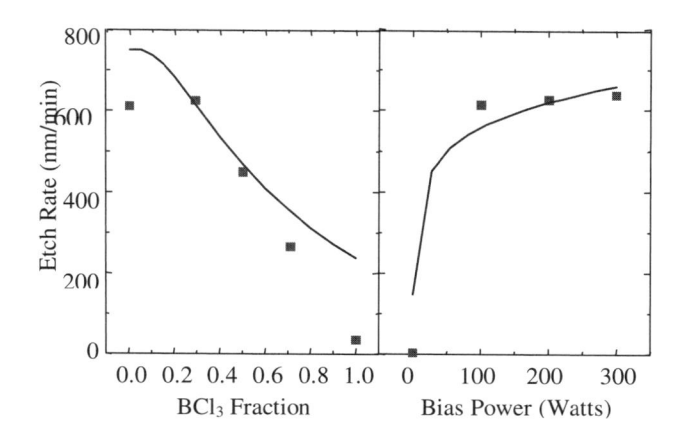

Figure 7. Agreement AURORA model predictions and experiment for etch rate versus key plasma reactor operating parameters using BCl₃/Cl₂/Ar chemistry of Meeks et al. (1998).

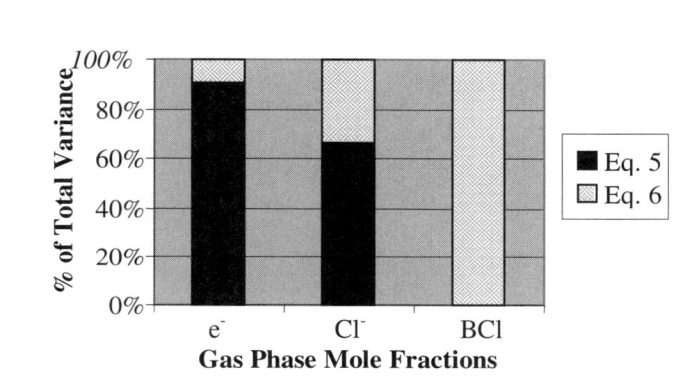

Figure 8. Percent contribution to total variance in predicted mole fractions resulting from ±20% uncertainty in Eq. (5) and (6) rate constants.

The quality and depth of knowledge delivered by the model can be improved by using the DEMM uncertainty analysis methods described above. Klipstein et al. (1997) conducted a preliminary study of the mechanism to analyze the relationship between two of the reaction steps believed to be key to predictive accuracy. While the analysis confirmed the importance of the reactions, it also provided a basis for quantifying the degree to which their rate constants needed to be refined. Fig. 8 shows the impact of ±20% uncertainty in the rate constants for the following reactions:

$$BCl_3 + e^- \rightarrow BCl_2 + Cl^- \quad (5)$$

$$BCl_2 + e^- \rightarrow BCl + Cl^- + e^- \quad (6)$$

Sensitivity analysis alone indicates that these reactions are important, but the uncertainty analysis quantifies the accumulated error in key outputs. In Fig. 8, it is clear that improvement in the confidence of predicted electron density and Cl⁻ concentration is dependent on both reactions, but that time spent improving the Eq. 5 rate constant will lead to less uncertainty in the results and better decision-making potential. Conversely, Eq. 6 is clearly the place to spend effort to improve confidence BCl concentration predictions. Since the DEMM analysis can also link these variables to ultimate process outputs like etch rate, it can aid process engineers and model developers in focusing their efforts. Furthermore, it gives the process engineer insight into how reliable the model will be as a decision-making guide.

Summary of Examples: Partial RPM Capability

Many other examples like these are known which demonstrate the value of such partial implementations of the RPM capability. Unfortunately, many of the most convincing examples are proprietary applications in specific industrial equipment or process development efforts, and are therefore not available for general publication, presentation and discussion. Nevertheless, the cases presented provide examples of how chemical mechanism generation, low-order process models and high-order process models can be profitably applied to create industrially useful knowledge. Given the projected growth and emergence of a knowledge-based economy, those who can most easily generate and access this kind of physico-chemical knowledge will be in the best position to make the decisions necessary to win out over their competition. Easy generation and access will only be accomplished through an integrated reactive process modeling approach such as that described conceptually above.

Conclusion

The key points of discussion are summarized as follows:

1. Knowledge is the dominant commodity in the economy of the future, and it is mobile.
2. More reliable knowledge leads to better decisions which improve technology and business plans and efficiency of operations.
3. Models, mental, empirical or rigorous, must exist to analyze data and information to create reliable, decision-worthy knowledge.
4. The increasing speed and complexity of technology and business processes are more often overwhelming mental models or exceeding the scope of empirical ones.
5. Rigorous institutional models and knowledge must therefore be created and then evolve continuously to replace the less reliable mental and empirical models.
6. Integrated schemes for physico-chemical modeling will be a part of the most successful institutional modeling processes.

The proposed RPM structure captures the key elements needed to create, disseminate and maintain an organizational resource for technology decision-making. As the presented case studies indicate, partial implementation of a stand-alone scheme provides significant value. However, the full power of physico-chemical modeling cannot be realized without complete implementation of such a scheme and its adoption by the organization as a key component of the institutional model.

Acknowledgements

The author would like to thank Drs. David Klipstein and Ellen Meeks of Reaction Design, Prof. Greg McRae of MIT, and Dr. Dan Coronell of Motorola for support in preparing the manuscript, and Prof. Klavs Jensen of MIT for his many years of contribution to reactive process modeling and his valued mentorship.

References

Amsden, A. A. 1993. *KIVA-3: a KIVA program with block-structured mesh for complex geometries*. Los Alamos, NM: Los Alamos National Laboratory LA-12503-MS.

Banerjee, S., Cole, J. V. and Jensen, K. F. 1998. Nonlinear model reduction strategies for rapid thermal processing systems. *IEEE Transactions on Semiconductor Manufacturing*, 11 (2), 266-275.

Bindra, A. 1998. Low-k dielectrics being characterized for key role in copper interconnects. *Electronic Design*, 46 (9), 35-36.

Cale, T. S. and Mahadev, V. 1996. Feature scale transport and reaction during low-pressure deposition processes. *Thin Films*, 22, 175-276.

Coronell, D.G., Egan, E.W., Hamilton, G., Jain, A., Venkatraman, R. and Weitzman, B., 1998. Monte Carlo simulations of sputter deposition and step coverage of thin films. *Thin Solid Films*, to appear.

Deane, A. E., Kevrikidis, I. G., Karniadakis, G. E., and Orzag, S. A. 1991. Low-dimensional models for complex geometry flows: Application to grooved channels and circular cylinders. *Physics of Fluids A*, 3 (10), 2337-2354.

DeJule, R., 1997. CMP challenges below a quarter micron. *Semiconductor International*, 20 (13), 55-60.

Drucker, P. F., 1997. The future that has already happened. *Harvard Business Review*, 75 (5), 20-24.

Edelstein, D., Heidenreich, J., Goldblatt, R., Cote, W., Uzoh, C., Lustig, N., Roper, P., McDevitt, T., Motsiff, W., Simon, A., Dukovic, J., Wachnik, R., Rathore, H., Schulz, R., Su, L., Luce, S and Slattery, J. 1997. Full copper wiring in a sub-0.25 µm CMOS ULSI technology. *In: International Electron Devices Meeting 1997, Washington, D. C. 7-10 December 1997*. New York: IEEE, 773-776.

Hebb, J. P. and Jensen, K. F. 1996. The effect of multilayer patterns on temperature uniformity during rapid thermal processing. *Journal of the Electrochemical Society*, 143 (3), 1142-1151.

Hierlemann, M., Kersch, A. Werner, C. and Schafer, H. 1995. A gas-phase and surface kinetics model for silicon epitaxial growth with SiH_2Cl_2 in an RTCVD reactor. *Journal of the Electrochemical Society*, 142 (1), 259-266.

Hutcheson, G. D. and Hutcheson, J. D. 1997. Technology and Economics in the Semiconductor Industry. *Scientific American Special Issue: Solid-State Century*, October.

I300I, 1998. *International 300 mm Initiative Presentation*. http://www.sematech.org/public/division/300/mission.htm: SEMATECH

Jensen, K. F., Simka, H., Mihopoulos, T. G., Futerko, P. and Hierlemann, M. 1996. Modeling approaches for rapid thermal chemical vapor deposition: Combining transport phenomena with chemical kinetics. *In F. Roozeboom, ed. Proceedings of NATO Advanced Study Institute*, Acquafredda di Maratea, Italy 3-14 July 1995. Dordrecht: Kluwer Academic Publishing, 305-331.

Kee, R.J., Rupley, F. M, Miller, J. A., Coltrin, M. E., Grcar, J. F., Meeks, E., Moffat, H. K., Lutz, A. E., Dixon-Lewis, G., Smooke, M. D., Warnatz, J., Larson, R. S., Mitchell, R. E., Petzold, L. R., Reynolds, W. C., Caracotsios, M., Stewart, W. E., and Glarborg, P., 1998. *The CHEMKIN Collection*. Version 3.03. San Diego: Reaction Design.

Meeks, E. and Ting, I. 1996. *PROTEUS: Plasma Reactor Overall-Transport Engineering-Use Simulator*. Livermore: Sandia National Laboratories Preliminary Report.

Meeks, E., Ho, P., Ting, A and Buss, R. J. 1998. Simulations of $BCl_3/Cl_2/Ar$ Plasmas with Comparison to Diagnostic Data. *Journal of Vacuum Science and Technology*, to appear July.

Merchant, T. P., Cole, J. V., Knutson, K. L., Hebb, J. P. and Jensen, K. F. 1996. A systematic approach to simulating rapid thermal processing systems. *Journal of the Electrochemical Society*, 143 (6), 2035-2043.

Oh, I.H. and Takoudis, C.G. 1991. Modeling of epitaxial silicon growth from the SiH_2Cl_2-H_2-HCl system in an RF-heated pancake reactor. *Journal of Applied Physics*, 69 (12), 8336-8345.

Saffo, P., 1997. Are you machine wise? *Harvard Business Review*, 75 (5), 28-30.

Salinger, A.G., Shadid, J. N., Hutchinson, S. A., Hennigan, G. L. Devine, K. D. and Moffat, H. K., 1997. *Massively Parallel Computation of 3D Flow and Reactions in Chemical Vapor Deposition Reactors*. Albuquerque: Sandia National Laboratories SAND97-3092.

Seligson, D., 1998. The economics of 300 mm processing. *Semiconductor International*, 21 (1), 52-58.

Semiconductor Industry Association, 1994. *1994 National Technology Roadmap for Semiconductors*. San Jose: SIA.

Semiconductor Industry Association, 1997. *1997 National Technology Roadmap for Semiconductors*. San Jose: SIA.

Senge, P. M., 1997. Communities of leaders and learners. *Harvard Business Review*, 75 (5), 30-32.

Shunk, D. L., 1991. *Integrated Process Design and Development*. Burr Ridge, IL: Irwin Professional Publishing.

Simka, H., Hierlemann, M., Utz, M. and Jensen, K. F. 1996. Computational chemistry predictions of kinetics and major reaction pathways for germane gas-phase reactions. *Journal of the Electrochemical Society*, 143 (8), 2646-2654.

Singer, P., 1996. 1996: A new focus on equipment effectiveness. *Semiconductor International*, 19 (1), 70-74.

Singer, P., 1997. Copper goes mainstream: Low k to follow. *Semiconductor International*, 20 (13), 67-70.

Tandon, S., 1998. Challenges for 300 mm plasma etch system development. *Semiconductor International*, 21 (3), 75-83.

Tatang, M. A., 1995. *Direct Incorporation of Uncertainty in Chemical and Environmental Engineering Systems*. Ph.D. Thesis. Cambridge, MA: Massachusetts Institute of Technology Department of Chemical Engineering

Tatang, M.A., Pan, W., Prinn, R.G. and McRae, G.J., 1997. An efficient method for parametric uncertainty analysis of numerical geophysical models. *Journal of Geophysical Research*, 102 (D18), 21925-21932.

Venkatesan, S., Gelatos, A. V., Hisra, S., Smith, B., Islam, R., Cope, J., Wilson, B., Tuttle, D., Cardwell, R., Anderson, S., Angyal, M., Bajaj, R., Capasso, C., Crabtree, P., Das, S., Farkas, J., Filipiak, S., Fiordalice, B., Freeman, M., Gilbert, P. V., Herrick, M., Jain, A., Kawasaki, H., King, C., Klein, J., Lii, T., Reid, K., Saaranen, T., Simpson, C., Sparks, T., Tsui, P., Venkatraman, R., Watts, D., Weitzman, E. J., Woodruff, R., Yang, I., Bhat, N., Hamilton, G. and Yu, Y. 1997. A high performance 1.8 V, 0.20 µm CMOS technology with copper metallization. *In: International Electron Devices Meeting 1997, Washington, D. C. 7-10 December 1997*. New York: IEEE, 769-772.

Willis, B. G. and Jensen, K.F. 1998. An evaluation of density functional theory and *ab initio* predictions for bridge-bonded aluminum compunds. *Journal of Physical Chemistry A*, 102 (15), 2613-2623.

SESSION SUMMARY: PLANNING AND SCHEDULING

Stephen P. Lombardo
The Coca-Cola Company
Atlanta, GA 30313

Summary

The Coca-Cola Company Enterprise is a complex network of diverse manufacturing and distribution operations. Concentrates and Beverage Bases, in liquid and dry parts forms, are manufactured in the company's 34 Concentrate plants around the globe. These are then sold and shipped to our partner bottlers who convert them to finished bottles and cans as well as syrups for fountain outlets. In the United States, our own plants manufacture syrups. Manufacturing is accomplished through a combination of batch and continuous processing. Key unit operations include mixing/blending, pasteurization, homogenization, and filling. A low level of process automation characterizes the facilities, and many of the tasks are labor and QA intensive. The most costly steps in the supply chain are in distribution since orders are manually assembled for delivery to bottler and retail customers.

Today, scheduling in the Coca-Cola System is primarily a manual operation. Several attempts have been made to implement automated tools that utilize rule/constraint-based approaches. The inconsistent success of these tools has prevented their widespread acceptance for use in everyday operations. Manual scheduling by its very nature leads to an operation, which is make-to-stock rather than JIT or make-to-order. This leads to operations which are characterized by: 1) high inventories, 2) inefficient utilization of capital assets, and 3) increased operating costs for existing operations. Also, The Coca-Cola Company has successfully utilized constraint-based tools for infrastructure planning of bottler bottling/canning and warehouse locations as well as supply logistics from manufacturing operations through warehouse and customer locations. Implementation of new MILP-based tools for infrastructure planning is currently scheduled for 1999.

Inconsistency in implementation successes is not limited to The Coca-Cola Company. High technology industries, such as oil, chemicals, and pharmaceutical have also had varied results in their attempts at automated scheduling and ERP/supply point optimization. There are several contributing factors which have led to this situation. First, many companies inadequately test prospective software tools prior to making purchase commitments. Normally, the primary criterion for selection is ease of use and flexibility for customization. Solution quality is evaluated with oversimplified models due to the cost and time required to solve a real operational problem. Second, all operations have unique features which pose significant complexity in model formulation. The resulting models become quite large, and often require long solution run-times yielding solutions which are operationally infeasible. Third, the lack of internal expertise leads to heavy reliance on external consultants for model development. This leads to an added cost to build, change, and maintain operational models beyond the original software cost.

The search is continuing by companies across all industries for powerful PC-based tools, which are easy-to-use, and can solve high-level and detailed operational problems across the supply chain. At The Coca-Cola Company, there is an effort currently underway to examine the long-term applicability of scheduling & planning tools for a number of different applications. First, to maximize the production efficiency of existing facilities. By increasing the capacity & labor efficiency of these operations, they can be operated in more of a JIT philosophy rather than make-to-stock. This would enable a significant reduction in inventories. Second, to maximize the efficiency of capital expenditures. For expenditures in existing operations, how the new equipment will be integrated into the existing labor constraints is often overlooked. For new facilities which often include new technologies, the scheduling feasibility of the operation to take full advantage of the innovations is not considered in the design. Normally, the new operation is managed at start-up using old paradigms. Third, to quantify cost/benefits of new technologies and systems. In many cases, the major hurdles to implement new technology are that the capital costs are too high as well as verification of benefits since the new technology normally is not operationally proven. Fourth, high level planning to optimize supply point logistics across the supply chain and contingency planning in the event that a link in the normal supply chain is expectant interrupted. Tools which give quick, accurate solutions will enable companies to quickly respond to changing business conditions and generate financial advantage over their competitors.

Today, given the rapid evolution of hardware and software, the level of technology in scheduling and planning is rapidly approaching the point where they will meet the future needs of industry. However, considerable work still needs to be accomplished. Particularly, there are currently no tools that are capable of uniform application to scheduling, design, and planning/supply chain. The implication of this is that separate tools are required with their uniquely built models to solve a particular detailed problem within the supply chain. This creates difficulties within industry to manage interfaces and databases for problems at various levels in the supply chain. The best of all possible worlds would be to have an integrated tool which utilizes a common detailed model to solve the desired problem, and an algorithm which extracts the required data according to the solution objective. The achievement of this goal will require increased collaborations between industry, academic, and supplier communities. By synergizing global research efforts to develop common, realistic test problems to evaluate software and new research developments, the effort will be facilitated and accelerated. Not to mention the fact that this work will be accompanied by the development of a performance database for industry by which to make software purchase decisions. The new criterion will be based on solution of their industry-specific problem rather than ease of use or simplified problems, which have been used historically.

The Scheduling and Planning session at FOCAPO'98 consisted of 3 papers; 2 academic and 1 industrial; representing an international perspective on scheduling and planning. They reviewed the current state of affairs in the scheduling and planning area, from detailed scheduling to high-level ERP, and the future challenges in academic research and application to industry. The first paper is entitled "Synthesizing Enterprise-Wide Optimization with Global Information Technologies : Harmony or Discord?" by Conor McDonald from Dupont. It focuses on the use of a MINLP solution methodology to solve a real-world industrial problem. It identifies as the key challenge for the use of optimization technology in everyday business planning as not the building of the complex models to solve the problem, but the integration of the models into the decision-making processes of the supply chain. The second paper is entitled "Single and Multi-Site Planning and Scheduling: Current Status and Future Challenges" by Nilay Shah from Imperial College. It specifically reviews techniques for production scheduling of individual and multi-site facilities. It identifies shared capacity between competing products, and scheduling under uncertainty as the key research challenges. The third paper is entitled " Towards the Convergence of Theory and Practice: A Technology Guide for Scheduling/Planning Methodology" by Joe Pekny and Rex Reklaitis from Purdue University. It examines the various methodologies which are utilized in the solution of industrial scheduling problems. It rates them according six key features, which includes their ability to accommodate the concept of NP-completeness.

All three papers seem to agree that mathematical programming techniques are the methodology of choice to best solve industrial scheduling and planning, and address the current and future research challenges. However, they also agree that much work is necessary to create software tools which can reliably solve industrial problems, and be utilized as an everyday tool by industry. Research of the next 3-5 years will tell us whether the prophecies of the authors is correct. The opinion of the session chair is that collaboration between academia and industry will be the critical factor which will determine the outcome of this technology debate.

SYNTHESIZING ENTERPRISE-WIDE OPTIMIZATION WITH GLOBAL INFORMATION TECHNOLOGIES: HARMONY OR DISCORD?

Conor M. McDonald
E.I. DuPont de Nemours & Company
Wilmington, DE 19880–0101
Conor.M.McDonald@usa.dupont.com

Abstract

Chemical companies continue to drive improvements in asset productivity for businesses characterized by complicated infrastructures, wide geographical spread, a broad distribution in asset capabilities, and products in varying stages of their life–cycle. In addition, the metrics and goals of business management, finance, manufacturing, research and marketing may be different and can frequently compete with one another. These factors provide compelling motivation to utilize optimization techniques to evaluate the relative economic effects of supply chain decisions.

To this end, an extensive discussion of the employment of optimization for the solution of important classes of problems encountered in practice will be provided. The models are deterministic in nature and traverse the full optimization spectrum. In particular, a specific example of a lot–sizing, asset allocation problem that is formulated as a Mixed Integer Non Linear Programming problem will be provided. This challenging, nonconvex problem is solved using a novel successive approximation approach, and the manner in which the results of the model were used will be discussed.

The two principal modes of utilization of these models will be explored: the first mode amounts to using sophisticated mathematical programming techniques to solve a particular problem instance a supply chain manager or scheduler might encounter. In this case, implementation is relatively straightforward. The second mode involves "live" or ongoing planning and scheduling models, which are fully integrated into the supply chain planning activities of the business – a desired state. In this environment, there remain significant challenges in deploying models for business gain despite the strong motivations noted above and the availability of advanced optimization tools and techniques. The high levels of information technology investment that continue to be made do not necessarily ameliorate the difficulties in integrating optimization technologies into the decision–making process. These challenges, which increase as one descends the supply chain decision hierarchy, are often more daunting than the development of the model itself. Examples from industrial implementations will be used to highlight these points.

Keywords

Supply chain management, Enterprise resource planning, Production planning, Production scheduling, Lot sizing problem, Asset productivity, Data integration.

Introduction

This paper will explore the challenge of employing advanced optimization technologies for the solution of large and complex business supply chain problems, and integrating these methods into the operational decision making process. The chemical industry in general, and DuPont in particular, has very high levels of capital investment. Due to the difficulty faced in obtaining returns on investment that exceed the cost of capital, there is immense pressure to improve asset productivity. Therefore, the main focus in this paper lies in finding ways to utilize these precious manufacturing assets in the most effective manner for maximum generation of profit.

Let us focus our area of investigation. First, large scale continuous chemical processes are the dominant mode of operation in most DuPont businesses. A large fraction of these units must manufacture multiple products and grades of products, and therefore semicontinuous operation is required. This involves expensive transitions between products or grades, and these are a major determinant of financial performance. Even for plants that contain batch units, these are often upstream or downstream of semicontinuous units. It is for this reason that multipurpose batch processes, such as those that might be found in DuPont's Agricultural Products division, are not considered here. Even though this class of problems has received extensive attention in the academic chemical engineering literature, this paper will focus mainly on production planning and scheduling for large semicontinuous units.

One of the most common manufacturing scenarios involves a continuous unit feeding a collection of downstream (semi)continuous or batch processors. Management of critical resources becomes a key issue due to the pressure on reducing operating costs and improving overall financial performance. There are two aspects of the problem. In one case, the downstream units are competing for a critical resource which is the material being produced by the upstream process. In the other, multiple products can be considered to be competing for capacity on one or more of the downstream processes, meaning the downstream operating units are the critical resources.

The models developed use deterministic data and therefore stochastic mathematical programming approaches are not considered. They are typically operational support models, and do not address the issue of design and capacity expansion. As has been stated, DuPont is tightly controlling capital investment, and therefore, the need for such models has been limited.

Drivers for Optimization Technology

It is important to understand the different components of the business organization, and how they interact. A key determinant of behavior is the set of performance metrics chosen by an organization. These can be very different and often conflict. For manufacturing, typical performance measures are yield, uptime and throughput. For finance, the metrics include return on investment, along with standard cash flow and profitability measures. For marketing, they will be market share, product pricing and new product penetration. For research and development, it will be new product portfolio and product development cycle time. For business management, it will be shareholder return, as well as the plethora of standard business measures of success.

The supply chain organization is responsible for operational decision making and coordinates planning and scheduling in cooperation with these other groups of the corporation. Therefore, it has a significant impact on business performance by virtue of its control over the manufacturing assets. Marketing will supply a forecast to the supply chain, and agreement on a final forecast may involve several iterations until convergence. This process drives the production goals, raw materials supply management and future asset plans. Finance will want

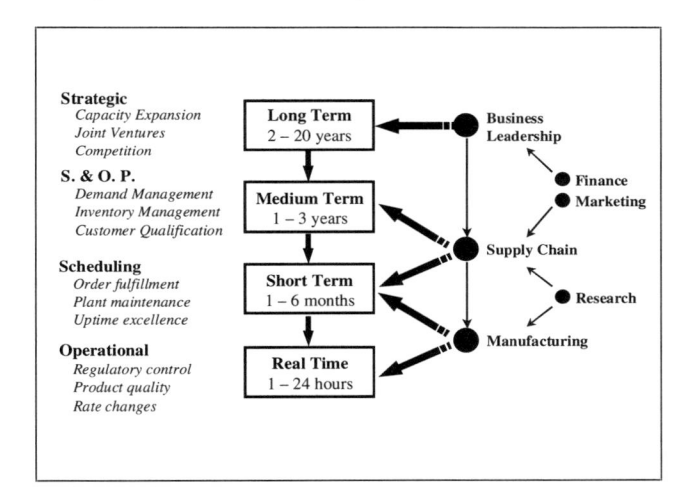

Figure 1. Typical functional hierarchies.

continual working capital and cost estimate projections from supply chain. They will also create financial targets. Research and Development will request production in order to run tests. Manufacturing will want rough cut schedules so that it may plan shop floor execution effectively. Fig. 1 shows a typical structure of these corporate functions and the areas of decision making where they have the most substantive impact. Managing a large portfolio of assets in the face of these conflicting objectives is a challenging task. For an interesting discussion of how the manufacturing environment must acknowledge its external environment to improve plant performance, see Backx *et al.* (1998).

A good example of how supply chain must manage competing goals is in relation to inventory levels. A recurring scenario is that business management desires a lowering of inventory levels to improve metrics like cash flow cycle time and inventory days of sales. However, manufacturing will push for longer run times in the process to improve yield and uptime, resulting in higher inventory levels. In addition, different plant personalities and cultures can lead to divergent operating policies. In one DuPont business, the European plants had much longer run lengths than their counterparts in the U.S. Their overall uptime was higher, but this led to high inventory levels, creating problems for the supply chain organization as they tried to meet financial working capital targets. Optimization can weigh these competing factors using objective, financially driven criterion, and converge operating and financial strategies across the organizational and regional divisions of the corporation.

There are many other reasons for making use of advanced optimization technology. First, there is often a broad distribution in the age and capability of our manufacturing assets. This complex asset mix has very high levels of capital investment, and maximizing profit per unit output is critical. Second, there is usually a complex product mix, with widely varying volumes and profit margins. Newer growth products and markets obviously receive considerable attention. They will typically have a higher margin, but will be relatively low volume, creating supply chain problems. A new product may have higher setup costs and lower yields compared to an established product, and this will reduce the productivity of the assets, despite the greater selling price. The more mature segment of the product portfolio will have higher levels of volume, but with limited or negative opportunities for volume expansion. However, economies of scale will make these products less expensive to manufacture.

In recent years, capacity utilization rates have climbed due to the intense pressure to reduce capital spending (arising from the struggle to outperform the cost of capital). This can create considerable customer supply problems. If a hurricane downs a plant for two weeks, or a customer triples his regular order size, then there may not be any safety capacity available to which supply chain can turn. The business can enter crisis mode very quickly, forcing itself to make drastic changes in production plans and schedules, without fully understanding their longer term effects on overall ability to supply customers.

The above represent compelling reasons to employ optimization in support of business operations planning.

The Role of Information Technology

It is clear from the last section that a wide array of data is required to make these kinds of decisions effectively. Manufacturing data is required to capture plant topology and capability. Financial information is required to measure the performance of the supply chain system itself. Marketing provides the forecast. These information flows are necessary to provide the ability to make operational decisions with all pertinent information available. The ultimate goal is to move from a set of competing, potentially antagonistic, fiefdoms to an organization coherently focused on satisfying customer needs while driving overall profitability.

There are two key elements for success in this area. The first is having cross functional data integration, and the second involves providing the ability to utilize this information for decision support. There are also two distinct classes of software to support these goals. The Enterprise Resource Planning (ERP) software providers constitute the first group, such as the market leader SAP (Waldorf, Germany), as well as others like Baan (Ede, Netherlands), and Oracle (Redwood, CA). Chemical companies have recognized the importance of ERP software as a key component of their business strategy. Some primary goals are to manage and integrate the flow and dispersal of information among the disparate groups of an organization and to link the flow of data from suppliers through manufacturing to customers. Converting transactional data to leverageable information across all functions of the corporation is the key motivation.

The ERP vendors are still perceived largely as suppliers of transaction and data management systems, with limited decision support capabilities. This defines the second broad category of software for advanced planning and scheduling, from vendors such as AspenTech (Boston, MA), i2 Technologies (Dallas, TX), and Manugistics (Rockville, MD). This kind of software uses the information gathered across a corporation for real–time decision support in order to improve supply chain management, providing functionality that is not currently available from the ERP vendors. It should be noted that recently, the ERP vendors are either developing their own software offerings (the Production Planning/Process Industry module – PPPI – from SAP, for example) or are initiating alliances (AspenTech and SAP) to expand their products into this critical area of decision support. There is also a broad class of modeling environments such as GAMS (Washington, DC), APC (West Lafayette, IN), ILog (France) and AspenTech (Boston, MA) that provide the capabilities of developing sophisticated optimization modeling for the supply chain.

Some enlightening examples that illustrate the importance of information to a business are now provided. Basic information like current inventory may only be available in paper form. In one business, two polymer plants on different continents were producing the same product simultaneously when the business was sold out in other products. Knowledge gaps such as these can lead to serious inefficiencies in effectively utilizing the manufacturing assets. It is also difficult to respond to changes in production costs and selling prices if they are only updated quarterly, meaning potential profit opportunities can be missed. Effective Available To

Promise (ATP) capabilities are necessary to make rapid price and delivery commitment and improve overall customer responsiveness. Achieving this requires an effective link between the customer service and production management software. Seamless integration of inventory, current orders and projected production plans which are available to a customer service representative is the desired state. Also crucial is the ability to communicate and deploy new information rapidly, much like the airline industry has done with yield management techniques, that update pricing strategies overnight. ERP software systems are again the enablers, but the chemical industry lags other sectors in its efforts to achieve these goals.

Current State of Optimization Modeling

There is no shortage of available optimization formulations for many classes of problems in the area of production planning and scheduling. A partial review is now supplied. See Shah (1998) for a more thorough literature review.

Production planning is concerned with optimizing the utilization of a manufacturing asset base over a relatively long time horizon. This horizon is usually set by the Sales & Operations Planning (S&OP) process, which generates ongoing plans for the business. Multiperiod models are typical: Bitran and Hax (1984) provide a review of the use of mathematical programming in this area. McDonald and Karimi (1997) developed a production planning model. This work emphasizes the importance of how inventory is costed, and the minimum run length constraints whose representation is critical for problems involving semicontinuous equipment. They present an analysis of how the ratio of time period length to minimum run length guides the type of model that should be selected. Sahinidis and Grossmann (1991) present an approach for allocation of resources (including capital) over a longer time horizon, incorporating capacity expansions into the formulation.

A drawback for these conventional multiperiod models in certain situations is that events are constrained to occur within a time period, without specification of the exact timing of the event. In an effort to address this limitation, Sahinidis and Grossmann (1991) formulate the cyclic scheduling problem and the optimal cycle time is an output of the model. Production slots are allowed to float over the time horizon. Pinto and Grossmann (1994) extend this formulation to a two–stage example, while allowing intermediate storage. However, it is not possible to measure the objective function at discrete points in time over the horizon in these formulations. Kondili et al. (1993) present a formulation which overcomes this drawback for the optimal planning of a cement mill. Karimi and McDonald (1997) discuss the application of a continuous time formulation to a short–term scheduling problem over a horizon of three months. One of its drawbacks is that instantaneous resource constraints cannot be included. Schilling and Pantelides (1996) present the most general approach to date, using the State Task Network (STN) framework. These formulations are very sophisticated and are capable of representing very complex manufacturing systems. However, they are combinatorially very challenging with large integrality gaps. The size of the formulations is intimidating, even for smaller systems, with concomitantly excessive computing requirements. This is one key area of research where the gap between the optimization approaches and their implementation remains large.

Kondili et al. (1993) present a state–task formulation (later extended and refined by Pantelides, 1993), which used a uniform discretization of time (UDM). This approach can generate a prohibitively large number of time periods. This has led to research activity in methods which could remove this drawback. Wilkinson et al. (1995) present an aggregated model which approximates the original UDM, and can provide tight upper bounds. The model has been used by Dimitriadis et al. (1997) in a rigorous decomposition scheme in which an aggregate formulation is used to define a series of detailed subproblems that can be solved to yield a feasible schedule. Zentner et al. (1994) develop a comparison between the UDM models and the Non Uniform Continuous Models (NUCM). which allow flexibility of modeling approaches over an extended time horizon. Bassett et al. (1996) discuss a decomposition approach for the solution of large–scale scheduling problems.

In the field of Operations Research, the economic lot scheduling problem (ELSP) has received much attention. Elmaghraby (1978) provides a review. These problems are important because any production planning and scheduling model for semicontinuous processes must embed this trade–off of inventory versus setup costs. Carreno (1990) develops a multi–machine model assuming an infinite horizon with a constant demand rate which must be met. Dobson (1987) allows lot sizes to vary.

Miller et al. (1993) present an approach for solving sequencing problems encountered at DuPont. The formulation has an underlying Traveling Salesman Problem (TSP) structure, and they present various deterministic and heuristic methods to solve these challenging NP–complete problems, in which sequence dependent transitions are critical. The deterministic algorithm of Miller and Pekny (1991) is used to generate bounds for the heuristic approaches. Pekny and Reklaitis (1998) provide an extensive discussion of the relationship between solution methodologies and NP–completeness.

Given the uncertainty associated with much of the data used in these models, formulations which include the stochastic nature of the problem are desirable. Liu and Sahinidis (1996) present a two–stage stochastic programming approach for planning, using a decomposition algorithm to help cope with the size of the model. Petkov and Maranas (1997) develop an interesting formulation for the scheduling of batch plants, involving the conversion of their stochastic formulation into a deterministic equivalent. Ierapetritou et al. (1996) discuss a two–stage stochastic programming formulation for production planning.

Fig. 2 presents a diagrammatic summary of the typical kinds of models and applications used in the area of production planning and scheduling.

	STRATEGIC PLANNING	OPERATIONS PLANNING	PRODUCTION SCHEDULING
Modeling Approach	•Stochastic •Multiperiod	•Multiperiod •Economic Lot-sizing	•CTF & UDM •Hybrid models •Heuristics
Typical Applications	•Capacity expansion •Competitive modeling	•Inventory standards •PSI Plans	•Resource Mgmt. •Transition Scheduling
Degree of Penetration	Low	Significant	Limited
I.T. Overhead	Low	Moderate → High	High

Figure 2. Typical modeling approaches.

The next section presents a multi–asset lot–sizing formulation. The motivation for this work was to extend the capabilities of a commonly used in house DuPont tool which essentially implements a single–machine lot–sizing algorithm. Due to a compelling business need, it was desired to extend the approach to multiple machines. In the next section, some of the drawbacks of the lot–sizing model are discussed, providing the motivation to develop more rigorous Mixed Integer Programming models. These sections dealing with current industrial practice in optimization provide the foundation for a discussion on how the models are actually implemented and used. A discussion of the successes and failures encountered will be provided, and the theme of interaction between optimization technology and enabling software will be revisited.

Multi–asset Lot–sizing Formulation

The motivation for this work was to extend the capabilities of a commonly used in house DuPont tool which implements a single–machine lot–sizing algorithm. This tool will henceforth be referred to as ELST (Economic Lot Sizing Tool). It evaluates the optimum balance between inventory and transition costs for semicontinuous machines on which multiple products are manufactured. Its main disadvantage is that it is not capable of performing asset optimization across multiple machines: the user must specify the allocation of product loads to machines. Another drawback is that it cannot make economic based decisions on what products to short if there is inadequate capacity.

A large DuPont business was using this tool to determine its inventory standard as it headed into a year where customer shortages were projected. The business manufactures several hundred products on multiple machines and locations. There is considerable variation in the capability of these machines. Each product is usually only made on a small subset of the total number of machines on which it could be manufactured. There are approximately 2000 possible allocations of products to machines, but with no independent subgroups of machines that can be considered separately. The approach was to manually adjust the product loading until the "best" solution was obtained. This manual process left them 1% short of the projected annual forecast and provided the motivation to consider the proposed optimization approach.

Optimization formulation

The optimal allocation of products to assets is determined while minimizing the following costs:

1. Inventory
2. Transition and waste
3. Costs of unmet demand

The product to machine loadings are an output of the model, which are then fed to the ELST. Before the problem is described, the variables and data used in the optimization formulation are now documented. Note that products are labeled $i \in I$, and machines are labeled $j \in J$. The horizon considered, labeled H, is a year.

Variables

1. T_{ij} :– total time in days over the horizon H spent manufacturing item i on machine j.
2. P_{ij} :– total production amount during H of item i on machine j.
3. n_{ij} :– number of campaigns over H for item i on machine j.
4. R_{ij} :– length of a campaign for item i on machine j.
5. Y_{ij} :– binary variable equal to 1 if item i is manufactured on machine j, 0 otherwise.
6. I_i^- :– amount of unmet demand for item i over the horizon H.
7. P_i^+ :– excess production of item i over H.

General Data

1. d_i :– daily unit demand for item i over the horizon H.

2. R_{ij}^e :– daily production rate for item i on machine j.
3. τ_{ij} :– transition time (in days) for item i on machine j.
4. MRL_{ij} :– minimum run length (in days) for a campaign of item i on machine j.
5. U_j :– uptime percentage for machine j.

Cost Coefficients

A key component of the problem formulation is the financial data used in the evaluation of the objective function:

1. v_{ij} :– allocated production cost (in \$/unit) of manufacturing item i on machine j.
2. h_{ij} :– inventory cost (in \$/H) for item i on machine j.
3. κ_{ij} :– transition cost (in \$) for item i on machine j.
4. μ_i :– unit selling price (in \$/unit) for item i.
5. ω_i :– unit waste price (in \$/unit) for item i.

The annual inventory cost is calculated as a fixed percentage (28%) of the allocated production cost so that the cost of stocking a unit of product i produced on machine j for a year is $0.28\ v_{ij}$ (\$/unit/year). The transition cost is obtained as the parts and labor cost and the difference between the value of first grade product and waste product made during the transition. The transition costs are sequence independent. No distribution costs were included in the model because the demand was not broken down by geographic area, although it would be easy to do so if this data was available.

Manufacturing Policy

The following assumptions are made:

1. Products are manufactured in regular cycles and each product campaign lasts the same length of time, R_{ij}.
2. Daily demand for product i, d_i, is constant over the planning period.
3. Daily supply of product i from machine j, labeled, s_{ij}, is also constant over the planning period.

The cyclic scheduling policy leads to the following fundamental nonlinear relation of importance:

$$R_{ij} = \frac{T_{ij}}{n_{ij}} \qquad (1)$$

For example, if a product is manufactured twice a year for a total of 20 days, then each campaign lasts 10 days ($R = 10$, $T = 20$, $n = 2$). s_{ij} is the daily supply of product i from machine j to customers, defined as follows:

$$s_{ij} = \frac{R_{ij}^e \cdot T_{ij}}{H} \qquad (2)$$

During each campaign, sufficient inventory must be built in order that demand may be satisfied until the next time product i is manufactured. This inventory is called cycle stock, labeled CS_{ij}. While the product is being manufactured, the excess inventory build–up is $R_{ij}^e - s_{ij}$. Therefore, the total amount produced during a campaign must be enough to supply customers daily with s_{ij} of product i over the cycle time. This observation leads to the following definition for the cycle stock:

$$CS_{ij} = \left[R_{ij}^e - \frac{R_{ij}^e \cdot T_{ij}}{H} \right] \cdot R_{ij}$$

$$= R_{ij}^e \cdot \left[1 - \frac{T_{ij}}{H} \right] \cdot \frac{T_{ij}}{n_{ij}}$$

The cycle time, labeled CT_{ij}, is defined as the time between production campaigns:

$$CT_{ij} = \frac{H}{n_{ij}}$$

If a product runs twice a year, then the cycle time is six months. Fig. 3 shows an example inventory plot for a single product i, where $n = 2$.

The primary variables used in the formulation are T_{ij} and n_{ij} (R_{ij} can be directly determined from Eqn. 1). The costs included in the objective function are inventory associated with cycle stock (Eqn. 3), transition costs (Eqn. 4), shortage costs (Eqn. 5), and excess production costs (Eqn. 6). Note how n_{ij} appears in the objective function. As it increases, the transition costs rise, while the inventory costs reduce. This constitutes the key cost trade–off in the face of the large number of product to asset choices. The objective function is given as:

$$\min OF = \frac{1}{2} \sum_{i,j} h_{ij} \cdot R_{ij}^e \cdot \left[1 - \frac{T_{ij}}{H} \right] \cdot \frac{T_{ij}}{n_{ij}} \qquad (3)$$

$$+ \sum_{i,j} n_{ij} \cdot \kappa_{ij} \qquad (4)$$

$$+ \sum_i I_i^- \cdot [\mu_i - \omega_i] \qquad (5)$$

$$+ \sum_i P_i^+ \cdot v_i \qquad (6)$$

The constraints are written as follows:

$$\sum_i \left[T_{ij} + n_{ij} \cdot \tau_{ij} \right] \leq U_j \cdot H \qquad (7)$$

$$T_{ij} \geq MRL_{ij} \cdot Y_{ij} \qquad (8)$$

$$T_{ij} \leq H \cdot Y_{ij} \qquad (9)$$

$$I_i^- - P_i^+ = d_i - \sum_j R_{ij}^e \cdot T_{ij} \qquad (10)$$

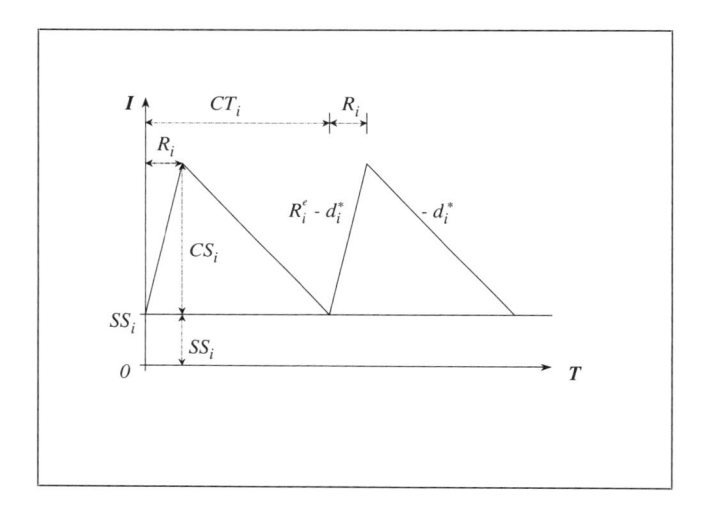

$$T_{ij}, n_{ij}, I_i^-, P_i^+ \geq 0; Y_{ij} \in \{0,1\} \qquad (11)$$

Figure 3. Example inventory profile for product i.

Eqn. 7 states that the total time spent producing and transitioning must be less than the available capacity on each machine. Eqns. 8 and 9 are required to ensure that the minimum run length constraint is satisfied. These constraints are generated for all possible item to machine combinations. Eqn. 10 calculates the difference between the demand over H and the total amount produced. The quantity on the right hand side of Eqn. 10 is either positive (when production falls short of demand) or negative (when production exceeds demand). Because I_i^- and P_i^+ are both constrained to be nonnegative, and because the objective function cost coefficients for both these variables are positive, then at least one of these variables must be zero. If both are zero, then production matches demand exactly, which is the desired solution. Excess production can occur for low volume items, where a single campaign lasting for the minimum run length produces more product than is demanded.

There are several difficulties with this formulation. First, the objective function is nonlinear and contains terms of the form:

$$\text{(i)} \ \frac{T_{ij}}{n_{ij}} \quad \text{and} \quad \text{(ii)} -T_{ij} \cdot \frac{T_{ij}}{n_{ij}}$$

The first function is a linear fractional term which is pseudoconvex by itself. However a summation of these pseudoconvex functions is nonconvex. In addition, recall that an item can run on several machines, but usually will run on a much smaller subset of machines. This implies that $n_{ij} \to 0$ for many item/machine combinations, collapsing the pseudoconvexity of this term as well as leading to numerical difficulties. The second term is a negative quadratic fractional term and is concave. In addition, binary variables are required because of the minimum run length constraints. This is therefore an extremely challenging Mixed Integer Non Linear Programming problem (MINLP). The sum of nonconvex and concave functions constitute the objective function to be minimized, subject to constraints containing binary variables. Given the size of the problem, it cannot be solved in reasonable time using a state–of–the–art MINLP optimization package such as DICOPT.

Recasting the Problem

Faced with these challenges, a different approach is required. A decomposition approach is taken:

1. Transform the decision variables of MINLP to T_{ij} so that these terms appear linearly in the formulation.
2. Approximate the remaining nonlinear variables using a lot sizing formula.

The transformations involved are summarized in Table 1.

Linearizing the nonlinear terms in the objective function yields an mixed integer linear subproblem (MIP). This is an approximation of the original problem but it is much easier to solve. By updating the critical variable (an estimate of the optimal number of campaigns) at each iteration, a set of MIP subproblems is solved successively until convergence occurs. We call this a successive MIP approach, or *s*MIP.

Note that the capacity consumed by transitions is approximated by Θ_{ij}. In this decomposition approach, updates for the allocation of products to assets are obtained from the solution of MIP as Y_{ij} and T_{ij}. At each iteration of *s*MIP, T_{ij}^* is updated using a economic order quantity (EOQ) formula. Once the above steps have been performed, then the adjusted formulation (denoted MIP) is as follows:

Table 1. Transformation of Terms.

	MINLP	MIP	Parameter Definition
Eqn. 3	$R_{ij}^e \cdot \left[1 - \dfrac{T_{ij}}{n_{ij}}\right]$	$R_{ij}^e - s_{ij}^*$	$s_{ij}^* = \dfrac{R_{ij}^e \cdot T_{ij}^*}{H}$
Eqn. 3	$\dfrac{T_{ij}}{n_{ij}}$	$\dfrac{T_{ij}}{n_{ij}^*}$	
Eqn. 4	$n_{ij} \equiv \dfrac{T_{ij}}{R_{ij}}$	$\dfrac{T_{ij}}{R_{ij}^*}$	$R_{ij}^* = \dfrac{T_{ij}^*}{n_{ij}^*}$
Eqn. 7	$T_{ij} + n_{ij} \cdot \tau_{ij}$	$\dfrac{T_{ij}}{\Theta_{ij}}$	$\Theta_{ij} = \dfrac{R_{ij}^*}{R_{ij}^* + \tau_{ij}}$

Adjusted Objective Function (*AOF*):

$$AOF = \frac{1}{2}\sum_{i,j} h_{ij} \cdot \left[R_{ij}^e - s_{ij}^*\right] \cdot \frac{T_{ij}}{n_{ij}^*}$$
$$+ \sum_{i,j} \frac{T_{ij}}{R_{ij}^*} \cdot \kappa_{ij}$$
$$+ \sum_i I_i^- \cdot [\mu_i - \omega_i]$$
$$+ \sum_i P_i^+ \cdot v_i$$

Eqn. 7 is adjusted as follows:

$$\sum_i \frac{T_{ij}}{\Theta_{ij}} \leq U_j \cdot H \quad (12)$$

Eqns. 8 – 11 are also required as before. Convergence is deemed to occur when the approximation associated with R_{ij}^* (labeled ε) is within tolerance, and the number of switches of products across assets (labeled N) is within limits:

$$\varepsilon = \sum_{i,j}\left|\left[R_{ij}\right]^K - \left[R_{ij}\right]^{K-1}\right|; \quad N = \sum_{i,j}\left|\left[Y_{ij}\right]^K - \left[Y_{ij}\right]^{K-1}\right|$$

Calculation of n^*

The approach described is predicated on the availability of estimates of T_{ij}^* and n_{ij}^*. Solution of MIP supplies an update of T_{ij}^*. However, a method of approximating the optimal number of campaigns for a given product and machine is required. For a single product and machine, the lot sizing problem is formulated, using the current value of T_{ij}^* with n_{ij} as the sole decision variable. The approach taken is to calculate an economic number of campaigns based on T_{ij}^*. Defining the coefficients of the lot sizing calculation as follows:

$$\bar{h} = h_{ij} \cdot \left[R_{ij}^e - s_{ij}^*\right] \cdot T_{ij}^*$$
$$\bar{\kappa} = \kappa_{ij}$$

The costs can then be constructed in terms of *n* as follows:

$$C = \frac{1}{2} \cdot \frac{\bar{h}}{n} + \bar{\kappa} \cdot n$$

Simple calculus delivers the minimum cost *n* as follows:

$$\frac{\partial C}{\partial n} = 0 \Rightarrow -\frac{1}{2} \cdot \frac{\bar{h}}{n^2} + \bar{\kappa} = 0$$
$$\Rightarrow n^* = \sqrt{\frac{\bar{h}}{2\bar{\kappa}}} \quad (13)$$

Safety Stock

In the update of n^*, it is also possible to consider the effect of safety stock levels, which are dependent on the variance associated with demand, $V[d_i]$, and the period of risk, POR_{ij}, defined as the time during which product *i* is not being manufactured:

$$POR_{ij} = CT_{ij} - R_{ij}$$
$$= \frac{H - T_{ij}}{n_{ij}}$$

If cycle times are large, then more safety stock will be required to last through the longer period of uncertainty during which the product is not being manufactured. Also, the more uncertain the demand for a product, the more safety stock will be needed:

$$SS_{ij} = \zeta_i \sqrt{V[d_i] \cdot \frac{H - T_{ij}}{n_{ij}}}$$

where ζ_i is the inverse probability norm for the selected service level, and $V[d_i]$ is an input derived from a forecasting system. Defining $\bar{s}$ as follows:

$$\overline{s} = h_{ij}\zeta_i\sqrt{V[d_i]\cdot\left(H-T_{ij}^*\right)}$$

allows the cost to be written in terms of n as follows:

$$C = \frac{1}{2}\cdot\frac{\overline{h}}{n} + \overline{\kappa}\cdot n + \frac{\overline{s}}{\sqrt{n}}$$

The minimum cost n is then given as follows:

$$\frac{\partial C}{\partial n} = 0 \Rightarrow -\frac{1}{2}\cdot\frac{\overline{h}}{n^2} + \overline{\kappa} - \frac{\overline{s}}{2n\sqrt{n}} = 0$$

$$\times\frac{n^2}{\overline{\kappa}} \Rightarrow n^2 - \frac{\overline{s}}{2\overline{\kappa}}\sqrt{n} - \frac{\overline{h}}{2\overline{\kappa}} = 0$$

$$\Rightarrow \left(\sqrt{n}\right)^4 - s'\left(\sqrt{n}\right) - h' = 0$$

This can be rewritten as follows:

$$\left(\sqrt{n}\right)^4 - s'\left(\sqrt{n}\right) - h' = 0 \quad (14)$$

where $s' = \overline{s}/2\overline{\kappa}$ and $h' = \overline{h}/2\overline{\kappa}$. Eqn. 14 is a quartic equation. It is possible to obtain the 4 roots of this equation explicitly. There is a complex conjugate pair of roots, one real negative root, and the required real positive root, calculated as follows[1]:

$$\sqrt{n^*} = \alpha + \frac{1}{2\sqrt{3}}\sqrt{\beta + \gamma} \quad (15)$$

where:

$$\alpha = \frac{\rho_2}{6}; \quad \beta = \frac{4h'}{\rho_1} - 3\rho_1; \quad \gamma = \frac{18s'}{\rho_2}$$

$$\rho_1 = \left\{\frac{1}{2}(s')^2 + \frac{1}{18}\sqrt{768(h')^3 + 81(s')^4}\right\}^{\frac{1}{3}}$$

$$\rho_2 = \left\{9\rho_1 - \frac{12h'}{\rho_1}\right\}^{\frac{1}{2}}$$

The complete procedure for solving the original MINLP is now supplied. The optimization approach supplies s_{ij}^* for input to the single machine lot sizing tool, which in turn generates the operating and inventory policy.

Step 0: Initialization

Select ε^L and N^L. Set $K = 0$.

---

[1] With gratitude to Stephan de la Veaux

Set $s_{ij}^* = 0$, $n_{ij}^* = 1$, $R_{ij}^* = MRL_{ij}$, $\Theta_{ij} = MRL_{ij}/(MRL_{ij} + \tau_{ij})$

Step 1: Solve MIP

Obtain T_{ij}^*, $Y_{ij}^* \equiv \left[Y_{ij}\right]^K$

Step 2: Update parameters

Update s_{ij}^* (Eqn. 2) and n_{ij}^* (Eqns. 13 or 15).
Calculate $n_{ij}^* = \max\left\{1, n_{ij}^*\right\}$; $\Theta_{ij} = R_{ij}^*/\left(R_{ij}^* + \tau_{ij}\right)$;
and $R_{ij}^* = \left[R_{ij}\right]^K = \max\left\{MRL_{ij}, T_{ij}^*/n_{ij}^*\right\}$

Step 3: Calculation of error

if ($\varepsilon \leq \varepsilon^L$) **and** ($N \leq N^L$) **then** Output s_{ij}^*.
else Set $K = K + 1$ and return to **Step 1**.

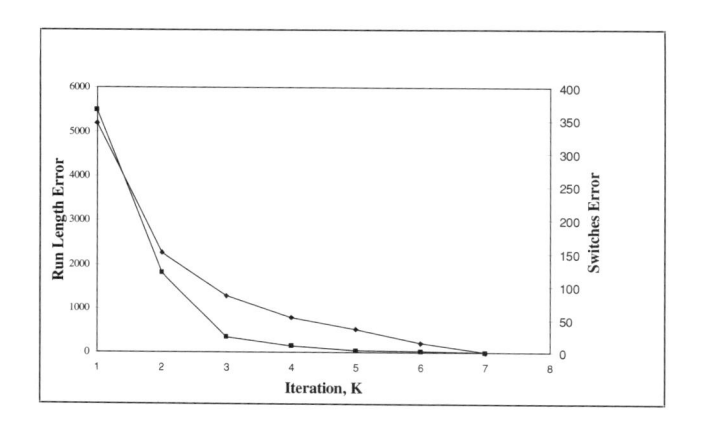

Figure 4. Convergence characteristics for lot sizing problem.

The model was run for a single fiscal year, and the convergence characteristics are shown in Fig. 4. It took 7 iterations, with each MIP taking approximately 3 minutes on an IBM Power2 SP node[2]. The results from this model generated an extra 1.5% of capacity from the existing assets, simply by adjusting the allocation of products to assets, alleviating the projected shortages. Even though the percentage value was small, it still represented a substantial multi-million dollar stake. The inventory standard was 40% higher than their existing one, with the model indicating that 16% fewer transitions should be carried out. Business management allowed these levels of inventory to be built.

Obviously, only the short-term capacity problem was fixed. The business must decide to either invest in new

---

[2] 120 MHz; 233 Mflops; 16.6 Spec95fp

capacity, or remove itself from its lowest margin businesses. These are questions that the current model does not address. In addition, the model was only used once for a single year end analysis. In the ideal case, a business would run this kind of model to establish the inventory standard on a regular basis. Because of the isolated nature of the project, no analysis of the supply chain performance was undertaken after the initial recommendations were implemented. This highlights the need to make these kinds of sophisticated optimization models an integral component of the business process for planning and scheduling if they are to be of continuing value. This issue will be discussed further in the section on Model Utilization and Implementation.

Extended MIP Formulation

Even though the MINLP lot sizing model enabled a step change advance in supply chain modeling, there are a number of drawbacks attached to these kinds of cyclic scheduling models:

1. Demand is assumed to be constant over the planning horizon. Seasonality and normal variations in the business cycle weaken this assumption.
2. Even though the output of the model yields a capacity feasible plan, it may not yield a feasible schedule. However, this was not a problem as the results were submitted to the in-house tool which always generated a feasible cyclic schedule.
3. If there are additional classes of transitions (a family of products which incurs a major transition) then the current model is not capable of solving this particular problem.
4. There is no guarantee that the global solution will be obtained, or even that the sMIP approach will converge. Also, no lower bounds on the optimal solution can be supplied.

In essence, the lot sizing model is a steady state model for the supply chain, but dynamic models are clearly more desirable. It is for this reason that multiperiod models are typically recommended which can remove these disadvantages. This kind of model has been developed in McDonald and Karimi (1997). Due to the large size of the problems, these kinds of formulations are usually decomposed. There are typically many hundreds of products, and tens of machines and time periods. These kinds of models allow production campaigns to vary from one time period to the other, removing one of the major drawbacks of the cyclic scheduling models. Inventory standards can be back calculated from the average inventory levels over the set of time periods. It should be noted that the model of Sahinidis and Grossmann (1991) avoids Drawbacks 2 and 3 presented above, although it does not accommodate multiperiod demand patterns.

Model Utilization and Implementation

In this section, we will focus on the two principal modes in which optimization models are utilized at DuPont. The first, and least problematic, involves solving a specific instance of a particular supply chain problem. The lot–sizing problem presented earlier is a good example of this. Experience with these problems has been overwhelmingly positive: results can be obtained and communicated quickly. However, the major drawback is that the results provide insight for a limited period of time only. As soon as any of the model assumptions or data have changed, the results are invalidated.

The second mode involves the implementation of planning or scheduling models, which are fully integrated into the daily supply chain activities of the business, providing ongoing decision support. In this arena, the challenges are quite different. Management of data – both its integration across business functions and its flow – are critical. The user interface – how the data and model results are presented to the user – is also critical. The experiences with implementation in this area are now described.

Successes and Failures

The discussion is divided into the areas of planning and scheduling. Obviously, many DuPont businesses use well established software like SAP for transaction management, MIMI for supply chain decision support, and Manugustics for forecasting. Even though there is considerable corporate pressure to use standardized tools, each business can (and often does) individually design and maintain its own software programs for planning and scheduling. This is only done if the business has a unique need which cannot be accommodated by the commercial software companies. The goal of our group is to provide a modular optimization modeling component which fits underneath these other software systems.

The first success story relates to a large business unit which has several manufacturing sites spread throughout the world. They needed to develop a global planning tool for the S&OP process. The time horizon for the S&OP process is usually 18 months using a monthly forecast. Note that this horizon is easily changed to accommodate other scenarios. The model implemented is very similar to the midterm planning model described in McDonald and Karimi (1997). It is a large planning model with many hundreds of products and is implemented in GAMS running on an IBM Power2 SP node. Fig. 5 shows the information technology structure implemented for this project. It consists of three layers, and is therefore labeled a 3–tier application. The bottom layer is the database server, where all the data is stored. The middle layer (or the application layer) contains the business application code used to manage the flow of data between the various components of the application. The communication between the layers is enabled by the Distributed Component Object Model (DCOM) of Microsoft®. It also manages the communication with the optimization model and this is strictly controlled by flat file specifications. A key attribute of the strategy is that the model must be merely a modular component of the tool so it can be ripped out easily, or alternatively, can still function if the software or business model changes. For example, when a web interface was mounted on the current system, this did not affect the model in any way. The top layer (presentation layer) consists of the actual user interface. The main advantage of these 3–tier systems is that they are scalable (it is easy to add multiple servers in the middle or bottom layers), and the top layer is a light client (most of the execution requirements reside on the lower layers) with vast improvements in performance for the user. From the IT perspective, it is evident that the system is very sophisticated. This tool is now being used for global S&OP, and there have also been enormous gains in supply chain productivity. For example, in one region, the planning process used to take several people weeks to complete. Now, one person can create a plan for the whole region in an afternoon. These are ERP benefits, arising from data integration and the management and communication of information. The model is used to create plans (and even schedules for some regions) over the S&OP horizon, achieving improved asset management and inventory control.

Similar planning and scheduling models were also developed for another business which featured considerable complexity despite its small size. It was a multisite business with some of the plants providing product for the other plants, as well as supplying customers directly. As with the previous example, the goal was to build a tool to support the S&OP process, as well as providing schedules over a four month horizon. The project lasted three years and was eventually abandoned. Production planning and scheduling models were implemented and validated. These models remained relatively static over the project's lifetime. The interface was developed in a crude 2–tier system. There was a succession of efforts to improve the interface which all failed, due to the inability to meet challenges such as cross–platform data networking and managing the integration of data across functions in a multiuser environment. Because the business itself was so complex, the software developers did not understand the real needs of the project. In addition, there was an absence of internal technical IT leadership to direct the project. The key learning is that the IT aspects of a project are the driver and determinant of success. Without this phase being implemented effectively, there is no chance of utilizing optimization technology.

In the area of short–term scheduling, which typically means time horizons of three or four months for DuPont businesses, there has not been much success in employing advanced optimization models such as the continuous time formulations discussed earlier. Even though there are a number of sophisticated formulations to choose from, problems remain with implementing models of such complexity and size. In addition, the user requirements are severe in terms of speed of performance and quality of results. There has been virtually no success in integrating sophisticated optimization models with the scheduling activity. For example the continuous time formulation of Karimi and McDonald (1997) was prototyped in a commercial supply chain management software package. To incorporate this kind of model for this particular business within the software package would take considerable resources and expertise and the effort could not be justified.

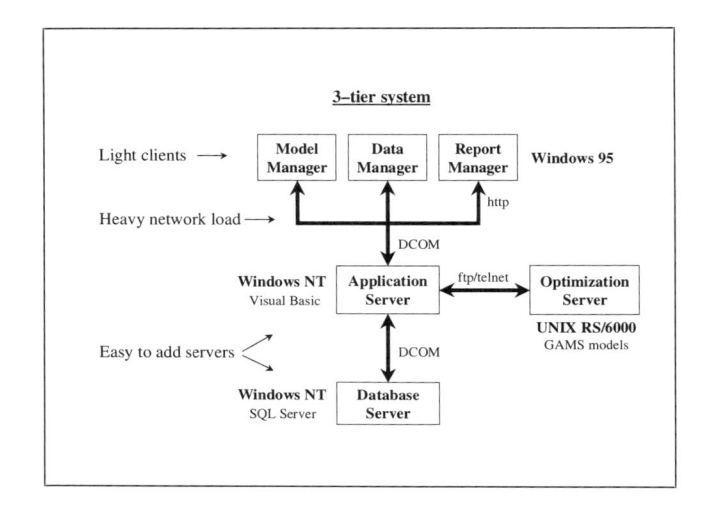

Figure 5. IT structure for S&OP application.

However, it is instructive to talk briefly about one encouraging development in the area of real–time online scheduling. Wilkinson and McDonald (1997) present a problem where a critical resource (a monomer in this case) feeds a set of downstream semicontinuous and batch units. The problem is formulated as a Mixed Integer Program (MIP), with the goal of maximizing the consumption of the critical resource over a period of 24 hours. The rate of downstream monomer consumption is currently less than the maximum possible levels of monomer production in a sold–out market. It is a low level model from the operational standpoint. In order to implement this in the control room, it is not feasible to build an interface between the DCS and a MIP model. Therefore, the model has been used to generate a set of rules which are implemented in GENSYM, a software system that facilitates real–time communication with the DCS and which incorporates expert system based reasoning. The model provides a basis for comparing the performance of various heuristic rule sets to this ideal case. These rule sets incorporate tried and tested shopfloor operational procedures. Additionally, if the plant topology or operations change, then the model will be used to provide new insight and provide guidance for the rule set. In this case, the optimization model provides scheduling insight. The employment of this insight is enabled by the software which interfaces higher level planning information (from MIMI) and lower level DCS information from the downstream process areas. The installation of the GENSYM software has already led to gains in productivity. At present this is due solely to increased visibility across the manufacturing areas, and forcing each area to commit to a schedule at the start of every day. Further gains are anticipated when the rule based coordination strategies are fully implemented. This illustrates the point that often as much business gain can be effected by improving information technology as from implementing optimization modeling.

Synthesizing Optimization and Information Technology

As can be seen, the common thread for success (failure) in these applications has been the presence (absence) of sophisticated enabling information technology capability, which enables the optimization modeling effort to be a seamless component of the total application. The optimization modeling effort must follow the data integration effort. There is no room for modeling arrogance. The formulation and development of a model consumes a relatively small percentage of the total project effort (usually 10 – 15%). The other activities of defining business processes, collating and cleaning data, managing the data flow across networks and presenting data to the user community consume far more resources than the modeling effort itself.

From the IT perspective, continued improvement in data integration is a key factor in motivating the application of optimization modeling. As data transparency and integration improve due to sustained IT investment, there are additional opportunities for optimization models to gain even more prominence and importance. It is clear that there is a wide portfolio of formulations in the area of planning and scheduling. For example, continuous time formulations are of great interest and are available but problems persist in implementing them as integrated components of a decision support tool set. This is due to problems related to large integrality gaps, high computing requirements and poor overall performance. The uncertainty in relation to demand forecasting and future raw material and product prices provides the natural motivation for the consideration of stochastic programming approaches. Implementation of these kinds of models is not currently planned at DuPont, but it still represents an important area of study. Due to the large size of the typical industrial problem, methods that employ techniques such as model aggregation and decomposition represent a promising area for research. Usually, these techniques are employed to some extent when solving large industrial applications, but the general frameworks for solution are not as formalized as they could be. There has been some debate over the role of generalized frameworks versus special purpose methods. The reality is that due to problem size and computational performance pressures, each model will be individually tuned for best performance. However, the model generalizations and formalisms provide the framework for successful implementations, and are therefore critically important.

Conclusions

The paper has emphasized that financial performance should drive the decision making process across functions, and that it is especially critical to align these goals with the utilization of the manufacturing assets. The role of IT is crucial in providing cross functional data integration, a critical step in supplying the pertinent information for effective decision making. Despite the availability of sophisticated ERP software, the gap between the utopian vision and everyday reality remains large. Examples of this are the absence of availability of basic but critical information like inventory or demand, or difficulty in obtaining reliable financial data. Cross functional links between customer service, manufacturing, finance and supply chain are often absent, hampering the organization's ability to handle rapid changes. However, once these business processes and supporting IT structures are in place, this can create step change improvements in supply chain productivity and performance. Achievement of these gains provides further impetus to integrate real–time optimization planning and scheduling technologies into supply chain management. As has been discussed, this has been achieved for some DuPont businesses.

However, there are still challenges in integrating optimization models in industrial practice especially as one descends the supply chain hierarchy. It is clear that there is a wide spectrum of production planning and scheduling algorithms available. However, there has been little progress in effectively integrating rigorous mathematical programming scheduling formulations with shopfloor execution due to poor computational performance, inability of the models to capture all the idiosyncrasies of the plant, and the absence of interfaces that are easy to work with. Links between the Distributed Control System (DCS) and higher level functions are even more rare, even though these may be critical in certain instances like the real–time scheduling example described earlier. Therefore, the main challenge is to integrate this sophisticated optimization arsenal into the decision making process of the supply chain.

Acknowledgements

The author is indebted to W. David Smith for his leadership, support and encouragement; as well as colleagues Iftekhar A. Karimi (who worked on the lot–sizing problem with the author), and Stephen J. Wilkinson for his helpful comments.

References

Backx, T., O. Bosgra, and W. Marquardt (1998). Towards intentional dynamics in supply chain conscious process operations. In *Proceedings of FOCAPO III*.

Bassett, M.H., J.F. Pekny, and G.V. Reklaitis (1996). Decomposition techniques for the solution of large–scale scheduling problems. *AIChE J.*, **42**(12):3373.

Bitran, G.R., and A.C. Hax (1984). The role of mathematical programming in production planning. In K.D. Lawrence and S.H. Zanakis, (Eds.), *Production Planning and Scheduling: Mathematical Programming Applications*, page 21. Industrial Engineering and Management Press.

Carreno, J.J. (1990). Economic lot scheduling for multiple products on parallel identical processors. *Mgmt. Sci.*, **36**(3):348.

Dimitriadis, A.N. (1997), N. Shah, and C.C. Pantelides. RTN–based rolling horizon algorithms for medium term scheduling of multipurpose plants. *Comput. Chem. Eng.*, **21S**:1061. In European Symposium on Computer Aided Process Engineering – 6.

Dobson, G. (1987). The economic lot scheduling problem: achieving feasibility using time–varying lot sizes. *Oper. Res.*, **35**(5):764.

Elmaghraby, S.E. (1978). The economic lot scheduling problem (ELSP): review and extensions. *Mgmt. Sci.*, **24**(6):587.

Ierapetritou, M.G., E.N. Pistikopoulos, and C.A. Floudas (1996). Operational planning under uncertainty. *Comput. Chem. Eng.*, **20**(12):1499.

Karimi, I.A. and C.M. McDonald (1997). Planning and scheduling of parallel semicontinuous processes. 2. Short–term scheduling. *Ind. Eng. Chem. Res.*, **36**:2701.

Kondili, E., N. Shah, and C.C. Pantelides (1993). Production planning for the rational use of energy in multiproduct continuous plants. *Comput. Chem. Eng.*, **17S**:123.

Liu, M.L. and N.V. Sahinidis (1996). Optimization in process planning under uncertainty. *Ind. Eng. Chem. Res.*, **35**:4154.

McDonald, C.M. and I.A. Karimi (1997). Planning and scheduling of parallel semicontinuous processes. 1. Production planning. *Ind. Eng. Chem. Res.*, **36**:2691.

Miller, D.L. and J.F. Pekny (1991). Exact solution of large asymmetric traveling salesman problems. *Science*, **251**:754.

Miller, D.L., H. Singh, and K.A. Rogers (1993). A modular system for scheduling chemical plant production. In *Proceedings of FOCAPO II*, page 355.

Pantelides, C.C. (1993). Unified frameworks for optimal process planning and scheduling. In *Proceedings of FOCAPO II*, page 1.

Papageorgiou, L.G. and C. Pantelides (1996). Optimal campaign planning/scheduling of multipurpose batch/semicontinuous plants. 2. A mathematical decomposition. *Ind. Eng. Chem. Res.*, **35**:510.

Papageorgiou, L.G. and C. Pantelides (1996). Optimal campaign planning/scheduling of multipurpose batch/semicontinuous plants. 1. Mathematical formulation. *Ind. Eng. Chem. Res.*, **35**:488.

Pekny, J.F. and G.V. Reklaitis (1998). Towards the convergence of theory and practice: A technology guide for scheduling/planning methodology. In *Proceedings of FOCAPO III*.

Petkov, S.B. and C.M. Maranas (1994). Multiperiod planning and scheduling of multiproduct batch plants under demand certainty. *Ind. Eng. Chem. Res.*, **36**(11):4864.

Pinto, J.M. and I.E. Grossmann (1994). Optimal cyclic scheduling of multistage continuous multiproduct plants. *Comput. Chem. Eng.*, **18**(9):797.

Sahinidis, N.V. and I.E. Grossmann (1991). MINLP model for cyclic multiproduct scheduling on continuous parallel lines. *Comput. Chem. Eng.*, **15**(2):85.

Sahinidis, N.V. and I.E. Grossmann (1991). Multiperiod investment model for processing networks with dedicated and flexible plants. *Ind. Eng. Chem. Res.*, **30**:1165.

Schilling, G. and C.C. Pantelides (1996). A simple continuous–time process scheduling formulation and a novel solution algorithm. *Comput. Chem. Eng.*, **20S**:1221.

Shah, N. (1998). Single– and multisite planning and scheduling: Current status and future challenges. In *Proceedings of FOCAPO III*.

Wilkinson, S.J. and C.M. McDonald (1997). Scheduling in mixed batch and semicontinuous processes. Presented at AIChE National Meeting, Los Angeles.

Wilkinson, S.J., N. Shah, and C.C. Pantelides (1995). Aggregate modelling of multipurpose plant operation. *Comput. Chem. Eng.*, **19S**:583. from *European Symposium on Computer Aided Process Engineering – 5*.

Zentner, M.G., J.F. Pekny, G.V. Reklaitis, and J. Gupta (1994). Practical considerations in using model based optimization for the scheduling and planning of batch/semicontinuous processes. *J. Proc. Cont.*, **4**(4):259.

SINGLE- AND MULTISITE PLANNING AND SCHEDULING: CURRENT STATUS AND FUTURE CHALLENGES

Nilay Shah
Centre for Process Systems Engineering
Imperial College of Science, Technology and Medicine
London SW7 2BY
United Kingdom
Email: n.shah@ic.ac.uk

Abstract

This paper describes the problem of production scheduling in the process industries. First, a review of the techniques for optimising production schedules at individual production sites is presented, with an emphasis on formal mathematical methods. Recent extensions to such work have resulted in techniques for both long-term planning of individual facilities and the overall planning of production and distribution in multisite flexible manufacturing systems. A number of challenges arise in such integrated manufacturing problems. Mathematical models of overwhelming size must be avoided and it is very difficult to apply simplistic approaches based on straightforward capacity models when the capacity is flexible and is to be shared between competing products. The recent research in single- and multisite planning will be reviewed. Another important factor, especially when considering longer term planning, is the presence of uncertainty, both with respect to the process (e.g. rates/processing times and yields) and the market. There have been a number of recent developments in tackling uncertainty in planning and scheduling, either explicitly during planning, or by including a real-time, reactive scheduling capability within the planning function. These will be considered here. The paper will conclude with views on future research challenges in planning and scheduling.

Keywords:

Process operations, Supply chain, Planning and scheduling.

Introduction

The theme of production planning and scheduling has received great attention in the recent past. Initially, especially from the early 1980s to the early 1990s, this was due to the resurgence in interest in flexible processing either as a means of ensuring responsiveness or adapting to the trends in chemical processing towards lower volume, higher value-added materials in the developed economies (Reklaitis, 1991; Rippin, 1993; Hampel, 1997). More recently, the topic has received a new impetus as enterprises attempt to optimise their overall supply chains in response to competitive pressures or to take advantage of recent relaxations in restrictions on global trade.

It is widely recognised that that the complex problem of what to produce and where and how to produce it is best considered through an integrated, hierarchical approach which also acknowledges typical corporate structures and business processes. This type of structure is illustrated below. In the most general case, the extended supply chain is taken to mean the multi-enterprise network of manufacturing facilities and distribution points that perform the functions of materials procurement, transformation into intermediate and finished materials and distribution of the finished products to customers. Recently, the concept of a "holonic network" has been proposed as an extremely responsive example of such a system (McHugh et al., 1995), where a network is rapidly formed by a group of firms to exploit a short-term opportunity. Although there has been some recent work in the planning of activities across such multi-enterprise networks, this will remain outside the scope of this paper.

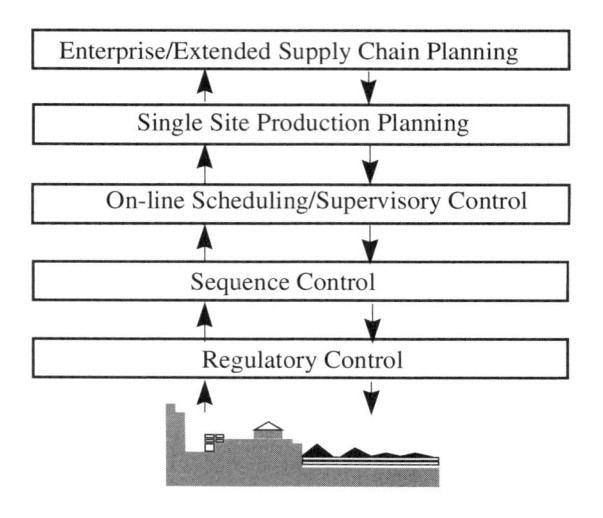

Figure 1. Process operations hierarchy.

The most common context for planning at the supply chain level is the co-ordination of manufacturing and distribution activities across multiple sites operated by a single enterprise (enterprise-wide or multisite planning). Here, the aim is to make the best use of geographically distributed resources over a certain time period.

The result of the multisite planning problem is typically a set of production targets for each of the individual sites, and rough transportation plans for the network as a whole. The production scheduling activity at each individual site seeks to determine precisely how these targets can be met (or indeed how best to compromise them if they cannot be met in whole). This involves determining the precise details of resource allocation over time.

Once a series of activities has been determined, these must be implemented in the plant. The role of the supervisory control system is to instantiate the correct sequences of control logic with the correct parameters at the correct time, making sure that conflicts for plant resources are resolved in an orderly manner. It is also useful at this level to create a schedule of planned operations over a short future interval using a model detailed enough to ensure that there are no anticipated resource conflicts. This "on-line" scheduling allows current estimates of the starting and finishing times of each operation to be known at any time. Although this capability is not essential for the execution of operations in the plant, it is vital if the hierarchical levels are to be integrated together so that production scheduling is performed in response to deviations in expected plant operation ("reactive scheduling").

Finally, the lowest levels of the hierarchy relate to execution of individual control phases and ensuring safe and economic operation of the plant. They shall not be considered further in this paper.

There is clearly a need for research and development in all the levels of the operations hierarchy. Two previous reviews in this area (Reklaitis, 1991; Rippin, 1993) summarised some of the main challenges as:

i. the development of efficient general purpose solution methods for the mixed integer optimisation problems that arise in planning and scheduling;
ii. the design of tailored techniques for the solution of specific problem structures which either arise out of specific types of scheduling problems or are embedded substructures in more general problems;
iii. the design of algorithms for efficient solution of general resource constrained problem; especially those based on a continuous representation of time;
iv. the systematic treatment of uncertainty;
v. the advancement of on-line techniques for rapid adaptation of operations;
vi. the development of methods for the integrated planning and scheduling of multisite systems.

Progress towards these challenges will be described. For reasons of presentation, the remainder of the paper is organised as follows. The next section considers work in the area of single-site production scheduling. The subsequent section will describe recent developments in multisite planning and medium and long-term planning at individual sites. The penultimate section will focus on the complications that arise out of uncertainty in process operations and means of dealing with them. The final section will make conclusions and propose future challenges in this area.

Single-site Scheduling

The scheduling problem at a single site is usually concerned with meeting fairly specific production requirements. Customer orders, stock imperatives or higher-level supply chain or long-term planning would usually set these, as described in subsequent sections. It is concerned with the allocation over time of scarce resources between competing activities to meet these requirements in an efficient fashion.

The key components of the scheduling problem are resources, tasks and time. The resources need not be limited to processing equipment items, but may include material storage equipment, transportation equipment (intra- and inter-plant), operators, utilities (e.g. steam, electricity, cooling water), auxiliary devices and so on.

The tasks typically comprise processing operations (e.g. reaction, separation, blending, packaging) as well as other activities which change the nature of materials and other resources such as transportation, quality control, cleaning, changeovers, etc.

There are both external and internal elements to the time component. The external element arises out of the need to co-ordinate manufacturing and inventory with expected product liftings or demands, as well as scheduled raw material receipts and even service outages. The internal element relates to executing the tasks in an appropriate sequence and at right times, taking account of the external time events and resource availabilities.

Overall, this arrangement of tasks over time and the assignment of appropriate resources to the tasks in a resource-constrained framework must be performed in an efficient fashion, which implies the optimisation, as far as possible, of some objective. Typical objectives include the minimisation of cost or maximisation of profit, maximisation of customer satisfaction, minimisation of deviation from target performance, etc.

As noted by Gabow (1983), all but the most trivial scheduling problems belong to the class of NP-hard problems; there are no known solution algorithms that are of polynomial complexity in the problem size. This has posed a great challenge to the research community, and a large body of work has arisen aiming to develop either tailored algorithms for specific problem instances or efficient general-purpose methods.

There are many different ways in which approaches to solving the scheduling problem may be classified. Here, the approaches are divided into heuristic/stochastic search techniques and mathematical programming techniques and then further subdivided into specific and general application domains. The latter division is intended to reflect the scope of the technique (in terms of plant structure and process recipes). Rippin (1983) classified different flexible plant structures as follows:

Multiproduct plants, where each product has the same processing network, i.e. each product requires the same sequence of processing tasks (often known as "stages"). Due to the historic association between the work on batch plant scheduling and that on discrete parts manufacturing, these plants are sometimes called "flowshops".

Multipurpose plants ("jobshops"), where the products are manufactured via different processing networks, and there may be more than one way in which to manufacture the same product. In general, a number of products undergo manufacture at any given time.

Heuristic/stochastic Search: Specific Processes.

Most scheduling heuristics are concerned with formulating rules for determining sequences of activities. They are therefore best suited to processes where the production of a product involves a prespecified sequence of tasks with fixed batchsizes; in other words variants of multiproduct processes. Often, it is assumed that fixing the front-end product sequence will fix the sequence of activities in the plant (the so-called "permutation schedule" assumption). Generally, the processing of a product is broken down into a sequence of jobs that queue for machines, and the rules dictate the priority order of the jobs.

Dannebring (1977), Kuriyan and Reklaitis (1985, 1989) and Pinedo (1995) give a good exposition on the kinds of heuristics (dispatching rules) that may be used for different plant structures. Typical rules involve ordering products (see, e.g., Hasebe et al., 1991) by processing time (either shortest or longest), due-dates and so on.

Most of the heuristic methods originated in the discrete manufacturing industries, and might sometimes be expected to perform poorly in process industry contexts. In process scheduling problems, most of the concerns with these approaches are associated with the divisibility of material in practice, which implies variable batch sizes, and batch splitting and mixing. In fact the last two activities are becoming increasingly popular as a means of effecting late product differentiation.

Stochastic search approaches are based on continual improvement of trial solutions by the application of an evolutionary algorithm which modifies solutions and prioritises solutions from a list for further consideration. The two main evolutionary algorithms applied to this area are simulated annealing and genetic algorithms. An early application of simulated annealing to batch process scheduling problems was undertaken by Ku and Karimi (1991), where they applied the algorithm to multiproduct plant scheduling. They concluded that such algorithms are easy to implement and tended to perform better than conventional heuristics, but often required significant computational effort.

Xia and Macchietto (1994) describe the application of simulated annealing and genetic algorithm techniques to the scheduling of multiproduct plants with complex material transfer policies.

More recently, Murakami et al. (1997) described a repetitive simulated annealing procedure which avoids local minima by using many starting points with fewer evolutionary iterations per starting point.

Sunol et al. (1992) describe the application of a genetic algorithm approach to a simple flowshop sequencing problem, and found the technique to be superior to explicit enumeration. As noted by Hasebe et al. (1996), the performance of a genetic algorithm depends on the operators used to modify trial solutions. They apply a technique that selects appropriate operators during the solution procedure for the scheduling of a parallel-unit process.

Overall, the stochastic search processes are best applied to problems of an entirely discrete nature where an objective function can be evaluated quickly. The classical example is the sequencing and timing of batches in multiproduct plant, where the decision variables are the sequence of product batches, and the completion time of any candidate solution is easily evaluated through recurrence relations or minimax algebra. The main disadvantages are that it is difficult to consider general processes, and inequality constraints and continuous decisions, although some recent work aims at addressing this. Xia and Macchietto (1997) use a stochastic search approach for the solution of the mixed integer non-linear programming model of Zhang and Sargent (1994, 1996) reviewed below.

Heuristic/stochastic Search: General Processes

The problem of scheduling in general multipurpose plants is complicated by the additional decisions of assignment of equipment items to processing tasks, task batchsizes and intermediate storage utilisation. It is difficult to devise a series of rules to resolve these, and there are therefore few heuristic approaches reported for the solution of this problem.

Kudva et al. (1994) consider the special case of "linear" multipurpose plants where products flow through the plant in a similar fashion, but potentially using different stages and with no recycling of material. They take account of limited intermediate storage, material receipts at any stage, soft order deadlines, changeover costs and pre-specified equipment maintenance times. A rule-based constructive heuristic is used, which requires the maintenance of a status sheet on each unit and material type for each time instance on a discrete-time grid. The algorithm uses this status sheet with a sorted list of orders and develops a schedule for each order by backwards recursive propagation. The schedule derived depends strongly on the order sequence. Solutions were found to be within acceptable bounds of optimality when compared with those derived through formal optimisation procedures.

Graells et al. (1996) presented a heuristic strategy for the scheduling of multipurpose batch plants with mixed intermediate storage policies. A decomposition procedure is employed where subschedules are generated for the production of intermediate materials. Each subschedule consists of a mini production path determined through a branch-and-cut enumeration of possible unit-to-task allocations. The mini-paths are then combined to form the overall schedule. The overall schedule is checked for feasibility with respect to material balances and storage capacities. Improvements to the schedules may be effected manually through an electronic Gantt chart or through a simulated annealing procedure.

As mentioned earlier, the application of heuristics to such problems is not straightforward. Although this effectively represents current industrial practice, most academic research has been directed towards the development of mathematical programming approaches for multipurpose plant scheduling. As will be described later, these approaches are capable of representing all the complex interactions present.

Mathematical Programming: Specific Processes

In this section, we shall first outline some of the features of mathematical programming approaches in general, and then consider their application to processes other than the general multipurpose one. The latter will be considered in the next section.

The application of mathematical programming approaches implies the development of a mathematical model and an optimisation algorithm. Most approaches aim to develop models that are of a standard form (from linear programming (LP) models for refinery planning to mixed integer non-linear programming (MINLP) models for multipurpose batch plant scheduling). These may then be solved by standard software or specialised algorithms that take account of problem structure.

The variables of the mathematical models will tend to include some or all of the following choices, depending on the complexity considered:

i. sequence of products or individual tasks;
ii. timing of individual tasks in the process;
iii. selection of resources to execute tasks at the appropriate times;
iv. amounts processed in each task;
v. inventory levels of all materials over time.

The dichotomic nature of some of the variables (sequencing and resource selection) implies that binary or integer-valued variables will be required.

The selection of values for all the variables will be subject to some or all of the following constraints:

i. non-preemptive processing–once started, processing activities must proceed until completion;
ii. resource constraints–at any time, the utilisation of a resource must not exceed its availability;
iii. material balances;
iv. capacity constraints–processing and storage;
v. orders being met in full by their due-dates.

Finally, optimisation methods dictate that an objective function be defined. This is usually of an economic form, involving terms such as production, transition and inventory costs and possibly revenues from product sales.

A critical feature of mathematical programming approaches is the representation of the time horizon. This is because activities interact through the use of resources and therefore the discontinuities in the overall resource utilisation profiles must be tracked with time, to be compared with resource availabilities to ensure feasibility. The complexity arises because these discontinuities (unlike discontinuities in availabilities) are functions of any schedule proposed and are not known in advance. The two approaches for dealing with this are:

1. Discrete-time–the horizon is divided into a number of equally spaced intervals so that any event that

introduces such discontinuities (e.g. the starting of a task or a due-date for an order) can only take place at an interval boundary. This implies a relatively fine division of the time grid, so as to capture all the possible event times, and in the solution to the problem it is likely that many grid points will not actually exhibit resource utilisation discontinuities.

2. Continuous time—here, the horizon is divided into fewer intervals, the spacing of which will be determined as part of the solution to the problem. The number of intervals will correspond more closely to the number of resource utilisation discontinuities in the solution.

In addition to the above, another attribute of time representation is whether the same grid is used for all major equipment items in the plant (the "common grid" approach) or whether each major equipment item operates on its own grid (the "individual resource grid"). Generally speaking, the former approach is more suitable for processes in which activities on the major equipment items also interact with common resources (materials, services etc.) and the latter where activities on the major equipment items are quite independent in their interactions with common resources. These distinctions will become clearer when individual pieces of research are discussed.

The simplest specific scheduling process is probably a single production line which produces one product at time in a continuous fashion. Work in this area has been directed towards deriving cyclic schedules (where the production pattern is repeated at a fixed frequency) which balance inventory and transition costs, by determining the best sequence of products and their associated run-lengths or lot-sizes. A review of this so-called "economic lot scheduling problem" is given by Elmaghraby (1978).

Sahinidis and Grossmann (1991a) consider the more general problem of the cyclic scheduling of a number of parallel multiproduct lines, where each product may in principle be produced on more than one line and production rates and costs vary between lines. They utilise a continuous time individual resource grid model, which turns out to be a MINLP. This includes an objective function that includes on combined production, product transition and inventory costs for a constant demand rate for all products. The assumption of cyclic operation simplifies the model and the Benders decomposition approach utilised to solve the problem.

Their work was extended by Pinto and Grossmann (1994) who considered the case of multiple production lines, each consisting of a series of stages decoupled by intermediate storage and operating in a cyclic mode. Each product is processed through all stages, and each product is processed only once at each stage. The model again uses a continuous time model, and it is possible to use the independent grid approach despite the fact that stages interact through material balances–this is due to the special structure of the problem. A separate grid is used for tasks as well, and constraints used to ensure correct anchoring of activities once tasks are assigned to units. A convex approximation to the inventory costs is used, and the problem solved to optimality using a Benders decomposition approach. Both of the approaches above were demonstrated on industrial-scale examples.

A number of mathematical programming approaches have been developed for the scheduling of multiproduct batch plants. All are based (either explicitly or implicitly) on a continuous representation of time.

Pekny et al. (1988) considered the special case of a multiproduct plant with no storage (zero wait, ZW) between operations. They show that the scheduling problem has the same structure as the asymmetric travelling salesman problem, and apply an exact parallel computation technique employing a tailored branch-and-bound procedure which uses an assignment problem to provide problem relaxations. The work was extended to cover the case of product transition costs, where the problem structure is equivalent to the prize-collecting travelling salesman problem (Pekny et al., 1990), and linear programming relaxations are used. For both cases, problems of very large magnitude were solved to optimality with modest computational effort. Gooding et al. (1994) augmented this work to cover the case of multiple units at each stage (the so-called "parallel flowshop" stage).

A more complete overview of the development of algorithms for classes of problems ("algorithm engineering") is given by Applequist et al. (1997) and a recent commercial development in this area is described by Bunch (1997).

Birewar and Grossmann (1989) developed a mixed integer programming model for a similar type of plant. They show that through careful modelling of slack times, and by exploiting the fact that relatively large numbers of batches of relatively few products will be produced (which allows end-effects to be ignored), a straightforward LP model can be used to minimise the makespan. The result is a family of schedules, from which an individual schedule may be extracted. They extend the work to cover simultaneous long-term planning and scheduling, where the planning function takes account of scheduling limitations (Birewar and Grossmann, 1990).

Pinto and Grossmann (1995) describe a MILP model for the minimisation of earliness of orders for a multiproduct plant with multiple equipment items at each stage. The only resources required for production are the processing units. The interesting feature of the model is the representation of time. Two types of individual time grids are used: one for units and one for orders. For each unit, a number of intervals of unknown duration are defined, which represent the possible sequence of tasks (one per interval). For each order, the time interval represents a processing stage. Processing times are unit-dependent, so these interval durations are also unknown.

A series of mixed integer constraints are then require to ensure that when a stage of an order is assigned to a unit, the start times on both grids are equal. The material balances are handled simply by precedence relations and hence do not constitute common resource interactions. Pinto and Grossmann (1997) then augmented the model to take account of interactions between processing stages and common resources (e.g. steam). Rather than utilise a common grid, they retain the individual grids, and account for the resource discontinuities through complex mixed integer constraints which weaken the model and result in large computational times. They therefore proposed a hybrid logic-based/MILP algorithm where the disjunctions relate to the relative timing of orders. This dramatically reduces the computational effort expended.

Moon et al. (1996) also developed a MILP model for ZW multiproduct plants. The objective was to assign tasks to sequence positions so as to minimise the makespan, with non-zero transfer and set-up times being included.

The extension of the work to more general intermediate storage policies was described by Kim et al. (1996) who proposed several MINLP formulations based on completion time relations.

The case of single-stage processes with multiple units per stage has been considered by Cerda et al. (1997) and Karimi and McDonald (1997). Both describe continuous-time based MILP models. Cerda et al. focus on changeovers and order fulfilment, while Karimi and McDonald focus on semicontinuous processes and total cost (transition, shortage and inventory) with the complication of minimum run lengths. A characteristic of both approaches is that discrete demands must be captured on the continuous time grid.

The work described above all relates to special process structures, which means that mathematical models can be designed specifically for the problem class. This ensures that, despite the typical concerns about computational complexity of discrete optimisation problems, solutions are available with reasonable effort. The drawback of the work is its limited applicability. Nevertheless, several models appear to have been developed with specific industrial applications in mind (e.g. Sahinidis and Grossmann, 1991a; Pinto and Grossmann, 1995; Karimi and McDonald, 1997).

Mathematical Programming: Multipurpose Plants

A large portion of the most recent research in planning and scheduling undertaken by the process systems community relates to the development of mathematical programming approaches applied to multipurpose plants. As intimated earlier, in this case the application domain tends to imply the solution approach—mathematical models are the best way of representing the complex interactions between resource allocations, task timings, material flows and equipment capacities.

The work in this are can be characterised by three different assumptions about plant operation.

i. the unique assignment case - each task can only be performed by a unique piece of equipment, and there are no optional tasks in the process recipe and batchsizes are usually fixed;
ii. campaign mode of operation - the horizon is divided into relatively long campaigns, and each campaign is dedicated to one or a few products;
iii. short-term operation - products are produced as required and no particular scheduling pattern may be assumed.

The first assumption is particularly restrictive. The second relates to a mode of operation that is becoming relatively scarce as it implies a low level of responsiveness. One sector in which campaign operation is still prevalent is in the manufacture of active ingredients for pharmaceuticals and agrochemicals. The short-term mode of operation is tending to become the most prevalent elsewhere as it best exploits operational flexibility to meet changing external circumstances.

Rich and Prokopakis (1986) developed a MILP model based on short-term operation and unique assignment for scheduling of multipurpose plants to meet orders, assuming that each order is met from its own production run.

Mauderli and Rippin (1979) developed a procedure for campaign planning which attempts to optimise the allocation of equipment to tasks. An enumerative procedure (based on different equipment-to-task allocations) is used to generate possible single-product campaigns which are then screened by LP techniques to select the dominant ones. A production plan is then developed by the solution of a MILP that sequences the dominant campaigns and fixes their lengths. The disadvantages of this work are the inefficiency of the generation procedure and the lower level of resource utilisation implied by single-product campaigns. Wellons and Reklaitis (1991a, 1991b) address this through a formal MINLP method to generate campaigns and production plans in a two-stage procedure, as do Shah and Pantelides (1991) who solve a simultaneous campaign generation and production planning problem.

An early application of mathematical programming techniques for short-term multipurpose plant scheduling was the MILP approach of Kondili et al. (1988). They used a discrete representation of time, and introduced the State-Task Network (STN) representation of the process (see figure 2; circles are material states and rectangles are tasks).

The STN representation has three main advantages:

i. it distinguishes the process operations from the resources that may be used to execute them, and therefore provides a conceptual platform from

which to relax the unique assignment assumption and optimise unit-to-task allocation;
ii. it avoids the use of task precedence relations which become very complicated in multipurpose plants–a task can be scheduled to start if its input materials are available in the correct amounts and other resources (processing equipment and utilities) are also available, regardless of the plant history;
iii. it provides a means of describing very general process recipes, involving batch splitting and mixing and material recycles, and storage policies including ZW, no-intermediate storage, multipurpose storage tanks and so on.

The formulation of Kondili et al. (1988) (described in more detail in Kondili et al., 1993) is based on the definition of binary variables that indicate whether tasks start in specific pieces of equipment at the start of each time period, together with associated continuous batchsizes. Other key variables are the amount of material in each state held in dedicated storage over each time interval, and the amount of each utility required for processing tasks over each time interval.

Their key constraints related to equipment and utility usage, material balances and capacity constraints. The common, discrete time grid captures all the plant resource utilisations in a straightforward manner; discontinuities in these are forced to occur at the predefined interval boundaries. Their approach was hindered in its ability to handle large problems by the weakness of the allocation constraints and the general limitations of discrete-time approaches such as the need for relatively large numbers of grid points to represent activities with significantly different durations.

Their work formed the basis of several other pieces of research aiming to take advantage of the representational capabilities of the formulation while improving its numerical performance. Sahinidis and Grossmann (1991b) disaggregated the allocation constraints and also exploited the embedded lot-sizing nature of the model where relatively small demands are distributed throughout the horizon. They disaggregate the model in a fashion similar to that of Krarup and Bilde (1977) are were able to improve the solution efficiency despite the larger nature of the disaggregated model. This was due to a feature particular to mixed integer problems: other things being equal, the computational effort for problem solution through standard procedures is dictated mainly by the difference between the optimal objective function and the value of the objective function obtained by solving the continuous relaxation where bound constraints rather than integrality restrictions are imposed on the integer variables (the so-called "integrality gap"). The formulation of Sahinidis and Grossmann (1991b) was demonstrated to have a much smaller integrality gap than the original.

Shah et al. (1993a) modified the allocation constraints even further to generate the smallest possible integrality gap for the type of formulation. They also devised a tailored branch-and-bound solution procedure which utilises a much smaller LP relaxation and solution processing to improve integrality at each node. The same authors (Shah et al., 1993b) considered the extension to cyclic scheduling, where the same schedule is repeated at a frequency to be determined as part of the optimisation. This was augmented by Papageorgiou and Pantelides (1996a, 1996b) to cover the case of multiple campaigns, each with a cyclic schedule to be determined.

Elkamel (1993) also proposed a number of measures to improve the performance of the STN-based discrete-time scheduling model. A heuristic decomposition method was proposed, which solves separate scheduling problems for parts of the overall scheduling problem. The decomposition may be based on the resources ("longitudinal decomposition") or on time ("axial decomposition"). In the former, the recipes and suitable equipment for each task are examined for the possible formation of unique task-unit subgroups which can scheduled separately. Axial decomposition is based on grouping products by due dates and decomposing the horizon into a series of smaller time periods, each concerned with the satisfaction of demands falling due within it. He also described a perturbation heuristic, which is a form of local search around the relaxation. The proposed techniques worked well on the class of test examples.

Yee and Shah (1997, 1998) and Yee (1998) also considered various manipulations to improve the performance of general discrete-time scheduling models. A major feature of their work is variable elimination. They recognise that in such models, only about 5-15% of the variables reflecting task-to-unit allocations are active at the integer solution, and it would be beneficial to identify as far as possible inactive variables prior to solution. They describe an LP-based heuristic, a flexibility and sequence reduction technique and a formal branch-and-price method. They also recognise that some problem instances result in poor relaxations and propose valid inequalities and a disaggregation procedure similar to that of Sahinidis and Grossmann (1991b) for particular data instances.

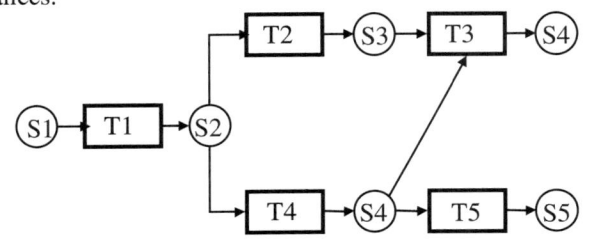

Figure 2. State-task network representation.

Gooding (1994) considers a special case of the problem with firm demands and dedicated storage only. The scheduling model is described in a digraph form where nodes correspond to possible task-unit-time allocations and arcs the possible sequences of the activities. The explicit description of the sequence in this form addresses one of the weaknesses of the discrete-time formulation of Kondili et al. (1998, 1993), which was that it did not model sequence-dependent changeovers very well. Gooding's (1994) model therefore performed relatively well in problems with a strong sequencing component, but suffers from model complexity in that all possible sequences must be accounted for directly.

Pantelides et al. (1995) reported a STN-based approach to the scheduling of pipeless plants, where material is conveyed between processing stations in movable vessels. This requires the simultaneous scheduling of the movement and processing operations.

Pantelides (1994) presented a critique of the STN and associated scheduling formulations. He argued that despite its advantages, it suffers from a number of drawbacks:

i. the model of plant operation is somewhat restricted—each operation is assumed to use exactly one major item of equipment throughout its operation;
ii. tasks are always assumed to be processing activities which change material states—changeovers or transportation activities have to be treated as special cases;
iii. each item of equipment is treated as a distinct entity—this introduces solution degeneracy if multiple equivalent items exist;
iv. different resources (materials, units, utilities) are treated differently, giving rise to many different types of constraints, each of which must be formulated carefully to avoid unnecessarily increasing the integrality gap.

He then proposed an alternative representation, the Resource-Task Network (RTN), based on a uniform description of all resources. In contrast to the STN approach, where a task consumes and produces materials while using equipment and utilities during its execution, in this representation, a task is assumed only to consume and produce resources. Processing items are treated as though consumed at the start of a task and produced at the end. Furthermore, processing equipment in different conditions (e.g. "clean" or "dirty") can be treated as different resources, with different activities (e.g. "processing" or "cleaning") consuming and generating them - this enables a simple representation of changeover activities. Pantelides (1994) also proposed a discrete-time scheduling formulation based on the RTN which, due to the uniform treatment of resources, only requires the description of three types of constraint, and does not distinguish between identical equipment items (which results in more compact and less degenerate optimisation models). He illustrated that the integrality gap could not be worse than the most efficient form of STN formulation, but that the ability to capture additional problem features in a straightforward fashion made it an ideal framework for future research.

The review above has mainly considered the development of discrete-time models. As argued by Schilling (1997), discrete-time models have been able to solve a large number of industrially-relevant problems (see, e.g. Tahmassebi, 1996), but suffer from a number of inherent drawbacks:

i. the discretisation interval must be fine enough to capture all significant events—this may result in a very large model;
ii. it is difficult to model operations where the processing time is dependent on the batchsize;
iii. the modelling of continuous and semi-continuous operations must be approximated, and minimum run-lengths give rise to complicated constraints.

A number of researchers have therefore attempted to develop scheduling models for multipurpose plants which are based on a continuous representation of time, where fewer grid points are required as they will be placed at the appropriate resource utilisation discontinuities during problem solution.

Zentner and Reklaitis (1992) describe a formulation based on the unique assignment case and fixed batchsizes. The sequence of activities as well as any external effects can be used to infer the discontinuities and therefore the interval boundaries. A MILP optimisation is then used to determine the exact task starting times.

Reklaitis and Mockus (1995) detail a continuous-time formulation based on the STN formulation, and exploiting its generality. A common resource grid is used, with the timing of the grid points ("event orders" in their terminology) determined by the optimisation. The model is a MINLP, which may be simplified to a mixed integer bilinear problem by linearising terms involving binary variables. This is solved using an outer-approximation algorithm. Only very preliminary findings are reported, but the promise of such models is evident.

Mockus and Reklaitis (1996) then reported an alternative solution procedure. They introduce the concept of Bayesian heuristics, which are heuristics that can be described through parameterised functions. The Bayesian technique iteratively modifies the parameters to develop a heuristic that is expected to perform well across a class of problem parameters. They illustrate the procedure using a material requirements planning (MRP) backward scheduling heuristic which outperforms a standard discrete-time MILP formulation solved using branch-and-bound.

Zhang and Sargent (1994, 1996) presented a continuous time formulation based on the RTN representation for both batch and continuous operations, with the possibility of batchsize-dependent processing times for batch operations. Again, the interval durations are determined as part of the optimisation. A MINLP model ensues; this is solved using a local linearisation procedure combined with what is effectively a column generation algorithm.

A problem with continuous time models of the form described above arises out of the inclusion of products of binary variables and interval durations or absolute starting times in the constraints. The linearisation of these products gives rise to terms involving products of binary variables and maximum predicted interval durations or starting times. The looser these upper bounds, the worse the integrality gap of the formulation and, in general, the more difficult it becomes to solve the scheduling problem. Furthermore, it is difficult to predict good duration bounds a priori. The poor relaxation performance of the continuous time models is the main obstacle to their more widespread application.

Schilling and Pantelides (1996) and Schilling (1997) attempt to address this deficiency. They developed a continuous-time scheduling model based on the RTN. They propose a number of modifications to the formulation of Zhang and Sargent (1996) which simplify the model and improve its general solution characteristics. A global linearisation gives rise to a MILP. They then developed a hybrid branch-and-bound solution procedure which branches in the space of the interval durations as well as in the space of the integer variables. For a given problem instance, this can be viewed as generating a number of problem instances, each with tighter interval duration bounds. The independence of these new instances was recognised by Schilling (1997), who implemented a parallel solution procedure based on a distributed computing environment. The combination of the hybrid and parallel aspects of the solution procedure resulted in a much improved computational performance on a wide class of problems.

Multisite and Long-term Planning

It might be expected that large benefits would ensue from co-ordinated planning across sites, in terms of costs and market effectiveness. Most business processes dictate that a degree of autonomy is required at each manufacturing and distribution site, but pressures to co-ordinate responses to global demand while minimising cost imply that simultaneous planning of production and distribution across plants and warehouses should be undertaken. This would result in the most efficient utilisation of all resources. A target-setting approach, where central plans set achievable production targets without imposing operational details is compatible with operational details being determined at each site.

The need for such co-ordinated planning has long been recognised in the management science and operations research literature. For example, Williams (1981) surveyed a series of heuristics for production-distribution scheduling in multisite systems for different network structures. There are two general weaknesses with research in this domain:

i. steady-state demands are assumed;
ii. simple expressions or even constants are used for plant capacity.

In practice, demands are usually time-varying and the capacity of a flexible manufacturing facility cannot be known a priori, but is rather a function of the product mix and the details of scheduling.

It is only very recently that researchers in the process systems engineering community have started to address this topic.

Wilkinson et al. (1994) showed how the RTN can be used to represent a variety of distribution options. The multisite planning problem can therefore be directly posed using the RTN representation and the discrete-time model of Pantelides (1994). Wilkinson et al. (1994) recognised that a potential problem with this approach is the very large problem sizes that will ensue. A secondary issue is that the development of a central plan to a very fine level of detail is probably unnecessary. This led to the development of an aggregation procedure (Wilkinson et al., 1995). The aim is to capture production and distribution capacities accurately without considering detailed scheduling. The method involves aggregating the many discrete intervals into fewer, longer intervals known as aggregated time periods (ATPs). The summation of the detailed variables and constraints over the detailed time periods in the ATP gives rise to surrogate aggregated variables and constraints which are much fewer in number. Detailed variables corresponding to activities that cross ATP boundaries are retained as "linking" variables. A feature of the formulation is that in addition to the basic summation, higher "orders" of aggregation are also defined, in a manner akin to moments of a distribution. This gives rise to a family of aggregate formulations. By increasing the order or the number of ATPs, the accuracy of the formulation is increased, and the aggregate formulation tends exactly to the original detailed formulation.

Wilkinson et al. (1996) applied this technique to a continent-wide industrial case study. This involved optimally planning the production and distribution of a system with three factories and fourteen market warehouses and over a hundred products. A great deal of flexibility existed in the network which in principle enables the production of products for each market at each manufacturing site. It was found that the ability of the technique to capture effects such as multipurpose operation, intermediate storage and changeovers gave rise to counter-intuitive results, such as producing materials further away from demand points than would be expected. This balances the complexity associated with producing many products in each factory with the extra distribution costs incurred by concentrating the manufacture of specific products at specific sites.

McDonald and Karimi (1997) describe a similar problem for multiple facilities which effectively produce products on single-stage continuous lines for a number of geographically distributed customers. Their basic model is of multiperiod LP form, and takes account of available processing time on all lines, transportation costs and shortage costs. An approximation is used for the inventory costs, and product transitions are not modelled. The model is extended to include minimum run-lengths (which requires the addition of binary variables and the linking variable concept). They include a number of additional supply chain related constraints such as single sourcing, internal sourcing and transportation times. The formulation was capable of solving realistic problems but is inadequate when the period durations are small relative to run-lengths.

The area of long- or medium-term planning at an individual site can be thought of as a special case of supply chain planning. The work in this area can be broadly divided into two types:

i. planning models based on lower-level scheduling models;
ii. independently-derived planning models.

The latter case shall be considered first. Again, a great deal of work has been undertaken in the OR/MS field. Gabbay (1979) presented a simple multiperiod model (based on a multiproduct structure) where each period has different demands and amounts processed at each stage are subject to aggregate capacity constraints, based on demand factors placed on the stage equipment by each product. This type of aggregate model is prevalent in the literature.

Other planning models of this type do not consider each product in isolation, but rather group products that place similar demands on resources into families, and base the higher level planning function on these families. This forms the basis of many hierarchical production planning systems (see, e.g., Bitran et al., 1982).

More sophisticated models exist in the process systems literature. Sahinidis et al. (1989) describe a model which selects processes to operate from an integrated network, and ensures that the network capacity constraints are not exceeded. Sahinidis and Grossmann (1991b) and Liu and Sahinidis (1995) describe means of improving the solution efficiency of this class of problems. Most other models of this type are used for planning under uncertainty and will be considered in a later section.

As pointed out by Reklaitis (1991), the distinction between planning and scheduling is a somewhat artificial one, and the more consistent the models across these levels, the better the scope for integration. Recent research has therefore aimed at developing planning models which are consistent with lower level scheduling models. These result in a better representation of flexible plant capacity and work in the full product space; it is often difficult to perform product grouping in the general multipurpose case.

Bassett et al. (1996a) give a detailed exposition on a number of decomposition techniques for the solution of large scale scheduling problems. They build upon the group's previous work in this area (Bassett et al., 1996b; Subrahmanyan et al., 1996) and focus mainly on long-term planning and scheduling at a single site, although a multisite motivating example is presented where the sites are effectively decoupled. The main objective is to combine aggregate and detailed models ultimately to develop detailed schedules over long horizons. Their approaches are described below.

First, an aggregation and disaggregation technique using a series of heuristics is presented. An aggregate multiperiod model is generated from a detailed STN-based discrete-time model by summing variables and constraints over the time periods and replacing these sums by aggregate variables and constraints in a manner somewhat similar to that of Wilkinson et al. (1994). The solution to the aggregate model effectively fixes the production requirements within each time period. In order to assess plan feasibility, a detailed short-term scheduling problem must then be solved for each period with the objective of meeting these requirements. If all individual problems are feasible, then the procedure terminates. Otherwise, some means of revising the solution must be implemented. Two techniques are described. First, a heuristic is used which attempts to move or merge batches, or to effect improved splicing between the individual scheduling problems. A more formal procedure is also described, where slack variables are introduced which indicate capacity or inventory shortfalls at the detailed level. These are then used to impose restrictions at the aggregate level, and the procedure continues. One problem identified with this approach arises out of the fact that activities are not allowed to cross aggregate interval boundaries. This results in significant end-effects, where processing chains between storage must be initiated at the beginning of each interval.

A more formal approach is also defined (based on that of Subrahmanyam et al., 1996) where the high-level model aggregate model is modified if any of the detailed scheduling problems in an interval are infeasible. In this case, the part of the model corresponding to the infeasible interval is expanded in full detail, and the high-level model becomes an aggregate-disaggregate model. In the worst-case, one would have to solve the detailed problem for the whole time horizon.

This then forms the basis for a single-level backwards rolling horizon heuristic. Here, there two main modifications. First, linking variables (Wilkinson et al., 1994) are introduced to allow activities to span aggregate intervals, and secondly a single level problem is solved in an evolutionary fashion. The algorithm proceeds as follows:

1. The planning horizon is divided into intervals (time blocks). All but the last are modelled in an aggregate form and the last is modelled in detailed form.
2. The resulting MILP is solved to optimality. If the all intervals are modelled at a detailed level, the algorithm stops.
3. Once the MILP is solved, the variables associated with the detailed intervals are fixed to their optimal values.
4. The latest aggregate interval is modelled in detail and the procedure is repeated from step 2.

The procedure is efficient because only one interval is solved in detail at any time (the variables corresponding to later detailed periods are fixed). This heuristic procedure appeared to work well on selected problems, and was enhanced by including an interval of intermediate detail between the detailed and aggregate ones. This is achieved by modelling the intermediate interval in detail but not enforcing integrality conditions on the binary variables.

Dimitriadis et al. (1997a) describe two rolling horizon procedures for medium-term planning and scheduling, based on the more general RTN formulation. They take advantage of the unique properties of Wilkinson et al.'s (1994) aggregation in this context. In the forwards rolling horizon algorithm, the horizon is divided into two time blocks. The first is relatively short and modelled in detail, while the second is relatively long and modelled using the aggregate scheduling formulation. The solution of this MILP gives rise to a detailed solution for the first period and an aggregate one for the second. Dimitriadis et al. (1997a) recognised that, rather than fix all the variables in the first period at the next iteration of the procedure, it makes sense only to fix the complicating integer variables and leave the continuous ones free for further optimisation. At the next iteration, there are three time blocks, the first one with fixed integer variables, the second one modelled in detail and the third (the remainder of the horizon) modelled at an aggregate level. The algorithm proceeds until a detailed solution is obtained for the entire horizon. Although a heuristic, the procedure was shown to generate very good solutions. There are no problems with end-effects as the aggregation procedure automatically includes linking variables. A backwards rolling horizon algorithm is also presented which proceeds in an opposite fashion.

Dimitriadis et al. (1997b) presented a rigorous method to develop such schedules, based on an iterative decomposition procedure. It requires the solution of a higher level aggregate scheduling problem which generates a number of independent detailed scheduling problems which can be solved in parallel. Although similar in spirit to the procedure of Subrahmanyam et al. (1996), it has one key difference. It exploits the specific feature of the aggregation procedure of Wilkinson et al. (1994) that the aggregate formulation tends to the detailed one as the number of ATPs increases. In this method, if a detailed scheduling problem is found to be infeasible, instead of modelling the interval in full detail at the aggregate level, the interval (ATP) is split into two intervals (ATPs) of equal duration. This immediately improves the accuracy, while only modestly increasing the computational effort. A great improvement in performance over a single-level formulation is possible, especially if the inherent parallelism is exploited.

Uncertainty in Planning and Scheduling

The area of uncertainty in planning and scheduling is worthy of a review in its own right, and will be considered somewhat briefly here.

Sources of uncertainty (which tend to imply the means for dealing with them) can crudely be divided into:

i. short-term uncertainties such as processing time variations, rush orders, failed batches, equipment breakdowns, etc.;
ii. long-term uncertainties such as market trends, technology changes etc.

Traditionally, short-term uncertainties have been treated through on-line or reactive scheduling, where schedules are adjusted to take account of new information. Longer-term uncertainties have been tackled through the solution of some form of stochastic programming problem. These two areas are considered below.

Reactive Scheduling

A major requirement of reactive scheduling systems is the ability to generate feasible updated schedules relatively quickly. A secondary objective is often to minimise deviations from the original schedule. As plants become more automated, this may become less important.

Cott (1989) presented some schedule modification algorithms to be used in conjunction with on-line monitoring, in particular to deal with processing time variations and batch size variations.

Kanakamedala et al. (1994) presented a least impact heuristic beam search for reactive schedule modification in the face of unexpected deviations in processing times and resource availability. This is based on evaluating possible product re-routings and selecting that which has least overall impact on the schedule.

Rodrigues et al. (1996) modify the discrete-time STN formulation to take account of due-date changes and equipment unavailability. They use a rolling horizon (rolling out a predefined schedule) approach which aims to look ahead for a short time to resolve infeasibilities. This implies a very small problem size and fast solution times.

Schilling (1997) adapts his RTN-based continuous time formulation to create a hierarchical family of MILP-based reactive scheduling formulations. At the lowest level, the sequence of operations is fixed as in the original schedule and only the timing can vary. At the top-most level, a full original scheduling problem is solved. The intermediate levels all trade off degrees-of-freedom with computational effort. This allows the best solution in the time available to be implemented on the plant.

Planning under Uncertainty

Most of the work in this area is based on models in which product demands are assumed to be uncertain and to differ between a number of time periods. Usually, a simple representation of the plant capacity is assumed, and the sophistication of the work relates to the implementation of stochastic planning algorithms to select amounts for production in the first period ("here and now") and potential production amounts in different possible demand realisations in different periods (see, e.g., Ierapetritou et al., 1996).

In relatively long-term planning, it is reasonable to introduce additional degrees of freedom associated with potential capacity expansions. Liu and Sahinidis (1996a, 1996b) and Iyer and Grossmann (1998) extended the MILP process and capacity planning model of Sahinidis and Grossmann (1991b) to include multiple product demand scenarios in each period. They then propose efficient algorithms for the solution of the resulting stochastic programming problems (formulated as large deterministic equivalent models), either by projection (Liu and Sahinidis, 1996a) or by decomposition and iteration (Iyer and Grossmann, 1998). A major assumption in their formulation is that product inventories are not carried over from one period to the next. This has the advantage in ensuring that the problem size is of $O(np \cdot ns)$ where np is the number of periods and ns is the number of demand scenarios, rather than $O(ns^{np+1})$. However, if the periods are too short, this will compromise the solution from two perspectives:

i. all products must be produced in all periods if demand exists for them - this may be suboptimal;
ii. plant capacity must be designed for a peak demand period (particularly constraining for seasonal products).

Clay and Grossmann (1994) address this issue. They considered the structure of both the two-period and multiperiod problem for LP models and derived an approximation method based on successive repartitioning of the uncertain space with expectations being applied over partitions. This has the potential to generate solutions to a high degree of accuracy in a much faster time than the full-scale deterministic equivalent model.

The approaches above are based on relatively simple models of plant capacity. Petkov and Maranas (1997) treat the multiperiod planning model for multiproduct plants under demand uncertainty. Their planning model embeds the planning/scheduling formulation of Birewar and Grossmann (1990) and therefore accurately calculates the plant capacity. They do not use discrete demand scenarios but assume normal distributions and directly manipulate the functional forms to generate a problem which maximises expected profit and meets certain probabilistic bounds on demand satisfaction without the need for numerical integration. They also make the no inventory carry-over assumption, but show how this can be remedied to a certain extent at the lower level scheduling stage.

Alternative, more recent approaches have attempted to characterise the effects of some sources of uncertainty on detailed schedules.

Rotstein et al. (1994) defined flexibility and reliability indices for detailed schedules. These are based on data for equipment reliability and demand distributions. Given a schedule (described in network flow form), these indices can be calculated to assess its performance.

Dedopoulos and Shah (1995) used a multistage stochastic programming formulation to solve short-term scheduling problems with possibilities of equipment failure at each discrete-time instant. The technique can be used to assess the impact of different failure characteristics of the equipment on expected profit, but suffers from the very large computational effort required even for small problems.

Sanmarti et al. (1995) define a robust schedule as one which has a high probability of being performed, and is readily adaptable to plant variations. They define an index of reliability for a unit scheduled in a campaign through its intrinsic reliability, the probability that a standby unit is available during the campaign, and the speed with which it can be repaired. An overall schedule reliability is then the product of the reliabilities of units scheduled in it, and solutions to the planning problem can be driven to achieve a high value of this indicator.

Mignon et al. (1995) assess schedules obtained from deterministic data for performance under variability by Monte Carlo simulation. Although a number of parameters may be uncertain, they focus on processing time. Performance and robustness (predictability) metrics are defined and features of schedules with good indicators are summarised (e.g. introducing an element of conservatism when fixing due-dates).

Honkomp et al. (1997) build on this to compare schedules generated by discrete-time and continuous-time algorithms and two means of ensuring robustness in the face of processing time uncertainties, namely increasing the processing times of bottleneck stages and increasing all processing times at the deterministic scheduling level. They found that the latter heuristic was better, and that the rounding effect of the discrete time model results in marginally better robustness. Robustness is defined with respect to variance in the objective function. Strictly speaking, penalising the variance of a metric to ensure robustness assumes that the metric is two-sided (i.e. "the closer to nominal the better" in the Taguchi sense). Since economic objective functions are one-sided ("the more the better") robustness indicators such as these should be used with caution. This has been noted recently by Ahmed and, Sahinidis (1998).

Gonzalez and Realff (1998a) analyse MILP solutions for pipeless plants that generated by assuming lower level controls for detailed vehicle movements and fixed, nominal transfer times. The analysis performed using stochastic simulation with variabilities in the transfer times. The system performance was found not to degrade considerably from its nominal value. They extended the work (Gonzalez and Realff (1998b) to consider the development of dispatching rules based on both general flexible manufacturing principles and properties of the MILP solutions. They found that rules abstracted from the MILP solutions were superior, and could be used in real-time.

Bassett et al. (1997) contrast aggregate planning and detailed scheduling under uncertainties in processing times and equipment failure. They argue that aggregate models that take these into account miss critical interactions due to the complex short-term interactions. They therefore propose the use of detailed scheduling to study the effects of such uncertainties on aggregate indicators such as average probabilities in meeting due dates and makespans. They also use Monte Carlo simulation, but use each set of sampled data to generate a detailed scheduling problem instance, solved using a reverse rolling horizon algorithm. Once enough instances have been solved for statistical significance, a number of comparisons can be made. For example, they conclude that long, infrequent breakdowns are more desirable, with obvious implications for maintenance policies.

Conclusions and Future Challenges

Revisiting the challenges posed by Reklaitis (1991) and Rippin (1993), it is clear that considerable progress has been made towards meeting them.

Overall, the emerging trend in the area of short-term scheduling is the development of techniques for the solution of the general, resource-constrained multipurpose plant scheduling problem. In order to capture the generality of resource-task interactions, the common grid approach is prevalent. More intervals are needed than the individual resource grid, but the formulation of common resource constraints is very difficult and is likely to be restricted to special cases (e.g. Pinto and Grossmann, 1997).

Many algorithms have been developed to exploit the tight relaxation characteristics of discrete-time formulations, in particular seeking to avoid having to solve the full problem in a single instance. There remains work to be done in this area, in particular to exploit the sparsity of the solutions. Direct intervention at the LP level during branch-and-bound procedures (e.g. column generation and branch-and-price) seems a promising way of solving very large problems without ever considering the full variable space. Rigorous decomposition techniques based on the time domain have been proposed; a new challenge is the development of such techniques which operate in the task/resource domain by identifying weakly connected components that can scheduled independently in an iterative procedure.

The anticipated continuous time formulations have started to emerge, but can still be considered to be in their infancy. The main challenge here is in continual improvement in problem formulation and preprocessing to improve relaxation characteristics, and tailored solution procedures (e.g. branch-and-cut, and hybrid logic-continuous variable-integer variable branching) for problems with relatively large integrality gaps.

An important contrast between early and recent work is that the early algorithms tended to be tested on "motivating" examples (e.g. to find the best sequence of a few products), while recent algorithms are almost always tested on (and often motivated by) industrial or industrially-based studies.

The multisite problem has received relatively little attention, and is likely to be a candidate for significant research in the near future. A major challenge is to develop planning approaches that are consistent with detailed production scheduling at each site and distribution scheduling across sites. An obvious stumbling block is problem size, and a resource-task based decomposition based on identifying weak connections should find promise here as the problems tend to be highly structured. It is important to retain an industrial focus to develop relevant techniques.

Researchers have attacked the problem of planning and scheduling under uncertainty from a number of angles, but have tended to skirt around the fundamental problem of multiperiod, multiscenario planning with realistic production capacity models (i.e. embedding some scheduling information) in the case of longer-term uncertainties. Issues that must be resolved relate mainly to problem scale. A sensible way forward is to try to capture the problem in all its complexity and then to explore rigorous or approximate solution procedures, rather than develop exact solutions to somewhat idealised problems. Process industry models are complicated by having multiple stages (periods) and integer variables in the second and subsequent stages, so most of the classical algorithms devised for large scale stochastic planning problems are not readily applicable.

The treatment of short-term uncertainties through the determination of characteristics of resilient schedules and then to use on-line monitoring and rescheduling seems eminently sensible. Further work is required in such characterisation and in the design of rescheduling algorithms with guaranteed real-time performance.

A final challenge relates to the seamless integration of the activities at different levels - this is of a much broader and more interdisciplinary nature.

References

Ahmed, S. and N.V. Sahinidis, "Robust Process Planning under Uncertainty", *Ing. Eng. Chem. Res.*, **37**, 1883-1892 (1998).

Applequist, G., O. Samikoglu, J. Pekny and G.V. Reklaitis, "Issues in the Use, Design and Evolution of Process Scheduling and Planning Systems", *ISA Trans.*, **36**, 81-121 (1997).

Bassett, M.H., J.F. Pekny and G.V. Reklaitis, "Decomposition Techniques for the Solution of Large-Scale Scheduling Problems", *AIChE J.*, **42**, 3373-3387.

Bassett, M.H, F.J. Doyle III, G.K. Kudva, J.F. Pekny, G.V. Reklaitis, S. Subrahmanyam, M.G. Zentner and D.L. Miller, "Perspectives on Model-Based Integration of Process Operations", *Computers chem. Engng.*, **20**, 821-844 (1996b).

Bassett, M.H, J.F. Pekny and G.V. Reklaitis, "Using Detailed Scheduling to Obtain Realistic Operating Policies for a Batch Processing Facility", *Ind. Eng. Chem. Res.*, **36**, 1717-1726 (1997).

Birewar, D.B. and I.E. Grossmann, "Efficient Optimization Algorithms for Zero-Wait Scheduling of Multiproduct Batch Plants", *Ind. Eng. Chem. Process Des. Dev.*, **28**, 1333-1345 (1989).

Birewar, D.B. and I.E. Grossmann, "Simultaneous Production Planning and Scheduling in Multiproduct Batch Plants", *Ind. Eng. Chem. Res.*, **29**, 570-580 (1990).

Bitran, G.R., E.A. Haas and A.C. Hax, "Hierarchical Production Planning: A Two-Stage System", *Opns. Res.*, **30**, 232-251 (1982).

Bunch, P., "A Simplex-Based Primal-Dual Algorithm for the Perfect B-Matching Problem–A Study in Combinatorial Optimisation", Phd Thesis, Purdue University (1997).

Cerda, J., G.P. Henning and I.E. Grossmann, "A Mixed Integer Linear Programming Model for Short-Term Scheduling of Single-Stage Multiproduct Batch Plants with Parallel Lines", *Ind. Eng. Chem. Res.*, **36**, 1695-1707 (1997).

Clay, R.L. and I.E. Grossmann, "Optimization of Stochastic Planning-Models", *Chem. Eng. Res. Des.*, **72**, 415-419 (1994).

Cott, B.J., "An Integrated Computer-Aided Production Management System for Batch Chemical Processes", PhD Thesis, University of London (1989).

Dannebring, D.G., "An Evaluation of Flowshop Sequencing Heuristics", *Man. Sci.*, **23**, 1174-1182 (1977).

Dedopoulos, I.T. and N. Shah, "Preventive Maintenance Policy Optimisation for Multipurpose Plant Equipment", *Comput. chem. Engng*, **S19**, S693-S698 (1995)

Dimitriadis, A.D., N. Shah and C.C. Pantelides, "RTN-based Rolling Horizon Algorithms for Medium-Term Scheduling of Multipurpose Plants", *Computers chem. Engng.*, **S21**, S1061-S1066 (1997a).

Dimitriadis, A.D., N. Shah and C.C. Pantelides, "A Rigorous Decomposition Algorithm for Solution of Large-Scale Planning and Scheduling Problems", paper presented at AIChE Annual Meeting, Los Angeles (1997b).

Elkamel, A., "Scheduling of Process Operations using Mathematical Programming Techniques", PhD Thesis, Purdue University (1993).

Elmaghraby, S., "The Economic Lot Scheduling Problem. Review and Extensions", *Man. Sci.*, **24**, 587-598 (1978).

Gabbay, H., "Multi-Stage Production Planning", *Man. Sci.*, **25**, 1138-1148 (1979).

Gabow, H.N., "On the Design and Analysis of Efficient Algorithms for Deterministic Scheduling", *Proc. 2nd Intl. Conf. Foundations of Computer-Aided Process Design*, Michigan, USA, 473-528 (1983).

Gonzalez, R. and M.J. Realff, "Operation of Pipeless Batch Plants - I. MILP Schedules", *Computers chem. Engng.*, **22**, 841-855 (1998).

Gonzalez, R. and M.J. Realff, "Operation of Pipeless Batch Plants - II. Vessel Dispatch Rules", *Computers chem. Engng.*, **22**, 857-866 (1998).

Gooding, W.B., "Specially Structured Formulations and Solution Methods for Optimisation Problems Important to Process Scheduling", PhD Thesis, Purdue University (1994).

Gooding, W.B., J.F. Pekny and P.S. McCroskey, "Enumerative Approaches to Parallel Flowshop Scheduling via Problem Transformation", *Computers chem. Engng.*, **18**, 909-927 (1994).

Graells, M., A. Espuña and L. Puigjaner, "Sequencing Intermediate Products: A Practical Solution for Multipurpose Production Scheduling", *Computers chem. Engng.*, **S20**, S1137-S1142 (1996).

Hampel, R., "Beyond the Millenium", *Chemistry and Industry*, **10**, 380-382 (1997).

Hasebe, S., I. Hashimoto and A. Ishikawa, "General Reordering Algorithm for Scheduling of Batch Processes", *J. Chem. Eng. Japan*, **24**, 483-489 (1991).

Hasebe, S., S. Taniguchi and I. Hashimoto, "Automatic Adjustment of Crossover Method in the Scheduling Using Genetic Algorithm", *Kagaku Kogaku Ronbunshu*, **22**, 1039-1045 (1996).

Honkomp, S.J., L. Mockus, G.V. Reklaitis, "Robust Scheduling with Processing Time Uncertainty", *Computers chem. Engng.*, **S21**, S1055-S1060 (1997).

Ierapetritou, M.G., E.N. Pistikopoulos and C.A. Floudas, "Operational Planning under Uncertainty", *Computers chem. Engng.*, **20**, 1499-1516.

Iyer, R.R. and I.E. Grossmann, "A Bilevel Decomposition Algorithm for Long-range Planning of Process Networks", *Ind. Eng. Chem. Res.*, **37**, 474-481 (1998).

Kanakamedala, K.B., G.V. Reklaitis and V. Venkatasubramanian, "Reactive Schedule Modification in Multipurpose Batch Chemical Plants", *Ind. Eng. Chem. Res.*, **33**, 77-90 (1994).

Karimi, I.A. and C.M. McDonald, "Planning and Scheduling of Parallel Semicontinuous Processes. 2. Short-term Scheduling", *Ind. Eng. Chem. Res.*, **36**, 2701-2714 (1997).

Kim, M., J.H. Jung and I.-B. Lee, "Optimal Scheduling of Multiproduct Batch Processes for Various Intermediate Storage Policies", *Ind. Eng. Chem. Res.*, **35**, 4048-4066 (1996).

Kondili, E., C.C. Pantelides and R.W.H. Sargent, "A General Algorithm for Scheduling of Batch Operations", *Proc. 3rd Intl. Symp. on Process Systems Engng.*, Sydney, Australia, 62-75 (1988).

Kondili, E., C.C. Pantelides and R.W.H. Sargent, "A General Algorithm for Short-Term Scheduling of Batch Operations - 1. Mixed Integer Linear Programming Formulation", *Computers chem. Engng.*, **17**, 211-227 (1993).

Krarup J. and O. Bilde, "Plant Location, Set Covering and Economic Lot Size: an O(mn) Algorithm for Structured Problems", *Int Ser. Num. Math.*, **36**, 155-180 (1977).

Ku, H. and I.A. Karimi, "An Evaluation of Simulated Annealing for Batch Process Scheduling", *Ind. Eng. Chem. Res.*, **30**, 163-169 (1991).

Kudva, G., A. Elkamel, J.F. Pekny and G.V. Reklaitis, "Heuristic Algorithm for Scheduling Batch and Semicontinuous Plants with Production Deadlines, Intermediate Storage Limitations and Equipment Changeover Costs", *Computers chem. Engng.*, **18**, 859-875 (1994).

Kuriyan, K. and G.V. Reklaitis, "Approximate Scheduling Algorithms for Network Flowshops", In. *Chem. Eng. Symp. Ser.*, **92**, 79-90 (1985).

Kuriyan, K. and G.V. Reklaitis, "Scheduling Network Flowshops so as to Minimise Makespan", *Computers chem. Engng.*, **13**, 187-200 (1989).

Liu, M.L. and N.V. Sahinidis, "Computational Trends and Effects of Approximations in an MILP Model for Process Planning", *Ind. Eng. Chem. Res.*, **34**, 1662-1673 (1995).

Liu, M.L. and N.V. Sahinidis, "Long-range Planning in the Process Industries - a Projection Approach", *Computers Ops. Res.*, **3**, 237-253 (1996a).

Liu, M.L. and N.V. Sahinidis, "Optimization in Process Planning under Uncertainty", *Ind. Eng. Chem. Res.*, **35**, 4154-4165 (1996b).

Mauderli, A.M. and D.W.T. Rippin, "Production Planning and Scheduling for Multi-Purpose Batch Chemical Plants", *Computers chem. Engng.*, **3**, 199-206 (1979).

McDonald, C.M. and I.A. Karimi, "Planning and Scheduling of Parallel Semicontinuous Processes. 1. Production Planning", *Ind. Eng. Chem. Res.*, **36**, 2691-2700 (1997).

McHugh, P. and G. Merli, *Beyond Business Process Reengineering*, John Wiley and Sons, London (1995).

Mignon, D.J., S.J. Honkomp and G.V. Reklaitis, "A Framework for Investigating Schedule Robustness under Uncertainty", *Computers chem. Engng.*, **S19**, S615-S620 (1995).

Mockus, L. and G.V. Reklaitis, "Continuous-Time Representation in Batch/Semicontinuous Process Scheduling - Randomized Heuristics Approach", *Computers chem. Engng.*, **S20**, S1173-S1178 (1996).

Moon, S., S. Park and W.K. Lee, "New MILP Models for Scheduling of Multiproduct Batch Plants under Zero-Wait Policy", *Ind. Eng. Chem. Res.*, **35**, 3458-3469 (1996).

Murakami, Y., H. Uchiyama, S. Hasebe, and I. Hashimoto, "Application of Repetitive SA Method to Scheduling Problems of Chemical Processes", *Computers chem. Engng.*, **S21**, S1087-S1092 (1997).

Pantelides, C.C., "Unified Frameworks for Optimal Process Planning and Scheduling", *Proc 2nd Conf. Foundations of Comp. Aided Proc.Op.*, CACHE Corp., 253-274 (1994).

Pantelides, C.C., M.J. Realff and N. Shah, "Short-Term Scheduling of Pipeless Batch Plants", *Trans. IChemE Part A*, **73**, 431-444 (1995).

Papageorgiou, L.G. and C.C. Pantelides, "Optimal Campaign Planning/Scheduling of Multipurpose Batch/Semicontinuous Plants. 1. Mathematical Formulation", *Ind. Eng. Chem. Res.*, **35**, 488-509 (1996a).

Papageorgiou, L.G. and C.C. Pantelides, "Optimal Campaign Planning/Scheduling of Multipurpose Batch/Semicontinuous Plants. 2. A Mathematical Decomposition Approach", *Ind. Eng. Chem. Res.*, **35**, 510-529 (1996b).

Pekny, J.F., D.L. Miller and G.J. McCrae, "Application of a Parallel Travelling Salesman Problem to No-Wait Flowshop Scheduling", paper presented at AIChE Annual Meeting, Washington D.C. (1988).

Pekny, J.F., D.L. Miller and G.J. McCrae, "An Exact Parallel Algorithm for Scheduling when Production Costs Depend on Consecutive System States", *Computers chem. Engng.*, **14**, 1009-1023 (1990).

Petkov, S.B. and C.D. Maranas, "Multiperiod Planning and Scheduling of Multiproduct Batch Plants under Demand Uncertainty", *Ind. Eng. Chem. Res.*, **36**, 4864-4881 (1997).

Pinedo, M., *Scheduling. Theory, Algorithms and Systems*. Prentice Hall, New York (1995).

Pinto, J.M. and I.E. Grossmann, "Optimal Cyclic Scheduling of Multistage Continuous Multiproduct Plants", *Computers chem. Engng.*, **18**, 797-816 (1994).

Pinto, J.M. and I.E. Grossmann, "A Continuous Time MILP Model for Short-Term Scheduling of Multistage Batch Plants", *Ind. Eng. Chem. Res.*, **34**, 3037-3051 (1995).

Pinto, J.M. and I.E. Grossmann, "A Logic-Based Approach to Scheduling Problems with Resource Constraints", *Computers chem. Engng.*, **21**, 801-818 (1997).

Reklaitis, G.V. "Perspectives on Scheduling and Planning of Process Operations", *Proc. 4th Intl. Sym. Process Systems Engng.*, Montebello, Canada (1991).

Reklaitis, G.V. and L. Mockus, "Mathematical Programming Formulation for Scheduling of Batch Operations based on Non-uniform Time Discretization", *Acta Chimica Slovenica*, **42**, 81-86 (1995).

Rich, S.H and G.J. Prokopakis, "Scheduling and Sequencing of Batch Operations in a Multipurpose Plant", *Ind. Eng. Chem. Res. Process Des. Dev.*, **25**, 979-988 (1986).

Rippin, D.W.T., "Batch Process Systems Engineering: a Retrospective and Prospective Review", *Computers chem. Engng.*, **S17**, S1-S13 (1993).

Rodrigues, M.T.M., L. Gimeno, C.A.S. Passos and M.D. Campos, "Reactive Scheduling Approach for Multipurpose Batch Chemical Plants", *Computers chem. Engng.*, **S20**, S1215-S1226 (1996).

Rotstein, G.E., R. Lavie, D.R. Lewin, "Synthesis of Flexible and Reliable Short-Term Batch Production Plans", *Computers chem. Engng.* **20**, 201-215 (1994).

Sahinidis, N.V., I.E. Grossmann, R.E. Fornari and M. Chathrathi, "Optimisation Model for Long-Range Planning in the Chemical Industry", *Computers chem. Engng.*, **15**, 255-272 (1991a).

Sahinidis, N.V. and I.E. Grossmann, "MINLP Model for Cyclic Multiproduct Scheduling on Continuous Parallel Lines", *Computers chem. Engng.*, **15**, 85-103 (1991b).

Sahinidis N.V. and I.E. Grossmann, "Reformulation of Multiperiod MILP Models for Planning and Scheduling of Chemical Processes", *Computers chem. Engng.*, **15**, 255-272 (1991).

Sanmarti, E., A. Espuña and L. Puigjaner, "Effects of Equipment Failure Uncertainty in Batch Production Scheduling", *Computers chem. Engng.*, **S19**, S565-S570 (1995).

Schilling, G. and C.C. Pantelides, "A Simple Continuous Time Process Scheduling Formulation and a Novel Solution Algorithm", *Computers chem. Engng.*, **S20**, S1221-S1226 (1996).

Schilling, G.H., "Algorithms for Short-Term and Periodic Process Scheduling and Rescheduling", PhD Thesis, University of London (1997).

Shah, N. and C.C. Pantelides, "Optimal Long-Term Campaign Planning and Design of Batch Plants", *Ind. Eng. Chem. Res.*, **30**, 2308-2321 (1991).

Shah, N., C.C. Pantelides and R.W.H. Sargent, "A General Algorithm for Short-Term Scheduling of Batch Operations - 2. Computational Issues", *Computers chem. Engng.*, **17**, 229-244 (1993a).

Shah, N., C.C. Pantelides and R.W.H. Sargent, "Optimal Periodic Scheduling of Multipurpose Batch Plants", *Ann. Ops. Res.*, **42**, 193-228 (1993b).

Subrahmanyam, S., J.F. Pekny and G.V. Reklaitis, "Decomposition Approaches to Batch Plant Design and Planning", *Ind. Eng. Chem. Res.*, **35**, 1866-1876 (1996).

Sunol, A.K., M. Kapanoglu and P. Mogili, "Selected Topics in Artificial Intelligence for Planning and Scheduling Problems, Knowledge Acquisition and Machine Learning", *Proc. NATO ASI on Batch Processing Systems Engineering*, 595-630 (1992).

Tahmassebi, T., "Industrial Experience with a Mathematical Programming Based System for Factory Systems Planning/Scheduling", *Computers chem. Engng.*, **S20**, S1565-S1570 (1996)

Wellons, M.C. and G.V. Reklaitis, "Scheduling of Multipurpose Batch Plants. 1. Formation of Single-Product Campaigns", *Ind. Eng. Chem. Res.*, **30**, 671-688 (1991a).

Wellons, M.C. and G.V. Reklaitis, "Scheduling of Multipurpose Batch Plants. 2. Multiple-Product Campaign Formation and Production Planning", *Ind. Eng. Chem. Res.*, **30**, 688-705 (1991b).

Williams, J.F., "Heuristic Techniques for Simultaneous Scheduling of Production and Distribution in Multi-Echelon Structures", *Man. Sci.*, **27**, 336-352 (1981)

Wilkinson, S.J., N. Shah and C.C. Pantelides, "Scheduling of Multisite Flexible Production Systems", paper presented at AIChE Annual Meeting, San Fransisco (1994).

Wilkinson, S.J., N. Shah and C.C. Pantelides, "Aggregate Modelling of Multipurpose Plant Operation", *Computers chem. Engng.*, **S19**, S583-S588 (1995).

Wilkinson, S.J., A. Cortier, N. Shah and C.C. Pantelides, "Integrated Production and Distribution Scheduling on a Europe-Wide Basis", *Computers chem. Engng.*, **S20**, S1275-S1280 (1996).

Xia, Q. and S. Macchietto, "Routing, Scheduling and Product Mix Optimization by Minimax Algebra", *Chem. Eng. Res. Des.*, **72**, 408-414 (1994).

Xia, Q. and S. Macchietto, "Design and Synthesis of Batch Plants–MINLP Solution Based on a Stochastic Method", *Computers chem. Engng.*, **S21**, S697-S702 (1997).

Yee, K.L., "Efficient Algorithms for Multipurpose Plant Scheduling", PhD Thesis, University of London (1998).

Yee, K.L. and N. Shah, "Scheduling of Fast-Moving Consumer Goods Plants", *J. Opl. Res. Soc.*, **48**, 1201-1214 (1997).

Yee, K.L. and N. Shah, "Improving the Efficiency of Discrete-Time Scheduling Formulations", *Computers chem. Engng.*, **S22**, S403-S410 (1998).

Zentner, M.G. and G.V. Reklaitis, "An Interval-based Mathematical Model for the Scheduling of Resource-Constrained Batch Chemical Processes", *Proc. NATO ASI on Batch Processing Systems Engineering*, 779-807 (1992).

Zhang, X and R.W.H. Sargent, "The Optimal Operation of Mixed Production Facilities–a General Formulation and some Approaches for the Solution", *Proc. 5th Intl. Symp. Process Systems Engineering*, 171-178 (1994).

Zhang, X and R.W.H. Sargent, "The Optimal Operation of Mixed Production Facilities–Extensions and Improvements", *Computers chem. Engng*, **S20**, S1287-S1292 (1996).

TOWARDS THE CONVERGENCE OF THEORY AND PRACTICE: A TECHNOLOGY GUIDE FOR SCHEDULING/PLANNING METHODOLOGY

J. F. Pekny and G.V. Reklaitis
Computer Integrated Process Operations Consortium
School of Chemical Engineering
Purdue University
West Lafayette, IN 47907-1283

Abstract

This paper discusses the nature and characteristic features of planning/scheduling problems arising in the chemical processing and related industries. The concept of NP-completeness is reviewed as it applies to this challenging family of combinatorial problems. It is shown that this fundamental property has key implications for the solution methodology for process scheduling problems and sets practical limits on the performance of this methodology. The three known strategies for addressing NP-completeness are outlined and their advantages and disadvantages contrasted. An overview is given of the essential elements of available solution methods which implement these strategies, including randomized search, rule-based methods, constraint guided search, simulation based strategies, as well as mathematical programming formulation-based approaches using conventional and engineered solution algorithms. An assessment of this technology is made in terms of five key features: solution quality, usability, extensibility, robustness, and strategy for accommodating NP-completeness. Based on available but preliminary evidence, engineered mathematical programming approaches are found to offer the best prospects for automatic solution of problems of industrial scope. However, the algorithm engineering approach is shown to require considerable intellectual investment on the part of the technology developer/provider. The business implications of the nature of scheduling technology both on the users and the technology providers are explored. The paper concludes with projections for the future evolution of this technology and the industry that services it.

Keywords

Process scheduling, Heuristic methods, Mathematical programming, Complexity theory, Engineered algorithms.

Introduction

The business environment of the early 1990's focused on driving profitability through cost reductions. These were achieved through lowering inventories, reducing staff, and shedding marginal business lines or excess manufacturing facilities. Since the resulting lean organizations and processes could no longer be made more profitable by sheer cost cutting, further gains could only be achieved through better coordination in the use of corporate, especially, manufacturing resources, and tighter integration of sales, resourcing, and manufacturing functions. It was soon widely accepted that coordination and integration require accurate and synchronized information on the state of the enterprise. Thus, corporations have energetically pursued the collection and computerization of, first, manufacturing- and, then, enterprise-wide information. This has proven to be a resource intensive process from organizational and information technology perspectives and thus, not surprisingly, has spawned a multi-billion dollar service industry. Unfortunately, more and more of the companies who have heavily invested in information systems, are reporting that the technology has not delivered all the benefits claimed and, thus, are abandoning or curtailing some of these efforts (Wysocki, [1998]). This is unfortunate because the next logical step is to use that expensively collected information to derive economic benefit by making superior business decisions, especially those related to manufacturing and supply functions. A sufficient number of companies have, however, recognized the need for this next step and this has given rise to an accelerating interest in planning and scheduling technology. The result has been not only a stimulation of research in new methodology but also the spawning of another multi-billion dollar industry that provides and services scheduling/planning tools. The ambitious goal of this paper is to provide a perspective on the nature of these tools and the industry that services them. We hope to forestall in the scheduling domain the consumer disillusionment that appears to be developing with respect to the information technology domain because of an incomplete understanding of what existing technology can, in fact, deliver.

Specifically, in this paper we will seek to characterize the properties of scheduling problems in the CPI and examine the implications of these properties. In particular we will discuss the implication of NP-completeness on solution technology, provide a high level overview of available solution technology, examine the business implications arising from that technology perspective, and offer some projections for the future evolution of the technology and the technology providers. Since our review of scheduling problem solution technology will be at the conceptual level, the reader interested in a more detailed technical exposition is invited to consult any of several reviews, including introductory (Reklaitis et al, 1997), practical (Applequist et al 1997; Zentner et al 1994b) and most recent (McDonald, 1998; Shah, 1998) treatments.

Nature of Planning/Scheduling Problems

In general, the planning and scheduling of production facilities is a decision making process that determines what, when, where, and how to produce a set of products given known requirements defined at specific points in time over a specific time horizon, a set of limited production resources, and an unambiguous description of the operations that must be executed to make each product.

- What refers to the amounts of each product to be made, defined in terms of lot sizes, batch sizes, or product run amounts.

- When denotes the timing of specific operations: start and end times, run lengths, etc.

- Where refers to specific sites, process units, or equipment items assigned to execute specific operations.

- How relates to allocation of specific amounts or levels of resources, such as labor, utilities, feedstocks, and intermediates, necessary to execute the production operations.

The distinction between planning and scheduling problems lies principally in the level of aggregation at which the issues of what, when, where, and how are addressed. Typically, in planning, a coarse aggregation is employed, thus requiring a loss of manufacturing detail such as the sequence or order in which specific manufacturing steps are executed. However, as manufacturing systems are driven more aggressively and thus planning and scheduling solutions are required to track resource constraints tightly, planning models must increasingly be disaggregated so that critical manufacturing details, such as sequence dependent effects are included. Thus, the distinction between the planning and scheduling levels are diminished to the point where the resource allocation decisions become inherently the same. The distinction between the two levels becomes largely a matter of computational and organizational convenience rather than one of different decision theoretic content. Hence, for purposes of the subsequent discussion, we will treat scheduling and planning as equivalent decision problems.

One of the key characteristics of scheduling problems is that the information, which is required to describe the manufacturing operations, the resource requirements and the product demands, is extensive and dynamic. Since the information spans technical, financial, and commercial domains, the sources of information are diverse and extend outside of the boundaries of the manufacturing organization itself. Moreover, the data changes rapidly over time as customer orders, resource availability, and the manufacturing processes themselves undergo changes. The resulting data complexity makes efficient management of information resources a necessary prerequisite for effective scheduling.

Furthermore, the determinations of when, where and how are inherently combinatorial in nature. The assignment of equipment and resources to operations and the determination of the order and hence timing of the execution of operations are discrete decisions which can be represented in a binary (yes /no) fashion. For instance yes, assign equipment 1 to operation A or, no, do not; yes, execute operation A on equipment 1 before operation B, or no, do not. Of course, as the number of such individual discrete decisions increases, the number of possible scheduling solutions grows in an exponential fashion. A typical manufacturing schedule can readily involve making hundreds of thousands of such decisions. Thus, most manufacturing scheduling problems necessarily will be information intense and highly combinatorial in nature.

A solution or instance of the what, when, where, and how's resulting from a scheduling process is feasible if it meets all of the imposed resource limitations, manufacturing constraints, and product requirements. Different feasible solutions are usually not equivalent: some are better than others for technical or economic reasons. Thus, a schedule will have associated with it an agreed-upon measure of merit, such as the manufacturing cost, profit, or number of late orders. A feasible schedule that attains the best possible value of the measure of merit is said to be optimal. Normally, a scheduling process seeks to generate optimal solutions, but sometimes that is impractical and thus good but suboptimal, feasible solutions must suffice. In such instances, it is desirable to obtain estimates or bounds of how close the candidate schedule actually is to the optimal. As will be noted later in this paper, scheduling procedures differ in their abilities to guarantee the attainment of optimal, suboptimal and bounded, or even feasible solutions.

Scheduling Problem Characterization

In the manufacturing setting, scheduling problems can be defined in terms of three basic entity types: equipment, resources, and tasks. Equipment are the physical facilities used to execute specific manufacturing operations. Resources are all other measurable entities employed or produced in manufacturing, such as materials, human labor, and utilities. A task is any manufacturing operation that transforms resources. In the process and related industries, a task typically consists of a contiguous set of chemical or physical operations, which is performed entirely within one equipment item or a closely coupled set of equipment items. The manufacture of CPI products normally requires that a set of tasks, e.g., reaction, separation, packaging, drying, etc., be performed in a certain order. Thus, associated with each product will be a directed network of tasks called a recipe, where the directed arcs in the network indicate the precedence order among tasks and at the same time represent the direction of material flow between tasks. In general a manufacturing facility will produce multiple products, each with a distinct recipe.

CPI scheduling problems have several distinct features that differentiate them from other manufacturing domains. These are associated with the nature of the materials being processed, the structure of the recipe network, and the complexity of the interactions between resources. The feedstocks, intermediates, and products arising in the process industries are mixtures of fluids and solids. That is, they are continuous materials and, as such, infinitely divisible and not expressible in some irreducible quantity, such as an object or an assembly of objects. The recipe structures characteristic of CPI products are general networks, with multiple inputs and outputs from each task as well as material recycles. By contrast in assembly processes the recipe structures are convergent while in packaging facilities the recipe structures are divergent. Finally typical tasks, such as reaction or separation, require and generate multiple resource types that are shared among tasks. Since these resources are normally limited in availability, there arises a competition for these resources among the tasks to be executed. For example, the refrigeration utility used in one task may be required by several other competing tasks and thus total refrigeration capacity will limit the tasks that may be concurrently executed. As a result of these features CPI scheduling decisions require consideration of complex interactions between equipment and resource constraints, material and resource balances to describe the generation and depletion of materials and other conserved resources, and complex precedence relations between tasks. As will be seen in subsequent sections of the paper, these features impose requirements on scheduling methodology that does not arise in other domains of manufacturing.

Basic Concepts of Complexity Theory and NP-Completeness

The fundamental computational features of combinatorial problems, of which process scheduling problems are prime examples, are characterized by complexity theory. The concept of NP-completeness (Garey and Johnson [1979], Harel [1987]), which is a cornerstone of the theory, sets the boundaries of what is achievable with scheduling technology from the solution methods up through the business models of software and service vendors that operate in the marketplace. This section reviews the concept of NP-completeness as it applies to scheduling and subsequent sections will address the implications for tools and business. For the next several sections the perspective is confined to that of solving problems since the capabilities of the solution approach largely control the properties of a scheduling tool.

In abstract terms a scheduling problem can be defined as determining

$$x \in S = (F, I) \qquad (1)$$

where x is a solution (schedule) from the set of possible solutions, S, and the set of possible solutions has been partitioned into those that are feasible, F, and those that are infeasible, I. Of course the principal goal in scheduling problems is not simply to find a solution but find a solution which optimizes a selected measure of merit. In symbolic terms the desire to obtain optimal solutions can be expressed as

optimize objective(x)

such that

$$x \in S = (F, I) \quad (2)$$

where objective(x) is the measure of merit of the solution. In fact the simple definitions (1,2) can be used to define two additional classes of problems which are useful for understanding scheduling technology. In particular, consider the scheduling problems implied by

$$x \in F \quad (3)$$

and

optimize objective(x)
such that

$$x \in F \quad (4)$$

These latter two classes of scheduling problems are considerably more difficult than class (1) and (2) because they require finding solutions which satisfy all the constraints. In fact, the degree of difficulty in solving problems in class (3) and (4) can be formalized. To this end, let description(F) represent a list of data used to describe the feasible set of solutions. For example, description(F) may include a list of demand amounts by product and time, bills of materials, processing times, manufacturing costs, equipment capabilities, etc. When written on a computer disk or piece of paper, description(F) will have a certain length, say L, for example the number of bytes or sheets of paper used to describe the problem. For a given algorithm that solves problem (3) or (4) the worst case amount of time required can be determined as a function of L, e.g. the solution time could scale as L, L^2, L^3, e^L, etc. Any algorithm whose worst case solution time is not bounded by a polynomial in L is said to be inefficient. Practically, an algorithm is considered efficient only if its solution time is bounded by a low order polynomial in L, otherwise a user may have to wait an unacceptable amount of time.

In complexity theory (Garey and Johnson, [1979]), the set of all problems for which algorithms exist whose worst case solution time is bounded by a polynomial in the length of its problem description is called set P, the set of problems solvable in polynomial time. There is also a set of problems E whose solution times are guaranteed to scale with a function that is exponential in the length of the description of the problem. For obvious reasons this class of problems is known to be provably intractable. In this class, for all but the smallest of problems, there is little hope of effective solution. Almost all scheduling problems in class (3) and (4) are not as easy as those in class P, but are not as difficult as those in class E. In particular, scheduling problems of class (3) and (4) are in the class of Non-deterministic Polynomial time Complete (NP-complete) problems. The worst case solution time of any algorithm for problems in this class is known to be exponential in the length of problem description. Practically this means that some scheduling problems can require an unreasonable time to solve, although there is no reason that the worst case execution time has to be realized for problems of interest. This observation forms the foundation of viable strategies for scheduling problem algorithms and will be explored in the next section.

To facilitate the subsequent discussion, we adopt the following terminology and conventions. An <u>instance</u> of a scheduling problem is defined to be a particular choice of values for the parameters, i.e. numerical values specified for all due dates, costs, process times, etc. A given scheduling problem is then considered to be the set of all possible instances. For example a scheduling problem might consist of a single unit on which orders are processed and in which a sequence dependent changeover cost is incurred. A particular instance of this scheduling problem is a set of numerical values specified in the changeover cost matrix between orders. The set of all possible changeover cost matrices taken together comprise the scheduling problem.

The Practical Implications of NP-Completeness for Process Scheduling

There are a number of practical implications to the ideas presented in the previous section. A number of these implications arise by considering the relationship between the various classes of scheduling problems (1), (2), (3), and (4). At first glance problem classes (1) and (2) do not appear particularly useful given that they include solutions that are infeasible. However, if problem class (1) is pursued instead of problem class (3) and problem class (2) is pursued instead of problem class (4) then the difficulties associated with addressing NP-complete problems can be avoided. In particular, by engineering the infeasible set I properly, an approach can always be used whose execution time is bounded by a low order polynomial, that is, by allowing the approach to produce an infeasible solution the execution time can be guaranteed to be kept reasonable. As discussed below, this is the approach of many heuristic methods for solving scheduling problems. Of course, another related strategy when confronted with having to solve problems in class (3) or (4) is to deploy resources to avoid having to address difficult scheduling problems. In practice this involves spending money to increase capacity or inventory or reducing customer service levels. Given the pressures for economic competitiveness, there is increasing unwillingness to deploy resources to avoid complex scheduling issues. This unwillingness is the principal driving force behind the expanding market for scheduling tools. However, approaches designed to allow infeasible solutions may often not be sufficiently sophisticated to allow a significant reduction in capital (inventory, capacity) or improvement in customer service.

A comparison of problem class (3) to (4) provides an interesting insight about algorithms that are only designed to find feasible solutions. Consider the class of problems related to problem class (4)

$$x \in F \text{ and objective}(x) \leq V \qquad (5)$$

If V is chosen suitably large, then problem class (5) is identical to problem class (3). However, repeated solution of a sequence of problem class (5) using a bisection procedure on V enables the use of algorithms designed only to find feasible solutions to **determine** optimal solutions with respect to arbitrary objectives. This simple argument indicates that the search for feasible solutions can be as difficult as the search for optimal feasible solutions. The implications for solution technologies are clear. The development of high performance algorithms for finding feasible solutions to scheduling problems is as difficult and valuable as the development of approaches for finding optimal solutions.

For economic reasons, the desired design objective for algorithms should be to address scheduling problems from class (3) and (4). However in practice the set of feasible solutions for a given problem instance is often empty due to excessive demand requirements, boundary conditions imposed on the solution by fixing a portion of a schedule, and/or infeasible inventory constraints. In the case of inventory constraints, minimum inventories can be specified to be excessive or maximum inventory constraints may preclude problem feasibility. In cases where the feasible set is empty, the goal of approaches becomes solving a related problem (or its optimization version)

$$x \in S = (F, I') \qquad (6)$$

where I' is a set of solutions infeasible to the original problem specification but possessing special properties. For example, if F is empty due to excessive demand then I' contains only those infeasible solutions where the amount of lateness in the production of orders is minimized or orders are shipped with the minimum amount of shortfall by the demand due date. In the case of restrictive inventory constraints, I' may only contain those infeasible solutions that minimize the amount of inventory violation. However, the key idea is that scheduling algorithms are not just required to be capable of solving problems in classes (3) or (4), but related problems that involve finding only certain kinds of solutions that are infeasible to the original problem. Indeed one highly desirable feature of any algorithm for scheduling problems is that it be able to determine when operating requirements are truly infeasible and then to propose a good, or best possible, response. In fact problems in class (5), for almost every practical specification of I', are in the class of NP-complete problems, that is the solution of problems in class (5) is as difficult as that of solving problems in class (3) or (4).

Strategies for Addressing the NP-completeness of Scheduling Problems

NP-completeness can be thought of as providing boundaries that must be obeyed by approaches for solving scheduling problems. This means that all scheduling approaches must be based on some strategy for dealing with the consequences of NP-completeness and that users of such approaches should be aware of the resulting practical limits on performance. In particular, there are three strategies for approaching the solution of the scheduling problems in class (3) and (4), see Applequist et al [1997] for a more detailed discussion:

Strategy 1 - Change the Problem to Be Solved to Make It Easier

This is by far the most popular strategy for addressing the NP-complete nature of scheduling problems, and much commercially available scheduling software is designed solely around it. Quite simply the strategy involves changing a given scheduling problem by modifying the constraints, parameters, and objective until a given scheduling algorithm can solve the problem, that is permit solutions infeasible to the original problem as in problem classes (1) and (2). Typically, such changes involve postponing due dates for demand fulfillment, allowing storage buffers to overflow, ignoring labor constraints, and disregarding the objective. The popularity of this strategy stems from the fact that it can be made to work regardless of the simplicity of the underlying scheduling algorithm. However, the chief disadvantage of this strategy is that substantial economic and performance penalties can result when a problem that has feasible solutions is changed to accommodate a solution scheme. From an engineering standpoint, methods can be differentiated by how the strategy is applied. For example, postponing due dates is often a preferred way of relaxing problem difficulty rather than allowing storage buffers to overflow. One design principle underscored by this strategy is that methods should be developed to minimize changes to scheduling problems. In this way, the economic impact of limits in scheduling technology is kept to a minimum.

Strategy 2 - Use an Exact Algorithm Which May Require Unreasonable Time

The theory of NP-completeness specifies that the worst case execution time for an exact algorithm on a given scheduling problem from class (3) and (4) is exponential in some measure of problem size. Practically this means that an exact algorithm may require an unacceptably long time on some problem instances. However, the theory of NP-completeness also does not require an exact algorithm to require an unreasonably long time. In fact, the notion of exact algorithms being slow for scheduling problems has developed primarily because general purpose optimization or search methods demonstrate unreasonably long behavior on most classes of industrial scale scheduling problems. Although the development of highly customized exact methods for large scale scheduling problems is still in its infancy, past research has shown the feasibility of such efforts (Zentner et al., 1994a; Zentner et al., 1994b). Of course in practical applications, even an occasional unreasonable execution time is unacceptable. Thus from a design perspective, exact methods must be paired with those methods using Strategy 1 or Strategy 3. The quality of the resulting systems can then be judged on the basis of how frequently the backup strategy must be invoked.

Strategy 3 - Use a Heuristic Algorithm That May Get a Poor Answer

Methods can be devised for process scheduling problems that will always have a reasonable execution time by pursuing problem class (1) or (2). However, the theory of NP-completeness then dictates that such methods can sometimes obtain poor answers or even fail to find a feasible solution. Since failure is unacceptable in practical applications, this requires that Strategy 3 also be combined with Strategy 1. Note that Strategy 3 is not mutually exclusive with Strategy 2 in the sense that exact algorithms may often be used in a heuristic mode.

Taken together, Strategy 2 and Strategy 3 highlight the uncertainty principle embodied in the theory of NP-completeness. That is, either the execution time or the feasibility/solution quality can be predicted with theoretical certainty, but not both. However, this uncertainty principle does allow room for considerable success. Customized approaches can significantly improve the probability of obtaining high quality solutions in a reasonable time. This state of affairs makes the development of scheduling methods a quintessential engineering activity. Real applications as well as advancing computer technology determine the appropriate tradeoffs between the three strategies. Note that the central idea is that clever application of problem specific knowledge offers the ability to achieve high quality answers in reasonable times without obscuring the details of the problem.

Observations About Individual Problem Instances

Many applications (e.g. marketing and sales demonstrations, journal publications, etc.) are concerned with the solution of a single problem instance over a period of time that is long relative to the amount of time that would be available to solve similar or related instances in practice. In these applications the goal is to develop some intuition as to the effectiveness of a given approach. At the conclusion of significant work on a highly targeted approach for such an instance, measures of solution time and quality are asserted as to the effectiveness of a set of ideas. The implications of NP-completeness have significant bearing on the interpretation of such assertions. In fact, nothing in the theory of NP-completeness precludes the construction of an effective approach dedicated to a single instance. In fact observations about prolonged study of individual instances arising from practical applications suggests that high quality solutions can virtually always be derived given enough time, cost, human interaction, and judicious relaxation of constraints. This is not surprising since instances derived from real world applications generally must be solved in practice through some means. A more complete test of a given scheduling approach is to record measures of effectiveness over a set of related instances, a fraction of which are available during the development of the approach and the remainder of which are used for analysis after the approach is complete. In fact this testing idea can be formalized as is done in the next section.

The Use of Instance Generators to Develop and Test Scheduling Approaches

An instance generator is an algorithm whose input is seed information, output is an instance of a scheduling problem, and execution time is bounded by a polynomial in the length of the output. A particular scheduling approach can then be applied to the output of the generator. By changing the seed information, an instance generator may be used iteratively to generate an entire family of instances. The effectiveness of a scheduling approach can then be measured according to the fraction of generator instances solved and the amount of resources required to do so. The generator can be used during the development of an approach using one set of representative seed information and then tested using a similar set of seed information. Nothing about NP-completeness precludes the development of approaches that successfully solve a very large fraction of the instances output by a given generator in reasonable time. Indeed, as discussed below, for the asymmetric Traveling Salesman Problem with uniformly generated random costs and other generators, the solution fraction for existing approaches is close to one even for very large problems (Miller and Pekny [1991]). The importance of using generators for rigorous testing arises from the observation that the performance of many approaches is very sensitive to what would appear to be minor instance changes. Indeed there are examples in the literature where minor changes in instance structure increases solution time by several orders of magnitude (Kudva and Pekny [1993]). The largest impediment to the use of generators to increase the reliability of scheduling approaches is the significant degree of manual intervention required in most approaches and the level of effort required to construct industrially meaningful generators. Practically, the construction of generators offers several advantages. For example, definition of a generator forces consideration of those features that are likely to arise in practice, provides an unlimited number of test cases, and provides the assumptions that may be used in algorithm development. That is a properly designed generator is an unlimited source of test instances that have a realistic combination of problem features/parameter values and can be used to guide and verify algorithm development.

How Effective Can Scheduling Approaches be Given NP-Completeness?

The most important implication of NP-completeness is that there can be immense differences in performance between two methods for a given problem, both in terms of solution time and quality. This implication underscores the importance of testing approaches on a representative set of instances. However, the philosophical and business implications of NP-completeness are also significant. Essentially there are two interpretations on the development of approaches that are consistent with what is known about NP-completeness. In the first and largely prevailing interpretation, methods must be developed on a rather narrow case by case basis using a toolbox type approach that requires significant intellectual and manual effort. In a second interpretation, an algorithm can be engineered to be effective on a given set of instance generators, that is,, the probability of the algorithm solving each of the instances produced by the generators is close to one while using reasonable computing resources. Whenever a new generator is developed the existing algorithm may be confounded, but with a research and development expenditure the algorithm may be adapted to be effective. This second interpretation of what is possible using NP-completeness admits the possibility of methods that are quite automatic and effective on a wide range of practical problems. These two interpretations imply quite different business models as will be discussed below.

The issues in the previous paragraph ultimately bring up the question of how effective a given algorithm can be on a new family of instances without any additional development having to be expended. The extreme possibilities are that a given algorithm is only effective on a single instance or that a given algorithm is effective on almost all instances of practical interest. Experimental evidence (Miller and Pekny [1991], Kudva and Pekny [1993]) suggests that neither extreme is true and in fact the effectiveness of an approach to a change in the structure of an instance depends on the nature of the change. In particular, changes to a family of scheduling problem instances can be classified as parametric or structural. A parametric change involves simply modifying some of the numbers in a given instance, for example changing the demands or initial inventory levels for a given process. A structural change involves adding new combinations of features, for example adding management of constrained labor to a scheduling problem or having to manage shared storage vessels rather than dedicated ones. In general a given approach is much more likely to fail when encountering instances with structural changes that have not been encountered before than when encountering a parametric change. Furthermore, the experimental evidence suggests that whether a particular change to instance structure significantly alters performance is highly dependent on the technology underlying the approach (Pekny, Miller, Stodolsky [1991]). In short, the significance of a change in instance structure can only be measured relative to a given technology. For one type of solution technology a specific change to problem structure might be significant, while for another technology the same change might be inconsequential.

Engineered Approaches to Scheduling Problems

The discussion in the previous sections clearly indicates that NP-completeness makes the development of scheduling tools an engineering activity where compromises must be made. As an engineering activity, the cost/benefit of developing sophisticated scheduling approaches must be traded against the cost/benefit of using the approaches in practice. The ideal scheduling tool would be able to address problem class (4), or equivalently problem class (3), and in the absence of any feasible solution be able to determine solutions infeasible to the original problem but possessing desirable properties. The implication of NP-completeness is that this ideal can only be achieved for well defined scheduling problems (e.g. ones for which generators are identified) and that as new problems are encountered additional research and development must be done to extend an approach. Furthermore, as engineered systems, scheduling tools must also be developed taking into account more than the underlying scheduling approach. Other issues involved with developing scheduling tools include graphical user interfaces, database connectivity, modeling aids, debugging tools, and process model libraries. Subsequent sections will provide an overview of technology from the standpoint of analyzing scheduling tools as engineered systems.

A Detailed Algorithm Engineering Example

The ideas in the previous sections can be made concrete by considering an algorithm engineering approach to a well known and simple to describe scheduling problem. In particular, consider a single unit operation on which a set of n orders is to be processed where each order requires a specified time and where a sequence dependent changeover time exists between orders. This problem is easily mapped to a Traveling Salesman Problem (TSP), where the number of orders to be processed directly maps to the number of cities (Pekny and Miller, 1991). Assuming an answer of a guaranteed quality is an objective, the preferred starting point for algorithm engineering is a mathematical formulation of the problem since this allows relaxation bounds to be readily obtained:

$$\text{minimize} \sum_{i=1}^{n} \sum_{j=1}^{n} c_{ij} x_{ij} \quad (A)$$

$$\sum_{i=1}^{n} x_{ij} = 1 \ \forall j \in \{1,...,n\} \quad (B)$$

$$\sum_{j=1}^{n} x_{ij} = 1 \ \forall i \in \{1,...,n\} \quad (C)$$

$$\sum_{i \in S} \sum_{j \in S} x_{ij} \leq |S|-1 \ \forall S \subset \{1,...,n\}, S \neq 0 \quad (D)$$

$$x_{ij} \in \{0,1\} \ i,j = 1,...,n \quad (E)$$

The objective function (A) seeks to find a minimum makespan solution where the c_{ij} are a function of the processing time and the changeover time. The equations (B) and (C) guarantee that each order is only processed once. Equation (D) guarantees a physically realizable schedule, and (E) enforce the integrality of the decision variables x_{ij}.

The first step to solving non-trivial problem sizes is to implicitly handle constraint set (D) since the number of these constraints scales exponentially with n. The most direct way is to ignore (D) and enforce any violated constraints through the branching rules. Relaxation of (E) to allow continuous variable values then leads to a tractable linear program which generic simplex or interior point methods can address for values of n up to several hundred. To accommodate dramatically larger problem sizes, the structure of (A), (B), (C), and the continuous relaxation of (E) can be exploited through specialized linear programming techniques. For this particular structure, shortest augmenting path methods may be used based on Dijkstra's algorithm (Balas, Miller, Pekny, Toth [1991]). A several order of magnitude speedup results from exploiting this linear programming structure and the speedup grows as n increases. Iterative solution of the linear programming relaxation through classic branch and bound methods may be used to obtain optimal solutions. In fact direct enforcement of the violated constraints in (D) through the branching rules can lead to on the order of a factor of two speedup in performance on many instance generators (Pekny and Miller [1992]) over simple binary branching. For a randomly generated asymmetric cost matrix with c_{ij} drawn uniformly from the integer interval [0,M] the resulting algorithm can routinely and reliably solve instances with on the order of 1,000 cities as long as M is greater than n (Miller and Pekny [1989]). For values of M approximately equal to or less than n, the uniform random generator routinely produces linear programming relaxations with 100% bound strengths but with a great deal of degeneracy in the linear programming solution. This degeneracy severely degrades the performance of branch and bound methods. However, the degeneracy can be mitigated through the use of an exact directed Hamiltonian cycle (DHC) algorithm applied to the support graph of each node in the branch and bound tree (Pekny, Miller, and Stodolsky [1991]). In fact the DHC augmented algorithm typically solves highly degenerate problems at the root node of the branch and bound tree.

For the uniform random generator, further dramatic improvements in problem size are possible by more detailed engineering of the linear programming solution. In particular, sufficient conditions can be developed that allow much of the cost matrix to be ignored during branch and bound solution (Pekny and Miller [1992]). Failure to satisfy the sufficient conditions simply results in resolving the problem while ignoring fewer matrix elements. Once the cost matrix is made sparse, d-heap based sorting techniques can be used to maintain the labels in Dijkstra's algorithm (Tarjan [1983]). This results in a speedup of nearly a factor of n for instances constructed using the uniform random generator for sizes in the range of tens of thousands of cities. Other enhancements to the linear program solver are possible (e.g. parametric solution of non-root node linear programs), but the net result is an algorithm that reliably performs on instances involving tens of thousands of cities. The addition of an upper bounding heuristic that "patches" constraint violations in (D) leads to an algorithm (summarized in Table 1) that has been proven to be reliable in obtaining provably optimal solutions to instances from the uniform random generator possessing up to 500,000 cities or 250 billion variables (Miller and Pekny [1991]). In particular, the probability of the engineered algorithm failing to solve in a time bounded by a polynomial in n on instances from the uniform random generator appears to vanish as n increases without bound. Intuition for this statement results from contemplating the interaction of the various algorithm features with respect to the structure of the instances that are likely to be generated. Surprisingly, although the algorithm sketched above was engineered with the uniform random generator in mind, computational evidence shows that is performs very well on other generators (Miller and Pekny [1991]). Thus the engineered algorithm is robust over several classes of generators.

Predictably, because of NP-completeness, there are several known generators that confound the algorithm, which performs so well on the uniform random generator.

For example, if the cost matrix is symmetric uniform random the approach sketched above only works to optimality for about 30 cities. One is tempted to conclude that the uniform random generator is somehow an easy case and that the symmetric random generator is harder. However, the above approach is extensible to the symmetric uniform random generator if the above formulation is modified to exploit symmetry and a two-matching linear programming relaxation is used (Pekny and Miller [1994]). An algorithm that does well on either uniform random or symmetric uniform random generators can then be developed by decomposing the cost matrix into symmetric and asymmetric parts (Fischetti and Toth [1989]). In fact a random Euclidean generator that involves placing cities randomly at points on a plane confounds this more sophisticated algorithm. However, the literature provides details of an approach that has been specifically engineered for Euclidean random generators and has proven effective for several thousand cities (Padberg and Rinaldi [1991]). Again, the evidence suggests that highly reliable algorithms can be engineered for specific types of TSP generators. If additional constraints types are appended to the formulation (A)-(E) above, the performance of the approach described above badly degrades. Such a structural change to the TSP necessitates the use of dramatically different types of algorithm engineering techniques from those described above (Kudva and Pekny [1993]). Intuitively, the addition new constraint families to a formulation seems to greatly change the algorithm engineering required for an effective approach relative to those used in the absence of the constraints. The example in this section outlined the algorithm engineering required for an exact approach to a specific NP-complete problem class. A similar discussion applies to the engineering of heuristics for NP-complete problems (Lin and Kernighan [1973], Kudva et al. [1994]), although heuristics do not provide the performance guarantees of mathematical programming based approaches. For this reason heuristics do not provide as much feedback, which can complicate the algorithm engineering.

The lessons of the TSP example are clear: (1) Algorithm robustness is more likely in the face of parametric changes to a generator, while generators that add new constraint families are much more likely to require significant algorithm engineering changes. (2) For mathematical programming formulations the majority of dramatic algorithm speedups occur due to exploitation of the structure of the underlying linear program. (3) The lack of a highly structured LP relaxation and the interaction of multiple constraint families significantly complicate algorithm advances on more general scheduling problems. (4) The observation of a given algorithm on an instance for which it is unsuccessful can lead to insights that can ultimately and dramatically improve algorithm performance.

Overview of Scheduling Technology

In this section we will highlight the key features of available scheduling technology, focusing specifically on how this technology addresses the underlying combinatorial character of these problems, namely their NP-completeness. We begin by recapitulating the architecture of scheduling systems and the desired properties of the solution technologies that constitute the heart of such systems.

Table 1. *Summary of Algorithm Engineering Features for Example TSP Discussion.*

Algorithm Engineering Feature	Observation
Ignore constraint (D)	Required for solving non-trivial problem sizes
Use Dijkstra's algorithm	Multiple order of magnitude speedup over generic LP solution
Parametric LP solution at non-root nodes in branch and bound tree	Order of magnitude speedup
Use sparse matrix methods on objective function	One to two order of magnitude speedup by ignoring large cost matrix value
Use d-heaps in Dijkstra's algorithm, etc.	Tailor LP solution for exploiting cost matrix sparsity, order of magnitude speedup
Directed Hamiltonian Cycle application at each node in search tree	Mitigate degeneracy in LP relaxation
Branching rules customized for constraint (D)	Approximate factor of two speedup

Scheduling System Architecture

A scheduling system must provide an effective framework for representing, generating, and reporting the solution to scheduling problems. As noted by Applequist et al [1997], this is most effectively done through a multi-layered approach consisting of four basic components:

- a graphical user interface
- a representation layer
- a formulation layer, and
- a computational object layer

The graphical representation layer offers point-and-click user input and navigation and supports graphical displays of recipe structures, equipment networks, "intelligent" Gantt charts, inventory and resource utilization plots, etc. The representation layer contains a complete and unambiguous description of the problem to be solved. In (Zentner et al [1998]) use of a structured language is advocated for this purpose as it allows convenient data interpretation and checking of consistency and logical correctness. The formulation layer consists of the translation steps required to convert the problem representation to a description that can be operated upon by the computational layer. For instance, at this level the problem may be cast into a specific equation based model formulation suitable for solution by an appropriate solver. The computational object layer contains the specific algorithms or procedures that are invoked to generate the solution schedule.

Good system design practice suggests that these layers should be highly modular and largely independent, using the object oriented paradigm. The GUI should be designed to represent the problem in terms natural to the plant scheduler. The representation layer should cast the problem into a structured form which is independent of the particular formulation and solution method selected and can be unambiguously understood by a technical support person. Generally, but not necessarily, the formulation and computational object layers may be more closely coupled. However, a comprehensive system will offer several alternative formulation - solution paths. Given current user expectations of applications, all commercial scheduling systems offer the first layer. Indeed some focus exclusively on this layer, offering only very limited tools for alternative schedule generation. Others intertwine the functionalities of the last three layers so tightly that a change in representation requires reconstruction of the entire solution process. As will be elaborated in the third section of this paper, the degree to which modularity of layers is offered will dictate the level of expertise which is required to maintain plant applications as the information and structure of the application change over time. For instance, the more tightly merged are the functionalities of the representation, formulation, and computational layers, the more the user organization will be tied to the vendor's consultants.

Scheduling Technology Features

Each of the layers described in the previous section serves an important role and contributes understanding during the process of collecting all of the relevant information and agreeing on the essential elements to be incorporated into the problem representation. However, the largest value is added at the computational object layer, since it is there that the solution schedules are generated. Indeed, it is the nature and capabilities of this layer that ultimately dictate the character of the solution and the effectiveness of the entire scheduling system. The features that are due solely to the nature and capabilities of the computational layer are:

- solution quality
- usability
- extensibility
- robustness
- complexity strategy

The quality of the solution generated by a solution methodology refers to its feasibility and optimality. A candidate schedule generated by the computational layer may be infeasible, feasible, suboptimal, suboptimal with bounds, or optimal. The optimal solution, that is, the solution which is provably the best feasible solution to the problem is, of course, the most desirable.

The usability of the solution methodology refers to its overall computational performance, including, clock time to obtain a solution, the degree to which the user must or can be involved in steering the solution process, and the manner in which failures are treated. Clearly, solution times as well as the predictability of solution times are an essential aspect of user acceptance of the system and its solver. Other usability issues include the degree to which the user has to steer the solution process and the controls that are provided to the user to impose solution preferences. Failure handling is concerned with the manner in which the system addresses premature termination of the solution process and the degree to which it provides help in identifying causes of failure, whether these are due to input inconsistencies or actual computational pathologies.

The extensibility of a methodology relates to its ability to accommodate incremental changes in the problem being solved and still be able to generate high quality solutions in reasonable solution times. These changes may be parametric, such as changing product demand levels, or structural such as adding equipment items or another category of limited resources. An extensible methodology will require modest additional effort to add parametric or structural extensions to an existing successful application.

In the context of scheduling methodology, robustness refers to the ability of a methodology to be successfully used to generate high quality solutions on another application of the same general type but structurally and parametrically distinct from the application type for which it was originally developed. A robust methodology will, by definition, require only modest additional effort to accommodate a new class of CPI applications once it has been proven to be successful on a given class.

Finally, the complexity strategy refers to the manner in which the solution methodology addresses the underlying NP-completeness of the scheduling problem. As noted earlier, the three basic strategies are to (1) change the problem, (2) use an exact method which may require unreasonable solution times, or (3) offer solutions of unknown quality.

In light of the fact that virtually all instances of scheduling problems of relevance to the process industries are NP-complete, it should be clear that there can be no single methodology or algorithm which is sufficiently robust to accommodate all classes of process industry problems. However, solution technologies do differ substantially in solution quality, usability, extensibility, robustness, and complexity strategy. In the next section we review existing scheduling methodology to assess the degree to which they possess these features.

Types of Scheduling Technologies

Scheduling methodologies can be grouped into two broad classes: heuristic methods and equation based methods. The former class includes randomized search methods, rule-based approaches, constraint guided search, and simulation based strategies. Equation based methods involve the representation of scheduling problems as mathematical programs and include two basic strategies: formulation-focused approaches which employ general implementations of mathematical programming algorithms and engineered mathematical programming approaches which rely on highly customized algorithms. In this section we will highlight the essential elements of these methods with the primary purpose of a comparative assessment of their features. The reader interested in further details and representative examples is directed to Applequist et al [1997] and the references cited therein as well as to the review papers by Shah [1998] and McDonald [1998].

Randomized Search Methods

This class of methods, which is among the most actively applied and commercially most widely disseminated, includes several prominent members: genetic algorithms, simulated annealing, and Bayesian Heuristic Methods. Of these the first two have been most actively pursued: hence, we will confine our discussion to them. For details on BHS, the reader is directed to (J. Mockus et al, 1997) and for a specific applications to scheduling to (Mockus and Reklaitis, 1997). Genetic algorithms and simulated annealing rely on representation of the solution of the scheduling problem as some form of ordered list of entities where the position in the list denotes the order in which the entities are to be processed on specific equipment items. A simple example is the list of product grades that are to be processed on a single equipment item. In this case, the position of the grade in the list indicates its order in the sequence. The two members of this class differ principally in the manner in which the fixed encoding of the solution structure (in the above example, the list of grades) is manipulated in order to obtain schedules with desired properties. Typically, the key property is a cost or performance based measure of merit.

Genetic algorithms modify the encoding using two basic change operators: randomized combination of sublists drawn from available preferred solutions into new lists and randomized rearrangement of the components of the list associated with a specific preferred solution. Solutions that have suitably good performance measures are retained among the preferred population and the process continued until some suitable stopping criterion is met. Solutions that violate resource limitation or other problem constraints are either penalized and thus rejected or else problem specific procedures are devised to modify the encoding to attempt to correct infeasibilities. For instance, certain types of constraints may be implicity incorporated so as to reduce the problem degree of freedom. The key algorithm design choices associated with this family of methods include the choice of solution encoding, the details of the change operators, the mechanism used for constraint handling, and a number of tuning parameters, such as the definition and size of the preferred solution set, the relative frequencies at which the change operators are applied, and termination criteria. (see Cheng [1996]; Sunol et al.[1996])

Simulated annealing may be viewed as a randomized hill climbing search which accepts non-improving solutions with some probability. The acceptance criterion is adjusted in a step-wise fashion as the search progresses so that nonimproving solutions are less likely to be accepted. Again the key components of this approach include a fixed encoding of the solution structure, a set of allowable change operators (for instance, pairwise interchange of positions in the list) for improving the solution, a randomization of the selection of change operators, an acceptance criterion with adjustable parameter, and a schedule for tightening the parameter of the acceptance criterion.. Infeasibilities are typically addressed via the same mechanisms as employed in the case of genetic algorithms. Solution quality improves asymptotically as the search is continued, providing that the change operators can be guaranteed to span the space of all feasible solution candidates (see Ku and Karimi [1991]; Murakami [1997]).

The randomized search methods exhibit the following common qualitative features:

Solution quality: Since both approaches must be seeded with feasible solutions, the issue of feasibility is passed to the user. Attained improvements in solution quality will strongly depend on how tightly constrained the application is and the length of time allowed for the solution process. With tightly constrained applications, the probability of finding a feasible solution through the random application of change operators can be vanishingly small. Hence, at best, suboptimal solutions will be found. Inherently there is no systematic way of obtaining good bounds on the attainable optimum values of the performance function.

Usability: This family of methods can have very reasonable solution times but these times often scale poorly with problem size, even with problems which are effectively unconstrained (see for example [Miller and Pekny, 1991]). The user has some control over the solution process through the selection of the initial seed solutions and by fixing portions of the encoding. There is very limited information available in the event of solution failure, e.g., no improvement of the seed solution is obtained within some allowed cpu time.

Robustness: The common features of randomized search methods are that the solution structure encoding, choice of change operators, and various tuning parameters are very problem specific. As a result, this family of methods scores poorly in robustness: a different set of recipe structures will require a different encoding and certainly will require retuning of parameters and change operators.

Extensibility : It is difficult for the user to modify these solution strategies as structural elements are added to an existing application since changes in structural elements generally require changes in the solution encoding, modifications of the change operators, and retuning of the algorithm parameters.

Complexity strategy : It is evident that complexity is addressed by shifting to the user the problem of generating an initial feasible solution (strategy 1). The issue of generating solution improvements is handled through strategy 3: providing a solution procedure that offers no assurance of quality or feasibility or both.

In general, robustness and usability of these methods are poor principally because the representation, formulation, and solution layers are very tightly coupled: a change in problem structure must be propagated through the entire solution approach.

Rule-based Methods

The motivation for rule-based scheduling methods is to emulate the decision making behavior of human schedulers by attempting to express it in terms of a set of rules. Typically rules are stated in the form of assignment or dispatching principles, such as, "assign the next task to the earliest available unit". Generally, multiple rules are required, each relevant in some context, and, thus, the system must include some mechanism for selecting appropriate rules and for resolving conflicts among rules. In the presence of constraints, additional mechanisms and rules must be employed to correct infeasibilities. For instance, as in the case of random search methods, certain types of constraints may be implicitly incorporated in the rule processing schema so as to reduce the problem degree of freedom (Mockus and Reklaitis, 1998). However, these constructions invariably are highly application dependent. Because of this complication, rule based methods are often combined with constraint based search. (see Kudva et al [1994]; Stephanopoulos and Han [1996])

Solution Quality: Neither feasibility nor optimality can be assured. Constraint satisfaction is particularly problematic. If no feasible solution is found there is no assurance that one does not in fact exist. Moreover, there is no way of assessing how close any particular candidate solution is to the optimal solution.

Usability: Solution times can be quite modest, especially since feasibility is not assured. The user is, of necessity, fully engaged in the process and, indeed, serves as part of the solution algorithm. User preference are typically imposed by choice of specific rules and by manual modifications/editing of the system proposed solutions to correct infeasibilities. There are no systematic ways of identifying causes of failures other than of presenting instances of infeasibilities.

Extensibility: Even parametric changes can significantly affect the effectiveness of the previously selected rules. Moreover, there is no mechanism for determining the extent to which the solution quality will deteriorate with parametric, much less structural, changes.

Robustness: Successful solution of one class of process applications with a given set of rules provides no assurance of success with another class. New rules will in general need to be created for different applications since no *a priori* set of rules which covers all situations exists.

Complexity strategy: In this family of methods strategy 1 is employed: the original problem is simplified and the burden of insuring feasibility is shifted to the user.

Rule-based systems typically are envisioned as a means of replicating the actions of human experts experienced in scheduling specific facilities. The very nature of such systems dictates considerable development and tuning effort for each new application since there is no expert to emulate. In general, there is no structured way to carry the insight gained from one application to the next.

Constraint Guided Search

This family of scheduling methods focuses on attaining feasible solutions that balance various constraints or schedule requirements. Satisfactory schedules are generated via the iterative refinement of an incumbent schedule which involves successively identifying and attacking the most limiting or most important constraints. Each successive constraint violation is corrected by the application of some suitable set of schedule modification rules. If it is not possible to satisfy all constraints, the constraints are selectively relaxed by making trade-offs. Such trade-offs of course will be controlled through user specified preferences (See Fox[1996]; Chiodini [1996]).

Solution Quality: Since the focus of the search is on obtaining feasible solutions, direct optimization of a selected performance measure is not an issue. However, by focusing on balancing constraint infeasibilities, acceptable solutions can be obtained. Since the search is heuristic, there is no guarantee that if no feasible solution is found none, in fact, exists.

Usability: Again solution times can be reasonable because feasibility is not assured. However, performance degrades significantly as the scheduling application becomes more tightly constrained and the constraint interaction become more complex. Yet, it is in those instances that scheduling methodology offers the most value. With this family of methods, the user must be actively involved in the solution process and can control the search by manipulating constraint priorities. Failure resolution is also left to the user.

Extensibility: As in the rule based case, even parametric changes can significantly affect the degree to which different constraints limit the solution. Structural changes can totally alter the constraint structure and interaction and thus extensions of successful application typically require substantial changes in search strategy.

Robustness: Very little learning can be carried from one application to the next as the constraint structure, relative importance, and preferences will be quite different.

Complexity strategy: This family of methods embodies strategy 3, namely, use of a heuristic approach that offers no assurance of quality or feasibility.

As with rule based systems, constraint based tools involve tight coupling between the representation, the solution operators, and the user. Thus, the development of applications is expert/consultant intense.

Simulation-based Methods

This approach to scheduling uses a discrete event or combined continuous-discrete simulation model of the production facility as a vehicle for testing alternative schedules. Typically, the user must provide a partial schedule in the form of an event stack or a task execution sequence along with some dispatching rules to resolve any timing and resource utilization conflicts which might arise in the course of simulating the imposed partial schedule. Since the simulation progresses by moving forward in time, if conflicting demands for resources or equipment arise, the dispatching rules only have available the information about the status of the simulation at that point in simulated time in order to resolve these conflicts. Thus, these scheduling actions can only be local in character. However, simulation as a tool has the advantage that highly complex recipes can be accurately represented and Monte Carlo sampling can readily be incorporated to assess the impact of stochastic problem parameters. Typically, these types of simulations are developed using any of a variety of simulation languages or packages. (See Kuriyan et al [1987]; Reklaitis et al. [1997]).

Solution Quality: Since the partial schedule imposed by the user drives the simulation the determination of optimal or even feasible schedules is largely left to the user. The simulation can only correct any conflicts by applying the dispatching rules or by delaying the start of events requiring overbooked resources. Clearly, it is not possible in general to provide bounds on the optimum performance function value.

Usability: The execution time of simulation models is typically proportional to the number of tasks or events that must be processed. Thus, the execution times are usually modest and predictable. The user is largely responsible for generating candidate schedules and is fully in charge of selecting dispatching rules and their parameters. Failures in this context normally are confined to inability of the simulated plant to meet production demands. Resolution is left to the user.

Extensibility: Parametric or structural changes to an application require both changes to the simulation model as well as changes to the partial schedule and dispatching rules. The effort involved in the former is modest as most packages make changes in simulation models relatively easy. However, the empirical development of good partial schedules necessarily will need to be repeated by the user in toto, as will the retuning of the dispatching rules employed.

Robustness: New applications, of course, require the *ab initio* construction of new simulation models as well as the identification of new partial schedules.

Complexity strategy: Inherently simulation based methods shift the burden of dealing with NP-completeness almost entirely to the user. This is therefore a clear instance of strategy 1.

The most appropriate role for simulation models is to serve as a detailed validation tool for schedules developed systematically using other scheduling techniques.

Equation-Based Methods

The representation of scheduling problems as mathematical programs, that is, optimization problems expressed in terms of binary (0-1) and continuous variables, has been proposed in the operations research literature nearly forty years ago. The basic form of these models is to maximize/minimize a suitable objective function subject to equality and inequality conditions, representing material balances, equipment allocation constraints, storage constraints, and resource utilization constraints. As summarized in [Shah, 1998], for most scheduling applications arising in the process industries, it is possible to choose the variables and formulate the objective function and constraints such that the resulting optimization problem is a mixed integer linear program. Although a number of different formulations have been proposed over the past decade, for applications of practical scope, all lead to very high dimensionality problems with tens or hundreds of thousands of 0-1 variables and constraints. The most important differentiation between the proposed approaches which use equation based mathematical programming formulations arises in the degree to which the solution strategy and the structure of the formulation are integrated. In what we will term the conventional mathematical programming strategy, a general-purpose commercial MILP solver is used to solve the formulation. By contrast, in the engineered mathematical programming strategy highly customized MILP solution methods that exploit the detailed problem structure are employed.

Conventional Strategy. It is now well understood that the brute-force mating of an MILP formulation with an MILP solver is highly unlikely to yield successful solution to problems with over a few hundred 0-1 variables. Rather, in the best of conventional practice, attempts are made to construct the MILP formulation so as to best exploit the commercial MILP solver of choice. These solvers, which invariably employ the classical branch and bound strategy wherein linear programming relaxations are solved at each node of the enumeration tree, offer the user little control over the solution path and limited facility to exploit either the problem physics or the knowledge gained from solving similar problems. Thus, the focus of research on MILP approaches to scheduling has been on seeking formulation improvements which lead to tighter LP relaxations, reduced 0-1 variable dimensionality, or more elegant mathematical structures. Indeed, considerable progress has been made in recent years in understanding how to do this (Grossmann et al 1996). However, major gains in computational efficiency are missed because the solver is treated as a black box.

Solution quality: Optimal solutions are guaranteed upon completion of the branch and bound process. Bounds on the optimal objective function value are available if the solution process is terminated prior to completion.

Usability: Solution times can be quite long and unpredictable, even from what appear to be minor parameter changes to an instance. Solution degeneracy is a major source of solution time unpredictability. User intervention in the solution process is neither required nor conveniently possible. However, the user can impose solution preferences by fixing or limiting variable values. Solver failures are difficult to diagnose. If no feasible solution can be found, auxiliary problems can be used to generate solutions that only violate the least important constraints.

Robustness: While formulations exist which cover broad classes of scheduling applications, the performance of general purpose MILP solvers can be very sensitive to problem structure and thus the composite robustness of the formulation-solution path can be quite poor, as expected from the NP-completeness analysis.

Extensibility: The use of mathematical models to represent the application readily allows extension of the formulation to accommodate new parameters and structures. For instance, additional problem structures simply involve the addition of additional sets of variables and constraints to the original set. However, the effect of these new model elements on the effectiveness of the solver is unpredictable and in general can considerably extend solution times.

Complexity strategy: Mathematical programming approaches inherently employ strategy 2 by providing optimal solutions at the cost of possibly unreasonable solution times. Once a feasible solution is found, the user has the option of terminating further search and using the relaxation bounds to estimate solution quality.

Engineered Strategy: The basic goal of this strategy is to obtain the highest overall effectiveness of the solution process by seeking the best match of formulation with a solution strategy which exploits the detailed formulation structure. This requires that the MILP solver is not treated as a black box but instead is structured to exploit the detailed constraint connectivity and the physical interpretations of variables and constraints. The work of Pinto and Grossmann (1997) in modifying the branch and bound logic to accommodate disjunctive constraint sets represents a step in this direction. In general, various algorithm engineering elements have been under development and use for decades to attack classical problems such as forms of the traveling salesman problem, the knapsack problem, the b-matching problem, and to some extent the quadratic assignment problem. In these applications, devices such as problem reduction methods, primal rounding heuristics to expedite the branching process, specially ordered set constructions, customized pivot selection methods, dynamic matrix factorization, and column generation methods are employed to take advantage of the known constraint structures present in the specific applications. This was the case with the engineering of the solution algorithm for the asymmetric TSP described earlier. It is important to note, however, that these algorithm engineering devices do not necessarily compromise the rigor of the solution process: attainment of the optimum may still be guaranteed. This is analogous to the difference between the steepest edge and partial pricing strategies in large-scale linear programming. Both are guaranteed to terminate at the optimum but the latter is much more efficient computationally (Martin, 1999).

Initial applications of algorithm engineering strategies to process scheduling are in progress but published applications are few. As a consequence of such algorithm engineering techniques, the solver is adapted to the characteristics of the problem at hand and substantial improvements in solution speed attained. By way of anecdotal example, the successive reduction in cpu time for a particular bulk process scheduling application is shown in Fig.1 (Zentner, [1997]). In this instance, the learning curve for the engineered approach extended over a five-year period because literally both the tool development and the research on the algorithmic enhancements had to be undertaken.

The engineered approach also lends itself to learning strategies under which techniques developed for specific constraint structures can be engaged when those structures are recognized in a new application. This suggests the use of iterative learning strategies to improve algorithm performance as problem instances are solved. Indeed, an iterative learning approach to enhance the effectiveness of the branch and bound solution process has been reported by Realff and Stephanopoulos (1998). However, the key disadvantage of the algorithm engineering approach is that it requires the development of sophisticated tools and approaches of high intellectual fixed cost and implementation complexity.

Solution quality: As in conventional case.

Usability: Solution times can be long, however, the engineered approach results in lower and more predictable solution times. The exploitation of structure facilitates avoidance of degeneracy and identification of model inconsistencies and thus facilitates failure handling.

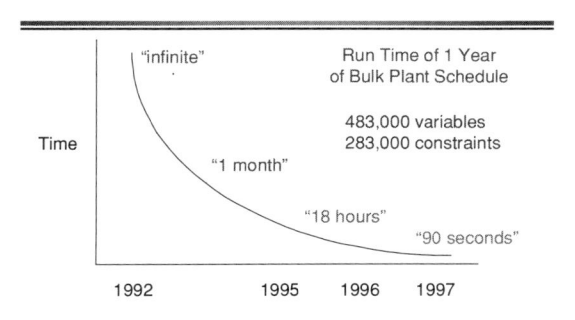

Figure 1. Effects of algorithm engineering on bulk scheduling application.

Extensibility: The engineered strategy shares the formulation extensibility advantages of the conventional equation based strategy but has the additional advantages provided by a solution approach that can take advantage of the details of problem structure. This capability inherently extends the adaptability of the strategy to both parametric and structural changes in the application. The advantages of the engineered strategy lead to the notion of <u>inductive algorithm improvement</u>, which is defined as the requirement that a given algorithm be extended to handle a new set of problematic instances while retaining the capability of solving all existing instances of interest. While such a requirement seems natural, only the ready extensibility of mathematical formulations promotes the development of an algorithm that can be manageably extended to a wide variety of problem structures.

Robustness: The engineered approach offers the hope of improved robustness through the recognition of types of constraint structures combined with a knowledge base of successful techniques for those structures. At present there are no automated procedures that facilitate the learning process.

Complexity strategy: The engineered mathematical programming approaches does employ strategy 2 but it relies on the characteristic of NP-complete problems that through careful exploitation of structure acceptable performance can be obtained.

The comparative features of existing scheduling methodology are summarized in Table 2. The obvious conclusions one can draw from this status assessment are that none of the technologies meet all of the desired characteristics. However, the engineered math

Table 2. Assessment of Features of Scheduling Technologies.

Technology	Quality	Usability	Extensibility	Robustness	C-Strategy
Random Search	Poor	Fair	Poor	Poor	1/3
Rule-Based	Poor	Fair	Poor	Poor	1
Constraint-guided	Poor	Fair	Poor	Poor	3
Simulation-based	Poor	Fair	Fair	Poor	1
Conventional MP	Good	Fair	Fair	Poor	2
Engineered MP	Good	Good	Good	Fair	2

programming approach seems to offer the best prospects for delivering a methodology that directly addresses the combinatorial complexity of the process scheduling problem while also providing technology features that are important in practice.

Business Issues

Of course the most important, and obvious, business issue associated with scheduling tools is that their use must result in more profitable operations. Use of scheduling tools may increase profitability in one or more of the following ways:

Increase Customer Service. The ability to schedule more complex process operations results in a decrease in the amount of time that a customer has to wait for a product or makes the delivery date more reliable. Increased service levels translate to improved competitiveness, both in terms of attracting new customers and retaining existing ones.

Lower Inventories. Decreasing inventories generally requires that upstream and downstream processes be managed with a greater degree of coordination. Scheduling tools may facilitate such coordination by timing upstream and downstream processes to operate in such a way that less inventory is required.

Reduce Need for Excess Capacity. Complex scheduling issues may always be avoided by retaining sufficient excess capacity. Conversely, one of the advantages of using effective scheduling tools is the ability to operate processes with less excess capacity. In a similar fashion, scheduling tools can help determine appropriate resource levels (e.g. staffing).

Provide Mechanistic Understanding of Process. In a changing business environment, processes are subjected to new and different combinations of stresses. For example, price competition, the need to manufacture new products, changes in raw material supply, environmental pressures, labor issues, opportunistic business, etc. Changes in business environment to some extent obviate experience and require a mechanistic understanding of how changes will effect the profitability of operations. Scheduling tools offer the opportunity to develop a mechanistic understanding of processes to show how changes in details translate into outcomes. If used properly, this insight can result in better strategic management and can provide an earlier indication of potential problems.

Identify Key Data. For purposes of strategic management, not all process and business data is equally important. However, the cost of data acquisition and maintenance scales with the volume and required level of accuracy. Indeed, a portion of the cost and time overruns of many large scale enterprise wide data system projects can be attributed to making them comprehensive in that all data is treated uniformly. Scheduling tools offer the potential of identifying which data is most important, often a fraction of the data that could be categorized. Thus scheduling tools offer the potential of directing enterprise wide data system efforts, reducing their scale, and generating cost savings/system simplification in the process.

Whether a particular scheduling tool is able to deliver any of these potential improvements in profitability very much depends on the type of technology employed and the business model of the customer and vendor. In order to explore the role of technology in delivering scheduling tool value, consider the usage model depicted in Figure 2. The attention of many companies is currently focused at the lowest level, namely, on providing systematic data availability through enterprise wide data systems. At the next level shown in Figure 2, the process model provides structure and organization to data. For purposes of scheduling, a process model consists of an objective (e.g. minimize cost, maximize profitability, etc.) and a specification of constraints (e.g. material balances, maximum storage levels, etc.). A process model is goal oriented in that the objective rates alternative schedules and clearly indicates the properties required of a feasible schedule. The function of the scheduling tool, the third level in Figure 2, is to produce a schedule that is of high quality with respect to the objective function while meeting all constraints. A user interprets the schedule to determine if strategic concerns are properly met and may iterate through many possible answers before accepting one for routine implementation. An effective tool may also be used for off-line analysis to help in making capital decisions such as how much new capacity should be added or ascertain the effect of labor or environmental rule changes.

Since production schedules coordinate much of a facility's resources, schedules have a significant impact on the efficiency of a facility and on whether strategic business goals are met. As scheduling technology becomes more effective, the scope of impact will increase as additional upstream and downstream processes are integrated with a given site. As this occurs, business processes must be restructured to reflect the corresponding impact. In particular, scheduling models can be used to answer strategic questions such as the effect of a marketing strategy on production assets, how to classify customers according to their marginal value, etc. As scheduling tools, their underlying models, and the large amounts of data they marshal become faithful software copies of processes how scheduling tools are used within the decision making process must be re-evaluated.

The role of the vendor business model also has a significant impact on the potential of a scheduling tool. A key aspect of the business model is whether tools are designed to function as automatically as possible or are toolboxes that require significant user direction. Another key aspect of the business model is the amount of consulting required to produce useful results and how much consulting is required to maintain tool functionality across normal process changes. Both of these aspects are directly controlled by the underlying scheduling technology and ultimately by the vendor's approach to addressing NP-completeness.

In considering Figure 2, note that the standard scheduling practice in most facilities is based on an experienced user who employs either a spreadsheet or paper based approach. Before moving on to discuss the potential impact of more sophisticated technology, the usage model of this standard practice will be considered to identify what improvements are possible. Also, consideration of the standard practice is important since a spreadsheet based approach often sets the expectations that must be considered when a more sophisticated technology is going to be deployed. Indeed expert schedulers are able to use spreadsheets to obtain more insight and handle considerably more complexity than would be possible by hand. Thus scheduling tools are expected to offer an even greater improvement in capability. As such, increasing use and familiarity with common applications, such as word processors, spreadsheets, personal databases, web browsers, etc. is changing the interface requirements and what is expected of scheduling algorithms.

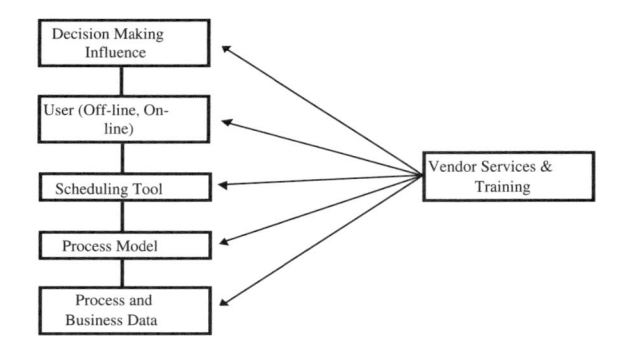

Figure 2. Usage model for scheduling tool.

With respect to Figure 2, a spreadsheet based approach generally requires a manual or semi-automatic feed of data through a file interface (e.g. comma separated values) into a spreadsheet structure that has been constructed by a present or past scheduler. Except for the simplest of processes or in the most elaborate application of spreadsheets, a schedule is not directly generated through the spreadsheet. Rather the spreadsheet calculations determine the material resource requirements and the numbers of tasks (activities) required to meet demands. Actual equipment assignment, sequencing, and timing is done manually with the spreadsheet, Gantt Chart tool, or paper Gantt Chart serving as a display mechanism. Because the user has to make so many interacting decisions, the spreadsheet approach tends to be tedious and time consuming. Furthermore, the amount of expended effort scales strongly with process complexity. In the spreadsheet approach to scheduling, the process model is implicit as is part of the scheduling function. In terms of the technology features defined in the previous section, the spreadsheet approach to scheduling is not robust in that it typically fails to deliver feasible solutions and systematic process optimization is almost never possible. Spreadsheet based approaches also are not very extensible in that process changes may involve considerable effort to reconfigure and re-interpret the spreadsheet analysis. With respect to business issues and despite its shortcomings, the spreadsheet approach to scheduling does indicate a very important point. Namely, spreadsheet based approaches establish that companies are able to organize and provide the data needed for any scheduling technology without requiring elaborate information integration efforts. In short, usage of scheduling technology is facilitated by integration but does not require it to be used. This raises some interesting possibilities for how scheduling tools might be employed in the future (see below). A compelling advantage for spreadsheet based approaches relative to more sophisticated technologies is that users or company support people can provide help and maintenance without recourse to outside consulting. Although expense is one reason for concerns about the need for outside consulting to maintain scheduling capability, the principal objection is the need for timely response. Unless a consultant is based on site or dedicated, at least a several day response time is often required for moderate changes. Thus the amount of consulting required to install and maintain a tool is a principal concern of the marketplace and is something that is directly controlled by the underlying technology.

The single most important attribute controlling ease of installation and maintenance is the separation of the process model from the solution technology used to produce a schedule. Explicitly constructing a proper model requires developing an understanding of what is important to consider in developing a schedule and what detail can be ignored or relegated to manufacturing execution. The development and understanding of facility scheduling models can result in a significant corporate resource. The most valuable technologies for expressing process models may require some training or vendor expertise in constructing the first few, but over time the libraries of models that companies build up mitigate the need for substantial outside consulting resources. Furthermore, the process models can serve as the implementation means for business plans.

As pointed out in the previous paragraph, separating the construction of process models from the underlying solution technology has several advantages. However, such separation also greatly increases the burden on solution technology. When the solution technology is intertwined with the process description, the complexity can more easily be controlled and the chance of a technology being exposed to situations in which it cannot properly function is more limited. Indeed, many of the technologies described in the previous section can only work when the user directs their application in an iterative fashion, with graphical feedback dictating how the user proceeds. Thus most technologies are used in a toolbox environment where the user decides which tools are applied and when to achieve the desired result. The presumption in separating the process description from the solution of the scheduling problem defined by that description is that the technology will automatically determine a solution without requiring tedious interaction on the part of the user. The user interface challenge with more automatic solution technology is the definition of parameters which allow the user to "sculpt" the shape of a schedule to meet strategic guidelines without requiring any significant direction of the solution technology. The degree to which a user must direct a solution is of more significance than just convenience. The more automated the scheduling process the more complexity that can be handled and the more likely that the benefits listed at the beginning of this section will be realized.

The discussion of the previous paragraph and in the solution technology section above suggests classifying the business usability of scheduling tools according to their degree of automation in producing a high quality or at least feasible result. At one end of the spectrum are electronic Gantt chart tools that are often applied in conjunction with a spreadsheet to help a user develop a schedule. At the other end of the spectrum are fully automatic scheduling tools which input a process description and "sculpting" parameters and output a schedule which can be analyzed and then refined if necessary. In between these extremes are the toolbox environments which allow the users to apply relatively straightforward operations to partially automate schedule construction. Virtually all existing technology lies within the electronic Gantt chart to toolbox environment part of the spectrum with fully automatic scheduling tools still not available for full scale deployment. The challenge in producing fully automatic tools derives from the NP-complete nature of scheduling problems. With time, research and development will be directed towards the typical scheduling structures that show up in practice (e.g. packaging, extrusion, etc.). As more of these structures and combined structures are successfully addressed, scheduling tools will be able to become more automated. Although because of NP-completeness fully automated scheduling tools are not really possible, research and development can make the likelihood of failure very small. If process modeling is conducted with some knowledge as to the capabilities of solution technology then the probability of failure can be kept near zero. Whether all the benefits listed at the beginning of the section can be achieved depends on the capabilities of the solution technology. Although the scheduling industry is currently very young, over time the capabilities are almost sure to grow.

Roughly speaking the investment in a scheduling tool can be classified as being apportioned to the interface or the underlying solution technology. As discussed in the previous section the interface includes graphics required for specifying a problem and understanding the solution, and the machinery for inputting process models, interfaces to enterprise wide data systems and common personal computer tools. Although interface work requires significant creativity, increased investment in the interface usually yields corresponding improvements in capability. In other words, the payoff for interface investment is predictable. However, as long as a reasonably functional interface is available, most of the benefits and ease of use stem directly from the investment in the solution technology. Unfortunately, the payoff for solution technology investment is not predictable. Large amounts of resources can be devoted to technology that turns out to neither robust nor extensible. In fact, the development of solution technology is expertise intense and relatively independent of the scope of investment, while the development of interfaces is more investment intense and more of a development commodity. In light of the rapidly growing market for scheduling tools, this state of affairs has resulted in a fairly predictable research and development strategy among scheduling vendors. In particular, most tools are the result of significant interface investment and solution technology strategies that require substantial user input to avoid many of the difficulties imposed by NP-completeness. This strategy allows vendors to capture market share, increase shareholder value, and position themselves to incorporate any advances in solution technology that they discover or become widely known. Furthermore, the lack of market understanding about the use of scheduling tools in practice and the fundamental challenge of NP-completeness has been consistent with a large investment in interfaces and marketing and less of a need to develop fundamentally new solution capability.

As a measure of the need for market education and a consumer report function, a web search on scheduling tools yields dozens of web pages that claim technology capable of routinely delivering feasible or even optimal solutions for complex processes. Such claims are almost never accompanied by caveats about specificity of details for which the technology can work or the process by which the technology is improved, yet NP-completeness disallows effective and completely general solution techniques. The best which can be hoped for is an extensible technology that can be iteratively improved. A significant opportunity exists for researchers and educators to develop aids to illustrate the practical consequences of NP-completeness. Improvements in market understanding will reduce the chance of failure in scheduling tool installation, properly frame the role of enterprise wide data systems, and help guide scheduling project management.

The Future of Process Scheduling Tools

The fundamental tradeoff present in the development of solution technology is the type of expertise deployed and where it is deployed, either in the field or in development efforts. A simplistic framework for solving scheduling problems requires a substantial amount of in the field customization in order to deliver satisfactory results. Furthermore, maintenance of a simplistic solution technology requires substantial on site effort to accommodate minor process changes. Scheduling tools that compensate for shortcomings in solution technology by requiring significant user direction still tend to be consulting intense. Hence vendors that build tools around relatively simplistic solution technology choose a business model that involves substantial consulting. Conversely, vendor resources can be deployed in an attempt to dramatically improve solution technology. If successful, a large investment in solution technology may dramatically reduce the amount of internal and external consulting that customers require to make and keep scheduling tools effective. Because of NP-completeness, a significant advance in solution technology is defined to be one that greatly facilitates automatic generation of more complex schedules. Obviously such an advance would eventually confer a significant competitive advantage for a vendor.

As long as the market does not discriminate among solution technologies, the deployment of consulting intense tools will persist. A major advance in solution technology that dramatically shifts the amount of effort required for tool deployment will spur a major change in the economics of the scheduling market. Given the enormous potential savings from improved scheduling capability and that approaches for addressing NP-complete problems are still in their infancy, the likelihood of dramatic advances in solution technology are significant.

A key research challenge in the area of process scheduling is the development of solution algorithm architectures that allow effort expended in addressing one class of instances to be usable in a more general context. The goal is to develop an algorithm architecture that directly supports *inductive development* whereby an algorithm that is capable of successfully solving instances from a given set of generators with high probability is readily extended to solve a new type of generator that arises in practice as well as the existing generators. One aspect of the power of an algorithm architecture is then measured by the ability of an algorithm at a given time to handle generators for which direct development efforts have not been expended. Another aspect of the power of an algorithm architecture is then measured by the amount of effort which is required to extend an algorithm to successfully handle a problematic generator. With an appropriate algorithm architecture, the inductive development process leads to an accumulation of scheduling knowledge within a given solution algorithm. As more practical cases are encountered, the algorithm becomes more capable and less consulting and development effort is required to deploy scheduling tools to new processes. The inductive development process does not avoid the issues of NP-completeness, rather there will always exist generators for which a given algorithm does poorly. However, nothing about NP-completeness prevents a given algorithm from being able to be developed for most, if not all, practical process scheduling applications. The choice of how much capability to engineer in a given algorithm is then solely an economic question that depends on the structure of the algorithm architecture, the number of applications to which an algorithm is to be subjected, the business model of the owner company, and market competition between different products.

Significant advances in solution technology will accelerate the trend towards using scheduling tools in conjunction with existing personal computer utilities such as spreadsheets, databases, reporting tools, web browsers, etc. The asymptote to this trend is an open scheduling environment in which users can download data from central databases, manipulate it in a spreadsheet or local database, generate a series of solutions to what-if questions, manipulate the results, develop customized presentations in familiar reporting tools, and upload results as appropriate for corporate wide publication. Such a usage mode for scheduling tools leverages the considerable investment that has been made in learning to use common utilities and frees vendors from having to duplicate common information management functionality. Of course for data to flow freely among tools from different vendors, open standards must be in place for data exchange. Fortunately, the natural evolution of corporate information technology projects seems to be towards such an open model.

References

Applequist, G., G., O. Samikoglu, J. Pekny, and G. Reklaitis, "Issues in the Use, Design, and Evolution of Process Scheduling and Planning Systems", *ISA Transactions*, Vol. 36, No. 2, pp. 81-121, 1997.

Balas, E., D. L. Miller, J. F. Pekny, and P. Toth, "A Parallel Shortest Augmenting Path Algorithm for Fully Dense Linear Assignment Problems", *Journal of the Association for Computing Machinery (JACM)*, Vol. 38, No. 4, pp. 985-1004, 1991.

Cheng, R., M. Gen, and Y. Tsujimura, "tutorial survey of Job-Shop Scheduling Problems using Genetic Algorithms, *Computers and Industrial Engineering*, Vol. 30, No. 4, pp. 983-997, 1996.

Chiodini, V., "Configurable On-Line Scheduling", *Integrated Computer-Aided Engineering*, Vol.3, pp.225-243, 1996.

Fischetti, M., and P. Toth, "An Additive Bounding Procedure for Combinatorial Optimization Problems", *Operations Research*, Vol. 37, pp. 319-328, 1989.

Fox, M. S., "Constraint Guided Scheduling: A Short History of Research at CMU", *Computers in Industry*, Vol.14, pp.79-88, 1990.

Garey, M. R., and D. S. Johnson, *Computers and Intractability: A Guide to the Theory of NP-Completeness*, Freeman, New York, 1979.

Grossmann, I.E., I. Quesada, R. Raman, and V. T. Voudouris, "Mixed Integer Optimization Techniques for the Design and Scheduling of Batch Processes", in *Batch Processing Systems Engineering* (G.V. Reklaitis, A.K. Sunol, D.W.T. Rippin, and O. Hortacsu (eds.), NATO ASI Series, Vol. 143, pp.451-494, 1996.

Harel, D., *Algorithmics: The Spirit of Computing*, Addison-Wesley, Reading, 1987.

Ku, H., and I. Karimi, "Evaluation of Simulated Annealing for Batch Process Scheduling", *Ind.Eng.Chem.Res.*, Vol. 30, pp.163-169, 1991

Kudva, G. K., and J. F. Pekny, "A Distributed Exact Algorithm for the Multiple Resource Constrained Sequencing Problem", *Annals of Operations Research*, Vol. 42, pp. 25-54, 1993.

Kudva, G., A. Elkamel, J. F. Pekny, and G. V. Reklaitis, "A Heuristic Algorithm for Scheduling Multi-Product Plants With Production Deadlines, Intermediate Storage Limitations, and Equipment Changeover Costs", *Computers and Chemical Engineering*, Vol. 18, No. 9, pp. 859-875, 1994.

Kuriyan, K., G.V. Reklaitis, and G. Joglekar, "Multiproduct Plant Scheduling Studies using BOSS", *Ind.Eng.Chem.Res.*, Vol. 26, pp.1551-1558, 1987.

Lin, S., and B. W. Kernighan, "An Effective Heuristic Algorithm for the Traveling Salesman Problem", *Operations Research*, Vol. 21, pp. 498-516, 1973.

Martin, R.K., *Large Scale Linear and Integer Optimization: A Unified Approach*, Klower Academic Publishers, Dordrecht, the Netherlands, 1999.

McDonald, C., "Synthesizing Enterprise Wide Optimization With Global Information Technologies: Harmony or Discord", *Proceedings of the Foundations of Computer Aided Process Operations Conference*, Snowbird, Utah, 1998.

Miller D. L., and J. F. Pekny, "Results from a Parallel Branch and Bound Algorithm for the Asymmetric Traveling Salesman Problem", *Operations Research Letters*, Vol. 8, No. 3, pp. 129-135, 1989.

Miller, D. L., and J.F. Pekny, "Exact Solution of Large Asymmetric Traveling Salesman Problems," *Science*, Vol. 251, pp. 754-761, 1991.

Mockus, J., W.D. Eddy, A. Mockus, L. Mockus, and G.V. Reklaitis, *Bayesian Heuristic Approach to Discrete and Global Optimization*, Kluwer Academic Publishers, Dordrecht, the Netherlands, 1997.

Mockus, L., and G.V. Reklaitis, "A Bayesian Approach to Batch Process Scheduling", *Intl. Trans. Operational Research*, Vol. 4, pp. 55-65 (1997).

Mockus, L., and G.V. Reklaitis, "Continuous Time Representation Approach to Batch and Continuous Process Scheduling - II. Computational Issues", *Ind.Eng.Chem.Res*, (in press).

Murakami, Y., H. Uchiyama, S. Hasebe, and I. Hashimoto, "Application of Repetitive SA Method to Scheduling Problems in Chemical Processes", *Computers and Chemical Engineering*, Vol. S21, No. 9, pp. S1087-S1092, 1997.

Padberg M., and G. Rinaldi, "A Branch and Cut Algorithm for the Resolution of Large Scale Symmetric Traveling Salesman Problems", *SIAM Review*, Vol. 33, pp. 60-100, 1991.

Pekny, J. F., D. L. Miller and D. Stodolsky, "A Note on Exploiting the Hamiltonian Problem Substructure of the Asymmetric Traveling Salesman Problem", *Operations Research Letters*, Vol. 10, No. 3, pp. 173-176, 1991.

Pekny, J. F., and D. L. Miller, "A Parallel Branch and Bound Algorithm for Solving Large Asymmetric Traveling Salesman Problems", *Mathematical Programming*, Vol. 55, pp. 17-33, 1992.

Pekny, J. F., and D.L. Miller, "A Staged Primal-Dual Algorithm for Finding a Minimum Cost Perfect Two-Matching in an Undirected Graph", *ORSA Journal on Computing*, Vol. 6, No. 1, pp. 68-81, 1994.

Pinto, J.M., and I.E. Grossmann, "A Logic-Based Approach to Scheduling Problems with Resource Constraints", *Computers and Chemical Engineering*, Vol. 21, 1997.

Realff, M.J., and G. Stephanopoulos, "On the Application of Explanation-Based Learning to Acquire Control Knowledge for Branch and Bound Algorithms", *INFORMS Journal on Computing*, Vol.10,No.1,pp.56-71, 1998.

Reklaitis, G. V., J. Pekny, and G. Joglekar, "Scheduling and Simulation of Batch Processes", *in Handbook of Batch Process Design*, (P.N. Sharratt,(ed.)), Blackie Academic & Professional, London, pp. 24-60, 1997.

Shah, N., "Single- and Multisite Planning and Scheduling: Current Status and Future Challenges", *Proceedings of the Foundations of Computer Aided Process Operations Conference*, Snowbird, Utah, 1998

Stephanopoulos, G., and C. Han, "Intelligent Systems in Process Engineering: a Review", *Computers and Chemical Engineering*, Vol. 20, No. 6/7, pp. 743-791, 1996.

Sunol, A. K., M. Kapanoglu, and P. Mogili, "Selected Topics in Artificial Intelligence for Planning and Scheduling Problems, Knowledge Acquisition, and Machine Learning", in *Batch Processing Systems Engineering* (G.V. Reklaitis, A.K. Sunol, D.W.T. Rippin, and O. Hortacsu (eds.), NATO ASI Series, Vol. 143, pp.595-630, 1996.

Tarjan, R. E., *Data Structures and Network Algorithms*, Society for Industrial and Applied Mathematics, Philadelphia, 1983.

Wysocki, Jr., B., "Pulling the Plug", *Wall Street* J., April 30, 1998.

Zentner, M., Advanced Process Combinatorics, Inc., West Lafayette, IN , private communication, November, 1997.

Zentner, M. G. Pekny, J.F., Miller, D.L. and Reklaitis, G.V., "RCSP++: A Scheduling System for the Chemical Process Industry", *Proceedings of the 5th International Symposium on Process Systems Engineering (PSE) 94*, Kyongju, Korea, 1994a.

Zentner, M. G., Pekny, J., Reklaitis, G., and Gupta, J., "Practical considerations in using model-based optimization for the scheduling and planning of batch/semicontinuous processes", *J. Proc. Control*, Vol. 4, No. 4, 1994b.

FROM WASTE TREATMENT TO POLLUTION PREVENTION AND BEYOND: OPPORTUNITIES FOR THE NEXT 20 YEARS

Henry T. Kohlbrand
The Dow Chemical Company
Midland, MI 48674

Abstract

Movement of business focus from a regional to a global basis is redefining how companies organize their activities. During the past 25 years, the chemical industry has been gradually moving from a focus on "end-of-pipe" treatment to source reduction, recycling, and reuse. Pollution Prevention, once viewed primarily as an environmental issue, is now becoming a critical business opportunity. This change is not coming easily because it requires institutional transformation in organizations where such change is difficult. The emphasis on Pollution Prevention is now broadening to include tools such as: life-cycle analysis, full-cost accounting, sustainable development, and eco-efficiency. As we look twenty years into the future, process design and optimization will incorporate more elements of the environment into the objective function that defines success. Modeling and simulation will play a key role in making these changes occur more rapidly. This paper reviews the drivers, change, and tools that can be used to accelerate this change.

Keywords

Pollution prevention, Waste elimination, Eco-efficiency, Industrial ecology, Full-cost accounting, Life-cycle analysis, Environment, Institutional change.

Introduction

During the last twenty-five years, the chemical industry has been in the process of moving from a focus on "end-of-pipe" waste treatment to source reduction, recycle and reuse. Recently, more effort has gone into developing full-cost accounting schemes and defining the life cycle and sustainability impact of processes and products. This evolution has its origins in the increasing regulatory climate of the 1970's and finds itself in the global competitiveness framework of the 1990's.

A recent review of several large chemical companies indicated that the cost associated with meeting environmental regulations represented an average of 3% of sales. In addition, a review of an oil refinery indicated that when emissions were measured (rather than estimated) significant opportunities were found. They also found that their environmental compliance costs were 3-5 times higher than their original estimates when carefully measured (AMOCO/USEPA, 1992). In a report to the U.S. Congress in 1986, the Environmental Protection Agency noted:

"Incentives of waste minimization are already strong, so EPA must capitalize on them. Most lacking is access by generators to the information that will demonstrate the economic benefits of waste minimization to industry, overcome logistical problems, and help develop creative new approaches." (U.S. EPA, 1986)

Unfortunately, standard methodologies and measurements have not developed as quickly in the past twelve years as has the opportunity to apply them. Waste elimination is a key lever to optimize the process. Although "pollution prevention" has been defined as a key activity by various publications and regulations, the real value generation comes from a focus on the reduction of process waste. *It is now becoming clear that we can not consider waste elimination activities solely as an environmental issue – waste generation and treatment is a critical business issue and needs to be viewed as a business opportunity.* Long term cost of ownership needs to be considered in addition to short term cash flow optimization.

The Evolving Business Environment

This change in focus on the business value of pollution prevention and industrial ecology has its roots in the continuously evolving global business climate. We live in a world with a constantly changing social and business environment. Over the past several years, we have witnessed the end of the Cold War, the breakup of the Communist Bloc, and the move to free-market economies around the world. This move from a regional to a world economic base is really effecting the chemical and pharmaceutical industry. The desire of many third world countries to improve their position in the world order has effected the way some companies look at cost of capital and ultimately effects the way we look at our business. Today the battlefield is based on economics and commercial activities, not guns and tanks.

As a result of these conditions, we are seeing trends that effect the way that we operate. We are seeing:

- Projects with shorter timetables
- Competitive cost and timing pressures
- An increase in the purchase or licensing of technology rather than internal development
- Large companies coordinating businesses on a global basis
- An increase in the number of joint ventures
- Globalization of communications using collaborative computing and information technology
- A move toward judging new products and processes using life cycle, eco-efficiency or sustainability metrics

In his book *"Post Capitalist Society"*, Peter Drucker (Drucker, 1993) has described the change that we are going through as one equivalent to the industrial revolution or the Renaissance. We are becoming more globally focused and dependent on an information-based economy. This is destroying old paradigms and fundamentally changing the way we can conduct business.

These trends provide us with a number of opportunities when examining how to best use modeling technology. We need to be able to move quickly in our development and optimization activities while maintaining a high degree of accuracy. The understanding of our systems needs to go beyond technical operation into product quality, performance and competitive assessment. Often, we need to be able to quickly evaluate the safe operation of processes designed and operated by other companies as we acquire technology through license or joint venture agreements. Indeed, at times we need to be able to fit technology developed at another company into our equipment in an efficient and timely manner. This environment requires us to take a look at our work processes and find ways to continuously adapt and improve them. Without this type of "re-engineering", this rapid change can strain the ability of our hazard evaluation and environmental compliance systems to keep pace with the business driven needs of our organizations. The effective use of technical and business models can help us to successfully cope with this changing business environment.

We have found that a partnership must be established between our manufacturing, engineering, research, technical service and commercial organizations. Often this partnership is extended to key customers or joint ventures as we evaluate our business needs and their effect on the design, operation and requirements of our manufacturing facilities. It is not enough to understand the technology alone – the impact of the process on product performance, economics and reliable operation must also be understood. Environment, health and safety issues can no longer be looked at as functional add-ons but must be integrated into the overall business strategy. Models developed to solve these problems must be capable of incorporating not only the technical inputs and outputs but the product and commercial aspects as well.

Making the Shift from Environmental to Business Focus – One Company's Journey

The Dow Chemical Company has a strong historical commitment to environmental awareness and minimization of environmental impact of our operations. Treatment technology was employed by Dow in the 1920's – long before it was required by government regulation. In the last few years, we have recognized that an increasingly proactive focus on waste elimination, in addition to regulatory compliance, was necessary. We have had a historical focus on process efficiency including yield improvement and energy conservation.

In 1986, Dow established the Waste Reduction Always Pays (WRAP) program, which generated an average of $20MM/year in cost savings. In 1995 a waste elimination effort was started, sponsored jointly by our manufacturing, environmental, and R&D functions. The following material is presented in summary of what we learned from our effort to incorporate pollution prevention planning into our business processes.

Where to focus

The synthesis and improvement of chemical processes can be very challenging. Properly done, it requires a balance of safety, reliability, economics and quality, while having an acceptable impact on the environment and society. Add to that the changing nature of customer needs (and the rapid changes in the global economy) and the development process becomes even more complicated. Woven into all of these activities are the government regulatory requirements that exist where the process will be developed and ultimately where the process will be carried out.

The basic stages used in process development are illustrated in Figure 1. The earlier that principles for waste elimination can be incorporated into the process development cycle, the easier and less expensive it is. If no attention is paid to this until the "detailed design" stage, practical options are much fewer than during "process synthesis" or "conceptual design".

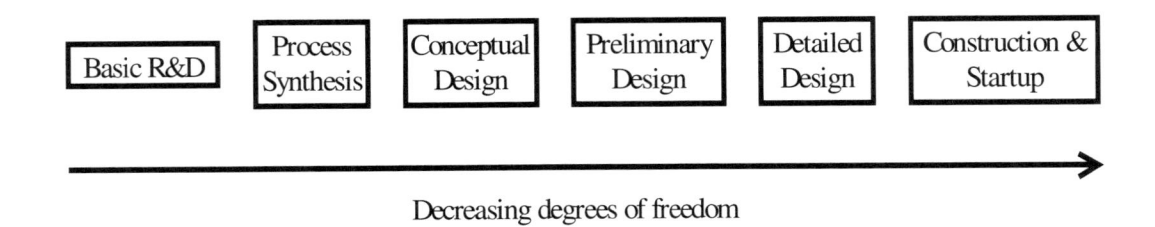

Basic R&D → Process Synthesis → Conceptual Design → Preliminary Design → Detailed Design → Construction & Startup

Decreasing degrees of freedom

Figure 1. Stages in developing a process.

Levers for Change

In order for an organization to make the shift from focusing on environmental compliance to asset opportunity, the following levers must be in place:

- Focus on business success
- Value system that identifies the real cost (present and future) of generating the waste/emissions
- Measurement and reporting system to communicate progress and share success
- Clearly defined roles and responsibilities within the organization
- An integrated look that goes beyond a single unit operation or process unit

Focus on Business Success

Business success can be measured in many different ways. Ultimately, the business needs to serve its key stakeholders which include employees, customers, stockholders and society. Each of these groups has its own measure of how the business is successful. A focus on success which looks at the impact of decisions on these stakeholders and balances both the short and long-term success factors for each will enable prosperity. In any case, the waste elimination opportunities must be framed in a way that the business can evaluate them side-by-side against other opportunities.

Full-cost Accounting

One of the key success factors involves having a value system that identifies the real cost associated with a process or product. Full cost accounting is a methodology which tries to organize different levels of cost. An understanding of full cost accounting must include direct, indirect, associated, and societal costs. Incorporating as many of these as possible into the business decision process is critical to the development of processes and products that will be sustainable in the future (UNEP, 1996; PCSD, 1996).

Most of those involved in manufacturing are comfortable with measuring the direct and indirect costs. Associated costs are more difficult because they include costs such as avoided fines where you are estimating the probability of having been fined vs. hard costs such as operating capital and expenses (White, 1993). Societal costs are even more difficult to include since there are no standard, agreed-upon methods to estimate them. The U.S. Environmental Protection Agency has introduced four tiers of full cost accounting. They include:

- Tier 0 - Usual costs equipment, labor and materials
- Tier 1 - Hidden costs compliance and permits
- Tier 2 - Liability costs penalties, fines and future liabilities
- Tier 3 - Less tangible costs consumer responses, employee relations, cost to society and the world

Measurement and Reporting System

Another success factor is the availability of data and information. Progress is difficult unless there is a measure of where you are starting from and where you want to go. In addition to detailed material balances and material flow data, it is also important to have a mechanism to share successes and opportunities within the organization. Often, an idea from another type of process or product can be creatively applied to solve other challenging problems. Good communication also helps to make sure that alternative uses for waste streams from one area can be brought to the attention of another. This leverages know-how to create value.

Clearly Defined Roles and Responsibilities

It is important that each organization define how they want to address the issues surrounding waste generation, treatment and elimination. Kraft (1992) has outlined a method to incorporate environmental reviews into the development and design of processes. He notes that during the early stages of new facility design, or process development, ample opportunity exists to implement design modifications that reduce the need for waste treatment via source reductions or reuse. His 10 steps are outlined below:

1. Perform initial assessments
2. Assign leadership responsibility
3. Define environmental objectives
4. Identify permit needs
5. Determine compliance requirements
6. Analyze waste minimization overall - note that waste minimization is more likely to be hindered by attitudes based on limited information and experience than on lack of effective technology. The means to reduce waste are imbedded in all aspects of production
7. Apply best environmental practices
8. Determine treatment/disposal options
9. Evaluate options
10. Summarize the results

Any management system that defines roles and responsibilities also needs to link these into the value creation process for the business as well as the balanced needs of the stakeholders.

Integrated View

It can be tempting to look at a single unit operation or even a single process when evaluating waste elimination opportunities (Spriggs, 1994b). For example, one reactor design produces a 95% pure product in a solvent which is easy to recover and recycle. Another reactor design produces a 99% pure product with and 1% undesired isomer which is very difficult to separate from the product. Even though the second reactor design appears to be more efficient, an integrated look at the separations section and reactor together might yield a different decision. Sometimes, one process' waste may become another process' raw material. Looking at integrated economics across a site, company or even between companies can yield surprising results.

Approaches – Principles and Heuristics

There are many different approaches that can be used to look at developing and existing processes. All of them look for a way to determine opportunity, then to value alternatives and finally to incorporate the results into a prioritization matrix. The approaches tend to be organized into two types – those focusing on the stage of the development of the process (basic R&D, process synthesis, design, etc.) and those focusing on process functions (chemistry, unit operations, maintenance, type of waste, etc.). One approach that incorporates both methodologies is outlined in Figure 2:

Each business has a project identification and prioritization process in place. This methodology is designed to fit into and not replace existing schemes for determining the most attractive areas for the business to work in. The level of detail called for in evaluating waste elimination options is usually dictated by the perceived economic return available.

General Rules

Hierarchy for Pollution Prevention

Consider using the accepted hierarchy for pollution prevention represented by:

Source reduction
Recycle/reuse
Treatment
Safe Disposal

The first objective is to look for opportunities to minimize waste at the source. If you don't make it in the first place, you don't need to worry about it. If source reduction is not feasible, the next option is to recycle the material or reuse it. This reuse may be by a unit in a different company. The next option is to treat the waste and the final option is to dispose of it.

Prioritization

Waste can be defined as any material coming from the process that the business does not desire to market, sell, or use. When evaluating opportunities, different factors for prioritization must be used. The weighting of such factors may be slightly different depending on geography and business. Among those that can be considered are:

- Toxicity
- Environmental impact (including environmental persistence)
- Cost

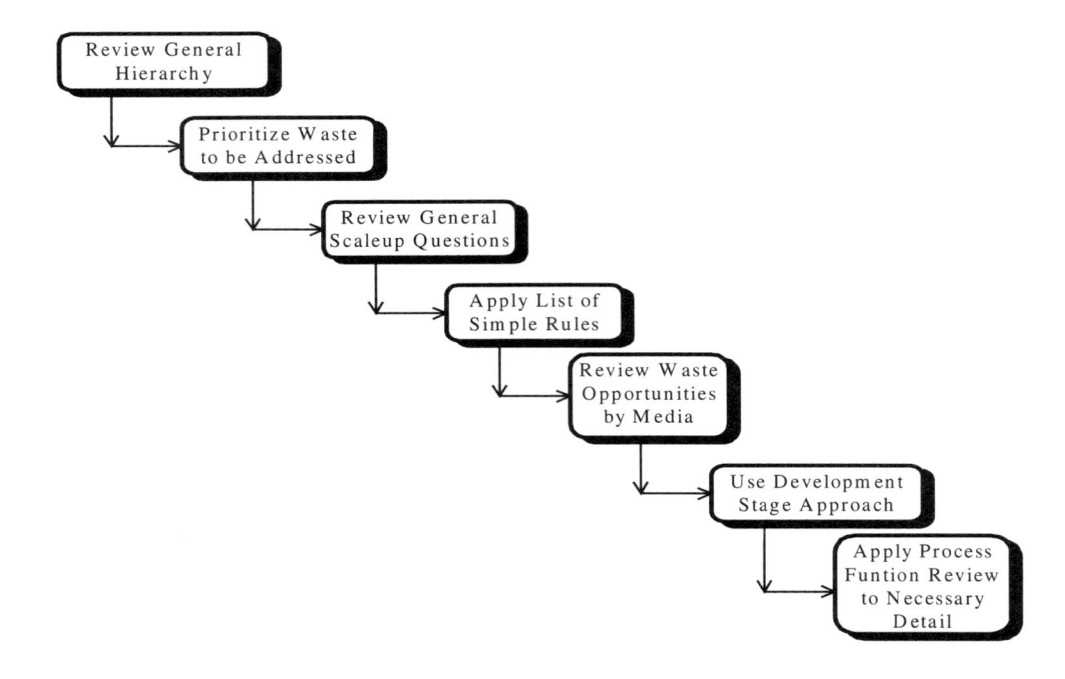

```
Review General
  Hierarchy
    ↓
  Prioritize Waste
  to be Addressed
    ↓
  Review General
  Scaleup Questions
    ↓
  Apply List of
  Simple Rules
    ↓
  Review Waste
  Opportunities
  by Media
    ↓
  Use Development
  Stage Approach
    ↓
  Apply Process
  Funtion Review
  to Necessary
  Detail
```

Figure 2. Overview of waste elimination process.

- Government regulations
- Corporate commitments
- Long-term risk
- Business risk
- Image/issues with the public
- Value to someone else

Waste Elimination and Process Scale-up

The purposes of scale-up studies include gaining a fundamental understanding of the process and the implication of changing scale on the process to be carried out, generating small amounts of material for customer evaluation, learning about the levers that impact the economics, and providing data for further developing the process. The actual approach used depends on several factors:

- Have we done this (or something similar) before?
- Has anyone else done this before?
- How much do we really understand the fundamentals?
- What is the scale (quantity)?
- What is the appropriate type of process operation? (multi-purpose batch to large continuous?)
- How much time do we have?
- How much can we afford to invest?
- How much business risk is involved?

These factors will influence the scale and types of activities involved. In addition to the classical experimental approaches (bench scale, mini-plant, pilot plant, market development plant), the use of modeling and simulation tools is becoming increasing popular (and powerful). In addition to the benefits of time and cost, such techniques also force the developer to consider the key fundamental factors and thereby organize a more effective experimental program.

Simple Rules

The first step in any screening process is to apply simple rules and evaluate the results. A sample set of considerations for the first stage in waste elimination follows:

- Wherever possible eliminate waste materials at their source. Can sometimes be done by switching raw materials.
- Recycle waste materials within the process. Where not possible, explore the possibility of using within other processes.
- If wastes formed reversibly, recycle to extinction.
- Look at the product - can another product which generates less waste fill the same customer need.
- Review chemistry alternatives including catalysis.
- For all heating duties use the utility with the lowest practical temperature.
- Minimize the total number of main equipment items and minimize piping connections.
- Look at changing set points or tightening control variations of key variables. Modifying single equipment items (like changing column

internals) can also yield significant improvements with little capital expenditure.
- Generally more difficult in batch than continuous processes.
- The inherent flexibility of batch plants makes raw material product substitutions simpler.
- Don't buy waste - in raw materials, packaging, etc.

Waste-focused Review for Evaluating Opportunities by Media

Once an initial review of the waste elimination opportunities has been done, the next step is to look at the media in which waste appears, to construct the next set of opportunities.

General

- It is important at this stage to characterize the types of waste being generated.
- Do we have an accurate material balance around the process?
- It is important to try and get a clear understanding of the environmental and health issues associated with each waste (present and future).
- Do we have a clear understanding of the variability of the composition and rate of waste generation? - Will the emissions change as the plant utilization or capacity changes?
- Are there community or regulatory issues associated with any wastes or emissions?
- Are fugitive emissions considered when selecting equipment?
- Would an environmental review be helpful in identifying issues and opportunities?

Water

- Do we understand the total oxygen demand, biological oxygen demand, metals, toxicity, etc. of each aqueous stream being generated?
- Is it possible to recycle process wastewater?

Air

- Do we understand the composition of the emissions and the toxicity of the streams?
- What are the best controls available to control emissions?
- Consider measuring from time-to-time to verify estimating techniques typically used for air emissions.

Incineration

- How well does the material burn?
- Are there any hazards associated with mixing these and other wastes prior to incineration?
- Does the emission from incinerating this waste contain any persistent bioaccumulative toxic materials?
- Does the material cause operability problems with the incinerator?
- Are there any concerns about transporting the material to an incinerator?
- What are the characteristics of any residual solids?

Organized Review Processes

Rossiter and Klee (Chapter 12 – Rossiter, 1995) have outlined a Hierarchical Process Review for Waste Minimization. This method provides a systematic approach to the problem of identifying appropriate process modifications to minimize waste generation. A general outline of some of their considerations is given below:

Detailed Process Function Approach

Level 1 Processing Mode - Continuous vs Batch

- Waste streams from batch are generally intermittent and often include effluents from cleaning between batches'
- Compositions and flow rates from batch processing generally vary vs those from continuous processes
- More startup and shutdown generally cause more wastes
- Production rates - under 500T./yr batch; between 500 and 5000 batch is common - at higher rates continuous preferred
- Product life - batch better suited for products with short lives where rapid response to market is preferred
- Multiproduct capabilities - If unit required to produce several similar products using the same equipment batch is usually preferred
- Process reasons - Batch preferred when process related factors require like cleaning requirements that necessitate frequent shutdowns, difficulty in scaling up lab data or complicated process recipes
- If potentially serious environmental problems are anticipated, continuous is preferred
- If batch preferred, consider buffering to smooth operation, etc.

Level 2: Input/output Structure

- Look at overall mass balance of in and out - look for items to eliminate, recycle, etc
- Can the feed quality be improved at low cost?
- Should impurities be removed before processing or be allowed to pass through the process
- Can any input streams be eliminated
- Can any waste output stream be used beneficially with the plant or sold to an external customer

Level 3: Recycle and Product Formation Considerations

- Break down to major component sections
- Do waste or output streams contain feed or product material that could be recovered or recycled
- Can reaction conditions be altered to minimize formation of waste or by-products
- Can waste or byproducts be recycled to extinction

Level 4: Separation Systems

- Detailed analysis of separation systems
- Look at alternative separation techniques or rearranging existing ones
- Are any waste streams the result of poor or inappropriate separations
- Can any wastes be removed from process effluents by adding new separations
- Are there alternative techniques/technologies that could replace or supplement existing separations and reduce releases

Level 5: Product Drying

- Alternatives that cause less degradation
- Can waste materials be removed from dryer off-gasses
- Can dryer wastes be recycled

Level 6: Energy Systems

- Pinch analysis
- How far can energy consumption be reduced economically
- Can temperature levels for heat delivery be reduced
- Are different fuels available

Level 7: Equipment and Pipework

- Evaluate and eliminate fugitive emissions
- Designing out problematic equipment
- Can total number of items and connections be reduced
- Are there alternative equipment items that result in fewer emissions
- Use welded vs flanged pipe

Waste Minimization in Design and Development

R.A. Jacobs (Jacobs, 1991) outlines his approach to waste minimization in process development and design. He defines the following categories of sources of waste:

- unrecovered raw materials
- unrecovered products
- useful by-products
- useless by-products
- impurities in raw materials
- spent process material
- packaging and container wastes

The following then represent questions and issues associated with each phase of process design:

Developmental Stage Approach

Product conception

- What are raw materials; are there any new chemicals for company to handle?
- Are there any toxic or hazardous chemicals generated?
- What are performance specs of product?; can they be widened to include impurities?
- How reliable is the process?; are all steps proven?; does company have experience in this area of technology?
- What types of wastes are generated?

Laboratory Research

- Typically CHE's focus on mass balances toward final product - need to expand focus to encompass how much does not end up in the final product? Where does it go? Meticulous attention to material balances is the key.
- Process simplification.

Process Development

- Now we verify what we have done so far. Pilot plants can be used to gauge accuracy of assumptions.
- Look at flexibility in selections of raw materials to minimize waste volume and toxicity.
- Methods of improving process reliability
- Ability to track and control all waste streams
- Potential impacts of the process on public
- Any impurity that cannot be included in the product ends up as waste.
- What kind of process measures are necessary to really run the scaled-up plant?

Mechanical Design

- Appropriately incorporate all of the waste-elimination measures into the final plant design.
- Reduction/control of fugitive emissions.
- Taking measurements to control and quantify wastes.
- Concentrate waste streams to maximum extent possible.
- Reduction of interim storage and container management.
- Utilization of waste treatments that allow recovery and reuse of rate materials and intermediates.

The methodology outlined above looks at opportunities based on developmental stage. Another option is to look at which part of the process might contribute to a waste elimination opportunity. In his article on pollution prevention strategies, Chadha (1994), outlines a method which divides opportunity into four categories: process chemistry, engineering design, operating practices, or maintenance practices. Although the purpose of this discussion is to center around the process development stage, it is well to note that decisions made during the development stage can have serious effects on process operability and maintenance practices.

Hybrid Development Stage and Process Function Approach

Engineering Design-based Strategies

- Storage and Handling Systems
- Process Equipment
- Process Controls and Instrumentation
- Recycle and recovery equipment

Process Chemistry and Technology Based Strategies

- Raw Materials
- Plant Unit Operations

Operations-Based Strategies

- Inventory Management
- Housekeeping Practices
- Operating Practices
- Cleaning Procedures

Maintenance-based Strategies

- Existing Preventative Maintenance
- Proactive Strategies (Monitoring and Inspecting vs. Reacting)

Ken Nelson formerly of Dow (1990) has published extensively in this area. He has introduced a another list of areas to look at for opportunity in waste reduction or elimination. They include:

Raw Materials

- Improve quality of feeds
- Use off-spec material
- Use inhibitors to prevent unwanted side reactions
- Change shipping containers (that can be reused)
- Reexamine need for each raw material - small additives that "help" but you are not sure of effectiveness

Reactors

- Improve physical mixing
- Distribute feeds better
- Improve catalyst
- Improve ways in which reactants are added
- Examine heating or cooling techniques
- Provide separate reactor for recycle streams
- Consider a different reactor design
- Improve control

Heat Exchangers

- Desuperheat steam - improves heat transfer coefficient
- Use lower pressure steam
- Install a thermocompressor
- Use on-line cleaning techniques
- Use scraped wall exchangers
- Use staged heating
- Monitor exchanger fouling
- Use non corroding tubes

Pumps

- Recover seal flushes and purges
- Use sealless pumps

Furnaces

- Replace the coil
- Replace furnace with intermediate exchanger
- Use existing steam superheat

Distillation

- Increase reflux ratio
- Add section to column
- Retray or repack
- Change feed tray
- Watch where you control - pick the most sensitive area in the column
- Insulate
- Improve distribution feed and reflux
- Preheat column feed
- Remove overheads from tray near top of column
- Increase size of vapor line
- Modify reboiler design

- Reduce reboiler temperature
- Lower column temperature
- Lower column pressure
- Improve condenser
- Forward vapor to next column (don't condense and revaporize)

Piping

- Watch for combining waste streams and having a mess to treat or recycle - keep separate as long as possible
- Eliminate leaks
- Consider metallurgy
- Use lined pipes or vessels
- Monitor vent and flare systems

Process Control

- Improve on-line control
- Optimize daily operation
- Automate start-ups, shutdowns and product changeovers
- Program plant to handle unexpected upsets and trips

Miscellaneous

- Avoid unexpected trips and shutdowns
- Use waste streams from other plants - use brine stream to do an extraction; reduce organics in extract
- Reduce the number and quantity of samples
- Recover product from tank cars and tank trucks
- Use removable insulation
- Maintain external painted surfaces
- Find a market for waste products

General Considerations

Batch vs. Continuous – Small vs. Large

Slightly different approaches must be used when evaluating small batch processes vs. large continuous processes. Although the basic principles remain the same, the actual process (and options) for reducing or eliminating waste are usually different. Often, batch processing units are built to be flexible and it tends to be easier to adapt them to new products and processes. The large continuous systems tend to be built for specific products and processes and are more difficult (and expensive) to adapt for other uses. The investment in a large continuous process unit is usually significant and the anticipated lifetime long, so it is very important that sustainability issues regarding the process and the product(s) be addressed for the expected lifetime of the investment.

Physical Limitations

Unfortunately, it is not always possible to minimize waste at the source or reuse it in an existing facility to the extent that would be desirable. The physical structure of an existing facility is already in place, and this inherently limits the flexibility and options available to reduce or reuse wastes.

Cultural Barriers

Since waste has been viewed historically as an environmental issue, it takes some effort to shift the focus to business opportunity. It is important to include long-term business risk and long-term environmental costs in the discussions of business opportunity. Incremental cost savings very seldom justify all of the real high impact projects.

Regulatory Barriers

In many geographic areas, environmental regulations are very restrictive on how to handle waste. In the case of a refinery, two projects to reduce volatile organic emissions were evaluated. Each reduced the same amount of a similar material. Project 1 was mandated by law and the emission source was in the center of the refinery. Project 2 was not required by law and was located at the edge of the facility much closer to a populated area. When an integrated solution is sought, it is important to include regulatory drivers with the economic, health, and environmental drivers.

Comfort Zone

End-of-pipe treatment is often low risk and uses very well known technology. Waste elimination opportunities can often involve higher risk because of unproved or less-used technology.

Logistics – Don't Buy Waste

There are times when a look at the raw material supplier can really make a difference. The 3M company has had a significant effort in working with their suppliers on packaging as a way of greatly reducing their solid waste. Another opportunity involves carefully looking at the purity or nature of impurities in raw materials. In one plant changing raw material suppliers (same absolute purity, different composition of impurities) resulted in a 75% reduction in waste generated in the reactor and distillation system. The purchasing function had been using the original supplier because of a very small price difference. When the total economics were evaluated, the decision to use the second supplier was made.

Helpful Tools with a Future Focus

Several technology developments are providing additional support to waste elimination efforts. They include:

- Modeling and Simulation Tools
- Mass Exchange Networks
- Life Cycle Analysis
- Brainstorming – Thinking "Out of the Box"
- Advances in Reactor Technology

Modeling and Simulation Tools

This area continues to advance at a rapid rate. Tools incorporating artificial intelligence are beginning to look promising but are dependent on a good set of starting heuristics like those listed in the screening tools outlined above. In addition to knowledge based approaches, numerical optimization, and process simulators incorporating environmental as well as cost optimizers are becoming more available. Combinatorical optimization (used to solve large scale optimization problems characterized by the "traveling salesman scheduling problem") will become much more important for comparing process alternatives in the future.

In any of these systems, the result is only as good as the input assumptions. Careful consideration of full costs and other factors must be included when depending on these tools to quantify alternatives. In addition to engineering tools, reaction synthesis and modeling tools are becoming more robust. These are typically used in the synthesis of environmentally benign reaction paths and in review "green chemistry" alternatives. The ultimate use of these tools depend on the careful selection of the "optimum" and definition of possible pathways to reach it.

Pinch Technology

Over the past few years, pinch technology, a systematic technique used for analyzing heat flows and heat exchange networks, has been extended to include mass exchange networks (Rossiter, 1995 pp 53-110). This form of pinch technology is most useful in evaluation of options in large, integrated continuous plants but can also be very useful in integrating material use across several batch processes. Much as one would look at temperature/ enthalpy relationships in heat exchanger design, the mass exchange network approach looks at mass/concentration relationships to optimize the exchange of mass in an integrated operation. One of the real barriers to more effective use of this tool is the presence (or lack) of small levels of impurities in streams where cross exchange is desirable and where the impurity can negatively impact on product quality or performance.

Life Cycle Analysis (LCA)

Life cycle analysis (Bailey, 1990) looks holistically at the environmental consequences associated with the cradle-to-grave life cycle of a process or product. How can waste be reduced or eliminated starting with the point of generation to its processing, treatment or ultimate disposal as a residual hazardous waste? Computer-based tools (such as the one designed by the U.S. Department of Energy) can be used very effectively to compare alternative choices in process route or product selection.

Brainstorming – Thinking "Out of the Box"

As we begin the initial stages of process scale-up the following question should be posed:

What would it take to have no waste?

Waste can be defined as any material coming from the process that the business does not desire to market, sell, or use. While discussing this question we should consider alternative:

- Unit operations
- Chemistry
- Equipment (reliability)
- Product mix

In most cases some waste streams need to be generated. When defining these streams we should understand:

- What are our alternatives?
- What are the long-term cost, health, and environmental issues associated with this stream?
- Can we calculate the investment (raw material and processing cost) for each lb of waste?
- Is there a "net present penalty" for the wastes being generated?

By asking a project team to consider zero-waste alternatives, often significant creativity is stimulated.

Reactors for the 21st Century

As we look for waste reduction and productivity there are many new developments of reactor technology (and extensions of existing technology) which will aid in the development of minimum waste processes in the future. Two recent articles discuss them in more detail (LeRou, 1996; Ondrey, 1996). They include:

- Reactive distillation
- Membrane reactors
- Ultrasound
- Microwave

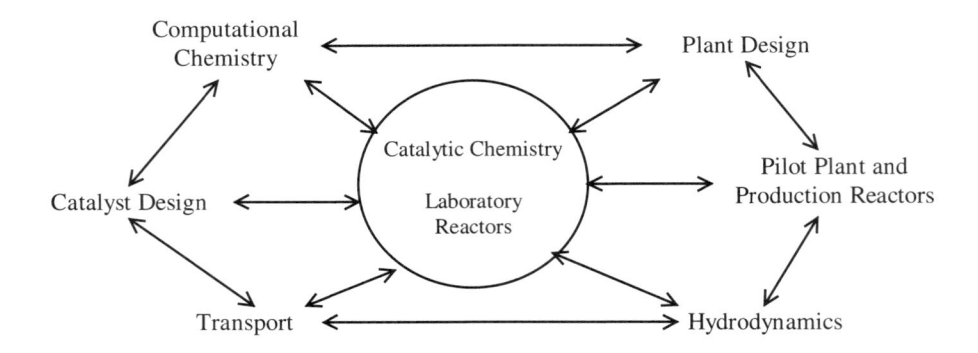

Figure 3. Reaction engineering now calls for a multidisciplinary approach.

- Falling film reactor
- Adsorptive reactor
- Continuous reactive chromatography

LeRou notes that in addition to kinetics and reactor modeling, the overall process economics should be built into any reactor design effort. Catalysis and fluid mechanics are now an integral part of reaction engineering. Priorities (listed in order of importance) include:

1. Process and product safety
2. Environmentally compatible products
3. Minimization of waste
4. Minimization of investment
5. Minimization of energy consumption
6. Operability and control

The reactor tends to be the most severe and complex operation in the entire plant and although it is not the only unit operation to optimize, often optimizing the reactor is a very significant part of optimizing the process. LeRou outlines time and space frames and suggests a multi-scale approach to optimizing reactor design and operation. Reaction engineering now calls for a multidisciplinary approach:

In the conventional paradigm for reactor development we tended to go from microreactor → bench-scale reactor → pilot reactor → commercial reactor. This process, more and more, is being compressed from a sequential mode to a highly parallel mode. It is not unusual to find laboratory reactor data that can not really be applied to the operating conditions of the plant when all of the characteristics of time and scale outlined in figure 3 (LeRou, 1996) are taken into account.

The ability to characterize reactor environment and incorporate issues such as micromixing, local energy density and fundamental kinetics is greatly changing the control we can exercise over the reactor output.

Selection of the Objective Function

One of the most difficult tasks in optimization is the choice of the appropriate objective function and associated constraints. Most chemical plant design uses economic factors in selecting the objective for design and operation. The true optimum is often best designated as a space rather than a single point. Various key parameters (such as raw material cost, energy cost, and product selling price) can vary and substantially change the final result. There can also be other variables (such as contingent costs associated with waste disposal) where future values are unknown. An alternative to the selection of a single objective function is the selection of two or three.

We have found that using three approaches involving economic factors, environmental impact (waste and emission production), and energy use yield a more complete picture of the optimum. By evaluating the process alternatives using these three criteria, a different set of heuristics tend to be used that result in a set of design and operating alternatives that provide a family of solutions to the problem. As long as the final result fits into the defined optimum for economic operation, designs which consume less energy or generate less waste can be defined (which may not have been considered when the economic objective was the only one used). This approach can also be used to stimulate the creativity of the professionals performing the synthesis and analysis.

The Future

In the next twenty years, the focus on pollution prevention will evolve into a balance between societal needs and ecological impact (Anastas, 1997; Oldenburg, 1997). Techniques such as life cycle analysis, sustainability metrics, and total cost accounting, will create a new view of process synthesis and product development. Multiple objective functions will be used to define success. A combination of economic (cost, yield, long-term cost of ownership) and environmental (life cycle, sustainability, contingent cost analysis) objective functions will be used to develop and improve products and processes.

Modeling (including process, business, and environmental) will play a key role in defining the "optimum". Much development is needed to be able to combine the results of models from these three areas effectively.

Examples

Oxidative Coupling

In this case, an old process based on the use of a chlorinated aromatic in a high pressure reactor system to make an organic intermediate was becoming less economically viable. The process generated about 1 kg. sodium chloride per kg. of product. The resulting aqueous waste stream was treated and discharged to an inland river system. A new process was evaluated which eliminated the use of chlorine, eliminated the aqueous salt stream and made a purer product at higher yield. The capital investment barrier was reduced when a full cost accounting approach was used to characterize the real savings of going to the new process. It also reduced volatile organic emissions of about 2 kg. VOC/kg. Product. The new process is also closer to raw material sources and customers reducing cost and environmental impact of transportation.

Choice of Neutralizing Agent

In a fine chemicals synthesis, the research chemist was using a process that generated sodium acetate as a result of neutralizing an intermediate. After discussing "waste reduction" opportunities with the process engineer assigned to the area a slight process change was made which generated a smaller amount of sodium carbonate. This had no material impact on the quality of the intermediate but did reduce organic loading to the waste water treatment plant. Once the researcher understood the challenge it was not difficult to adjust her thinking.

In 1995 the U.S. Chemical Manufacturers Association published a booklet containing a number of successful case studies in pollution prevention. *Table 1* represents a summary of some of the projects and results from that study.

Conclusion

The focus on process waste has gone from being an environmental compliance issue to one of key business importance. In most cases, technology is available to reduce, or in some cases, eliminate process waste, but full cost accounting and a change in business cultural mindset needs to take place. The opportunity is great to increase business profitability, reduce environmental impact of products and processes, and create a sustainable future.

Although many opportunities to reduce or eliminate waste will not be economically attractive using a short-term time window, proper identification of real, long-term costs will result in the best project list. Creative use of optimization technology using multiple objective functions will have a positive impact on process synthesis and optimization. These models will incorporate economic, business, and environmental models to define the "optimum" space for products and processes.

Keeping in mind the changing global pressures in all of our businesses, we need to have Wellington's image of an engineer in our minds as we move forward.

"Engineering is the art of doing that well with one dollar, which any bungler can do with two after a fashion", Arthur Mellen Wellington, 1847-1895

References

Anastas, P.T., Breen, J.J., 1997. Design for the Environment and Green Chemistry. *J. Cleaner Prod*, V.5, N 1-2, pp. 97-102

Bailey, P.E., 1991. Life Cycle Costing and Pollution Prevention. *Pollution Prevention Review*, Winter 1990-91, Vol 1, No 1, pp. 27-39

Berger, S.A., 1994. The Pollution Prevention Hierarchy as an R&D Management Tool. *Pollution Prevention via Process and Product Modifications*, AIChE Symposium Series #303 Volume 90 1994

Blau G.E., 1995. Introduction to Green Trends in Design. *4th Inter. Conf. On Found. Of Comp. Aided Proc. Design* AIChE Symp. Ser. #304

Chadha, N., 1994. Develop Multimedia Pollution Prevention Strategies. *Chemical Engineering Progress*, Nov 1994, Vol. 90 Issue 11, pp. 32-39

CMA, 1995. Profiles in Prevention, Case Histories of Pollution Prevention in the Chemical Industry. *Chemical Manufacturers Association*, Arlington, VA 22209

Crabtree, E.W., El-Halwagi, 1994. Synthesis of Environmentally Acceptable Reactions. *Pollution Prevention via Process and Product Modifications*, AIChE Symposium Series #303 Volume 90 pp. 117-127

Diwekar, U.M., 1994. A Process Analysis Approach to Pollution Prevention., *Pollution Prevention vis Process and Product Modifications*, AIChE Symposium Series #303 Volume 90, pp. 168-179

Doerr, W.W., 1993. Plan for the Future with Pollution Prevention. *Chemical Engineering Progress*, Jan 1993 Vol 89 Issue 1 pp. 24-29

Drucker, P., 1993. *Post Capitalist Society*. Harper Business, ISBN 0-88730-620-9

El-Halwagi, M.M, Petrides, D.P. (ed.), 1994., *Pollution Prevention via Process and Product Modifications*, AIChE Symposium Series #303 Volume 90

Forman, A.L., et. al., 1994. Environmentally Proactive Pharmaceutical Process Development. *Pollution Prevention via Process and Product Modifications*, AIChE Symposium Series #303 Volume 90 pp. 36-45

Freeman, H, et. al. 1992. Critical review of the literature on pollution prevention. *Journal Air Waste Manage. Assoc.* Vol. 42 No 5 pp. 618-656

Frosch, R.A., 1995. Industrial Ecology. *Environment.* V.37, N.6, Jul Aug 1995, pp. 16-37

Jacobs, R.A., 1991. Waste Minimization-Part 2: Design Your Process for Waste Minimization. *Chem Eng Prog.* Vol 87 No 6 pp. 55-59

Joback, K.G., 1994. Solvent Substitution for Pollution Prevention., *Pollution Prevention via Process and Product Modifications.* AIChE Symposium Series #303 Volume 90 pp. 98-104

Kraft, R.L. 1992. Incorporate Environmental Reviews into Facility Design. *Chemical Engineering Progress.* Aug 1992, Vol 88, No 8, pp. 46-52

Lerou, J.J., Ka, M. Ng, 1996. Chemical Reaction Engineering: A Multiscale Approach to a Multiobjective Task. *Chemical Engineering Science.* Vol 51 No 10 pp 1595-1614

Linninger, A.A., et.al., 1994. Synthesis and Assessment of Batch Processes for Pollution Prevention. *Pollution Prevention via Process and Product Modifications.* AIChE Symposium Series #303 Volume 90 pp. 46-58

Nelson, Ken, 1990. Use These Ideas to Cut Waste. *Hydrocarbon Processing,* Vol. 69, No. 3

Newton, J., 1990. Setting Up a Waste Minimization Program. *Poll Eng.* Apr 1990, pp. 75-80

Oldenburg, K.U., Geiser, K. 1997. Pollution Prevention and ... or Industrial Ecology. *J. Cleaner Prod.* V.5, N.1-2, pp. 103-108

Ondrey, G., 1996. Reactors for the 21st Century. *Chemical Engineering.* June 1996 pp. 40-45

PCSD, 1996. Sustainable America - Report by The President's Council on Sustainable Development. Doc # 061000008578, *Government Printing Office,* Washington, D.C. (or www.whitehouse.gov/pcsd)

Pojasek, R.B., 1991. Contrasting Approaches to Pollution Prevention Auditing. *Pollution Prevention Review* Vol.1 N. 3 1991

Rittmeyer R.W., 1991. Waste Minimization-Part 1: Prepare An Effective Pollution Prevention Program. *Chem Eng Prog* May 1991 pp. 55-62

Rossiter, Alan, 199.5 *Waste Minimization Through Process Design.* McGraw Hill 1995

Smith, R., 1995. Industrial Transition to Sustainable Development, *SRI Consulting.* Report D96-1994,

Spriggs, H.D., 1994a. Design for Pollution Prevention. *Pollution Prevention via Process and Product Modifications,* AIChE Symposium Series #303 Volume 90 pp. 1-11

Spriggs, H.D., 1994b, Integration, The Key to Pollution Prevention. *Waste Management,* V 14 Nos 3-4, pp. 215-219

UNEP, 1996. Eco-efficiency and Cleaner Production - Charting the Course to Sustainability. *United Nations Environment Programme* (UNEP)

U.S. EPA, 1986. Report to Congress-Minimization of Hazardous Wastes, Washington D.C., EPA/530/SW/86/033, October, 1986

U.S. EPA, 1992. *AMOCO/USEPA Pollution Prevention Project.*

White, A.L., Becker, M., Savage, D., 1993. Environmentally Smart Accounting: Using Total Cost Assessment to Advance Pollution Prevention. *Pollution Prevention Review,* Summer 1993, pp. 247-259

Please note: Although not all references are cited, all of these references were used in the development of the concepts presented in the paper

Table 1. CMA (1995) Examples.

Company	Description	Solution
Occidental Chemical	83 tons of hazardous waste generated during ethylene production	Simple filtration system reduced waste by 65% and saved $125,000/yr due to reduced emissions and labor costs
Ciba-Geigy	3700 tons of calcium carbonate generated during herbicide production would be disposed of in landfills	New filtration process strengthened and purified the calcium carbonate allowing it to be used for farming. $595,000 in savings for disposal ($115,000/yr investment). Savings to farmers was $400,000/yr.
Chevron Chemical	Unreacted phosphorus pentesulfide had to be filtered and removed from the process	Adding a settling step reduced waste and cleaning time but ultimately it was determined that this material could remain in the final product
Rhone-Poulenc	Acid regeneration plant produced acid wastewater	Fundamental understanding of furnace operation was developed. Operating procedures modified to minimize excess oxygen and raise combustion temperature. Waste acid reduced 75%. Material costs reduced $500,000
Dow Corning	Neutralization of waste stream produced 3,000,000 lbs of TDS waste and 6700 cubic yards of landfilled waste/yr.	Separation and recovery captured 90% of the waste stream, which was useable product. Less frequent batch pumpouts of distillation column, separator installation and recovery of silane. Cost $3.8MM - savings of $1.7MM/yr.
DuPont Chambers Works	In specialty polymer production, sticky polymer must be periodically removed from the reactors using flammable organic solvent. 5,000 lbs/wash 24-32 hours.	Water jet cleaning system installed. Cleaning time is now 4 hours, waste generation reduced 98% and manufacturing capacity increased 8%.
Ciba Geigy	Di-amino stilbene di-sulfonic acid (DAS) made using 3 reactions (sulfonation, oxidation, reduction). Solid waste produced in all three.	Continuous sulfonation replaced dilute oleum sulfonation (+30% cap.), air oxidation replaced bleach oxidation (+17% yield), hydrogen reduction replaced iron reduction (+20% cap.).
Monsanto	3-step process to 4-aminodiphenylamine starting with chlorobenzene generated salt, xylene, organic residue	New chemistry eliminated solvents, only by-product water. Organic waste reduced 74%, inorganic waste 88.9%. New process uses 50% less raw material/lb. product
Phillips Petroleum	Atactic polypropylene made during manufacture of isotactic polypropylene. 5-8MM lbs/yr of flammable waste.	Licensed new high activity catalyst which eliminated formation of atactic polymer. 225M lbs emissions also eliminated.

COMPUTATION AND COMPETITIVENESS: MANAGING TECHNOLOGY IN THE INFORMATION AGE

Michael P. Ramage
Mobil Oil Corporation
Fairfax, VA 22037

Introduction

The petroleum industry is riding a wave of innovation generated by the co-evolution of two major forces of change: The emergence of knowledge as a competitive business advantage and the rapidly advancing power of information and computational technology.

Over the past decade, the increased ability to acquire and process information has impacted every aspect of the way we find, produce, and process petroleum. It has made it possible to adopt global strategies by coupling market changes with oil field development, to synthesize and study the interaction of materials at molecular and atomic levels, and to model physical phenomena with detail that was not possible just a few years ago.

In exploration, the search for oil and gas in frontier areas has made the cost of drilling a dry hole high and the cost of missing a find unacceptable. To improve the odds, neural nets on parallel computers turn seismic and well log data into precise underground maps that help point out where the oil is and where it isn't.

On the manufacturing side, advances in computer hardware and algorithms have enabled rigorous on-line optimization of large subsets of petrochemical plants and refineries, resulting in significant value enhancement and improved response to market changes.

In the research lab, molecular-level modeling is helping evaluate new material and templates to synthesize catalysts. This cost-effective prescreening permits evaluation of more alternatives and shortens the R&D cycle by obviating time-consuming experiments in the lab.

Data from exploration, producing, and refining sites are immediately available to scientists and engineers in their offices, permitting immediate responses to problems around the globe. Powerful mathematical algorithms using the latest computer hardware are being developed to integrate and solve models spanning the entire petroleum value chain.

The quest to gain competitive advantage from knowledge has changed the industry. It has spurred development that challenges the way we manage technology. We need to be careful not to go too far and to evaluate the added value and business impact of these efforts. The challenge is to focus the technology where it has the most benefit to the business.

Changes in information technology continue to affect the skills required for every technical professional. Engineers, scientists, plant operators and technicians all have a rapidly escalating need to have technical computing as a core skill. Academic institutions need to further reinforce this core skill in the professionals of tomorrow.

The revolution has only begun. There are more many more challenges ahead for both industry and academics. We need to develop new environmentally noninvasive techniques for acquiring the data required to model reservoirs with the increasingly fine detail needed for the optimal development of oil fields. We will need to predict the molecular composition of refinery streams as a function of how the refinery is operated. In order to optimize the integrated operation of our business, we will have to track the progress of molecules from below the ground to the gas pumps while accounting for equipment design and reliability.

Editor's Note: The text, slides, and a video clip associated with this presentation may be found at the World Wide Web location:

http://che.www.ecn.purdue.edu/FOCAPO98

CULTURE SHIFT: POSITIONING TECHNICAL COMPUTING TO ENABLE SUSTAINED PROFITABILITY IN THE SPECIALTIES BUSINESS

Gary E. Blau
Purdue University
W. Lafayette, IN 47907-1283

Kay E. Kuenker
Dow AgroSciences
Indianapolis, IN 46268

Introduction

In recent years, the chemical process industries, agrochemical and pharmaceutical industries have demonstrated their abilities to increase productivity and earnings through mergers, one-time restructuring, downsizing and outsourcing of various work processes. In many cases, this has simultaneously reduced their diversification and focused efforts on primary or core business units. As we approach the new century, companies are faced with the challenge of continuing to demonstrate increased productivity, but with a leaner organization. Key decisions have to be made in terms of capital expenditures, research directions, organizational structure and product supply chain. These decisions will have a long-term impact on these companies well beyond the year 2000.

Is there technology that can help companies not only make these important decisions, but also ensure effective operation of leaner organizations? The answer is a resounding yes! As we have seen from this conference, never before has the rate of computer based technical development been so great. Rapid advances in algorithms, hardware, software and languages have allowed the development of quantitative decision making tools which make it possible to formulate and solve seemingly intractable problems in acceptable time frames. Globalization, ubiquitous networks, and telecommunications have made it possible for the lean organization to operate in a virtually "boundariless" world. Quantitative methods incorporating mathematical and statistical analysis concepts have matured to the point where they can help improve the productivity of virtually every work process within the company, Blau (1997). Unlimited amounts of data are available at the push of a button transforming companies from their traditional venues into "information" companies. However, all of this splendid technology comes with an entry fee. To effectively incorporate this advanced technology into corporate decision making, companies must experience a new paradigm in the way they operate. In fact, to be truly effective, the culture of companies must undergo a dramatic shift such that information technologies and quantitative thinking become institutionalized throughout the entire product life cycle.

This paper addresses some of the issues faced by companies trying to get these new technologies adopted as a way of life or a part of culture within the business environment. Although there are both hard and soft issues to overcome, particular attention will be centered on the human factor as well as the risks associated with new technology implementations. After these issues are presented, the paper concludes with some ideas and strategies for proper positioning of technical computing in a company to help mitigate the difficulties in their adoption process. By understanding the changes that have occurred, it is hoped that both vendors and academics will better be able to position their new tools for introduction into the company structure.

The Challenge

Some of the issues facing management in growth companies are listed in Table 1.

Table 1. Key Issues Faced by Management.

- How to prioritize R&D projects to maximize the impact on long-term company profits
- How to gain the most bang from limited capital bucks
- How to become more effective in the new product development process
- How to beat competitors in getting new products to the marketplace
- How to deploy resources to maximize the impact on company profits
- How to maximize the efficiency of the supply chain
- How to insure product integrity/How to assess the risks in doing so
- How to insure successful scale-up of prospective manufacturing processes
- How to insure existing processes are most efficient and environmentally benign

Managers are desperate for help in making these complex decisions not only to gain a competitive advantage, but also to survive. Information technologies (in this paper a term which encompasses both technical computing and information sciences) has the potential to help management with these decisions. In fact, many expensive programs are sold to upper management with this expectation in mind. However, frequently they do not meet expectations, but take on a life of their own, since they are not regulated by the same checks and balances that control other technologies more familiar to the company and the decision makers. A recent article in the Wall Street Journal (1998) quoted the consulting firm CSC Index in Cambridge, Massachusetts and the American Management Association saying, "Fifty percent of all (information) technology projects fail to meet the chief executive's expectations". In the same article, Howard Selland, president of Aeroquip Corporation, talked about his company's conversion from Microsoft Corporation's Windows® 3.1 to Windows® 95 saying, "We converted every person in a 50-person research lab at a cost of $20,000 per person or $1 million without one single piece of paper justifying the expense being circulated through our normal approval chain". As a result of these types of experiences, management is losing its' interest in spending large sums of money on new systems without guarantees that they will deliver on their promises.

What are some of the newer enabling technologies that have been dealt with in this conference and can have the potential to impact the issues in Table 1? A condensed list is shown in Table 2.

Motivation for each of the areas of computational model development has been addressed in other papers in this conference proceedings and will not be repeated here. Although it has not been dealt with specifically at this conference, other enabling technologies also under the

Table 2. Enabling Technologies to Improve Corporate Decision Making.

Computational Methods
Planning & Scheduling algorithms
Multivariate statistical analysis
Plant-Wide Optimization procedures
Steady-State/Dynamic Equation Solvers
Hardware Platforms
Networking Enhancements
Comprehensive Software Solutions (e.g. SAP*)

general umbrella of information technologies are having a major impact on company decision making; this includes hardware platforms and integration between hardware and software environments. Networks are advancing to the point that they are ubiquitous with the eventuality that data, voice, Internet, etc. will all merge into one. In addition, industry is moving to comprehensive software solutions rather than the niche solutions of the past. Software, such as SAP*, provides a solution to cross-functional business processes and is helping not only to integrate functions within a company, but also to effect standardization across the industry.

These technologies have the potential to change and in many cases have already changed the product development process as well the entire value chain. However, it is still difficult to get buy-in and sponsorship for the adoption of many of these technologies. The challenge addressed in this paper is to elucidate the reasons for this problem and to suggest solutions, so that these technologies can be brought into a company in a cost effective manner and meet heralded expectations.

Evolution of the Business Information Technology (IT) Relationship

It is instructive to study the historical change in the relationship between information technologies and various company business units. A Venn Diagram for technical computing in the 1970's is shown in Figure 1. There was little intersection between the users of technical computing across a business.

R&D used technical computing primarily for engineering and computational research. In manufacturing, the primary use was for plant design, simulation and control. In both R&D and Manufacturing, engineers and the occasional physical chemist were the focal point for the use of the technology. In the financial arena, there was a closer integration between the business work processes and the Cobal programmers in the information systems departments. Here, the focus of computing was on data entry and transactional systems.

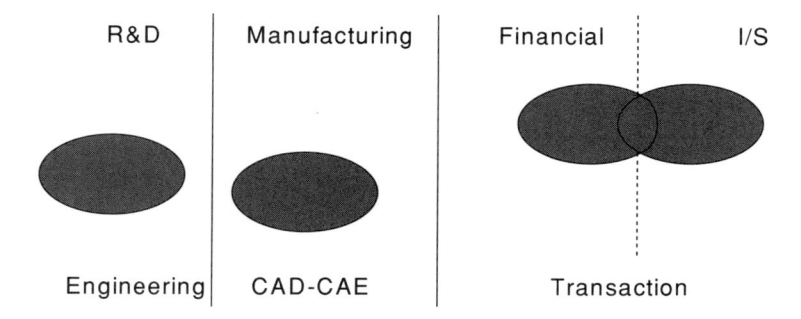

Figure 1. Technical computing in the 1970's.

Computational methods, hardware and application areas were all separated and little benefit could be realized from even the possibility of integration. Consequently, differences of technology and culture fostered an environment which encouraged freedom and flexibility to adopt technologies that benefited one department or work group without consideration for the other company functions. Despite this freedom, a power struggle always existed at the upper levels of companies for control of technical computing.

It will be helpful to share a personal experience from the Dow Chemical Company. In 1968 one of the authors, Gary Blau, joined Dow Chemical Company's Computational Research Laboratory. He entered an exciting research environment:

- chemical engineers were developing Dow's own process simulator while another group worked on the same thing in Dow's Texas division
- COBAL programmers were working on Information Search Engines
- computer literate physical chemists were developing algorithms for predicting physical property relations for new various molecules
- math modelers were developing statistical and operations research tools for modeling various work processes

All of this work was done on punch cards in ALGOL on a Burroughs computer. The operations were housed in Dow's independent research lab under the direction of the vice-president of R&D. Two years later the R&D director retired. The lab was disbanded and the employees were distributed throughout R&D or to the "Global" Information Systems (I/S) group. The move was justified by efficiency. The Burroughs mainframe andALGOL vanished. FORTRAN became the language of the "techies".

Just three years later, the R&D community in total frustration over the lack support from the "business" controlled computing community, forced management to reform the lab. This time, it was called the Systems Research Lab. The scenario repeated itself. The lab grew and prospered. The group was so innovative in its use of computing, they mistakenly outperformed the business controlled I/S function. When the protecting R&D vice-president departed, the lab was disbanded again. The same "efficiency" reasons were given.

A few years later, the lab was brought together once again when the needs of R&D were not met. However, this current version of the lab met a different fate, because of the winds of change in the business. All previous versions of these organizations were given financial support from the R&D coffers to work on truly innovative projects. In the 90's, when downsizing was implemented, this source of funding was essentially eliminated. Support from various business units was needed for all technical computing projects. Consequently, innovation was stifled and many of the scientists in these organizations decided to join various business entities where thjey could operate more freely. The result was a distribution of technical computing expertise throughout the company.

One very positive aspects of distributing technical computing to the various business units is having specialists focused on the needs of a specific organization. However, the lack of a supported, central technical computing function has significantly impacted the way in which new technologies, such as those described in this conference, are brought into a company. Under the 1970's/80's scenarios, the scientists in the technical computing function became internal champions for new technologies. They either developed these technologies themselves, became the focal points for small software vendors or new tools emerging from a university setting. For example, the development of the ASPEN process simulator in the mid 70's was shepherded in by a "who's who" of technical computing from the industrial setting. The introduction of new technologies in the 1990's will follow a different track.

Information Technology, a concept of the 1990's, has emerged as the common denominator for technical computing regardless of the application; be it R&D, manufacturing or financial. The move to a horizontal integration of computer technology across the functions versus vertically within the functions has been well documented, Grove (1996.). This trend towards the integration of information technology across a company is one of the keys toward implementation of the enabling technologies described in this conference.

R&D　　Manufacturing　　Financial

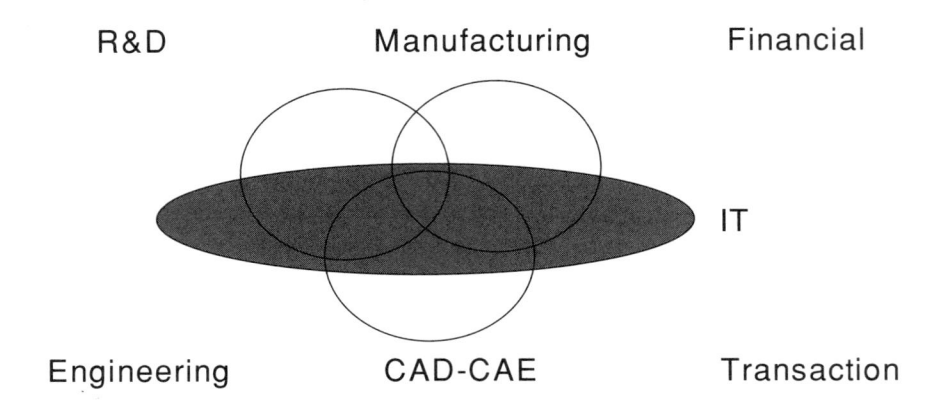

IT

Engineering　　CAD-CAE　　Transaction

Figure 2. A Venn diagram for the 1990's reveals a remarkably different environment.

Additionally, this emergence of information technologies has become a tremendous driver for change in the business models of the company. It is essential that everyone engaged in the technical computing area understand how IT is changing, or can change, the way business is being conducted, because today's existing business partners may not exist - or will exist, but very differently than they do today.

Information Technology Evolution

So what are some of the characteristics of Information Technology and computing that have changed the look of the Venn diagram? Computing has changed from being mainframe based with time sharing to having networks of personal workstations. Historically, problems were solved on a functional, localized level instead of solving problems on a global level with remote decision making. A substantial time lag existed in gaining access to information to the current state of being overloaded with information in real time. IT functions were housed in financial departments separate from R&D, but today IT departments are globally positioned to serve all the business functions.

COBOL was the main force for financial computing in contrast to today's thoughts that COBAL programming is passé, with the very real exception of the year 2000 problem. People with strong computer skills were needed for computer based problem solving, but today end-users with minimal computer skills have been empowered through the use of desktop tools such as FileMaker Pro® and Microsoft Excel®, etc. to do their own "programming" to gain personal efficiencies.

Another example, familiar to the authors which demonstrates the positioning of technical computing within Dow, was the internal development of the Simusolv[1] computer program. During one of the cycles when technical computing existed as an organization, the need to develop a nonlinear parameter estimation capability to build mechanistic models was recognized. Internal resources were used to write a program to meet this need for the other research functions. When resources dried up to support this product, a period when technical computing was organizationally distributed, the internally developed product was turned into a product for sale outside of Dow. This provided the funds to support development, and the internal version of the product became a tool used by over 500 scientists within the company. However, based on what has been alluded to above, such an internally developed and supported product can not exist in the 90's, especially one that is poorly understood by a technically challenged management structure. Consequently, the product was withdrawn from the market, and an attempt is being made to replace it with a standard one supported by outside resources.

Not only has there been a change in the technology and the organization of information technology within a company, but there has also been a change in the culture of the organization as well. In the 70's and 80's, there was a much more hierarchical based organizational structure with "top-down" decision making... the classical pyramid model. Today, there is a much more flattened organizational structure. Within Dow Chemical, for example, the CEO of the company has mandated no more than six layers of reporting structure between himself and the plant operator. It would appear that this would move decision making to a lower level, but the reality is that it has just minimized the number of independent decision makers. Consequently, upper level management is being forced to deal with decisions that they previously passed

---

[1] Trademark of Dow Chemical

onto middle managers who were frequently better trained to deal with them.

Since decisions involving technical computing frequently require the expenditures of large sums of dollars and cannot be handled by the individual business themselves, upper level management has asked for help with these decisions. The creation of the CIO position has been one attempt to deal with this problem. However, the attrition rate for these individuals has taken on staggering proportions. To be successful in this position, the CIO must have a unique relationship with the CEO and others on the management board. They must understand the demands and the uncertain expectations of meeting the needs of diverse internal business areas with minimal internal technical resources available to help with the decision making.

What has happened to the internal technical resources? Why is upper level management reaching out to external consultants and occasionally software/hardware vendors themselves to help them with their information technology decisions? The answer is simple. The technical resource are gone! In the 70's and 80's, when graduates signed up with a company in many cases it was for life. There was considerable loyalty both by the company and by the employee. Today, the norm is self loyalty, and it is expected that an employee will work for several employers during his or her career. This is particularly true for information technology specialists. The CPI cannot compete with consulting firms in attracting, acquiring and retaining information technology professionals. It is not too difficult for such a professional to double or triple his or her salary by quitting a traditional company to work for a consulting company; and self-loyalty makes this decision rather simple. One associated problem with loosing these professionals is that it's increasingly difficult to deal with proposals from outside consulting firms and vendors. It should come as no surprise that many companies are sold systems with all conceivable bells and whistles, while a much simpler, less sophisticated system with proven technology would work as well.. This contributes to the resulting termination of 40% of all information technology projects reported in the Wall Street Journal Article. On the bright side, this is creating an opportunity for a another set of independent consultants that can teach management the following:

1. the basics of information technology,
2. what to expect from large information technology firms,
3. the level of internal information technology resource base required (regardless of salary levels), and
4. how to achieve a reasonable return from their investment in information technology.

Impact of Information Technology on Business

The previous section has illustrated how changes in corporate decision making and culture have reshaped the technical computing. Before moving forward with suggestions for dealing with the changes, it will be worthwhile to emphasize the importance of the information revolution on the way business is being conducted. The first illustration is a business familiar to the authors: the Agricultural Products Industry. The changes that have occurred as a direct result of advances in information technology are shown in Figure 3.

Technology Impact to the Ag Industry

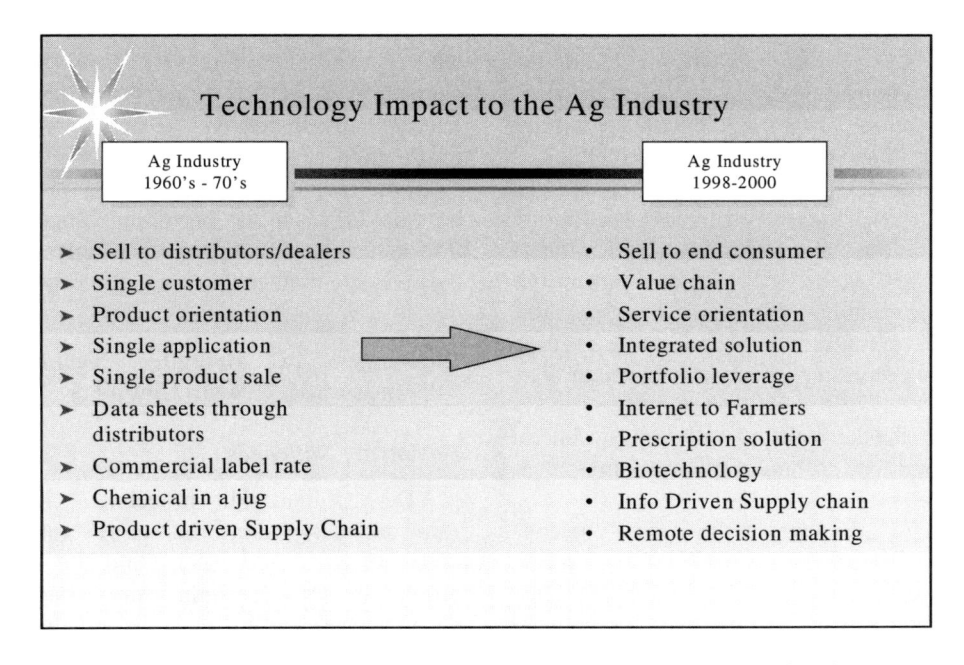

Ag Industry 1960's - 70's	Ag Industry 1998-2000
► Sell to distributors/dealers	• Sell to end consumer
► Single customer	• Value chain
► Product orientation	• Service orientation
► Single application	• Integrated solution
► Single product sale	• Portfolio leverage
► Data sheets through distributors	• Internet to Farmers
► Commercial label rate	• Prescription solution
► Chemical in a jug	• Biotechnology
► Product driven Supply Chain	• Info Driven Supply chain
	• Remote decision making

Figure 3. Impact of information technology on the agrochemical industry.

Information Technology Evolution

Transaction Based 1960 - 1973	Operating Efficiencies 1973-1980's	Virtual Corporation 1990's
➤ Accounting/Finance ➤ Payroll ➤ Sales Statistics ➤ Cost Distribution ➤ Inventory Accounting	• Materials Management • Shop Floor Control • CAD-CAE • Analytical Computation • Word Processing	• Direct Customer Access • Product Customization • Fast Response • JIT • EDI-E-Commerce • Data Warehousing

Value of IT to Business

Direct Savings → Indirect Savings → Growth

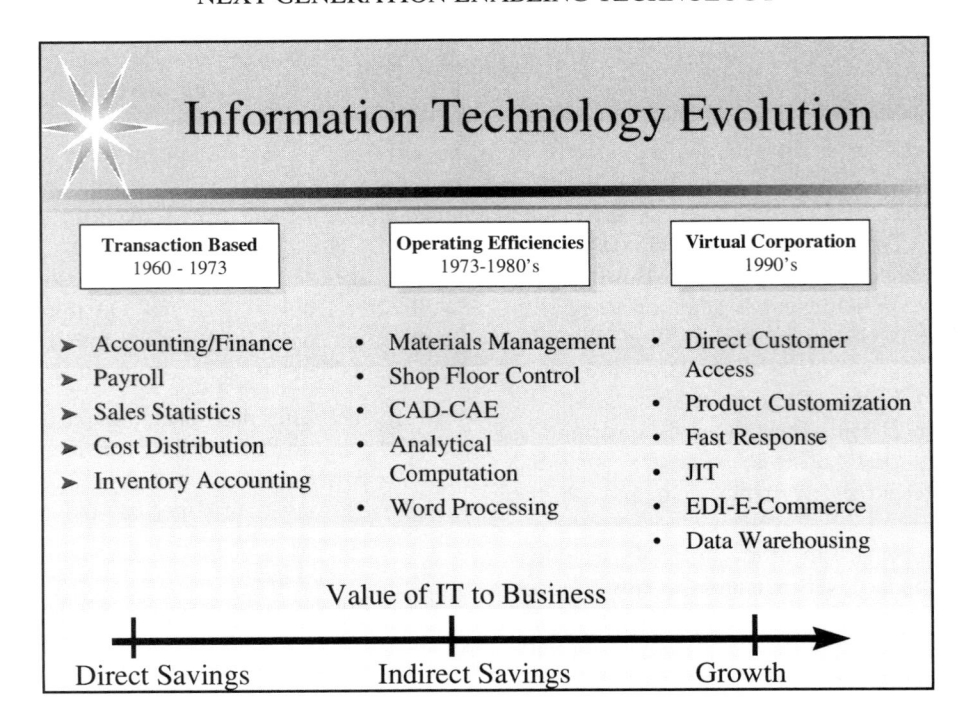

Figure 4. Evolution of information technology.

From the 60's to the 80's, the primary customers for agricultural products were basically distributors and dealers, rather than the end-user -- the grower. Today, technology has changed the value chain and made it possible to sell directly to the end consumer, the farmer, the golf course operator or the home owner, by compiling key information such as customer profiles.

Historically, the industry had a product orientation characterized by selling "chemicals in a jug" to the dealers and distributors. Today, information technology has given the industry a service orientation. For example, Dow AgroSciences now sells a product called Sentricon that is a complete service for termite control including the monitoring of bait stations and the resulting information generated. Single product, single application sales have given way to selling integrated solutions to problems. In the past a single, commercialized label rate characterized Agricultural Products whereas today Globally Positioning Satellite technology allows for prescription application of the products to grower's fields. All of these concept use information from the customer to permit a personalized service orientation. The future will be even more exciting and dramatic as this industry enters the biotechnology revolution.

Moving beyond the agricultural industry, a general evolution of information technology in commerce is depicted in Figure 4.

In the 60's and 70's, industry realized direct savings from information technology by focusing on accounting, finance, sales and cost distribution inventories. Then some of the technical advances improved operational efficiencies through materials management, computer engineering, shop floor control and word processing. Today, technical computing has revolutionized how business is conducted by providing direct access to customer information, such that the entire value chain is transformed to meet the customized needs of the user. All of this is being achieved with fast response as just-in-time manufacturing and electronic communication are completely changing the distribution channel. Companies that capitalize on changes in the available information will be successful and achieve that elusive competitive advantage.

Examples of this new paradigm in electronic commerce are everywhere. Books are ordered on-line with a charge card instead of in bookstores. Do-it-yourself investing has diminished the role of the broker in making invest decisions and providing investment information. The impact of the SMART card on consumer buying will dramatically influence the offering by suppliers. It is becoming abundantly clear that information technology is not only changing the way business is conducted, but also the way people live.

Challenges in Bringing Technical Computing Advances into a Growth Company

Technical Challenges

Despite the rapidity of technical change, the wealth of value added by the information revolution and the merger of Information Technology and the business, it is still difficult to introduce significant technology advances into an industrial setting. Some of the reasons for this difficulty are summarized in Figure 5.

Challenges (continued)

➤ Globalization
➤ Rapid advancement of technology change
➤ Paradigm shift of IT's position in business
➤ Strategic Use of IT to drive competitive advantage (for survival)
➤ Getting control of information overload

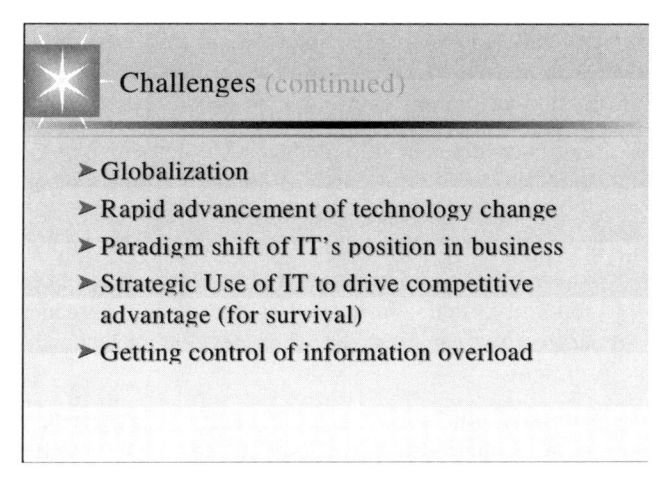

Figure 5. Challenges to introducing technology advances.

Globalization is becoming the norm in industry today. Area or country focus is becoming only a historical reference point. Operating and competing as global organizations is only possible through advances in computing technology, particularly network technology. However, globalization can only be achieved by standardization. Unfortunately, standardization is an anathema to rapid advancement of technology and innovation.. It is a major decision to determine not only when to advance to a new technology, such as those described in this conference, but also when to upgrade versions of an existing system. It is no longer possible to take one work group or even a function to a new technology as shown in Figure 1. Now the global nature of the business forces everyone to a much broader scope in the decision making process, since everyone must adopt the new technology.

As described in Figure 2, there has been a major shift in the role of information technology within the business. This has created internal contests for ownership, sponsorship and dealing with the "not invented here" syndrome. This friction must be eliminated. The I/T department and R&D must become much more closely aligned if they are both going to obtain strategic advantage of information technology in order to allow a company to survive and perhaps even have a competitive advantage.

One final issue to deal with, which will be addressed more completely in subsequent paragraphs, is the issue of information overload. Once it is known that a decision on new computer technology is imminent, the professionals within a company are soon deluged with mountains of e-mail messages spewing claims for and against the new technologies. It is almost impossible for someone to sort through this morass of information without the benefit of some unique skill sets.

The Human Factor

There are a lot of technical challenges when bringing in new technologies, but the greatest challenge is the human factor. Upper-level managers are forced to make decisions for which they are technically challenged. They are often not familiar with the technology or its capabilities. They frequently make decisions in ignorance or delegate this responsibility to staff groups or lower level managers. As has been pointed out earlier, the size of staff groups and middle management positions have been "readjusted" through various downsizing programs. As a result of these depleted ranks, managers who need to make these important decisions fall victim to the claims of "airline tabloids" or the rhetoric of vendors who use "computer lingo" to overwhelm their audiences. The problem is further compounded, because management does not think of themselves as working for an "information" company, but rather have maintained the traditional product mentality. As a result, they look at computer technology as a tool rather than as a core competency that will give them a competitive advantage. Consequently, they find it difficult to attract and train in-house information and technical computing professionals who can help them effectively navigate through the rapidly changing world of information technology.

Potential Solutions to the Problem

Despite all the challenges identified in the previous section, new technologies will find their way into growing companies, or the company will cease to exist. The authors propose various suggestions involving the various groups present at this conference to help industry expedite the introduction of new technologies in a cost-effective manner. Many of these suggestions are of the "soft" variety, but frequently these "soft" issues are the most difficult.

Education

Upper level management and operations personnel need to learn the benefits of the tools described at this conference. If management appreciates how this technology can help them answer the issues of Table 1, they will take the time to learn. Vendors, academics and independent consultants will have to familiarize themselves with the challenges of industry and develop materials and presentations which will teach these individuals in a language that they understand. This is another way of saying that all parties should attend the same language school before they decide to communicate. An example of this is a course that one of the authors is developing to assist upper-level management in using information technology to select new products for development. In this course, management is shown how to extract meaningful information from the wealth of instantly available data in a useful form for decision making. Rather than ramming probabilistic concepts down the throats of an audience , a set of examples showing how traditional economic approaches can be augmented with the real-time, globally consistent information available via company computing networks.

Any remaining in-house experts also need to keep abreast of the latest technologies. As their importance is realized by industry, special efforts will be needed to afford these individuals time to learn about futuristic advances by attending conferences, such as this one, rather than "fire-fighting" within the latest quadrant one crisis.

A New Skill Set for the Technical Computing Professional

Many industrial representatives at this conference may considered to be technical computing professional. In order to function successfully in the companies of the 1990, it is no longer sufficient to be a technical professional in an area.. In order to bridge the knowledge gap, the technology expert must become a communicator and a salesman to management. To this end, a knowledge of the business is required along with a command of business jargon as well as their familiar "techie" speak. It is no longer possible for such individuals to work in isolation. They must develop excellent collaboration and networking skills. They must become team leaders instead of team members as the importance of their technology becomes increasingly recognized by the corporation. This is an area where universities can help. Today, there is little encouragement to develop these "soft skills" beyond what is needed to communicate with the instructor and to interview for a job. The demand for students with these "soft skills" will grow astronomically as the importance of both bringing in their technologies and having them utilized by the company are realized.

Positioning Technical Computing within a Company

Because of the ubiquitous nature of information technology, it is essential to establish a high-level, corporate-wide steering team within a company to set direction for technology acquisition. This team could then network with localized, functional level teams (e.g. area/R&D/manufacturing, etc.). These internal organizations should then strengthen and formalize interactions with academic organizations to ensure that future human resource needs with the appropriate "soft skills" are met as well as the newer tools required for the next generation of applications are developed. Another forum might be the establishment of a consortium across companies to design and establish various technical information guidelines. Examples might include an electronic notebook consortium, a consortium for the electronic registration of new products by government agencies, risk assessment models for government regulated industries, standardized scheduling test problems for new algorithm developments, an agenda for software advances for 2020, etc. Such consortia have several attractive features conducive to the current information technology environment in industry:

- eliminates the need for a dedicated in-house expert
- standardizes the information technology environment; e.g. analogous to ASTM
- opens case studies to industry, academia and vendors
- externalizes a voice, which frequently has more credibility than in-house sources, and broader perspectives can be taken when developing technical solutions

Naturally, this is not a total panacea. Proprietary aspects of business can be an issue as well as the slower progress in gaining consensus and implementing solutions. It is also essential that the academic community, as well as vendors, be willing to participate in such a venture. Once again, competitive issues stimulated by a "not-invented-here" syndrome and traditional funding sources need to be resolved.

Positioning Technical Computing with the University

A very attractive venue for positioning novel technical computing advances developed in the university environment is the start-up company. Several examples of this innovative style of marketing have been described at this conference. From industry's point of view, technical credibility is frequently guaranteed if the name of a university is associated with the a technical computing enterprise. Coincidentally, industry is more comfortable with obtaining software from a company rather than the academic strength software of a university. The start-up company will be faced with two major challenges: the lack of a technically strong internal champion in a corporation, and the need to make the offerings compatible with the existing standardized information technology infrastructure. Both of these issues will promote the stimulate the eventual acquisition of these start-up companies by vendors who have established reputations within the various industry segments.

Conclusions

Spectacular advances in computer technology offer the potential for dramatic changes in the way decisions are made and business is conducted. These technical changes and the cost of implementation have dramatically shifted the way they can be introduced and integrated into a company. Each of the groups represented at this conference, industry, vendors and academia, has a stake in positioning technical computing, so that it becomes the driving force to lead productivity increases in the next century. Those companies and individuals who have the visionary skills to anticipate the technical and cultural shifts will emerge as the winners.

Acknowledgments

The authors would like to thank Armin Pressler and Jerry Ytzen of Dow AgroSciences for their vision in applied information technology.

References

Blau, G.E., "A Systems Engineering Approach to New Product Development", CAST Communications, Vol. 20, No 2, 1997.

Wysoki, B., "Some Firms Let Down by Costly Computers, Opt to De-Engineer", Wall Street Journal, Vol. CI, No 84, 1998.

Grove, A.S., "Only the Paranoid Survive", Currency Doubleday, 1996.

SESSION SUMMARY: CONTRIBUTED PAPERS

Ignacio E. Grossmann
Department of Chemical Engineering
Carnegie Mellon University

Venkat Venkatasubramanian
School of Chemical Engineering
Purdue University

Introduction

The session on Contributed Papers at the FOCAPO98 Meeting was very successful, and greatly exceeded our expectations. The first major surprise for us was to see that more than 80 abstracts were submitted to the session. From past experience we expected no more than 30 or so. The second major surprise was to see that the interest and quality of the papers was very high. And thirdly, the other surprise was that a significant number of papers were submitted by industry. This is in stark contrast with many meetings on process systems engineering which are mainly attended by academics.

Table 1. Paper Contributions by Country and Sector.

COUNTRY	UNIVERSITY	INDUSTRY	COLLABORATION	TOTAL
U.S.	16	10	5	31
U.K.	6	1	2	9
Germany		1	3	4
Austria			2	2
Switzerland		2		2
Argentina	2			2
Spain	2			2
Australia	1		1	2
Japan	1	1		2
Korea	2			2
Canada	1	1		2
Denmark	1			1
Finland	1			1
Norway			1	1
Portugal	1			1
Romania	1			1
South Africa	1			1
Sweden	1			1
Taiwan	1			1
TOTALS	38	16	14	62

As seen in Table 1, the total number of papers that was accepted was 62 from 19 different countries. About one half of the papers came from the US. Except for the United Kingdom and Germany, all other countries had one or two papers. It is interesting to see that 16 papers came from industry, and that 14 were collaboration papers co-authored by industrialists and academics. Therefore, about 44% of the papers had industrial participation. From the above statistics, it should be clear that process operations is currently an area of great interest by both academia and industry.

The session was divided into two major sections. One dealt with Planning, Scheduling and Control/Dynamics, which represent core areas of process operations. The other section dealt with Identification/Diagnosis, Design and Information Systems, which represent the supporting and enabling technologies for process operations. Below we briefly give an overview of the papers in each of these sessions. Our intent is not to give a detailed description, but simply to classify and highlight major points of each paper. As the reader will realize, there is a very rich variety in the content of the contributed papers.

Planning, Scheduling and Control

Although the division of Planning and Scheduling is often non-trivial, we tried to divide the papers according to the emphasis on their time scale: long term for planning, and short term for scheduling.

Planning.

There were a total of 13 papers on Planning. Papers that were more *practical* in nature, can be classified as follows:

Supply Chain

"Issues In Next Generation Supply Chain Management" by Bala Ramachandran, which explores the effect of dynamics of supply chains through simulation.

"An Industrial Perspective of Supply-Chain Optimization and Simulation" by Oscar Rosen, which discusses the need of integrating mathematical programming with discrete event simulation techniques for modeling scheduling problems arising in supply chain problems.

New Product Development

"Pilot Plant Operations in Pharmaceutical Research and Development in the Next Century" by Jonathan M. Vinson, Owen D. Keck, Prabir K. Basu, R. Brian Houston and Linas Mockus. This paper outlines an information system to support the planning of pre-registration of products.

"Bringing New Specialty Chemicals to Market - A Nightmare in Planning and Scheduling" by Karl Schnelle and Gary Blau, which discusses the evaluation of uncertainties in the market and yields through the use of spreadsheets.

Operations Optimization

"Total Site Optimization of a Petrochemical Complex" by M. Turkay, T. Asakura, K. Fujita, C. W. Hui, Y.Natori, Y. Masaiwa, H. Oonishi and I. B. Tjoa. This paper describes the solution of a very large MILP model with millions of variables through the use of disjunctive programming which yields very tight relaxations.

"Improving Batch Manufacturing Process Operations Using Mathematical Programming Based Models" by Paul R. Bunch, Doug L. Watson, and Joseph F. Pekny, which describes a hierarchical decomposition scheme for integrating MILP planning and scheduling models.

Papers that were *theoretical or computational* in nature in Planning can be classified as follows:

Batch Operations

"A New Approach to Batch Process Planning", by Yang Gul Lee and Michael F. Malone, which described the application of simulated annealing for the NPV optimization of flowshop plants.

"Operational Planning for Chemical Processes using Geographic Information and Environmental Modeling", by Derya B. Ozyurt and Matthew J. Realff, which describes a multiperiod batch scheduling model in which environmental concerns are taken into account.

"A New Conceptual Approach for Enterprise Resource Management Systems" by M. Badell and L. Puigjaner, which presents a novel framework for simultaneous production-cash flow planning in batch plants.

Large-scale decomposition

"Auction-Driven Coordination for Plantwide Optimization", by Rinaldo A. Jose and Lyle H. Ungar, which presents an intriguing slack resource auction approach in contrast to the traditional Lagrangian decomposition used in plantwide optimization.

"Disjunctive Multiperiod Optimization Methods for Process Networks" by Susara Van Den Heever and Ignacio E. Grossmann, which describes a combined disjunctive optimization and decomposition method to greatly reduce the expense in multiperiod MINLP problems.

Uncertainty

"Reliability and Maintenance Considerations Process Operations Under Uncertainty" by E.N. Pistikopoulos and C.G. Vassiliadis, which addresses the problem of planning under uncertainty through the incorporation of maintenance scheduling.

"Planning under Uncertainty: A Parametric Optimization Approach" by E.N. Pistikopoulos and Vivek Dua, which describes novel algorithms for multiparametric MLP and MINLP, that can be used in planning under uncertainty.

Scheduling

There were a total of 12 papers on Scheduling. Papers that were more *practical* in nature, can be classified as follows:

Mixed-integer programming

"Scheduling of a Multi-Product Polymer Batch Plant" by Christian Schulz, Sebastian Engell and Rudiger Rudolf, that compares continuous and discrete time models using a branch and bound method for solving the underlying MINLP problem.

"Integrated Scheduling in Steel Plants", by Vipul Jain and Ignacio E. Grossmann, which describes a continuous time MILP model of a job shop using an equivalent representation in terms of a multistage plant.

"Large Scale Multi-Facility Planning Using Mathematical Programming Methods" by Paul R. Bunch, Rex L. Rowe and Michael G. Zentner, which describes an aggregate MILP multifacility problem involving 2000 pharmaceutical products.

"Solving Trim-Loss Problems with Variable Raw Paper and Trim-Loss Widths", by Iiro Harjunkoski and Tapio Westerlund, which describes an MILP reformulation of a non-convex integer nonlinear program.

Uncertainty

"Minimizing Production Cycle Time at a Formulations and Packaging Plant", by Matthew Bassett, which addresses the minimization of uncertain changeover times by applying Monte-Carlo simulation to a traveling salesman optimization model for changeovers.

"Using Dynamic Modeling to Improve Scheduling Capability" by Bernard McGarvey and Bruce Bickle, which analyzes schedule robustness from varying times in a solvent recovery system.

Papers that were *theoretical or computational* in nature in Scheduling can be classified as follows:

Optimization Strategies

"Maximum Dispersion Algorithm for Multi-Site Delivery Scheduling" by S.D. Mokashi and A.C. Kokossis, that describes an algorithm based on graph theory that can handle very effectively problems up to 1000 vertices.

"An Optimal Control Approach for Scheduling Mixed Batch/Continuous Process Plants with Variable Cycle Time" by H.P. Nott and P.L. Lee, which describes an intriguing approach for a single-product job-shop problem through a two-stage strategy based on a relaxation.

AI techniques

"Machine Learning Approaches for Enhancing the Optimization of Batch Production Scheduling and Process Operation Problems" by Matthew J. Realff and Kenneth J. Kirschner, which describes learning procedures for heuristics in branch and bound methods.

"A Task-Resource Based Framework for Process Operations Modeling" by Gabriela S. Mannarino, Horacio P. Leone and Gabriela P. Henning, which describes an object oriented approach for modeling language for process operations.

Batch processing

"Integration of Production Scheduling and Activity Based Costing Models" by V.Lionis, G.E. Rotstein and E. Mohamed, that describes an accounting approach to cost products and manufacturing operations.

"Batch Production Scheduling with Flexible Recipes: The Single Product Case" by Moises Graells, Espen Loberg, Antonio Delgado, Enrique Font and Luis Puigjaner, which describes the application of a genetic algorithm to optimize batch times based on quality models.

Control/Dynamics

There were 5 papers on control and dynamics. Of these, they could be broken down as being oriented toward applications or being theoretical/computational in nature.

Applications

"CLIFFTENT: Determining Full Financial Benefit from Improved Dynamic Performance" by Pierre R. Latour, which describes a precursor tool to real-time optimization for evaluating its potential benefits.

"Real-Time Optimization of FCC Recovery Section", by John Brydges, Andrew Hrymak and Thomas Marlin, which describes the application of advanced NLP sensitivity analysis for ensuring model fidelity.

"Dynamic Simulation and Control Strategy Evaluation for MTBE Reactive Distillation" by Douglas A. Bartlett and Oliver M. Wahnschafft, which describes the use of steady-state and dynamic simulation for designing a control system on a reactive distillation column.

Theoretical/computational

"Model Predictive Control and Identification: A New Approach to Closed-Loop Identification and Adaptive Control" by Alexander Schwarm, S., Alper Eker and Michael Nikolaou, which describes a novel chance-constrained optimization model for this problem.

"Operational Implications of Optimality" by John Bagtrep Jorgenson and Sten Bay Jorgensen, a paper that confirms the empirical observation that complex process behavior is often found at operational optima.

Identification and Diagnosis

There were a total of 11 papers on Identification and Diagnosis. They can be grouped into the following broad categories:

Multivariate Statistical Analysis in Process Monitoring and Diagnosis

"Process Operation Improvement Based on Multivariate Statistical Analyses of Process Data" by Chonghun Han. The work presents a methodology of combining various multivariate statistical analyses towards process improvement and a case-study application on an industrial blast furnace.

"Fault Isolation in Large Scale Industrial Processes Using Fisher Discriminant Analysis" by Evan L. Russell and Richard D. Braatz. This paper presents the Fisher Discriminant Analysis as an alternative to PCA for dimensionality reduction. The capabilities are demonstrated using both simulated and real life industrial data.

"Faster Development of Fermentation Processes. Early Stages Process Diagnosis" by Joao P. Lopes, Jose' C. Menezes presents a method for combining PCA and PLS for fault isolation that includes information on product quality.

"Process Monitoring Using the Clustering Method and Functional-Link-Associative Neural Network" by Kyung Joo Mo, Dongil Shin, Sooyoung Eo, and En Sup Yoon. This paper presents a new process monitoring methodology that is based on the modified adaptive k-means clustering algorithm and nonlinear principal component analysis (NLPCA).

Optimization Techniques

"Adaptive Random Search and Short-Cut Techniques for Process Model Identification and Monitoring" by Gheorghe Maria. The paper describes a adaptive random strategy that includes a short-cut estimator to overcome poor conditioning in optimization problems in process monitoring.

"One Step Collective Gross Error Identification and Compensation in Linear Dynamic and Steady State Data Reconciliation" by Miguel Bagajewicz and Qiyou Jiang. This paper presents a method for simultaneous identification and estimation. A spanning tree approach is used to capture the gross error sets which is then solved using an MINLP based method to obtain the minimum.

Fault Detection and Abnormal Situation Management

"Improved Statistical Process Control Using Wavelets" by Bhavik R. Bakshi and Sermin Top presents a unified framework for commonly used SPC methods. The approach uses multiscale wavelets to decouple deterministic and stochastic components and decorrelate the measurement data.

"Optimal Alarm Logic Design for Process Networks" by Chuei-Tin Chang and Chii-Shang Tsai. The paper presents a systematic approach to identify data redundancies in the process network and synthesize the corresponding alarm generation logic.

"Applying a Procedural Reasoning and Reactive Approach for Abnormal Situations in Refinery Control" by David J. Musliner. The Abnormal Event Guidance and Information System (AEGIS) is described and the various components and lessons from their implementation are discussed.

"Performance Assessment and Diagnosis of Refinery Control Loops" by N.F. Thornhill, M. Oettinger, and P. Fedenczuk. This paper presents an approach to assign default values to parameters in control loops refinery wide for use in performance assessment algorithms. Operational signatures are examined as a cause for closed loop impulse response oscillations.

"Enhancement of Gross Error Detection When Data Are Serially Correlated" by Ruth Kongsjahju and Derrick K. Rollins. An extension to the unbiased estimation technique is presented in this work coupled to a new pre-whitening technique to transform the data.

Design

There were a total of 13 papers on Design. They were along the following general themes:

Design and Performance Improvement via Modeling and Simulation

"Operational Design and its Benefits in Real-Time Use" by Yuji Naka, Rafael Batres, and Tetsuo Fuchino. An approach to better management of change and clearer operational design is presented that couples the operational aspects of design with the manufacturing execution system.

"The Use of Dynamic Event Driven Simulation to Size Buffer and Storage Systems" by Joerg Krames, Dirk Wollenweber, and Christine Edinger describes the application of a discrete event simulation tool and compares it to the traditional thermodynamic used more commonly in the process industry.

"An Advanced Process Analysis System for Pollution Prevention" by Xueyu Chen, Kedar Telang, and Ralph W. Pike. This work describes an integrated system that combines reactor and pinch analysis, on-line optimization and pollution indices for waste reduction. The system has been successfully applied to a Monsanto/IMC Agro contact process for sulfuric acid.

"The Cost of Crossing Reaction Equilibrium in a System that is Overall Adiabatic" by W. Nicol, D. Glasser, and D. Hildebrandt. This work extends an earlier approach by the authors that demonstrated the possibility of equilibrium crossing in a single adiabatic reactor. The current approach includes cost effects of such an event.

"Flexibility Analysis of Natural Gas Plants" by A.M. Eliceche. This work uses plant flexibility analysis to identify active design constraints and an NLP based approach to extract the maximum allowed variations. The aim is to perform debottlenecking of the design using a cost-flexibility trade-off analysis.

"On Combining Operating Features in the Design of a Batch Process" by Girish Joglekar and Steven M Clark presents a simulation based feasibility study of an industrial solvent recovery operation that is environmentally conscious.

"On the Efficiency of Material Flow Simulation for Process and Multipurpose Plant Optimization" by P. Solot Phillipe presents different models in use for these simulations and describes two practical case-studies to which they have been successfully applied.

"A New Concept to Improve Operation and Performance of an Industrial Size Tube Reactor by Means of CFD Modeling" by H.J. Warnecke, M. Schaeffer, J. Pruess. This paper presents a hybrid approach that combines basic CFD simulation with the concepts of reactor modeling to describe the complex absorption of gaseous components into the liquid-phase in a tubular reactor.

"Design and Optimization of Chromatographic Processes in Production Scale" by Jochen Strube and Henner Schmidt-Traub. This article explores the possibilities of separating racemates via kinetic resolution through preferred crystallization or preparative chromatography on chiral stationary phases (CSP), especially in its application form as Simulated Moving Bed (SMB) chromatography.

Theoretical/Optimization Techniques in Process Design

"Using Homotopy Continuation and Grobner Bases for Finding all Solutions of Steady State Process Design Problems" by D. A. Harney and N. L. Book. This paper describes an approach to generate the Grobner basis for a system of design equations and using it in tandem with Homotopy continuation to extract all solutions of the system.

"Nonlinear Parameter Estimation Using Interval Analysis" by Chao-Yang Gau and Mark A. Stadtherr. This paper outlines an interval-based nonlinear parameter estimation method with potential global optimum capabilities and demonstrates its application in estimating the parameters for the Wilson VLE equation.

"A Systematic Procedure for Optimal Operation of Heat Exchanger Networks" by Bjornn Glemmestad and Truls Gundersen presents a comprehensive framework and methodology that involves an initial degrees of freedom analysis followed by a suite of different techniques from simple qualitative methods to on-line periodic optimization to minimize energy cost.

"Process Optimization and Process Robustness using Experimental Design: STAVEX" by Y.L. Grize and W. Seewald. This paper describes the STAVEX expert system and two industrial case-studies addressed using the system for process performance and robustness optimization.

Information Systems

There were a total of 8 papers on Information Systems. The papers contributed, addressed the following major topics:

Representation, Architecture and Modeling

"Information Models for Batch and Real Time Chemical Process Data" by Neil L. Book and Arvind Sharma describes the use of information models to capture transition data between snapshots in different types of process operation i.e. batch, semi-continuous etc.

"A Process Plant Knowledge Repository for Multiple Applications" by M.J. Elsass, D.C. Miller, J.R. Josephson, J.F. Davis describes the use of the device centered functional representation to model heterogeneous forms of knowledge. This paper describes the sharable framework based on the above representation.

"An Initiative for Integrated Computer-aided Process Engineering by Andreas A. Linninger, Helmut Krendl and Helmut Pinger" presents a framework for integrating different pieces of the modeling process using the combination of a phenomena-driven language and a general mathematical language.

"An Open Software Architecture for Process Modeling and Model-Based Applications" by A.I. Kakhu, B.R. Keeping, Y. Lu, and C.C. Pantelides presents some mechanistic basis for an openness in modeling tools and describes the concept of a foreign object which is a construct for achieving openness in the gPROMS modeling tool.

Applications and Development Tools

"Application Programming Interfaces for EXPRESS Information Models" by Milind M. Khandekar and Neil L. Book. This paper describes an API tool developed to produce the interface from information models written using the EXPRESS modeling language.

"A Process Engineering Information Management System using Word-Wide Web Technology" by R Andrews and J. W. Ponton. This paper presents an platform independent object-oriented Java-based architecture for information inter-communication between different modeling and design tools as well as manual analysis, with advantages and future potential.

"From Engineering Analyses to Operators Training" by Peter Stanley summarizes the wide-ranging capabilities of state of the art training systems that include accurate simulations based on first principle engineering models.

"OLE Automation: Bringing the Power of Process Simulation to Plant Operations Personnel" by James McGill presents a protected OLE environment that enable plant operations personnel to access process simulations. The environment is customized to ensure integrity of the results as well as ease of understanding.

Summary

From this brief introduction and overview it can be seen that all the contributed papers have something interesting to offer. Whether it is reporting experience with methodologies by industrialists, or proposing new algorithms for solving basic problems in process operations, the papers cover a large number of topics. We hope that the reader will enjoy the rich variety of contributed papers.

Acknowledgment.

We would like to acknowledge the great help of George Applequist in managing the submission of manuscripts through the FOCAPO98 website. The authors are also most grateful to all the reviewers of these manuscripts who promptly reviewed the papers.

ISSUES IN NEXT-GENERATION SUPPLY CHAIN MANAGEMENT

Bala Ramachandran
Unilever Research
Port Sunlight Laboratory
Bebington, Wirral L63 3JW, U.K.

Abstract

This paper raises some important challenges to tactical supply chain management and control from the point of view of a fast moving consumer goods industry. Factory utilizations are higher than ever before as a result of strategic regional and global sourcing owing to pressures from increased asset utilization which has resulted in increased strains in the supply chain. For example, planning marketing promotions with major retailers would require detailed factory scheduling in addition to planning. Information regarding sales and marketing intelligence tend to be very uncertain which has to be taken into account while considering tradeoffs between reallocating production in time or to other sites in the regional sourcing network; otherwise, this could lead to high noise amplifications in the supply chain from the point of sales through the distributor stocks to factory production, resulting in higher costs and lower customer service. Practical examples are numerous where the consumer offtake is relatively flat while the material transactions upstream in the supply chain undergo wild fluctuations. Hence sales forecasting, detailed factory scheduling, logistical constraints and reallocation of demand to different points in the sourcing network along with the associated costs need to be considered in an integrated framework to facilitate effective decision making. Sales, Marketing, Manufacturing and Logistics considerations should be unified in a decision support system capable of functioning in real time. In this paper, our initial efforts in this area are discussed with practical examples from the consumer goods industry to stimulate further work in this direction.

Keywords

Supply chain management, Systems dynamics, Consumer goods industry.

The Emerging Business Environment

The fast moving consumer goods industry has been undergoing significant restructuring owing to pressures from increased asset utilization. Customer perceptions and expectations have also considerably changed in this decade due to a number of factors, notably initiatives such as Efficient Consumer Response, Continuous Replenishment Planning etc. As a result of this, the global business environment is highly interlinked. Maintaining a competitive edge in such an environment requires a lean and flexible supply chain and the emphasis on this is stronger than ever before. Moreover, we are on the verge of a breakthrough in electronic commerce and undoubtedly, this will manifest itself in new opportunities for supplying products to consumers. This would entail a further reduction in lead times and place a larger emphasis on high customer service at lower costs.

Research Issues

There are a number of research issues to be addressed in order to develop a supply chain capable of functioning effectively in the emerging business environment.

Supply Chain Structure and Connectivity

The information technology revolution in the past decade has spawned a number of opportunities manifesting in considerable changes in the supply chain. Foremost among these, is the surge of interest in enterprise resource planning, which facilitates handling of information across the entire supply chain, beginning from the suppliers through to the retailers concurrently to facilitate better decision making. This entails an increased level of interaction between different parts of the supply chain. According to complex adaptive systems theory (Kauffman, 1993), there is an optimum level of interconnectivity in the system for it to be most effective in its objective while maintaining stability. A higher interconnectivity than the optimum level might lead to successive bifurcations of the equilibrium behavior ultimately leading the system to chaotic behavior. On the other hand, a lower interconnectivity than the optimum level would imply that the system is below its Pareto optimal surface, thus suggesting that resources are not utilized most effectively. It has been postulated that systems need to be at the edge of chaos to be able to quickly adapt to the needs of a changing environment. This raises very interesting questions regarding what is the optimal level of supply chain interconnectivity so that the system is adaptive while at the same time, lead times can be reduced providing customized customer service at low costs. What should be its geographical structure and how much autonomy should each constituent part have ? What kinds of parameters in the supply chain would manifest in chaotic behavior and what are the key performance indicators to monitor its performance and take corrective action?

Supply Chain Amplification

Practical examples abound where the consumer offtake is relatively flat while the material transactions upstream in the supply chain undergo wild fluctuations. This effect is known as the supply chain amplification effect or the Forrester effect. Fig. 1 shows a real life example of this phenomenon in the consumer goods industry. This is primarily due to the tendencies of different stages of the supply chain trying to protect themselves from adverse situations by holding high stocks, leading to wild fluctuations on the factory loads. This problem is only bound to become worse when asset utilizations are high. This raises interesting possibilities of whether it is possible to design supply chain networks which inherently have good dynamical properties. Towill and Del Vecchio (1994) have modelled the production-distribution system using a simple dynamical representation to study the frequency response of different regions of the design space. More work is necessary on realistic representations of supply chain networks to determine how the amplification is dependent on different supply chain parameters and what factors constitute a stable and adaptive supply chain design.

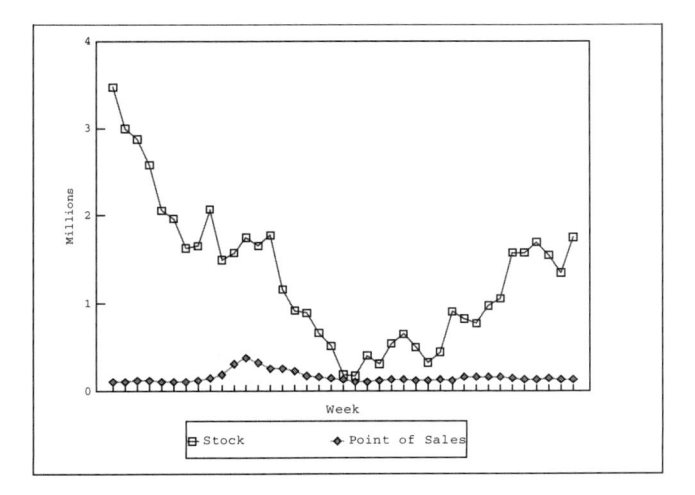

Figure 1. Supply chain amplification - illustrative example.

The effects of supply chain amplification become more severe as the number of global interactions increase as it is happening currently. Moreover, this also has an impact on raw material and packing material suppliers. Systems dynamics (Forrester, 1961) was originally used to illustrate the implications of the Forrester effect. However, all the work until now in this area has concentrated on unconstrained systems. Factory capacity constraints are going to play a critical role in the emerging business environment of much fewer factories making a wider range of stock keeping units (SKUs). Other forms of constraints include limited transportation availability within a geographical region, minimum order quantities etc. The primary objective is to maintain high customer service levels in all geographical regions subject to capacity constraints in different parts of the sourcing network at a low cost. This raises both strategic and tactical issues. What is the most adaptive and cost effective supply chain in this capacitated environment and what kinds of organizational changes are necessary to support such an operation? On the tactical side, capacity constraints demand a number of functional changes in the roles assigned to personnel supporting different aspects of the supply chain as a result of the increased interactions between different parts of the network. For example, planning marketing promotions with major retailers would require detailed factory scheduling in addition to planning. The high uncertainties in the information relating to sales and marketing intelligence should be considered as well. Hence, sales forecasting should be integrated with the manufacturing aspects in order for the retailer accounts to be managed smoothly. The primary tradeoff here is whether to postpone production due to capacity constraints or reallocate production to other parts of the sourcing network subject to issues relating to cost and transportation constraints. Below is a real example to illustrate some of the issues mentioned above.

Illustrative Example

The supply chain under consideration is among the simplest possible, but still some of the effects mentioned above have already begun to manifest themselves in the supply chain operations. It is a production-distribution system comprising of a factory and a sales organization which supplies the market retailers. There is considerable strain on the factory because the market demand is almost equal to the maximum factory capacity and also exceeding it in a number of weeks. In such cases, future forecasts are used to advance production when the market demand significantly exceeds the factory capacity. Safety stock levels are set based on desired customer service levels, forecast accuracies and factory production reliabilities. For this market segment, the point of sales data accounted for promotional effects is almost flat while the factory production and stocks undergo wild fluctuations as shown in Fig. 1. This leads the retailers to believe that they have to hold high stock buffers to supply their stores when the factory is encountering problems which in turn places an additional burden on the factory, leading to a supply chain exhibhiting positive feedback effects. There are additional constraints on the supply chain such as minimum run lengths for brand orders, discounts for full truck loads etc. We have constructed a simple model to simulate the dynamics of the supply chain. Annual stock profiles from the model for a representative set of products shown in Fig. 2 show clear evidence of the fluctuations seen in reality.

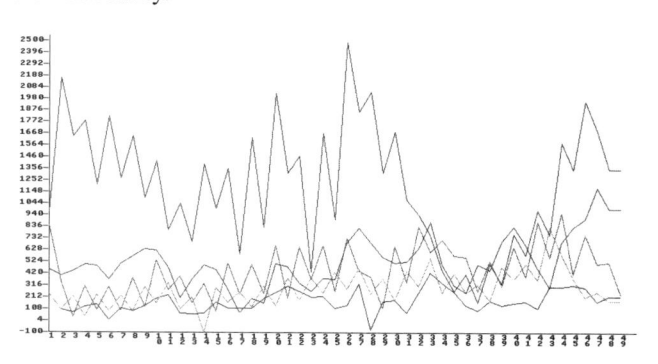

Figure 2. Model stock profiles - initial scenario.

The model also reproduces the capacity constraints in the factory. Fig. 3 is a plot of the weekly factory utilization (in shifts) compared with the nominal work load. Added on top of the factory capacity constraints is the uncertainty regarding sales and marketing intelligence. Hence, the supply chain tends to build up stock at different stages to safeguard against stockouts. However, even this fails to act as a sufficient insurance during times of considerable strain resulting in high costs and customer dissatisfaction. We have analyzed the effects of various parameters such as forecasting accuracy, production reliability, factory

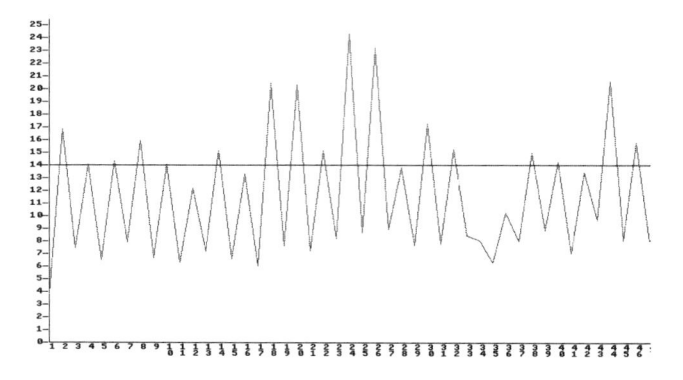

Figure 3. Factory utilization profile - initial scenario.

production policies on the supply chain noise to determine an improved set of parameters. Fig. 4 shows the stock profiles for the representative brands chosen in Fig. 2 using the optimized supply chain. What is noteworthy is that the effects of the capacity constraints are also far less stringent in the optimized case as shown in Fig. 5.

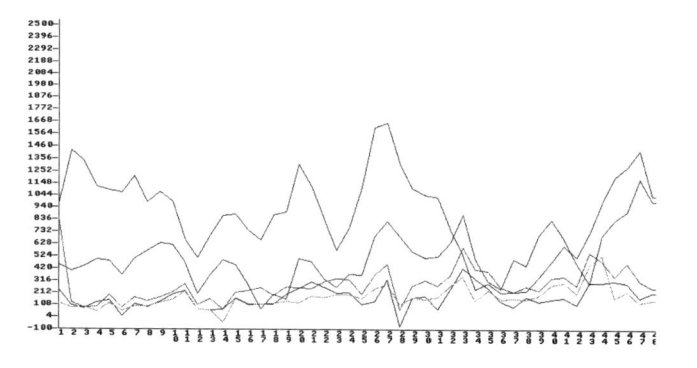

Figure 4. Model stock profiles - improved supply chain.

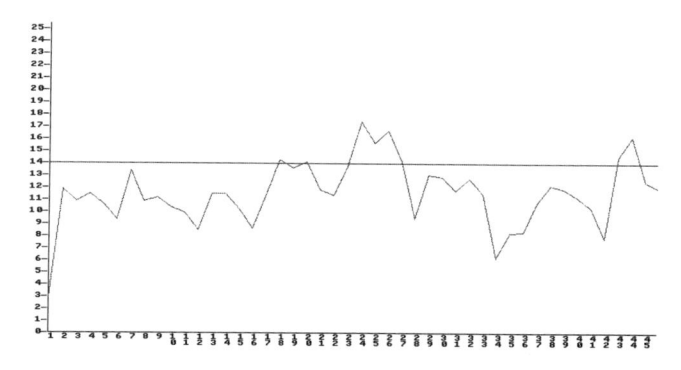

Figure 5. Factory utilization profile - improved supply chain.

Concluding Remarks

The model presented in this paper is basically an open loop simulation to optimize the supply chain policies. The next step is to develop a closed loop model of the supply chain with the objective of maximizing the customer service levels at low cost. In control theoretic terms, the set point is the desired customer service levels and the source of noise is the marketing uncertainties. It is conceivable that advanced control concepts can be exploited in supply chain dynamics. The underlying model has to account for logistics, manufacturing and marketing considerations. It is hoped that the evidence cited in this paper will stimulate further interest in the academic community in this direction.

References

Kauffman, S.A. (1993). The Origins of Order : Self-Organization and Selection in Evolution, Oxford University Press.

Towill and Del Vecchio, A. (1994). The Application of Filter Theory to the Study of Supply Chain Dynamics. *Prod. Plan. & Cont.*, **5**, 82-96.

Forrester, J.W. (1961). *Industrial Dynamics*, Systems Dynamics Series, Productivity Press.

A NEW APPROACH TO BATCH PROCESS PLANNING

Yang Gul Lee* and Michael F. Malone
Department of Chemical Engineering
University of Massachusetts
Amherst, MA 01003

Abstract

In this study a new approximate algorithm for batch process planning using a simulated annealing technique is described. The versatility of simulated annealing allows us to handle a broad range of batch process planning problems involving a general profit function. This new approach is designed to deal with discrete demands and due dates. That is, it is targeted to the planning of a batch process running in *open shop* mode (Graves 1981). Hence, it seeks to find a feasible production plan that can meet all the demands and keep as close as possible to the due dates given by customers. From another point of view, we can estimate if a batch process can meet given due dates or not, i.e., whether it is underutilized or overutilized and for what products and periods. We demonstrate how to incorporate to handle constraints such as intermediate products, compatibility between products and units, etc. This can be done without algorithm engineering or tailoring to handle the constraints. In examples, we study a parallel network flowshop (*multiplant*) structure with zero-wait policy between stages. The effects of setup time and setup cost on the capacity of an open shop will be illustrated. We compare the results to those found by sequencing rules often used for production scheduling such as: *FCFS* (First Come First Serve), *SPT* (Shortest Processing Time), and *EDD* (Earliest Due Date). Significant improvements in solutions are found compared to the results from these typical heuristics.

Keywords

Batch process planning, Open shop, Closed shop, Due date, Inventory, Setup cost, Setup time.

Introduction

Batch process scheduling and planning have recently received greater attention in the literature as chemical and related processes evolved for the production of many different types of high value-added products (Birewar and Grossmann, 1990; Kudva et al., 1994; Lazaro et al., 1989; Papageorgiou and Pantelides, 1996). However, the general scheduling problem is known to be *NP-hard*. Because of the computational difficulty, approximate methods using heuristics have been developed (Kudva et al., 1994). These methods are fast but sometimes end up with a solution far from the global optimum. Mathematical programming approaches often adopt *a multiperiod LP model* (Birewar and Grossmann, 1990; Papageorgiou and Pantelides, 1996). This divides the production horizon into several fixed-length periods and the production level is decided in each period to optimize an objective function. The formulation for mathematical programming is not easy to generalize and each problem with different features may be formulated separately. The formulation for large-sized problems must often be tailored in order to get an optimum or near-optimum solution in reasonable times.

A new approximate method, simulated annealing has been introduced in the batch process scheduling (Das et

---

* To whom all correspondence should be addressed. Present address: Simulation Sciences Inc., 601 Valencia Ave. Brea, CA 92823.

al., 1990; Ku and Karimi, 1990; Malone, 1989; Tricoire, 1992). Thanks to dramatic reductions in the cost of computing, it is often true that the engineering time in reformulating a problem or tailoring an algorithm in a mathematical programming approach is more costly than the calculation time of simulated annealing. Particularly when the problem size is large, simulated annealing offers more advantages although there are certain problems for which more rapid and exact algorithms are available, e.g., Miller and Pekny (1991). Another advantage of simulated annealing is its versatility in handling fairly general objective functions and constraints. Hence, one simulated annealing code can be applicable to a broad range of problems without significantly tailoring the algorithm.

In most of the previous studies dealing with batch process planning, a multiperiod planning model was used. It was implicitly assumed that batch processes operated in a *closed shop* mode where customer requests are fulfilled by inventory and production tasks are results of inventory replenishment decisions (Graves, 1981). However, there is an incentive to operate batch plants closer to the *open shop* mode in which production orders are made directly to the batch plant by customer requests. The major benefit of an open shop is much less inventory than in a closed shop. In an open shop, due dates are a key factor because all production orders come with discrete due dates from customers and no inventory is stocked. Therefore, meeting due dates of every customer orders is a major consideration of production planning and scheduling of an open shop. So far, most studies for batch chemical processes have dealt with the closed shop and only a few studies have focused on the importance of keeping due dates (Musier and Evans, 1991; Ku and Karimi, 1990; Lee, 1997).

In this paper, we describe a new approach to the planning of batch processes whose structure is parallel network flowshops (*multiplant*) operated in the open shop mode. The method relies on simulated annealing for solutions. The versatility of simulated annealing allows the use of a general objective function consisting of the net present value (*NPV*) of revenue, plus due date penalties, inventory cost, and the setup costs. This approach can accommodate discrete demands with due dates and can solve a planning problem in which intermediate products are involved.

Problem Definition

The multiplant structure (Fig. 1) is a set of network flowshops (*production lines*). Each production line has a sequence of stages, and each stage involves one or more pieces of processing equipment (*units*) separated by intermediate storage. For simplicity, zero-wait intermediate storage is used in this study. The number of stages can be different in each production line. The set of units in each stage can process one or more operations (*tasks*). This 'task-to-unit assignment' doesn't change during the production horizon and the sequence of stages in each production line is fixed. A product is processed by passing through the sequence of stages in a production line.

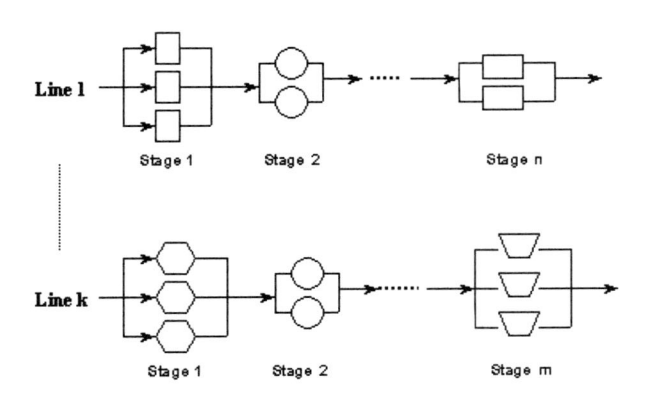

Figure 1. Multiplant batch process structure.

Input Information

The goal of batch process planning is to determine a feasible production plan that maximizes an objective function, given the:

number of production lines, number of stages in each production line, and number, types, and sizes of parallel units in each stage
task-to-unit assignment
product structure of intermediate and final products, and product recipe
processing times, setup times, transfer times, and size factors of each task
product values, raw material and processing costs, setup costs of each product
demands and due dates for each product, and production horizon

The *size factor* is the required size of a unit to produce a unit amount of a product. The size factor for each task is required to estimate the maximum batch size of a product in each production line.

Objective Function

The objective function of production planning optimization is a profit function, which consists of the product value as well as the manufacturing costs such as raw materials, processing and labor. In the planning of an open shop, it is crucial for an algorithm to find a feasible production plan, which produces all demands on time (if possible) and to consider the inventory cost. To find feasible plans, we use due date penalties in the objective. If not all of the due dates can be met, the inventory cost for holding finished products and the shortage cost for late production are also included in the objective. Since the equipment changeover may be a frequent with significant costs, and since it depends on the production sequence, the setup cost is considered in a separate term, not in the manufacturing cost term. The capital investment is not considered in this study.

The resulting objective function is to maximize a profit penalized by due date (lateness) penalties and inventory (earliness) costs and including setup costs. For a production line l, the net present value (NPV_l) and the net penalized present value ($NPPV_l$) are:

$$NPV_l = \sum_{i=1}^{N_{c_l}} \sum_{j=1}^{N_{b_i}} V_p B_{p,l} \exp[-r \max(t_{f_j}, t_{d_p})] - \sum_{i=1}^{N_{c_l}} \sum_{j=1}^{N_{b_i}} PRM_p B_{p,l} \exp[-rt_{s_j}] \qquad (2)$$

$$NPPV_l = NPV_l - \sum_{i=1}^{N_{c_l}} C_{S_i} + C_{V_i} + C_{D_i} \qquad (3)$$

where N_{c_l} and N_{b_i} are the number of campaigns in production line l and the number of batches in campaign i, respectively. The subscript p represents the product produced in the campaign i and $B_{p,l}$ is the maximum batch size of product p on production line l. V_p is the product value and PRM_p is the manufacturing cost which is the sum of the raw material costs, processing costs, and labor costs for product p produced in campaign i. The completion time and the starting time for a batch j are t_{f_j} and t_{s_j}, respectively. Product p has a due date of t_{d_p} and the discount rate is r for continuous compounding. C_{S_i} is the setup cost for campaign I; this is imposed only when the product produced in the previous campaign, $i-1$, is different from that of the present campaign. C_{V_i} and C_{D_i} are the inventory cost and the due date penalty charged during campaign i, respectively. The profit for a production plan is the sum of $NPPV_l$ for all production lines N_l.

$$Profit = \sum_{l=1}^{N_l} NPPV_l \qquad (4)$$

Estimation of Inventory Cost

The inventory costs charged on holding finished products can be simplified into two categories: *capital cost* and *storage cost*. The capital cost is the loss of opportunity to invest funds tied up in inventory and is mainly affected by the annual discount rate. The storage inventory cost stands for all other costs required to keeping finished products in a physical facility. The capital inventory cost is often the most significant component of holding costs (Love, 1979). However, high value-added products often need special treatment such as refrigeration, humidity control and etc. In that case, the storage cost may become a dominant component of the inventory cost.

To estimate the inventory cost of each product, it is necessary to calculate the completion time of each campaign in a production plan. The minimum batch size of each product in each production line is calculated from its size factor and the size of units in each stage of that production line. Then, the minimum batch size is used to estimate the number of batches produced in a campaign. The campaign length is estimated from the number of batches. More detailed derivation for the completion time calculation can be found in Tricoire (1992).

Simulated Annealing Parameters

In this study, the Kirkpatrick cooling schedule (Kirkpatrick et al. 1983) is used.

$$T_{k+1} = \alpha T_k \qquad (5)$$

The parameters determining Kirkpatrick schedule are the initial temperature (T_0), the cooling rate (α), the length of Markov Chain, and a stopping criterion. The initial temperature should be high enough to accept most of the generated solutions. This means that the initial acceptance ratio (the ratio between the number of accepted solutions and the total number of the randomly generated solutions) at T_0 should be close to 1.0. In contrast, the cooling process should proceed until the acceptance ratio is close to zero. We find T_0 empirically so that the acceptance ratio is close to 1.0. The cooling stops if the average profits at 4 consecutive temperatures are below a prefixed value (Aarts and van Laarhoven, 1985). The prefixed value for the termination is decided empirically so that the acceptance ratio becomes close to zero.

Another requirement in a cooling process is that for the Markov chain constructed at each cooling temperature, the distribution of importance-sampled data is close to the equilibrium distribution. In other words, the process remains in quasi-equilibrium during the cooling process. To maintain quasi-equilibrium, the cooling process should be slow or the length of Markov chain should be long. This condition is realized by setting the cooling rate parameter (α) to between 0.90 to 0.98 and the length of Markov chain is decided by precalculations.

Move Set

Another component of a simulated annealing algorithm is the generation of new solutions (a 'move') from the previous solution. In order to guarantee asymptotic convergence of a simulated annealing algorithm, it must be *possible* to generate any feasible solution. Among all of the possible moves that could be used for given types of problems, only some sets of moves yield good solutions

We have designed and tested a set of moves for batch process planning problems (Lee, 1997). To generate a random solution during the calculation, each type of move is used in equal proportion. These proportions don't change the final solution very much but in certain problems, removing some move types increases the calculation efficiency. For example, if a process has dedicated production lines only, we can reduce the proportion of line assignment moves to zero and increase the others.

Examples

The relationship between the inventory and the setup costs has been a big issue in production planning for closed shops (Aggrawal and Park, 1993; Karmarkar, 1987; Trigeiro et al., 1989). In the planning of an open shop, keeping due dates is the most important issue and how to find a feasible production plan which produces all the demands on time with a given batch process is the most important goal. Reducing inventory costs may be the next most important goal. The inventory cost and the setup cost are intimately related to the lot-sizing of production campaigns. That is, there is a trade-off between the setup cost and inventory cost. When setup costs are not large, the inventory cost can be minimized by many short campaigns and by running all production lines simultaneously before due dates. By doing this, the length of time when finished products are stored as inventory is minimized. But if setup costs are significant, the number of changeovers should be reduced and the longer campaigns are preferred to reduce setup costs. The setup cost also plays with due date (lateness) penalty in the planning of an open shop. Especially, when an open shop is in a capacity-limited condition, some of the demands cannot be produced on time due to the longer campaigns caused by the large setup costs.

In the following examples, we will show how the due date penalties, the inventory costs, the setup times, and the setup costs interplay in the production planning of an open shop. The base case example is as follows. A multiplant has 3 identical production lines, which consist of 5 processing stages and the zero-wait storage policy between stages is used. Each stage has one processing unit that has the same size, 10 m^3, and performs a single task. There are four products to be produced and they all have the same recipe, which is composed of 5 tasks in series. All the units are compatible with each product. Therefore, all the products are compatible with any production line. There are no intermediate products and the production horizon is 3000 hours.

Case 1: High Due Date Penalties

In reality, it is often critical for a batch plant to keep due dates. In other words, due date penalties are typically severe. To study this situation, we use a large due date penalty, 5.0 ($/kg/week). The capital inventory cost for a production plan was also estimated.

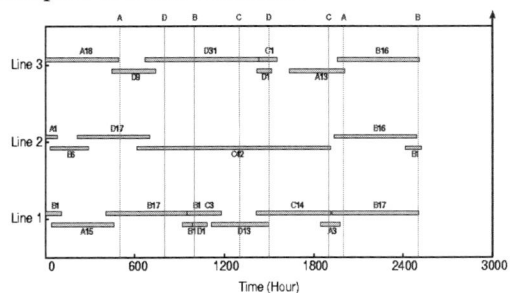

Figure 2. The best production plan in Case 1 when large due date penalties are charged on all products.

Figure 2. shows the production plan found. In this diagram, each horizontal bar represents a production campaign. The character and the following number above or below a production campaign indicate the product and the number of batches produced in that campaign, respectively. The vertical line shows a due date time for a product, whose label is shown on top of each due date line. The thicker vertical line at 3000 hours represents the production horizon. An inventory diagram corresponding to this production plan is shown in Fig. 3 in which the solid line is the inventory level of each product. The vertical line shows a due date for each product. The numbers at the top and bottom of a due date line represents the due date time and the amount due at that time, respectively. Note that the negative inventory level represents a shortage. For Fig. 3 there is no shortage throughout the production horizon.

Once all of the demands are satisfied on time, then the next target is to minimize the inventory cost. Figure 2 shows that in order to minimize the inventory cost, we should use all of the production lines simultaneously near the due dates of each product, which minimizes the duration that the finished products are stored as inventory.

For given product demands and due dates, it is often asked if we can get a feasible solution in which all the demands are produced on time. This is equivalent to deciding if a batch process will be overutilized or underutilized with the given demands and due dates. Simulated annealing is a good tool to check the utilization

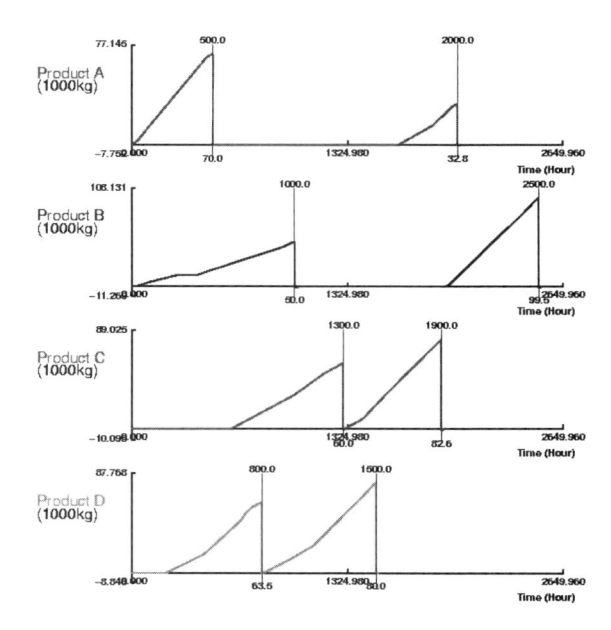

Figure 3. Inventory diagram shows that every product can be produced on time in Case 1.

of a batch process; as Fig. 2. shows, the process is fully utilized. This batch process produced a total of 583,400 kg of 4 products on time. The total capital inventory cost is $31,970.

Case 2: Large Setup Times and Large Setup Costs

For this example, all of the setup times for products are 10 times larger than those in Case 1 and, otherwise the same values for other parameters were used. Figure 4 shows the resulting plan. Because of the large setup times, there are a smaller number of campaign changeovers than in Case 1. Many changeovers will make some of products violate their due dates because the long setup times reduce the available production time within the horizon. To make up for the low time utilization and to reduce late production, the number of changeovers should be minimized. In the plan shown in Fig. 4., 4.65% (25,050 kg) of the demands are produced late.

The effect of setup costs is similar to the setup times in that it leads to a few long campaigns in the production plan. The production planning was also carried out with setup costs of $75,000 per changeover between all of the products. Reducing the number of changeovers and keeping a few long campaigns increase the inventory cost and the late production. In contrast, the total setup cost increases when many small campaigns are used. In the inventory diagram (not shown here for reasons of space) some of products are in shortage because longer campaigns are used to minimize the setup cost. In this case, 14.1% (76,140 kg) were not produced by the due dates and the capital inventory cost was increased by 20.05% ($38,380).

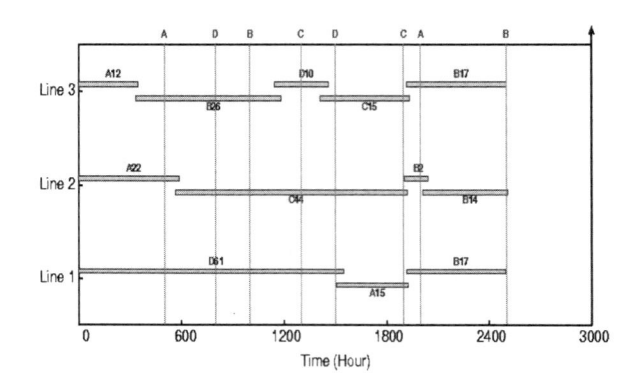

Figure 4. In Case 2, the number of changeovers is reduced because of longer setup times.

These two examples demonstrate the trade-off between the inventory cost and the setup cost or revenue loss during setup times. It is also shown that the effective capacity of a batch process changes according to its setup times and costs. The simulated annealing approach is useful to check variations in the process capacity under the given conditions.

Heuristics

For comparison, we tested the sequencing rules (Nahmias, 1997) that are the most frequently used for real production scheduling: *FCFS* (First Come First Serve), *SPT* (Shortest Processing Time), and *EDD* (Earliest Due Date). In FCFS, demands are processed in the sequence in which they are ordered. In this case, we consider that the sequence is A1, A2, B1, B2, C1, C2, D1, D2, where A1 and A2 are the demands of product A at its first and second due dates and so forth. With the SPT, we produce the product demands having the shortest processing time first and imagine that all orders are made in time to do this. In this example, the sequence by the SPT rule is A2, C1, B1, D1, A1, D2, C2, and B2. The EDD rule processes the demand in increasing order of their due dates. The sequence of demands by the EDD rule is A1, D1, B1, C1, D2, C2, A2, and B2.

The input data for Case 1 was used for this test and each demand in the sequence obtained by each rule is assigned to a single campaign on a production line by turns. The profits, *NPVs*, due date penalties and inventory costs for the solution obtained by the three rules and the best solution found by simulated annealing (Case 1) are summarized in Table 1. All the rules have much lower profits than the best solution of Case 1. The plan of FCFS produces all of the demands for product D very late and

Table 1. The Profits, NPVs, Due Date Penalties and Inventory Costs of Product Plans Obtained by Heuristics.

Heuristics	Profit ($10^6)	NPV ($10^6)	Due Date Penalty($10^6)	Inventory Cost ($10^5)
FCFS	-0.175	3.783	3.900	0.580
EDD	2.901	3.805	0.877	0.264
SPT	2.596	3.805	1.181	0.283
Simulated Annealing	3.771	3.804	0.000	0.320

also carries larger inventories of product A, B, and C than the other. The plans from EDD and SPT have the same *NPVs* as the best solution of simulated annealing and lower inventory costs. However, both of them produce some product demands very late. In contrast, the simulated annealing solution produces all of the demands on time but carries the more inventory than the EDD and SPT.

Conclusions

We have developed a new approximate algorithm for batch process planning using a simulated annealing technique. The algorithm seeks a feasible solution which satisfies all of the demands, meets the due dates from customers, and also maximizes a profit function including the net present value (*NPV*), the inventory cost, and the setup cost.

Case studies showed that setup times and setup costs change the effective capacity of a batch process. The simulated annealing algorithm can be used to estimate the capacity as well as the degree of utilization of the batch process. Comparing to typical heuristics, the simulated annealing algorithm found a much better solution than those of the heuristics. It wasn't shown here but a batch process-planning problem having intermediate products could be solved by this algorithm. For cases with a large number of due dates as well as a complicated product structure, the algorithm could find good feasible solutions even in the face of incompatibility between products and production lines (Lee, 1997).

References

Aarts, E. H. L. and P. J. M. Van Laarhoven (1985). Statistical Cooling: a General Approach to Combinatorial Optimization Problems. *Philips J. Res.*, **40**, 193-226.

Aggarwal, A., and J. K. Park (1993). Improved Algorithms for Economic Lot Size Problems. *Operations Research*, **41**, 549-571.

Birewar, D. B. and I. E. Grossmann (1990). Simultaneous Production Planning and Scheduling in Multiproduct Batch Plants. *Ind. Eng. Chem. Res*, **29**, 570-580.

Das, H., P. T. Cummings, and M. D. Levan (1990). Scheduling of Serial Multiproduct Batch Processes via Simulated Annealing. *Computers Chem. Engng*, **14**, 1351-1362.

Graves, S. C. (1981). A Review of Production Scheduling. *Operations Research*, **29**, 646-675.

Karmarkar, U. S. (1987). Lot Sizes, Lead Times and In-Process Inventory. *Management Science*, **33**, 409-423.

Kirkpatrick, S., C.D. Gelatt Jr., and M.P. Vecchi (1983). Optimization by Simulated Annealing. *Science*, **220**, 671-680.

Ku, H., and I. A. Karimi (1990). Scheduling in Serial Multiproduct Batch Processes with Due Date Penalties. *Ind. Eng. Chem. Res.*, **29**, 580-590.

Kudva, G., Elkamel, A., Pekny, J. F., and G. V. Reklaitis (1994). Heuristic Algorithm for Scheduling Batch and Semi-Continuous Plants with Production Deadlines, Intermediate Storage Limitations and Equipment Changeover Costs. *Computers Chem. Engng*, **18**, 859-875.

Lazaro, M., Espuna, A., and L. Puigjaner (1989). A Comprehensive Approach To Production Planning in Multipurpose Batch Plants. *Computers Chem. Engng*, **13**, 1031-1047.

Love S. F. (1979). *Inventory Control*. Chapter 8, McGraw-Hill, New York.

Lee, Y. G. (1997). A New Approach to Batch Chemical Process Planning. Ph.D. Thesis, University of Massachusetts, Amherst.

Nahmias, S. (1997). *Production and Operations Analysis*. Richard D. Irwin, a Times Mirror Higher Education Group, Inc., Chicago.

Malone, M. F. (1989). Batch Sequencing by Simulated Annealing. AIChE Annual Meeting, San Francisco, CA.

Miller, D. L. and J. F. Pekny (1991). Exact Solution of Large Asymmetric Traveling Salesman Problems. *Science*, **251**, 754-760.

Musier, R. F. H., and L. B. Evans (1989). An Approximate Method for the Production Scheduling of Industrial Batch Processes with Parallel Units. *Computers Chem. Engng*, **13**, 229-238.

Papageorgiou, L. G., and C. C. Pantelides (1996). Optimal Campaign Planning/Scheduling of Multipurpose Batch/Semi-Continuous Plants. 1. Mathematical Formulation. *Ind. Eng. Chem. Res*, **35**, 488-509.

Trigeiro, W. W., Thomas, L. J., and J. O. McClain (1989). Capacitated Lot Sizing with Setup Times. *Management Science*, **35**, 353-367.

Tricoire, B. (1992). Design and Scheduling of Multiproduct Batch Plants with Application to Polymer Production. Ph.D. Thesis, University of Massachusetts, Amherst.

AUCTION-DRIVEN COORDINATION FOR PLANTWIDE OPTIMIZATION

Rinaldo A. Jose and Lyle H. Ungar
University of Pennsylvania
Philadelphia, PA 19104

Abstract

Model predictive control strategies generally focus on controlling plant outputs to setpoints; in industry, however, a more desirable goal is maximizing a plant's profitability. In principle, this can be done by creating a plant model and maximizing profit with respect to the market prices of the plant's inputs and outputs, but in practice, such centralized approaches often cannot effectively be applied at the operations time scale due to the size and complexity of the problem. One solution is to use decentralized optimization at the unit operations level by tearing process streams and coordinating the resulting pieces. Such optimization, however, requires that unit inputs and outputs be priced. We show that a traditional Lagrangean-based approach to this pricing fails for simple systems. Instead, we define slack resources over the torn process streams and price them using auctions. Unlike Lagrange multipliers, slack resource prices contain useful information and can be used to make decisions regarding capital improvements, thus providing a strong tie between the operations and management layers in chemical plants.

Keywords

Auctions, Distributed optimization, Resource prices, Process decomposition, Optimal coordination, Penalty-based methods, Lagrangean decomposition.

Introduction

In principle, one can construct a detailed model of a plant and use it to centrally maximize the plant's profit by finding the "best" or most profitable setpoints for the process streams. In reality, such large-scale problems are quite difficult, and, at least for the foreseeable future, intractable, especially for large refineries. The assumption that one could assemble and use all the information from each unit is also fairly strong. Often, units are designed and run separately by different sets of people in different organizations; even the control systems are fine-tuned individually for each unit operation.

Plantwide optimization problems have a definite structure which can be exploited for decomposition: chemical plants are composed of separate units linked together by process streams. This component-wise structure can be used to attack these optimization problems, both by simplifying them and by solving them in their "natural form". In other words, unit operations are essentially separate, though interacting, pieces and often have separate control systems and operators; there are benefits to preserving this as much as possible.

One approach to problems with this type of structure is process decomposition followed by some form of penalty-based coordination (Findstein, 1980; Mesarovic, 1970; Morari, 1980). Here, one tears process streams and locally optimizes the units assuming they are all independent; this generates stream value proposals from each unit. If the stream value proposals across each tear are all the same, there is no need for coordination—the units' locally optimal solutions taken together are globally optimal as well. If, on the other hand, there are differences (which is most often the case), the units have conflicts, and the local solutions are not feasible. In order to generate feasible solutions, one applies a penalty to the units so that they alter their proposals. The goal of this approach is to reach a globally optimal, system-wide solution by repeatedly adjusting the applied penalty.

If the system-wide objective is to maximize profit, it is natural to view the applied penalties in terms of prices. Starting from an optimization perspective, one might try to apply some form of Lagrangean relaxation to the problem and use Lagrange multipliers as penalties to coordinate units. We show, however, that this is inadequate for the typical case where units in one part of a plant are indifferent to what happens in another part. We claim, instead, that starting with a clear definition of system resources is a better strategy. This leads to what we call a slack resource auction: a price-based penalty method that can optimally coordinate a broad class of systems including those for which Lagrangean-based methods are inadequate.

Problems with Lagrangean-based Coordination

Consider a plant consisting of two units, A and B, connected by a process stream (Fig. 1). Each unit has a set of local constraints and a local objective or utility function. In addition, suppose each unit is indifferent to the stream's state except for the temperature. After decomposing this system by tearing the stream (Fig. 2), the goal of the plant manager is to solve the following problem:

$$\text{Maximize } u_A(T_A) + u_B(T_B)$$
subject to :
$$T_A = T_B \quad (1)$$
A's constraints
B's constraints

where T_A and T_B are the proposed stream temperatures coming from Unit A and going into Unit B, and where u_A and u_B are their utility functions. Applying Lagrangean relaxation to the linking constraint moves it out of the constraint set and introduces it into the objective as a soft constraint:

$$\text{Maximize } u_A(T_A) + u_B(T_B) - \lambda(T_A - T_B)$$
subject to :
A's constraints
B's constraints
$$(2)$$

This suggests the following decomposition:

$$\text{Maximize } u_A(T_A) - \lambda T_A \qquad \text{Maximize } u_B(T_B) + \lambda T_B$$
subject to :   subject to: (3)
A's constraints   B's constraints

The Lagrangean-based procedure is then

1. Set λ to zero.
2. Have subproblems submit their proposed temperatures T_A and T_B.
3. If T_A and are T_B sufficiently close, stop.
4. If $T_A > T_B$, increase λ and go to Step 2.
5. If $T_A < T_B$, decrease λ and go to Step 2.

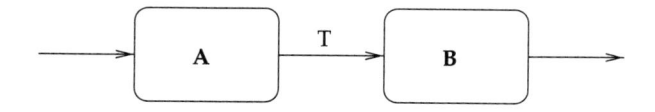

Figure 1. Simple system.

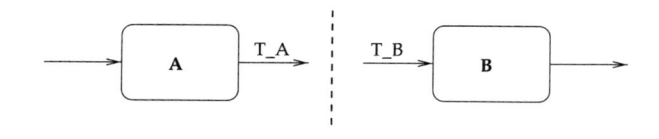

Figure 2. Decomposed system.

Although the Lagrangean-based method seems reasonable, it is unable to coordinate a system in which one of the units is indifferent to the temperature.

Example 1: One Indifferent Unit

Consider the pair of units shown in Fig. 2, and suppose that their utility functions over the stream temperature are defined by Fig. 3. Because B is indifferent to the temperature, the system-wide solution will lie at A's locally optimal point of 30 °F. Applying a Lagrangean-based decomposition results in the system described by Eqn. 3. The goal of this method is to vary λ until $T_A = T_B$ at the system solution. This approach, however, does not converge (see Fig. 4).

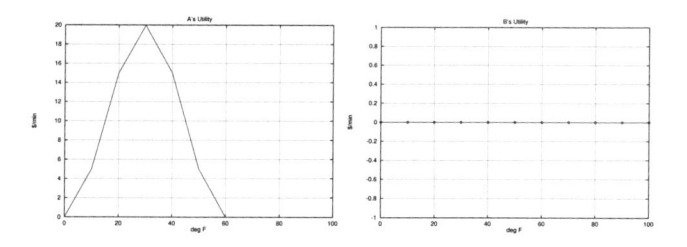

Figure 3. Utility functions of units A and B.

We can explain this failure as follows. At the beginning of the procedure, λ is set to zero and the agents submit their stream temperature requests. While A submits its locally optimal temperature, B is indifferent and simply picks one of its own feasible points. In general, the proposed temperatures will differ and so λ

will be adjusted. If λ becomes positive, then Unit B will select its maximum feasible temperature according to

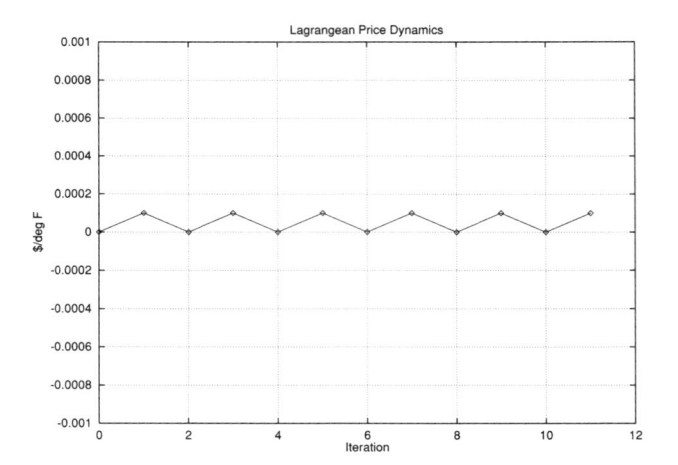

Figure 4. Failure of Lagrangean procedure to converge.

Eqn. 3; likewise, if λ becomes negative, Unit B selects its minimum feasible temperature—these are the only temperatures that B will ever propose. Because, in general, neither of these will match A's locally optimal temperature, the method will not converge.

There are other problems with Lagrangean-based methods. Not only are Lagrangean methods unsuitable for coordinating systems where some of the units are indifferent, they are also unable to coordinate systems where at least one of the units has a piecewise linear utility function—the indifferent unit being a special case of this. In addition, prices in one part of a plant cannot be meaningfully compared with prices in another part. Thus, the Lagrange multipliers provide no useful information with regard to making capital investments. Consider, for instance, a plant manager who has the option of placing a heat exchanger somewhere in the plant. If temperature prices could be meaningfully compared, the manager would place the heat exchanger at the point where the system would benefit the most—at the stream with the highest temperature price. Lagrangean-based methods cannot help here.

A Better Approach to Coordination: Slack Resource Auctions

The problems with Lagrangean-based coordination stem directly from failing to define precisely what the system resources are. In Ex. 1, the system resource appears to be the stream temperature. This is awkward at best. For instance, what does it mean to have a supply of temperature? In principle, this resource is unbounded—the units can select any stream temperature they both agree to. In order to construct an effective coordination mechanism with meaningful prices, it is crucial that we define meaningful resources.

System resources, in the usual sense, are quantities like steam, electricity, or cooling water. These resources have a definite supply, and there is a definite way of dividing them between units. From the system's point of view, divisible resources can always be expressed in the form

$$D_1(\mathbf{x}_1) + D_2(\mathbf{x}_2) + \cdots + D_n(\mathbf{x}_n) \leq S \quad (4)$$

where x_i is Agent i's decision variable vector, D_i maps Agent i's activities into a demand for system resources, and S is the supply. The left-hand-side of this constraint represents the aggregate demand for resources, and each of its terms is one of the units' individual demands.

There are system resources, however, that are not divisible. The temperature of a linking stream, for instance, cannot be divided between units; it is an indivisible value which must be shared. The associated linking constraints have the form

$$T_A = T_B \quad (5)$$

One way of re-expressing this constraint so that it looks like a divisible resource constraint is

$$|T_A - T_B| \leq S \quad (6)$$

or

$$\tfrac{1}{2}|T_A - T_B| + \tfrac{1}{2}|T_A - T_B| \leq S \quad (7)$$

where each term on the left-hand-side of Eqn. 7 represents one of the units' demand for the slack in temperature, and S is the supply of "temperature slack" or the system's tolerance for a difference in the temperatures of the torn linking stream. Taking the temperature slack to be our resource leads to the following decomposition of the above problem:

$$\begin{array}{ll} \text{Max } u_A(T_A) - p|T_A - T_B^0| & \text{Max } u_B(T_B) - p|T_A^0 - T_B| \\ \text{subject to:} & \text{subject to:} \\ \quad A\text{'s constraints} & \quad B\text{'s constraints} \end{array} \quad (8)$$

where T_A^0 and T_B^0 are parameters from the auctioneer defining the "slack poles" of the stream temperature slack, and p is the slack resource price. One finds the optimal price by applying the following slack resource auction procedure:

1. Decompose the system by breaking the appropriate linking streams.
2. For each relevant stream variable x which joins Units i and j, define the slack resource R_x.
3. For each slack resource R_x, define a price p_x and a slack supply S_x.

4. Set all prices to zero.
5. Request proposed stream states from units.
6. For each slack resource R_x, compute the slack demand D_x and the excess demand $E_x = D_x - S_x$.
7. For each resource, if the excess demand is negative, lower the price and go to Step 5.
8. For each resource, if the excess demand is positive, raise the price and go to Step 5.
9. If the excess demand is zero for all slack resources, stop.

Example 2: One indifferent unit revisited

Consider the system of units described in Example 1 and apply a slack resource auction to it by introducing $|T_A - T_B|$ as a resource. This yields a coordination problem in the form of Eqn. 4. Unlike the Lagrangean procedure, the slack resource auction converges to the system-wide solution at an equilibrium price of 0 (see Fig. 5). Since prices measure conflict, this is correct—Unit B is indifferent to the temperature, and so there is no conflict at 30°F.

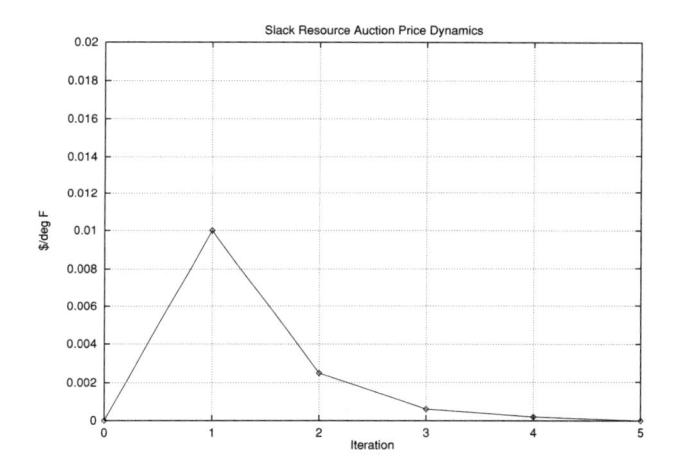

Figure 5. Slack resource auction price dynamics.

Example 3: Three Units with Recycle

Now consider a slightly more complicated system consisting of three units joined by process streams and a recycle loop (Fig. 6). For the purposes of discussion, suppose that each linking stream has a product stream draw which defines the value of that stream as a function of temperature (Fig. 7).

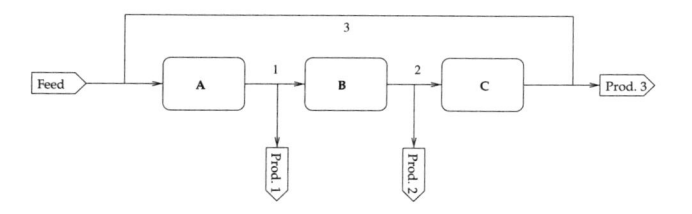

Figure 6. Three units with recycle.

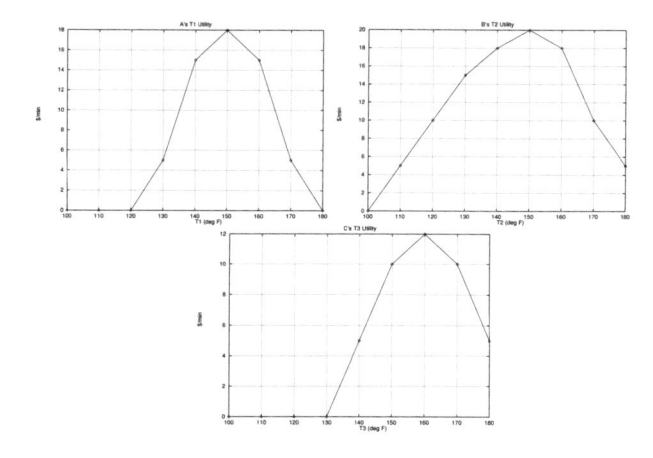

Figure 7. Product stream values.

Suppose further that each unit has an ideal temperature drop between its inlet and outlet for which there is no cost of heating or cooling. This ideal temperature drop and the cost of heating and cooling for each unit are shown in Table 1. These parameters give rise to utility functions, for each unit, over the inlet and outlet stream temperatures (Fig. 8).

Table 1. Heating Parameters.

Unit	Ideal ΔT (°F)	Heat Cost ($/°F)	Cool. Cost ($/°F)
A	10.0	0.10	0.10
B	10.0	0.20	0.20
C	20.0	0.15	0.15

The goal of this problem is to determine the triple $(T_1; T_2; T_3)$ which maximizes the sum of the units' utility functions. Following the slack resource auction procedure, one tears the process streams, defines a slack temperature resource over each of them, and adjusts prices based on demand. The final, system-wide optimal result is

Stream	Temp. (°F)	Slack Price ($/°F)
Stream 1	150.0	0.41
Stream 2	150.0	0.30
Stream 3	160.0	0.20

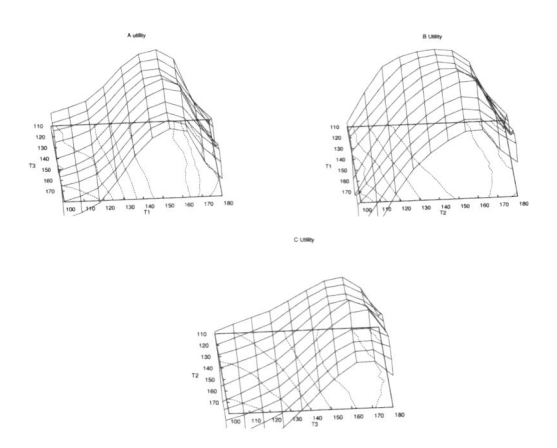

Figure 8. Unit's utility functions over inlet and outlet temperatures.

Discussion

It is natural to ask what the relationship is between the equilibrium prices of a slack resource auction and the dual prices from the original centralized problem. There is a well-known market-based interpretation of the primal and dual problems in optimization: given a primal problem, the dual problem can be viewed as a "malevolent market" whose goal is to drive the net profit of the primal problem to as low a value as possible by manipulating resource prices (Bradley, 1977). This malevolent market is assumed to be complete—composed of an infinite number of infinitely varied agents. Now consider a plant decomposed into separate units. The market here is made up of only a finite number of unit operations. Therein lies the difference between dual prices and auction prices: coordinating auction prices result from an imperfect market. Auction prices are therefore generalizations of dual prices. As the number of units increases, auction prices will approach the corresponding dual prices. This same difference can be seen when coordinating LPs using Dantzig-Wolfe decomposition—the prices generated by the master problem are never equal to the dual prices of the corresponding constraints.

Slack resource auctions are not limited to the simple coordination problems treated here. In particular, they can coordinate several unit operations with recycle and units with certain types of non-convex utility functions. With a slight modification to the pricing, we can guarantee that slack resource auctions will always coordinate units with concave objectives and convex feasible regions. The idea is this: augment the amount normally charged each unit, $p|T_A - T_B|$, with a small quadratic term, $q(T_A - T_B)^2$, where q is some small positive constant. The existence of an optimal resource price vector can be demonstrated (Jose, 1997) by casting this augmented slack resource auction as a nonlinear complementarity problem:

Find p such that:
$$-\Delta(p;q) \geq 0 \quad , \quad (9)$$
$$p^T(-\Delta(p;q)) = 0$$

where $\Delta(p;q)$ is the excess demand vector and p is the slack price vector.

The price adjustment mechanism we have described here is very simple. Viewed from a control perspective, we are essentially using PI controllers to drive the demand for slack resources to the level of their supply. This works well for simple examples but can be slow for larger systems. Faster convergence may be obtained by viewing the price adjustment procedure as following a homotopy path along some implicitly defined demand curve towards a point where the excess demand is zero. With appropriate normalization of the resource prices, standard fixed-point algorithms might also prove useful in adjusting price vectors in a more sophisticated and more efficient manner.

Conclusion

If the plantwide objective is to maximize profit and if it is necessary to decompose a chemical plant into pieces, it is natural to consider price-based penalty methods for coordinating the units. Standard Lagrangean-based approaches are ineffective not only because the Lagrange multipliers lack meaningful interpretation as prices, but also because they fail to coordinate simple, but common, systems. Defining meaningful resources suggests slack resource auctions as an alternative which yield meaningful prices and, more importantly, are able to coordinate systems for which Lagrangean auctions are unsuitable.

When we move beyond plantwide optimization to *inter-plant* optimization all of the arguments for decomposition-based strategies, and hence auction-based coordination, are strengthened. Inter-plant optimization problems are much more complex than those for a single plant. Accessing and retrieving complete information from each plant is practically impossible. Slack resource auctions, however, can be applied to inter-plant optimization in the same way that they were applied to plantwide optimization. Slack resource prices have the same interpretation as before, and under the convexity conditions stated above, there will be some set of prices which optimally coordinates all the plants in the system.

Although the focus on coordination was motivated by the difficulty of solving plantwide optimization problems centrally, there are other benefits. Prices from a slack resource auction provide plant managers with useful information regarding the "hot spots" or bottlenecks in a plant. Slack prices not only help identify bottlenecks, but they also give a measure of their severity, or equivalently, a measure of how much the system as a whole would benefit by allowing more slack there. For instance, when dealing with stream temperatures, slack resource prices can help answer the question, "How much would we benefit by adding a heat exchanger there?" Seeing where these bottlenecks are and being able to evaluate their seriousness are important tools for tying together the operations and management layers in chemical plants.

References

Bradley, S. P., Hax, A. C., and Magnanti, T. L., 1977. *Applied Mathematical Programming.* Addison-Wesley.

Findstein, W., Bailey, F., Brdys, M., Malinowskip, K., Tatjewski, P., and Wozniak, A., 1980. *Control and Coordination in Hierarchical Systems.* vol. 9 of *International Series on Applied Systems Analysis.* John-Wiley and Sons.

Jose, R. A., Harker, P. T., and Ungar, L. H., 1997. Auctions and Optimization: Methods for Closing the Gap Caused by Discontinuities in Demand. *In: Third International Conference on Computing in Economics and Finance, Stanford.*

Mesarovic, M., Macko, D., and Takahara, Y., 1970. *Theory of Hierarchical Multilevel Systems.* vol. 68 of *Mathematics in Science and Engineering.* Academic Press.

Morari, M., Arkun, Y., and Stephanopoulos, G., 1980. Studies in the Synthesis of Control Structures for Chemical Processes—Part 1. *AIChE Journal,* **26**, 2, 220-231.

RELIABILITY AND MAINTENANCE CONSIDERATIONS IN PROCESS OPERATIONS UNDER UNCERTAINTY

Efstratios N. Pistikopoulos and Constantinos G. Vassiliadis
Department of Chemical Engineering
Centre for Process Systems Engineering
Imperial College, London SW7 2BY, U.K.

Abstract

In recent FOCAPO Meetings the importance of reliability and maintenance in process operations has been emphasized. This has motivated the need for the development of contemporary techniques and tools for availability assessment of process systems, which go beyond traditional practices, by focusing on the interactions of reliability and maintenance optimization with the detailed process operation in its dynamic, continuously changing environment. This work describes recent research developments towards this objective - in particular, we propose a formal optimization based framework, which links maintenance scheduling to process operation in the presence of uncertainty.

Keywords

Maintenance optimization, Process operation, Process uncertainty.

Introduction

Uncertainty in a chemical process is, widely, classified in two main categories: *continuous,* (feedstream availability, kinetic constants, prices, etc.) and *discrete* (equipment availability). The need for integrating both types of uncertainty in the study and optimization of process operation has been repeatedly pointed out (Van Rijn, 1987; Grievink et al., 1993; Thomaidis and Pistikopoulos, 1994; Pistikopoulos, 1995) and general mathematical frameworks have started to appear (Straub and Grossmann, 1990; Thomaidis and Pistikopoulos, 1995). However, maintenance considerations in the presence of uncertainty have not been explicitly accounted for. On the other hand, typical studies on maintenance optimization (Alkhamis and Yellen, 1995; Dedopoulos and Shah, 1995; Tseng, 1996; Tan and Kramer, 1997) do not take into account the interactions between maintenance planning/scheduling and process operation under uncertainty. This work presents a formal mathematical framework which links maintenance optimization to process operation under uncertainty. The specific problem addressed in this study is as follows.

Given:

- a mathematical model, describing the process system
- the reliability and maintenance characteristics of the plant equipment components
- the type and cost of different available maintenance policies
- quantitative information regarding the uncertainty involved in process parameters

the goal is to determine an optimal maintenance policy, i.e. the one which maximizes plant profitability over a specified time horizon.

Mathematical Foundations and models

Consider the chemical complex shown in Fig. 1, for the production of chemical C from chemical A. Such a system involves four units and sixteen system states. A system state is defined as one combination of the possible operating conditions (up or down) of all system components. The model that describes the process is given in Table 1. By defining a function $r(x_k, \theta)$ describing the revenue generated per unit time by each state as a function of process variables, (x), and uncertain parameters, (θ), the expected revenue rate ERR_k of a state k is given by

$$ERR_k = \int_{\theta \in FOR_k} \max_{x_k} r(x_k, \theta) j(\theta) d\theta \quad \text{(P1)}$$

s.t.

$$h_k(x_k, \theta) = 0, \forall k \in S$$

$$g_k(x_k, \theta) \leq 0, \forall k \in S$$

where

- h_k, g_k are the vectors of equality and inequality constraints for each state k of the process
- $\theta \in T$ is the vector of continuous uncertain parameters, assumed to follow a continuous probability distribution function $j(\theta)$ and $T = \{\theta, \theta^l \leq \theta \leq \theta^u\}$
- S is the set of operable states of the system.

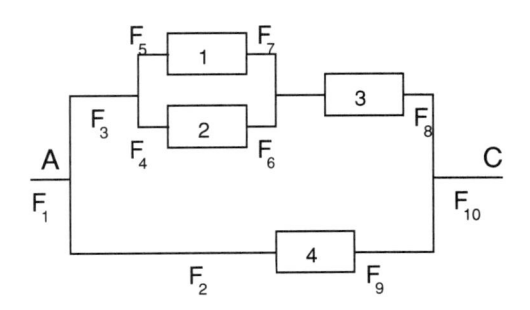

Figure 1. Chemical complex with four units.

Table 1. Model Equations.

Mass Balance	Specifications
$F_1 = F_2 + F_3$	$F_5 \leq 5$
$F_3 = F_4 + F_5$	$F_4 \leq 5$
$F_{10} = F_8 + F_9$	$(F_6 + F_7) \leq 7$
$F_7 = \alpha_1 \cdot F_5$	$F_2 \leq 9$
$F_6 = \alpha_2 \cdot F_4$	$F_1 \leq S$
$F_8 = \alpha_3 \cdot (F_6 + F_7)$	$F_{10} \geq D$
$F_9 = \alpha_4 \cdot F_2$	

Then, the expected revenue of a multistate system over a time horizon H is given by

$$ER = \int_H \sum_{k \in S} P_k(t) \cdot ERR_k \, dt \quad (1)$$

where $P_k(t)$ is the probability of occurrence of state k as a function of time. In general, the state probabilities can be evaluated as a function of time when the availability characteristics of the units are given as a function of time. Note, from Eqn. (1), that the probability of the system being in each state k of the system at any specific time t, $P_k(t)$, is determined by the probability of each component j being up (or down) at time t which in turn depends on the implemented maintenance policy. This suggests that a model is required to describe the probability of each component being up as a function of time *during and after maintenance actions,* which is based on a maintenance policy model. This is presented next.

Maintenance Policy Model

Consider a system with M components. Let $R_{1,j}(t)$ be the initial (before any maintenance actions are performed) reliability function of component j and $t_i, \forall i \in [2 \cdots N-1]$ the time instants at which we perform maintenance to any of the components, according to a maintenance schedule, while t_1 and t_N are assigned the values of 0 and H respectively. The maintenance policy in the time horizon (t_1, t_N) is based on the following assumptions:

- each time maintenance is performed, it is performed on only one of the units in the following way:
1. *corrective*, if the unit is down
2. *preventive*, if the unit is up

    This assumption is common in the literature (see, for example, Tseng (1996)) and valid in many real cases.
- all maintenance is of an AGAN (as good as new) type
- all failure events are independent

Suppose that the system is in the interval between maintenance actions t_{i-1} and t_i and let $A_{i-1,j}(t)$ denote the availability function of unit j (i.e. the probability of unit j being up) during this interval. We introduce two types of optimization variables: t_i, which are *continuous variables*, representing the time instant of maintenance action $i-1$, and $u_{i,j}$, which are *0-1 variables* taking the value of 1 if maintenance at time t_i is performed to component j, or 0 if maintenance at t_i is performed to a component other than j. The availability of component j after t_i will then be

$$A_{i,j}(t) = (1-u_{i,j}) \cdot A_{i-1,j}(t) + u_{i,j} \cdot A_{1,j}(t-t_i') \quad (2)$$

where $t_i' = t_i + \tau_i$, τ_i being the expected duration of maintenance action $i-1$. Equation (2) describes the following two cases:

- $u_{i,j} = 0$: maintenance is performed to a component other that j and, hence, j retains the same availability characteristics as in the previous interval
$$A_{i,j}(t) = A_{i-1,j}(t)$$
- $u_{i,j} = 1$: maintenance (corrective or preventive) at t_i is performed to j, which according to our assumptions, becomes AGAN
$$A_{i,j}(t) = A_{1,j}(t-t_i')$$

From Eqn. (3) we can, recursively, establish analytical expressions for the availability functions of all units after each maintenance action and, an a similar way, during each maintenance action. The availability of unit j after the $(\lambda-1)^{th}$ maintenance action, $A_{\lambda,j}(t)$, is given by

$$A_{\lambda,j}(t) = \sum_{k=1}^{\lambda} [(\prod_{i=k+1}^{\lambda}(1-u_{i,j})) \cdot u_{k,j} \cdot R_{1,j}(t-t_k')] \quad (3)$$

while the availability of a unit being up during the $(\lambda-1)^{th}$ maintenance action, $\overline{A}_{\lambda,j}(t)$, is given by

$$(1-u_{\lambda,j}) \cdot \sum_{k=1}^{\lambda-1} [(\prod_{i=k+1}^{\lambda-1}(1-u_{i,j})) \cdot u_{k,j} \cdot R_{1,j}(t-t_k')] \quad (4)$$

With units failing independently, the probability of being in state k of the system is given by

$$P_k(t) = \prod_{i \in OP} A_i(t) \cdot \prod_{j \notin OP}(1-A_j(t)) \quad (5)$$

where $A_{i(t)}$ is the availability of unit i at time t and OP is the set of operable units in state k. From Eqns (3)-(5) analytical expressions can be found, describing the probability of system being in a particular state k at time t as a function of the maintenance policy. Furthermore, the *expected preventive maintenance cost* is given by

$$\cos t_{\text{preventive}} = \sum_{\lambda=2}^{N-1} \sum_{j=1}^{M} u_{\lambda,j} \cdot A_{\lambda-1,j}(t_\lambda) \cdot C_{prev,j} \quad (6)$$

where $C_{prev,j}$ is the preventive maintenance cost of unit j, and the *expected corrective maintenance cost* is given by

$$\cos t_{\text{corrective}} = \sum_{\lambda=2}^{N-1} \sum_{j=1}^{M} u_{\lambda,j} \cdot (1-A_{\lambda-1,j}(t_\lambda)) \cdot C_{corr,j} \quad (7)$$

where $C_{corr,j}$ is the corrective maintenance cost of unit j. Similarly, the expected duration of maintenance task $\lambda-1$ is given by

$$\tau_\lambda = u_{\lambda,j} \cdot [(1-A_{\lambda-1,j}(t_\lambda)) \cdot \tau_{corr,j} + A_{\lambda-1,j}(t_\lambda) \cdot \tau_{prev,j}] \quad (8)$$

where $\tau_{prev,j}$, $\tau_{corr,j}$ are the durations of preventive and corrective maintenance tasks, respectively, for component j (see Vassiliadis and Pistikopoulos (1998a) for a more detailed analysis).

Optimization Model

Based on the previous analysis and on the expressions derived, the problem of identifying the optimal maintenance policy of a process operating in the presence of uncertainty is described by the following optimization model.

$$\max_{u_{\lambda,j},t_\lambda,N} \sum_{\lambda=1}^{N-1} \int_{t_\lambda+\tau_\lambda}^{t_{\lambda+1}} \sum_{k \in S} P_{\lambda,k}(t) \cdot ERR_k \, dt$$
$$+ \sum_{\lambda=2}^{N-1} \int_{t_\lambda}^{t_\lambda+\tau_\lambda} \sum_{k \in S} \overline{P}_{\lambda,k}(t) \cdot ERR_k \, dt \quad (P2)$$
$$-m.\cos t(u_{\lambda,j},t_i,N)$$

s.t.
$$\sum_{j=1}^{M} u_{\lambda,j} = 1, \forall \lambda \in [2 \cdots N-1] \quad (9)$$
$$t_\lambda + \tau_\lambda \leq t_{\lambda+1}, \forall \lambda \in [1 \cdots N-1] \quad (10)$$

N is an integer variable denoting that the total number of maintenance actions is $N-2$. The third term of the objective function represents the total expected maintenance cost and is the sum of expressions (6) and (7). Equation (9) states that only one maintenance action is performed at a time, while Eqn. (10) states that maintenance action $\lambda-1$ occurs before maintenance action λ. The objective function also consists of two terms, one describing the revenue generated between maintenance actions and another for the revenue generated by the process during maintenance actions ($\overline{P}_{\lambda,k}(t)$ is the probability of the system being in state k during the $(\lambda-1)^{th}$ maintenance action). Note that both uncertainty and maintenance characteristics are incorporated in the objective function. Reliability and maintenance characteristics directly determine the probability of the system being in a particular state k and, on the other hand, uncertainty influences the expected revenue rate of this state.

Problem (P2) is a mixed-integer nonlinear optimization formulation which, additionally, involves integral terms with integrands defined implicitly, as a function of 0-1 variables. In the next section, we propose an efficient two-step approximate solution strategy, which avoids the direct solution of (P2).

Maintenance Optimization Algorithm

Step1: Determining an Initial Sequence

In the first step of the algorithm we assume that we perform maintenance to each component periodically, every t_j time units. The availability function of component j after the nth maintenance action is given by

$$A_j(t) = R_j(t - n \cdot t_j), \quad (n-1) \cdot t_j \leq t \leq n \cdot t_j, \quad \forall j. \quad (11)$$

These maintenance time instants $t_j, 2 \cdot t_j, \cdots, n \cdot t_j$ correspond to a specific (yet unknown) availability value for component j, $A_{th,j}$, which is called the *availability threshold value of j*. Therefore, an alternative way to identify the optimal maintenance time instants for each component and, hence, the sequence and number of maintenance actions is to determine a set of optimal availability threshold values that maintain each component availability, in average, to such a level that maximizes process profitability. This is illustrated in Fig. 2, for a two component system. Whenever the availability of the first unit falls below its threshold value $A_{th,1}$, maintenance is performed to the first unit and the same holds for the second unit. In essence, it is the availability thresholds that determine the number of maintenance actions (three for unit one, six for unit two) and their sequence (212212212). These constant (over time) threshold values are determined from the solution of an optimization problem derived from (P2) by expressing all the terms as functions of the threshold values (Vassiliadis and Pistikopoulos, 1998b). In (P3) the availability levels of each component are optimized so as to maximize the expected profit and balance the trade-offs between process availability and maintenance costs.

$$\max_{A_{th,j}} \sum_{k \in S} H_k(A_{th,j}) \cdot \Pr_k(A^*_j) \cdot ERR_k$$

$$-m.\cos t(A_{th,j}) \quad (P3)$$

$$\text{s.t.} \quad 0 \leq A_{th,j} \leq 1, \quad \forall j \in [1 \cdots M]$$

where

- $H_k(A_{th,j})$ is the maximum amount of time that the system can spend in state k as a function of the availability thresholds. This is equal to the time horizon H subtracting the expected duration of maintenance tasks for this state.

- A^*_j is the average availability of component j in the time period to the first maintenance action t_j. Since periodic maintenance has been assumed, A^*_j corresponds to the average availability for the whole time horizon and is given by (Lewis, 1996)

$$A^*_j = \int_0^{t_j} R_j(t) dt$$

where $R_j(t)$ is the reliability function of component j.

- $\Pr_k(A^*)$ is the probability of the system being in state k as a function of the average availability of the components.

- $m.\cos t(A_{th,j})$ is the expected maintenance cost expressed as a function of the availability threshold values.

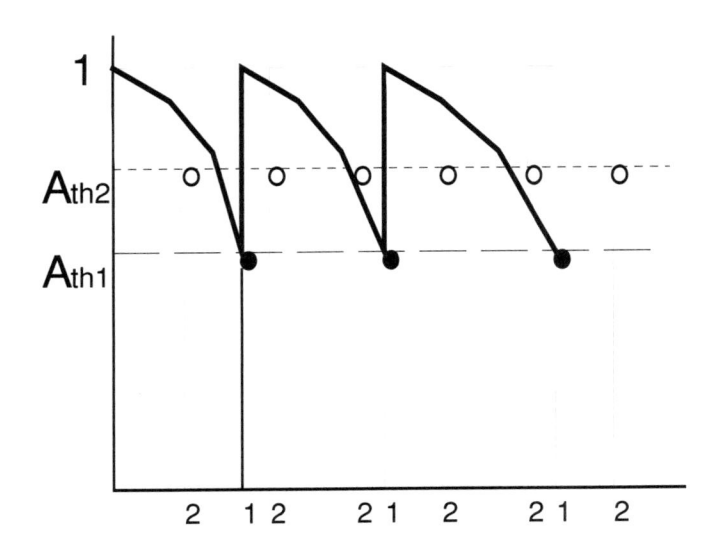

Figure 2. Availability thresholds.

In the case of exponentially distributed reliability functions for the components,

$$R_j(t) = e^{-\frac{t}{\mu_j}}$$

where μ_j denotes the mean time to failure of component j, problem (P3) is recasted as

$$\max_{A_{th,j}} \sum_{k \in S} H_k(A_{th,j}) \cdot \Pr_k(A^*_j) \cdot ERR_k$$

$$-m.\cos t(A_{th,j}) \qquad \text{(P3a)}$$

$$\text{s.t.} \quad 0 \le A_{th,j} \le 1, \quad \forall j \in [1 \cdots M]$$

$$m.\cos t(A_{th,j}) = -\sum_j \frac{H}{\mu_j \cdot \ln(A_{th,j})} \cdot$$
$$\cdot \left((1 - A_{th,j})C_{corr,j} + A_{th,j}C_{prev,j}\right), \quad \forall j$$

$$H_k(A_{th,j}) = H + \sum_{j \in OP} \frac{H}{\mu_j \cdot \ln(A_{th,j})} \cdot \tau_j, \quad \forall j,k$$

$$\tau_j = (1 - A_{th,j}) \cdot \tau_{corr,j} + A_{th,j} \cdot \tau_{prev,j}, \quad \forall j$$

$$\Pr_k(A^*) = \prod_{i \in OP} A^*_i \prod_{j \notin OP}(1 - A^*_j), \quad \forall j$$

$$A^*_j = \frac{\mu_j}{t_j}\left(1 - e^{-\frac{t_j}{\mu_j}}\right), \quad \forall j$$

where τ_j is an approximation of the expected duration of maintenance tasks performed to component j. Should the reliability characteristics of the components be described by other distributions (e.g. Weibull), problem (P3) is modified accordingly.

Problem (P3a) is a standard nonlinear mathematical programming problem the solution of which provides the optimal number of inspections $\overline{N}$ and an initial maintenance sequence $\overline{u}_{\lambda,j}$.

Step 2. Finding an optimal solution

1. With fixed $\overline{N}$ and $\overline{u}_{\lambda,j}$, problem (P2) reduces to a standard nonlinear optimization problem (P4):

$$\max \sum_{t_\lambda}^{\overline{N}-1} \int_{t_\lambda+\tau_\lambda}^{t_{\lambda+1}} \sum_{k \in S} P_{\lambda,k}(t) \cdot ERR_k \, dt$$

$$+ \sum_{\lambda=2}^{\overline{N}-1} \int_{t_\lambda}^{t_\lambda+\tau_\lambda} \sum_{k \in S} \overline{P}_{\lambda,k}(t) \cdot ERR_k \, dt \qquad \text{(P4)}$$

$$- m.\cos t(\overline{u}_{\lambda,j}, \overline{N}, t_\lambda)$$

$$\text{s.t.} \quad t_\lambda + \tau_\lambda \le t_{\lambda+1}, \forall \lambda \in [1 \cdots N-1]$$

The solution of (P4) yields the exact maintenance time instants t_λ, $\forall \lambda \in [2 \cdots N-1]$.

2. If some of the equality constraints in (P4) are active, i.e. $t_\lambda + \tau_\lambda = t_{\lambda+1}$, we go back to step 2.1 and check whether changing the sequence of the maintenance actions involved in this constraint will improve the solution of (P3).

3. If there are no active constraints or no better solution is obtained, stop.

Note that the proposed algorithm eliminates the combinatorial complexity of the problem, offering enough hope for the solution of large scale industrial problems.

Example

Table 2. *Reliability and Operating Characteristics.*

Unit	MTTF	Cprev	Ccorr	a
1	5000	15000	50000	.92
2	5000	16000	56000	.9
3	8000	18000	48000	.85
4	13000	20000	60000	.75

Table 3. *Expected Revenue Rates.*

State	ERR_k
1111	1147
0111	1125
1011	1133
1101	343
1110	114
0011	343
0101	343
1001	343
0001	343

The chemical complex shown in Fig. 1 converts species A to C, either through the production of the intermediate product B or by direct conversion of A to C. The supply of raw material A and the demand of product C are considered as continuous uncertain parameters, described by normal pdf's, with mean value of supply $S^N = 12$ and demand $D^N = 7$ and a standard deviation $\sigma_S = \sigma_D = 1$. The cost of raw material is 40 per unit and the profit from end product is 200 per unit. The model for this process is shown in Table 1. The operating, reliability and maintenance characteristics are provided in Table 2. The NLP formulations were solved using GAMS modelling package (Brooke et al., 1988). A time horizon of H=2000 units has been considered. The integration in order to evaluate the expected revenue rate of each system state (problem (P1)), is achieved using a Gaussian integration scheme, similar to the one presented in Acevedo and Pistikopoulos (1998). The expected revenue rates are summarized in Table 3. In this process, there are nine states with a non-zero revenue rate.

The objective is to identify the optimal maintenance policy, i.e. the optimal number of maintenance actions required, the optimal sequence of actions and the exact maintenance time instants.

Step 1: The optimal availability thresholds for each component are first determined (from the solution of (P3a)), with which the number of maintenance actions required and the corresponding maintenance sequence can be derived, as shown in Table 4.

Table 4. Results of Step 1.

Unit	A_{th}	Instants (initial values)	M. Actions
1	.774	1280	1
2	.748	1453	1
3	.916	699,1398	2
4	.939	821,1642	2

Note that unit 4 has the largest availability threshold, which means it has to be maintained more frequently. This is expected, since it 'participates' in more profitable states of the process.

Step 2: From the initial maintenance time instants obtained in Step 1, the initial maintenance sequence is (3,4,1,3,2,4). By fixing this sequence, the solution of (P4) provides the optimal maintenance instants, as shown in Table 5. At the solution, the sequence constraint involving the 4th and the 5th maintenance actions becomes active. This is an indication that the 5th maintenance action should have taken place before the 4th. Indeed, by reversing the order of these two actions and solving (P4) again we obtain the optimal maintenance policy.

Table 5. Results of Step 2.

It.	Sequence	Instants	Profit (10E6)
1	3,4,1,**3,2**,4	642,649,749,**1262,1262**,1303	2.05
2	3,4,1,2,3,4	661,667,725,1086,1313,1319	2.06

Conclusions

The maintenance optimization algorithm, presented in this paper, identifies an optimal maintenance policy (number of maintenance actions required per unit, their sequence and the exact maintenance instants) for processes operating in the presence of uncertainty. Its application to a small process example highlights the important interactions between reliability, maintenance and uncertainty characteristics.

References

Acevedo, J. and E.N. Pistikopoulos (1998). Stochastic optimization based algorithms for process synthesis under uncertainty. *Computers Chem. Engng.*, in print.

Alkhamis, T.M. and J. Yellen (1995). Refinery units maintenance using integer programming. *Applied Mathematical Modelling*, **19**, 543.

Brooke, A., D. Kendrick and A. Meeraus (1988). *GAMS: A User Guide...*, Redwood City, CA.

Dedopoulos, I.T. and N. Shah (1995). Optimal short-term scheduling of maintenance and production for multipurpose plants. *Industrial & Engineering Chemistry Research*, **34**, 192

Grievink, J.K., R.D. Smit and C.F.H. Van Rijn (1993). Managing reliability and maintenance in the process industry. *Conf. On Foundation of Computer Aided Process Operations, FOCAP-O*

Lewis, E.E. (1996). *Introduction to reliability engineering*. John Wiley and Sons, Inc.

Pistikopoulos, E.N. (1995). Uncertainty in process design and operations. *Computers Chem. Engng.*, **19**, 553.

Straub, D.A. and I.E. Grossmann (1990). Integrated stochastic metric of flexibility with discrete states and continuous uncertain parameters. *Computers Chem. Engng.*, **29(3)**, 967.

Straub, D.A. and I.E. Grossmann (1993). Design optimization of stochastic flexibility. *Computers Chem. Engng.*, **17(4)**, 339.

Tan, J.S. and M.A. Kramer (1997). A general framework for preventive maintenance optimization in chemical process operations. *Computers Chem. Engng.*, **21(12)**, 1451.

Thomaidis, T.V. and E.N. Pistikopoulos (1994). Flexibility, reliability and maintenance in process design. *Conf. On Foundation of Computer Aided Process Design, FOCAP-D*

Thomaidis, T.V. and E.N. Pistikopoulos (1995). Optimal design of flexible and reliable process systems. *IEEE Transactions on Reliability*, **44(2)**, 243.

Tseng, S.T. (1996). Optimal preventive maintenance policy for deteriorating production systems. *IIE Transactions*, **28**, 687.

Van Rijn, C.F.H. (1987). A system engineering approach to reliability, availability and maintenance. *Conf. On Foundation of Computer Aided Process Operations, FOCAP-O*.

Vassiliadis, C.V. and E.N. Pistikopoulos (1998a). On the interactions of chemical process design under uncertainty and maintenance optimization. *Annual Reliability and Maintainability Symposium, Anaheim, CA*.

Vassiliadis, C.V. and E.N. Pistikopoulos (1998b). On the incorporation of reliability and maintenance in optimal process design and operation under uncertainty. *IRC report*.

PLANNING UNDER UNCERTAINTY: A PARAMETRIC OPTIMIZATION APPROACH

Efstratios N. Pistikopoulos and Vivek Dua
Imperial College
Centre for Process Systems Engineering
London SW7 2BY, U.K.

Abstract

In this work, we propose a parametric optimization approach for addressing planning problems under uncertainty. For linear models, a multi-parametric mixed-integer linear program (mp-MILP) is formulated. The solution procedure involves an iterative strategy between a multi-parametric linear program, for a fixed planning strategy, through the solution of which valid parametric upper bounds are established and corresponding regions of optimality in the space of uncertain parameters are identified, and master sub-problems updating the planning strategy in each parametric region. For convex nonlinear planning models, a multi-parametric mixed-integer nonlinear program (mp-MINLP) is formulated. An outer-approximation iterative solution strategy is presented, in which valid parametric upper bounds are established through the solution of multi-parametric nonlinear problems. An approximation is constructed by considering multi-parametric LPs at a number of points, and solving them to derive linear parametric profiles and corresponding parametric regions. New points are generated by comparing these linear approximations to the nonlinear solution at the vertices of the parametric regions. A master mp-MILP sub-problem is then generated and solved to update the vector of binary variables (planning strategy) and the corresponding sub-regions of the uncertain parameter space.

Keywords

Planning under uncertainty, Outer-approximation, Parametric mixed-integer optimization.

Introduction

Planning typically involves decisions regarding the selection of new processes, expansions and/or shut-down of existing facilities, optimal operating patterns for production chains, etc. In making such decisions, it is imperative to account for and assess the impact of uncertainties in the process and business environment, for example regarding prices, demand volumes and technology. In this way robust strategies can be developed and contingency plans put into effect.

The problem of planning in the presence of incomplete information and uncertainty (for long range and short term production planning) has received a lot of attention in the literature, with multi-period (Sahinidis et al., 1989; Sahinidis and Grossmann, 1991; Liu and Sahinidis, 1996a; Iyer and Grossmann, 1998) and stochastic (Bloom, 1983; Borison et al., 1984; Modiano 1987; Bienstock and Shapiro, 1988; Ierapetritou and Pistikopoulos, 1994; Ierapetritou et al., 1996; Liu and Sahinidis, 1996b) approaches being the most popular. In the multi-period formulation, for a given forecast of demands and prices over a time horizon, an optimal planning strategy is identified. One drawback of this approach is that if the forecasted values are given by ranges, rather than a number of discrete points (which usually is the case) then, the planning problem has to be resolved for all the expected values that lie within the forecasted range. In the stochastic approach, the aim has been to identify the single "most preferred" planning policy, so as to maximize an "average" profit criterion (or minimize economic risk) over possible future outcomes or scenarios (due to the uncertainty involved). One drawback of this approach is that the identified single planning strategy may not be optimal for all scenarios, i.e., it may be more useful to have information about the values of uncertain parameters for which some other planning strategy is more profitable rather than obtain a single strategy which tries to capture all the scenarios based upon an "average" performance criterion.

The objective of this paper is to propose an alternative, parametric optimization approach for addressing planning under uncertainty problems. In this approach, the uncertain parameters, such as demand and supply, are described by certain upper and lower bounds and the solution is given by:

1. Linear parametric profiles, as a function of uncertain parameters, which are valid in the sub-regions of the space of uncertain parameters, and
2. Planning strategies which are associated with those sub-spaces of uncertain parameters.

In the next sections, we describe a mathematical formulation of the planning model under uncertainty, and we propose a solution approach for solving such multi-parametric optimization models.

Mathematical Formulation

The planning problem under uncertainty can be stated as follows. Given,

- A superstructure of (i) processes and capacity expansion options, and (ii) chemicals
- Investment and operating costs
- Lower and upper bounds on the uncertain parameters (e.g., supply and demand of raw materials and products respectively),

determine:

- The optimal objective value (cost) as a function of uncertain parameters
- The complete set of optimal planning strategies (process selection and capacity expansion decisions) for the entire space of uncertain parameters.

Mathematically, this problem can be represented as the following multi-parametric mixed-integer optimization model:

$$z(\theta) = \min_{x,y} d^T y + f(x)$$
$$s.t. \quad Ey + g(x) \leq b + F\theta \quad (1)$$
$$\theta_{min} \leq \theta \leq \theta_{max}$$
$$x \in X; y \in \{0,1\}$$

where,

x is a vector of continuous variables, which describes the operating conditions, for example the flow-rates of chemicals,

y is a vector of 0-1 binary variables, which describes the selection or not of the processes and possible capacity expansions,

θ is a vector of continuous uncertain parameters, bounded between lower and upper bounds given by θ_{min} and θ_{max} respectively; we assume that the terms involving uncertainty are separable and present on the right hand side of the constraints,

f is a scalar objective function, which describes the operating cost,

g is a vector of constraints, which describes the input-output relation associated with the processes, material balances, etc.,

d is a constant vector, which describes the investment cost of the processes,

b is a constant vector; E, F are constant matrices.

For the case when f and g are linear functions of x, (1) corresponds to a multi-parametric mixed-integer linear programming problem (mp-MILP), whereas for the case when they are (convex) nonlinear, (1) is a multi-parametric mixed-integer nonlinear programming formulation (mp-MINLP). In the next section, we describe algorithms for solving such mp-MILP and mp-MINLP planning problems.

Solution Approaches

Multi-Parametric MILP Planning Models

For linear planning models (1) can be recasted as follows:

$$z(\theta) = \min_{x,y} c^T x + d^T y$$
$$s.t. \quad Ax + Ey \leq b + F\theta \quad (2)$$
$$\theta_{min} \leq \theta \leq \theta_{max}$$
$$x \in X; y \in \{0,1\}$$

where c is a constant vector and A is a constant matrix. The algorithm for the solution of (2) is based upon decomposing the problem into a multi-parametric LP (mp-LP) and a deterministic MILP sub-problem. The solution of the mp-LP, which is obtained for a fixed integer solution, provides a parametric upper bound, whereas the solution of the deterministic MILP, which is obtained by treating θ as a free variable, provides a new integer solution. The parametric solutions corresponding to two different integer solution are then compared, using a procedure proposed by Acevedo and Pistikopoulos (1997), to keep tighter (lower of the two) bounds. The algorithm terminates when the solution of the deterministic MILP is infeasible. The detailed steps of the algorithm are as follows.

Initialize an upper bound $\hat{z}^*(\theta) = \infty$. An initial planning strategy is fixed, $y = \bar{y}$, in (2) to obtain a multi-parametric LP problem of the following form:

$$\hat{z}(\theta) = \min_x c^T x + d^T \bar{y}$$
$$\text{s.t.} \quad Ax + E\bar{y} \leq b + F\theta$$
$$\theta_{\min} \leq \theta \leq \theta_{\max} \quad (3)$$
$$x \in X$$

The solution of (3) is given by a set of linear parametric profiles, $\hat{z}(\theta)^i$, and the corresponding critical regions[2], CR^i, (Gal and Nedoma, 1972). $\hat{z}(\theta)^i$ is then compared to $\hat{z}^*(\theta)$ to update the current upper bound. For each critical region, CR^i, (regions obtained from the solution of (3) and the regions which are infeasible where $\hat{z}(\theta) = \infty$) a deterministic MILP sub-problem is formulated as follows:

$$\bar{z} = \min_{x, y, \theta} c^T x + d^T y$$
$$\text{s.t.} \quad Ax + Ey \leq b + F\theta$$
$$c^T x + d^T y \leq \hat{z}(\theta)^i \quad (4)$$
$$\sum_{j \in J} y_j - \sum_{j \in L} y_j \leq |J| - 1$$
$$\theta \in CR^i; x \in X; y \in \{0,1\}$$

where, θ is treated as a free variable; $J = (y|y = 1)$ and $L = (y|y = 0)$, and $|J|$ is the cardinality of J, to identify the next set of integer solutions. Note that the inequality, $c^T x + d^T y \leq \hat{z}(\theta)^i$, prohibits the identification of those integer solution which are higher than the current upper bound, $\hat{z}(\theta)^i$; and the inequality, $\sum_{j \in J} y_j - \sum_{j \in L} y_j \leq |J| - 1$, corresponds to integer cuts which prohibit previous integer solutions from appearing again. The integer solution and corresponding critical regions obtained from (4) are then returned back to (3) to obtain another set of parametric profiles and critical regions. If there is no feasible solution to the problem in (4) for some critical regions, then the algorithm terminates in those regions with the current solution in those regions as the final solution.

---

[2] A critical region is defined as the feasible region of θ associated with an optimal basis.

Multi-parametric MINLP Planning Models

For the case when f and g are convex nonlinear functions of x, (1) corresponds to a multi-parametric mixed-integer convex nonlinear programming problem - note that although here we assume that nonlinear equalities can be relaxed as inequalities, the theoretical framework presented in this work can be extended to incorporate nonlinear equalities (similar for example to the work of Acevedo and Pistikopoulos, 1996). The solution approach for this problem is based upon an iterative outer-approximation decomposition strategy (Duran and Grossmann, 1986). The problem is decomposed into (i) a Primal and (ii) a Master Sub-problem, representing parametric upper and lower bounds respectively, and then iterating between these two sub-problems.

First the best upper bound is initialized as $\hat{z}^*(\theta) = \infty$. Then the Primal Sub-problem is obtained by fixing the vector of integer variables, $y = \bar{y}$, which transforms (1) into the following multi-parametric NLP problem:

$$\hat{z}(\theta) = \min_x d^T \bar{y} + f(x)$$
$$\text{s.t.} \quad E\bar{y} + g(x) \leq b + F\theta$$
$$\theta_{\min} \leq \theta \leq \theta_{\max} \quad (5)$$
$$x \in X$$

The solution of the multi-parametric NLP (5) is approached by outer-approximating the nonlinear terms f and g at x^*, resulting in a multi-parametric LP of the form as follows:

$$\bar{z}(\theta) = \min_x d^T \bar{y} + f(x^*) + \nabla_x f(x^*)(x - x^*)$$
$$\text{s.t.} \; E\bar{y} + g(x^*) + \nabla_x g(x^*)(x - x^*) \leq b + F\theta$$
$$\theta_{\min} \leq \theta \leq \theta_{\max} \quad (6)$$
$$x \in X$$

The multi-parametric LP (6) is solved to obtain a set of linear parametric profiles, $\bar{z}(\theta)$, and the corresponding regions of optimality, CR. These linear profiles are then compared to the nonlinear solution at a number of points so as to identify new points for linear approximation. The points for comparison are obtained by using the convexity property of the objective function value: the maximum discrepancy between the linear and nonlinear function will lie at one of the vertices of the critical region. Thus, it suffices to compare the linear and nonlinear functions at the vertices of the critical regions. At the point where the difference between the linear and nonlinear function is maximum, another mp-LP is formulated and solved to obtain another set of parametric profiles. Sets of parametric profiles, obtained from solving mp-LPs at, say, two different points, are then compared to retain the tighter of the (two) solutions using a comparison procedure described by Acevedo and Pistikopoulos (1997). Such linear approximations are continued until the difference between the linear function and the nonlinear function at the vertices of the critical region is within a certain tolerance. $\tilde{z}(\theta)$ is then compared to $\hat{z}*(\theta)$ to update the current upper bound. The final solution of Primal Sub-problem is given by linear parametric profiles, $\hat{z}(\theta)^i$, and their corresponding critical regions, CR^i (Dua and Pistikopoulos, 1998).

A Master Sub-problem is then formulated to identify the next set of integer solutions by formulating a deterministic MINLP, by introducing constraints similar to that in (4), of the following form:

$$\bar{z} = \min_{x,y,\theta} d^T y + f(x)$$
$$s.t. \quad Ey + g(x) \le b + F\theta$$
$$d^T y + f(x) \le \hat{z}(\theta)^i \quad (7)$$
$$\sum_{j \in J} y_j - \sum_{j \in L} y_j \le |J| - 1$$
$$\theta \in CR^i; x \in X; y \in \{0,1\}$$

The planning strategy (y) obtained from the solution of (7) is returned back to the Primal Sub-problem. The algorithm terminates when (7) is infeasible in each critical region.

An alternate Master Sub-problem can be formulated by linearizing the nonlinear functions, f and g, at the vertices, (x^v), of the critical region, to obtain an mp-MILP problem of the following form:

$$\bar{z}(\theta) = \min_{x,y} \mu$$
$$s.t. \quad \mu \ge d^T y + f(x^v) + \nabla_x f(x^v)(x - x^v)$$
$$Ey + g(x^v) + \nabla_x g(x^v)(x - x^v) \le b + F\theta \quad (8)$$
$$\sum_{j \in J} y_j - \sum_{j \in L} y_j \le |J| - 1$$
$$\theta \in CR^i; x \in X; y \in \{0,1\}$$

where, μ is a free variable. The solution of mp-MILP (8), $\bar{z}(\theta)$, is given by parametric profiles, their critical regions, and, corresponding integer solutions. Note that the formulation in (8) is obtained by outer-approximating the nonlinear functions, which underestimates the objective function and overestimates the feasible region, and hence, represents a lower bound. In the regions where the lower bound exceeds the current upper bound, the algorithm stops, and for the rest of the regions the corresponding y vector is returned to the Primal Sub-problem.

Papalexandri and Dimkou (1998) proposed an alternate Master Sub-problem formulation, based upon Generalized Benders Decomposition principles, of the following form:

$$\bar{z}(\theta) = \min_y \mu$$
$$s.t. \mu \ge d^T y + f(x^v) + \lambda^T [Ey + g(x^v) - b - F\theta] \quad (9)$$
$$\theta \in CR^i; x \in X; y \in \{0,1\}$$

where, λ, is a vector of Lagrange Multipliers. Note that (9) is an mp-MILP problem and represents a parametric upper bound. The solution of (9) is analyzed similar to that of (8) to terminate the algorithm and return the vector of 0-1 variables.

In the next sections, we present two examples to illustrate the significance of obtaining parametric solutions of planning problems.

Example 1 - Linear Planning Model

Consider the superstructure of planning problem as shown in Fig. 1, where a chemical C can be produced from chemical B through two different process alternatives, i.e., process 4 and process 5, and chemical B can be produced from chemical A through three alternatives, i.e., process 1,2 and 3. Only processes 1,2 and 3 have an option of capacity expansion and at the most two of the three processes 1,2 and 3 can be selected. The uncertain parameters, θ_1 and θ_2 are given by the supply of A and demand of C respectively. The mathematical model is given in Table 1 and the solution is given in Table 2 and Fig. 2.

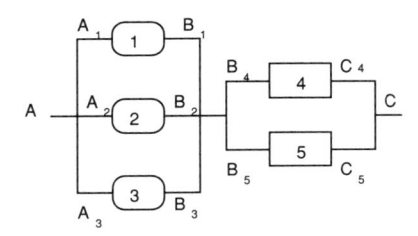

Figure 1. Planning Superstructure - Example 1.

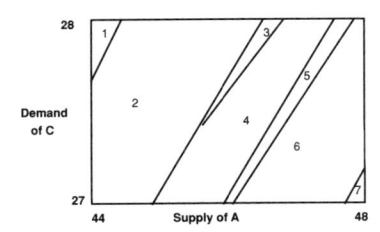

Figure 2. Parametric Regions - Example 1.

The first few solution steps are as follows: (i) Starting with y-1 = 01111000, the solution of mp-LP is given by the first two solutions in Table 2; y-1 is infeasible in the rest of the region, CR^{inf}. (ii) No new y is found in the first two regions and y-new = 01111001 is obtained in $CR^{\inf}$. (iii) The solution of mp-LP corresponding to the y-new is given by the third solution in Table 2 and the corresponding CR^{new} is given by: $\theta_1 - 1.47\theta_2 \geq 5.35$, $\theta_1 - 1.47\theta_2 \leq 6.35$; y-new is infeasible in the rest of the region, $CR^{\inf-1}$. (iv) A new integer solution, y = 11011000, is found in both the regions, CR^{new} and $CR^{\inf-1}$. (v) Solution of the mp-LPs corresponding to this new integer solution and comparison to the current solutions results in 3rd, 4th and 5th solutions in Table 2, and after another set of iteration 6th and 7th solutions are also identified.

The parametric solution (Table 2 and Fig. 2) provides a complete picture of the solution space, showing that:

- Processes 4 and 5 have to be selected for all scenarios
- Only Process 3 will require a capacity expansion,
- Except for a small region of the uncertain parameter space, process 2 will always be selected,
- While for low values of supply of A process 3 is preferred, for higher values process 1 will be installed.

Table 1. Mathematical Model for Example 1.

$$z(\theta) = \min 200A + 5A_1 + 15A_2 + 5A_3$$
$$+ 15B_4 + 12B_5 + 6A_1^{CE} + 17A_2^{CE} + 7A_3^{CE}$$
$$- 700C + 80y_1 + 130y_2 + 150y_3 + 100y_4$$
$$+ 80y_5 + 50y_1^{CE} + 75y_2^{CE} + 90y_3^{CE}$$

s.t.
$B_1 = 0.6(A_1 + A_1^{CE})$
$B_2 = 0.8(A_2 + A_2^{CE})$
$B_3 = 0.7(A_3 + A_3^{CE})$
$C_4 = 0.85B_4$
$C_5 = 0.8B_5$
$A = \theta_1 = A_1 + A_1^{CE} + A_2 + A_2^{CE} + A_3 + A_3^{CE}$
$B_4 + B_5 = B_1 + B_2 + B_3$
$C = \theta_2 = C_4 + C_5$
$A_1 \leq 25y_1; A_2 \leq 25y_2; A_3 \leq 25y_3;$
$B_4 \leq 30y_4; B_5 \leq 30y_5;$
$A_1^{CE} \leq 8y_1^{CE}; A_2^{CE} \leq 8y_2^{CE}; A_3^{CE} \leq 8y_3^{CE};$
$y_1 \geq y_1^{CE}; y_2 \geq y_2^{CE}; y_3 \geq y_3^{CE};$
$y_1 + y_2 + y_3 \leq 2$
$44 \leq \theta_1 \leq 48$
$27 \leq \theta_2 \leq 28$

where the superscript CE corresponds to capacity expansion.

Table 2. Parametric Solution of Example 1.

	Cost $z(\theta)$	Operating Region CR	Process Selection Decision y	Capacity Expansion Decision y^{CE}
1	$135\theta_1 - 560\theta_2 + 340$	$\theta_1 - 1.5625\theta_2 \leq 0.78125$	01111	000
2	$186.2\theta_1 - 640\theta_2 + 300$	$\theta_1 - 1.5625\theta_2 \geq 0.78125$ $\theta_1 - 1.47\theta_2 \leq 5.35$	01111	000
3	$186.2\theta_1 - 640\theta_2 + 354.8$	$\theta_1 - 1.47\theta_2 \geq 5.35$ $-\theta_1 + 2.076\theta_2 \geq 1.26$	01111	001
4	$175\theta_1 - 622.5\theta_2 + 363.75$	$\theta_1 - 1.47\theta_2 \geq 5.35$ $-\theta_1 + 2.076\theta_2 \leq 1.26$ $\theta_1 - 1.47\theta_2 \leq 6.35$	11011	000
5	$175\theta_1 - 622.5\theta_2 + 363.75$	$\theta_1 - 1.47\theta_2 \geq 6.35$ $\theta_1 - 1.5625\theta_2 \leq 3.90625$	11011	000
6	$186.2\theta_1 - 640\theta_2 + 320$	$\theta_1 - 1.5625\theta_2 \geq 3.90625$ $-\theta_1 + 2.083\theta_2 \geq 8.625$	11011	000
7	$183.4\theta_1 - 640\theta_2 + 397.2$	$-\theta_1 + 2.083\theta_2 \leq 8.625$	10111	001

Table 3. Mathematical Model for Example 2.

$$z(\theta) = \min 3.5y_1 + y_2 + 1.5y_3 + B_2 + 1.2B_3 + 1.8(A_2 + A_3)$$
$$+ 7B_P - 11C_1$$

s.t.
$-0.9B_1 + C_1^2 \leq 0$
$-A_2 + EXP(B_2) \leq 0$
$-A_3 + EXP(B_3/1.2) \leq 0$
$A = \theta_1 = A_2 + A_3$
$B_1 = B_P + B_2 + B_3$
$B_P \leq 1.95$
$C = \theta_2$
$A_2 \leq 20y_2; A_3 \leq 20y_3;$
$B_2 \leq 20y_2; B_3 \leq 20y_3; C_1 \leq 20y_1;$
$y_1 + y_2 + y_3 \leq 2$
$0.50 \leq \theta_1 \leq 0.75$
$5.5 \leq \theta_2 \leq 6.00$

Looking at the results in more detail, for different scenarios of demand and supply, the operating region has been divided into 7 operating regions (Table 2 and Fig. 2). Associated with each region is (i) a cost, (ii) process selection decisions, and (iii) capacity expansion decisions. The cost is given by a linear function of supply and demand (θ_1 and θ_2 respectively) and can be calculated by merely substituting the value of demand and supply without any further optimization. Also, the sensitivity of the cost to uncertain supply and demand in different operating regions is given by the their coefficients in the linear parametric expressions. The selection and rejection decisions for different operating regions are also depicted in Table 2, where a value of 0 corresponds to the rejection, whereas a value of 1 corresponds to the selection of a process. Similarly, for the operating regions where a capacity expansion is required.

Example 2 - Convex nonlinear planning model

Consider a chemical complex, shown in Fig. 3, in which a chemical C is produced from chemical B through process 1, while chemical B can either be purchased from the market or can be produced from chemical A through two different alternatives, process 2 and process 3. Considering the supply of chemical A and demand of chemical C as uncertain parameters, the corresponding multi-parametric mixed integer nonlinear programming mathematical model is given in Table 3. The application of the solution procedure described earlier leads to the results shown in Table 4 (for $\varepsilon \leq 0.3\%$). Note that all the three master sub-problems, (7), (8) and (9), were tested on this example - see Dua and Pistikopoulos, 1998, for details.

Conclusions

A parametric optimization strategy for solving planning under uncertainty problems has been presented and applied to two planning examples. The key feature of solving these problems parametricaly is that the solution provides an overview of all the possible outcomes. One distinct advantage of such parametric optimization approach is that the complete map of optimal planning solutions as a function of the uncertainty involved is derived, while critical (important) uncertain factors can be systematically identified, regions of unrealistic/infeasible operations depicted, and the economic benefits (and risks) due to uncertain parameters realizations consistently evaluated without any further optimization calculation. From a practical viewpoint, a key advantage of a parametric optimization approach to planning under uncertainty is the fact that "what-if" scenarios and their impact are in essence displayed one next to the other, providing the decision maker with the complete "roadmap" of optimal actions. While a mathematical framework for solving planning problems involving uncertain parameters has been outlined, the current research mainly focuses on implementation of these algorithms in order to solve large scale problems. The computational complexity of these problems is expected to increase with increase in the number of binary variables and uncertain parameters. However, the key advantage of these approaches, in the context of solving large scale planning models, is that these approaches are based upon efficiently characterizing the space of uncertain parameters and hence solving of a large number of planning problems for fixed values of uncertain parameters is avoided.

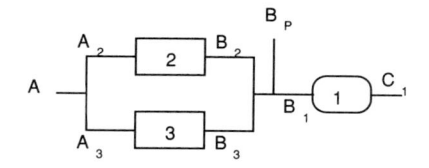

Figure 3. Planning Superstructure – Example 2.

Table 4. Parametric Solution of Example 2.

	Cost	Operating Region	Process Selection Decision
1	$z(\theta)$	CR	y
2	$-2.178\theta_1 - 4.8322\theta_2 - 14.26$	$-0.686\theta_1 - 4.8322\theta_2 \leq 4.73$ $0.63\theta_1 + 0.21\theta_2 \geq 1.63$	101
3	$-2.8105\theta_1 - 5.0422\theta_2 - 12.63$	$-0.795\theta_1 + 0.8515\theta_2 \leq 4.48$ $0.63\theta_1 + 0.21\theta_2 \leq 1.63$ $-0.608\theta_1 + 0.851\theta_2 \leq 4.502$	101
4	$-2.8105\theta_1 - 5.0422\theta_2 - 12.63$	$0.63\theta_1 + 0.21\theta_2 \leq 1.63$ $0.66 \leq \theta_1$	101
5	$-1.85\theta_1 - 5.0422\theta_2 - 13.26$	$-0.608\theta_1 + 0.851\theta_2 \geq 4.502$ $\theta_1 \leq 0.66$	110
6	Infeasible Region	$-0.686\theta_1 - 4.8322\theta_2 \geq 4.73$	***
7	Infeasible Region	$-0.686\theta_1 - 4.8322\theta_2 \leq 4.73$ $-0.795\theta_1 + 0.8511\theta_2 \geq 4.48$	***

References

Acevedo, J. and E.N. Pistikopoulos (1996). A parametric MINLP algorithm for process synthesis problems under uncertainty. *Ind. Eng. Chem. Res.*, **35**, 147-158.

Acevedo, J. and E.N. Pistikopoulos (1997). A multi-parametric programming approach for linear process engineering problems under uncertainty. *Ind. Eng. Chem. Res.*, **36**, 717-728.

Bienstock, D. and J.F. Shapiro (1988). Optimizing resource acquisition decisions by stochastic programming. *Management Sci.*, **34**, 215-229.

Bloom, J.A. (1983). Solving an electricity generating capacity expansion planning problem by generalized Bender's decomposition. *Oper. Res.*, **31**, 84-100.

Borison, A.B., P.A. Morris, and S.S. Oren (1984). A state-of-the world decomposition approach to dynamics and uncertainty in electric utility generation expansion planning. *Oper. Res.*, **32**, 1052-1068.

Dua, V. and E.N. Pistikopoulos (1998). Algorithms for the solution of multi-parametric mixed-integer optimization problems. *IRC Report*.

Duran, M.A. and I.E. Grossmann (1986). A mixed-integer nonlinear programming algorithm for process systems synthesis. *AIChE J.*, **32**, 592-606.

Gal, T. and J. Nedoma (1972). Multi-parametric linear programming. *Management Science*, **18**, 406-422.

Ierapetritou, M.G. and E.N. Pistikopoulos (1994). Novel optimization approach of stochastic planning models. *Ind. Eng. Chem. Res.,* **33**, 1930-1942.

Ierapetritou, M.G., E.N. Pistikopoulos and C.A. Floudas (1996). Operational planning under uncertainty. *Computers Chem. Engng.,* **20**, 1499-1516.

Iyer, R.R. and I.E. Grossmann (1998). A bilevel decomposition algorithm for long range planning of process networks. *Ind. Eng. Chem. Res.,* **37**, 474-481.

Liu, M.L. and N.V. Sahinidis (1996a). Long range planning in process industries: a projection approach. *Computers Oper. Res.,* **23**, 237-253.

Liu, M.L. and N.V. Sahinidis (1996b). Optimization in process planning under uncertainty. *Ind. Eng. Chem. Res.,* **35**, 4154-4165.

Modiano, E.M. (1987). Derived demand and capacity planning under uncertainty. *Oper. Res.,* **35**, 185-197.

Papalexandri, K.P. and Dimkou, T.I. (1998). A parametric mixed integer optimization algorithm for multi-objective engineering problems involving discrete decisions. *Ind. Eng. Chem, Res.,* **37**, 1866-1882.

Sahinidis, N.V., I.E. Grossmann, R.E. Fornari and M. Chathrathi (1989). Optimization model for long range planning in the chemical industry. *Computers Chem. Engng.,* **13**, 1049-1063.

Sahinidis, N.V. and I.E. Grossmann (1991). Reformulation of multiperiod MILP models for planning and scheduling of chemical processes. *Computers Chem. Engng.,* **15**, 255-27

PILOT PLANT OPERATIONS IN PHARMACEUTICAL RESEARCH AND DEVELOPMENT IN THE NEXT CENTURY

Jonathan M. Vinson, Owen D. Keck, Prabir K. Basu,
R. Brian Houston, Linas Mockus and Alan R. Noren
Searle, a Monsanto Company
Skokie, IL 60077

Abstract

We present our vision of the chemical pilot plant for research and development at Searle as it affects production of pharmaceutical compounds for pre-registration activities. The heart of this vision is an information system that integrates the disparate functions of research and development into a flexible and efficient network. The flow of product from bulk substance to the formulations, based on clinical demand and toxicology needs, must be fully integrated and continuously coordinated. While project management establishes overall strategy, each group must be sensitive to one another so that internal conflicts do not compromise these objectives.

Keywords

Batch processing, Data integration, Computer-aided engineering, Planning, Scheduling, Computer integrated manufacturing.

Introduction

A common problem in pharmaceutical development is the frequent changes in planning; not everyone learns of the changes in a timely fashion. As a result, the clinic might find out about delays in production only when there is not enough drug compound to meet their demands. In this vision, the information system keeps all parties up to date regarding the latest changes in planning and the latest data from engineering, quality and purchasing so that they can make informed decisions. High-level business plan changes disseminate quickly through the network, and the affected functions can coordinate their activities appropriately. The schedule of all pilot plant activities is part of the information system as are all the engineering, quality, purchasing, material and facilities databases. Everyone involved in pre-registration activities has access to the information; thus routine activities, such as preventive maintenance, are not surprise delays to a pilot run. Another example is that there are no shocks in materials management caused by surprise requests for raw materials that the development team has known about for weeks. The information system ties engineering to chemical development, analytical, quality, materials and facilities.

In the pilot plant, this vision impacts many of the current activities undertaken by the engineering and operations staff. Preparation of production runs in the pilot plant absorbs an inordinate amount of time; much of it spent collecting data and disseminating information throughout various functions in the company. As a result of this information system and the use of integrated engineering tools, the time required for documentation of pilot plant runs will decrease dramatically. Time-saving arises from the automation of repetitive tasks, such as generation of operating instructions and gathering data for analysis of hazards, operability and quality concerns.

Chemical production also links directly to this information system. Electronic, rather than paper, operating instructions guide the operators and provide a framework for electronic approvals. The information system provides a mechanism to inform analytical services when samples are waiting for analysis and electronically report the results back to operations. Materials management links to the production plan, so they transfer raw materials to the plant as needed and collect products and waste materials in a timely fashion. Electronic notebooks tie directly to the generation of operating instructions, creating an electronic paper trail from the chemist's laboratory to the pilot plant.

All the information ties back to the higher functions, so formulators and the clinical pharmacists know when their supplies are coming and can plan accordingly. In this new vision the drug pre-registration activities become streamlined and the potential pipeline of new drugs expands.

Since the pilot plant is a research and development facility, there are many needs in addition to the requirement for drug substance. The development team needs to collect data on the effectiveness of the process without negatively affecting the quality or quantity of the product. The final route for generating the drug compound must be robust and well characterized. These and other complicating factors emphasize the necessity of an integrated collection of databases, software tools and information in such an environment.

Electronic Batch Records

The driving force for this vision of our pilot plant is the development of a paperless pilot plant. By the year 2000 we expect to be using Electronic Batch Records (EBR) to direct and record all the pilot plant activities involved in production of a batch of chemical. A paperless pilot plant will help ensure consistency and repeatability from run to run, particularly in combination with the intelligent master operating record tool described below.

The full implementation of an EBR system will force the integration of many of the core information systems. All material and movable equipment (e.g., scales and pumps) will have bar-coding for check-in and check-out purposes. The Laboratory Information Management System (LIMS) will feed its results directly to the EBR system, reducing wait time and clerical errors. The EBR system will rely on Documentum® for access to standard operating procedures, material safety data sheets, and even P&ID's. The EBR will set control-system parameters and collect sensor data and attach them to the operating record. Planning and scheduling systems will have real-time access to the current state of the campaigns in the pilot plant and will be able to reactively reschedule if necessary.

Another benefit of an electronic batch record includes the use of electronic signatures for operators, supervisor sign-off and chemical analyses. Since an EBR utilizes procedural and data interlocks, cGMP compliance will be enhanced.

Planning and Scheduling

A major part of the central information system is a schedule that incorporates the high-level plans of the company with the day-to-day activities in the pilot plant. The high-level plans dictate the gross demands for research chemicals along with setting priorities on the projects. The actual demands for chemicals from the pilot plant come from the pre-clinical and clinical research groups whose demands stem from their needs and, ultimately, by the corporate plan. Day-to-day activities in the pilot plant also affect its ability to meet demands of its customers.

Our current schedule consists of many separate schedules that do not necessarily correlate. The long-range plan stems from clinical research & development goals. In practice the pilot plant development team updates this schedule weekly, however it is never completely in synch with the pilot plant. The pilot plant operations team creates the short-term, eight-week plan. The team updates this schedule at a weekly planning meeting, based on the status of the projects in the pilot plant. The schedule is available on the network for anyone to access. However, even this detailed plan does not explicitly include scheduled cleaning and preventive maintenance.

An example is the planning and preparation required for the start of a new campaign in the pilot plant (Basu, 1997). Documentation preparation must begin approximately six weeks before the start of a new campaign. This preparation includes reviews for process hazards and operability; design and review of special setups; ensuring that raw materials are available and released for use. However, six weeks is occasionally not enough time to complete complicated setups or source highly unusual raw materials. In these cases, the people involved must look forward into the site-wide plan to ensure coverage of this kind of detail. With an integrated, networked information system this information is available at one's computer. Alternatively, the schedule can capture time-sensitive issues and inform the appropriate parties.

Figure 1 shows a flowchart of an integrated planning system, such as described here; a single master schedule holds all the details of the pilot plant operations and the higher-level goals. While only a few users can modify and update the schedule, anyone may examine it to see what is happening in the pilot plant or check for future activities. When major changes or delays affect project deadlines, then the system must automatically inform the affected parties.

In the early development of a compound, the decision as to which of several possible synthesis routes to select is a scheduling problem that we frequently address in research and development. Most often we select the route that minimizes costs assuming fixed external factors. However, the pilot plant operates several syntheses at once and equipment limitations and availability must be part of route selection. Given the current schedule of plant activities, the best route may be one that appears to cost more but uses idle equipment. Of course, one must consider the impact at the manufacturing facility as well. The development team conducts this route selection in the early phases of development. Once the route is fixed, the development team cannot easily change it due to drug registration procedures. Thus one must know the basic schedule of activities and have access to the latest information.

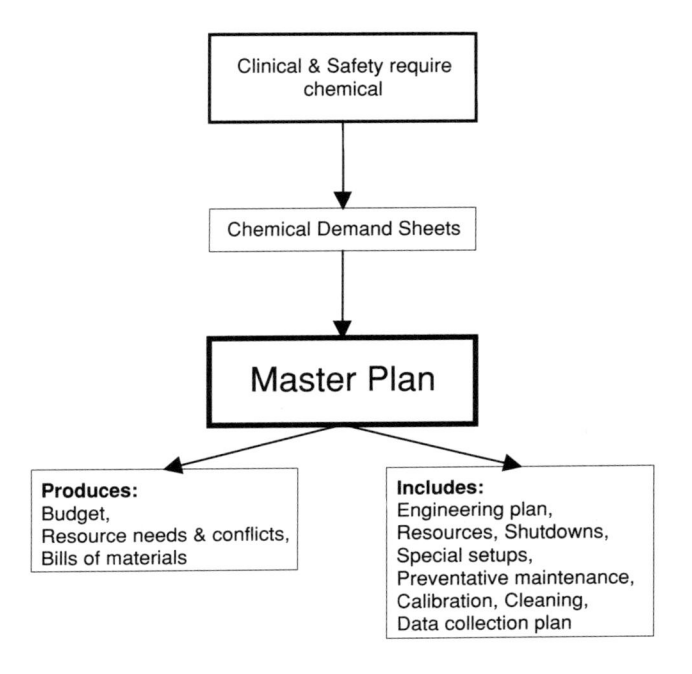

Figure 1. Vision of an integrated plan.

While the main purpose of our foray into planning and scheduling is the integration of disparate plans in the pilot plant operation, these tools bring much more capability to the table. When making long-range decisions regarding development of drugs, corporate planners consider the likelihood of a compound surviving each development phase, the perceived market for the drug and other global factors. Planning decisions at each level impose constraints on the lower levels. For instance, long range plans regarding which therapeutic areas and compounds to develop constrain the drugs in the pipeline. With global planning systems in place, one can examine the effect of decisions and uncertainty on the more detailed plans. Blockbuster drugs are virtually guaranteed development, but there are borderline drugs whose development is less certain. With software, one can adjust the uncertainty parameters to find the compounds that are sensitive to the uncertainty. The same kind of process applies to assessment of the human and capital resource needs of the company: When and how many new people to hire? Considering current understanding of the marketplace and technology, will our existing facilities meet future production needs (Applequist, 1997; Honkomp, 1997; Lombardo, 1997)?

In addition to these questions, we are currently developing an algorithm to automatically generate the MILP formulation for the optimizer based on the engineering information. Once a schedule has been optimized, one may want to simulate it to determine if there are unexpected interactions that were not captured by the MILP formulation. This work is currently under development.

Databases

Along with the schedule of pilot plant activities, another primary need is a central repository of data. A materials database includes safety and health information as well as standard physical properties. Equipment data include exact measurements of the unit, operating boundaries, heat transfer parameters, and material of construction. A database of chemical synthesis information includes the characteristics of the reaction, special considerations of the synthesis, and time required for the unit operations (i.e., reaction, crystallization, drying). For this database, there must also be connections to reports and lab analyses through Documentum® and electronic notebooks. Hazard analyses require a database of material interactions that indicates the reactivity of all binary combinations of materials. Currently, old reports and expertise are the only source of this information rather than a globally accessible matrix. As a result, the engineer must regenerate the matrix with help from the team, and this may lead to inconsistencies between the reports.

While most of the databases are straightforward collections of information, some of the data are located in multiple sources such as books, catalogs and software. It will be beneficial to speed and accuracy to have this information stored in a network-accessible format. This is particularly true of equipment specifications and material properties. Current data sources do not contain all the necessary data and some conflict with one another due to out of date information.

These databases are also central to the needs of many of the other tools that are part of the vision for pilot plant operations. Although packaged software usually comes with proprietary databases, in-house software needs to have access to data as well.

Engineering Tools

Development of manufacturing process requires a variety of tools, from basic spreadsheets to simulation of complex flow sheets. We have developed material balance spreadsheets to help calculate exact material charge quantities. We need this flexibility since the yields of upstream transformations and the downstream requirements change in a research and development setting. We have created similar spreadsheets to assist in scaling to different size equipment. Scale-up requires more information from the development team regarding necessary processing conditions. Important data for these calculations include material properties and equipment capacities. The process assessment tool described below is an attempt to standardize these calculations and improve reporting consistency.

We have developed a workbook template to help the engineers generate the process hazards analysis (PHA) report. This report is a collection of lab analyses, reaction characterizations, spreadsheet calculations, the results of the "what if" analysis described below, and meeting minutes. This "electronic PHA" acts as an activity checklist as well as a place to hold the appropriate information. Data needed for the PHA report include safety and health summaries, material interactions, waste profiles, equipment data and reaction information.

As teams develop new production routes, it goes through many reviews before the process can run in the pilot plant (Basu, 1997). Throughout this review process, the engineer must modify the detailed instructions and update special equipment setup drawings. The intelligent master operating record tool described below is the first step towards unifying and consolidating these necessary, but tedious, activities. There are several potential future developments that integrate these tasks and the other pilot plant activities. One example is an expert system that captures the typical concerns raised during a safety review to ensure the engineers think about critical questions in every process. This would also guide the engineers toward best-practice solutions to common difficulties.

Intelligent Master Operating Record (MOR) Tool

For every new process in the pilot plant there must be a MOR that specifies exactly how to run that process. The responsible engineer generates the MOR for a new synthesis route, based on the chemist's laboratory recipe. In addition, the engineer must generate a new MOR any time there are major changes to the process, such as different equipment or significant deviations from the original instructions. When the synthesis runs in the pilot plant, the document control department creates a copy of the MOR, which will be the pilot plant Operating Record (OR). The engineer completes the OR with exact charge amounts and other details by hand. The MOR and OR are subject to Good Manufacturing Practice (GMP) and are auditable by the FDA. As such, the MOR must be accurate and consistent, no matter who has created it.

Besides the documentation requirements for the operators and quality assurance, instructions must represent a safe and operable process. Also, the team involved in the synthesis will want to collect data during the run to optimize and create the best process when it transfers to manufacturing (Basu, 1998). To this end, there is a series of process reviews which are guided by the Process Sequence Diagram (PSD), a flowchart of the instructions. After each review, the engineer may need to modify the PSD, adding significantly to the time burden of the engineer if done by hand. In the current structure, the average time from initial preparation of the PSD to the final MOR and a run in the pilot plant is six weeks. The potential time-savings and frustration reduction represented by automating some of this process is quite significant.

We have developed software in collaboration with Purdue University that automates the MOR-writing process and saves time preparing for the safety and operability reviews (Viswanathan, 1998a, 1998b). The engineer enters the basic instructions to the program and adds details as necessary. For each type of instruction the program generates a standard "boiler-plate," or macro, of text based on information entered by the user. The program generates both the PSD and MOR from the same information, so it incorporates any additions, deletions or modifications to the PSD directly into the MOR. In the past, the PSD was a template for the MOR, which the engineer wrote near the end of the review process. Simple clerical errors could lead to mistakes in the MOR, requiring more reviews and changes to the MOR document.

In addition to the instructions, the MOR must include a variety of attachments: safety and health summaries of all the raw materials and intermediates; the process sequence diagram; a dispensing log that indicates material and quantity charged in each charge instruction; a list of equipment used; a waste profile summary for each waste stream; and several other informative documents. The MOR must also contain drawings for unique equipment set-ups and unusual instructions. All of these pages must be sequenced properly in the MOR. The tedious pagination and instruction numbering tasks become quite simple with the intelligent MOR tool. The tool must access the equipment, material and personal protective equipment databases to generate the full MOR.

We are currently working with Mitchell Scientific to develop an interface to Emission Master®, which calculates vapor emissions of standard batch operations. The EPA has approved Emission Master® for estimates of vapor emissions. The interface will read text output from the MOR tool and automatically generate the emissions report without additional contributions from the user.

This tool may also be able to simulate the operating record. This simulation might take the form of a dynamic simulation of each operation to predict the material and energy balances. Alternatively, the simulation might draw on historical plant data to predict the time spent in each operation. The scheduling program could then use this time information to improve scheduling efforts for future campaigns.

Future improvements to this program include adding more intelligence to the generation of instructions. The program could suggest when to use purges, recommend when to clean and what solvent to use, and suggest charging setups. When the electronic batch record system is up and running, the MOR generated by this tool will export directly to the pilot plant over the network. In combination with electronic laboratory notebooks, a tool may be developed to translate a chemist's recipe directly into a set of operating instructions (Miller, 1997). Finally, the tool could generate the MOR for manufacturing, which has a different format but contains the same information. This can save a great deal of time in the transfer of production technology.

Automatic Process Hazards Analysis (PHA)

One of the major pre-run reviews is the Process Hazards Analysis. The engineer analyzes the process to ensure that appropriate controls are in place to avoid potential hazards. Other tasks include ensuring that the process "fits" the selected equipment and that there are no untoward interactions between the compounds in the process and with the equipment materials of construction. If there are new compounds involved, the team must conduct corrosion tests. All of this information is part of the Electronic PHA mentioned previously.

The largest and most time consuming part of the PHA is the "what if" analysis. The development team splits the planned synthesis route into groups of operations, or nodes. The team examines each node for typical upsets, such as high or low temperature, high or low pressure, utility failures, etc. The current system requires that the team identify and discuss possible hazards due to these upsets and record preventive measures. This review consumes much time and requires the presence of several key people.

In collaboration with Purdue University, we are developing a tool that will automate the "what if" analysis in the PHA (Srinivasan, 1998a, 1998b). We are making significant changes from the continuous version of this program to properly handle batch operations. The program is based on a grafchart, modular framework.

Instead of analyzing groups of operations for the effect of upsets as is done by hand, the program examines the effect of upsets in each instruction. The charge operation, for example, would generate questions about upsets in the temperature, charge amount, agitation, etc. The engineer can test these upsets one-by-one through the graphical interface, or the program can run through them sequentially. For each upset the program steps through the instructions to determine its potential downstream. The automatic PHA also looks for potential upstream causes to indicate possible sources of problems. For each upset tested, the program generates a table of potential causes, the upset, and hazards and quality issues. The engineer and development team must then use their judgment to filter unlikely hazards. For the remaining hazards the team must identify the controls necessary to reduce their severity. This approach is more thorough and the team can use its expertise to evaluate the results rather than generating the common potential hazards. Rather, she can focus on controlling these hazards when they arise. In addition, the development team has the opportunity to devote more time to the special concerns associated with the synthesis route that the automatic PHA program cannot capture.

Besides upsets, the automatic PHA tool also has the capacity to reason about the effect of maloperation. This includes such errors as a reversal of operations, a missed operation or an incorrect material. The engineer can conduct these tests through the graphical interface.

The automatic PHA tool will automatically pull the appropriate material interaction information from the database. The tool requires a detailed material database that includes such properties as the flash point, upper- and lower-explosive limits and temperature sensitivity along with the standard material properties. The PHA tool also needs equipment information such as capacity and operating limits to check against the results of process upsets or maloperations.

Both the PHA and MOR tools use Genysm's G2 environment, and they use similar internal representations of the synthesis route. They are both based on the ISA S88 standard for representation of batch processes. As a result, there is a natural synergy between the tools, and we are combining them. This integrated tool will incorporate more of the PHA report requirements. For example, the report requires a chart comparing the volume, temperature and pressure of the process to the rating of the equipment. A material interaction matrix points out any severe reactions between chemicals in the process and chemicals that are common to the pilot plant. The PHA report must also include vent-size calculations for worst-case scenarios such as a runaway reaction or an external fire. With access to the necessary data, the engineer merely sets up the scenario and the calculations are automatic.

Finally, the current PHA review system focuses on the chemical synthesis and the main equipment used in the campaign. Another review examines operability issues, such as long stretches of piping for slurries and the effects of extreme weather or unusually long hold times, etc. The framework of the auto PHA program is applicable to this kind of analysis as well.

Process Assessment Tool (PAT)

To speed development of new drug compounds, we have created software to conduct initial cost estimates based on the preliminary chemical route of a new compound. The user sets up the basic order of operations and the chemicals required for and created by the synthesis route in a graphical interface. PAT then generates a material balance sheet that shows materials fed to the major operations of the synthesis route. As the user gathers more information, this material balance grows in complexity and detail.

At any level of detail PAT generates cost estimates to achieve a first-pass of the per kilogram costs of a compound. The user can then compare costs between a variety of possible routes to help select the best one. PAT calculates raw material costs by a proprietary routine that estimates the cost based on quantity required and historical purchase data. The fixed and variable costs are based on the latest accounting information for the pilot plant. These cost data along with material property data are available from the network databases.

Once the development team selects a synthesis route, the PAT format is a template for improving and tweaking the process. For example, the engineer can conduct sensitivity analyses on the variables in the process. When the process moves from one scale to another, the material balance is the basis of scale-up calculations.

In the future, a single package may inherit the capabilities of PAT and the PHA and MOR programs. An integrated program of this sort has the benefit of retaining information throughout the development of a synthesis route. Figure 2 shows the information flow in such an integrated system. In this case, the preliminary analysis of PAT feeds directly to a block PSD (process sequence diagram) which the user expands with detailed instructions. This PSD is the basis of both the hazard analysis and the final MOR.

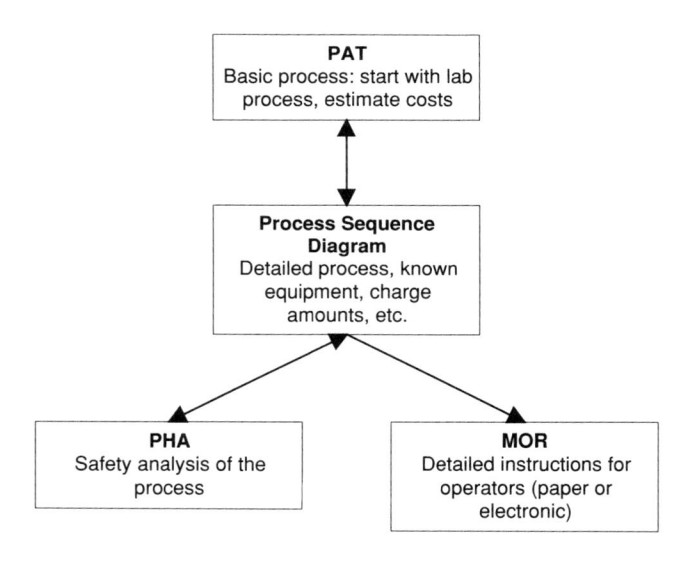

Figure 2. Interaction between PAT and the automatic PHA and MOR tools.

Conclusions

With information available to everyone at the click of their mouse, the job of gathering important data will become easier. Managers and scientists will be able to do their jobs more effectively when they know the most recent plans for the next week or the next year. People will evaluate their needs based on accurate estimates of their future activities.

This information system is currently being created at our research and development facility. The materials and equipment databases exist and are being used for hand calculations and in a variety of engineering tools. The process assessment and the intelligent master operating record (MOR) tools are already in use for all new MOR's. The automatic PHA tool is still under development, with a planned release by the end of 1998. The implementation of the first phase of electronic batch records is scheduled for 1999 with the paperless pilot plant to follow in the year 2000. The planning and scheduling systems are under development and should be in place by the end of 1998.

References

G. Applequist, O. Samikoglu, J. Pekny, G. Reklaitis (1997). Issues in the Use, Design, and Evolution of Process Scheduling and Planning Systems. *ISA Transactions*, 36: (2) 81-121.

P. K. Basu, A. R. Noren and R. A. Mack (1997). Pharmaceutical Pilot Plants are Different. *Chemical Engineering Progress*, June, 66-77.

P. K. Basu, J. Quaadgrass, R. A. Mack and A. R. Noren (1998). Achieve the Right Balance in Pharmaceutical Pilot Plants. *Chemical Engineering Progress*, February, 67-74.

S. Honkomp, J. Pekny, G. Reklaitis (1997). Robust Planning and Scheduling of Process Development Projects under Stochastic Conditions. *Fall Annual AIChE Mtg., Los Angeles, CA,* Paper 211a.

S. P. Lombardo and M. G. Zentner (1997). Improvement of Process Operations through Integrated Optimization of Design, Planning, and Scheduling. *Fall Annual AIChE Mtg., Los Angeles, CA,* Paper 56f.

D. C. Miller, J. F. Davis, J. R. Josephson and B. Chandrasekaran (1997) A Process Design Decision Support System for Developing Process Chemistry. *Fall Annual AIChE Mtg., Los Angeles, CA,* Paper 198i.

R. Srinivasan and V. Venkatasubramanian (1998a). Automating HAZOP Analysis of Batch Chemical Plants: Part I. Knowledge Representation Framework. *Comput. Chem. Eng.*, Accepted for Publication.

R. Srinivasan and V. Venkatasubramanian (1998b). Automating HAZOP Analysis of Batch Chemical Plants: Part II. Algorithms and Application. *Comput. Chem. Eng.*, Accepted for Publication.

V. Venkatasubramanian and R. Vaidhyanathan (1994). A Knowledge-Based Framework for Automating HAZOP Analysis, *AIChE Journal*, 40, 496-505.

S. Viswanathan, C. Johnson, R. Srinivasan, V. Venkatasubramanian and K.E. Arzen (1998a). Operating Procedure Synthesis for Batch Processes: I. Knowledge Representation and Planning Framework. *Comput. Chem. Eng.*, Accepted for Publication.

S. Viswanathan, C. Johnson, R. Srinivasan, V. Venkatasubramanian and K.E. Arzen (1998b). Operating Procedure Synthesis for Batch Processes: II. Implementation and Application. *Comput. Chem. Eng.*, Accepted for Publication

AN INDUSTRIAL PERSPECTIVE OF SUPPLY-CHAIN OPTIMIZATION AND SIMULATION

Oscar Rosen
The Procter & Gamble Company
Cincinnati, OH 45232-1736

Abstract

This work gives an Industrial perspective of our experience with Simulation, Optimization and their combined use. Commercially available simulation packages typically incorporate metamodeling and optimization options that allow users to automate the design optimization process. Although this approach can be applied to complex Supply-chain problems, other Management Science (MS) and Operations Research (OR) techniques offer powerful alternatives. Yet, OR tools seldom incorporate aids to understand the impact of process and parameter variability. Solutions are prescriptive and feedforward open loop in nature. Robustness considerations, although known to be important, are an afterthought in the methodology that normally cannot be easily tested or automated. Here, a general overview of approaches for supply-chain analysis is presented. The relative strengths and deficiencies of these methods are then explored in some detail with emphasis on how they can be used in practice. Finally, it will be shown how these methods can be used in an integrated approach.

Keywords

Supply-chain modeling, Mathematical programming, Discrete-event simulation, System dynamics, Systems thinking, Theory of constraints, AI optimization.

Introduction

Supply-chain issues are increasingly critical to industrial competitiveness. Recently, the Government sponsored study on Technology Vision 2020 for the U.S. Chemical Industry (1996) recognized the value of logistics and recommended the development of better methods to manage the supply-chain. At Procter & Gamble (P&G) rigorous models are routinely used to refine plans and to support key decisions.

More than ever before, P&G has been working closely with some of our suppliers and customers. We consider problems that involve the efficient movement of materials all the way from suppliers to consumers and the streamlining of new initiatives (Fig. 1). The key objective is to balance the metrics of customer service, time to market, throughput, inventories and costs. Typical areas of interest are sourcing location, manufacturing production planning, project planning, transportation, and distribution networks.

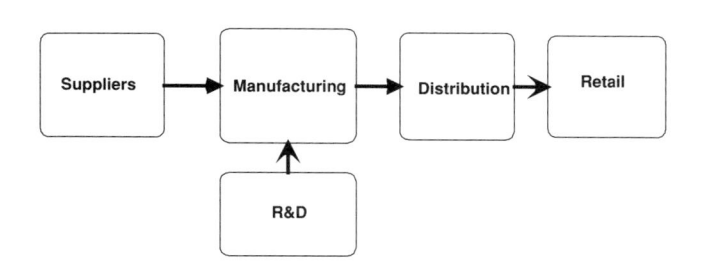

Figure 1. The supply-chain.

In this paper, we first review the factors that force us to use decision tools since they form the basis for model and tool selection. Then, a brief summary of our experiences with the various modeling approaches for supply-chain management is presented. The purpose is not to detail specific issues we have encountered, but to provide a brief overview of the area and to suggest possible research directions.

Bias, Complexity, Variability and Uncertainty

Decisions in supply-chain and enterprise analysis are usually complicated by several factors: *bias, detail complexity, dynamic complexity, adaptive complexity, variability and uncertainty.*

Bias is introduced in the process of model building. The model builder or model sponsor can incorporate the *bias*, deliberately or not, given selfish motives, authority relations and level of experience. Also, there is a strong tendency to seek evidence that confirms our beliefs and not for evidence that can discredit them.

Detail complexity is the result of systems with a large number of variables or components. For example, at P&G many of our manufacturing facilities are involved in the production of hundreds or thousands of products. Tractability and computational effort can then become a problem, in particular when the solution effort grows rapidly with the size of the problem. Fortunately, advances in computer technology and algorithmic development have steadily allowed us to tackle larger problems. Problems that 5 years ago forced us to group products by family, make simplifying assumptions and break into sub-problems, we can now routinely analyze as a whole.

Dynamic complexity arises from the presence of feedback loops, time delays, non-linearity, chaotic behavior, and multiple interactions between variables. To account for these factors, the decision-maker must obtain a more fundamental understanding of the system. Without this understanding, causal and empirical models can be invalid. Furthermore, analysis methods that rely on discretizing the time horizon may require smaller time steps to adequately model the behavior, thereby contributing to detail complexity.

Adaptive complexity is the tendency of systems under study to evolve under competition. Competitors, customers, suppliers and consumers all react as a result of our actions and other external factors. Our initial actions may change the system in such a way that our prediction of the future behavior may become obsolete.

Variability is the lack of repeatability in activities and behaviors at different times. It is the stochastic nature of the problems. Daily demand for our products as well as processing times, transfer times, lead times and human performance all vary over time to some extent. Since elements within the supply-chain tend to work only with local information, variation can be distorted and amplified as it propagates to upstream processes. For example, P&G found that the diaper orders received by manufacturing has a degree of variability that cannot be explained by consumer demand fluctuations alone (Lee *et al.*, 1997).

Variability in product demand and processes can be correlated. If the demand for a group of products goes up or down, the demand of competing products may go in the same or the complete opposite direction, i.e. the demands are cross-correlated. Furthermore, if the process reliability is poor during a particular shift, it is more likely to be poor the next shift unless strong corrective action is taken, i.e. processes are auto-correlated.

Uncertainty refers to our lack of knowledge about the value of the system parameters. As any practitioner knows, data gathering and validation is usually a formidable challenge. Data is usually hard to get and often it ends up being incomplete or inaccurate. This problem is more pronounced in the early stages of new initiatives and for activities with long lead times. The further out in the planning horizon, the more uncertain a forecast becomes. By the time we have reasonable estimates, many decisions will be made with a certain risk. In Industry, the data that you need will likely be outdated, biased or incomplete.

Some new decision tools embraced by companies have inadvertently amplified the issues presented in this section. In particular, we have found that the ease of use of current business simulation tools has resulted in an increase in decisions being made by new hires with little or no experience in supply-chain issues or in simulation methodology.

Types of Models

Based on their output, supply-chain models basically can be either **prescriptive** or **descriptive** (Sterman, 1996). **Prescriptive** models tell us what we need to do to achieve our objectives. Recommendations are based on *heuristics* or *optimization* techniques. **Descriptive** models tell us what will happen if we make a set of decisions. The response is characterized by means of *equation solving* or by *simulation*.

To estimate the impact of parameter uncertainty or risk, *Monte Carlo* and other *sampling simulation* methods can be used to generate sets of possible answers from static equation solvers, dynamic simulators and optimizers.

While, analysts usually use a combination of techniques to derive their recommendations, specific tools available in Industry tend to be intrinsically either more descriptive or more prescriptive. What are the respective advantages of *heuristics, optimization* and *simulation,* and how can we leverage all of the approaches?

Heuristics

Heuristics are rules and principles that have been derived from experience, by long trial and error, or based on logical arguments. Heuristic rules are very popular because people can easily comprehend them and can be extremely effective for very specific problems. Unfortunately, situations that occur infrequently are normally not accounted for by these rules. Most heuristic rules have to be derived case by case and offer limited reapplicability. Because they do not explicitly incorporate detail or dynamic complexity, they tend to be suboptimal. To get the most out of them, heuristic based approaches require a knowledgeable user, someone who understands well both the approach and the system being analyzed.

The following areas are important Industrial applications of heuristic methods:

Heuristics in Scheduling

Most of our operations at P&G are scheduled today with tools that rely heavily on heuristics. Schedules are created with spreadsheets or with specialized scheduling tools that are mostly rule based with multiple controlling parameters. Others have used expert systems to combine and direct the scheduling heuristic rules (Henning and Cerda, 1996).

Examples of heuristics that we have found useful are: the allocation rules that specify the preferred resources for given tasks, the sequencing rules that specify the preferred order to execute tasks, the routing rules that specify the preferred product flow management, the cycle rules that specify the frequency with which things ought to be done, and the conditional rules that specify future actions based on the history of the system.

In day to day operations, variability, dynamic complexity and uncertainty are handled by increasing the frequency of rescheduling. Also, Gantt Chart manipulation tools allow users to make minor adjustments to production plans.

Scheduling heuristics are fast and flexible for simple systems with little interaction between resources. It is more difficult to find reliable and effective rules for multi-stage processes with extensive equipment and people sharing. For these systems, mathematical programming methods have proven to be helpful. *However, heuristics must be added to mathematical programming methods to speed-up computations, to implement desired operating strategies and to gain the approval of the tool by the end user.*

Advisor Systems

Advisor systems are guided sets of heuristic rules, sometimes augmented with equation solvers and optimizers. Advisor Systems are good for monitoring processes and providing decision support for the less experienced operators and analysts. However, for rapidly changing environments and high adaptive complexity, pure advisor systems are difficult to maintain and support.

Systems Thinking

Made popular by Senge (1990), Systems Thinking, ST, is the ability to see and deduce behavior patterns, rather than single events. It involves the identification and classification of system behavior into *archetypes* that are recurrent in a wide variety of systems. The is to find a way to change the structure of a system to improve its performance and robustness (Richardson, 1996). For example, the archetype "Fixes that Fail", illustrates how feedback loops introduce unexpected system changes that can prevent the successful implementation of problem solutions.

Large system improvements can be obtained by identifying systems archetypes without the need of other techniques. However, ST nicely complements simulation and optimization. For example, simulation is used extensively in ST to identify archetypes, to demonstrate the resulting negative performance, and to test strategies that correct the behavior. Also, ST can supplement optimization for post-analysis to protect against dynamic complexity since one of the archetypes can limit the projected benefits.

Theory of Constraints

In a production system, the most heavily loaded piece of equipment is the resource constraint. Theory of Constraints, TOC, is based on 5 steps: (1) identify the system constraint, (2) exploit the constraint, (3) subordinate all the decisions for non-constraints, (4) elevate the constraint, and (5) identify the new constraint and start again. Goldratt (1990) developed sequencing heuristic rules to efficiently schedule and exploit the resource constraint. Subordination, or support of the decisions made for the constraint, is achieved by sizing buffers for equipment before and after the constraint and by scheduling material releases to support the constraint activities. Adding capacity to the system in the right place elevates, or eliminates, a constraint.

The recommended TOC scheduling sequencing rules work well for moderate levels of complexity and clearly identifiable constraints. For larger system with multiple interactions between resources, optimization with broader heuristics provides a more robust approach. Nevertheless, the principle of focusing on constraint resources can simplify many apparently complex problems by limiting the detail that needs to be considered at one time.

Initially, TOC was related to production scheduling (Goldratt, 1990b) and inventory management (Goldratt, 1986). Today TOC has become more generic and focused on conflict resolution (Goldratt, 1994), dealing with people as well as production systems. Conflicts are identified and eliminated by using special TOC influence tree diagrams and conflict resolution techniques. Therefore, TOC incorporates the human factor in ways that simulation and optimization cannot when policies and procedures constrain the throughput of a system.

Simulation

As a technique, simulation is one of the most widely used in operations research and management science. Given its flexibility, it is used regularly to address a variety of engineering problems at the design, planning and operational levels (Law and Kelton, 1991).

The reasons for the use of simulation are compelling. Equation solving, heuristic and optimization models often do not take correctly into account the dynamic complexity and the variability in systems. Sometimes what-if information is more important than knowledge of the optimal decision because it can be translated into actionable responses to specific situations. The current capability of many simulation tools to visualize an operation sometimes adds to the final understanding and aids in detecting design flaws.

Since simulation is a descriptive approach, modelers need to run multiple what-if scenarios to derive prescriptive recommendations. Until recently, metamodels based on design of experiments were a favored vehicle for optimization (Law and Kelton, 1991; Madu, 1990). More recently, several software simulation tools have added stochastic and AI based optimization techniques to automatically run simulations (Glover et al., 1996; Pierreval and Caux, 1996). Kalasky (1996) even goes as far to suggest that simulation techniques can replace, and perhaps eventually displace, existing static and deterministic optimization methods in the areas of planning, engineering design and scheduling.

Unfortunately, while it is true that we can add precision with simulation by including process variability, it is also true that the necessary data is usually unavailable. Data gathering, validation and interpretation are often difficult tasks. Moreover, as Law et al. (1994) show, using the wrong probability distributions can result in large output error. Yet, distribution-fitting software usually assumes uncorrelated and stationary IID data, in spite of the fact that processes in large supply-chain problems tend to be non-stationary and correlated. Trends, natural non-homogeneity and lack of statistical control are common. Even if the variability of inputs was well understood, the large uncertainty in some key parameters does not allow a gain in accuracy over other techniques for many problems.

The following areas are important types of supply-chain simulation methods:

Discrete-event Simulation

Discrete-event simulation is the representation of a system as it evolves over time with the state of the system changing only at discrete, but possibly random, set of times. Widely used because current tools can include generic queuing models, stationary Markov, Poisson and renewal processes, as well as non-stationary Poisson and ARIMA processes. Some tools even have the capability to model continuous-time processes. Models can be built quickly by connecting preset elements and by filling out dialog boxes. Although animation has been promoted as an important feature, the difficulty and costs to setup computer visualization or animation are not worth the benefits in a great number of applications. Simple graphs, charts and program traces can achieve the same results.

Because they are so flexible, there is a tendency for the less experienced modelers to include everything in detail in their models, together with unnecessary constraints. Since the structure usually hides the details, the assumptions that went into a model can be difficult to examine, particularly if it is poorly documented.

System Dynamics

Developed by Forrester (1961), System Dynamics (SD) models evolve continuously over time and make use of only few primitives: tasks, inventories, equations, graphical functions and flow controls.

Why use SD tools given that the simulation output derived from SD can be also obtained with a good discrete-event software tool that can model continuous-time processes?

Good for process reengineering, SD models clearly show the relationships between tasks in the decision process, as opposed to being equipment centered. Because SD models are essentially a logical map of processes and decisions, the tools are ideal for modeling and discussions in group settings. Model assumptions are not hidden. SD tools encourage the use of broad models with soft or qualitative factors such as the level of expertise of people, instead of including too much detail. Therefore, SD models are normally used to demonstrate long term trends rather than short term behavior. The use of qualitative factors is an acknowledgment that some important factors cannot be obtained or quantified. Currently, large detail complexity cannot be included in SD models.

Learning environments and role playing

Advances in multimedia combined with simulation form an excellent tool to train people on how to make better decisions. Good implementations are still not easy, but as software becomes more sophisticated, it will increase the effectiveness of this approach. The key is for the models to capture the desired behaviors.

Complex Adaptive Systems

In Complex Adaptive (CA) models, multiple evolving agents are used to represent parts of the system or natural phenomena. They are valuable because they can produce complex emergent phenomena out of a small set of relatively simple rules, constraints and relationships. (Wildberger, 1996; Holland, 1995). CA systems are similar to the genetic algorithms in optimization that are discussed in the next section in that they both use evolutionary computing techniques and fitness functions.

ince some CA models appear to imitate the complex interactions of people and businesses in competitive environments such as the stock market (Wan and Hunter, 1997), they may be explain one day why certain behaviors are observed.

Optimization

Optimization is the cornerstone of supply-chain management. It is used at all levels from operations planning and scheduling to design, sourcing and financial analysis. Novel and non-intuitive answers can be found with optimization tools because they do not need to introduce preconceived notions about the solution This reduces the likelihood of unintentional bias in the model. It is not uncommon to find that optimum solutions do not match what people were expecting it to be.

Getting the "right" objective function is probably the most important step with these approaches. Multiple goals and qualitative considerations make objective definition difficult. Many suboptimal recommendations are obtained because of a poor problem formulation rather than the optimization technique used.

The following areas are important types of deterministic optimization methods. If necessary, all the approaches can be combined with simulation to add stochastic behavior.

Mathematical Programming

Mathematical programming is a highly developed area based on expressions of the general form:

$$\begin{aligned} &\text{Minimize} \quad F(\mathbf{x}) \\ &\text{Subject to} \\ &\qquad h(\mathbf{x}) \leq 0 \\ &\qquad g(\mathbf{x}) = 0 \end{aligned} \quad (1)$$

Linear Programming (LP), where the **x** variables are continuous, and Mixed Integer Linear Programming (MILP), where **x** elements can continuous or discrete, are routinely applied to supply-chain problems. Dynamic complexity can be included by using Nonlinear Programming (NLP) or Mixed Integer Nonlinear programming (MINLP). Unfortunately NLP and MINLP can handle fewer variables and are therefore used only for specific applications.

Until recently, the problems in planning, design and scheduling that could be solved using deterministic optimization methods were too small, took very long to solve, and recommendations were impractical. Thanks to new algorithmic developments, models with thousands of variables can now be solved efficiently. With the help of computer technology and by exploiting domain specific heuristics, the number of variables can easily reach the millions.

The robustness of the decisions can be evaluated using sensitivity analysis techniques. Some types of parameter uncertainty can even be incorporated explicitly in the formulation of the problem. In day to day scheduling applications, variability, dynamic complexity and uncertainty are handled by increasing the frequency of rescheduling. Also, Gantt Chart manipulation tools allow users to make minor adjustments to the plans .for factors not included in the mathematical formulation.

Nevertheless, to really understand the impact of variable processing times and lead times, unexpected events and forecasting errors, simulation must be used. To add simulation on top of optimization, the concept of a model predictive controller can be used to represent the process of reoptimizing as illustrated in Fig. 2. Reoptimizing is done when the system significantly deviates from the planned performance, at periodic intervals of time and/or when new information becomes available.

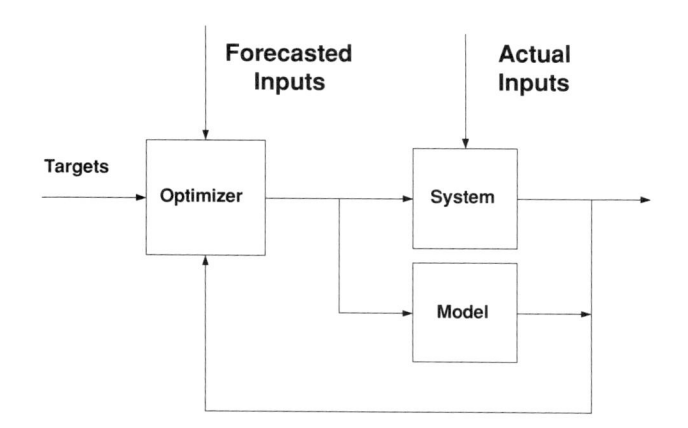

Figure 2. Rescheduling approach.

AI and Genetic Algorithms

AI techniques can be combined with mathematical programming to improve the computational efficiency and the robustness of solutions. With qualitative or declarative representation languages, domain specific knowledge and factors that cannot be expressed in the form a mathematical equation can be incorporated in the solution search.

Several population based methods are available to solve optimization problems directly (Cagan *et al.*, 1996). In Genetic Algorithms (GA), offspring candidate solutions are generated from parent sets and then the least fit are discarded. In Tabu Searches (TS), restrictions are imposed to discourage new candidate solutions to resemble solutions encountered earlier in the simulation. In Simulated Annealing (SA), a function is used as a degree of leniency in deciding to accept new trial points.

These techniques are popular because they can be applied to many problems, they relatively easy to program and they tend to converge better to global optimal solutions. In contrast, some mathematical programming techniques are difficult to program and tend to get stock in local optimum. AI techniques are not used more often because it takes very long for them to converge in large size problems, if they ever converge.

Integrated Approach to Supply-Chain Analysis

Our need to extend models to include suppliers, customers and the consumer has required that the analysis tools that we use have both breadth and depth. In Industry, we seldom have the time to master multiple tools and we would like to easily add detail to a model as required. Therefore, an integrated approach of the methods presented here is desirable.

One outage is the lack of mathematical programming solvers within commercially available simulation environments. In supply-chain problems, simulation formulations currently need extensive use of heuristics to represent the decision rules that guide the model. If we could replace some of the heuristics with optimization algorithms, we could explore more readily the impact of better synchronization of day to day operations at different stages and locations. In this case, each simulation run would make use of mathematical programming to guide the system and model parameters would be adjusted between simulations using stochastic optimization.

We also note that there is a lack of simulation capability within commercially available optimization environments. Simulation can help account for process variability and dynamic complexity to test for the robustness of optimum solutions. In this case, we would start by solving the problem using the deterministic optimization tool and then feed the solution to a simulation environment that would allow the dynamic reoptimization of a subset of the parameters during simulation runs.

Regardless of the approach taken, simulation and optimization must be combined. There should be no need to input the problem for optimization and then a second time for simulation. Therefore, the problem formulation should be decoupled from the solution approach. Given the nature of supply-chain problems, the problem formulation should permit hierarchical structures to represent the desired level of detail. The capability to incorporate heuristics is also necessary to test specific operating strategies.

Finally, although methodologies such as Systems Thinking and Theory of Constraints provide some qualitative perspective on global system behavior and process improvement, further research is needed to understand the impact of qualitative factors, the human behavior, and information availability throughout the supply-chain.

Conclusions

We have reviewed how heuristics, simulation and optimization are important for supply-chain analysis, as well as some of their shortcomings. Two new directions in analysis tools have been proposed. We can add mathematical programming within the current simulation tools, or we can add simulation to current mathematical programming tools. Research is required to determine which approach will be most valuable to practitioners in Industry and under what conditions.

References

American Chemical Society, 1996. *Technology Vision 2020. The U.S. Chemical Industry.* Wasington, DC.

Cagan J., I. E. Grossmann, and J. Hooker (1996). Combining artificial intelligence and optimization in engineering design: a brief survey. In J. F. Davis, G. Stephanopoulos, and V. Venkatasubramanian (Eds.), *International Conference on Intelligent Systems in Process Engineering.* CACHE Publications. pp. 110-118.

Forrester, J. W. (1961). *Industrial Dynamics.* MIT Press, Cambridge, Mass.

Glover, F., J. P. Kelly, and M. Laguna (1996). *New advances and applications of combining simulation and optimization.* Proceedings of the 1996 Winter Simulation Conf. Coronado, California. pp. 144-152.

Goldratt, E. M. (1990). *The Haystack Syndrome.* North River Press, New York.

Goldratt, E. M. (1994). *It's not Luck.* North River Press, New York.

Goldratt, E. M. and R. E. Fox. (1986). *The Race.* North River Press, New York.

Henning, G.P. and J. Cerda (1996). An expert system for the scheduling of multistage multiproduct plants manufacturing assorted products. In J. F. Davis, G. Stephanopoulos, and V. Venkatasubramanian (Eds.), *International Conference on Intelligent Systems in Process Engineering.* CACHE Publications. pp. 110-118.

Hollander, J. H. (1995). *Hidden Order: How Adaptation Builds Complexity.* Addison-Wesley Publishing Company, New York.

Kalasky, D. R. (1996). *Simulation-based supply-chain optimization for consumer products.* Proceedings of the 1996 Winter Simulation Conf. Coronado, California. pp. 1373-1378.

Law, A.M. and W. D. Kelton (1991). *Simulation Modeling & Analysis,* 2nd ed. McGraw-Hill, New York.

Law, A.M., M. G. McComas (1994). The crucial role of input modeling in successful simulation studies. *Industrial Engineering,* **26, No. 7**, 55-57.

Lee H. L., V. Padmanabhan, S. Whan (1997). Information distortion in supply chain: the bullwhip effect. *Management Science,* **43, No. 4**, 546-558.

Madu, C. N. (1990). Simulation in manufacturing: a regression metamodel approach. *Computers and Industrial Engineering,* **18, No. 3**, 381-389.

Pierreval, H., and C. Caux (1996). Using simulation in production management. *SIMULATION*, **66, No. 2**, 73-74.

Richardson, G. P. (1996). *Modelling for Management: Simulation in support of systems thinking,* Vols. I and II. Dartmouth, Cambridge.

Senge, P. (1990). *The Fifth Discipline: The Art and Practice of a Learning Organization.* Doubleday, New York.

Sterman, J. D. (1996). A skeptic's guide to computer models. In G. P. Richardson (Ed.), *Modelling for Management: Simulation in support of systems thinking*, Vol. I. Dartmouth, Cambridge. pp. 3-23.

Wan, H. A., and A. Hunter (1997). On artificial adaptive agents models of stock markets. *SIMULATION*, **68, No. 5**, 279-289.

Wildberger, M. A. (1996). *Introduction & overview of "artificial life" – evolving intelligent agents for modeling & simulation.* Proceedings of the 1996 Winter Simulation Conf. Coronado, California. pp. 1373-1378.

TOTAL SITE OPTIMIZATION OF A PETROCHEMICAL COMPLEX

Metin Turkay, Tatsuyuki Asakura, Kaoru Fujita, Chi Wai Hui* and Yukikazu Natori
Development and Engineering Research Center
Mitsubishi Chemical Corp., Mizushima Plant
Kurashiki, Okayama 712, Japan

Yoshihisa Masaiwa and Haruyoshi Oonishi
Planning and Coordination Center
Mitsubishi Chemical Corp., Mizushima Plant
Kurashiki, Okayama 712, Japan

I. Bhieng Tjoa
MC Research & Innovation Center, Inc.
Mt. View, CA 94041

Abstract

This paper addresses the total site optimization of petrochemical complexes. A systematic optimization approach in the optimization of petrochemical complexes in which all of the processes are integrated can provide useful insights into to operation of the complex and detect operational and economical improvements. The total site optimization problem is modeled in the Generalized Disjunctive Programming (GDP) framework in which discrete behaviors of processes are represented naturally. The nonlinear process models are reformulated with piecewise linear functions, and then classified as convex and non-convex. The resulting mixed-integer optimization problem is solved by a novel decomposition algorithm. The proposed decomposition algorithm is used to optimize a petrochemical complex. It is shown that the modeling approach and the decomposition algorithm are robust and reliable in solving very large-scale optimization problems.

Keywords

Total site optimization, Generalized disjunctive programming, Mixed-integer optimization, Linear programming, Production planning.

Introduction

A petrochemical complex is a highly integrated combination of process plants that have certain level of interaction among each other. The literature in which the process plants in a petrochemical complex are considered integrated by one model is concentrated to specific problems. Long-range planning and capacity expansion of petrochemical complexes has been studied as a multi-period optimization problem (Sahinidis et al., 1989;

---

* Current address: Department of Chemical Engineering, The Hong Kong University of Science and Technology, Clear Water Bay, Kowloon, Hong, Kong

Sahinidis and Grossmann, 1991; Liu and Sahinidis, 1996; Iyer and Grossmann, 1998). The process models in these studies consist of linear material balances where only major chemicals are considered. The pinch analysis of petrochemical complexes has been reported to provide some energy savings (Linnhoff, 1993). The optimization of single process plants with detailed models has also attracted some attention in process control applications (Trieber et al., 1992). The real-time optimization of a complete ethylene plant has been studied and the advantage of optimizing complete plants compared to a collection of units was reported (Emoto et al., 1994). Recently, Tjoa et al. (1997) reported MINLP based ethylene plant scheduling and optimization. This study was extended for optimal feedstock planning and scheduling (Ota et al., 1998).

Although planning of petrochemical complexes and optimization of single process plants has been studied, there is no reported study in the optimization of petrochemical complexes in which process plants are treated as integrated systems rather than disconnected process plants. The main obstacles in the optimization of petrochemical complexes are the difficulties in modeling large-scale systems and lack of reliable large-scale optimization algorithms.

In this paper we describe the total site optimization of a petrochemical complex as an integrated collection of processes. Optimal operation for a process plant may result in suboptimalities or even infeasiblities for other plants in the same complex due to strong interactions. A systematic approach in which petrochemical complexes are treated as integrated process plants can provide useful insights into the operation of the complex and detect economic improvements.

Problem Statement

The petrochemical complexes are composed of process plants that have discrete operational features. An important step in the optimization of a petrochemical complex is the model development. The optimization model should include important characteristics of the petrochemical complex so that approaches for the optimal operation, maintenance, scheduling and planning can be analyzed simultaneously.

Generalized Disjunctive Programming (GDP) has been used as the modeling framework (Turkay and Grossmann, 1996) in the total site optimization of a petrochemical complex. The GDP is a natural modeling framework in which discrete decisions and discrete behaviors of processes/units in a process network are represented. Another advantage of GDP modeling framework is the ability to represent processes/units as separate modules that allows flexibility in model generation, model analysis, and custom solution algorithm development.

The nonlinear process plants are reformulated with piecewise linear functions in the following form,

$$y_i = a_{ij} x_i + b_{ij} \quad j \in D_i \quad (1)$$

where the main variable x_i has lower and upper bounds for each section j of the piecewise linear function i. The function (1) can be represented in the GDP modeling framework as follows:

$$\bigvee_{j \in D_i} \begin{bmatrix} y_i = a_{ij} x_i + b_{ij} \\ x_{ij}^L \leq x_i \leq x_{ij}^U \end{bmatrix} \quad (2)$$

The functions in (2) are classified into two groups as shown in Fig. 1: convex and non-convex (Nemhauser and Wolsey, 1988). The convex piecewise linear functions have the property that the slope of the function increases with respect to sections as shown in Fig. 1a. In general convex piecewise linear functions correspond to piecewise linear formulation of convex nonlinear functions. Similarly, non-convex piecewise linear functions correspond to non-convex nonlinear functions as shown in Fig. 1b. The slope of the non-convex piecewise linear functions does not increase monotonically with respect to sections.

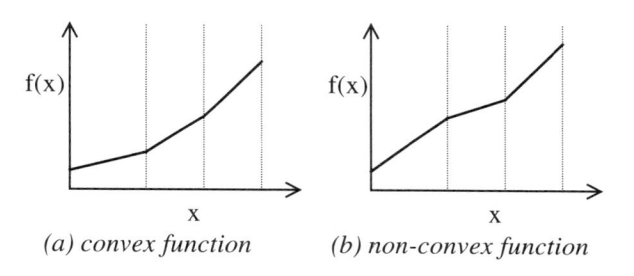

(a) convex function   *(b) non-convex function*

Figure 1. *Convex and non-convex piecewise linear functions.*

The non-convex piecewise linear equations can be reformulated with the convex hull formulation of disjunctions that was developed by Balas (1985) and Turkay and Grossmann (1996). The convex hull formulation has proved to provide a tight representation of disjunctive models (Turkay and Grossmann, 1998). This reformulation technique requires the introduction of a 0-1 variable for each section of piecewise linear function and disaggregation of continuous variables for each section. The function (2) is reformulated with convex hull formulation of disjunctions as follows:

$$y_i = \sum_{j \in D_i} y_{ij}$$

$$x_i = \sum_{j \in D_i} x_{ij}$$

$$\left.\begin{array}{r} y_{ij} = a_{ij}x_{ij} + b_{ij}z_{ij} \\ x_{ij}^L z_{ij} \le x_i \le x_{ij}^U z_{ij} \end{array}\right\} \; j \in D_i \quad (3)$$

$$\sum_{j \in D_i} z_{ij} = 1$$

$$z_{ij} = 0,1 \quad \forall j \in D_i$$

The reformulation of convex piecewise linear functions is made by disaggragating continuous variables, x_i and y_i, for each section of the piecewise linear function without the use of 0-1 variables. The number of 0-1 variables used to reformulate piecewise linear functions can be a major computational bottleneck in finding optimal solution of mixed-integer programming problems. Substantial reduction in the number of 0-1 variables can be obtained with this classification considering that in general 50% of the process models are convex.

The GDP model for the total site optimization of a petrochemical complex is given in (4). The objective function is the minimization of the operating cost that is a function of raw material and utility costs. In some case studies, the objective function can be substituted by the maximization of profit. The overall material and energy balances are included in the first inequality in (4). The process plant models are formulated with disjunctions such that when the plant/unit k is operated (i.e., P_k) piecewise linear and linear equations are included in the model. When the plant/unit k is not operated due to maintenance or any other reason the variables involved in plant k are set to fixed values (most of the time zero, but under special conditions non-zero). The propositional logic representation of processes/units in the petrochemical complex are included ensuring that when a process/unit stops the operation of all related processes/units also stops.

$$\min \; cost = cy + dx$$
$$\text{subject to}$$
$$\mathbf{E}x \le e$$

$$\bigvee_{j \in D_i} \begin{bmatrix} P_k \\ y_i = a_{ij}x_i + b_{ij} \\ x_{ij}^L \le x_i \le x_{ij}^U \end{bmatrix} \\ y_k = a_k x_k + b_k \\ y_k^L \le y_k \le y_k^U \\ x_k^L \le x_k \le x_k^U \end{bmatrix} \vee \begin{bmatrix} \neg P_k \\ y_i = y_i^N \\ x_i = x_i^N \\ y_k = y_k^N \\ x_k = x_k^N \end{bmatrix} \; k \in K \quad (4)$$

$$\Omega(P_k) = \text{True}$$
$$P \in \{\text{True, False}\}$$

GDP Decomposition Algorithm

GDP decomposition algorithm consists of a series of LP subproblems and a final MILP problem that includes all characteristics of the total site model. LP subproblems are used to determine the optimal operating regions for process plants/units and identify corresponding part of the piecewise linear functions. The LP subproblems have two shortcomings in handling these functions. When a variable has a value of 0, the function value has a non-zero value due to the fact that the constant b_{ik} in equation (1) cannot be eliminated (Fig. 1). This is a characteristics of the LP models that may be avoided either by the use of 0-1 variables or changing the process models to eliminate constants. The use of 0-1 variables converts the LP problem into MILP problem. The second shortcoming is a rather specific one that is related to the form of the piecewise linear functions. If the functional form is similar to the one in Fig. 2 that usually correspond to linearization of s-shaped nonlinear functions, then the LP subproblems may cycle.

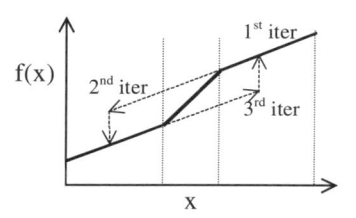

Figure 2. Illustration of cycling in LP subproblems.

MILP problem includes all of the features of the total site optimization of a petrochemical complex and eliminates shortcomings of the LP subproblems. The decomposition algorithm decomposes the problem into manageable subproblems while preserving the features and not compromising from the global optimality of the overall system.

The 0-1 variables in the total site optimization model are:

i. each section of the non-convex piecewise linear functions
ii. alternative feedstocks for certain processes
iii. the decisions for stopping/starting each process
iv. capacity expansion for certain processes
v. addition of new processes
vi. first day of regular maintenance period for each process

Example

The total site optimization study has been applied on the Mizushima Plant of Mitsubishi Chemical Corp. The optimization model included 53 major process plants. A process unit is treated as a process plant if that unit can be operated independently. There are 9 such process units in the Mizushima Plant. The optimization model includes 1202 chemicals among which 32 are major products and 16 are major raw materials. An important feature of the model is the ability to select different raw materials for the same process depending on the season, availability and price. The ethylene process provides most of the raw materials for the process plants in the Mizushima complex. This process uses heavy or light naphtha as major raw material and produces the following major intermediate products: ethylene, propylene, residue gas, C4 fractions and gasoline. The ethylene process consists of 19 furnaces in 4 types, quenching section and separation section. The ethane and propane by-products are separated and recycled back to the furnaces for co-cracking with naphtha.

The utility plant consisting of three boilers and five steam turbines supplies electricity and various levels of steam to other process plants in the complex. Each boiler uses a different type of fuel and produce high-pressure steam. The high-pressure steam from the boilers is combined with the steam from the ethylene process and distributed to turbines in the utility plant. Some of the single-stage turbines are operated when there is a need for additional electricity. These compressors have lower and upper bounds on the steam flowrate; therefore, models for the operation of the single-stage compressor requires the use of 0-1 variables. The operation of the multi-stage compressors is optimized to satisfy the steam requirement of the process plant at various levels of pressure and maximize the electricity generation. If the amount of electricity generated by the utility plant does not satisfy the demand of the process plants, external purchase of electricity is possible.

Other process plants in the complex include polyvinyl chloride, polyethylene, polypropylene, acetone, and aromatics plants. There are important interactions among these processes, since their operation is strongly affected by the operation of ethylene plant. The operation of the ethylene plant is also strongly affected by the demand of final products. The utility plant that provides steam and electricity to all of the process plants is strongly affected by operating conditions of every major process plant. All of the process plants require a yearly regular maintenance that lasts from 20 days to 2 months depending on the complexity of the process. When a process plant that is in maintenance provides a raw material for a downstream plant, enough inventory of that material has to be built over time or transferred from another complex if the downstream plant is not in maintenance. The optimization model should include the transportation and inventory costs to determine the best policy for coordinating inventory management and maintenance of process plants.

Table 1. Summary of Problem Size with Conventional Techniques.

	MILP
0-1 variables	2,036,700
cont. variables	13,263,735
constraints	13,572,525

Discrete time representation of the multi period optimization problem is used in the model. The optimization period is 365 days and each day consists of three shifts with different operating hours. The number of hours in each shift depends on the electricity purchase contract with a local utility company. When the total site optimization model is formulated for solving with conventional solution techniques the problem size becomes impossible to solve as shown in Table 1.

A reduction in the problem size is possible by further analysis of the process models and cost data. However, the problem size remains a bottleneck for solving this problem with conventional optimization algorithms. A GDP decomposition algorithm has been developed for solving the total site optimization of a petrochemical complex problem as described in the previous section. An important feature of the algorithm is that it decomposes the large scale MILP problem into a sequence of smaller scale LP subproblems and reduces the problem size and computational effort in the MILP problem. We also analyzed the process models and cost data to reduce the problem size during the execution of the steps in the GDP decomposition algorithm.

The optimization model is generated in the GAMS modeling language (Brooke et al., 1992) and the LP subproblems and MILP problem are solved with the interior point method of CPLEX solver. The LP subproblems determine the optimal sections in non-convex piecewise linear functions and the number of 0-1 variables for the MILP problem is reduced substantially. The 0-1 variables for MILP problem includes sections for the unconverged piecewise linear functions, alternative feedstocks, stopping or starting of certain process plants, capacity expansions, and maintenance periods for process plants. The operating budget of a petrochemical complex has been completed with successful solution of this model for 1998 financial year. The problem is solved on a 266 MHz Pentium II PC with 384 MB RAM in about 4 hours of computing time. It is observed that half of this time is spent in generating equations in the LP subproblems and the final MILP problem.

The optimization system is integrated with a model manager and database system by a graphical user interface to make the system user friendly. The model manager collects the data from process plants and creates input data for the optimization model in the proper format after interaction with the database. The optimization results are passed to the database in order to create easy to understand reports and production, and inventory profiles for major chemicals.

Conclusions

A GDP model and decomposition algorithm was developed for the total site optimization of petrochemical complexes. The model integrates all of the process plants in a petrochemical complex to capture interactions among these plants in one model. It is shown that conventional optimization techniques are not sufficient to solve this problem. A novel GDP decomposition algorithm was developed that consists of a series of LP subproblems and a MILP problem. The LP subproblems determine the optimal sections for piecewise linear functions and MILP problem optimizes the complete system. The robustness and efficiency of the approach described in this paper is tested on an actual petrochemical complex.

References

Balas, E. (1985). Disjunctive programming and a hierarchy of relaxations for discrete problems. *SIAM J. Alg. Disc. Meth.*, **6**, 466-486.

Brooke, A., D. Kendrick, and A. Meeraus (1992). *GAMS-A user's guide*, release 2.25, Boyd and Fraser Publ. Co, Danvers, MA.

Emoto, G., Y. Ota, H. Matsuo, M. Ogawa, D.B. Raven, R.F. Preston, and J.S. Ayala (1994). Integrated advanced control and closed-loop real-yime optimization of an olefins plant. *Proceedings of IFAC Symposium ADCHEM 94'*, Kyoto, Japan.

Iyer, R.R and I.E. Grossmann (1998). A bilevel decomposition algorithm for long-range planning of process networks. *Ind. Eng. Chem.*, **37**, 474-481.

Linnhoff, B. (1993). Pinch analysis-A state-of-the-art overview. *Chem. Eng. Res. Des.*, **71**, 503-522.

Liu, M.L. and N.V. Sahinidis (1996). Long range planning in the process industry. *Comput. Oper. Res.*, **23**, 237-253.

Nemhauser, G.L. and L.A. Wolsey (1988). *Integer and Combinatorial Optimization*, John Wiley & Sons Inc., New York, 11-12.

Ota, Y. and I.B. Tjoa (1998). Optimal naphtha selection for olefins plant feedstock planning and scheduling. *1998 AIChE Spring National Meeting*, paper 62e, *New Orleans, LA*, (March, 1998).

Sahinidis, N.V., I.E. Grossmann, R.E. Fornari, and M. Chathrathi (1989). Optimization model for long range planning in the chemical industry. *Comput. Chem. Eng.*, **13**, 1049-1063.

Sahinidis, N.V. and I.E. Grossmann (1991). Reformulation of multiperiod optimization model for long range planning in the chemical industry. *Comput. Chem. Eng.*, **15**, 255-272.

Tjoa, I.B., Y. Ota, H. Matsuo, and Y. Natori (1997). Ethylene plant schedulin system based on a MINLP formulation. *Comput. Chem. Eng.*, **21**, S1073-S1077.

Trieber, S., R.S. McLeod, T.J. Boyle, G. Powley, and S. Lee (!992). Closed-loop plant wide optimization. *Control System '92 Conference, Whistler, BC, Canada*, (September 1992).

Turkay, M and I.E. Grossmann (1996). Disjunctive programming techniques for the optimization of process systems with discontinuous investment costs-Multiple size regions. *Ind. & Engr. Chem. Res.*, **35**, 2611-2623.

Turkay, M and I.E. Grossmann (1998). Tight mixed-integer optimization models for the solution of linear and nonlinear systems of disjunctive equations. *Comput. Chem. Eng.*, in print.

OPERATIONAL PLANNING FOR CHEMICAL PROCESSES USING GEOGRAPHIC INFORMATION AND ENVIRONMENTAL MODELING

Derya B. Özyurt and Matthew J. Reallf
School of Chemical Engineering
Georgia Institute of Technology
Atlanta, GA, 30332-0100

Abstract

In this research we study a system consisting of time varying industrial processes and a river as the ecosystem process. We use geographic location information to define the various subsystems and to present the effect of operational changes on environmental quality. A Geographic Information System (GIS) is interfaced to a standard optimization package, GAMS (Brooke, Kendrick, Meeraus, 1992) that performs optimal decision making based on cost and environmental quality constraints. This can lead to more rational assessment of the environmental impact and enable novel solutions to environmental problems.

As a case study, we consider poultry facilities located along a river. The GIS is used to physically locate the various poultry facilities and their discharge points to the river. It is also used to define the river in term of the different "reaches", sections in which the flow is regarded as constant, since tributaries define reach boundaries. The flow and other hydrological data are also extracted from a GIS database. The production scheduling problem is solved using the RTN formulation (Pantelides, 1994), coupled with a model for generation, removal and discharge of BOD, and a simple model for the discharge in the river (Loucks, Revelle, Lynn, 1967; Schnoor, 1996).

Keywords

Geographic information systems, RTN, Environmental modeling, BOD, Operational planning.

Introduction

Operational planning of single- and multi-site processes has been studied in some depth (Pantelides,1994; Reklaitis,1996; Wilkinson, Shah, Pantelides, 1994). However, the effects of the operational changes on the environmental quality have only recently been the subject of study (Sharratt, Kiperstok, 1996; Stefanis, 1997). This is for several reasons.

First, defining an environmental cost which will be placed in the objective of the optimization is challenging. Second, a systematic framework for assessing the impact of operational changes on environmental quality is lacking. Finally, the parameters for mathematical formulation and optimum actions depend on the geographical location of the process, since this will determine the environmental media which are affected, the dispersion rates within those media, and the ecosystems impacted. Thus, it is necessary to describe the system in its broadened environmental dimension, and a reasonable representation requires an information source like **G**eographical **I**nformation **S**ystem (GIS) (Huxhold, 1991), integrated with the overall framework (Daniel, Diakoulaki, Pappis, 1997).

Recently, two approaches to extending process analysis and synthesis to consider environmental impacts have been proposed. Stefanis (1996) embedded Life Cycle Analysis into the design, synthesis and operation of chemical processes and product design. Sharratt and Kiperstok (1996) developed a method to consider the combined effect of several plants on a common receiving body. In the former approach no attempt was made to consider multiple plants, and in the latter there was minimal modeling of either the dynamics of the receiving body or the economic objectives of the facilities. In both cases, the impact of the actual geographical location data or environmental dynamics were left unmodeled. Thus in these research efforts, the environment-plant and plant-plant-environment interactions and role and importance of the real geographical location are omitted.

To illustrate these and our initial approach, suppose that we have two facilities along a watershed. The river has several natural and industrial inflows and also outflows. If we think on a facility level and try to minimize the harmful outflow of one facility to the environment (consisting of the river and the ecological system it contains), we can do so by modeling the operations in the facility and their effect on the environmental indicators. This modeling can be performed without any concern of the world outside the plant borders. However, in most cases the overall optimum will depend, for example, on the location of the discharges, on the dynamics of the river and ecological system, and the optimized operations of the other facility. Therefore, we should include these other factors and develop a conceptual framework which will allow us to formulate appropriate models of multiple process facilities and their geographical and ecological environments. Then using this framework we will introduce a mathematical representation of the system and show a simple example problem.

Framework

The framework (Fig. 1) which we will demonstrate, contains several modules. These modules describe the different components of the problem. However, they are connected to a decision module so that we can apply a common objective to this multi-perspective problem definition.

The plant module will mostly contain information about the industrial facilities, their input, output, operations, etc., whereas the Environment module will contain receiving media models, fate and transport information, and ecological system descriptions. The GIS module will coordinate the other two through its data structures organized around geographic locations.

Geographic Information Systems (GIS)

GIS is a computer-based data base management system for capture, storage, retrieval, analysis, and display of locationally defined (spatial) data (Huxhold, 1991) with the unique visualization and geographic analysis benefits offered by maps.

GIS, which is a combination of spatial and attribute data, can identify spatial relationships, for example, we can get population and income information about the neighbors of a given country. The spatial problem solving, such as choosing a location to meet some requirements, is also one of the many features of geographic information systems (Laurini, Thompson, 1992).

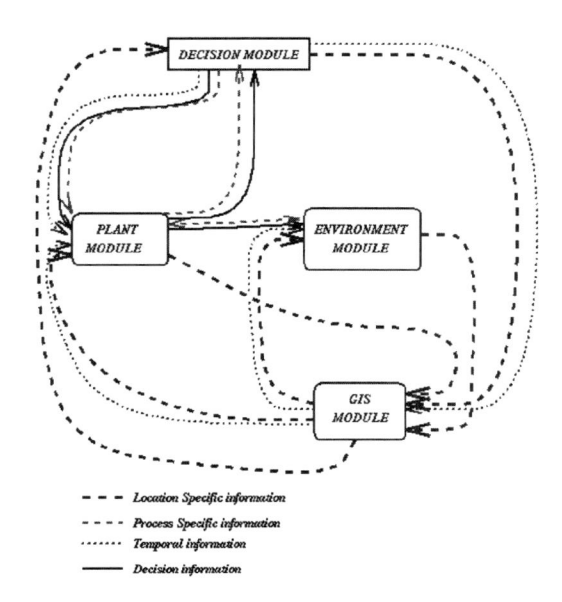

Figure 1. The representation of the framework and its information flows.

The single most distinguishing characteristic of a geographic information system is its ability to integrate information from many different sources and at many different levels of detail. This characteristic enables GIS to have a wide range of applications from map updating to finding solid waste collection routes. These applications can be enhanced by combining the GIS with other tools. For instance, the integration of GIS and spatial process models enables the models to deal with large volumes of spatially oriented data that geographically anchor processes occurring across space (Nyerges, 1993).

These modeling exercises usually do not contain any optimization and are in the form of descriptive analysis. They can for instance help planners identify, analyze and simulate the impacts of alternative management policies and practices on some predefined factors (Dikshit, Loucks, 1995-1996). GIS has been used to interface the user to an optimization environment and to display and manipulate the results of this optimization (Camm *et al.*, 1997). Here we will use a similar approach to tackle the operational planning problem with environmental considerations.

Methodology

Our methodology is to employ general adaptable mathematical formulations which will represent the plant, environment and decision modules of our framework. Then we will interface this modules with a GIS system to connect the abstract representation to physical reality using geographical maps and location based data.

Model Formulation

Mathematical formulation of the facilities

To see the time-and-location-dependent environmental impact of a system of plants, we will use a generic mathematical model to describe the behavior of each plant. This model needs to be simple enough to permit system optimization without sacrificing too many degrees of freedom. We will use a Resource-Task Network model (Pantelides, 1994) for the production operations. For every facility a separate RTN model is built.

Facility to environmental output model

If we know for every task the corresponding BOD level generated by this task in a given facility **f**, and assume additivity of BOD levels, we can write :

$$BOD_{f,t} = BOD_{f,t-1} + \sum_k \sum_{\theta=0}^{\tau_k} v_{f,k,BOD,\theta} \xi^{BOD}_{f,k,t-\theta} - BOD^{dis}_{f,t^*} \quad \forall f,t \quad (1)$$

where BOD^{dis}_{f,t^*} stands for the amount of BOD which is accumulated in a tank and ready to be discharged after BOD removal process. t denotes the time intervals which span the time horizon, and t^* is the time point when the wastewater tank is emptied. At every time of discharge we empty the wastewater tank, i.e.

$$BOD_{f,t^*} = 0 \quad \forall f, t^* \quad (2)$$

after reducing some of the BOD content with a waste treatment technology the remaining BOD level (BW_{f,s_f,t^*}) is mixed into the river

$$BW_{f,s_f,t^*} = BOD^{dis}_{f,t^*} - BOD^{rem}_{f,t^*} \quad \forall f, t^* \quad (3)$$

The amount of wastewater which is discharged into the river can also be given as an excess resource constraint similar to the BOD level constraints.

Mathematical formulation of the river

The model we are using will be a simple linear model adapted from the work of Loucks and coworkers (1967).

Total Mass Balance

In a given reach **s** the total amount flows into the reach, $QS_{s,t}$, is equal to the sum of stream coming from the previous reach, $QS_{s-1,t}$, the tributary flow entering the reach, $QT_{s,t}$, and the wastewater flow discharged into the reach, $QW_{s,t}$.

$$QS_{s,t} - QS_{s-1,t} - QT_{s,t} - QW_{s,t} = 0 \quad \forall s,t \quad (4)$$

Total BOD at the beginning of the reach

Assuming complete mixing, the BOD concentration at the beginning of each reach, $BB_{s,t}$, times total flow, is equal to the sum of the BOD concentrations at the end of the previous reach, $BE_{s-1,t}$, in the tributary, $BT_{s,t}$, and in the wastewater effluent, $BW_{s,t}$, times their respective flows.

$$BB_{s,t}QS_{s,t} = BE_{s-1,t}QS_{s-1,t} + BT_{s,t}QT_{s,t} + BW_{s,t}QW_{s,t} \quad \forall s,t \quad (5)$$

While the total and tributary flows are constant at their steady state values, $QW_{s,t}$ can vary with time and result in a bilinear constraint requiring special treatment (Al-Khayyal, 1992).

Total BOD at the end of the reach

The rate of change in the BOD concentration with time is proportional to the concentration of BOD present, B, and to the rate of the BOD addition, Rr, due to runoff and scour. In other words,

$$\frac{dB}{dt} = -(K_1 - K_3)B + Rr \quad (6)$$

where K_1 and K_3 represent rate constants for deoxygenation and sedimentation, respectively (Loucks, Revelle, Lynn, 1967).

By integration of this differential equation, we can get the BOD concentration at any point corresponding to a time, downstream from an initial BOD concentration. If we assume that we know the BOD concentration at the beginning of each reach, the parameter values K_1 and K_3 and Rr and the average time the stream travels from the beginning to the end of the reach, τ_s, then we can write the following for each reach **s**.

$$BE_{s,t} = \lambda_s(\tau_s) BB_{s,t-\tau_s} + \sigma_s(\Delta t) \quad \forall s,t \quad (7)$$

Here $\lambda_s(\tau_s)$, $\sigma_s(\Delta t)$ are time dependent parameters resulting from the integration, they are calculated using K_1, K_3 and Rr. The parameters of these equations can be input into the model using the GIS. Although, in general they are not available for all streams in a geographic region, they can be generated and linked to the GIS database. On the other hand some estimations for them are available (Roesner, Shubinski, 1978; Schnoor, 1996).

Total oxygen deficit at the end of the reach

The rate of change in the dissolved oxygen deficit concentration with time is proportional to the concentration of BOD present, B, the existing oxygen deficit, D, and the rate of oxygen production or reduction, A, due to plant photosynthesis and respiration. In other words,

$$\frac{dD}{dt} = K_1 B - K_2 D - A \qquad (8)$$

where K_2 is the reaeration constant. Using the same method as the determination of BOD at the end of the reach, we can evaluate the dissolved oxygen deficit at the end of a reach, $DE_{s,t}$, with the following formula

$$DE_{s,t} = \alpha_s(\tau_s) DB_{s,t-\tau_s} + \gamma_s(\tau_s) BB_{s,t-\tau_s} + \rho_s(\Delta t) \quad \forall s,t \quad (9)$$

Dissolved oxygen concentration at the end of the reach

At the end of every reach the dissolved oxygen concentration, $CE_{s,t}$, is the difference between the saturated oxygen concentration, $CS_{s,t}$, and the total oxygen deficit, $DE_{s,t}$.

$$CE_{s,t} = CS_{s,t} - DE_{s,t} \quad \forall s,t \qquad (10)$$

Dissolved oxygen concentration at the beginning of the reach

Similar to Eqn. (5) we can find the dissolved oxygen concentration at the beginning of the reach, $CB_{s,t}$;

$$CB_{s,t} QS_{s,t} = CE_{s-1,t} QS_{s-1,t} + CT_{s,t} QT_{s,t} + CW_{s,t} QW_{s,t} \quad \forall s,t \quad (11)$$

Total oxygen deficit at the beginning of the reach

It, $DB_{s,t}$, is the difference of the saturated oxygen concentration, $CS_{s,t}$, existing dissolved oxygen level, $CB_{s,t}$.

$$DB_{s,t} = CS_{s,t} - CB_{s,t} \qquad (12)$$

Objective

Our objective is to minimize the total production cost and BOD reduction cost for the given horizon. One possible representation, for a given time horizon H, is as follows

$$\min \sum_f \sum_{t=1}^{H} \xi_{f,k,t} C_{f,k} + \sum_f \sum_{t=1}^{H} C_f^{rem} BOD_{f,t^*}^{rem} QW_{s_f,t^*} \quad (13)$$

where $\xi_{f,k,t}$ is the *extent* of the task k (Pantelides, 1994), $C_{f,k}$ is production and C_f^{rem} BOD removal cost.

Implementation of GIS

The GIS application in our framework should accomplish at least four tasks. The first task is to locate the river which we are interested in. Second task will obtain a buffer zone around a river, which defines the boundary of our system. The third task locates the manufacturing sites within the constructed buffer coverage. The analysis require the corresponding Standard Industrial Classification (SIC)[3] code (Standard Industrial Classification Manual, 1987) of the mentioned facilities. The attribute data used contains SIC codes for primary and secondary products which allowed us to locate the industrial sites. The final task connects the GIS application with another application. Here the idea is to transfer the currently available data to another application program, which can either generate additional data or use the transferred data as a decision aid.

A general GIS system can perform the above mentioned tasks by using its spatial analysis and interfacing tools.

The minimal GIS data that can be used will be a coverage containing information about the hydrology and location of rivers and another which enables us to locate the facilities and get information about them.

Example

Problem Definition

Allen Creek is a small stream near Atlanta, Georgia. Different types of industries are located around this receiving body. We are interested in the effect of the production in the poultry plants to the Allen Creek. And we will minimize the total cost involved to maintain a sustainable stream.

---

[3] SIC is a four digit code which classifies the establishments by type of activity they are engaged. It facilitates the collection, tabulation, presentation and analysis of data relating to establishments.

GIS Data and Implementation

The basemaps of our application are two coverages[4]. The data sets are downloaded from GIS Clearinghouse (Georgia GIS Data Clearinghouse, 1996), namely the Georgia Manufacturing Directory (a data set containing points which represent the locations of all manufacturers with their production information and SIC codes) and Major Streams of State Georgia (a GIS data, which contains statewide information for the major streams).

For this study we used a UNIX-based GIS system by the name of ARC/INFO (1997). It enables us also to build user interfaces, and can be connected to other applications with network programming tools (Comer, Stevens, 1996).

The resulting map from the GIS application can be seen on Fig. 2.

The data required for the river model is given in Table 1. Although Allen Creek consists of 18 reaches, we will only use the information for the last four reaches, since the facilities in the system boundary are located around these reaches (reach number 15 and 17).

Process Information

For illustrative purposes the processes in the facilities are simplified. First, both facilities are assumed to be exactly the same, which have same production time (Table 2), demand profile for five successive days (Table 3). Second, every product is produced using all plant resources in a certain amount of time and generate a BOD level (Table 2). Third, both facilities discharge the same amount of wastewater (100 m^3) in the middle and at the end of the day.

Results

The simple production model coupled with the river model is solved for two different cases. First, the discharged amount from any facility is limited to 100 g/m^3. Obviously, both plants having the same demand profile and production line spent the same amount for production ($ 15,330) and removal ($ 10,018). Table 4 shows the amount of BOD generated (in kg) and removed, in successive rows for each plant.

The same problem, solved with environment dictating the water quality criteria, namely the maximum total oxygen deficits are introduced for every reach (in our case this limits are set to 3.5, 1.8, 1.7 and 1.0 g/m^3, respectively). Although there is only a small change in the cost of BOD removal ($10,045 and $10,051 for plant 1 and 2 respectively), the flexibility of the production is used to accommodate feasible BOD generation and removal. In this case both facilities tried to remove most of the BOD content in their wastewater. First plant discharged approximately 20 g/m^3 BOD, whereas the second plant is allowed to discharge only 1-2 g/m^3 BOD in wastewater (Table 5). This is because the second facility is located downstream of the first one and depends on the actions of the first facility.

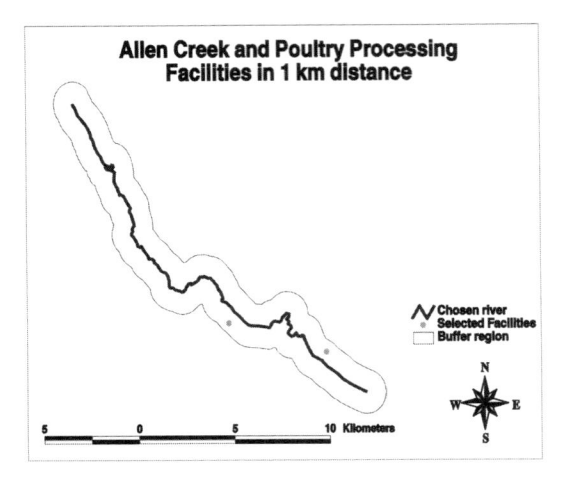

Figure 2. Result of the GIS application for the case study.

Conclusions

This poultry example is the embodiment of a general framework to represent operational aspects of environmental planning using several interacting process models. The methodology can be expanded to include air and soil as the media using relevant descriptive models. In this way, a geographic location based environmental impact assessment is made possible, where the source (production processes), the sink (environment) and their interaction are individually described within the same framework.

Moreover, the impact caused by a new plant construction, for instance, can be studied in the same framework. Also, multimedia modeling (Cohen, Ryan, 1985) can be used in environmental module to incorporate the transport of pollutants in other media besides water.

Future work will be done in using the formulation with more detailed production models. Some of the assumptions (e.g. constant wastewater flowrate) will be relaxed. The data from the GIS will be standardized.

---

[4] **coverage** : A set of thematically associated data considered as a unit. A coverage usually represents a single theme such as soils, streams, roads, or land use (ArcDoc, 1995).

Table 1. Stream and Wastewater Data for Allen Creek

Reach (s)	T_s (days)	QT_s (m³/day)	BT_s (g/m³)	CT_s (g/m³)	K_{1s} (day⁻¹)	K_{2s} (day⁻¹)	K_{3s} (day⁻¹)	Rr_s (g/m³/day)	A_s (g/m³/day)	CS_s (g/m³)
1	0.301	245000	4	6	0.31	1.02	0.02	0.15	0.5	10.20
2	1.168	0	0	0	0.36	0.8	0.04	0.14	0.85	9
3	0.358	0	0	0	0.34	0.73	0.05	0.11	0.39	9
4	0.0121	0	0	0	0.35	0.2	0.06	0.05	0.07	8.35

Table 2. Time Required to Process 1000 lb. of Product and BOD Level Generated.

Products	Production Time (in days)	BOD (mg/lt)
p1	1/16	3122
p2	3/32	3177
p3	1/32	3069
p4	1/32	3027

Table 3. Demand Profiles for the Products in Five Following Days.

Products	Demands for days (in 1000 lb.)				
p1	3	5	5	5	5
p2	-	5	3	4	3
p3	9	-	2	2	2
p4	10	5	10	10	8

Table 4. Generated and Removed BOD for Maximum BOD Discharge Constraint Problem.

Plants	Day 1	Day 2	Day 3	Day 4	Day 5
PL1	39.764	21.864	24.706	27.828	15.775
	39.664	21.764	24.606	27.728	15.675
PL2	30.840	42.707	33.922	27.570	27.788
	30.740	42.607	33.822	27.470	27.688

Table 5. Generated and Removed BOD for Maximum Oxygen Deficit Constraint Problem.

Plants	Day 1	Day 2	Day 3	Day 4	Day 5
PL1	43.012	24.761	30.707	24.706	30.654
	42.990	24.739	30.685	24.684	30.632
PL2	33.794	24.706	21.814	27.830	30.734
	33.792	24.704	21.813	27.829	30.732

References

Al-Khayyal F. A. (1992). Generalized bilinear programming- Part I. Models, applications and linear programming relaxation. *Eur. J. Oper. Res.* **60**, 306-314.

ARC/INFO -- Professional GIS (1997). URL: http://www.esri.com/base/products/arcinfo/arcinfo.html

Brooke, A., D. Kendrick, and A. Meeraus (1992). *Release 2.25 : GAMS : A User's Guide*. Boyd & Fraser Publishing Co., MA, USA.

Camm, J. D., T. E. Chorman, F. A. Dill, J. R. Evans, D. J. Sweeney, G. W. Wegryn (1997). Blending OR/MS, Judgment, and GIS : Restructuring P & G's Supply Chain. *INTERFACES*, **27**, 1, 128-142.

Cohen, Y., P. A. Ryan (1995). Multimedia Modeling of Environmental Transport: Trichloroethylene Test Case. *Environ. Sci. Technol*, **19**, 5, 412-417.

Comer, D. E. and Stevens, D. L. (1996). *Internetworking with TCP/IP :Client-Server Programming and Applications*, Vol. III, 3rd ed. Prentice Hall, NJ, USA.

Daniel, S. E. and Diakoulaki, D. C. and Pappis, C. P. (1997). Operations research and environmental planning. *European Journal of Operational Research*, **102**, 248-263.

Dikshit, A. K. and Loucks, D. P. (1995-1996). Estimating Non-Point Pollutant Loadings - I: A Geographical-Information- based Non-Point Source Simulation Model. *Journal of Environmental Systems*, **24**, 4, 395-408.

Georgia GIS Data Clearinghouse (1996). URL: http://www.gis.state.ga.us.

Huxhold, W. E. (1991). *An Introduction to Geographic Information Systems*. Oxford University Press, NY, USA.

Laurini, R. and D. Thompson (1992). *Fundamentals of Spatial Information Systems*. Number 37 in The A.P.I.C. Series. Academic Press, U.K.

Loucks, D. P., C. S. Revelle and W. R. Lynn. (1967). Linear Programming Models for Water Pollution Control. *Management Science*, **14**, 4, B166-B181.

Nyerges, Timothy L. (1993). Understanding the Scope of GIS: Its Relationship to Environmental Modeling. In Goodchild, M. F. and Parks, B. O. and Stegaert, L. (Eds.), Environmental Modeling with GIS, Chapter 8, pp. 75-93, Oxford University Press.

Pantelides, C. C. (1994). Unified frameworks for optimal process planning and scheduling. *Proc. Second Conf. on Foundations of Computer Aided Operations*. (Editors: D.W.T. Rippin and J. Hale). CACHE Publications. pp. 253-274.

Reklaitis, Gintaras V. (1996). Overview of Scheduling and Planning of Batch Process Operations. In G. V. Reklaitis and A. K. Sunol and D. W. T. Rappin and Öner Hortaçsu (Ed.), *Batch processing systems engineering: Fundamentals and Applications for Chemical Engineering*. Springer Verlag. pp 660-705.

Roesner, L. A. and R. P Shubinski (1978). *Modelling Procedures Manual*. Water Resources Engineers, Springfield, Virginia.

Schnoor, J. L. (1996). *Environmental Modeling*. John Wiley & Sons, Inc., USA.

Sharratt, P. N. and A. Kiperstok (1996). Environmental Optimisation of Releases from Industrial Sites into a linear receiving body. *Computers and Chemical Engineering*, **20**, Suppl., S1413-S1418.

Standard Industrial Classification Manual (1987). Executive Office of the President; Office of Management and Budget.

Stefanis, Stavros K. (1996). *A Process Systems Methodology for Environmental Impact Minimization*. PhD Thesis, University of London.

Stefanis, S. K., A. G. Livingston and E. N. Pistikopoulos (1997). Environmental impact considerations in the optimal design and scheduling of batch processes. *Computers and Chemical Engineering*, **21**, 10, 1073-1094.

Wilkinson, S. J., N. Shah and C. C. Pantelides (1994). Scheduling of Multisite Production Systems, AIChE Annual Meeting, San Francisco.

BRINGING NEW SPECIALTY CHEMICALS TO MARKET - A NIGHTMARE IN PLANNING AND SCHEDULING

Karl Schnelle*
Dow AgroSciences
Indianapolis, IN 46268

Gary Blau
Purdue University
W. Lafayette, IN 47907-1283

Abstract

In the low volume, high value added world of specialty chemical production, the job of a supply chain manager is extremely complex because of the use of multiple facilities (some third party subcontractors) for producing the intermediates and final products, shortened lead times, and the increasing global character of suppliers and customers. Additional complexities are encountered when one or more new products are introduced into the marketplace at once. In addition to the normal uncertainties in the day-to-day running of a production facility, new product introductions include uncertainties in capacities, yields, demand levels, demand timings, and product pricing.

One area most critical to improved decision making is supply chain planning and campaign scheduling. Supply chain planning is required to ensure that each intermediate is available at the right time, in the right place, and in sufficient quantities. To help shorten the time required to make supply chain decisions, a two-step procedure is used that includes the significant uncertainties. Planning which campaigns were to be run in which facilities and in which order is accomplished first. Next, detailed campaign scheduling is done to determine campaign start and stop times and monthly production amounts at each facility. Using the agrochemical industry as a real-world example, this paper examines the use of a supply chain model to organize the required information and to enhance the decision making process. With this supply chain model, the nightmares can be avoided and the effect of uncertainties on the supply chain can be understood by all parties involved.

Keywords

Supply chain, Intermediates, Actives, Formulated products, Supply, Demand, Uncertainties, Ramp-up.

Introduction

Many specialty chemicals such as pharmaceuticals, agrochemicals, and biotech products are manufactured in low volume, multi-product batch facilities. Different intermediates and formulated products may share production facilities but use distinct processing trains, or they could even share certain equipment units. If equipment sharing is the case, typically long production campaigns (on the order of months) of single intermediates are run and then careful turnarounds (cleanouts) are performed. Because many of these

---

* Author to whom correspondence should be addressed. E-mail: kschnelle@dowagro.com

chemicals that are new to the market are structurally complex, the number of processing steps (the chemical route) typically is becoming longer and more complex. Thus, manufacturing becomes more sophisticated as the number of required intermediates for each new active compound increases.

As a result, the job of a supply chain manager is extremely complex because of the use of third party subcontractors for some of the intermediates (or final products), shortened lead times, and the increasing global character of suppliers and customers. Additional complexities are encountered when one or more new products are introduced into the marketplace at once. In addition to the normal uncertainties in the day-to-day running of a production facility (equipment breakdowns, unavailability of required labor and utilities, raw material shortages, etc.), new product introductions include uncertainties in demand levels, demand timings, and product pricing. Software systems or modeling tools are needed to include many of the uncertainties associated with planning and to shorten the time required to make supply chain decisions.

A critical area for improved decision making is supply chain planning and campaign scheduling. Supply chain planning is required to ensure that each intermediate is available at the right time, in the right place, and in sufficient quantities so that the next intermediate can be produced.

Rapid Ramp-Up

In recent years, an added complexity to supply chain management is the emphasis on decreasing the time required to push new products into the marketplace. Product and process development are now being overlapped to produce initial quantities of an active chemical; these small runs are typically required for environmental and toxicological studies and field trials prior to the product being registered and approved for sale. Pilot plant studies are also being skipped. The driver for this rapid ramp-up is that less financial risk occurs when process development is delayed as long as possible. Also, the probability for regulatory approval and commercial success are better known later in the development.

Specialty chemicals are generally produced in batch facilities. When they are first manufactured in a regular production facility, ramp-up is the period of time to reach normal production levels, which may be on the order of months. Ramp-up occurs because operators become better acquainted with the process and better trained, because process problems caused by new construction are identified and fixed, and because chemistry problems that were not identified in the accelerated process development are fixed. Many of these improvements are completed in the ramp-up phase and not in process development because of lack of time in development and greatly reduced pilot plant work. Even if the time to market had not been accelerated, many potential improvements could not have been identified until production started because the many interactions are too hard to define beforehand. Instead of solving all problems on the benchtop or in the pilot plant, the "learn by doing" mode (Pisano, 1997) is employed in the full scale production facility.

Timing is crucial as the first major campaign of the product should be ready for the first season after a product becomes approved for sale. If the product becomes registered before expected, then ramp-up must occur even faster. With quicker ramp-up and registration, the product can penetrate the market faster and reach higher sales numbers. Then, the development costs are recouped quicker. Fast ramp-up signifies both fast gains in productivity (kgs. out the door) and high yields (less raw material consumed per kg. of product). The result is less need for investment in capital.

Critical Issues

Because of rapid ramp-up, production parameters and uncertainties are hard to quantify. In addition to uncertainty in process productivity and yields, the other end of the supply chain - the market - is also hard to predict. Both product demands and prices are uncertain to a high degree; their estimates may be updated as often as several times per year near the end of a 10-15 year development cycle. Many individual estimates are done by market segment (usually by location or crop) and are then rolled-up into global demands by a product manager. Because of the numerous sources of the various demand and price estimates, most values are given as "best guess" and any predictions of uncertainty are hard to establish.

Modeling Methods

It is essential to build and use a supply chain model to organize the fundamental information and to enhance the decision making process. Basically, the model should consist of a material balance that takes into account essential information starting with the first major intermediate through the active product to the final formulated product. The material balance is constrained on the supply side by yield and capacity estimates. Because the products are new, these parameters must ramp-up over time to their mature value. On the market side, the model is also driven by demand amounts and timing. However, product sales price is not a model input because price is used at a higher level of analysis to determine NPV of the product. If the NPV is recalculated and falls too low, then the product may not be viable and, therefore, removed from the supply chain model. The demands also increase over time to their mature amounts. Depending on the distance between the final manufacturing site and the market, the timing of the demands can be critical as well.

The time step of the model is a compromise between the interval between required demands and the quickest interval that production parameters can be changed. For example, the global agrochemical business is very cyclic in nature as products are required for planting in the Spring in the northern hemisphere and Fall for the southern hemisphere. In many cases, a month time step is used for agrochemical planning and scheduling because campaigns are on the order of months, demands may occur more than once a year, and production changes or equipment additions may take weeks or months to plan and implement.

Practical Use of the Model

The model fills a niche between the simple supply chain issues that can be decided by the "back-of-the-envelope" method and those that require many inter-related decisions to be made. Typically in this case, many siting decisions have already been made (as in a retrofit situation), or there are only two, or possibly three, options to consider. In other words, the problem is constrained enough that only a limited number of tradeoffs need to be evaluated, and a feasible solution is all that is required. However, complexity can still occur for several reasons:

- length of supply chain (many intermediates),
- campaign turnarounds (shared equipment),
- uncertain production and demand levels (new markets or products),
- rapid ramp-up over time (new processes),
- complex market patterns (global markets, seasonal demands)

For instance, the model is needed to determine when to increase capacity of a certain intermediate, or when mature yields of an intermediate are needed. These planning questions can be handled with the spreadsheet scenario approach. The example problem given below illustrates the scope of the decisions that can be solved.

A mathematical programming approach, based on supply chain design work similar to Subrahmanyam, *et. al.* (1994) or on production planning by Bassett, *et. al.* (1997), could be used for this task. However, a simpler heuristic method, with the material balance setup to handle multiple scenarios, is all that is required because of the type of problem described above. The amount of work needed to write and solve a mathematical program, such as McDonald and Karimi (1997) used to solve much larger industrial problems, is not required for the key decisions that are needed for this application. Nor is the detailed, industrial-scale batch scheduling needed as reported by Sanmarti, *et. al.* (1997) and Richards, *et. al.* (1997).

This model is a constrained monthly material balance programmed in a spreadsheet environment for easy modification and communication. So that the model output is understood by most key industrial customers (such as production supervisors, marketing managers, and product managers), a general desktop tool is advantageous. Inventory amounts, production amounts, amount consumed by the succeeding step, capacities, and yields are calculated and tracked on a monthly basis. These five variables are grouped by intermediate; intermediates and products are grouped by production facility along rows. Each column of the spreadsheet is one time step. Therefore, the spreadsheet has the look and feel of a Gantt chart.

Acquisition of the various data, organization of the data, and interpretation of the model results are complex enough to require a dedicated modeling expert during the startup phase of a supply chain. When production is established and the first sales season finished, ownership of the model can be transferred from the modeler to a production planner.

Because of the structure of the model, several different modes of use are possible. Various scenarios may be constructed based on the type of issue being studied:

- supply chain planning vs. campaign scheduling
- demand vs. supply perspective
- short term vs. long term planning
- demand vs. production uncertainty

In supplying new specialty products to market, a two-step procedure is used to make the production decisions. First, *planning* which campaigns are to be run in which facilities and in which order is accomplished. This may be a complex issue if shared equipment is required. In planning, two perspectives are used. The *demand perspective* involves the backward flow of information; demand data is used to set production capacities for each intermediate regardless of capacity constraints. As a first pass, this scenario sets the initial monthly goals for production personnel.

The bottle necks are subsequently identified by looking at *the supply perspective*, or forward flow of information. The realistic capacities and yields are used to determine the maximum possible production for all products, which is compared to the estimated market demands. Then, the gaps between the supply scenarios and the demand scenarios can be identified and corrected. Depending on the size of the gaps, correction may involve laboratory analysis to increase yields, small production changes to solve the issue, or larger process engineering work to modify or add to the process.

The next step, detailed *campaign scheduling,* is performed to determine campaign start and stop times and monthly production amounts. For the *short term* look, capacity and yield are typically ramped-up month-to-month in the first few campaigns. For the *longer term* view of the supply chain plan, the production parameters for the campaigns are assumed to be at mature levels. Even though some specialty chemicals may take only five or six years to reach mature demands, the long term plan may extend ten years or more because of overlapping product introductions. Because the framework is already setup for planning, the same procedures can be followed for this scheduling step.

Uncertainties can be included in any of the above modes of use. Both supply and demand side uncertainties can be used to construct base case, best case, and worst case scenarios. The number of scenarios depends on the size of the uncertainties and where bottlenecks are identified. For instance, if campaign scheduling identifies a certain intermediate as a bottleneck, then both worst-case and best-case scenarios can be constructed to bracket the predicted production amounts.

An Example Problem

To illustrate the benefits of the supply chain model, an example problem based on an industrial agrochemical case is given. The current regulatory and market circumstances require new agrochemicals to be more environmentally friendly and more efficacious; therefore, more complex processing steps are required to bring new products to market. In this example, two newly discovered products, A and B, are being introduced to the market over several years and manufactured in different shared facilities. Four major intermediates, I1 to I4, are needed to manufacture A and B. Initial plans call for some equipment for I1 and I2 to be shared with an existing product, C. I3 shares a processing train with another product, D, as well. I4, active A, and active B all share the same equipment and therefore must be campaigned. After active compounds A and B are manufactured, they must be formulated at another facility into final product A and B.

Market Uncertainties

Expected demands and sales prices for new products are estimated by the individual product managers several times during the process development and production scale-up phases. For this example, the forecasts are re-done once a year. Demand indices, are given in Figures 1 and 2 for products A and B, respectively. Both product demands increase for five or six years until they reach mature sales in 2002 to 2004. Uncertainty in the forecast is reflected in different scenarios: base case, upside, and downside. For these two products, no downsides are envisioned.

In the case of product A, the base case mature demand increased between the 1995 and the 1997 forecasts. The rate of increase to mature levels also increased. Added complexity also results because the forecast done in 1996 contained some 1997 production requirements, which were not anticipated in 1995. In other words, the facilities would have to be up and

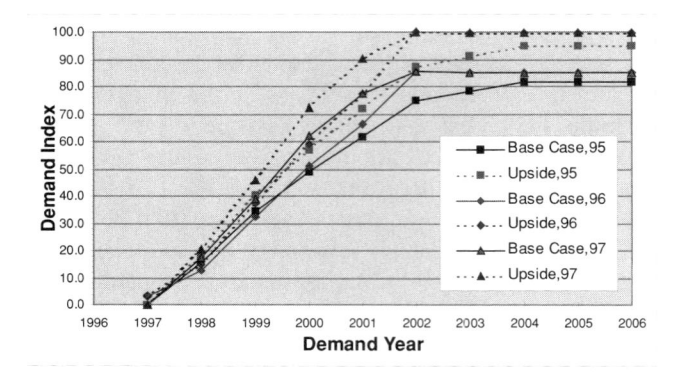

Figure 1. Demands over time for product A.

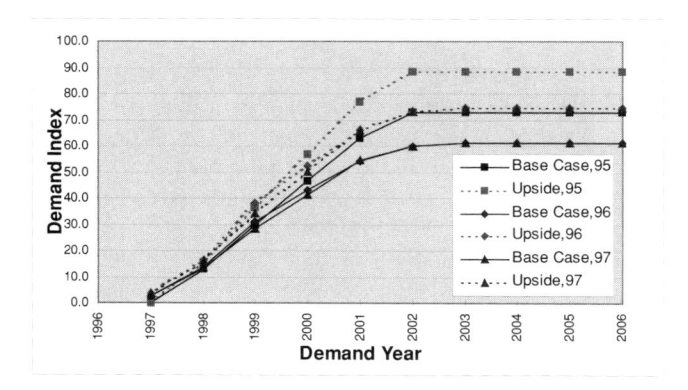

Figure 2. Demands over time for product B.

running one year earlier. A major change in planning resulted because of anticipation of an earlier registration. However, this change did not materialize as the early 1997 forecast returned to 0.0 for 1997.

For product B, the forecast dropped between 1995 and 1997, so that any gains in product A are somewhat offset by B. In this case, upsides continue in the same proportion as the forecasts were updated.

As these market forecasts are updated, the supply chain model must be updated to keep production aligned to current market conditions. If a large enough change occurs between forecasts, the decision must be made to add major projects to the existing facilities to increase capacities to the new requirements. Because of the lead time from project identification to start-up, it may be impossible to increase capacity in time to meet the new demands.

Supply Uncertainties

In addition to market uncertainties, both production amounts and yields are uncertain in any new processes. In addition, turnaround times (shutdown, cleanout, and startup) for shared equipment are estimated in a conservative manner until actual times can be substituted.

The first estimate of production capacities for both intermediates and final products was made in conjunction with the 1996 demand forecast. No ramp-ups were considered for this first estimate. Values, shown as yield indices and monthly capacity indices (unit-less processing train capacity), are in Table 1. The yield estimates are aggressive as the plan calls for best expected yields at the beginning of production for four of the six chemicals.

The first iteration of the model uses Table 1 as inputs. After better estimates for ramp-up of capacities and yields were available, the model is re-iterated. The initial estimates in Table 1 can be compared to the later ramp-up estimates for capacities, both for base case and upside values. An example of the base case is shown in Figure 3. In a similar manner, later yield estimates are also ramped-up to the maximum index over time.

Table 1. Initial Yield and Monthly Capacity Estimates.

Product	Yield Index	Capacity Index
I1	100	22
I2	96	23
I3	99	63
I4	100	24
A	100	19
B	100	21

To double check the accuracy of the monthly schedule, other required inputs include cycle times and batch sizes. The number of days of production added to any turnaround times is then used as a check on the monthly production plan. If production + turnarounds > 30 or 31 days for any month, then the model flags that month as infeasible.

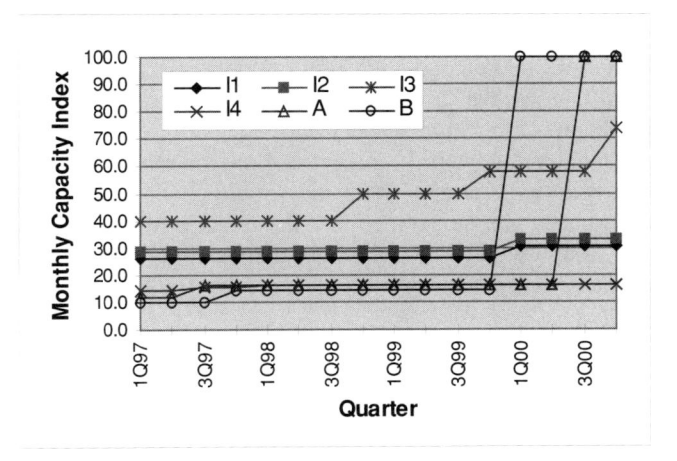

Figure 3. Ramp-up of monthly capacity.

Results for this Example

For the industrial case upon which the example is based, more than sixty scenarios were produced over an eighteen month period. Many of these individual scenarios were updated over time as well, as better market and production data became available.

The Work Process

The work process used in industry for this example is first to evaluate planning scenarios, followed by more detailed campaign scheduling. The first planning model constructed is for both the short and long term time frame from the demand perspective, as shown as bubble 1 in Fig. 4. Then a short term supply perspective scenario (2) is built to ensure that the first major market demands can be met.

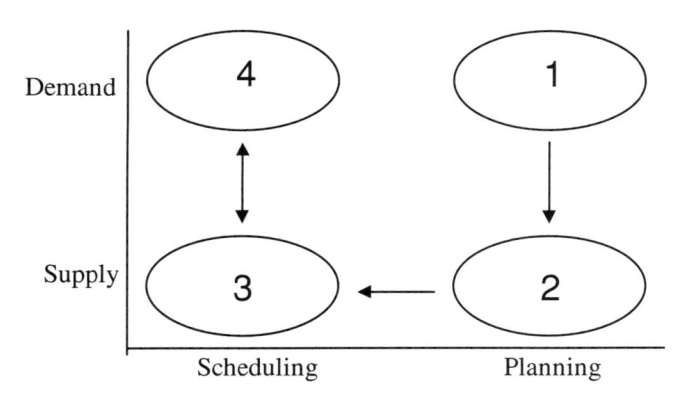

Figure 4. Typical work process.

When a feasible plan is found, the short term supply perspective is used for campaign scheduling (3). Then the campaigns are scheduled out into the long term horizon. When the supply perspective does not satisfy the market demands, the demand perspective (4) is built to identify where the gaps exist. As new production information is gathered and new chemistry or process ideas are generated, the supply perspective is used to ensure that any potential improvements to the supply chain will succeed in satisfying the forecasted demands.

The demand and supply scenarios (3 and 4) are used in an iterative fashion until a feasible solution is found. Uncertainties in demands are included at any point to identify gaps by running base case and upside scenarios. If a major change in the supply chain occurs, then the work process would jump from (4) to (1) and repeat. For example, if a new facility is being evaluated to manufacture one or more of the intermediates or products, then the planning phase may have to be repeated.

The supply chain model can help address many key questions. The following questions are based on the actual industrial implementation of the model:

Can we satisfy the market for the first year of demands?

The first results from (1) indicated that, for the short term, the plan was very tight for the first major demands of A and B. The term tight refers to the fact that no slack exists in the plan, *i.e.* no extra production time for reprocessing bad batches, for large equipment failures, etc. Different planning scenarios were run that included the capacity and yield uncertainties, as shown in Table 2. To satisfy 100% of both A and B demands for the first major sales year, both capacity and yield has to reach the upside potential. As a result of the model scenarios, resources were applied to the supply chain to ensure that the upsides were achieved.

Table 2. Demand Satisfied Based on Short Term Planning Scenarios.

Capacity Index	Yield Index	A (%)	B (%)
base case	base case	80	85
base case	upside	90	90
upside	base case	90	100
upside	upside	100	100

Do we have capacity to produce C and I2 (and D and I3) in shared equipment?

The campaign scheduling model also demonstrated that C could not be made in shared equipment unless the capacity index of I2 went from 23 to 27. Because of the downside potential of C's demands and the cost to increase I2 capacity, C was removed from A and B's supply chain. The model supplied the basis for this decision.

The long term plan is also very tight like the short term. Although enough capacity exists to make I3 and D in the same facility, the timing is off to satisfy demands when they are needed. Therefore, the D campaign is split in two, and extra turnaround times are added to the plan. The detailed campaign scheduling confirmed the feasibility of the new plan.

Where are the major bottlenecks?

Mature demands were impossible to meet under the original assumptions for the model. By examining the entire supply chain for I1-I4, active A and B, and formulated A and B, bottlenecks were identified at both I2 and I4 production. After several scheduling iterations from the supply perspective, active A and B were removed from the shared equipment with I4. A new processing train was the most cost effective method to relieve the I4 bottleneck.

As soon as the plan was updated, the I2 bottleneck emerged. A sensitivity analysis was performed on the model to determine how much I2 capacity was needed and at what time as a function of I4 yields, because of opportunities to increase one or both. Figure 5 depicts the tradeoff over time for four levels of I4 yield. As I4 yields increase, less I2 capacity is required to meet the increasing demands. As a result, both process chemistry and production resources were used to increase I2 capacity and I4 yield concurrently.

Conclusions

The need exists for an uncomplicated solution to supply chain planning problems that can be rapidly implemented. A spreadsheet based supply chain model is a consistent tool that can do this. Because of this model, over ten times the number of scenarios as usual can be evaluated in a short time. The model accounts for key uncertainties and enables key bottlenecks to be

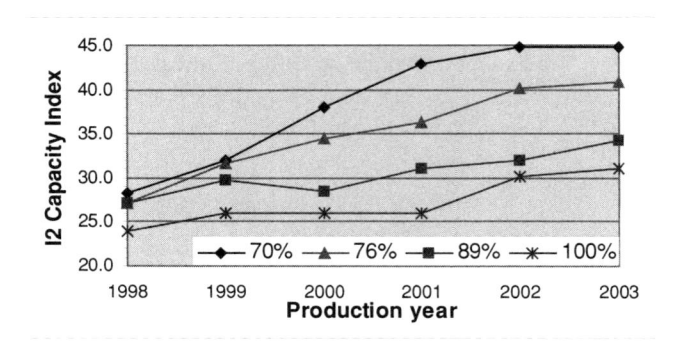

Figure 5. Effect of I4 yields on I2 capacity.

discovered quickly. With product managers usually emphasizing the short term supply chain, this model allows them to study the longer term without much more effort.

The supply chain model can also be used as a communication tool for the business. Because of the nature of a new product ramp-up, both production supervisors and marketing managers must estimate what the future will hold (production rates and market demands, respectively). The use of the model can help the supply chain manager quantify the effects of these uncertainties and help both sides understand when and where gaps in the supply chain exist. Because of the uncomplicated construction of the model, the information flow can evolve from a demand perspective into the production or supply perspective. Therefore, both production and business leaders can see the effects of their uncertainty estimates. With the supply chain model, the nightmares can be avoided and the effect of uncertainties and ramp-up on the supply chain can be understood by all parties involved.

References

Bassett, M. H., J. F. Pekny, and G. V. Reklaitis (1997). Obtaining realistic production plans for a batch production facility. *Computers Chem. Engng.*, **21**, S1203-S1208.

McDonald, C. M., and I. A. Karimi (1997). Planning and scheduling of parallel semicontinuous processes 1. Production planning. *Ind. Eng. Chem. Res.*, **36**, 2691-2700.

Pisano, G. P. (1997). *The Development Factory*, Harvard Business School Press, Boston.

Richards, K. M., M. G. Zentner, and J. F. Pekny (1997). Optimization-based batch production scheduling. AIChE Spring Meeting, Houston.

Sanmarti, E., A. Espuna, and L. Puigjaner (1997). Batch production and preventive maintenance scheduling under equipment failure uncertainty. *Computers Chem. Engng.*, **21**, 1157-1168.

Subrahmanyam, S., J. F. Pekny, and G. V. Reklaitis (1994). Design of batch chemical plants under market uncertainty. *Ind. Eng. Chem. Res.*, **33**, 2688-2701.

IMPROVING BATCH MANUFACTURING PROCESS OPERATIONS USING MATHEMATICAL PROGRAMMING BASED MODELS

Paul R. Bunch and Doug L. Watson
Eli Lilly and Company
Indianapolis, IN 46285

Joseph F. Pekny
Department of Chemical Engineering
Purdue University
West Lafayette, IN 47907

Abstract

We describe the development and implementation of a hierarchy of process models used to improve operations in a batch manufacturing facility. Specifically, models have been developed to facilitate long-range planning and shop floor scheduling in a dedicated multiple stage batch processing facility. The objective of this work was to develop a systematic means of setting target production rates at each stage of the process to balance production resources and while meeting product demands and allowing the plant to be shut down for routine maintenance.

A high level planning model was first developed to set target production rates at each stage of the process. Detailed operations taking place within each stage were not included in the high level planning model. Rather, each stage of the operation was considered as a task or small group of tasks with an average production rate and resource availability. This high level approach reduced the computational burden required to determine planned production rates over a long time horizon (3 months - 3 years). Since the planning model summarized the details of the relationships among the operations within a stage, it periodically set short-term target production rates for a particular stage that were quite aggressive. For this reason, detailed models were developed for the bottleneck stages within the operation. This allowed the bottleneck stages to achieve higher throughput so that production rates determined by the planning model could be met.

Since the modeling and solution frameworks for the planning and scheduling models are common, the transition from the planning model to the scheduling model required a minimal amount of work. By combining the high level planning model with the detailed models for the bottleneck stages, we have developed a means of smoothing resource usage through the plant. Better planning for unusual conditions such as plant shutdowns and scheduling complex stages of the operation for the targets set by the planning model have been achieved. Furthermore, since a mathematical programming framework has been used, the feasibility of meeting a set of proposed demands can be directly verified.

Keywords

Planning, Scheduling, Modeling, Mathematical programming, Optimization, Batch manufacturing.

Introduction

Manufacturing firms are operating in an intensely competitive environment in which it is imperative to optimize the benefit derived from their scarce resources, including capital and manpower. A highly informed and demanding consumer base requires that a large variety of products of high quality and safety be immediately available at a reasonable cost. This state of affairs places extreme pressure on those accountable for resource management in manufacturing firms. Their challenges include developing optimal strategies for meeting customer demand while reducing inventories, or increasing throughput with minimal capital expenditures.

The ability to systematically and efficiently evaluate the performance of a set of manufacturing facilities under a range of conditions is critical for strategic management of the facilities. This ability would allow one to explore the impact on operations of varying product mixes, demand profiles, manpower availability, shift changes, yield and cycle time improvements, and process changes, among other things. Such information could be used to rationalize the purchase of capital equipment to add production capacity or to coordinate the production of a range of products at multiple production facilities.

We have been exploring the use of mathematical programming based models to evaluate the capabilities of our manufacturing facilities to meet demand for products under a range of operating assumptions. In particular, mixed integer linear programming (MILP) models have been developed for several of our manufacturing facilities individually, and a model has been developed which considers several manufacturing sites simultaneously.

In this paper, we describe the development and implementation of a hierarchy of process models used to improve operations in a batch manufacturing facility used to produce one of Eli Lilly and Company's most successful products. This manufacturing facility, which consists of seven sequential stages, has been difficult to operate for several reasons. In particular, inventory cannot be stored between many of the stages since intermediates are unstable. Additionally, cycle times at each stage are long, batch sizes and processing times vary among stages, and each stage is run by different operating personnel. Coordination between the stages in the operation has been an ongoing challenge since operating policies within a given stage have been based upon local objectives such as maintaining a minimum production rate or inventory level. Using local decision rules causes the process to be run out of balance – some stages could not meet demand while other stages were idle – causing large swings in inventories. Since process times were long (days), there was a significant time delay between when a local decision was made and the impact of that decision was "felt" by upstream and downstream stages. This facility was particularly difficult to manage when faced with unusual conditions such as swings in demand, equipment failure within a given stage, and planned shutdown periods which are necessary to perform maintenance.

Specifically, models have been developed to facilitate long-range planning and shop floor scheduling in a dedicated multiple stage batch processing facility. The objective of this work was to develop a systematic means of setting target production rates at each stage of the process to balance production resources and while meeting product demands and allowing the plant to be shut down for routine maintenance.

Background

Consider a batch manufacturing operation that produces a single product and consists of a set of N sequential stages, as shown in Figure . In this system, we allow for the possibility that material produced in manufacturing stage j may be stored before proceeding to stage $j+1 \ \forall \ j \in 1,...,N-1$. Assuming that each stage corresponds to a unique operating unit, this operation can be thought of as a *flowshop* since each batch uses the same operating units in the same order. Consider the following conditions for $j \in 1,...,N-1$:

a) storage of materials is allowed in processing units within stage j;
b) storage of materials is allowed between stages j and $j+1$.

In a real system represented by Figure , it is possible that these conditions are true simultaneously or exclusively. Furthermore, it is possible that neither of these conditions is true, which implies that delays are not allowed between stages. In this case, which is referred to as the *no-wait flowshop*, the total processing time is the sum of the processing times in each individual stage.

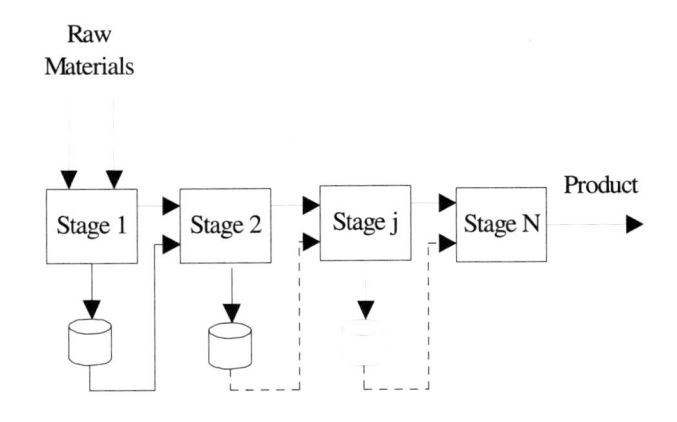

Figure 1. A sequential manufacturing line with intermediate storage.

Suppose that $\forall j \in 1,\ldots N$, stage j requires a fixed time t_j to process a single batch of material of fixed size $f_j Q$, where $f_j > 0$ is the batch size multiplier for stage j and Q is a nominal batch size. The maximum average rate R, at which the manufacturing line in Figure , can supply product is given by

$$R = \left[\min_j \left(\frac{f_j}{t_j} \right) \right] \times Q \qquad (1)$$
$$= \frac{f'}{t'} \times Q$$

Furthermore, stage j is referred to as a bottleneck stage *iff*

$$\frac{f_j}{t_j} = \frac{f'}{t'} \qquad (2)$$

Bottleneck stages limit the productivity of the manufacturing line since stages that are not bottlenecks are idle. In practice, the bottleneck in a process may change due to varying processing times, batch sizes, and product quality among other things. Efforts at increasing the throughput of flowshops are focused on increasing the throughput of bottleneck stages and can consist of adding capital equipment or increasing utilization of equipment in the bottleneck stages with improved scheduling.

Flowshop scheduling has received a significant amount of research attention, beginning with the work of Johnson (1954) and has been studied under a variety of assumptions -- Baker (1974) and Reklaitis (1982, 1991). Much of the research on flowshop scheduling has focused on developing optimal schedules with respect to some time of objective function. Examples of objectives include minimizing the time required to process all jobs (makespan), minimizing machine idle time, minimizing the maximum or mean tardiness of jobs with deadlines, or minimizing the cost of processing all jobs -- Graham *et al.* (1979), Pekny and Miller (1991).

Inventories are used in manufacturing operations to protect against swings in average production rates caused by variability in process times, batch sizes and demands, discrepancies in the average production rates between the stages in the line, and unplanned equipment failure. Most operations strive to maintain inventories at the minimum level possible while still providing adequate protection against the conditions mentioned. No-wait processing (and thus zero inventory) is always necessary when the intermediates are unstable.

Model Development

The description given of our process is quite simplistic. Specifically, the operation has been described as a manufacturing line consisting of a number of sequential stages. In reality, each stage performs a series of complicated chemical transformation and separation processes. The division of the process into stages is based on function and physical facilities. Personnel in each stage of the operation are responsible for managing the production processes in their particular area. Ideally, one could coordinate production activities for the entire facility by developing a large scheduling model that captures the details of all processing steps in all stages. This approach is not practical, however since the solution of the resulting model would be intractable. Furthermore, since the separate stages operate with a degree of autonomy, the model should not assume, for instance, that the start time of a reaction in stage one should have a significant influence on when a separation process should start in stage seven since these start times will be offset by weeks. Both from a computational perspective and an intuitive perspective, it made sense to develop the model within a hierarchical framework.

A high level representation of our manufacturing facility is depicted in Figure . The process is represented as a series of sequential processing (P) stages with intermediate storage between some of the stages. Storage is not allowed prior to completing processing on stage P4 since the intermediates produced from stages P1, P2, and P3 are unstable. Thus the no-wait processing condition is imposed on stage P1-P4. There are several points in our manufacturing process where the quality of intermediates must be checked. To make the model more realistic, these steps were considered as well as the no-wait processing condition. Figure 2 is a high level representation of the processing steps in our manufacturing operation in which only material flows from one stage to the next. Both material and information flow are depicted in Fig. 3. This reflects that material produced by means of a manufacturing process must then be tested in a laboratory (L) and then approved (A). For example, stage P7 cannot begin processing until processing in stages P6 is completed and the intermediate produced by stage P6 has been laboratory tested (L2) and approved (A2). This is an important feature to capture in the modeling process since testing and approval of intermediates and final products can be more time-consuming that making them. Laboratory testing and approval are included in the model by noting that material may be available from a process at a time that is later than that at which the processing is completed. The start times of production lots are adjusted to allow for testing and approval lead times.

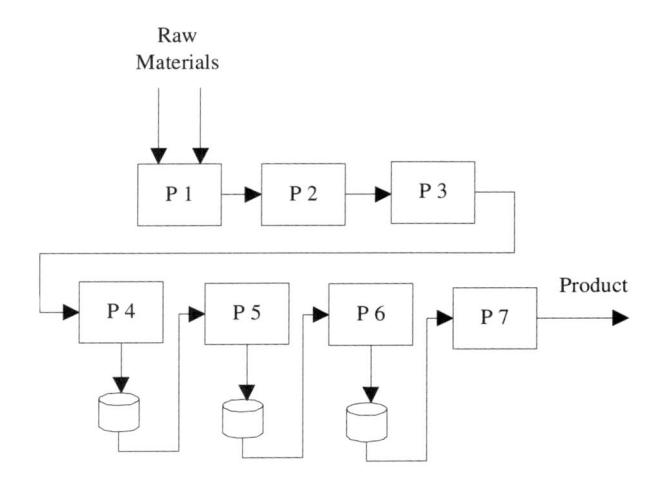

Figure 2. Manufacturing operation with flow of material only (processing steps).

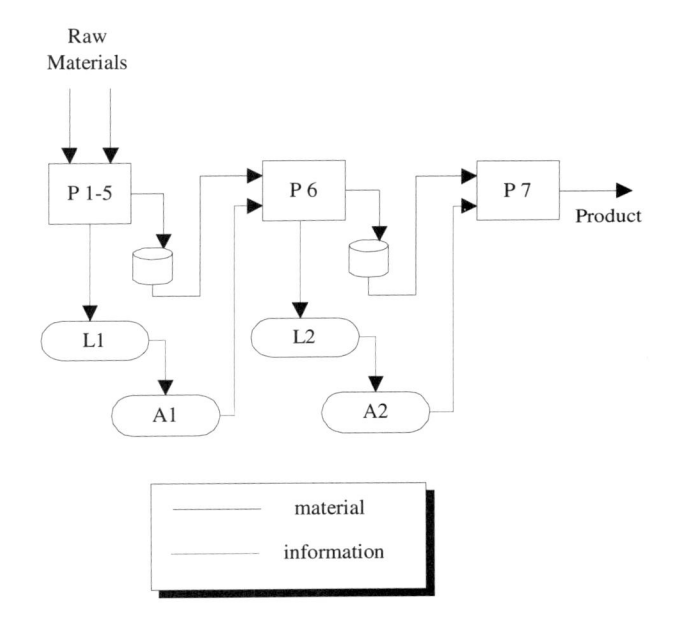

Figure 3. Manufacturing operation with flow of material and information.

Using VirtECS planning and scheduling software, we created an MILP model of the process represented by Figure 3. Input to the model consisted of the bill of materials for each step, the time required complete each step, initial inventories, equipment downtime, and a time varying demand profile. The solution methodology determined a production plan for the facility over a time horizon. In most cases, we were interested in generating plans using a horizon of 3 months to one year. The production plan determined the number and timing of batch starts for each of the seven stages so that the overall demands could be met within the scheduling horizon. The plans set target production rates for each stage over a specified time period. Personnel in each of the individual stage of the process were then responsible for meeting the production rate for their area. Since the planning model uses average production rate data, the target production rates for individual stages could be rather aggressive and difficult to execute. In other cases, the production rates required of an particular stage were low relative to the capability of the stage, implying that the plan could be easily executed.

If the production plan indicated that the production rate required of a particular stage was aggressive, then there was a potential that the overall production line would not meet its demand requirements. Based on historical production rates for each stage, it was a simple matter to determine which stages, if any, would have trouble meeting the production rates specified by the plan. The stage that would fall short of the rate required by the plan would be the bottleneck. In order to increase throughput in bottleneck stages, detailed models were developed. The detailed models are used to gain an understanding of the causes of bottlenecks. Note that the interaction between the planning model and the detailed scheduling models requires human intervention and is based on experience with the process. This is because the assumptions built into the planning model are based on experience and the results make sense only in the context of experience as well.

As mentioned, each stage in our process consists of chemical transformation and separation processes and their corresponding models are quite complex. As an example, consider Figure 4, which shows material flow through operating units (U1-U9) within stage P2 of the operating line. Personnel responsible for this stage must manage the assignment of equipment to tasks and the timing of tasks so that the production rates required by the plan could be met. That is, they must solve a detailed scheduling problem for their stage. These problems are quite difficult for bottleneck stages, necessitating the development of detailed models for each of the manufacturing stages in Figure . Thus, the overall approach for managing production in this manufacturing facility is hierarchical in nature. First, a planning problem is solved to determine the overall production rates over a time period for each stage in Figure . Next, if the production rates required for a given stage were close to the demonstrated capacity for the stage, a detailed scheduling model was run to help develop a feasible schedule so that the production plan could be met.

This approach makes sense from an operational perspective since the various stages operate somewhat independently and the computational expense of solving scheduling problems is incurred only for those stages in which capacity problems would be encountered.

All models were constructed using commercially available scheduling software which allowed us to represent the salient features of our problem with intuitive constructs. In particular, models were developed by populating spreadsheets with relevant data or by developing process flow diagrams and specifying attributes, such as cycle and recipe details, of the objects within the diagrams. Data from these intuitive constructs were used to generate and solve MILP models representing our manufacturing operations. The mathematical programming formulations used are based on uniform discretization models described by Elkamel (1993), and Pekny and Zentner (1993). Solutions were also available in intuitive formats such as Gantt charts or data tables that specified task start times. Thus we were able to make use of powerful mathematical programming formulations and solution methods without knowing the arcane details. This has greatly facilitated the model development and learning process by allowing us to focus on the details of our manufacturing operations.

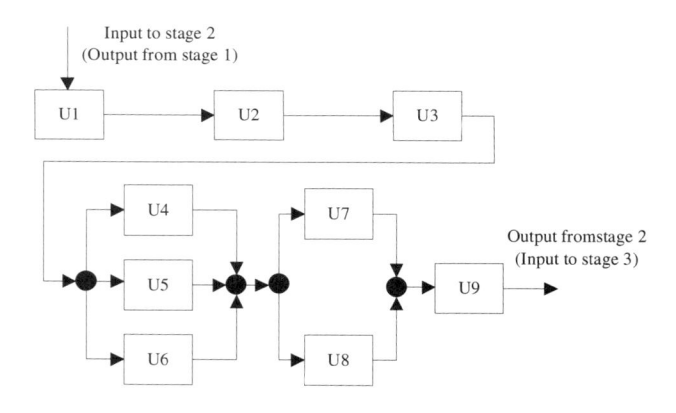

Figure 4. Detailed material flow for stage P2.

Results and Conclusions

Planning models were most often developed for time horizons ranging from three months to one year. Solution of models with long time horizons was possible since the planning model represented the operation at a high level of abstraction. Solution times for one year and three year planning model are approximately three minutes and 10 minutes, respectively, using a Pentium 200 MHz processor on a Windows NT server. Although the scheduling models of the individual stages contained much more detail than the planning models, we were still able to generate feasible schedules in reasonable time since the scheduling horizon of the detailed models was much shorter than for the planning models. Generating a one month schedule for a specific stage typically required on the order of 5 minutes of computational time.

The model described in this paper has been in use for approximately 9 months. Prior to implementation, the model was validated by comparing the production rates from the plan to production rates achieved historically. The comparison indicated that both the model assumptions and data were quite good. Implementation of the model has allowed us to set optimal target production rates over long time horizons while considering the impact of planned shutdowns, unplanned shutdowns due to equipment failures, and improvements in processing rates as well as laboratory analysis times. Furthermore, the model has provided a systematic means for developing coordinated start-up and shutdown strategies (production rates and inventory levels) for planned plant downtimes. In essence, implementation of this model has allowed us to explore the performance of our manufacturing facility under a wide range of actual and proposed operating conditions. Furthermore, given the computational performance, this exploration can proceed quickly, allowing one to gain greater insights into the operational performance of the facility.

In addition to being used to evaluate operating alternatives, the model has been used to drive operational changes. It was determined by running the model that we would have difficulty meeting demand over a certain horizon. In order to meet the demand, ideas were generated to increase throughput. Since the throughput in most of the processing steps could not be increased without sacrificing product yield or quality, we explored decreasing laboratory and testing times. It was found that over a short time window, certain approval times could be cut by as much as 50%, allowing us to meet demand for this time horizon. The model allowed us to quantify the impact of decreasing the approval time and communicating the urgency of the need to do so.

We have discussed the development of a hierarchy of mathematical programming models to improve the performance of a batch manufacturing facility. In particular, the models have provided the ability to quickly evaluate the performance of the manufacturing facility under a wide range of operating conditions. The hierarchical approach was useful since it allowed us to quickly develop a model at a high level of abstraction that accurately represented our process. Developing the model at the proper level of abstraction allowed us to focus only on the details that were necessary to develop the planning model. Furthermore, this approach was consistent with the operational policies of the plant – each stage of the process operated somewhat autonomously by executing a production plan. Finally, the high level planning models could be solved quite efficiently. When necessary, individual detailed scheduling models could be solved after the production plans were generated. Using this approach allowed us to coordinate a series of production stages without the unnecessary overhead that would have be required of a large monolithic scheduling model of the entire production process.

References

Baker K. R. (1974). *Introduction to Sequencing and Scheduling*. Wiley, New York.

Graham R. L., Lawler E. L., Lenstra J. K., and Rinnooy Kan A. H. G. (1979). Optimization and approximation in deterministic sequencing and scheduling: a survey. *Ann. Discr. Math.*, **5**, 287-326.

Hopp W. J. and Spearman M. L. (1996). *Factory Physics – Foundations of Manufacturing Management*. Irwin, Chicago.

Johnson S. M. (1954). Optimal two- and three-stage production schedules with set-up times included. *Naval Res. Logist. Q.* **1**, 61-68.

Pekny J. F. and Miller D. L. (1991). Exact solution of the no-wait flowshop scheduling problem with a comparison to heuristic methods. *Computers chem. Engng.*, **Vol. 15, No. 11**, 741-748.

Reklaitis G. V. (1982). Review of scheduling process operations. *AIChE Symp. Ser.* **78**, 119-133.

Reklaitis G. V. (1991). Perspectives on scheduling and planning of process operations. *Fourth Int. Symp. On Process Systems Engineering*.

DISJUNCTIVE MULTIPERIOD OPTIMIZATION MODELS FOR PROCESS NETWORKS

Susara A. van den Heever and Ignacio E. Grossmann
Department of Chemical Engineering
Carnegie Mellon University
Pittsburgh, PA 15213

Abstract

Multiperiod optimization models for design and planning in the chemical industry have received considerable attention in recent years. In this paper, we present a general multiperiod Mixed Integer Nonlinear Programming (MINLP) model, which incorporates design, as well as operation and expansion planning, and takes into account the investment costs, operating costs and expansion costs in each time period. We present a hybrid algorithm for solution of the model and compare the solution time to that of DICOPT++, a commercial solver. The proposed algorithm makes use of disjunctive programming, as well as a bilevel decomposition technique proposed by Iyer and Grossmann (1997). A numerical example is presented and results show that the proposed method shows a significant decrease in total solution time compared to DICOPT++. Results also show that problems with a significant number of time periods could be solved in reasonable time.

Keywords

Multiperiod optimization, Disjunctive programming, Non-linear, Design, Planning.

Introduction

Multiperiod optimization models for design and planning in the chemical industry have received considerable attention in recent years (Dedopoulos and Shah, 1996; Iyer and Grossmann, 1996; Iyer and Grossmann, 1997; Papalexandri and Pistikopoulos, 1994; Sahinidis et al., 1989; Varvarezos et al., 1992). Multiperiod plants are process plants where costs and demands typically vary from period to period due to market or seasonal changes (Fig. 1). Examples of multiperiod plants include refineries, utility systems and oil production platforms.

Models for multiperiod optimization typically have an objective, such as minimizing cost, subject to constraints in the form of equations and usually involve both continuous and discrete variables. Continuous variables can be either state variables representing operating conditions, such as time dependent flows and

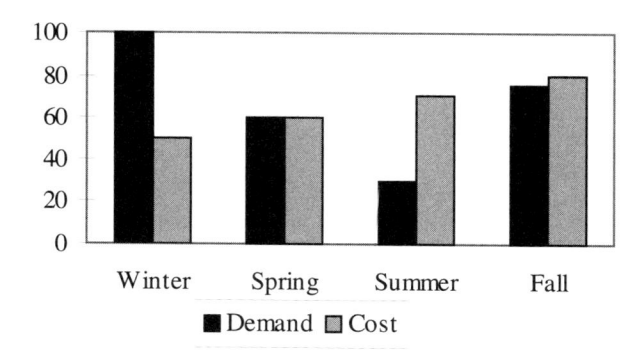

Figure 1. Seasonal changes in demand and cost.

temperatures, or design variables representing equipment sizes.

Discrete variables can be binary (0 or 1) or Boolean (true or false) and represent discrete decisions, for instance to invest in a process or to operate a process in a given period. Equations representing these models can be linear, for example mass balances, or nonlinear, for example process performance equations. Constraints can be valid for all periods or for an individual period and some variables and/or constraints may link the time periods (Fig. 2), preventing a decomposition solution where time periods are solved independently. Models involving 0-1 and continuous variables with nonlinear constraints are classified as Mixed Integer Nonlinear Programming (MINLP) problems. MINLP problems are known to be NP-complete (Garey and Johnson, 1978), meaning they require exponential solution times in the worst case. For multiperiod MINLP models, solution times quickly become intractable due to the large increase in the number of variables and constraints with each additional period. This difficulty becomes particularly acute when binary variables are involved at each time period. Therefore, there is a clear need for developing more efficient algorithms and models.

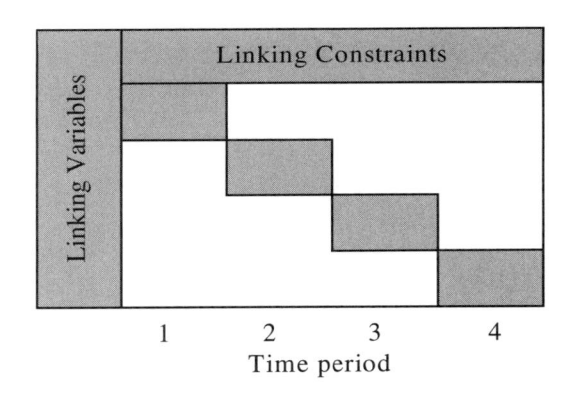

Figure 2. Block-diagonal structure with linking constraints and variables.

In this paper we propose a general disjunctive multiperiod MINLP model for design, operation planning and expansion planning in the process industry. We first show how this model reduces to several major classes of multiperiod problems. Motivated by the use of generalized disjunctive programming (Turkay and Grossmann, 1996) and a bilevel decomposition technique proposed by Iyer and Grossmann (1996), we propose a hybrid disjunctive bilevel decomposition algorithm for solution of the proposed model. The effectiveness of this algorithm is demonstrated with an example and computational results.

Problem Statement

The problem considered in this paper can be stated as follows. Given is a process network superstructure operating over T time periods. Different time periods can result from physical time periods, such as months, seasons or years, from discretized stochastic problems or from multiple scenarios relating different operating conditions. Demands and costs vary from period to period resulting in a different problem for each time period. Deterministic data for costs, demands and other parameters are assumed for the model. The problem involves selecting and optimizing the process over all time periods, with the objective of minimizing cost or maximizing profit. Constraints include mass balances, heat balances, unit performance equations, linking constraints and logical relationships between units and decision variables. We consider three main decisions in the optimization problem:

a. Selection of a network structure
b. Operation/shutdown of process in period t
c. Expansion of a process in period t

Decision (a) is a one-time decision involving investment costs, while the other two decisions are valid for every time period and involve both fixed and variable operation and expansion costs. Constraints are both linear and nonlinear, while variables are discrete (binary decision variables) and continuous (flows, equipment sizes, etc.). Therefore, the model falls under the general class of multiperiod MINLPs. A general disjunctive model for design and operation/expansion planning is presented next.

Disjunctive model

Consider optimizing a given process network superstructure over time periods $t = 1...T$. Based on the work by Raman and Grossmann (1994), who demonstrated the representation of mixed-integer logic through generalized disjunctive programming for process networks, we derive a novel general disjunctive model for multiperiod design and operation/expansion planning.

i) Sets

    I   set of streams
    J   set of processes
    T   set of time periods

ii) Indices

    i stream in set I
    j process in set J
    t time period in set T

iii) Parameters

- α_{jt} variable expansion cost for process j at time t
- β_{jt} fixed expansion cost for process j at time t
- γ_{jt} fixed operating cost for process j at time t
- c_{it} cost associated with stream i at time t
- U valid upper bounds for corresponding variables

iv) Variables

Binary decision variables

- y_j Selection of investment in process j
- w_{jt} Operation of process j at time t
- z_{jt} Capacity expansion of process j at time t

Continuous decision variables

- Q_{jt} Capacity of process j at time t
- QE_{jt} Capacity expansion of process j at time t
- x_{it} State/operating variables at time t
- CO_{jt} Operating cost process j at time t
- CE_{jt} Expansion cost process j at time t

Model (P)

$$\min \sum_t \sum_j CO_{jt} + \sum_t \sum_j CE_{jt} + \sum_t \sum_i c_{it} x_{it} \qquad (1)$$

subject to

$$g_t(x_t) \le 0 \quad \forall t \qquad (2)$$

$$\left[\begin{array}{c} y_j \\ \left[\begin{array}{c} w_{jt} \\ h_{jt}(Q_{jt}, x_t) \le d \quad (3) \\ CO_{jt} = \gamma_{jt} \quad (4) \\ \left[\begin{array}{c} z_{jt} \\ Q_{jt} = Q_{j,t-1} + QE_{jt} \quad (5) \\ CE_{jt} = \alpha_{jt} QE_{jt} + \beta_{jt} \quad (6) \end{array} \right] \vee \left[\begin{array}{c} \neg z_{jt} \\ Q_{jt} = Q_{j,t-1} \quad (7) \\ CE_{jt} = 0 \quad (8) \end{array} \right] \end{array} \right] \vee \left[\begin{array}{c} \neg w_{jt} \\ B^{jt} x_t = 0 \quad (9) \\ CO_{jt} = 0 \quad (10) \end{array} \right] \right\} \forall t \end{array} \right] \vee \left[\begin{array}{c} \neg y_j \\ B^{jt} x_t = 0 \\ \forall t \end{array} \right] \forall j$$

$$y_j \Rightarrow \bigvee_{t=1}^{T} z_{jt} \quad \forall j \qquad (11a)$$

$$\neg y_j \Rightarrow \neg z_{jt} \quad \forall j,t \qquad (11b)$$

$$w_{jt} \Rightarrow \bigvee_{\tau=1}^{t} z_{j\tau} \quad \forall j,t \qquad (11c)$$

$$z_{jt} \Rightarrow \bigvee_{\tau=t}^{T} w_{j\tau} \quad \forall j,t \qquad (11d)$$

$$\Omega(y) = 0 \qquad (12)$$

$$CE, CO, Q, QE, x \ge 0$$

$$y, w, z \in \{True, False\}$$

Model (P) is a new general representation incorporating three levels of decision making, namely design, operation planning and expansion planning. These hierarchical decisions are represented by the Boolean variables, y_j, w_{jt} and z_{jt}, respectively.

Equation (1) is the objective function minimizing total cost, where CO_{jt} and CE_{jt} represent operating and expansion costs, respectively. The third term represents costs in terms of stream variables, x_{it}. Constraints valid for all periods, such as mass balances over mixers, are represented by Eqn. (2). Constraints represented by

Eqn. (3) are valid for a given process that is selected (y_j = true) and operating in a particular period (w_{jt} = true), for example process input-output relationships. Linking constraints are represented by Eqn. (5) and are enforced when expansion takes place (z_{jt} = true). These equations state that the capacity at the current period equals the capacity at the previous period plus the capacity expansion. Equation (9) sets all state variables associated with process j to zero if it is not operated in period t (w_{jt} = false), while Eqn. (7) ensures that capacity expansion of process j is zero if it is not expanded in period t (z_{jt} = false). Equations (8) and (10) set expansion/operating costs to zero if process j is not expanded/operated in period t.

Equations (11a)-(11d) represent logical relationships between the binary variables. Equation (11a) states that a process is expanded in at least one period if it is selected, while Eqn. (11b) states that a process is never expanded if it is not selected. Equation (11c) states that a process can only be operated if it is already expanded beyond zero capacity and Eqn. (11d) states that a process is only expanded in a certain period if it is used in a future period. Equation (12) represents logic propositions relating Boolean design variables, y, for the topology of the network (combinations of processes that are permitted).

Model (P) can easily be reduced to a more specific model by ignoring the appropriate terms and variables. This diversity is illustrated for three particular cases.

i) Multiperiod design (only y). This problem involves only discrete decisions y_j for the selection of the topology of a process network (Papalexandri and Pistikopoulos, 1994):

$$\min \quad Z = \sum_j CI_j + \sum_t \sum_i c_{it} x_{it}$$

subject to

$eqn.(2)$

$$\begin{bmatrix} y_j \\ eqn.(3) \\ Q_j \ge Q_{jt} \\ CI_j = \beta_j + \alpha_j Q_j \end{bmatrix} \vee \begin{bmatrix} \neg y_j \\ eqn.(9) \\ CI_j = 0 \\ \forall t \end{bmatrix} \quad \forall j \quad (P1)$$

$eqn.(12)$
$CI, Q, x \ge 0, \quad y \in \{True, False\}$

ii) Multiperiod design and capacity planning (y and z). This problem involves the selection of the topology (y_j) of a process network, as well as the potential capacity expansion at each time period (z_{jt}) (e.g. see Sahinidis et al., 1989; Varvarezos et al., 1992):

$$\min \quad Z = \sum_t \sum_j CE_{jt} + \sum_t \sum_i c_{it} x_{it}$$

subject to

$eqn.(2)$

$$\begin{bmatrix} y_j \\ eqn.(3) \\ \begin{bmatrix} z_{jt} \\ eqns.(5),(6) \end{bmatrix} \vee \begin{bmatrix} \neg z_{jt} \\ eqns.(7),(8) \end{bmatrix} \forall t \end{bmatrix} \vee \begin{bmatrix} \neg y_j \\ eqn.(9) \\ \forall t \end{bmatrix} \quad \forall j \quad (P2)$$

$eqns.(11a),(11b),(12)$
$CE, Q, QE, x \ge 0, \quad y, z \in \{True, False\}$

iii) Multiperiod Planning (only w). This problem arises in process networks with fixed topology and fixed capacities in which decisions involve the start up/shutdown of processes (w_{jt}) (Iyer and Grossmann, 1997):

$$\min \quad Z = \sum_t \sum_j CO_{jt} + \sum_t \sum_i c_{it} x_{it}$$

subject to

$eqn.(2)$

$$\begin{bmatrix} w_{jt} \\ eqns.(3),(4) \end{bmatrix} \vee \begin{bmatrix} \neg w_{jt} \\ eqns.(8),(10) \end{bmatrix} \quad \forall j, t \quad (P3)$$

$\Omega(w) = 0$
$CO, Q, x \ge 0, \quad w \in \{True, False\}$

Note that (P1), (P2) and (P3) are all particular cases of model (P).

Disjunctive OA Algorithm

Turkay and Grossmann (1994) proposed a logic-based OA algorithm based on the OA method by Duran and Grossmann (1986) which involves iteration between an NLP subproblem where all binary variables are fixed and an MILP master problem where the nonlinear equations are linearized at the NLP solution points. In the logic-based algorithm, an initial set covering problem is solved to determine the least number of possible configurations, N, to cover all processes. These N configurations are used in the N initial NLP subproblems to ensure initial linearizations for all equations. The NLP subproblems only include equations for existing units (i.e. equations of disjunctions with true value). The algorithm used in this work is an extension to multiperiod problems of Turkay and Grossmann's method, which was restricted to a single period.

In applying this algorithm, we decompose model (P) into an NLP sub-problem and MILP master problem. Both problems are converted to mixed-integer form through the convex hull formulation (Balas, 1985; Turkay and Grossmann, 1996). The advantages of this formulation are that it reduces the dimensionality of the problem by only considering disjunctions for which the Boolean variable is true, it avoids singularities due to linearizations at zero flows, and it eliminates non-convexities of non-existing processes.

Bilevel Decomposition Algorithm

Iyer and Grossmann (1996) developed a method for linear multiperiod problems where the design and planning problem is decomposed into an upper level design problem (DP) and a lower level operation planning problem (OP). The solution to (DP) yields a lower bound and fixed configuration to be used in (OP). An upper bound is obtained from (OP), as well as values of z, QE and x used to formulate cuts. Integer cuts in y are used in (DP) to exclude subsets and supersets of previously obtained feasible configurations and to exclude infeasible configurations from future calculations. Design cuts in QE, x and y are used to force values of state variables in (DP) to be greater than or equal to their values in (OP) if the same configuration is chosen in both problems. The solution of (OP) with the current upper bound is the final solution after convergence is achieved. Iyer and Grossmann's work was restricted to linear problems and the sub-problems were formulated in the full space as mixed integer problems and solved with a branch and bound method for MILPs.

Hybrid Disjunctive Bilevel Decomposition Algorithm

In this section, we briefly describe the proposed solution method for model (P) (Fig. 3). The proposed algorithm is a combination of the disjunctive OA algorithm and the bilevel decomposition algorithm. A complete description of the algorithm will be given in a future paper (Van den Heever and Grossmann, 1998).

We first apply the bilevel decomposition to problem (P) to obtain the upper level design problem (DP) and lower level planning problem, which we name (OEP) seeing that our problem includes both operation and expansion planning. These subproblems are then solved with the disjunctive OA algorithm for which the MILP master problems are derived by applying the convex hull to each disjunction. The basic idea is that an outer loop iterates between (DP) and (OEP), similar to the algorithm of Iyer and Grossmann, while both (DP) and (OEP) are solved through inner loops using the disjunctive OA algorithm. (DP) is a relaxation of the original problem, since w_{jt} and z_{jt} are relaxed. This ensures that the optimal objective value Z^{DP} corresponds to a lower bound. (OEP) is in a reduced solution space, seeing that a subset of fixed y variables are used as obtained from (DP), making it possible to ignore a large number of equations.

In this work, the operation and expansion planning are incorporated into one subproblem, (OEP), whereas Iyer and Grossmann considered these planning decisions separately. In addition, the method proposed here is applicable to nonlinear problems, and an optimal solution is guaranteed in a finite number of iterations if the problem is convex.

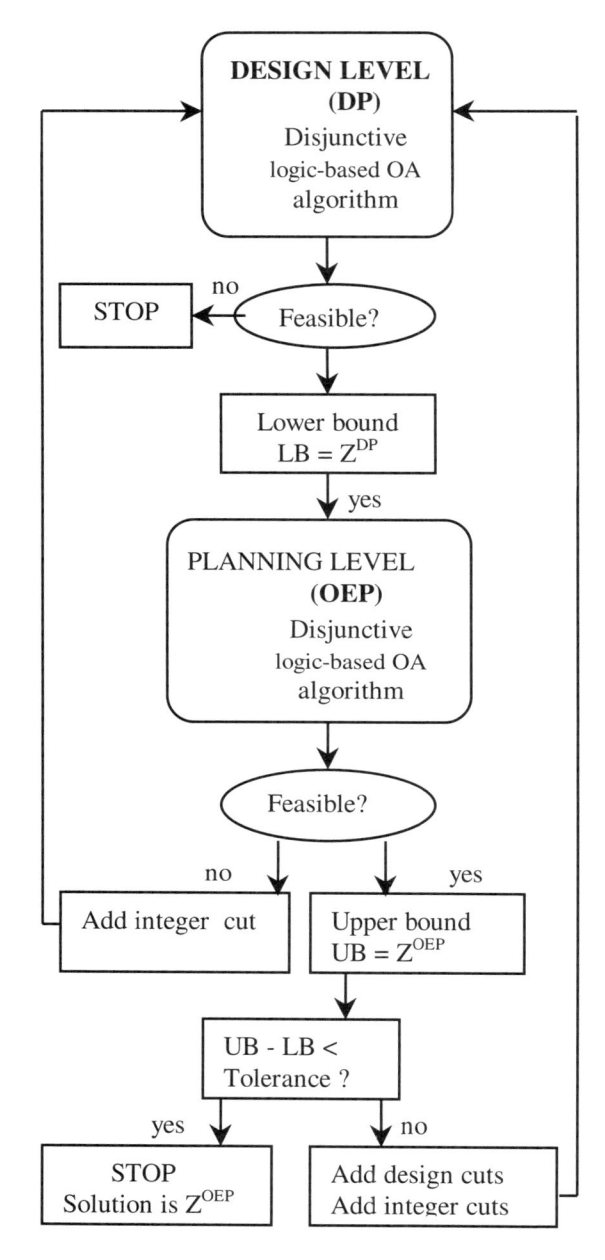

Figure 3. Hybrid algorithm.

Example

Consider the superstructure (Duran and Grossmann, 1986) in Fig. 4. By formulating it first as a generalized disjunctive program as in model (P), the problem was decomposed into (DP) and (OEP) and solved with the hybrid disjunctive bilevel decomposition algorithm outlined in Fig. 3. For comparison, the problem is formulated as an MINLP model, that is, without the use of disjunctions, and solved with DICOPT++. The latter algorithm is an MINLP solver used in the commercial optimization software package GAMS (Brooke *et al.*, 1992). Since the problem is convex, all three methods obtain the same objective value for problems with up to 25 time periods.

Table 1 shows a comparison of the total solution time. The hybrid disjunctive bilevel decomposition algorithm shows significant reductions in total solution times compared to DICOPT++. In particular, for the 21 period problem a reduction of nearly two orders of magnitude was achieved (156 sec. vs. 9341 sec.). All computations were done on HP/UX-9000. MILPs were solved with CPLEX, while NLPs were solved with MINOS5.

Conclusions

We have proposed in this paper a novel disjunctive optimization model for process networks, incorporating design, operation planning and capacity expansion. This model reduces to more specific multiperiod models, thus providing a very general model. Furthermore, a hybrid disjunctive bilevel decomposition algorithm has been proposed to solve this model. The proposed method has shown significantly reduced solution times compared to the commercial solver DICOPT++.

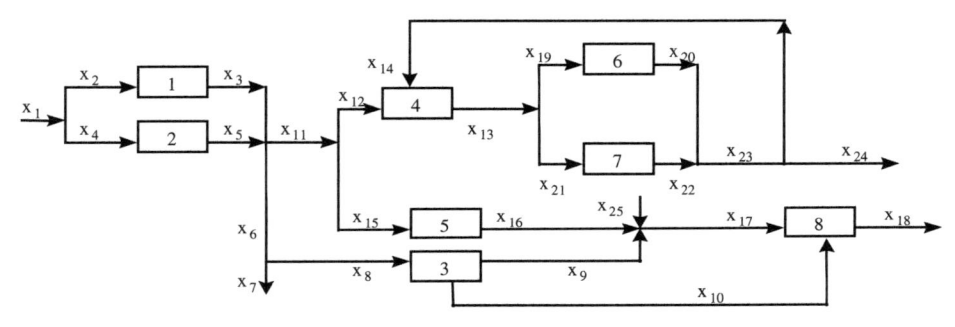

Figure 4. Eight process superstructure.

Table 1. Comparison of Results.

Number of time periods	Discrete variables	Continuous variables	Constraints	Total solution time (CPU sec.) Full space (DICOPT++)	Total solution time (CPU sec.) Hybrid
1	24	41	142	0.5	0.8
9	152	361	1086	43.5	13.9
13	216	521	1558	222.8	30.8
17	280	681	2030	730.8	46.7
21	344	841	2502	9341.2	156.3
25	408	1001	2974	>>10000	3841.7

References

Balas, E. (1985), Disjunctive programming and a hierarchy of relaxations for discrete optimization problems, *SIAM J. Alg. Disc. Meth.*, **6**, 466-486.

Brooke, A., Kendrick, D., and Meeraus, A. (1992), *GAMS: A User's Guide, Release 2.25*, The Scientific Press, South San Francisco.

Dedopoulos, I.T., and Shah, N. (1996), Long-term maintenance policy optimization in multipurpose process plants, *Trans IChemE*, **74**, **Part A**, 307-320.

Duran, M.A., and Grossmann, I.E. (1986), An outer approximation algorithm for a class of mixed-integer nonlinear programs, *Math. Program.*, **36**, 307-339.

Garey, M.R., and Johnson, D.S., *Computers and Intractability: A Guide to the Theory of NP-Completeness*, W.H. Freeman and Company, New York, 1978.

Iyer, R.R., and Grossmann, I.E. (1996), Synthesis and operational planning of utility systems for multiperiod operation, Work Paper.

Iyer, R.R., and Grossmann, I.E. (1997), Optimal multiperiod operational planning for utility systems, *Computers chem. Engng*, **21**, **no. 8**, 787-800.

Kocis, G.R. and Grossmann, I.E. (1989), A modeling and decomposition strategy for the MINLP optimization of process flowsheets, *Computers chem. Engng*, **13**, 797-819.

Papalexandri, K.P., and Pistikopoulos, E.N. (1994), A multiperiod MINLP model for the synthesis of flexible heat and mass exchange networks, *Computers chem. Engng*, **18**, **no. 11/12**, 1125-1139.

Raman, R., and Grossmann, I.E. (1994), Modeling and computational techniques for logic based integer programming, *Computers chem. Engng*, **18**, **no. 7**, 563-578.

Sahinidis, N.V., Grossmann, I.E., Fornari, R.E., and Chathrathi, M. (1989), Optimization model for long range planning in the chemical industry, *Computers chem. Engng*, **13**, **no. 9**, 1049-1063.

Turkay, M., and Grossmann, I.E. (1996), Logic-based MINLP algorithms for the optimal synthesis of process networks, *Computers chem. Engng*, **20**, **no. 8**, 959-978.

Van den Heever, S.A., and Grossmann, I.E. (1998), Nonlinear disjunctive methods for multiperiod optimization, In preparation.

Varvarezos, D.K., Grossmann, I.E., and Biegler, L.T. (1992), An outer-approximation method for multiperiod design optimization, *Ind. Eng. Chem. Res.*, **31**, **no. 6**, 1466-1477.

A NEW CONCEPTUAL APPROACH FOR ENTERPRISE RESOURCE MANAGEMENT SYSTEMS

M. Badell and L. Puigjaner
Department of Chemical Engineering
Universitat Politècnica Catalunya
Diagonal, 647. ETSEIB. DEQ, Barcelona 08028, Spain

Abstract

This work presents a new conceptual approach to develop Enterprise Resource Planning (ERP) systems for the batch industry, which greatly differs from the traditional planning concepts. ERP systems cannot be developed by making corrections to the MRP systems, which set the material requirements as a guiding force. Instead of planning first materials, then production, and if time permits, the cash flow, the finance is taken into account while the scheduling of production/materials is performed. The production management system requires a two-layered model, which allows the vertical integration of the enterprise system. An Enterprise Resource Management (ERM) system framework is proposed considering two economic times: the budget assignment and the production planning/performance.

Keywords

Enterprise Resource planning, Financial scheduling, Vertical integration, Autonomous order entry system, Networked enterprise.

Introduction

It is becoming clearer that manufacturing execution systems cannot be limited by the plant boundaries. Real time production and transactional business applications typically run in different computer applications. The potential for improved market response time, realistic schedules with more job tracking, on time development of new products, optimal treatment of the financial resources and better decision support tools cannot be achieved without the integration of the plant with the business level. The virtual groupings of manufacturing units known as Networked Enterprises foster changes in the traditional concepts. Autonomous agent-based systems capable of providing a quick response to orders through the optimization of price-time trade-off solutions are now necessary. The aim of this work is to create systems that enable to find the best solutions in order to reproduce money and to provide the systematic planning, monitoring and forecasting of the production and its economy.

An Approach to the Overall Business Management

The consideration of financial and production planning as one entire process appears to be the key that closes the loop of the business regulation cycle (Badell and Puigjaner 1998). Although the final objective of an enterprise is maximizing profit, production/financial decisions are usually made without a precise knowledge of the available economic resources and the consequences of utilizing them. Failed businesses are due mainly to cash flow management and indebtedness. In financial terms the liquidity problem is acquiring increasing importance in the fine chemicals industry because the raw materials form a significant weight of the overall production costs and can adversely affect the optimal performance in the cash flow balance. Cash flow management must also involve changing a production schedule. Day-to-day decisions in marketing, production scheduling, inventory

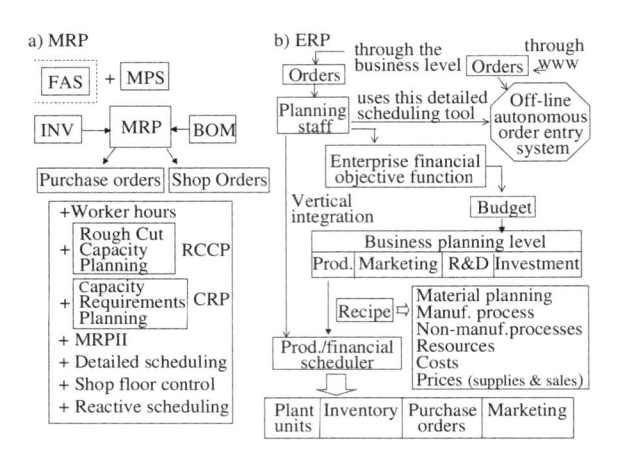

Figure 1. (a) The MRP and the versions evolution, (b) The ERP conceptual approach.

and material resource planning require a rigorous timing of financial resources.

Typical MRP systems lack scheduling functions and MRP II does not take into account the exact situation in the plant. From the several levels of planning usually required, the first ones are performed using manual planning tools. The use of hierarchical production planning frequently causes great discrepancies requiring many replanning activities because the top plan proposed is *blind* to the real production situations.

The core of MRP systems was material-oriented and still the same material logic remains in the current MRP II systems. Its aim is to bring the required materials to the right place in the right quantities at the right time (Baker, 1993). The timing of the requirements for labor force, equipment and resources was added to MRP framework to avoid planning conflicts, as shown in the Fig. 1(a). However, in this piece-worked assembly the most important strategic variable – enterprise finances – is not located at any hierarchical level and therefore the environment lacks the degrees of freedom needed for the overall management.

The use of the integrated systems developed under MRP logic cannot lead to an enterprise strategic decision support tool because they are management information systems, not decision-making systems, although its framework can provide the context to include specific decision-making models. Scheduling is a decision-making procedure, so the introduction of the material logic within the scheduling tool makes more sense than including the scheduling tool in the material logic framework. In the MRP-based systems material plans are made first, then capacity is considered giving as input fixed process lead times. With the proposed methodology the solution is radically different owing to the prevailing position given to the financial aspects in the decision-making system. Current plans often ignore the financial objectives since they are a result of the previously accorded production aims. The production and finance functionality forms a master/slave relationship where "production" decides – usually on the basis of a due date production policy – and "finance" obeys. The difficulty lies in the fact that improving cash planning affects production scheduling by altering payments for final products and billings of raw materials. High-level staff lacks time to frequently prepare/update short-term cash plans. This shortcoming is avoided by applying scheduling viewpoints not only to materials, manpower, equipment, and utilities, but also to cash flow, which is a resource. In addition to the intrinsic importance of liquidity, only the monetary units can offer a failure/success measure pattern of the firm's activities and a common denominator in the resource assignment.

In Fig. 1(b) is exposed the new ERP approach, with *custom-made* policy in the modeling activity. The system considers two basic economic times: the budget assignment and the financial/production planning and performance including its economic execution. The production recipe in the scheduling tool manages the supply timing and the explosion of materials. Taking into account the inventory status, the supplies are calculated at the same level of production in order to achieve the necessary synchronization. Supplies and production are scheduled guided by an adaptable-to-case objective function with financial constraints that also include the specific desired priority given to customer orders. The master/slave relationship between production and finance during planning is converted to a partnership one guided by a trade-off objective function that includes both viewpoints.

A two-layered planning model is considered, which consists in the real time production layer model and an autonomous order entry layer model. By means of the autonomous planning layer model the requested orders are attached at the end of the current plan creating a new *virtual* plan. The autonomous upper level system can be accessed through the Web or through the business level, which provides a quick feedback/response to the orders. A knowledge-based model updated with the day-to-day events supports a rule-based priority system for the determination of customer priorities. This architecture allows the vertical integration at the enterprise.

The proposed ERP system demands real time management of data on the state of the process, the state of payments and billings, the finance, the production accounting, the inventory and the market. Owing to the various regulation loops developed at the scheduling time when the proposed planning takes place, the supply and marketing functions are related to the global planning objective through the constrained distribution of the financial resources.

The Financial Scheduling Approach

The approach presented in Cantón (1997) is used to solve the multipurpose scheduling problem with financial considerations. A new type of recipe representation is given by the Process Material Network (PMN) where the nodes are materials, boxes symbolize *black box* processes and arcs represent precedence relations and transfers of mass, energy and information. The recipes represented at the top level by PMN are transformed successively until the conversion into the Events and Operations Network (EON). In the EON the events are the nodes of the net, boxes represent the operations as *black boxes* and arcs represent the precedence of the events and operations.

The problem statement with the EON format makes it possible to isolate the timing solution, thus achieving a reduction in computer time. Each operation has its operation time TOP and waiting time TW. In the graphical simulation a box sized by the operation and waiting time (TOP+TW) represents the operation. For perishable products the parameter TW^{max} defines the wait policy.

The nodes have associated a time value Tn. Each operation has a previous node, with its initial time associated T_{NI} and a successor node, with its associated final time T_{NF}. Links are included between nodes to represent the consecutive or simultaneous operations. When two nodes are connected, one is the source node and the other is the destination node. T_{ND} and T_{NO} are the times associated to destination and source nodes linked with inequality time constraints.

The problem is formulated by the objective function Z (Eqn. (1)) for minimum makespan *MS* in terms of the continuous time variable Tn subjected to several constraints. *N* represents the total nodes; *M* is the total operations; *K* is the total links; and C^1, C^2 and C^3 are the weight coefficients for each term of the objective function.

$$Z = C^1 \cdot MS + \sum_{m=1}^{M} C_m^2 \cdot TW_m + \sum_{n=1}^{N} C_n^3 \cdot T_n \quad (1)$$

$$MS \geq T_n \geq T_n^{max} \quad n=1,\ldots N \quad (2)$$

$$T_{NF_m} - T_{NI_m} - TOP_m = TW_m \quad m=1,\ldots M \quad (3)$$

$$0 \leq TW_m \leq TW_m^{max} \quad m=1,\ldots M \quad (4)$$

$$T_{ND_k} \geq T_{NO_k} \quad k=1,\ldots K \quad (5)$$

Equation (2) defines the makespan in order to bound the constrained problem among all the infinite feasible solutions. Eqns. (3) and (4) are the constraints related to operations that bound the time values of the initial and final nodes and the waiting time. Eqn. (5) refers to the links between nodes with inequality time constraints. The timing is solved using an algorithm following the heuristic rule of the earliest release operation.

In the adaptation of this approach to financial cases new items are included within the recipe context. Commercial operations are simulated using *virtual process units*. The sequences of non-manufacturing jobs (commercial events) leading to storable products (raw materials, final product) are subject to storage and precedence constraints. The schedules are gradually improved by a simulated annealing (SA) algorithm which considers a trade-off between customer satisfaction (meeting due dates) and the company liquidity (minimum net cash flow constraint). The quadratic deviation between the due dates and the obtained delivery time is used as the objective function to be minimized under the constraint of minimum cash. A safe figure for cash is calculated using the inventory control model by fixed order quantity.

The hard constraint of minimum cash creates a limited *liquidity stream* that connects and regulates in real time the inventory-production-marketing chain. Due to its finite nature in time, money can be represented by state profiles like any other resource or process utility. Owing to its discrete behavior, the cash flow balance can be manipulated on event vs. time diagrams with the timing of liquidity inflows and outflows. The inclusion of the net cash flow profile, a *Money Gantt Chart* superimposed to the production schedule, and the production costing of the process shows explicitly the money movement and the state of solvency at each moment.

Case Studies

The first case study illustrates the methodology proposed for optimum scheduling incorporating minimum cash flow constraints. The prices of the final products A to H per batch, the prices of raw materials, and the due dates are shown in Table 1. The equipment consists of five discontinuous stages (reactors R1-R3 and separators S1-S2), eight pumps (P1-P8), and six intermediate storage (IS) tanks (IS1-IS6). The recipes of products A to H are described in Table 2.

Table 1. Prices in Monetary Units (mu) of Final Products, Raw Materials and the Due Dates.

	F.Prod	Raw Material	Due Date (h)
A	200	120	35.5, 66.5
B	220	140	62, 99
C	170	100	37.5, 67.5
D	110	90	88.5, 109.5
E	150	100	92.5, 113.5
F	180	130	94, 103
G	170	110	83.5, 104.5
H	190	140	83, 114

The production cycle considered is a week, 120 h, during which it is necessary to meet orders of 16 batches, two batches/product, with the due dates given in Table 1. The following assumptions are made:

The supplier delivers raw materials and receives payment for them instantaneously. The customer receives the product ordered and pays for it instantaneously. Zero stock of raw materials and products is assumed. The production week analyzed does not coincide with the end of the month. Therefore the expenses during the week will only be of raw materials. The variability of cash outflow is assumed as a random variable with normal distribution and a mean value (440 mu/day) for weekly cycles. The estimation of minimum net cash flow calculated by the fixed order quantity model of inventory is 500 mu for a mean value of 440 mu and a 1% accepted risk of being left without cash. The minimum cash value is used as a hard constraint. This means that if the cash falls beneath the minimum, e.g., beneath 500 mu, a short-term loan must be requested. The initial liquidity is 1000 mu, considering it as twice the minimum cash.

Firstly, by means of a least squares objective function, a minimum quadratic deviation between the due dates and delivery dates is assured without the introduction of the minimum cash flow constraint. The random search of the SA algorithm required more than 6,000 iterations for each program run, beginning with an initial seed that consisted of the sixteen batches placed one after another totaling a makespan of 205.5 h. The resulting scheduling met the due dates requested by the customers. The cash flow profile is drawn for the 120 hours of a weekly production cycle. The solution obtained violates the restriction of 500 mu as minimum cash on the first day during 18 hours, leaving cash totaling only 290 mu (Fig. 2). Then, the scheduling program was modified to introduce the restriction of minimum cash flow in the least squares objective function to minimize the distance in time between the due dates and the delivery times. The maximum deviation of due dates is only 7.5 h when the minimum cash flow is not violated (Fig. 3).

This is explained by the fact that the intermediate storage immobilizes money when the process time is enlarged.

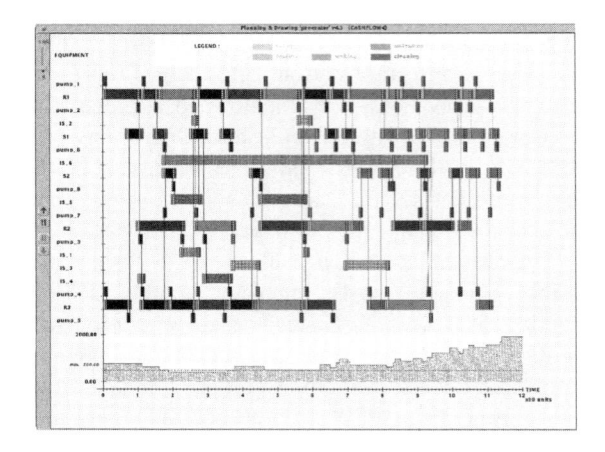

Figure 2. Schedule that meets the due dates but violates the minimum liquidity restriction.

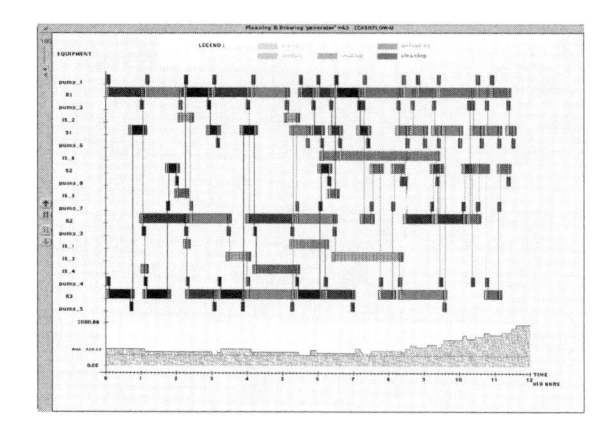

Figure 3. Schedule showing the best delivery dates without violation of minimum cash.

Table 2. Recipes of Products A to H (h).

A	B	C	D	E	F	G	H
P1(0.5)	P1(0.5)	P4(0.5)	P1(0.5)	P1(0.5)	P1(0.5)	P1(0.5)	P2(0.5)
R1(9)	R1(9)	R3(6)	R1(2)	R1(4)	R1(3)	R1(9)	R2(2)
P2(0.5)	P2(0.5)	P5(0.5)	P2(0.5)	P2(0.5)	P2(0.5)	P2(0.5)	P7(0.5)
R2(12)	IS2...	S1(3)	S1(3)	S1(1)	S1(2)	S1(2)	S2(2)
P3(0.5)	P7(0.5)	P3(0.5)	P6(0.5)	P6(0.5)	P6(0.5)	P6(0.5)	P4(0.5)
IS1...	R2(10)	IS4...			IS6...		R3(3)
P4(0.5)	P3(0.5)	P4(0.5)			P8(0.5)		P7(0.5)
R3(7)	IS3...	R3(6)			R2(7)		S2(2)
P5(0.5)	P4(0.5)	P7(0.5)			P7(0.5)		P8(0.5)
	R3(12)	S2(2)			S2(2)		
	P5(0.5)	P8(0.5)			P4(0.5)		
	S1(4)	IS5...					
	P6(0.5)	P1(0.5)					
		R1(6)					
		P2(0.5)					
		S1(2)					
		P6(0.5)					

In Fig. 4 the schedules are compared. The quality of the schedule without violation of minimum cash is greater due to the smaller use of intermediate storage (25 % less).

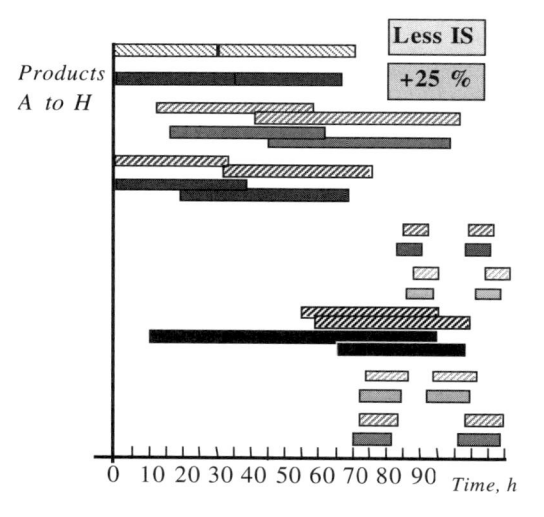

Figure 4. The schedule without violation of minimum cash (shadowed) uses less IS.

The second case study uses the same example to illustrate the methodology using the priority screening system. In Table 3 are detailed three orders of two batches of each product, the profit of each product and the two due dates for each batch ordered.

Table 3. Profit of Products Requested by Three Orders (2 batches/product) and Due Dates.

Product	A	B	C	D	E	F	G	H
Order	1	2	3	3	3	3	3	3
Profit (mu)	80	80	70	20	50	50	60	50
% total profit	17	17	15	4	11	11	13	11
Due date(1st)	35.5	62	37.5	88.5	92.5	94	83.5	83
Due date(2^n)	66.5	99	67.5	109.5	113.5	103	104.5	114

Figure 5 shows the knowledge-based model that supports the rule-based priority screening system with three intermediate objectives and a global objective defined by qualitative variables and linguistic terms. With the confirmed orders recorded in a daily orders database in the autonomous planning layer and the information retrieved at the business level, the autonomous agents access the rules. The system performs a priority screening setting the appropriate rules with the selected linguistic terms (Table 4) and establishing the priority.

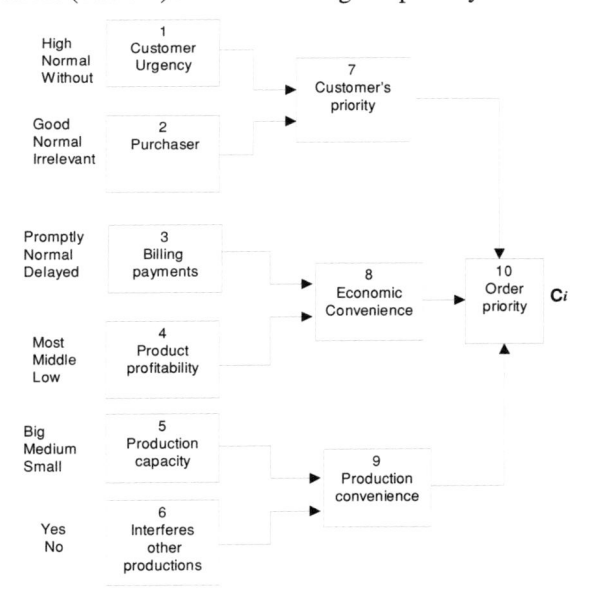

Figure 5. Rule-based priority screening system.

Thus the objective function of the quadratic deviation between due dates and the obtained deliveries for this case must have the following form:

$$\sum_i = (Tf_i - Di)^2 \cdot Ci \quad (6)$$

where Tf_i is the delivery time found for product i, Di is the due date for product i, and C_i represents the coefficient for product i obtained as a result of a priority analysis. To eliminate the influence on non-priority due dates in Equation (6) coefficients $C_i = 100$ are used for the production of A and B, and $C_i = 1$ for products C to H.

A plan with non-violated due dates for the prioritized products A and B can be obtained as shown in Fig. 6 and Table 5. A makespan of 113.75 h is obtained with a violation of the minimum cash flow for small periods of time, i.e., a cash flow of 410 mu during 16.37 h and of 290 mu during 0.2 h (12 min).

Table 4. Customer, Order and Activated Linguistic Terms in the Knowledge-based System.

Ord.	Quant	Urgency	Purchaser	Paym.	Profi	Cap.	Int.	Prio
1	2 of A	High	Normal	Prom	Most	Big	No	100
2	2 of B	Normal	Good	Prom	Most	Big	No	100
3	2 of C	Without	Irrelevant	Delay	Med	Small	Yes	1
3	2 of D	Without	Irrelevant	Delay	Low	Small	Yes	1
3	2 of E	Without	Irrelevant	Delay	Low	Small	Yes	1
3	2 of F	Without	Irrelevant	Delay	Low	Small	Yes	1
3	2 of G	Without	Irrelevant	Delay	Med	Small	Yes	1
3	2 of H	Without	Irrelevant	Delay	Low	Small	Yes	1

Table 5. Delivery Dates Giving Priority to A, B.

Product	Due date	Delivery date
A	35.5, 66.5	35.25, 66.25
B	62.0, 99.0	57.25, 94.25

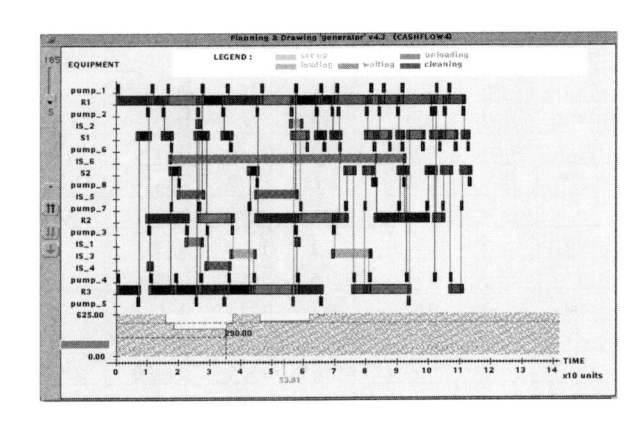

Figure 6. Schedule with priority given to A and B.

A third case study is presented to illustrate the schedule dependence of the product cost in multipurpose plants. It will be analyzed the situation where customer demand is lumpy, e.g., high demand periods are followed by periods of low or no demand. Two batches of each product A to H following the recipes of Table 2 are processed one after another. The Table 6 shows the variable costs, prices and process times per batch of 100 kg of final product. Considering the contribution to sales ratio (CSR) it seems as if the products A and C are the more promising to be produced.

Table 6. Variable Costs, Prices, and Process Time.

Product	Price	V.Cost	Differ.	(CSR)	P.Time
A	200	120	80	0.40	30.0
B	220	140	80	0.36	37.5
C	170	100	70	0.41	28.5
D	110	90	20	0.18	6.5
E	150	100	50	0.33	6.5
F	180	130	50	0.27	16.5
G	170	110	60	0.35	12.5
H	190	140	50	0.26	11.5

Table 7. Simultaneous Processing, Time (h), Produced Value, Variable, Fixed and Total cost, and Profit Obtained.

Simult.	Time	Value	V.Cost	F.Cost	T. Cost	Profit
(A)	10,00	6440	3864	4500	8364	-1924
A+B	21,00	25264	15584	9450	25034	230
(B)	14,00	7788	4956	6300	11256	-3468
(B+C)	4,50	5024	3074	2025	5099	-75
(C)	23,50	13090	7700	10575	18275	-5185
C+D	2,50	5629	4283	1125	5408	221
(D)	0,50	847	692	225	917	-70
D+E	3,50	13988	10215	1575	11790	2198
E	1,50	3465	2307	675	2982	483
E+F	1,50	5013	3425	675	4100	913
(F)	2,50	2556	1845	1125	2970	-414
F+G	11,00	26282	17853	4950	22803	3479
F+G+H+A	1,50	7018	4831	675	5506	1512
F+H+A	1,00	3339	2353	450	2803	536
H+A	9,00	20677	14429	4050	18479	2198
A+B	19,50	23448	14460	8775	23235	213
(B)	16,00	8910	5622	7200	12822	-3912
(B+C)	4,00	4449	2722	1800	4522	-73
(C)	24,00	13379	7866	10800	18666	-5287
C+D	2,50	5629	4283	1125	5408	221
(D)	0,50	847	692	225	917	-70
D+E	3,50	13988	10215	1575	11790	2198
E	1,50	3465	2307	675	2982	483
E+F	1,50	5013	3425	675	4100	913
(F)	2,50	2556	1845	1125	2970	-414
F+G	11,00	26282	17853	4950	22803	3479
F+G+H	1,50	6058	4255	675	4930	1128
F+H	1,00	2679	1956	450	2406	273
(H)	9,00	14877	10949	4050	14999	-122
Total	205,5	278000	185861	92475	278336	-336

However, the process time is longer, owing also to the possible intermediate storage, and therefore their fixed costs are higher. Products with shorter process times enlarge the potential capacity of the plant and it is possible to produce more thus generating more profit. Therefore A and C are far from being the best. Lack of economic information often lead to this type of misunderstandings. Even the production planning following the *Just in Time* policy can go against the business interests. In Table 7 are detailed the cost calculations necessary to draw the profit profile at every moment. The profile of production economics expressed by means of the process profit and the breakeven line can be explicitly drawn on-line by the scheduling program (Fig. 7) and thus avoid the difficulties.

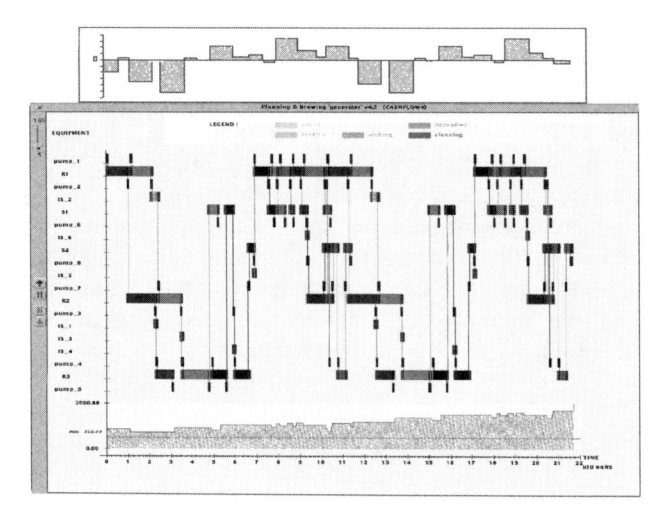

Figure 7. Real time profit profile.

Conclusions

This work provides a new approach for the development of ERP systems, which avoids the re-planning activities of MRP-based systems at the firm's hierarchical levels. The traditional planning concepts are greatly modified implementing a production/finance *partnership* relation during planning. First and foremost finances are considered simultaneously with production and materials. Money not only initiates business, it also can measure the efficiency and perform "on-line" control on all activities avoiding the "blind" financial decision-making usually present in the current practice. This framework can be used in ERM systems owing to the links created at functional levels between production, supply, marketing and financial management. The studies undertaken here indicate that these solutions point in the right direction but additional effort is required in its practical development.

Acknowledgements

The support of the European Community, Dir. Gral. de Invest. Cient. Técnica, Spain and CIRIT is appreciated.

References

Badell M. and L. Puigjaner, 1998. An ERP system prototype in the batch industry, *Xth Intern. Work. Seminar on Prod. Economics, Igls, 16-20 Feb.1998*.V.3, 17-27.

Baker, K.R., 1993. Requirements planning in logistics of production and inventory. *In:* S.C.Graves, eds. Vol. 4. U.K.: Elsevier Science Pub. B.V., 571-674.

Cantón, J. , 1997. "Temporizador para Plantas Multipropósito", Carrer Final Thesis, Univ. Politécnica Cataluña.

SCHEDULING OF A MULTI-PRODUCT POLYMER BATCH PLANT

Christian Schulz and Sebastian Engell
Process Control Group
University of Dortmund
D-44221 Dortmund, Germany

Rüdiger Rudolf
Technikum Joanneum,
Construction Engineering and Management,
A-8020 Graz, Austria

Abstract

In this paper, a real-world scheduling problem from the polymer industries is investigated. Special non-standard features of the problem are the high degree of coupled production where none of the different products can be produced separately but their relative proportion can be influenced by the choice of the recipes, and that the discontinuous and the continuous part of the plant are connected by a mixing stage which gives rise to nonlinear relationships between the batches. Two different mathematical models are presented: a continuous-time and a fixed-grid model. Both models give rise to large, nonconvex, mixed integer nonlinear problems (MINLP). The size of the problems makes it impossible to use general purpose algorithms. We present scheduling algorithms which take the specific properties of the problem into account and lead to good suboptimal solutions. The two problem formulations are compared wrt. the computational effort required to compute the schedules.

Keywords

Scheduling, Multiproduct batch plants, Mixed integer nonlinear programs (MINLP), Nonconvex optimization.

Introduction

In this contribution, we discuss a real-world scheduling problem from the polymer industries. In the plant, two types of expandable polystyrene are produced in several grain size fractions; each grain size fraction of the two types is sold as a single product. The reaction part of the plant is operated in batch mode whereas the final processing is done continuously. The two sections of the plant are connected by mixers.

In our modelling effort, we tried to develop compact, generic models for this type of problems. The crucial decision which has to be made is the choice of the representation of time, where, roughly speaking, two concepts can be distinguished. In the first, the planning horizon is divided into intervals of fixed length and the scheduling of task and resource usage is relative to this grid. The second approach is to use a continuous representation of time, where the duration of all intervals is determined by the optimisation algorithm. This approach may lead, depending on the process characteristics, to a smaller number of intervals and thus a smaller number of variables in the scheduling problem.

In this contribution we give a comparison of the two modelling techniques for the scheduling problem described briefly above. In the models, we exploit specific properties of the process in order to reduce the problem size. In particular for the continuous representation of time, modifications in the synchronisation mechanism and the use of a priori knowledge about the minimal period of time between events leads to a significant reduction of the problem size.

The resulting problems are large nonconvex MINLPs with nonlinearities which neither can be linearized nor convexified exactly. Therefore, solution algorithms which exploit the problem structure had to be developed to produce solutions within reasonable times (1-2 hours).

Process Description

The plant shown in Fig. 1 is used to produce two different types of polystyrene in several grain fractions. The production process is divided into the main steps preparation of raw material, polymerization, finishing of the polystyrene suspension in continuously operated production lines, and splitting into the different grain fractions for final storage. The process is of the flowshop type, i. e. all recipes have the same basic structure and differ only in their parameters and in certain steps.

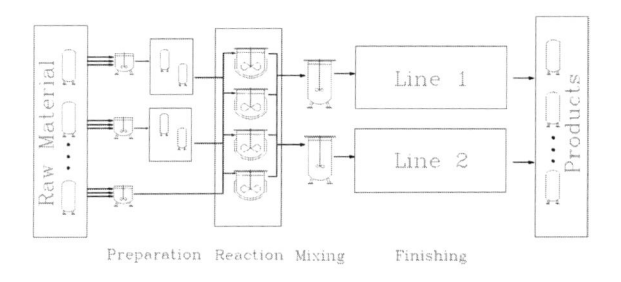

Figure 1. Polymerization process.

In the preparation step, batches of input material are mixed in vessels where they have to reside for a certain period ranging from half an hour to several hours. Then, the mixtures are pumped into one of several storage vessels the capacity of which is adapted to the batch size in the polymerization step. There, further additives are added. The storage vessels are grouped according to the types of polymer produced and this assignment cannot be changed during plant operation. One of the inputs to the polymerization is a mixture of styrene and some additives. The choice of the additives determines the grain size distribution of the product and the type of polystyrene. These parameters are not varied continuously but different fixed sets of parameters are used.

To start the polymerization, input material is pumped into one of the reactors. The batch size is constant in the polymerization step due to technological restrictions.

For safety reasons, the start of a polymerization run in one reactor is prohibited for a certain period of time after the start in any other reactor.

The continuously operating finishing lines are coupled with the reactors by two vessels in which the batches are mixed. Each line is assigned to one type of polystyrene.

A major problem for production planning results from the limited influence of the free process parameters on the particle size distributions. The choice of the recipe affects the distribution among the fractions but all fractions are always produced in significant amounts. Thus, the production of a certain grain size fraction cannot be performed by a certain batch exclusively and all batches are coupled.

During the scheduling horizon a number of orders have to be fulfilled. Each order consists of a due date and an amount of some grain fraction. The main objective is to produce the required grain fractions with minimum delay. On the other hand, there will always be a certain amount of production to storage and unwanted grain fractions are always produced for which only low prices can be obtained. The second objective thus is to minimize the amount of overproduction.

Continuous-time Representation

The main reason for developing mathematical models based on a continuous representation of time was the observation that for models with fixed interval duration the number of events in the model by far exceeded the number of possible events in the process, cf. e. g. Reklaitis (1995). Since each possible event is associated with several variables in the model, the reduction of the number of events can be expected to lead to a better performance of the solution algorithms.

Recently, several types of continuous-time models were suggested, e.g. (Zhang, 1995; Schilling, 1996; Grossmann, 1996). One main difference among the models is how events which occur in different parts or stages of the process are handled. One approach is to define a common reference grid with which all events are synchronized and where all resource balances are calculated. Another approach is to define one grid for each stage and to synchronize only the stages which are directly connected, e. g. via mass balances. An advantage of the latter approach is that one can easily compute the maximum number of events for each reference grid, whereas the first approach avoids the additional effort of synchronization. A common feature is, however, that external events as e.g. supplies, demands and changes in resource availability must be synchronized with the stages affected.

For the process discussed in this paper, it turned out to be effective to use two groups of stages each with a common reference grid.

The first group consists of the preparation stage together with the raw materials, the second comprises the polymerization stage, the mixers with the finishing lines and the final storage. Forming the second group, we exploit the fact that almost all events in this part are driven by the events in the reactor group where the polymerizations take place. Besides, almost all events in the stages after the polymerization are related by a constant shift because all operations in the polymerization stage have the same duration. This allows to use a single reference grid for this group where all internal events are inherently synchronized by a constant shift. The only events which have to be considered as external events are due dates for customer orders and changes in the resource availability. Changes in the mode of operation of the finishing lines are also referenced against this common grid, which does not restrict the degrees of freedom since the throughput is defined as an integral quantity.

Another key issue in continuous-time models is how to determine or bound the maximum number of events determined since this number has a major impact on the problem size, esp. on the number of binary (or integer) variables. A bound for one type of events, the starts of the polymerizations, is given by the minimum offset between two starts; this, however, does not limit the overall number of events if it is still required that the start *and* the end of all operations must coincide with points of the reference grid. This condition is, however, usually required in order to formulate resource (or capacity) balances in a uniform manner. For the specific process considered here, one can however drop this condition since the capacity constraints can be represented in an alternative manner (see below). The maximum number of – internal – events then equals the maximum number of polymerizations which can be started in the planning horizon.

Reactor Group

To state the mathematical formulation of the reactor group, we introduce the following variables:

t_n : time of event n in the reactor group
W_{ni} : equals 1 if polymerization i is started at event n, 0 otherwise
d_n : duration of interval n

The time of each event has an upper bound of H^{max} which is larger than the regarded planning horizon since we allow operations which are started at the end of the planning horizon. We enforce that at each event, one operation is started:

$$\sum_i W_{ni} = 1 \quad \forall n. \quad (1)$$

For each event, it must be ensured that the minimum offset q between the operations is fulfilled:

$$t_{n+1} = t_n + d_n, \quad q \leq d_n \quad \forall n. \quad (2)$$

Capacity (or resource) constraints usually are imposed by summing up the starting and finishing of operations at each point of the grid. For our problem, we use two different approaches, a simple one which is applied for constant resource availability and a more complex one which has to applied when the maximum availability, N_R^{max}, changes over the scheduling horizon. In the first case, the resource constraints can be fulfilled by the conditions

$$t_n - t_{n-N_R^{max}} \geq d_p \quad n = N_R^{max} + 1..N, \quad (3)$$

where d_p denotes the duration of the polymerisations. In the second case, we have to synchronize an external event, a change of the resource availability, against the internal events since it is not known a priori how the external and internal events are ordered relatively to each other. The resource availability is modeled similar to the approach in Zhang (1995), where a sequence of intervals is used during which the availability remains constant. We define a number of r intervals $[t_r^R, t_{r+1}^R[$ during which the maximum capacity C_r^R is given. Furthermore, the number of operations running at each event has to be calculated. Therefore, the following binary variables are defined:

$X_{nn'}$: 1, if $t_n - t_{n'} \leq d_p$, 0 otherwise
Y_r^R : 1, if $t_r^R \leq t_n$, 0 otherwise.

The capacity restriction then can be represented as

$$\sum_{\substack{n'<n \\ n-n'<d_p/q+1}} X_{nn'} \leq \sum_r C_r^R (Y_{rn}^R - Y_{r+1,n}^R) - 1 \quad \forall n, \quad (4)$$

where the left term denotes the number of running operations and the right term acts as a filter which calculates the capacity at the time of event n. In the sum on the left, the number of summands is reduced to the necessary minimum; the number of the binary variables Y_{rn}^R can also be restricted by calculating a maximum duration of each event: if a certain duration is exceeded, one (or both) of the finishing lines will run empty, regardless of the amount stored initially in the mixing stage. The definition of these bounds is omitted due to the limited space, as well as the set of equations defining the binary variables Y_{rn}^R and $X_{nn'}$ (Zhang, 1995).

Mixing Stage and Final Storage

In the mixing stage, each event occurs with a constant offset relative to the reactor group, we can thus use the same index n for each event. Before stating the equations for the mass balances, the necessary variables and parameters are introduced:

m_{ns} : Mass of each grain size fraction s
M_{nk} : Total mass in mixer k
f_{ns} : Feed of fraction s

F_{nk} : Integral feed into finishing line k
p_{ns} : Mass of product s
b : Batch size of each polymerization (const.)
ρ_{is} : Relative production of fraction s by polymerization i

The mass balances then follow directly as

$$m_{ns} = m_{n-1,s} + b\sum_{i}\rho_{is}W_{ni} - f_{n-1,s} \quad \forall n,s \quad (5)$$

$$M_{nk} = \sum_{s\in S_k} m_{ns} \quad \forall n,k \quad (6)$$

$$F_{nk} = \sum_{s\in S_k} f_{ns} \quad \forall n,k, \quad (7)$$

together with the bounds

$$M^{\min} \leq M_{nk} \leq M^{\max}$$
$$d_n F^{\min} \leq F_{nk} \leq d_n F^{\max} \quad (8)$$

In the mixer, the mass from several polymerizations is mixed; thus, we cannot assume that each output is directly transferred into the finishing lines. Instead, we must calculate the concentration of each fraction in the mixers and ensure that a feed with this concentration is fed into the finishing lines:

$$\frac{m_{ns}}{M_{nk}} = \frac{f_{ns}}{F_{nk}} \quad \forall n,k, s\in S_k. \quad (9)$$

Since the amount of each fraction s in the feed is added to the amount in the final storage after passing the finishing lines, it can be calculated by the expression

$$p_{ns} = p_{n-1,s} + f_{n-1,s} \quad \forall n,s \quad (10)$$

Preparation Stage

Each line in the preparation stage operates independently of the other lines, the only coupling occurs at the connection to the reactor where the intermediates are transferred. For the reasons mentioned above, each line j has its own reference grid t_l which is synchronized with the reference grid of the reactor group. Since all intermediates must be used up before the next batch of intermediates is produced, we filter out the polymerizations which take place between two such productions. This can be accomplished by introducing another binary variable:

Y_{jnl} : 1, if $t_{lj} \leq t_n$, 0 otherwise.

The mass balance of the intermediates then is

$$m_{lj}^I = m_{l-1,j}^I + B_{lj} - b\rho_j \sum_{n}(Y_{jnl} - Y_{jn,l+1}) \quad (11)$$

with

m_{lj}^I : Mass of intermediate j at event l
B_{lj} : Batch size of intermediate j at event l
ρ_j : Relative amount for polymerization.

The bounds of all quantities as well as the capacity restrictions for the vessels are straightforward. Furthermore, the number of the necessary grid points l and the range of the sum in Eqn. (11) can be restricted if we take the consumption of intermediates by the polymerizations into account.

Objective Function

To reflect the production goals, the fulfillment of the due dates together with minimum overproduction, we do not directly calculate the lateness of each order. Instead, for each order, the over- and underproduction is calculated at the due date and the objective then is to minimize a weighted sum of over- and underproduction.

Since each due date has to be regarded as an external event, it must be synchronized with the internal events. The synchronization follows the same principle as above; thus, for each of the m due dates t_m^L a binary variable is defined:

Y_{mn}^D : 1, if $t_m^L \leq t_n$, 0 otherwise.

The amount of each product at the due date can then be calculated by a filter similar to the one in Eqn. (4), but an additional complication has to be taken into account: the amount of each fraction in the final storage increases between two consecutive events because the finishing lines continuously feed into the final storage. Thus, the intermediate amount, denoted by p_m^D, has to be considered in the filter equation:

$$p_m^D = \sum_{n}\left(Y_{n+1,m}^D - Y_{nm}^D\right)\left(p_{ns} + f_{ns}\frac{t_m^D - t_n}{d_n}\right) \quad \forall m. \quad (12)$$

The objective function then can be stated as

$$\min \sum_{m} \alpha_m \max\left(0, D_m - p_m^D\right) + \beta_m \max\left(0, p_m^D - D_m\right) \quad (13)$$

which can easily be expressed by linear terms.

Fixed-grid Representation

Due to the limited space, we can only give an outline of this model and highlight the main differences to the continuous-time representation. It basically follows the ideas expressed in Kondili et al. (1993), i. e., one common reference grid is used for the whole plant. Then, at each point of the grid, any of the operations of the intermediate and polymerization stage can be started but no start is required. The mass balances, e. g. Eqns. (5)-(7), (9)-(11) and the objective function (13), then are formulated at all points of the grid.

The main difference to the previous model, however, is that no synchronization is necessary, since all events, internal and external ones, can be mapped on the grid; the remaining binary decision variables only represent the decision for the choice and the timing of the operations. The lack of synchronizations simplifies the solution strategy for this problem representation.

Solution Strategy

Both models presented above are large, nonconvex MINLPs whose nonlinearities ((9) for both models and (12) for the continuous-time representation) cannot be eliminated or transformed into convex representations.

The size of the problems, esp. the number of binary variables (cf. Table 1 below), makes it impossible to use general purpose algorithms like DICOPT++, Visvanathan and Grossmann (1990), or branch & bound algorithms. The application of branch & bound algorithms is problematic due to the nonconvexity of the problem, because good bounds cannot be derived or require a large computational effort for solving and tightening convex relaxations. In the algorithms presented here, however, the continuous relaxation is used as a basis for the scheduling decisions since its solution provides good hints for the scheduling decisions.

The strategy of our algorithms is to find an integer feasible solution of the problem by a depth-first search with repeated solutions of relaxed MINLP on each search level. The key question for a depth-first search is the choice and setting of the binary variables which differ for the two types of problems due to the different types of binary variables in the model, although they both represent the same scheduling decisions.

In general branch & bound algorithms, the choice of the branching variable usually depends on numerical properties, e. g. the fractional value or the reduced costs etc. This is inappropriate for our problem since we wish to reduce the likelihood of frequent and deep backtracking due to infeasibilities, because of the computational effort which is spent for the solution of each relaxed MINLP. Furthermore, we try to fix as many binary variables as possible in each step by rounding heuristics.

The core decisions of the scheduling algorithms are:

- to choose the polymerizations and their timing
- to schedule the operations in the intermediate preparation stage
- to determine the feed rates for the two finishing lines.

We first describe the algorithm for the fixed-grid representation and then the modifications which are necessary to set the binary synchronization variables of the continuous-time representation.

Instead of the criteria mentioned above, the choice of the binary variables starts with the beginning of the planning horizon by subsequently scheduling the polymerizations. The scheduling of the intermediate stage is performed after each of the scheduling decisions by setting the affected binary variables up to the starting time for the currently scheduled polymerization.

The choice of the polymerization which is scheduled next is based on the largest next nonzero value of the corresponding relaxed binary variables; "next" in this context means the time interval starting at the current scheduling time.

Before solving the next relaxed MINLP, all binary variables which can be fixed by applying the capacity restrictions of the affected resources are set. This is done as far forward in time as possible. Since the search starts with the beginning of the planning horizon, we can thus ensure that no capacity restrictions of the reactors are violated, not even for the intermediates. The only source of infeasible subproblems are violations of the capacity constraints of the mixers and the finishing lines which are, however, inevitable and cause backtracking.

In order to limit the amount of backtracking, the next choice after a backtracking step is determined by the reason for the infeasibility. The constraint violation reported by a NLP-solver can however not reveal this reason. Instead one gets a good guess from partially simulating the production plan which allows to detect e. g. overflows. This partial simulation is performed before solving the relaxed MINLP because the determination of infeasibilities by simulation is computationally much less expensive than by an NLP-solver.

When backtracking, one has the following choices:

- schedule a polymerization which produces into another line at the same point of time
- schedule the last polymerization earlier or later
- go back in time and perform the above possibilities for the previous polymerization.

At each stage, the scheduling of the production of intermediates has to be recalculated. The search is stopped when the end of the planning horizon is reached.

A similar algorithm is used for the solution of the continuous-time model with the exception that the second choice for backtracking is omitted because the event times are determined by the NLP-solver. In order to find a branching strategy for the additional binary synchronization variables, several strategies were tested: to branch on these variables along with branching on the variables denoting the start of the operations, to fix them before and to fix them after the scheduling choices were made. It turns out, however, that these strategies fail because they dramatically increase the amount of backtracking because infeasibilities introduced by branching on one of these variables are mostly detected far down in the tree. This is due to choices concerning conflicting external events.

The strategy which is currently used and which turned out to be effective branches on one group of synchronization variables before the scheduling decisions are made: all synchronization variables which belong to the reactor group, reflecting both internal and external events. The remaining synchronization variables belong to the due dates of the customer orders and are fixed along with the scheduling decisions.

In order to fix the first group, a linear substitute problem is formed which abstracts from the single fractions and takes into account only the overall masses in the mixers and the finishing lines. The integer values of a solution of this problem are always feasible values for the complete problem and are used as fixed values during the solution of the remaining nonlinear problem.

Numerical Results

To investigate the relative merits of the two models and algorithms, tests with several planning horizons were performed. For each of these tests, the capacity of the reactor group was assumed to change over time. Table 1 lists the size of the problems.

Table 1. Problem Sizes.

Scheduling Horizon in Days	8	10	12	14
Continuous-Time Model			
Variables	2656	3437	4465	5516
Binary Variables	1009	1385	1993	2639
Nonzeros	15438	20995	30334	39033
Fixed-Grid Model			
Variables	8095	10511	12927	15581
Binary Variables	1848	2424	3000	3654
Nonzeros	67965	88923	109911	130869

Due to the nonconvexity of the problem, a global optimum cannot be determined in general. To examine the quality of the solution, we used satisfiable demands in the objective function which were determined by simulation runs. Thus the optimal value of the objective function always equals 0. Furthermore, the overall sum of the demands is shown which serves as a relative measure of the delays in the schedule since overproduction is weighted by a factor of 1/10 relative to underproduction and, as an analysis has shown, the main part of the value of the objective function results from under-production.

The numerical results shown in Table 2 are only an excerpt, but give a good indication of the main tendencies. In this table, DS denotes the sum of demands, FG the fixed-grid and CT the continuous-time model. The results are of the same magnitude but towards longer scheduling horizons, the fixed-grid model provides the better results in most cases. The required solution times, however, are much longer than for the continuous-time model which is due to the larger problem size.

Table 2 Numerical Results.

Scheduling Horizon in Days		8	10	12	14
Objective Value	DS	33.4	52.6	99.1	129.3
	FG	1.1	2.8	2.4	2.9
	CT	1.0	1.0	3.6	4.3
Solution Time (sec)	FG	810	1685	3720	5208
	CT	109	248	456	878

Conclusions

We presented and compared two types of mathematical models for a complex real-world scheduling problem. It turns out that the solutions in both cases are of almost the same quality. Although the solution algorithm is more complicated, the computation times for the continuous-time model are considerably smaller. Thus, it might be possible to improve the quality of the solution by examining additional branches of the decision tree.

Acknowledgements

This research was funded by the Deutsche Forschungsgemeinschaft in the framework of the coordinated research program „Real-time Optimization" (Grant DFG En152/17). The 3rd author obtained financial support from the SFB F003 "Optimierung und Kontrolle" while he was with the Institute of Mathematics B, Technical University of Graz, Austria. The support of all these institutions is very gratefully acknowledged.

References

Kondili, E., Pantelides, C.C., and R.W.H. Sargent (1993). A general algorithm for short-term scheduling of batch operations. Part I - MILP formulation. *Computers & Chemical Engineering,* **17**, 211-227.

Pinto, J. M. and I. E Grossmann (1996). A continuous time MILP model for short term scheduling of batch plants with pre-ordering constraints. *Sixth European Symposium on Computer Aided Process Engineering ESCAPE-6*, Rhodes, Greece, pp. 1197-1202.

Reklaitis, G. V. (1995). Scheduling approaches for the batch process industries. *ISA Transactions* **34**, 349-358.

Schilling, G., and C. C. Pantelides (1996). A simple continuous-time process scheduling formulation and a novel solution algorithm, *Sixth European Symposium on Computer Aided Process Engineering ESCAPE-6*, Rhodes, Greece, pp. 1221-1226.

Viswanathan, J., and I. E. Grossmann (1990). A combined penalty function and outer approximation for MINLP optimization. *Computers & Chemical Engineering,* **14**, 769-782.

Zhang, X. (1995). Algorithms for optimal process scheduling using nonlinear models. PhD Thesis, University of London.

MAXIMUM DISPERSION ALGORITHM FOR MULTI-SITE DELIVERY SCHEDULING

S. D. Mokashi and A. C. Kokossis
Department of Process Integration, UMIST
P. O. Box 88, Manchester, UK

Abstract

The problem of optimal scheduling of product deliveries from multiple sites to customers in general is neither amenable to simple transportation models nor to algorithms on vehicle routing. Previous experience on single site problems shows the limitation of a pure mathematical programming approach in solving large scale industrial problems. This work provides a new formulation for multi-site delivery problems in the form of a weighted graph representation of various delivery options and their costs. The notion of dispersion of a graph is introduced to form the basis of the formulation and the optimization objective. The optimization objective is to search for a set of vertices, which form a dispersion, with the maximum aggregate weight. The maximum dispersion is shown to correspond to the optimal delivery schedule. An exact algorithm to achieve this objective is outlined. The efficiency of this algorithm is demonstrated by its relatively small solution trajectories and CPU time for graphs of reasonable size.

Keywords

Dispersion, Maximum dispersion, Graph, Delivery scheduling, Optimization.

Introduction

Scheduling of delivery lines forms an important component of the supply chain optimization as a whole. This is all the more true in the case of deliveries from multiple sites, since optimal delivery costs can significantly influence the other stages of the supply chain like production and inventory levels and schedules at the production sites as well as inventory at intermediate stock points.

The delivery problem in question can be summarized as follows:

Given:

1. ns production sites and/or inventory stock points (SPs).
2. nc customer orders received at a centralized place.
3. Location of customers and SPs in terms of distances from each other.
4. Delivery costs as a discrete function of distance of a trip and maximum load in a trip $DC(q_T, d_T)$.
5. Maximum of n customer orders can be combined in the scheduled trip of a carrier.
6. Capacity of carriers.

Objective: Find the optimal schedule of deliveries so as to minimize delivery cost, and the different orders delivered in each of the scheduled trips T

$$T = \{S_s, C_1, C_2, ..., C_m, S_d\} \quad (1)$$

Previous experience on optimization of single site delivery scheduling (Mokashi and Kokossis, 1997; Ronen 1992) has shown the inefficiency of using general purpose mathematical programming techniques for problems of industrial size. The only pragmatic solutions left to logistics engineers are then ad hoc simplifications or resorting to heuristics. The problem at hand is even larger and demands a fresh approach that can deliver optimal solutions in real time. It does not fall under the category of conventional transportation problems as the costs are not simply linear functions of the load transported from the source to the destination. Although the problem involves choice of different routes, vehicle routing algorithms existing in the literature (Laporte, 1992) cannot be directly applied because of the complex trade-offs between choice of a minimal route and options of combining different loads.

The philosophy of contextual optimization presented for single site optimization (Mokashi and Kokossis, 1997) can be practiced even for multiple sites by exploiting contextual information characteristic of multi-site problems to evolve a new problem representation and search method for optimizing delivery schedules.

Problem Formulation

The new representation is a graph whose vertices represent different trips or order combinations. Each trip is an ordered set of customers, starting and ending with an SP. A trip is constrained by the following considerations.
1. The quantity of product carried at the beginning of the trip should not exceed the carrier capacity.
2. The number of customers on a trip should not be more than n.

Any customer can be catered by one of several trips which has the customer as one of the destinations. Since each customer needs to be satisfied by just one of several trips, such trips conflict with each other. Any two vertices/trips with such a conflict are connected by an edge on the graph. The edge set of the graph representation can be formally defined as follows,

$$E = \{(v_1, v_2) | \exists C \in T_1, T_2, v_1 \equiv T_1, v_2 \equiv T_2\} \quad (2)$$

Hence no two vertices with a common edge can form a feasible schedule/ solution. This feature can be more generically defined as a dispersion set.

Dispersion Set

A dispersion set is defined as a set $D \subseteq G$ of vertices such that each vertex is connected to some vertex of the graph G and no two vertices of D are connected to each other.

This definition of a dispersion set is different from the continuous and discrete dispersion problems studied in the literature (Erkut, 1990; Ravi, Rosenkrantz and Tayi, 1990) These problems are concerned with locating facilities on a network such that they as far away as possible from each other in some sense. The dispersion set defined above however is concerned with locating vertices so that they are dispersed (not connected to) with respect to each other.

Optimization objective

The delivery cost for any trip may in general be any function of the load carried and the distance covered. For the purpose of the problem considered in this paper, the cost is obtained from a tariff table which gives the delivery cost as a function of the quantity q_T and the total distance of the trip d_T. Since any customer order can always be delivered as a single order trip $T=\{S_s, C, S_d\}$, the costs of trips in which this order is combined must be lower than the single delivery cost. The weight corresponding to any trip/vertex on the graph is calculated as follows,

$$w(T) = \sum_{C \in T}(DC(\hat{S}(C), C, \hat{S}(C)) - DC(T)) \quad (3)$$

Now, the total delivery cost is given by,

$$TDC = \sum_{T \in D \subseteq \overline{D}} DC(T) \quad (4)$$

The sum of weights of vertices which form a dispersion is given by,

$$w(D) = \sum_{T \in D} w(T) = \sum_{T \in D}(\sum_{C \in T} DC(\hat{S}(C), C, \hat{S}(C)) - DC(T)) \quad (5)$$

Eqn. (5) can be reduced to the Eqn. (6).

$$w(D) = \sum_{C \in C} DC(\hat{S}(C), C, \hat{S}(C)) - \sum_{T \in D} DC(T) \quad (6)$$

The first term in Eqn. (6) is constant and the second term is equivalent to Eqn. (4). This implies that the dispersion with the maximum aggregate weight corresponds to the scenario of minimum delivery cost.

Pre-processing

The construction of the graph introduced in the previous section is an important pre-processing step before the actual solution is attempted. The pre-processing stage can be considered in two steps
1. Generating trips/vertices from data on customer orders
2. Establishing the set of edges.

Generation of Vertices/trips

In general any customer can be catered by any SP. However if this was allowed in generating trips or order combinations then it will generate a large number of redundant vertices on the graph and defeat the purpose of having various SPs distributed in the vicinity of different customers. Hence in order to limit the generation of redundant vertices the idea of having neighborhoods around SPs is introduced. Each SP S will have a neighborhood radius r_s. Any customer lying outside this radius will not be catered by this SP. Hence for any trip T, each customer on this trip will be within this radius.

$$d(S,C) \le r_S \qquad (7)$$

In the case of trips which start and end with different stock points, the criterion can be as follows.

$$d(S_s,C) \le r_{S_s} \wedge d(S_d,C) \le r_{S_d} \qquad (8)$$

Once the neighborhoods are established, all possible trips in each neighborhood can be generated considering the capacity constraint of the carrier and the maximum number of allowable customers on a single trip. For each trip, the route with minimum distance is found. This will ensure minimum cost for that trip for the tariff policy considered in the previous section. This will set up the vertices for the graph to be processed. The weights of these vertices can be calculated using Eqn. (3).

The idea of neighborhoods can be understood with a simple example of two SPs (Fig. 1). Some of the valid trips, that can be formed are $T_1=\{S_1,C_1,C_2,C_3,S_1\}$, $T_2=\{S_1,C_4,S_2\}$ and $T_3=\{S_2,C_6,C_7,S_2\}$. Examples of invalid trips would be $\{S_1,C_3,C_4,S_2\}$ and $\{S_2,C_6,C_4,S_1\}$.

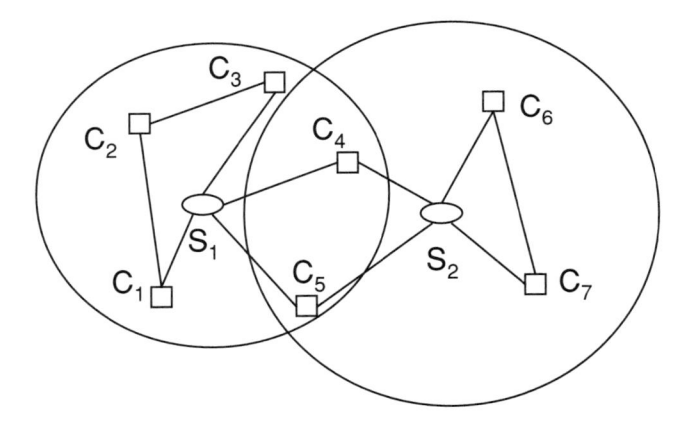

Figure 1. Example of neighborhoods.

Establishing the Edge Set

The edge set can be easily represented as a connectivity matrix from Eqn. (2). However this has two disadvantages

1. It would lead to a large sparse matrix
2. Such a representation will not efficiently support the operations in the maximum dispersion algorithm.

Hence the edge set is represented using the notion of a complete graph from graph theory. A complete graph (K) consists of vertices, all of which are connected to each other. Now, it is very easy to identify complete graphs in the multi site formulation discussed earlier. Since all trips containing any particular customer order as one of its members must conflict each other, these must be connected to each other forming a complete graph. Hence, corresponding to each customer order, there exists a complete graph. The graph as a whole is therefore made up of such overlapping complete graphs. The edge set E can now be represented as nc complete graphs, each K containing as many vertices as the number of trips or delivery options available for any particular customer order.

Figure 2 shows an example of trips/ vertices connected to each other through edges which represent conflicts. v_1 and v_2 are connected because they have customer 1 in common. The graph has 8 complete graphs each with cardinality 3. The trips are shown in braces without the SPs since conflicts are not related to SPs.

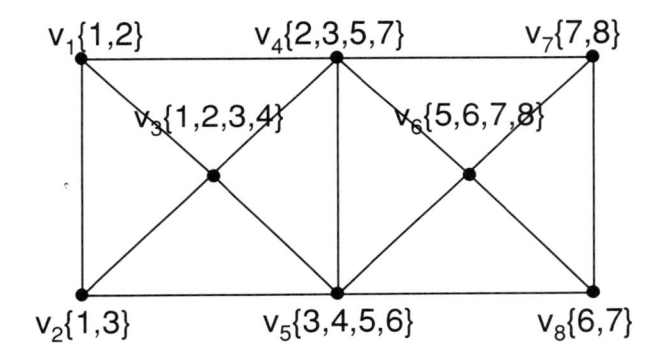

Figure 2. Conflicts and edges.

Maximum Dispersion Algorithm (MDA)

Before the maximum dispersion algorithm can be described, it is essential to define the dispersion operation which forms the basis of this algorithm.

Dispersion Operation

The dispersion operation can be performed on any existing dispersion of a graph. In general it can be defined as follows,

$$w(v) = w(v) - \sum_{K \in \overline{K}(v) \cap \overline{K}(D)} w(K) \qquad \forall v \in G \qquad (9)$$

such that,

$$\sum_{K \in \overline{K}(v)} w(K) = w(v) \qquad \forall v \in D \qquad (10)$$

It should be noted that this operation is not unique, in that the resultant values of the weights of vertices $w(v)$, depend upon the choice of the weights $w(K)$. The dispersion operation can be used as a useful criterion in the algorithm. If no vertex is left on graph G with a positive weight after performing the dispersion operation on the set D of dispersion vertices, then D is the maximum dispersion. This idea not only serves as a termination criterion, but also helps in identifying promising vertices which can form a better dispersion than the current one.

The idea of the dispersion operation can be better illustrated with an example. Figure 3 shows a graph with 8 vertices and an existing dispersion $D=\{v_1,v_5,v_7\}$. The dispersion operation is performed using the complete graphs $K_1=\{v_1,v_2,v_3\}, K_2=\{v_3,v_4,v_5\}$, $K_3=\{v_4,v_5,v_6\}$ and $K_4=\{v_6,v_7,v_8\}$. The operation using $w(K_1)=1$, $w(K_2)=0.6$, $w(K_3)=0.4$ and $w(K_4)=1$ generates the positive vertex v_6. The dispersion operation could also have been performed by using alternate complete graphs like $K_5=\{v_1,v_3,v_4\}$ and $K_6=\{v_2,v_3,v_5\}$.

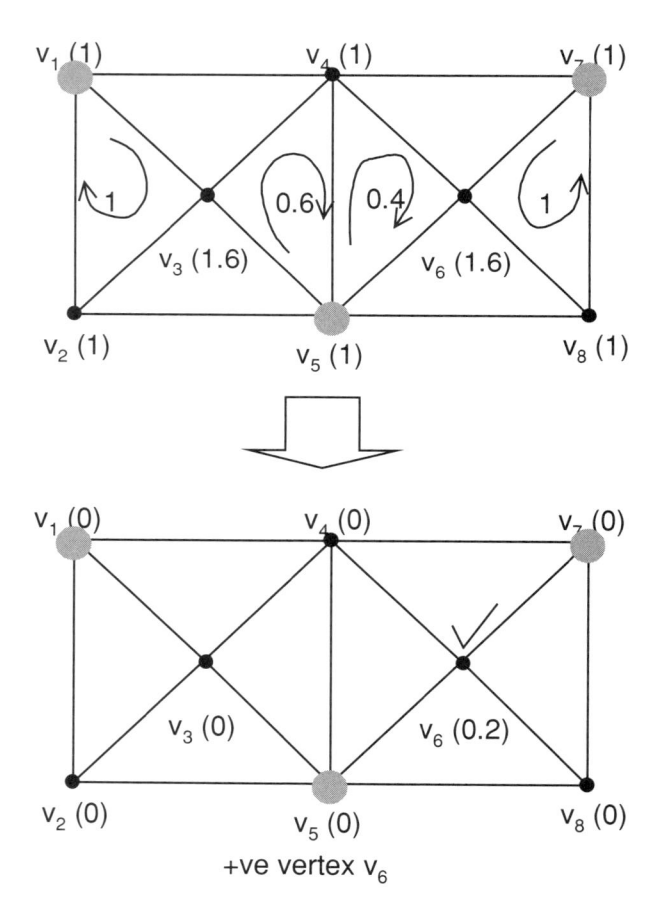

Figure 3. Dispersion operation.

Initialization

The algorithm can be initialized with any set of dispersion vertices, which can then be improved upon using the dispersion operation. However it would be a better to start with a good initial solution to accelerate the solution procedure. The celebrated greedy algorithm which is an exact algorithm for determining the minimum spanning tree on a graph (Biggs, 1985) could generate a good initial solution. The greedy algorithm for initializing the dispersion of a graph can be summarized as follows,

1. Choose a vertex $v \in G$ with maximum value of the weight $w(v)$
2. $G = G - \overline{K}(v)$
3. If $G=\emptyset$, then terminate, else goto 1.

Maximum Dispersion Algorithm

The MDA starts from the initial solution and finds better solutions using the dispersion operation and successively reducing the graph. The MDA can be outlined as follows,

1. Find initial dispersion D, $l=1$.
2. Perform dispersion operation on D.
3. Form a set of positive vertices P_l.
4. Select a positive vertex $v_p \in P_l$. If no +ve vertex is left and $l=1$ then terminate, else $l=l-1$, goto 4.
5. Remove vertices connected to v_p form G.
6. Calculate residual weight w_R given by,

$$w_R = \sum_{v \in D \cap \overline{K}(v_p)} w(v) - w(v_p) \quad (11)$$

8. If w is +ve then disperse on the reduced graph, $l=l+1$, goto 3.
9. If w=0 then goto 4.
10. If w is -ve then a better dispersion is obtained. Discard all the vertices removed from the graph and belonging to the earlier dispersion. Add all the newly selected positive vertices to form the new dispersion. $l=1$, goto 2.

Results

A sample run of the algorithm is shown in Fig. 5 on a problem (Fig. 4) with 20 vertices and 18 complete graphs for the purposes of illustration. Figure 4 shows the graph with the initial dispersion obtained from the greedy algorithm. The solution trajectory for this problem provides some insight into the performance of this algorithm. Whenever there are local parts of a dispersion that can be replaced by a better one, the algorithm does not have to dig deep to turn up with a new solution. Thus the algorithm will proceed replacing locally these better dispersion sub-sets and improving the current solution. Only in cases where a current dispersion is to be totally replaced by a better one, will the search depth be high. However in such a case the search breadth is expected to be short.

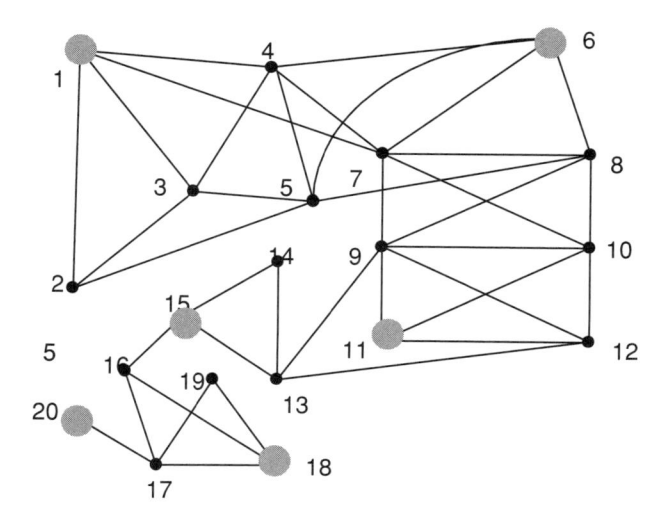

Figure 4. Example graph with initial dispersion.

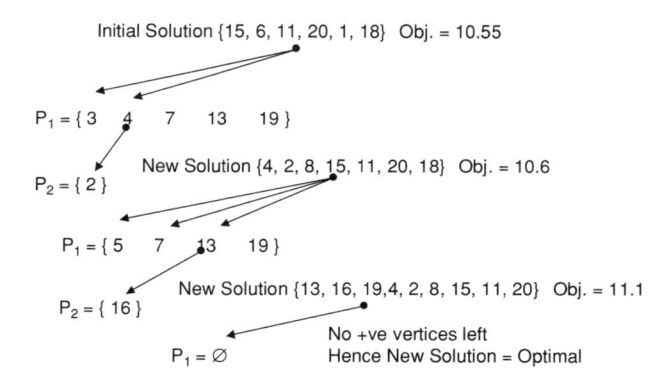

Figure 5. Solution trajectory.

The algorithm was also found to give solutions within seconds of CPU time for graphs with 1000 vertices (trips) and 100 complete graphs (customers).

Conclusion

A new representation for the formulation of multi-site delivery scheduling problems has been presented. The graph representation is generic enough to model different types of costing policies since it does not use any particular cost function. The notion of complete graphs from graph theory was exploited to efficiently support the operations of the maximum dispersion algorithm presented in this paper. The efficiency of the algorithm is evident from the CPU times for graphs of 1000 vertices.

The efficiency of this algorithm provides the motivation for its further extension to embrace more generic features of multi-site delivery problems such as multi-echelon deliveries and placement of intermediate stock points on a given network. These extensions can be realized by using the MDA as fathoming tool for tailored branch and bound algorithms so as to give tight lower bounds (for minimization) for the relevant sub-problems.

Nomenclature

- C Customer
- $\overline{C}$ Set of all customers
- d distance between locations in the argument.
- d_T distance of trip T.
- D dispersion set.
- $\overline{D}$ Set of all possible dispersion sets of a graph.
- DC delivery cost of trip.
- E set of edges
- G Graph (V,E)
- K complete graph
- $\overline{K}$ Set of all complete graphs of the vertex or set of vertices in the argument.
- l depth of search in MDA
- m no. of customers on a trip.
- n max. no. of customers allowed in a trip.
- P_l set of positive vertices at depth l.
- q_T load or amount of product in trip T.
- r_S radius of neighborhood of SP S.
- S SP (stock point or site)
- $\hat{S}$ nearest SP to customer.
- S_s, S_d source and destination SPs
- T trip $\{S_s, C_1, C_2, ..., C_m, S_d\}$
- TDC total delivery cost
- v vertex
- v_p *positive weight vertex*
- V set of vertices
- w weight of vertex or set of vertices
- w_R residual weight.

References

Biggs N. L. (1985) Trees, sorting and searching. *Discrete Mathematics*, Clarendon Press, Oxford. pp., 189-193.

Erkut E. (1990) The discrete p-dispersion problem. *European Journal of Operational Research*, **46**, 48-60.

Laporte G. (1992). The vehicle routing problem: An overview of exact and approximate algorithms. *European Journal of Operational Research*, **59**, 345-358.

Mokashi S. D. and A. C. Kokossis (1997). The maximum order tree method : A new approach for the optimal scheduling of product distribution lines. *Computers Chem. Engng.*, **21**, S679-S684.

Ravi S. S., D. J. Rosenkrantz and G. K. Tayi (1994). Heuristic and special case algorithms for dispersion problems. *Operations Research*, **42**, 299-310.

Ronen D. (1992), Allocation of trips to trucks operating from a single terminal. *Computers and Operations Research*, **19 No. 5**, 445-451.

AN OPTIMAL CONTROL APPROACH FOR SCHEDULING MIXED BATCH/CONTINUOUS PROCESS PLANTS WITH VARIABLE CYCLE TIME

H. P. Nott and P. L. Lee*
School of Engineering
Murdoch University, Perth WA 6150, Australia

Abstract

Effective scheduling of operations in the process industry has the potential for high economic returns. Process plants containing both batch and continuous units present a difficult scheduling problem, which is traditionally modelled by considering time over a uniform discretisation and then solved as a job shop scheduling problem. This generally results in a large mixed-integer linear programming problem, which is known to be extremely difficult to solve. Furthermore, when batch cycle times are decision variables in the process, the complexity of the problem is increased significantly. In this paper a new application area for optimal control is presented - the scheduling of mixed-batch/continuous system when batches have variable batch cycle times. The optimal control formulation is shown to enhance solution performance compared to the traditional job shop formulation. The motivation for this work is an existing scheduling problem in the sugar milling industry. However, a smaller problem is considered which has similar characteristics. This aids performance comparison.

Keywords

Scheduling, Mathematical modeling, Variable cycle time, Optimal control.

Motivation

This study is motivated by the need for an optimal scheduling policy in sugar milling (Fig. 1). Efficient scheduling of operations from crystallisation through to raw sugar production, incorporating various batch (eg pans) and continuous units (eg centrifugals and dryers) has the potential for high economic return.

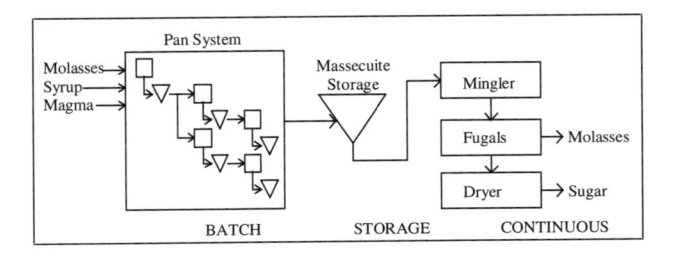

Figure 1. Simplified sugar milling system.

Although the pans in the sugar milling system (Fig. 1) process a fixed quantity of material, the length of time taken to process each batch is a decision choice. Thus, these batches are referred to as having *variable cycle times*. The physical system is constrained by bounds on the pan cycle times, as well as storage and flowrate limitations. A profit (turn over) is associated with the quantity of sugar produced, while there is a cost incurred from the processing of each pan operation. It also is considered costly to have a pan idle at any time and it is not advisable to make drastic changes to the continuous flowrate.

* Author to whom all correspondence should be addressed: peter@eng.murdoch.edu.au

Introduction

The scheduling of mixed-batch/continuous systems is a complicated problem and is generally accomplished by discretising time (Fig. 2) over the solution horizon. Moreover, the time discretisation size must be small enough to adequately represent the continuous units of the system. Thus, a much larger problem is produced than for the associated purely batch systems. Traditionally represented as a job shop scheduling problem, this then produces a large mixed-integer linear programming problem (MILP) as represented by Kondili *et al* (1993). Furthermore, although systems are generally restricted to having fixed batch cycle times, Nott (1998) has shown that incorporating variable cycle times in the model (where appropriate by the physical system) enables superior profit solutions. However, Nott (1998) has also gone on to show that representing such problems by the traditional job shop scheduling formulation is at severe computational expense - with the resultant MILP taking thousands of times longer to solve than the fixed cycle time formulation, in some instances.

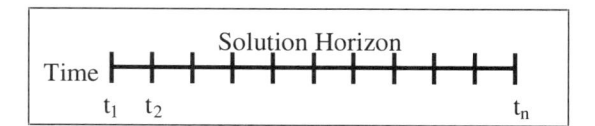

Figure 2. Discretised time.

An optimal control based strategy is presented that incorporates the benefits of variable cycle times, while minimising the resultant computational complexity.

Performance comparisons are made to the traditional job shop scheduling formulation by considering a simple problem which has similar characteristics to the sugar mill scheduling problem. Results are compared over a realistic parameter set as given in conjunction with the nomenclature at the end of this paper. Computations are performed on an IBM RS6000/380. MILPs generated are solved using the CPLEX 4.0 solver (CPLEX Optimization Inc, 1996), and AMPL (Fourer *et al* 1993, AMPL/CPLEX 1994) as a front end modelling language. CPLEX uses a branch and bound algorithm, (Williams, 1993) to solve MILPs.

Simple Model

The simple model used for performance comparison consists of two batch units 'dropping' into a single storage facility and then leading to a continuous production process, as seen in Fig. 3 (boxes, triangles and arrows represent batch, storage and continuous units respectively). A profit is associated with each unit leaving the system, and each batch begun incurs a cost. Batch units process batches with variable cycle times, within specified bounds, or they may be idle, although this is discouraged. The storage and flowrates are restricted, while changes in the flowrate are also deterred.

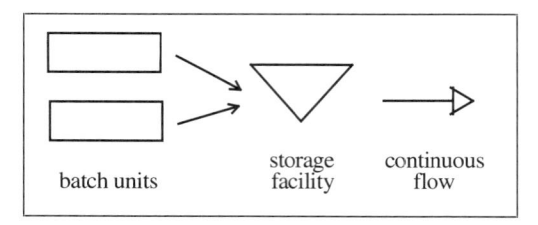

Figure 3. Simple batch-continuous model.

The system objective is represented by:

maximise {total system over solution horizon}
- {costs for batches}
- {penalty for idle periods}
- { penalty for changes in flowrate}.

Job Shop Formulation

For uniform time discretisations (Fig. 2), the job shop formulation is based on determining individual solution choices for each component of the system at every unit of time over the solution horizon. For each batch unit u, the decision $W_{u,t}=1$ represents the beginning of a new operational state, and states are defined as being either real batches: $Y_{u,t}=1$, or idle periods: $Y_{u,t}=0$ at each time unit t. As seen in the Appendix A, the complete formulation involves a large number (of the order of $6*num\_units*(TF-T0)$ or 300 for this case study) of boolean relationships/constraints as necessary to ensure cycle times lie within the feasible range, as well as to identify idle periods in the batch units.

Although deceptively simple in appearance, results presented later in this paper will confirm the computational effort required to solve this problem.

Optimal Control

Optimal control can be a useful modelling tool for complex scheduling problems (Kogan and Khmelnitsky, 1996). The optimal control problem is expressed as: find among all the acceptable control functions u, the one which minimises the performance criteria J, subject to the dynamic system constraints and all initial and final boundary conditions. The dynamic systems are described in terms of state variables $x(t)$ by:

$\dot{x} = f(x(t), u(t), t)$ for continuous time

or $x(k+1) = f(x(k), u(k))$ for discrete time

The state and control variables may both be constrained. The most general performance criterion is:

$$J = F(x(t_f)) + \int_{t_0}^{t_f} L(x(t),u(t),t)dt \text{ for continuous time}$$

or $J = F(x(N)) + \sum_{k=0}^{N-1} L(x(k),u(k))$ for discrete time

Fig. 4 shows the relationship between control vectors, **u** and state vectors, **x** for model P.

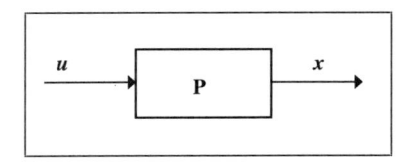

Figure 4. Dynamic model.

Often the state equations of model P are very complex, and the model is simplified or relaxed to P0. This incurs a modelling error of ΔP as is shown in Fig. 5.

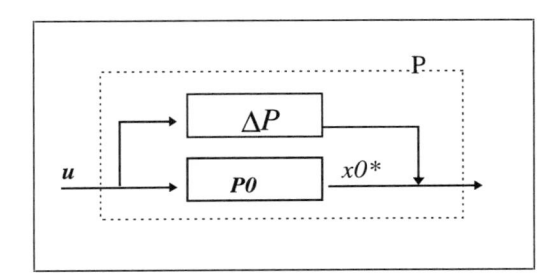

Figure 5. Relaxed model and modelling error.

In the past, optimal control has been applied to continuous time production in a hierarchical two-level splitting of the optimisation problem (Kogan and Khmelnitsky, 1996). The first level uses a relaxed model of the problem to produce some reference trajectory, while the second component attempts to track the reference trajectory using the actual model in real time. In this sense, optimal control has been able to deal with real-time disturbances. However, Brandimarte *et al* (1995) and Kim and Leachman (1993) identify that an optimal control strategy may assist in reducing the complexity of the off-line scheduling problem. In this paper a new application area for optimal control is presented - the scheduling of mixed-batch/continuous system when batches have variable batch cycle times. The structure of this optimal control implementation is to firstly determine an optimal trajectory for the relaxed model, P0, (Fig. 6).

The full process model, P then attempts to "track" this optimal trajectory. The success of this method depends on how close the optimal trajectory, x0* (from the relaxed model P0) approximates the true optimum, x* (which would result from a optimisation of the complete model, P).

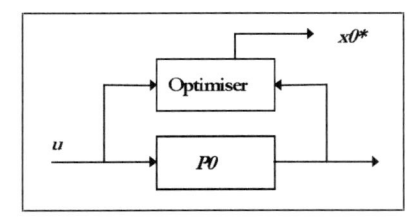

Figure 6. Optimal control stage 1: optimise relaxed model.

To reduce the effect of modelling error a closed-loop or feedback control may be applied where a controller C takes corrective action when a tracking error, x0*-x is detected, as seen in Fig. 7.

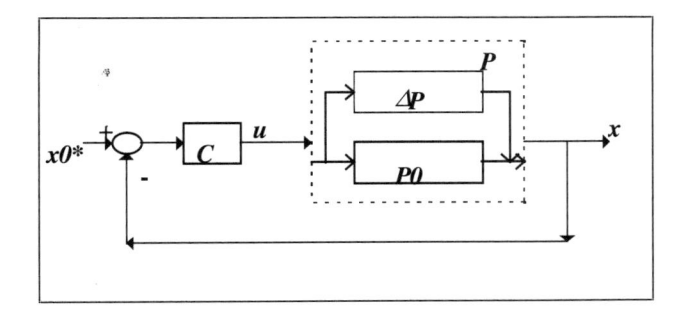

Figure 7. Optimal control corrective control / feed-back loop.

Optimal Control Implementation

The optimal control implementation maintains a discrete time representation and involves some feedback information, or corrective control measures. The optimal control algorithm is represented by:

1. Use a relaxed representation of the model to produce some reference trajectory.
2. Modify the reference trajectory to remove any implied idle periods. How this is achieved is explained in a subsequent section of this paper.
3. Track the reference trajectory using the complete model.
4. If the solution schedules an idle period:
   a) Let the reference trajectory = complete solution.
   b) Go back to step 2.
   c) Otherwise stop.

The complete optimal control formulation is shown in Appendix B.

The Relaxed Model and Reference Trajectory

Optimal control has most popularly been applied to manufacturing scheduling (Brandimarte *et al*, 1995) and this commonly involves a continuous flow model representation of discrete decisions. With the presence of both batch and continuous units responsible for a significant amount of the complexity of the mixed-batch/continuous scheduling problem, (Nott, 1998) the batch components of the system are initially relaxed to behave as continuous units, as seen in Fig. 8 (key as used in Fig. 3).

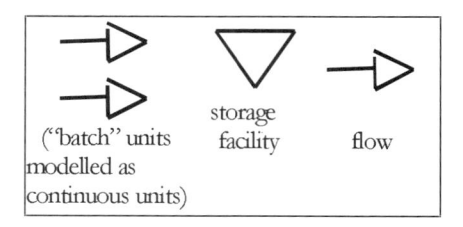

Figure 8. The relaxed model.

Thus relaxed model is expressed as a Linear Programming (LP) problem with two main decisions: $F_{u,t}$ - the flowrate through the 'batch' unit u at time t, and X_t - the flowrate through the continuous unit. The system is constrained by feasible flowrates (these are calculated for the 'batch' units by their minimum cycle time and batch size) and storage limitations and the objective is to:

maximise {system production over solution horizon}
 - {costs for individual batches}
 - {penalty for forced idles}
 - {penalty for changes in continuous flowrate}.

The solution produces a reference trajectory for the cumulative flow through each batch unit.

The success of the optimal control approach is dependent on how close the reference trajectory approaches the true optimal solution. Although the cumulative production curve from the relaxed model and the complete model may appear different (ie continuous curve vs a step function, as seen in Fig. 9), the impact on the solution may not be too large as the only objective function component of the complete model that is not present in the relaxed model is the deterrence of idle periods. The results obtained will confirm this hypothesis for this case study.

Corrective Control Procedures

The optimal control algorithm is unable to explicitly identify idle periods in the schedule. However, the reference trajectory produced from the relaxed model will highlight any tendency to schedule idle periods. Furthermore, the reference trajectory curve may be considered as *a priori* information to be used in solving the complete model in the next step. This information can then be manipulated to reduce any tendency towards scheduling idle periods while maintaining the overall shape of the reference trajectory (or production through the 'batch' units). This has analogies to the corrective control action applied in closed loop control.

Rounding the reference trajectory to the closest multiple of batch size for each unit, the solution is influenced to schedule an idle period when the rounded trajectory returns a cycle time larger than the maximum cycle time, *maxP*.

The tendency to schedule idle periods in the complete model is reduced by relaxing the penalty for the complete model solution deviating from the reference trajectory. This is implemented by modifying the reference trajectory. The easiest way to achieve this is to either lower the reference trajectory before the beginning of the offending batch or raise it towards the end of the batch. This is shown in Fig. 9. The shaded area indicates where the reference trajectory is modified, removing the preference to schedule idle periods.

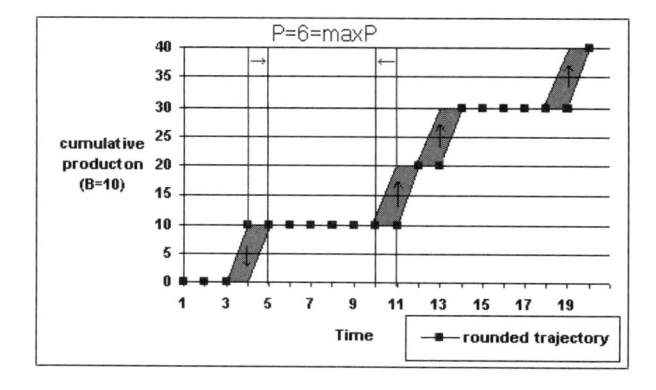

Figure 9. Reducing influence of scheduling idle periods.

Track Reference Trajectory with the Complete Model

The complete MILP model with batch units reinstated is used to track the reference trajectory for cumulative production through each batch that was produced from the relaxed model (Fig. 10). The is accomplished by penalising any deviations from the reference trajectory, T_u. Additional to the tracking objective, the complete model also maximises total production (the primary objective of the overall problem) and minimises changes in the continuous flowrate.

The batch component of the system is fully defined by binary variable: $E_{u,t}=1$ which indicates a batch ends on unit u at time t. Specifically, it is no longer necessary to explicitly represent the start of batches, or equivalently where idle periods are located. Thus, although the constraints present in the model are similar to those in the job shop scheduling formulation, many (of the order of 5*num_units*(TF-T0), 250 for this case study) of the boolean constraint relationships are no longer necessary. Issues relating to the presence of idle periods are taken up in the corrective control procedures.

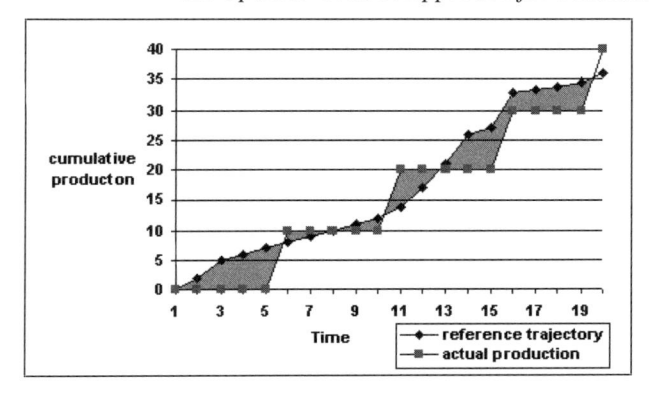

Figure 10. Tracking the reference trajectory.

Results

For the problem specifications given with the nomenclature, the optimal objective function value is found to be 1415.0833 by the job shop scheduling formulation. This involves 316,260 simplex iterations and traverses 22,448 branch and bound nodes. The MILP requires 585.46 seconds to solve, and a total of 598.20 seconds user time.

The optimal control method does find the known optimal objective function value of 1415.0833. The solution is found in a total of 0.63 seconds-user time, yielding an improvement of over 99%.

Conclusions

The optimal control approach presented in this paper has excellent performance for a problem that has been shown to be very difficult to solve. While the solution produced is not guaranteed to be optimal, a high quality solution is able to be returned.

Nomenclature & Case Study Parameters

Parameters Specifications Value

set U=1..num_units	batch units	1..2
set T=T0..TF	solution horizon	1..25
Z{U}	size of each batch	[8, 10]
minP{U}..maxP{U}	batch cycle times	[3, 4]:[5,6]
exp{U}	expected cycle time	[4, 5]
minX..maxX	feasible flowrates	2.5 .. 5.0
minS..maxS	feasible storage	2..15
S0	initial storage	10
sell	turn-over/produced	20
cost{U}	cost price per batch	[60, 50]
penI	idle penalty	100
penX	penalty X changes	1.0

Job Shop Formulation Variables

X {t}	flow in continuous unit time t to $t+1$
S {t}	amount in storage at time t
PR {t}	cumulative production up to time t
W {u, t} binary = 1	if a batch begins on unit u at time t
C {u, t} integer	# of batches on unit u up to time t
Y {u, t} binary:	type of batch: 0 = idle, 1 = real
B {u, t} binary=1	if a real batch begins on u at t
E {u, t} binary=1	if a real batch ends on u at t

Optimal Control Approach

Stage 1 Variables

F{u,t}	"flowrate" through unit, u at time $t-1$ to t
PRU{u,t}	cumulative prod through unit u up to time t
N{u}	number of batches scheduled of unit u
S{t}	amount in storage facility at time t
X{t}	flowrate through continuous unit time t to $t+1$
PR{t}	cumulative system production up to time t

Stage 2 Additional Parameters

T{u,t}	tracking curve (stage1) of production through unit u up to time t
penT	penalty for deviating from the tracking curve

Stage 2 Variables

E{u,t} binary =1	if batch ends on unit u at time t
PRU{u,t} integer	production through unit u up to time t
S{t}	amount in storage facility at time t
X{t}	continuous flowrate from time t to time $t+1$
PRU{t}	cumulative system production up to time t

References

AMPL/CPLEX (1994) *Using AMPL with CPLEX Optimizers on UNIX Systems,* CPLEX Optimization, Inc., Nervada.

Brandimarte P., Ukovich W. and Villa A. (1995) Continuous flow models for batch manufacturing: a basis for a hierarchical approach, *Int. J. Prod. Res.* 33(6):1635-1660.

CPLEX Optimization Inc. (1996), CPLEX-MIP, Suite 279, 930 Tahoe Blvd., Bldg. 802, Incline Village, Nervada.

Fourer R., D.M. Gay and B.W. Kernigham (1993) *AMPL : A modeling language for mathematical programming,* Boyd Fraser, Massachusetts.

Kim S-Y. and Leachman R.C. (1993) Multi-project scheduling with explicit lateness costs, *IIE Trans.* 25(2):34-44.

Kogan K. and Khmelnitsky E. (1996) An optimal control model for continuous time production and setup scheduling, *Int. J. Prod. Res.* 34(3):715-725.

Kondili E., Pantelides C.C. and Sargent R.W.T. (1993). A general algorithm for short-term scheduling of batch operations - I. MILP formulation. *Comp. Chem. Engng.* 17(2):211-227.

Kossik J.M., Miller G. (1994), Optimize cycle times for batch biokill systems, *Chem. Eng. Prog.* 90(10):45-51.

Nott H.P. (1998). Modelling alternatives for scheduling mixed batch/continuous process plants with variable cycle time, *PhD thesis submitted to school of Engineering, Murdoch University, Australia*

Williams H.P. (1993), *Model building in mathematical programming,* Wiley, England.

Appendix A. Job Shop Formulation

$$\text{maximise } sell * PR_{TF} - \sum_{\substack{u \in U \\ t \in T}} cost_u * B_{u,t} - penI * \sum_{\substack{u \in U \\ t \in T}} (1 - Y_{u,t}) - \sum_{t \in T: t > T0} penX * |X_t - X_{t-1}|$$

subject to

initial batch	$\forall u \in U$	$C_{u,T0} = 1$
	$\forall u \in U$	$E_{u,T0} = 0$
batch count	$\forall u \in U, \forall t \in T: t > T0$	$C_{u,t} = C_{u,t-1} + W_{u,t}$
min cycle times	$\forall u \in U, \forall t \in T,$	
	$\forall tt \in (t+1)..min(t + minP_u - 1, TF)$	$C_{u,tt} - C_{u,t} \leq (TF - T0) * (1 - B_{u,t})$
max cycle times	$\forall u \in U, \forall t \in T: t \geq T0 + maxP_u$	$C_{u,t} - C_{u,t-maxP_u} \geq 1$
unit idle periods	$\forall u \in U, \forall t \in T$	$W_{u,t} + Y_{u,t} \geq 1$
feasibility	$\forall u \in U, \forall t \in T: t > T0$	$Y_{u,t-1} - W_{u,t} \leq Y_{u,t} \leq Y_{u,t-1} + W_{u,t}$
assign batch begins	$\forall u \in U, \forall t \in T$	$W_{u,t} + Y_{u,t} - 1 \leq B_{u,t} \leq (W_{u,t} + Y_{u,t})/2$
assign batch ends	$\forall u \in U, \forall t \in T: t > T0$	$W_{u,t} + Y_{u,t-1} - 1 \leq E_{u,t} \leq (W_{u,t} + Y_{u,t-1})/2$
initial storage		$S_{T0} = S0$
update storage	$\forall t \in T: t > T0$	$S_t = S_{t-1} - X_{t-1} + \sum_{u \in U} E_{u,t} * Z_u$
storage feasibility	$\forall t \in T$	$S_t - X_t \geq minS$
initial production		$PR_{T0} = 0$
update production	$\forall t \in T$	$PR_t = PR_{t-1} + X_{t-1}$

VARIABLES :

$\forall t \in T: minX \leq X_t \leq maxX \qquad \forall t \in T: minS \leq S_t \leq maxS \qquad \forall t \in T: PR_t \geq 0$

$\forall u \in U, \forall t \in T: W_{u,t}, C_{u,t}, Y_{u,t}, B_{u,t}, E_{u,t} \in \{0,1\}$

Appendix B. Optimal Control Formulation

Stage 1 Formulation: Relaxed Model

$$\text{maximise} \quad sell * PR_{TF} - \sum_{u \in U} cost_u * N_u - \sum_{t \in T: t > T0} penX * |X_t - X_{t-1}|$$
$$- penI * \sum_{u \in U} (TF - T0 - N_u * \min P_u + 1) \quad \text{when } TF - T0 > N_u * minP_u - 1$$

subject to

initial batch flowrate	$\forall u \in U, \forall t \in T: t \leq minP_u$	$F_{u,t} = 0$
feasible batch flowrate	$\forall u \in U, \forall t \in T: t < TF$	$\sum_{tt \in t..min(t+minP_u-1,TF)} F_{u,tt} \leq Z_u$
initial batch production	$\forall u \in U$	$PRU_{u,T0} = F_{u,T0}$
batch unit production	$\forall u \in U, \forall t \in T: t > T0$	$PRU_{u,t} = PRU_{u,t-1} + F_{u,t}$
number of batches	$\forall u \in U$	$PRU_{u,TF} = N_u * Z_u$
maximum batches	$\forall u \in U$	$N_u \leq (TF - T0)/minP_u$
initial amount in storage		$S_{T0} = S0$
update storage	$\forall t \in T: t > T0$	$S_t = S_{t-1} - X_{t-1} + \sum_{u \in U} F_{u,t}$
feasible storage	$\forall t \in T$	$S_t - X_t \geq minS$
initial production		$PR_{T0} = 0$
update production	$\forall t \in T$	$PR_t = PR_{t-1} + X_{t-1}$

VARIABLES :

$\forall t \in T: minX \leq X_t \leq maxX \qquad \forall t \in T: minS \leq S_t \leq maxS \qquad \forall t \in T: PR_t \geq 0$

$\forall u \in U, \forall t \in T: F_{u,t}, PRU_{u,t}, N_{u,t}$

Stage 2 Formulation: Tracking Reference Trajectory

$$\text{maximise } sell * PR_{TF} - penT * \sum_{\substack{u \in U \\ t \in T}} |PRU_{u,t} - T_{u,t}| - \sum_{t \in T: t > T0} penX * |X_t - X_{t-1}|$$

subject to

initial batch	$\forall u \in U, \forall t \in T: t \leq minP_u$	$PRU_{u,t} = 0$
	$\forall u \in U, \forall t \in T: t \leq minP_u$	$E_{u,t} = 0$
batch unit production	$\forall u \in U, \forall t \in T: t > minP_u$	$PRU_{u,t} = PRU_{u,t-1} + E_{u,t} * Z_u$
minimum cycle time	$\forall u \in U, \forall t \in T$	$\sum_{tt \in t..min(t+minP_u-1,TF)} E_{u,tt} \leq 1$
initial storage		$S_{T0} = S0$
update storage	$\forall t \in T: t > T0$	$S_t = S_{t-1} - X_{t-1} + \sum_{u \in U} E_{u,t} * Z_u$
feasible storage	$\forall t \in T$	$S_t - X_t \geq minS$
initial production		$PR_{T0} = 0$
update production	$\forall t \in T: t > T0$	$PR_t = PR_{t-1} + X_{t-1}$

VARIABLES :

$\forall t \in T: minX \leq X_t \leq maxX \qquad \forall t \in T: minS \leq S_t \leq maxS \qquad \forall t \in T: PR_t \geq 0$

$\forall u \in U, \forall t \in T: E_{u,t} \in \{0,1\}, PRU_{u,t} \in Z$

INTEGRATED SCHEDULING IN STEEL PLANTS

Vipul Jain and Ignacio E. Grossmann
Department of Chemical Engineering
Carnegie Mellon University
Pittsburgh, PA 15213

Abstract

This paper deals with the scheduling of orders in the steel rolling mills and the machine shop for manufacturing steel of desired specifications in a steel plant. A representation that integrates the flow of orders through the rolling mills and the machine shop into a single flowsheet is presented. The proposed representation transforms partially structured scheduling problem into a completely structured scheduling problem that resembles the multistage problem considered by Pinto and Grossmann (1995,1996). However, there are two major differences: Firstly, an order can skip stages, and secondly, a given machine can be used in multiple stages. A continuous time MILP model, which is an extension of the model presented by Pinto and Grossmann (1996), is presented to solve this class of scheduling problems. Finally, a number of numerical examples are discussed to highlight the importance of the proposed representation and solution method.

Keywords

Scheduling, MILP, Steel rolling mills, Flexible flowshops.

Introduction

Over the past few years there have been significant developments in mixed integer programming based modeling techniques for scheduling problems (Kondili *et al.*, 1993; Shah *et al.*, 1993; Pinto and Grossmann, 1995; Mockus and Raklaitis, 1996; Zang and Sargent, 1996). However, the computational performance of these mixed integer models is far from satisfactory because the computational time for these models increases exponentially with the size of the problem. Recently, the development in the integer programming based heuristic scheduling algorithms by Pinto and Grossmann (1996) has proven to be very effective for solving large real life multistage scheduling problems. The key idea of this method is to use heuristics to fix or restrict some difficult decisions, like sequencing or assignment, in the mixed integer model. The resulting mixed integer program, which can be solved efficiently to determine remaining decisions, enforces the constraints and structure of the scheduling problem. In this paper, we generalize this method for a broader class of scheduling problems.

This paper is organized as follows. In the next section, we present a real life scheduling problem, which is motivated by steel manufacturing process. A representation that incorporates all allowable manufacturing sequences is then presented. This representation leads to a structure which broadly fits the structure of the problem considered by Pinto and Grossmann (1995,1996). However, there are two main differences. Firstly, an order can skip stages, and secondly, a machine is used in multiple stages. A continuous time MILP model that is an extension of the model proposed by Pinto and Grossmann (1996) is then presented to solve this class of problems. Finally, some examples are considered which are solved using the integer programming based heuristic method, and some conclusions are drawn.

Problem Description

The steel manufacturing plant considered in this paper consists of multiple rolling mills and a machine shop for manufacturing steel of desired specifications. The steel plant is required to complete a set of customer orders by certain due dates. These orders are translated into plant orders (quantity/specification) and distributed to different rolling mills (see Fig. 1). The output of each mill can go to one of the three destinations: Either it goes out of the system ("Finished Orders"), or it goes to the next rolling mill ("Rolling Mill i to Rolling Mill j, where i < j), or it goes to the machine shop.

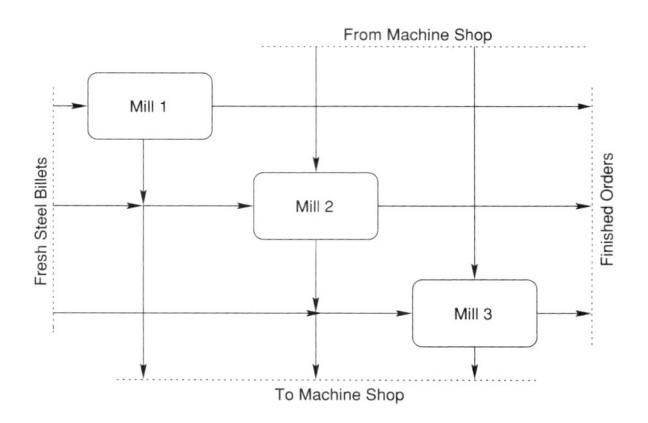

Figure 1. Flow of orders through steel rolling mills.

Orders are sent to the machine shop either to perform a pre-specified task or to perform corrective work. The machine shop consists of two stages. The first stage is the straightening stage and the second stage is the shearing stage. Each of these stages consists of multiple machines that operate in parallel. An order may use one or both the stages of the machine shop, but straightening is always done before shearing. The flow of orders in the machine shop with three straightening units and two shearing units is shown in Fig. 2. The output from the machine shop has one of two destinations: Either they are shipped out of the system, or they are sent to the next mill (Orders that came from Rolling Mill i will go to Rolling Mill j, where i < j). It should be noted that the requirement for corrective work is uncertain and cannot be predicted in advance. Therefore, rescheduling may be required for both rolling mills and machine shops to incorporate the additional step.

Each plant order is associated with the specified quantity and specification of steel to be manufactured, and the sequence of tasks in the rolling mills and the machine shop. The aim of this work is to develop a scheduling algorithm to find a workable schedule for the processing of a given set of orders in the rolling mills and the machine shop. The desired objective for the scheduling algorithm is to minimize the sum of completion times. This is because if the need for corrective work arises an attempt can be made to complete the order before the due date for the updated sequence of rolling mill and machine shop tasks. It is desired that the scheduling algorithm should be computationally fast because rescheduling of the whole system is required whenever corrective work is considered necessary for any order, at any instance. In the next section, we present a representation of the steel manufacturing process that integrates the rolling mills and the machine shop into one integrated flowsheet.

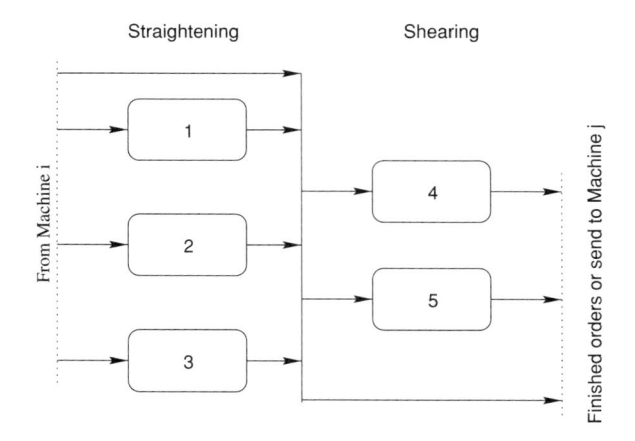

Figure 2. Flow of orders in the machine shop.

Representation

The flow of orders through the rolling mills or the machine shop resembles what in operations literature is referred as a flexible flowshop plant. However, if they are considered along with each other the statement is no longer valid, and the problem does not match any standard structure. This is because the machine shop can be used by an order more than once. However, there are two pieces of information in the problem description that can be exploited for developing a simple representation of the flow of orders in the steel plant. Firstly, the number of rolling mills determines the maximum number of machine shop related tasks for an order and secondly, the machine shop task follows a rolling mill task. Using this information, we can represent a steel plant with three rolling mills, three straightening units, and two shearing units as in Fig. 3.

The unified representation has nine stages. The stages 1, 4 and 7 correspond to the rolling mills 1, 2 and 3, respectively. The stages 2 and 3, 5 and 6, and 8 and 9 correspond to the machine shop. Arrows show the direction in which an order can move through the system. It should be noted that the machine shop is repeated thrice because there are three rolling mills. Furthermore, stages corresponding to the machine shop are exactly after each stage of rolling mill. The proposed representation does not impose any restriction on the sequence of rolling mill and machine shop tasks an order can undergo, except for the ones imposed by the description of the problem itself. This representation is very useful as it gives a structure to the flow of orders through the steel plant. Furthermore, this flowsheet representation resembles the structure assumed

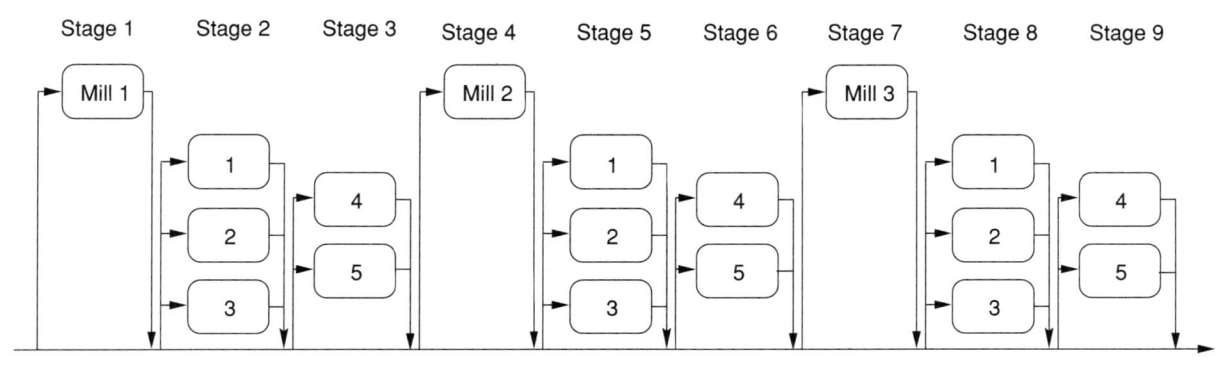

Figure 3. Unified representation of the flow of orders in the steel plant.

by Pinto and Grossmann (1995,1996), except for two major differences. Firstly, an order can skip stages and secondly a machine is assigned to more than one stage. In the next section we use this representation to develop a Mixed Integer Linear Programming (MILP) model to solve this scheduling problem.

MILP Model

The problem of finding a schedule in the context of unified representation can be summarized as follows. The steel plant is required to complete plant orders by certain due dates. Each order has a fixed processing time for every unit it might use and passes through a subset of stages of the unified representation. The aim of the problem is to determine the start time, the end time, and the machines used for each order in the stages required for completion of that order.

As mentioned earlier, the structure of the problem in current form resembles the one assumed by Pinto and Grossmann (1995,1996). In their first paper (Pinto and Grossmann, 1995) they proposed a comprehensive continuous time MILP model for scheduling such systems with the objective of minimizing earliness. The model proved to be computationally very expensive even for small test problems because of the combinatorics involved. To improve the computational efficiency they proposed the use of additional preordering constraints. The motivation behind using preordering constraints was to fix sequencing decisions a priori by using single machine scheduling algorithms and leave the decisions regarding the assignment of orders to machines as variables. By using the additional preordering constraints, they were able to increase the computational speed of the MILP model. Since it did not reach the desired level of efficiency, they proposed a two step solution algorithm for solving the MILP model with preordering constraints. In the first step, they used an alternate objective function to find an assignment of various units used in each stage for each order. In the second step they solved a linear programming problem to obtain the start time and end times of various orders in different units for the assignment obtained in step one. The method was very effective for solving large scheduling problems because for all the examples considered linear programming relaxation of the MILP in step one gave an integer solution. In their second paper (Pinto and Grossmann, 1996) they proposed an alternate MILP that had preordering as the underlying representation of the model, as opposed to including preordering as separate set of constraints. Since the problem size was much smaller, the computational performance of the two step method for solving the MILP scheduling model showed a significant improvement.

It is interesting to note that the initial comprehensive model of Pinto and Grossmann (1995) can be used to solve the scheduling problem at hand by making a small change in the objective function. However, the model will have the same limitation regarding the computational time. For the current scheduling problem computational speed is of prime importance because scheduling has to be done in real time. A computationally inexpensive feasible schedule is more desirable than the computationally expensive optimal schedule. In this section, we extend the alternate MILP model proposed by them to solve the current scheduling problem efficiently.

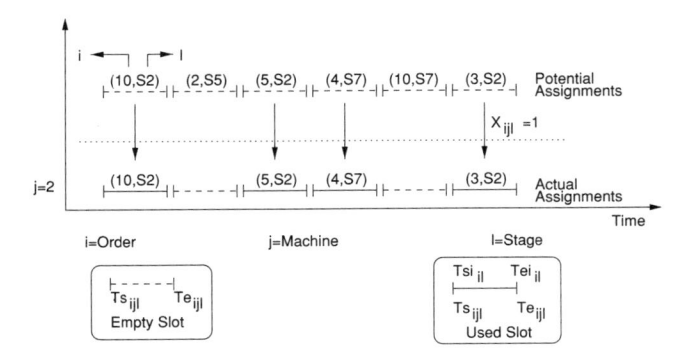

Figure 4. Preordering in a unit.

Before presenting the details of the model let us first present a brief nomenclature. Indices i, j and l denote an order, a machine and a stage, respectively. The set of orders ($\{1,2,3...\}$) is denoted by I, the set of stages ($\{S1, S2...\}$) by L, and the set of units ($\{M1, M2...1,2...\}$) by J.

The set of stages required to complete Order i is denoted by L_i ($L_i \subseteq L$). The set of machines that can be used by Order i is represented by J_i ($J_i \subseteq J$). The set of units belonging to Stage l is denoted by J_l ($J_l \subseteq J$).

The central idea of the model is the representation of time. Two different time coordinates are used to handle the assignment of an order to a machine. The first set of time coordinates is from the perspective of an order and is represented by variables Ts_{il} and Te_{il}. The variable Ts_{il} denotes start time of Order i in Stage l, and the variable Te_{il} denotes end time of Order i in Stage l. The second set of time coordinates is from the perspective of a machine and is represented through time slots. A time slot represents possible use of a machine by an order in a stage. The variables Ts_{ijl} and Te_{ijl} are used to denote start and end of a time slot corresponding to Order i, Machine j and Stage l. If Machine j is used by Order i for Stage l then the length of the time slot is equal to corresponding processing time, otherwise it is zero. The binary variable X_{ijl} is used to denote this decision. It is equal to one when Machine j is used by Order i for Stage l, and zero otherwise. The time slots, defined for a machine, are arranged in a sequence that is determined by a preordering heuristic. The two timing coordinates are mapped onto each other when an assignment is made to complete a stage of an order on a particular machine. Figure 4 illustrates the idea of time slots and associated variables. Using these ideas, we can now write the MILP model for the problem.

Objective Function: As discussed earlier, the desired objective of the problem is to finish all the orders as soon as possible. Hence, a reasonable objective for the MILP model is to minimize the weighted sum of completion times.

$$\text{Minimize} \sum_{i \in I} h_i Te_{i\tilde{l}_i} \quad (1)$$

Here parameter h_i is the weight for the completion time of Order i and index $\tilde{l}_i$ is the final stage of Order i.

Assignment Constraints: There are two sets of assignment constraints. The first set ensures that only one unit is used in stages with multiple units to complete the task associated with an order.

$$\sum_{j \in (J_i \cap J_l)} X_{ijl} = 1 \quad \forall (i \in I), (l \in L_i) \quad (2)$$

The second set of assignment constraints enforces an upper bound on the number of tasks (K_j) that can be performed on a machine. This constraint ensures that tasks are distributed among parallel machines (Pinto and Grossmann, 1996).

$$\sum_{i \in I_j} \sum_{l \in (L_i \cap L_j)} X_{ijl} \leq K_j \quad \forall (j \in J) \quad (3)$$

Definition of End Times: The end times in two time coordinates are related to the corresponding start times by the following constraints.

$$Te_{ijl} = Ts_{ijl} + X_{ijl}(T_{ijl} + SU_j)$$
$$\forall (i \in I), (l \in L_i), j \in (J_i \cap J_l) \quad (4)$$

$$Tei_{il} = Tsi_{il} + \sum_{j \in (J_i \cap J_l)} X_{ijl}(T_{ijl} + SU_j) \quad \forall (i \in I), (l \in L_i) \quad (5)$$

Here SU_j is the set time on Machine j and T_{ijl} is the processing time of Order i on Machine j for Stage l.

Timing Constraints: There are three sets of timing constraints. The first set of timing constraints fixes the relative sequence of time slots. By definition, a time slot represents the possible use of a machine by an order for a stage. By fixing the sequence of time slots, preordering of different tasks is ensured in the model definition itself.

$$Te_{ijl} \leq Ts_{i'jl'} \quad \forall (i \in I), (l \in L_i), (i' \in I), (l' \in L_{i'}),$$
$$j \in (J_i \cap J_l \cap J_{i'} \cap J_{l'}), Po_{i'jl'} = Po_{ijl} + 1 \quad (6)$$

Here Po_{ijl} is the position number of time slot of Machine j for Stage l of Order i. The second set of timing constraints ensures the direction of flow of order through the system.

$$Tei_{il} \leq Tsi_{il'} \quad \forall (i \in I), l \in (L_i - \{\tilde{l}_i\}), (l' \in L_i), (l' \geq l) \quad (7)$$

The third set of timing constraints ensures that each order (i) is completed before the due date of that order (d_i).

$$Tei_{i\tilde{l}_i} \leq d_i \quad \forall (i \in I) \quad (8)$$

Time Matching Constraints: These constraints relate the two time coordinates. They ensure that if Stage l of Order i is processed in Machine j ($X_{ijl}=1$) then Ts_{ijl} is equal to Ts_{il}.

$$-U(1 - X_{ijl}) \leq Ts_{ijl} - Tsi_{il} \quad \forall (i \in I), (l \in L_i), j \in (J_i \cap J_l) \quad (9)$$

$$U(1 - X_{ijl}) \geq Ts_{ijl} - Tsi_{il} \quad \forall (i \in I), (l \in L_i), j \in (J_i \cap J_l) \quad (10)$$

Variable Definition:

$$X_{ijl} \in \{0,1\} \quad \forall (i \in I), (l \in L_i), j \in (J_i \cap J_l) \quad (11)$$

$$Ts_{ijl}, Te_{ijl} \geq 0 \quad \forall (i \in I), (l \in L_i), j \in (J_i \cap J_l) \quad (12)$$

$$Tsi_{il}, Tei_{il} \geq 0 \quad \forall (i \in I), (l \in L_i) \quad (13)$$

The MILP model (M1) with Eqn. (1) as objective and Eqns. (2) to (13) as constraints can be used to solve the scheduling problem considered in this paper. However, before attempting to solve the model, preordering must be performed by assigning correct value to the parameter Po_{ijl}. Preordering can be done by using the heuristic presented by Pinto and Grossmann (1996), but special care has to be taken because an order can be processed in a machine more than once. The modified heuristic for this case is as follows.

1. For each Order i with due date d_i and processing time T_{ijl} in Machine j for stage l, evaluate:

$$Tsi_{il}^U = d_i - \sum_{l' \in L_i, l' \geq l} \min_{j \in (J_i \cap J_l)} \{T_{ijl}\} \quad \forall (i \in I), (l \in L_i) \quad (14)$$

Where Tsi_{il}^U is the upper bound on the start time of Order i in Stage l.

2. Sequence all the time slots corresponding to a machine j in increasing Tsi_{il}^U. Parameter Po_{ijl} is assigned the position number (1,2...) of the time slot corresponding to Stage l of Order i in the sequence for Machine j. The ties are broken using the release dates, which are defined as the lower bounds on start times.

$$Tsi_{il}^L = \sum_{l' \in L_i, l' < l} \min_{j \in (J_i \cap J_l)} \{T_{ijl}\} \quad \forall (i \in I), (l \in L_i) \quad (15)$$

The MILP model M1 differs from the one presented by Pinto and Grossmann (1996) in two respects. First, the objective function is different. Second, the definition of second time coordinate is different (Ts_{ijl} vs. Ts_{ij}). As shown in the next section, Model M1 is computationally not fast enough for real time scheduling of the industrial sized problems. It can however be solved using the two step method of Pinto and Grossmann (1996). In the first step, the MILP model is solved with an alternative objective of minimizing the total process time,

$$\text{Minimize} \sum_{(i \in I), (l \in L_i), (j \in J_i \cap J_l)} h_i X_{ijl} T_{ijl}. \quad (16)$$

to determine the assignment of tasks to machines. In the second step, model M1 is solved for fixed assignments to obtain the schedule for processing of orders in the steel plant.

The schedule, generated by the proposed method, has to be changed whenever an order needs additional processing steps. In such a situation, rescheduling can be done very easily by the proposed MILP model. However, the problem data has to be updated to account for tasks already completed, additional tasks, and tasks in progress. In context of the MILP model, the following updates are sufficient:

1. Identify the set of machines that are being used to process orders at the time of change in schedule. For all these machines, use the completion time of the tasks that are being completed as a lower bound for the start time of all the tasks on that machine. This will ensure that all the machines will finish the task they are doing before they are assigned to a new task.

2. For each Order i, update the set L_i by deleting all the stages that have been completed, and by adding new stages to incorporate additional steps required. It might be possible that a stage of an order might be under processing at the time of rescheduling. For such orders, remove that stage from the set L_i, and use its completion time as the lower bound for the start time of that order in any stage. This ensures that next stage of an order will not start before an earlier stage has been completed.

By solving the MILP model, with the updated sets and additional lower bounds, a schedule can be obtained for the remaining tasks. In summary, the MILP model M1 along with two step solution algorithm can be used to perform real time scheduling in the steel plants. In the next section, we present a number of example problems to compare the performance of different solution algorithms and highlight the importance of issues involved.

Computational Results

Let us first consider a small example problem. Given is a steel plant with three rolling mills, three straightening units and two shearing units (see Fig. 3). There are seven orders that have to be completed in this plant. The due dates and processing steps for these orders are summarized in Table 1. The aim of the problem is to find a schedule to complete these orders. The data regarding the processing time could not be presented here because of the restriction on the length of this paper.

Table 1. Due Dates and Stages for Orders.

Order	Due Date (d_i)	Stages Required
1	140	S1
2	110	S1,S2,S3,S4
3	140	S4,S5,S6
4	110	S1,S2,S3,S4,S5,S6,S7
5	110	S4,S5,S6,S7,S8,S9
6	110	S7
7	140	S1,S7,S8,S9

The problem can be modeled using the MILP model M1, which in turn can be solved either directly as an MILP or by using the two stage method presented in the previous section. It took 0.25 CPU seconds on a HP C110/9000 workstation to solve the problem directly as an MILP and 0.22 CPU seconds for the two stage method. . The GAMS modeling language (Brooke et al., 1992) was used to generate the model and CPLEX 4.0 (CPLEX, 1995) was used as the MILP solver. The computational times are not significantly different because of the small problem size. The Gantt chart corresponding to the solution obtained by using the two step method is shown in Fig. 5.

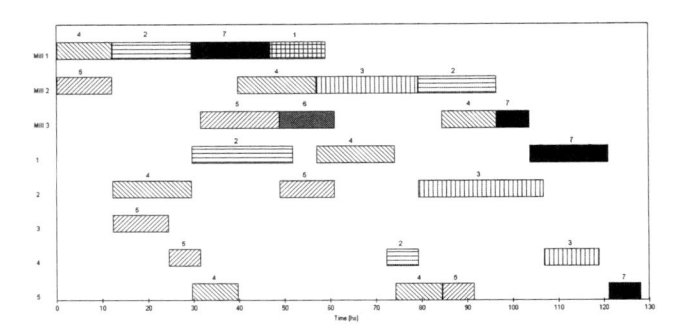

Figure 5. Schedule for seven orders.

Two other examples with the same plant topology but with higher number of orders were considered. The computational results for all the examples are summarized in Table 2.

Table 2. Computational Results.

Orders	Bin.Vars., Cont.Vars., Constraints	CPU Time (secs.) Direct MILP	2 Step Method	% difference in objective function.
7	47,201,299	0.25	0.22	3.7
12	68,288,415	178.92	0.33	6.3
15	71,311,443	12.09	0.42	6.4

The computational results indicate that the two step method is much faster for larger problems. Furthermore, for all the examples the difference in the value of objective function for the two solution methods is less than 7%. It should be noted that the computational time for solving the problem directly as an MILP is very unpredictable, for example, the smaller problem with 12 orders took more time than the larger problem with 15 orders. It was observed that the model had zero integrality gap in the step 1 of the two step algorithm. This conforms to the similar observation made by Pinto and Grossmann (1996). These results indicate that the Model M1 when solved using the two step algorithm is an efficient method to perform scheduling in steel plants.

Conclusions

The problem of scheduling orders in the steel rolling mills and the machine shop of a steel plant to manufacture steel of desired specifications has been addressed in this paper. The scheduling problem does not fit any structure in the literature. A representation was presented to transform it into a multistage scheduling problem similar to the one considered by Pinto and Grossmann (1995,1996), but with two major differences: Firstly, an order can skip stages, and secondly, a given machine can be used in multiple stages. An MILP model, which is an extension of the model presented by Pinto and Grossmann (1996), was presented to solve this type of scheduling problems. Finally, a number of numerical examples were presented to highlight the importance of the proposed representation and solution method.

Acknowledgements

The authors would like to thank Dr. Bruno Repetto of MSA-PASS for useful discussions regarding the working of the steel mills. The authors would also like to acknowledge the Benjamin Franklin Program for providing the financial support for this work.

References

Brooke, A., D. Kendrick, and A. Meeraus (1992). *GAMS: A users guide.* Scientific Press, Palo Alto.

CPLEX (1995). *Using the CPLEX callable library.* CPLEX Optimization Inc.

Kondili, E., C. C. Pantelides, and R. W. H. Sargent (1993). A general algorithm for scheduling of batch operations-I. *Computers Chem. Engng.*, **17**, 2, 221-227.

Mockus, L. and G. V. Reklaitis (1996). Continuous time representation in batch/semi-continuous process scheduling: randomized heuristic approach. *Computers Chem. Engng.*, **20**, Suppl., S1173-S1177.

Pinto, J. M. and I. E. Grossmann (1995). A continuous time mixed integer linear programming model for short term scheduling of multistage batch plants. *Ind. Eng. Chem. Res.*, **34**, 9, 3037-3051.

Pinto, J. M. and I. E. Grossmann (1996). An alternate milp model for short term scheduling of batch plants with preordering constraints. *Ind. Eng. Chem. Res.*, **35**, 1, 338-342.

Shah, N., C. C. Pantelides, and R. W. H. Sargent (1993). A general algorithm for scheduling of batch operations-II. *Computers Chem. Engng.*, **17**, 2, 229-234.

Zhang, X. and R. W. H. Sargent (1996). The optimal operation of mixed production facilities – A general formulation and some approaches for solution. *Computers Chem. Engng.*, **20**, 6/7, 897-904.

LARGE SCALE MULTI-FACILITY PLANNING USING MATHEMATICAL PROGRAMMING METHODS

Paul R. Bunch and Rex L. Rowe
Eli Lilly and Company
Lilly Corporate Center
Indianapolis, IN 46286

Michael G. Zentner
Advanced Process Combinatorics, Inc.
West Lafayette, IN 47906

Abstract

We describe the development of a mathematical programming based planning model to facilitate strategic asset management decisions in a manufacturing environment. In particular, a mixed integer linear programming (MILP) model was developed to find the lowest cost alternative using existing manufacturing assets to satisfy demand for approximately 2000 finished pharmaceutical items. The model was used to determine the production quantities over a time horizon of each item on each operating unit within a geographically distributed manufacturing network.

In addition to providing an optimal assignment of products to facilities over a time horizon, a model based approach has provided a systematic framework to evaluate alternative operating policies and assumptions. For instance, the impact of process improvements, addition of operating units, equipment failure, and operating costs on the optimal assignments was quickly assessed within the modeling framework. Furthermore, the modeling framework provided the structure to coordinate and standardize information among individuals within the various facilities within our geographically distributed network of manufacturing facilities. The modeling based approach exploring this problem has thus greatly facilitated process learning.

A general description of the process flow will be provided along with the model formulation. Additionally, the model development process will be discussed with emphasis placed on how significant computational performance improvements were realized through efficient problem representations. That is, as critical factors required for accurate problem representation were better understood, the model could be developed at the appropriate level of abstraction, reducing the number of binary variables in the formulation. We will discuss how the model was used to explore a range of possible operating scenarios. Finally, characteristics of the computational tool used which were critical to the model development process will be summarized.

Keywords

Planning, Scheduling, Modeling, Mathematical programming, Optimization, Supply chain management.

Introduction

Manufacturing firms are operating in an intensely competitive environment in which it is imperative to optimize the benefit derived from their scarce resources, including capital and manpower. A highly informed and demanding consumer base requires that a large variety of products of high quality and safety be immediately available at a reasonable cost. This state of affairs places extreme pressure on those accountable for resource management in manufacturing firms. Their challenges include developing optimal strategies for meeting customer demand while reducing inventories, or increasing throughput with minimal capital expenditures.

The ability to systematically and efficiently evaluate the performance of a set of manufacturing facilities under a range of conditions is critical for strategic management of the facilities. This ability would allow one to explore the impact on operations of varying product mixes, demand profiles, manpower availability, shift changes, yield and cycle time improvements, and process changes, among other things. Such information could be used to rationalize the purchase of capital equipment to add production capacity or to coordinate the production of a range of products at multiple production facilities.

We have been exploring the use of mathematical programming based models to evaluate the capabilities of our manufacturing facilities. In particular, mixed integer linear programming (MILP) models have been developed for several of our manufacturing facilities individually, and a model has been developed which considers several manufacturing sites simultaneously.

Consider a multi-facility planning problem, which consists of developing strategies for meeting customer demands for a range of products over a time horizon within a network of geographically distributed manufacturing facilities. Solution of this problem, which consists of the specifying the production quantities of each item on each machine within the network for each time period in the horizon, is difficult for several reasons. First, the number of alternative ways in which production operations can be assigned to machines explodes exponentially as the number of products, machines, and facilities increases. Second, the problem requires a tremendous amount of data such as demand orders, machine availability, transportation costs, allowable inventory levels, and product specific costs and production rates on each machine. Third, the problem constraints, which can include equipment and manpower availability, material balances, order due dates, storage and production capacities and equipment availability, can create complex interactions among the alternatives.

In this paper, we describe the development of a large-scale mathematical programming model for multi-facility planning. The model development process will be discussed with emphasis placed on how computational performance improvements were realized through efficient problem representations. That is, as the factors required to accurately represent our manufacturing system were better understood, the level of abstraction in the model was adjusted to reflect the tradeoff between model detail and computational performance. We will discuss how the model was used to explore a range of possible operating scenarios. Finally, characteristics of the computational tool used which were critical to the model development process will be summarized.

Problem Description

The manufacture of pharmaceuticals typically involves 1.) production of bulk (active) compounds, 2.) formulation of the active compound with inert materials, 3.) filling appropriate containers (vials, bottles, etc.) with formulated material, and 4.) finishing (packaging and labeling). The final products which finally make their way to customers are often referred to as stock keeping units (skus). A specific sku uniquely specifies 1.) the active compound, 2.) the dosage or concentration of the active, 3.) the presentation of the compound (e.g. tablets or solid solution), 4.) the container size (e.g. 250 or 500 ml vials), 5.) the market which will be served. In order to meet the needs of customers throughout the world for a large number of pharmaceuticals, Eli Lilly and Company produces thousands of skus, each of which must be processed by our filling and finishing manufacturing facilities. To meet the worldwide demand for our products, we operate approximately 25 fill/finish facilities throughout the world. The ability to systematically and efficiently coordinate our manufacturing facilities would allow us to satisfy the demand for our products in an economically optimal fashion and provide a framework for evaluating alternatives to expand the capacity of our manufacturing network.

As a first step, we chose to narrow the scope of this coordination effort by developing a decision support tool to analyze alternatives for supplying a single geographical market with product. This market, which serves our European customer base, receives products primarily from three fill/finish manufacturing facilities, which are located in Sesto, Italy; Fegersheim, France; and Geissen, Germany; and will be referred to as S, F, and G, respectively. The scope of the coordination effort was narrowed both to make the problem more manageable and to answer some immediate questions about capacity expansion alternatives that were under consideration. Specifically, the Fegersheim plant was forecasting a capacity problem and was having difficulty meeting demands so purchase of another line was being considered. Meanwhile, the Geissen plant appeared to have available production capacity and could possibly be used to help meet the demand for our products.

The manufacturing network considered is shown in Figure 1. The raw materials, which may originate from a number of our manufacturing plants throughout the world, are formulated products that must be filled in appropriate containers. These containers are then labeled and then are ready to exit the network. This network portrays a simplified view of the material flow since each manufacturing facility consists of a number of fill/finish lines, each of which can perform filling, finishing, or both. Products, which consist of finished pharmaceuticals, are shipped to distribution sites throughout the world. It was our goal to generate production plans to satisfy demand for approximately 2000 finished pharmaceutical items from this network of manufacturing facilities. The plans generated must span a time horizon of no less than one year since they might imply decisions that would be difficult or time consuming to implement.

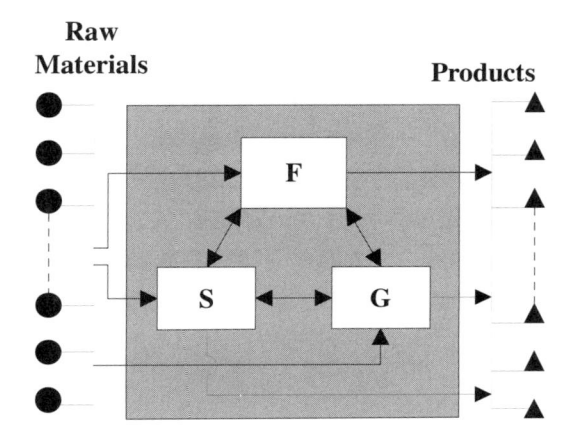

Figure 1. Manufacturing network consisting of three manufacturing facilities (F,S,G).

We developed a decision support tool with the capability of generating feasible if not optimal production plans. A feasible production plan is one which satisfies all problem constraints. A feasible production plan with a cost no greater than all feasible production plans is an *optimal production plan*. Constraints in this problem included:

- Demand due dates
- Mass balances
- Line production rates
- Available production time (considers shifts and shutdowns)
- Inventory levels
- Line to product pairings (some item could only be filled/finished on specific lines)

Costs important to comparing alternative solutions included:

- Line operating costs
- Inventory costs
- Transportation costs

The line production rates and operating costs were different for each machine and the available production time was allowed to vary for each manufacturing facility to conform to the current operating policy. For example, one plant staffs it operations 24 hours/day, 6 days/week whereas another plan only staffs its operations 8 hrs/day, 5 days/week. All production plans, whether feasible or optimal, provided the assignment of production steps to equipment over a time horizon to satisfy product demands.

Model Development

Production plans were generated by solving mixed integer linear programming models (MILP) representing the manufacturing network described. Commercially available scheduling software (VirtECS) was used for both problem representation and solution. The MILP model is based on an aggregate formulation (Basset, et al., 1996; Subrahmanyam, et al., 1996). This formulation is much sparser than those using a uniform discretization (Elkamel, 1993; Kondili, et al., 1988) which divides the time horizon into equal time slots based on the processing times and demand parameters. The aggregate model divides the time horizon of interest into a number of planning periods of equal size. The formulation sparseness, which results from suppressing the details of the scheduling problem except at the boundaries of the planning periods, provides significant computational efficiencies. The drawback, however, is that the timing and sequencing of production batches during the planning periods is lost. Furthermore, there is no guarantee that the scheduling problem implied by a feasible solution to the aggregate planning model will be feasible. To circumvent this problem, we used conservative estimates for equipment availability during the planning horizon.

The underlying structure of our model was determined by specifying costs, lines on which each product could be made, downtime, and demand profiles. The model building process proceeded in stages. First, a base model was constructed which included the equipment in our network, machine production costs, transportation costs, and a relatively lean demand profile. The initial model only specified demand for 10 products over horizon of one year using quarterly planning periods. These models were solved to optimality with less than a minute of solution time. As the number of products (demands) was incrementally increased, solution times increased as expected. The rate of increase in solution time with the number of demands indicated that the solution time of our entire problem (2000 products, > 1 year planning horizon) would be unsatisfactory.

To address this problem, we grouped individual products into *product families*, which we defined as a group of products using the same container size and going to the same market. This grouping reduced the number of products to be considered from 2000 to approximately 500. The initial model (base model) included constraints which had developed over time. For instance, based on strategic decisions made in the past, some products were made only at a specific site. Additionally, the model constrains the equipment availability in each facility based on the staffing policies of the corresponding facilities. One plant only runs 1 shift/day on weekdays whereas another plant runs 3 shifts/day for six days, for instance. Naturally, these constraints may have been limiting the effectiveness with which were operating our manufacturing network. By including these constraints, however, we were able to gain confidence in the model output and compare our current practices to alternative operating philosophies. After the base model was developed, alternative models were developed to explore a range of manufacturing scenarios.

The model development process was greatly facilitated by using a commercially available tool that allowed us to represent the important features of our problem with intuitive constructs. That is, we were able to make use of powerful mathematical programming formulations and solution methods without knowing the arcane details. In particular, models were developed by simply populating spreadsheets with the relevant data or by developing process flow diagrams and specifying attributes of the objects within the diagrams.

Model Results

For each planning period in the horizon, the model output consists of a list of production operations to be executed on each fill/finish line, and the amount of forced downtime and idle time on each fill/finish line. *Forced downtime* is specified in the problem description and accounts for the time that one expects a line to be down due to plant staffing policies, equipment failure, to perform maintenance, and to perform changeovers. *Idle time,* which is the time that a line is available but not needed to meet the demand for products, can be considered potential production capacity. An example of the output from the aggregate planning model is shown in Figure 2. The horizontal axis represents the time boundaries in the planning periods and the vertical axis represents the modeled operating lines at F, G, and S. Also available is a listing of the manufacturing steps that are required within each production block, equipment utilization, inventory levels, etc. Note that the specification of production steps within each production block does not include the timing of each production step since these details are lost when the formulation is aggregated. The aggregate formulation assumes that the available production time within each planning period can be used to meet production requirements and production tasks are planned based on the cycle time and production quantities. To verify the feasibility of plans generated by our model, then, detailed scheduling problems would need

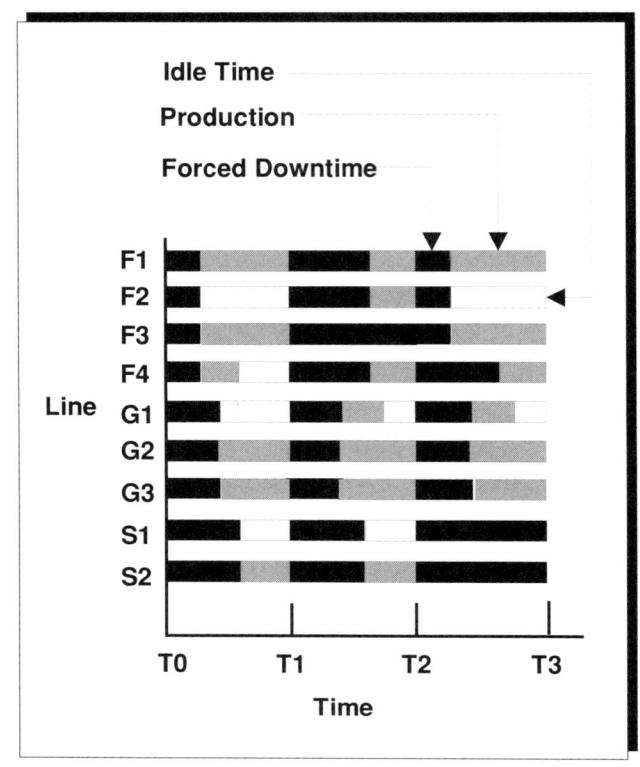

Figure 2. Output from aggregate planning model.

to be solved which could specify the timing and sequencing of the production steps in each planning period.

The models developed to represent our manufacturing network were extremely large. Even though an aggregate formulation was used, the number of planning periods was relatively small, and the products were aggregated into families, the mathematical programming formulations representing our manufacturing network were extremely large. Table 1 summarizes formulation statistics and computational results for a range of planning problems considered. Note that limited problem statistics were available since we used a commercially available package which generated the mathematical programming formulation directly from a spreadsheet. The software gives the user a very high degree of control over the problem specification and solution statistics but interrogation of the mathematical programming formulation is limited. In all cases, the planning horizon of one year was divided into four three-month planning periods. Our measure of problem size was the number of products in the demand file. The problem that we were interested in solving corresponds to the final entry in Table 1, which represents 2000 pharmaceutical products that have been reduced into 532 product families. Solution of these problems is computationally intensive – as expected, the time required even to find the first feasible solution increases rapidly with problem size as measured by the number of products in the demand file.

In most large scale cases, *good* feasible plans were obtained using the best feasible solution found after a fixed amount of computational time.
The ability to generate feasible solutions to industrial planning problem of this scale in less than one hour is of tremendous value. The mathematical programming framework allows one to measure the quality (compared to the optimal) of all feasible solutions generated. Plans generated using ad-hoc methods are likely to be difficult and time consuming to generate, may be infeasible, and provide no measure of solution quality.

Generating feasible plans quickly has allowed us to explore a wide range of manufacturing scenarios. For instance we considered the impact of adding new operating lines in specific sites, increasing the number of shifts the plants operated, and improving the productivity of specific manufacturing lines. The modeling framework established allowed us to quickly evaluate a large number of scenarios. This was very important to evaluate the best means to add production capacity and optimize existing capacity.

Conclusions

As manufacturing firms face increasing competition, they will be forced to operate their facilities in an efficient manner. This will require optimization both within individual manufacturing facilities and among multiple manufacturing facilities. When multiple facilities are considered, optimization becomes difficult due to the tremendous number of alternative ways of meeting demand for products and the interactions among the alternatives. Mathematical programming provides a framework for representation and solution of large-scale multi-facility planning problems. The advantage of this framework is that feasible solutions of measurable quality can be quickly generated. Although aggregate models, such as those we have discussed, generate sparser formulations than detailed uniform time discretization models, the scheduling problems implied by their solutions may be infeasible. Accordingly, detailed scheduling problems should be solved after plans are developed, and the plans should be modified if necessary.

We developed a mathematical programming based model to generate production plans for a manufacturing network consisting of three facilities which satisfy demand for approximately 2000 finished pharmaceuticals. Although optimal plans were desirable, feasible production plans were of tremendous value as well. By using an aggregate model and grouping products into families, we were able to obtain good feasible solutions within an hour of computational time. This performance has allowed us to systematically and efficiently explore the impact of a wide range of manufacturing alternatives under consideration. The use of commercially available software has allowed us to make use of mathematical programming formulations and solutions without requiring us to learn the arcane details. This has greatly facilitated the model development and learning process.

References

Bassett M., Pekny J., and Reklaitis G. (1996). Decomposition techniques for the solution of large-scale scheduling problems. Process Systems Engineering, Vol. 42, No. 12, 3373-3387.

Elkamel A. (1993). Scheduling of process operations using mathematical programming techniques: towards a prototype decision support system. Ph.D. Thesis, School of Chemical Engineering, Purdue University, West Lafayette, IN.

Kondili E., Pantiledes C., and Sargent R. (1988). A general algorithm for scheduling batch operations. Proc. Third Int. Symp. On Process Systems Engineering, pp. 62-75.

Subrahmanyam S., Pekny J., and Reklaitis G. (1996). Decomposition approaches to batch plant design and planning. Ind. Eng. Chem. Res., 35, 1866-1876.

Table 1. Formulation Statistics for Several Planning Problems.

Products	Variables	Constraints	Non-zeroes in Constraint Matrix (%)	LP Relaxation Value	Gap Between 1st Feasible Soln and LP Relaxation (%)	Time to 1st Feasible Soln (s)
10	1346	1094	0.3	-209201	30	9
20	2496	2024	0.1	-560201	28	16
100	9010	7153	0.02	-1784110	24	675
532	67550	53082	0.005	-13112800	2	2770

MACHINE LEARNING APPROACHES FOR ENHANCING THE OPTIMIZATION OF BATCH PRODUCTION SCHEDULING AND PROCESS OPERATION PROBLEMS

Matthew J. Realff and Kenneth J. Kirschner
School of Chemical Engineering
Georgia Institute of Technology
Atlanta, GA 30332-0100

Abstract

This paper describes an approach to learning from the results of rigorous branch and bound optimization. The goal is to develop a simple set of rules which can be used as heuristics in place of the optimization procedure for future cases.

This methodology has been applied to two distinct problems: a selection problem formulated as a multi-bin knapsack problem, and a cutting-stock problem. In the selection problem, both learning methods were correct in over 85% of the cases examined. Since a random selection procedure would only correctly classify 58% of the cases, it can be concluded that learning is occurring. On the other hand, both learning algorithms accurately classify only about 80% of the patterns examined in the cutting-stock problem. Since it is possible to correctly classify approximately 81% of the patterns by simply rejecting every pattern, learning does not appear to be occurring in this problem.

This paper provides a description of the two problems as well as the results of the learning on each problem. Additionally, a brief discussion of the similarities and differences in the problems is provided.

Keywords

Knapsack problem, Cutting-stock problem, Decision-tree learning, Nearest-neighbor learning.

Introduction

The modeling and optimization of chemical engineering operations has improved system performance and systematized the decision making process for a significant number of chemical processes. Problem types which have benefited from this type of work include scheduling (Pantelides, 1994) and operations planning problems (Lakshmanan and Stephanopoulos, 1990) (Dimitriadis, et. al., 1997). Unfortunately, many optimization problems are known to be NP-hard (Garey and Johnson, 1979). This means that is it unlikely that algorithms that require polynomial time in the size of the problem will be developed.

There has been a significant amount of work in developing heuristic methods to solve NP-hard combinatorial optimization problems without guaranteeing global optimality but in polynomial time. Complete optimization models have been developed for both the knapsack problem (Martello and Toth, 1988), (Martello and Toth, 1990a) and the cutting-stock problem (Winston, 1991). Since solving these formulations can, in some case, take a significant amount of time, a large number of heuristics have been developed for the knapsack problem (Eben-Chaime, 1996) (Pisinger, 1995) and the cutting-stock problem (Wascher, 1996). While these heuristics, and the heuristics for many other problems, give reasonable solution performance in many cases, most heuristic methods perform poorly in worst-case scenarios. Thus, the most reliable methods for obtaining high quality or optimal solutions are enumeration in a branch and bound framework (Nemhauser and Wolsey, 1988) or dynamic programming (Bertsekas, 1987). However, using enumerative approaches creates a new set of problems including how the procedure will be controlled to avoid excessive computation times.

The problems which occur in worst-case scenarios create a need for new solution techniques. Additionally, the optimization models and heuristics which are created do not always provide direct information about the problem. For example, in solving a selection problem, the optimization procedure will provide a list of items which should be selected and which items should be rejected. However, the reason for each decision is not included. Since the reason for making a decision is often as important as the actual decision, a method capable of providing both pieces of information would be very valuable.

Machine learning, the branch of artificial intelligence whose goal is to develop computer systems capable of acquiring and enhancing knowledge (Briscoe and Caelli, 1996), has been used successfully in the past to aid the solution procedures for optimization problems. The majority of this work, however, has focused on model formulation and selection (Dutta, 1996) (Cagan et al., 1996). In addition to these areas, machine learning can be used to develop models for a given problem in two distinct manners. The first method involves developing a predictive model from raw data. In this case, the goal is to use plant data to develop a model which, given information about one portion of the plant, can accurately predict information about another portion of the plant. For example, given the inputs to a system (flow rates, composition, etc.), the model can predict a system's outputs.

Models describing process operations have already been produced for many processes, however. Unfortunately, many of these models are complex and difficult to understand. Additionally, it may not be possible to apply the models quickly for a specific scenario. These circumstances create a need for the second model development method: develop a simple model or set of rules which can be used in place of the existing complex model. To do this, both the raw data (the data supplied to the complex model) and the results of the complex model are required. Learning from both sets of data, it may be possible to summarize the decision making in the complex model in a form that is both accurate and simple to understand. The work presented in this paper falls into this category of learning. It is unique in that machine learning is used to summarize the complex decision-making of an optimization procedure into simple decision rules.

These decision-rules can be generated in a number of ways. In this paper, the learning is applied in two manners: decision-tree learning and nearest-neighbor learning. Decision-trees are a knowledge representation scheme in which a body of data or a series of related decisions is represented graphically (Turban, 1992). At the top of a decision-tree is the root node which contains all of the possible decisions (or results) for a given system. From this node, a test is performed on one feature of the data to separate the data as effectively as possible into distinct classes. The result of the test for each case determines which branch from the node to follow. Once the first test has been completed, further tests are performed (on the same feature or a new feature). This iterative testing procedure continues until terminal nodes (or leaves) are reached and classification of the data is complete (Saraiva, 1995). Figure 1. provides an example of a decision-tree (Saraiva and Stephanopoulos, 1992).

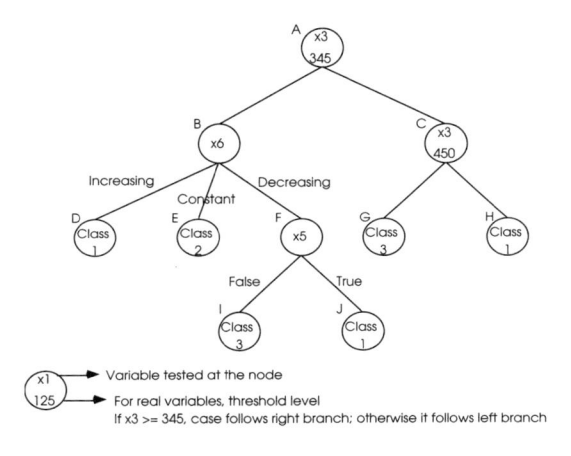

Figure 1. Example of a decision tree.

Many decision-tree systems have been developed including ID3 (Quinlan, 1986), C4.5 (Quinlan, 1992), and CART (Breiman, 1984).

In nearest-neighbor learning (Salzberg, et. al. 1995) (Atkeson, et. al., 1997), the characteristics of test cases are simply compared to the characteristics of the cases in the training set. The test cases are then classified based upon their proximity to the training cases.

These learning procedures have been applied to two distinct problems: a waste volume reduction problem and a one-dimensional cutting-stock problem. The next section provides information on the waste volume reduction problem. This includes a description of the problem, the tests that were performed, and the results. Next, information on the cutting-stock problem is provided. A comparison of the two problems is then performed and conclusions about the suitability of learning methods to problem types is drawn.

Waste Volume Reduction Problem

Problem Definition

Nuclear power plants create a significant amount of relatively low-level radioactive waste which can be disposed of in special secure landfills. The source of this waste is the resin material used to filter cooling water. After the resin has completed its useful life, it is removed and drummed ready to be further processed or placed in high integrity containers (HICs) for disposal. This waste is placed in HICs in order to protect the environment and people handling the waste material. In many cases, the volume of material in the containers which must be placed in a landfill is significantly less than the total volume of material in the container. This is because the organic component of the waste can be decomposed and separated from the rest of the material. The materials in the waste include water, sulfur, organic compounds, and condensable metals. Of these, only the radioactive metals must be placed in a landfill. The non-metal components can be removed and, after cleaning, released to the environment. The condensable metals which are not radioactive are very difficult to separate from the radioactive metals and are, therefore, not removed. Since the cost of placing radioactive waste in a landfill is high, this volume reduction can be a desirable and profitable process.

The major decisions involved in the operation of such a facility are:
- Given a limited processing capacity and a large potential supply of material, which subset of drums should be selected for processing?
- Given a limited capacity for any specific batch, that may contain the residuals from several drums, how should the drums be sequenced to minimize the number of batches required?

In order to solve this problem, a variation of the 0-1 Knapsack problem, the multi-bin knapsack problem, (Balas and Zemel, 1980) can be used. Two important constraints in the selection and grouping of the containers are the maximum volume for each batch and a limit on the total amount of accumulated radioactivity in each batch. The volume of a batch is constrained by the finite volume of the crucible in which the condensable metals accumulate. Similarly, the radioactivity is constrained in order to keep within safe levels. Mathematically, these constraints are:

$$\sum_k v_k * Y_{i,k} \leq v * X_i \quad \forall i \quad (1)$$

$$\sum_k C_k * Y_{i,k} \leq \overline{C} * X_i \quad \forall i \quad (2)$$

where v_k is the volume required by the material in container k, $Y_{i,k}$ is a binary value which states whether or not container k was assigned to batch i, v is the upper volume limit for a batch, X_i is a binary variable which states whether or not batch i exists, C_k is the curie load of container k, and $\overline{C}$ is the curie limit for a batch.

In addition to these, several other constraints are required for this problem. Two of these constraints are limits on the total amount of material which can be processed during the time available. Another requires that a batch must exist if one or more containers are placed into it. Additionally, each container can be placed in only one batch. Finally, there are constraints used to decrease the redundancy of the problem so that the problem solves more quickly than it might otherwise. These constraints can be represented as:

$$\sum_{i,k} Y_{i,k} * V_k \leq \overline{V} \quad (3)$$

$$\sum_{i,k} Y_{i,k} \leq \overline{H} \quad (4)$$

$$X_i \leq X_{i-1} \quad \forall i \quad (5)$$

$$X_i \geq Y_{i,k} \quad \forall k \quad (6)$$

$$\sum_i Y_{i,k} \leq 1 \quad \forall k \quad (7)$$

where V_k is the total volume of material in container k, $\overline{V}$ is the limit on the total volume which can be processed, and $\overline{H}$ is the limit on the total number of containers that can be processed.

The objective for this optimization problem is to maximize profits for processing the containers. If a container is selected (and therefore processed), revenues are received based upon the amount and type of material in the container. The revenues for each container can be calculated before the optimization procedure is executed. On the other hand, several costs are associated with processing the material. These costs include general processing costs for all of the material in the container, cleaning costs, and the cost to dispose of the processing batches. Mathematically, the objective function is represented as:

$$\text{Max} \quad Z = \sum_{i,k}(Y_{i,k} * \varphi_k) - \sum_i (X_i * \varphi_i) \quad (8)$$

where ϕ_k is the revenue received for processing container k and ϕ_i is the cost of processing batch b.

Data Generation

For this work, a problem is defined as one scenario in which a number of containers must be selected for processing. In this case, the only difference between two problems is the specific container data. Therefore, information must be generated for a number of containers. The data generated for this work was based upon information provided by an industrial partner. For each characteristic of a container (mass, weight percent sulfur, ppm iron, etc.), a range of expected values was determined.

The problem was then examined to determine the number of containers which should be included in each data set. The industrial partner operates the processing equipment on a one month schedule. In this time, a maximum of 35 containers can be processed. At the same time, for containers with characteristics reasonably similar to those provided, there will not be a high enough condensable metals content to prevent the processing of 35 containers. Therefore, 35 containers will always be processed in these tests. Since the data provided contained information on approximately 50 containers, each problem was generated with 50 containers. This means that, in general, 70% of the containers will be selected and 30% will be rejected for each problem solved.

Using the ranges on each characteristic provided, 40 training cases (2000 total containers) were created. Additionally, 30 test cases (1500 total containers) were created. To test the learning's robustness to changes in the data, test sets were also created using four data ranges modified from the standard data ranges. Each modification involved altering the acceptable ranges for each container characteristic. The four modifications were expanding the ranges by 25% of the magnitude of each range, contracting the ranges by 25%, shifting the ranges up 10%, and shifting the ranges down 10%.

Since the altered ranges were generated by making changes to the expected values, it was possible for a range to include negative values. A negative mass or amount of a component is not feasible. Therefore, all negative values were replaced with zero. This problem was only encountered with the ppm of iron in the container when the ranges were shifted down 25% or expanded 10%. Table 1 provides the data ranges used for this problem. In this table, the ranges have been scaled from 0 - 1 and the characteristics included in the column "All Other Characteristics" are: mass, weight percent sulfur, weight percent SO_4, weight percent water, ppm Na, ppm K, and ppm Zn. For each of the altered data regions, 30 test cases were created.

Table 1. Scaled Data Regions.

Scenario	Range for Fe	Range for Other Characteristics
Standard	0.03 - 0.89	0.1 - 0.9
Increase 25%	0 - 1	0 - 1
Decrease 25%	0.14 - 0.78	0.2 - 0.8
Shift Up 10%	0.12 - 0.98	0.18 - 0.98
Shift Down 10%	0 - 0.81	0.02 - 0.82

Tests and Results

Once the data sets were generated, the individual cases were solved using the formulation described above[5]. When each case was solved, a file was created containing all the characteristics of each container and whether the container was selected or rejected. The 40 training cases were combined into one large file to be used as the training set by the learning software.

The training set was then examined for trends using an ID3 decision-tree learning algorithm. Three of the rules that were developed are:

IF (S ≤ 0.397 AND total metals ≤ 0.203) THEN select
IF (S ≤ 0.456 AND total metals > 0.204) THEN select
IF (S > 0.767 AND total metals ≤ 0.695) THEN reject

Using these and the other developed rules, the containers in the test set were classified. Additionally, the test set containers were classified using a nearest-neighbor algorithm. With both algorithms, the parameters used were found through tests not included in this study. Table 2. provides a summary of the accuracy of the two classification methods.

Table 2. Classification Accuracy.

Data Region	Decision-Tree	Nearest Neighbor
Training Set	92.5%	---[6]
Standard	88.9%	86.9%
Decreased 25%	89.4%	86.4%
Shifted Down 10%	88.2%	86.1%
Increased 25%	87.6%	87.9%
Shifted Up 10%	85.6%	85.5%

If a random selection approach were utilized, only 58% of the containers would be correctly classified. Since the two learning methods correctly classify over 85% of the containers, it is evident that learning is occurring. Also, the learning can be considered robust to changes in the data since the classification performs well for all ranges examined.

The ability to classify these containers is very useful in understanding the actual process and can be used to determine which containers should be selected in the future. The rules which are created by the learning, and represented in the decision-tree, provide a simple way to

---

[5] These tests were solved on a DEC AlphaStation 400 running DEC/OSF 2. GAMS version 2.25 was used along with CPLEX version 4.0 to solve the optimization problem.

[6] Nearest-neighbor will always correctly classify an element of the training set.

perform this classification. This is especially helpful since decisions regarding which containers to select are often made without knowing the attributes of *all* the available containers. When this knowledge is not available, an optimization procedure cannot be used. The decision rules, however, can still be implemented.

Cutting-Stock Problem

Problem Description

In many industries, such as textiles and paper production, all manufacturing is performed to make large, standard length rolls. These rolls (or raws) are then cut to smaller final lengths to meet specific demands. While the raws are of a standard size, the final lengths can vary significantly. Since wasting material is costly, it is desirable to know the most economical manner for cutting the raws given a set of demands. This problem is known as the cutting-stock problem (Chvatal, 1983).

The objective in this problem is usually to minimize the number of standard rolls required to meet the given demand for finals. The only constraint is that the demand for each finals length must be met. The following equations provide the model used for this problem.

$$\text{Min} \quad Z = \sum_j X_j \qquad (9)$$

$$\text{ST} \quad \sum_j A_{j,i} * X_j \geq D_i \qquad (10)$$

In these equations, Z is the objective function value, X_j is the number of pattern j chosen, $A_{j,i}$ is the number of final i contained in pattern j, and D_i is the demand for final i.

Each of the possible combinations of cuts that can be made to a single raw is known as a pattern. In order to learn the features that determine whether a pattern should be selected or rejected, the characteristics of the pattern must be quantified. One of the characteristics is the number of a specific final in the pattern being examined, the $A_{i,j}$ coefficient. This varies from zero to two for the cases examined. However, it is also important to consider the demand for the final. If the demand for the final is low, the pattern is less likely to be required (there are usually several patterns which include each final). The demand for each final varied from approximately 1 to 100. In order to weight both the number of each final in a pattern and the demand approximately equal, the log of the demand for each final was multiplied by the number of that final in the pattern. This was performed for each possible pattern in the problem case.

Tests and Results

For this problem, 100 cases were generated using a software package called CUTGEN1 (Gau, 1995). In every case, the raw length was set to be 1000 and 25 finals lengths were allowed to vary randomly from 375 to 625. The demand for each final was normally distributed with a mean of 50. Of the 100 cases generated, the first 30 were considered the training set and the remaining cases were test sets.

Once the cases were generated, they were solved using the formulation provided above. After each problem was run, an output file was created containing the characteristics of each pattern (the log of the demand for each final times the number of that final in the pattern) and whether or not the pattern was selected. The 30 training cases were then used for learning while the 70 remaining cases were used as test cases. Table 3 shows how effectively the patterns were classified for both decision-tree learning and nearest-neighbor learning.

Table 3. Percent of Patterns Correctly Classified.

Learning Algorithm	Classification Percent
Decision-Tree	81.7%
Nearest-Neighbor	78.8%

If one were to simply reject all patterns (decide that none should be used), the "classification" would be correct 81.2% of the time. Therefore, it does not appear that a significant amount of learning is occurring in this problem.

Problem Comparison

The two problems discussed in this paper are similar in two important manners. First, in their simplest form, both problems can be considered selection problems. In the waste volume reduction problem, a portion of the available containers of low-level radioactive waste are being selected for processing while the remaining containers are being rejected and disposed of directly. On the other hand, in the cutting-stock problem, patterns of final cuts are being selected to minimize the number of standard length raws required to meet product demands. Second, key constraints for the two problems (equations 1, 2, and 10) are very similar.

While both of these problems can be considered selection problems, there are a number of differences that might lead to the significantly different results in the ability of the machine learning algorithms to correctly classify the items. One of these differences involves how well the characteristics chosen to represent a single instance (container or pattern) actually correspond to the selection problem. In the case of the volume reduction problem, it is clear that the amount and type of material in the containers is important. In the cutting-stock problem, however, the key features are not as evident. It would appear that whether or not a final is included in a pattern and the demand for each final are the key features.

However, the correct method for combining these two features, or if they should be combined, is not as evident.

A second important difference is how closely clustered in multi-dimensional space the individual instances are in the two problems. In the volume-reduction problem, there seems to be a significant amount of clustering. There are a large number of similar containers, the majority of which were classified in the same manner in different problems. On the other hand, there does not appear to be nearly as much clustering in the cutting-stock problem. Additionally, cases which are similar are not necessarily classified in the same manner. This makes it very difficult to perform decision-tree or nearest-neighbor learning. The clustering of similar cases is required for these learning methods to work effectively. Finally, the two problems differ drastically in the effect caused by small changes in the problem data. When small changes are made to the problem data in the volume reduction problem, the effect on the final solution is generally very small. In some cases, despite a moderate change in the problem data, the final solution turns out to be exactly the same. This is not the case in the cutting-stock problem. Small changes in the problem data can lead to drastic changes in the final solution. The features which are important in deciding which patterns will be required appear to change from case to case depending upon the specific scenario. Therefore, it will be difficult to develop a set of rules which can be used to determine whether a pattern should be selected or rejected.

Future Work

As described in the previous section, there are several differences between the two problems discussed in this paper which may lead to the drastically different performances of the learning algorithms. At this time, however, it is unknown which of these differences (or a combination of these and/or other differences) is causing the results to vary so significantly. One area of future work is to examine these and other variations in the two problems to see what might be the primary cause of difficulty. A second area of future work involves further examining feature selection in the problems. One possible reason for success in the knapsack problem and failure in the cutting-stock problem is the selection of the features to learn from. If the features are poorly selected, the learning will be inhibited. Therefore, further work into feature selection and/or deriving new features based on known information could be a valuable area of research.

Finally, these techniques will be applied to the pipeless batch plant scheduling problem. This problem has been examined in the past and patterns in the operation of the equipment have been observed. For this reason, it is believed that learning can be successfully applied to this problem.

Nomenclature

$A_{j,i}$ = number of cut i in pattern j
C_k = curie load of material in container k
$\overline{C}$ = maximum curie load of a batch
D_i = demand for cut i
$\overline{H}$ = maximum containers which can be processed
V_k = volume of material in container k
$\overline{V}$ = maximum total volume which can be processed
X_i = whether or not batch i exists (knapsack problem)
X_j = number of pattern j required (cutting-stock problem)
$Y_{i,k}$ = whether or not container k is in batch i
Z = objective function value
v = upper volume limit for a batch
v_k = volume of material in container k
i = set of batches (knapsack problem)
i = set of final cut lengths (cutting-stock problem)
j = set of patterns (cutting-stock problem)
k = set of containers (knapsack problem)

References

Alkeson, C. G., A. W. Moore, and S. Schaal (1997). Locally weighted learning, *Artificial Intelligence Review*, **11**, 11-73.

Balas, E. and E. Zemel (1980). An algorithm for large zero-one knapsack problems. *Operations Research*, **28**, 1130-1154.

Breiman, L., J. H. Friedman, R. A. Olshen, and C. J. Stone. *Classification and Regression Trees*. Wadsworth and Brooks, Belmont, CA, 1984.

Briscoe, G. and T. Caelli (1996). *A Compendium of Machine Learning Volume 1: Symbolic Machine Learning*, Ablex Publishing Corporation, Norwood, New Jersey.

Cagan, J., I. E. Grossman, and J. Hooker (1996). Combining artificial intelligence and optimization in engineering design: a brief survey, *International Conference on Intelligent Systems in Process Engineering*.

Chvatal, Vasek (1983). *Linear Programming*. W. H. Freeman and Company, New York.

Dimitriadis, V. D., N. Shah, and C. C. Pantelides (1997). Modeling and safety verification of discrete/continuous processing systems, *AIChE Journal*, **43**, 1041-1059.

Dutta, A. (1996). Integrating ai and optimization for decision support: a survey. *Decision Support Systems*, **18**, 217-226.

Eben-Chaime, M. (1996). Parametric solution for linear bicriteria knapsack models, *Management Science*, **42**, 1565-1575.

Garey, Michael R. and David S. Johnson (1979). *Computers and Intractability: A Guide o the Theory of NP-Completeness*, W. H. Freeman, San Francisco.

Gau, T. and G. Wascher (1995). CUTGEN1: a problem generator for the standard one-dimensional cutting stock problem. *European Journal of Operational Research*, **84**, 572-579.

Lakshmanan, R. and G. Stephanopoulos (1990). Synthesis of operating procedures for complete chemical plants. III. Planning in the presence of qualitative, mixing constraints, *Computers and Chemical Engineering*, **14**, 301-317.

Martello, S. and P. Toth (1988). A new algorithm for the 0-1 knapsack problem, *Management Science*, **34**, 633-644.

Martello, S. and P. Toth (1990a). An exact algorithm for large unbounded knapsack problems, *Operations Research Letters*, **9**, 15-20.

Nemhauser, George L. and Lawrence A. Wolsey (1988). *Integer and Combinatorial Optimization*, Wiley, New York.

Pantelides, C. C. (1994). Unified frameworks for optimal process planning and scheduling, *Proc. Second Conf. On Foundations on Computer Aided Operations*, (Editors: D.W.T. Rippin and J. Hale), CACHE Publications, 253-274.

Pisinger, D. (1995). An expanding-core algorithm for the exact 0-1 knapsack problem, *European Journal of Operational Research*, **87**, 175-187.

Quinlan, J. R. (1986). Induction of decision trees. *Machine Learning*, 1:81-106, 1986. Reproduced in *Readings in Machine Learning*, Shavlik and Dieterich, editors, pp. 57-69.

Quinlan, J. R. (1992). *C4.5: Programs for Machine Learning*, Morgan Kaufmann, San Mateo, CA, 1992.

Salzberg, Steven, Arthur L. Delcher, David Heath, and Simon Kasif (1995). Best-case results for nearest-neighbor learning, *IEEE Transactions on Pattern Analysis and Machine Intelligence*, **17**, 599-608.

Saraiva, P. M. (1995). Inductive and analogical learning: data-driven improvement of process operations, *in* G. Stephanopoulos and C. Han (eds), *Advances in Chemical Engineering; Intelligent Systems in Process Engineering Part II: Paradigms from Process Engineering*, Vol. 22, Academic Press, pp. 377-435.

Saraiva, P. M. and G. Stephanopoulos (1992). Continuous process improvement through inductive and analogical learning, *AIChE Journal*, **38**, 161-183.

Turban, E. (1992). *Expert Systems and Applied Artificial Intelligence*, Macmillan Publishing Company, New York.

Wascher, Gerhard and Thomas Gau (1996). Heuristics for the integer one-dimensional cutting stock problem: a computational study, *OR Spektrum*, **18**, 131-144.

Winston, W. L. (1991). *Introduction to Mathematical Programming Applications and Algorithms*, PWS-Kent Publishing Company, Boston.

USING DYNAMIC MODELING TO IMPROVE SCHEDULING CAPABILITY

Dr. Bernard McGarvey and Mr. Bruce Bickle
Clinton Laboratories
Eli Lilly and Company
Clinton, IN 47842

Abstract:

Every manufacturing operation requires assistance from support services in order to function properly, examples being maintenance, analytical laboratory support and utilities. Knowing the schedule of activities in the manufacturing area is critical for these support areas if they are to provide the best possible service. It is thus desirable to provide the support services with a manufacturing schedule so that they can schedule their own operations. In reality, if the schedule provided by manufacturing is subject to change, then it is of limited use by the support areas. Therefore the support areas normally place requirements on the robustness of the schedules provide by manufacturing. It is in manufacturing's own interest to provide such a robust schedule since this will improve the support service operations and this will in turn help the manufacturing operation. Since all operations contain variability, however, supplying such a robust schedule is not guaranteed.

In this paper, an example is presented where a manufacturing operation supplies a schedule to a maintenance organization. This schedule allows the maintenance organization to see when particular equipment is available so that maintenance tasks can be scheduled. A measure of schedule robustness was developed and customer valid requirements established. Measurement of data from actual operations showed that the requirements were not being met and so the system was incapable. It was not clear though how improvements might be made to make the system capable.

A dynamic model of the manufacturing operation had previously been constructed and this model was used to investigate whether the issue might be systemic or special cause. Having identified the issue as systemic, the model was used to identify a solution without the need to do experimentation in the field. This solution was implemented and an immediate and significant improvement in the schedule robustness was achieved so that the system is now capable.

This concept of using a dynamic model to measure schedule robustness is now being applied to other areas where dynamic models have been developed.

Keywords

Dynamic modeling, Robust operations, Scheduling, Simulation, Variability reduction.

Introduction

The ability of manufacturing operations to provide reliable schedules to support operations is very important so that the support operations can provide good service back to manufacturing operations. For example, in a pharmaceutical manufacturing operation, products are tested extensively to assure the quality of the product. Very often, it may be necessary to perform some initial setup on the laboratory testing equipment prior to the arrival of a sample from manufacturing. The manufacturing schedule can be used to determine when the initial activities should be started so that the testing equipment is ready when the sample arrives in the laboratory. If the sample is early, it must wait for the testing equipment to be ready. If it is late, then the testing equipment may have to be diverted to another operation and so the resources used in the initial set up are wasted.

Also, if the manufacturing schedule indicates that a piece of equipment will be down at a certain time, maintenance operations can allocate their resources so that the maintenance personnel are ready to work on the equipment as soon as it is available. If the equipment is not available at the scheduled time, maintenance personnel may be forced to go to another task. When finally the equipment does become available, the maintenance personnel originally allocated may no longer be available. Thus the effectiveness of the maintenance system is reduced, which will ultimately lead to more variability in the manufacturing system. On the other hand, if the manufacturing schedule is reliable, then the appropriate maintenance can be performed, which in turn helps reduce the level of manufacturing variability. Again, it is in manufacturing's own best interest to create reliable schedules for its maintenance support operations.

In practice, one of the main reasons why the actual equipment availability time may differ significantly from the scheduled time is due to variability in manufacturing operations. It is not difficult to imagine that if this operational variability is large, then the schedule will not be of much value to the support operations. Indeed, it will not be of much use to manufacturing operations itself, as manufacturing tries to synchronize its activities. Scheduling operations with significant variability is an area of current research interest (Schertz et al., 1997). It appears from this work that an approach to handling this variability is to construct scheduling policies that are robust to certain levels of variability. Such a policy is good only in an average sense. Common cause or systematic variability can be handled by the policy. Variability larger than that included in the policy must be dealt with as it occurs, so that in effect it is treated as special cause. Scheduling capability thus depends on creating a scheduling system which accounts for the inherent variability in the manufacturing system. The success of the scheduling system depends on our ability to understand the implications of this variability on operations. This leads naturally to the concept of scheduling robustness. A scheduling system can be considered robust if sources of common cause variability do not render the schedule of no value in planning the allocation of resources.

In this paper we discuss a scheduling system that has been developed for a solvent recovery component of a pharmaceutical manufacturing operation. This solvent recovery operation takes used solvent from the manufacturing operations, purifies (recovers) it and then makes it available again for the manufacturing operations. The manufacturing operation is a batch operation and so the used solvent arrives at the solvent recovery operations in batches. The solvent recovery operations are complicated because there are multiple solvents being recovered that share common equipment and solvents may have multiple customers, as well as multiple purity specifications. A dynamic model (simulation) of the solvent recovery system was created to provide insight into the solvent recovery operations (Bickle et al., 1996). Part of the output of this model is a schedule of equipment startups and shutdowns. Each week a two week forward schedule is generated. This schedule is then used by the solvent recovery operations and some support service operations to plan activities. In order to provide an indication of how good the schedule is, the solvent recovery operation measures the difference between the startup and shutdown times predicted by the schedule and the actual times. The schedule uncertainty is defined as the standard deviation of these differences. A small value for the schedule uncertainty would indicate a robust schedule.

The initial agreement between the solvent recovery operations and the main support groups was that the schedule uncertainty should be no greater than 1 hour for all important activities that would occur in a certain time horizon. This time horizon was defined as 7 days from the time the schedule was generated. After a period of time, the performance data was compiled and Fig. 1. shows the trend in schedule uncertainty as a function of days since the schedule was generated, for one of the key solvents being recovered.

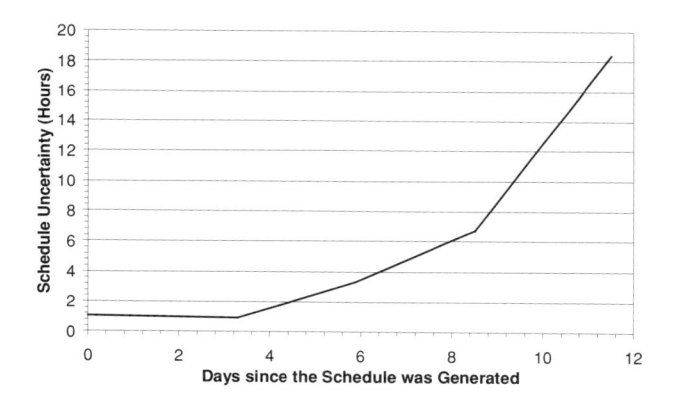

Figure 1. Schedule uncertainty performance.

It can be seen that up to a time horizon of three days, the schedule uncertainty was acceptable. However, after 3 days, the uncertainty increased dramatically. At 7 days, the uncertainty was about 4.8 hours. Thus significant improvement would be required if the original agreement was to be met. Also, it was evident to solvent recovery operations that by improving the schedule robustness, they could expect improved performance from their support groups.

Improving the Schedule Robustness

One of the important initial questions which had to answered was whether the sharp increase in schedule uncertainty after 3 days was a common cause issue or a special cause issue. From a management perspective, this is a crucial issue since the actions required to close the gap in performance will be very different depending on whether the root cause is common cause or special cause. In fact, Deming (1986) points out that "confusing special cause with common cause is one of the major issues with management". At this point it was realized that the dynamic model contained all known sources of common cause variability and so it should be possible to use the model to help answer the above question.

The way in which the dynamic model reflects variability is that variables such as flow rates, task times and flows of used solvent from manufacturing had probability distributions associated with them. The model samples the distributions each time step in the simulation and these sampled values are used to calculate the behavior of the system. The model was set up so that the sampling depended on a seed that varied from one run of the model to the next. Thus two runs of the model would give different values for the model outputs. Therefore, in order to look at the predicted effect of common cause variability on the schedule, a base case run was made and the important startup and shutdown times recorded. Then multiple runs of the model were made and the equivalent times recorded. A schedule difference was then calculated as the difference between the base time and the time given in the multiple runs for equivalent events (for example, the start up of a certain recovery operation). The schedule uncertainty is then the standard deviation of these differences. Figure 2. shows the comparison of schedule uncertainty as predicted by the model and the actual performance data.

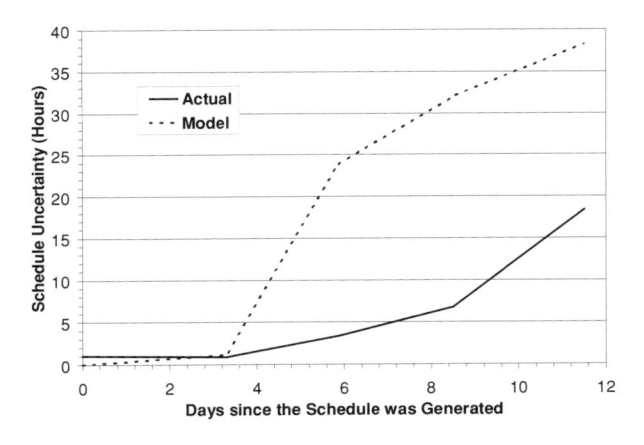

Figure 2. Comparison of actual vs predicted schedule uncertainty.

It can be seen that the model predicts that the schedule uncertainty should be acceptable up to a 3 day forward look but after this it degrades rapidly, in agreement with the actual performance data. The model predicts that the schedule uncertainty degrades faster than what is found in practice. However, this was not considered as important as the fact that the model predicts the degradation occurs at the same point in time as what is actually observed in practice.

This means that the degradation in schedule robustness after 3 days is a result of common cause variability and not special cause variability. Thus we have a systemic problem. In fact the solvent recovery operations are essentially doing "the wrong things well". Trying to improve performance by telling the solvent recovery operations to do their tasks better will not lead to any improvement in performance. In fact, since the operations personnel will be keenly aware of that they are carrying out their instruction set correctly, management credibility will almost certainly be hurt by such an approach to improvement. The key to improving performance is to recognize that the answer lies somewhere in the in the structure of the model. This structure consists of the connections, both physical and logical, between the components of the system and the decisions being made in the model.

Initially, the physical and logical connections of the model were analyzed but nothing could be inferred from this analysis. Then the decisions being made in the model were analyzed and it was found that there was a key decision being made that was causing the increase in schedule uncertainty. This decision related to the startup of one of the recoveries. As manufacturing processes solvent, it produces used solvent and sends this used solvent to a storage tank. In order for a recovery to start, there must be used solvent available in this storage tank. As soon as there is enough used solvent present and the equipment is available, then the recovery starts. It was found that "enough" was being defined as a specified volume of solvent. The reason why this definition was so important can be best understood by referring to the systems model shown in Fig. 3.

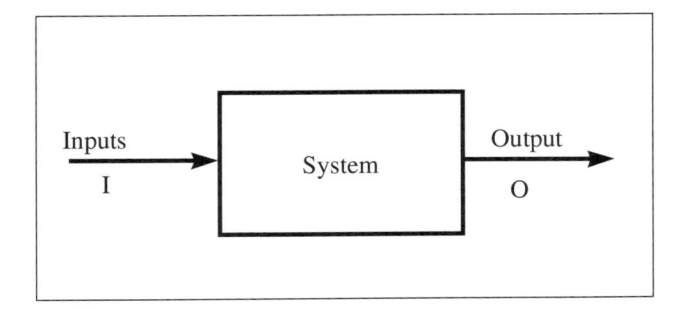

Figure 3. A systems model for the solvent recovery operations.

Considering the solvent recovery operations as a system, then the effect of this system is to map a set of inputs, I, to an output, O. This can be written as

$$O = f(I, S) \quad (1)$$

where S represents the system and the function f represents the mapping. In the present case, the output O is the schedule accuracy. The inputs, I, are the movements of used solvent from the manufacturing operations to the solvent recovery operations. The decision to startup the solvent recovery operation is a decision contained in S. If there is variability in the inputs I, then only one of two things can happen. On the one hand, the variability in the inputs I can be passed directly by the system S to the output O. On the other hand, the system S can absorb some of the variability in the inputs I so that the variability in the output O would be less than in the previous situation. It was found that by using the current startup decision, any variability in the receipts of used solvent from the manufacturing operations was being passed directly onto the startup time and hence the schedule.

If an improved decision was to be found, the volume in the storage tank would have to be varied in order to absorb the variability in receipts. The decision was modified so that instead of depending on a predetermined volume being in the storage tank, the startup decision depended on a predetermined number of batches of used solvent being received from the manufacturing operations. The startup volume in the storage tank would now vary in accordance with the variability in receipts from the manufacturing operations instead of being a fixed value. The variability in used solvent receipts would therefore be absorbed by the system and so would not be passed on to the recovery startup time and the schedule uncertainty. With this modification to the model, the schedule uncertainty was again predicted from the model.

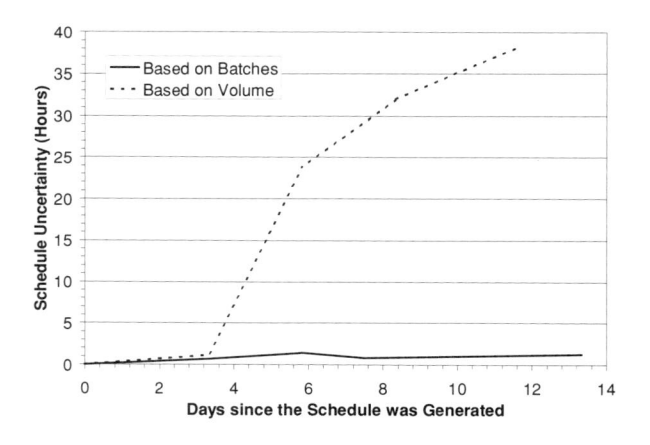

Figure 4. Comparison of schedule uncertainty for volume based startup decision vs batches based startup decision.

The results are shown in Fig. 4., which compares the schedule uncertainty based on volume to that based on number of manufacturing batches. The reduction in predicted schedule uncertainty is dramatic. The new decision criterion was incorporated into the model and the schedules generated from this model were used to plan activities. Figure 5. shows the improvement in the actual schedule uncertainty.

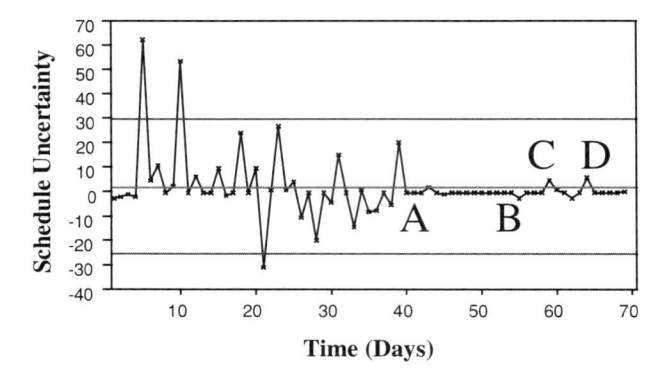

Figure 5. Improvement in actual schedule uncertainty using improved decision criterion.

Point A indicates the point at which the new decision criterion was introduced. The improvement in schedule robustness is enormous. The points B, C and D are due to special cause events where there were some equipment failures. The scheduling system capability had been improved dramatically with a simple change in a decision making criterion. It is important to note that trying to implement traditional scheduling systems based on combinatorial optimization (Morton et al., 1993; Zentner et al., 1994) could not succeed with the original amount of schedule uncertainty. With a robust schedule, the dynamic simulation can successfully schedule the operations. Also, without the dynamic model, we might still be trying to decide if the issue was one of special cause or common cause and it is difficult to see how we would have isolated the root cause of the scheduling issue without the model. Finally, the improvement in schedule uncertainty for the key solvent recovery studied also translated into improvements for other solvent recovery operations.

Order Policy Robustness

The idea of schedule robustness has also been applied to the ordering of a key raw material that is used by the whole site. Once ordered, it arrives on the site in rail cars and it subsequently unloaded into a main distribution tank. There is a significant lead time between ordering a rail car and its delivery on site. Both the daily use of the raw material and the rail car lead time have significant variability. This has lead to the site keeping a large number of rail cars on site and hence a large inventory. Also, the supplier has only a fixed number of rail cars available for all our company manufacturing sites. This means that keeping too many rail cars on site leads to further degradation in delivery lead time and so causes the site to want to hold even more rail cars on site. A further complication is that orders are placed one month in advance to help the supplier plan his deliveries. Once the orders have been placed, there is great reluctance to modify them.

It was thus decided to create an optimal order policy so that inventories could be reduced, rail cars freed up and the risk of running out of material kept below an acceptable level. The basis of the order policy was to order a rail car if the projected future site inventory, I_F, is less that some preset inventory threshold, I_{TH}. The future time is defined as the end of the order planning horizon which is defined as from the current date to the current date plus the mean value of the rail car delivery lead time, μ_{RCLT}. The reason for this is that this mean time is, on average, the earliest time that an order decision will affect the inventory level on site, I_{OS}. The future volume is calculated as

$$I_F = I_{OS} + \beta V_{IT} - U_A \quad (2)$$

where V_{IT} is the volume of material already in transit from previous orders, β is a supply line factor and U_A is the anticipated usage of the raw material over the order planning horizon. The factor β represents the fraction of material currently in transit that you estimate will be delivered within the order planning horizon, (Mosekilde et al., 1990). U_A is obtained from the equation

$$U_A = \mu_{UA} + \alpha \sigma_{UA} \quad (3)$$

where μ_{UA} is the average anticipated usage, σ_{AU} is the standard deviation of the anticipated usage and α is a factor to allow for the risk of inflated usage due to variability in the daily usage of the raw material. The raw material usage in the rail car lead time planning horizon, U_{RCLT}, is a random variable given by

$$U_{RCLT} = LT_{RC} * U_D \quad (4)$$

LT_{RC} is a random variable for the rail car lead time with mean μ_{RCLT} and standard deviation σ_{RCLT}. U_D is a random variable for the daily usage with mean μ_{UD} and standard deviation σ_{UD}. Using Eqn. (4) and the formulas from Taylor (1980) for calculating the mean and standard deviation of a function of independent random variables, the mean and standard deviation of U_{RCLT}, denoted μ_{UA} and σ_{AU} respectively, are given by

$$\mu_{UA} = \mu_{RCLT} \mu_{UD} \quad (5)$$

$$\sigma_{UA}^2 = \mu_{RCLT}^2 \sigma_{UD}^2 + \mu_{UD}^2 \sigma_{RCLT}^2 \quad (6)$$

The rail car lead time and raw material daily usage statistics were obtained from historical data. A simulation was set up to model the performance of this order policy over a one year period. The performance depends on the three parameters α, β, and I_{TH}. An optimization was then used to find an optimal set of values for these parameters. The formulation of the optimization problem was

$$Minimize[\mu_{IOS} + k\sigma_{IOS}] \quad (7)$$

where μ_{IOS} is the average daily on site inventory level and σ_{IOS} is the standard deviation of the daily site inventory level over the time period of the simulation. The factor k allows us to choose an optimal solution which places emphasis on minimizing the day to day variation in inventory as well as minimizing the average inventory level. A large value of k will tend to favor solutions where the day to day variation of the inventory is smaller but in so doing, it may favor solutions with a higher average inventory level. Similarly, a small value of k may give optimal solutions which have large day to day variation in inventory even though the average inventory is smaller. Thus k allows us to incorporate risk aversion into our optimal policy. Given the anticipated concern people might have about large day to day inventory variations, a value of k = 2 was used in the optimization. Along with Eqn. (7), the following constraints were used

$$\Pr obability[I_{MIN} \geq I_{TH}] \geq 0.995 \quad (8)$$

$$I_{MIN} > 0 \quad (9)$$

where I_{MIN} is the minimum site inventory level over the whole simulation.

The optimization technique used was simulated annealing (Rayward-Smith et al., 1996). This is a heuristic technique using a probabilistic hill climbing algorithm. The simulation with the optimized parameters is now used to predict when orders should be placed. Every month, the simulation was run to generate the orders. The simulation is also run on a daily basis to ensure that the orders that are currently placed do not need any modifications. This optimized order policy has allowed inventories to be reduced by approximately 50%.

One of the big concerns expressed by the people placing the orders was that there might be more need to modify the orders when the optimized policy was used. The large excess of inventory would no longer exist to cushion the system from variability. If significant and frequent modifications to the orders were required, it would undermine the utility of the optimized order policy. It was agreed that modifications of 2 days or less would not be important. The simulation was modified so that it could measure the order schedule uncertainty. A base case was run using the optimal values for α, β, and I_{TH}. This leads to a time ordered sequence of rail car order times.

Then a large number of simulations was run. Each run of the simulation would lead to a different prediction for the rail car order times. The timing for each order could be compared to the timing of that particular order for the base case run. Thus an order deviation time could be calculated as the difference between the base case order time and the order time for the multiple simulations. The order schedule uncertainty is then defined as the standard deviation of these order time deviations. Naturally, the further out in time the order, the larger the uncertainty. Figure 6. shows the variation of order schedule uncertainty with days into the future that the order is placed.

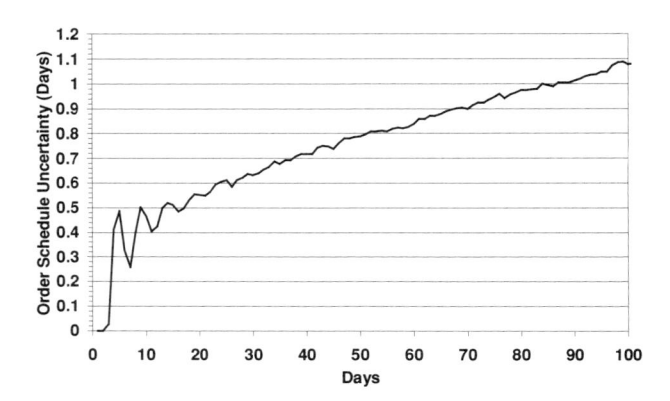

Figure 6. Variation of order schedule uncertainty.

At 30 days, the simulation predicts that the uncertainty in the orders is 0.6 days approximately. This assured the ordering personnel that the order schedule the simulation produced was robust. This greatly facilitated their acceptance of using the simulation to produce the orders. Note that this level of robustness is directly determined by the factor k in Eqn. (7). If the order robustness had not been acceptable, we could have increased the value of k and redone the optimization.

Finally, accounting for the effects of special cause events was accomplished by giving the ordering personnel the ability to modify the rail car lead time data and/or raw material daily usage data. For example, if it was known that the usage in the next month would be twice as much as normal, the user could increase the mean daily usage by a factor of two prior to running the simulation.

Conclusions

This paper has shown how an understanding of system dynamics via modeling and simulation can be used to increase the reliability of scheduling systems for those systems with significant variability. The concept of schedule robustness has been defined. The simulations themselves can be used to generate schedules and if the generated schedules are not robust, then the models can be used to help analyze opportunities for improvement.

The advantages of using this approach to scheduling is that variability is easily included and that complex decision criteria, reflecting real human decision making, can also be incorporated into the models. It also has the advantage that the simulations created have many more uses than just scheduling.

Currently, this approach to scheduling is limited to small systems, for example, systems with a small number of products that are not too interconnected. The approach cannot be guaranteed to give optimal solutions (due to the use of a heuristic technique), but it can find solutions which are better than the current situation.

In the future, it is planned to integrate the scheduling of the solvent recovery operations with the bulk pharmaceutical operations. This will necessitate linking a number of models which are now independent. The way the models will be linked is not yet clear. Also, with developments in the area of optimization of simulations, it will be possible in the future to create more optimal schedules.

References

Bickle, B., and B. McGarvey (1996). Application of Dynamic Modeling to a Solvent Recovery Area at a Pharmaceutical manufacturing Plant. *Systems Dynamics Conference*, July, 1996, held at Cambridge, Massachusetts.

Deming, W. E. (1986). *Out of the Crisis*. Published by the Massachusetts Institute of Technology.

Morton, T. E., and D. W. Pentico (1993). *Heuristic Scheduling Systems*. John Wiley & Sons, Inc.

Mosekilde, E., E. Larsen, and J. Sterman (1990). Coping With Complexity: Deterministic Chaos in Human Decisionmaking Behavior. In J. L. Casti and A. Karlqvist (Eds.) *Beyond Belief - Randomness, Prediction and Explanation in Science*. CRC Press. pp 199-229.

Rayward-Smith, V.J., I. H. Osman, and C. R. Reeves, and G. D. Smith, (1996). *Modern Heuristic Search Methods*. John Wiley & Sons, Inc.

Schertz, D., G. V. Reklaitis, G.V., and B. M. McGarvey, (1997) *Scheduling Under Uncertainty*. AIChE November Meeting, Los Angeles.

Taylor, W. A. (1991). *Optimization & Variation Reduction in Quality*. Mc-Graw-Hill, Inc.

Zentner, M. G., J. F. Pekny, G. V. Reklaitis, and J. N. Gupta, (1994). Practical considerations in using model-based optimizing for the scheduling and planning of batch/semicontinuous process. *J. Proc. Cont.*, 4, 4, 259-280.

MINIMIZING PRODUCTION CYCLE TIME AT A FORMULATIONS AND PACKAGING PLANT

Matt Bassett
Dow AgroSciences
Indianapolis, IN 46268

Abstract

An integer programming (IP) based optimization tool for generating a minimum clean-in-place (CIP) production cycle has been developed for the production planners at a liquid formulations packaging plant. A FORTRAN program generates the results by solving a traveling salesman problem (TSP) to optimality. Whereas the planners previously used their intuition to determine good production cycles (taking several hours per week), using this tool, the planners are now able to determine optimal production cycles in minutes. The planners can also determine which product changeover times should be given priority for reduction by determining the "critical" changeovers in the production cycle. Finally, the robustness of production cycles to variations in changeover times can be determined by performing Monte Carlo simulation.

Keywords

Optimization, Sequencing, Clean-In-Place, Traveling salesman problem, Scheduling

Introduction

Dow AgroSciences' newest formulations and packaging plant was designed to be a highly flexible, state-of-the-art facility. The plant formulates approximately 50 products using in-line mixing technology to directly feed the packaging line (Fig. 1). Different combinations of active ingredients are used for each formulated product. This means the equipment must be cleaned out between products to reduce the possibility of cross-contamination as well as to remain in compliance with our product integrity policy. The use of in-line mixing leads to straightforward clean-in-place (CIP) procedures for changing between products. Depending on the similarity of two products, the clean-out between them could vary from simply blowing the lines, to requiring a large number of solvent washes. For a given ordering of the products (production cycle), the total time spent cleaning the equipment (CIP time) can be calculated using existing plant data and/or educated guesses. By minimizing this CIP time, productivity can be increased and solvent expenses decreased for the plant.

Currently, the plant utilizes one master production cycle that is determined by an experienced scheduler. When production is not necessary for selected products, they are simply skipped over in the master production cycle. This may or may not give rise to the best production cycle for this reduced set of products. To

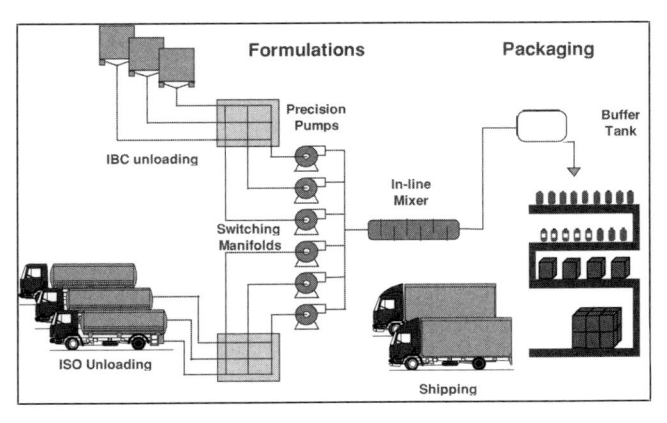

Figure 1. FP+ layout.

Table 1. Changeovers Times for Given Acceptable Contaminant Levels.

Acceptable Contaminant Level (ACL)	Changeover Time
>200.0	1
200.0	4
50.0	9
20.0	13
12.0	14
10.0	15
2.4	20
2.0	21
0.8	24
0.3	28

* if a phase change is necessary an additional 8 time units is required.

Table 2. Subset of Changeover Matrix for Example Problem.

From\To	1	2	3	4	5	6	7	8	9	10	11	12	13	14	15
1	0	9	24	24	24	24	1	1	0	8	9	17	17	17	32
2	1	0	1	1	1	1	1	1	1	9	9	9	9	9	9
3	13	13	0	1	1	1	1	1	1	21	9	21	21	21	32
4	13	13	1	0	1	1	1	1	1	21	9	21	21	21	32
5	13	13	1	1	0	1	1	1	1	21	9	21	21	21	32
6	13	13	1	1	1	0	1	1	1	21	9	21	21	21	32
7	24	13	24	24	24	24	0	1	1	21	9	21	21	21	32
8	24	13	24	24	24	24	1	0	1	21	9	21	21	21	32
9	13	13	24	24	24	24	1	1	0	9	9	21	21	21	32
10	21	21	32	32	32	32	21	21	9	0	1	1	1	1	24
11	21	21	32	32	32	32	21	21	9	1	0	1	1	1	24
12	21	21	32	32	32	32	21	21	9	1	1	0	9	9	24
13	21	21	32	32	32	32	21	21	21	13	13	13	0	1	24
14	21	21	32	32	32	32	21	21	21	13	13	13	1	0	24
15	17	21	21	21	21	36	17	17	21	13	13	13	13	13	0

guarantee that the best cycle is used at the plant, an optimal production cycle for this reduced set is found that minimizes the CIP time required. Determining the optimal production cycle is, however, a combinatorially complex problem. The combinatorial complexity arises from the fact that the required products can be produced using numerous different production cycles. Enumerating all of the possible production cycles to find the optimal (that is, checking all orderings) can take far longer than is reasonable or possible. By using mathematical programming techniques, it is possible to significantly reduce the time and effort of the planners to determine optimal production cycles.

This paper presents the tool developed for obtaining optimal production cycles given the products to be produced and the possible changeover times (or costs) between all the products. The tool consists of an integer programming (IP) based optimization algorithm linked to an Excel© worksheet. This allows the planners easy access to the results and the freedom to generate their own production cycles and compare them to the optimal. The user can also determine the sensitivity and/or the robustness of an optimal production cycle to variations in the changeover times. For clarity, a single example will be followed throughout the paper, although at times only a subset of the problem data will be presented due to space limitations.

Approach

To determine an optimal production cycle for a set of products, it is necessary to have a measure of the changeover time and/or the cost of changeover or solvent between all pairs of products. The decision to use time instead of changeover or solvent cost was arbitrary since for this problem they are proportional quantities. Plant personnel provided values for the Acceptable Contaminant Levels (ACL) between products. Using these ACLs, changeover times were calculated from plant CIP data to generate most likely changeover times. In Table 1, the changeover times utilized during the development of this tool are shown. Using this table, a changeover matrix is generated for all products. Table 2 shows a subset of the full changeover matrix. These values can be continually updated by the user as more data is collected on actual changeover times.

Given a changeover matrix, the problem of finding an optimal production cycle can be framed as a Traveling Salesman Problem (TSP) (Nemhauser and Wolsey, 1988). In its original form, the solution of the TSP finds a tour that visits every city in a given set exactly once while minimizing the total travel time needed to do this. This is exactly the problem we have, with products taking the place of cities and changeover times replacing travel times. The problem can be concisely stated using an integer programming (IP) mathematical formulation.

$$Min \sum_{i=1}^{n} \sum_{j=1}^{n} c_{ij} x_{ij} \qquad (1)$$

$$\sum_{i=1}^{n} x_{ij} = 1 \quad for\ all\ j = 1,\ldots,n \qquad (2)$$

$$\sum_{j=1}^{n} x_{ij} = 1 \quad for\ all\ i = 1,\ldots,n \qquad (3)$$

$$\sum_{i \in Q} \sum_{j \in Q} x_{ij} \leq |Q| - 1 \ for\ all\ subsets\ Q \\ of\ formulated\ products \qquad (4)$$

$$x_{ij} = 0\ or\ 1 \qquad (5)$$

By solving this mathematical formulation and interpreting the binary x_{ij} variables (1 if product j immediately follows product i in the production cycle and 0 otherwise), we are able to determine an optimal production cycle. Equation 1 states the objective of minimizing CIP time where c_{ij} is the changeover time

---

© Copyright 1985-1995 Microsoft Corporation.

from product *i* to product *j*. Equations 2 and 3 ensure that there is exactly one ancestor and one predecessor for each product, while Equation 4 ensures that optimal production cycles contain all required products. The optimal solution of this mathematical formulation is found using a FORTRAN program (Carpaneto *et al.*, 1990). This program uses Branch and Bound to determine the optimal solution without resorting to complete enumeration of all possible solutions (Salkin and Mathur, 1989).

Getting Results

For a more user-friendly interface, the FORTRAN program is linked via a dynamic link library (DLL) to an Excel worksheet which is shown in Figure 2. In the first column the user inputs the list of products that need to be produced during the next production cycle. This input can be in any order the user wishes -- random, alphabetically, possible production order, etc. Changeover times for all products must have been previously added to the changeover matrix.

The user has the choice of two buttons. If the user pushes the first button, it calculates the CIP time for the list assuming the given ordering. This allows the user to determine CIP times for user-defined production cycles. Next to each product is shown the time necessary to changeover when switching to the following product. The worksheet assumes that the first product on the list follows the last product on the list to form a cycle. Looking at the first column in Fig. 2, Product 1 follows Product 38 and requires a 21 time unit changeover.

If the user pushes the second button, it determines the minimum CIP time and outputs an optimal production cycle. It does this by collecting all the changeover times for the selected products and passing this information to the FORTRAN program which returns the answer. It is possible for there to be more than one production cycle that achieves the same minimum CIP time, but currently only a single solution is returned to the user. The output is very similar to that returned by the first button but now an optimal production cycle has been generated. There is a small possibility that, due to memory limitations, the answer returned will not be optimal and if this is true, the user will be informed that the answer is not optimal.

Using the optimal production cycle, or their own production cycle, the user can manually manipulate the ordering (through cutting and inserting cells on the worksheet) and determine what effect, if any, different orderings have on the CIP time. The user can also determine what effect a variation in changeover times has

User Defined Cycle	CIP		Optimal Cycle	CIP
1	9		1	9
2	1		2	1
3	1	Highlight User Defined Cycle and press button	3	1
4	1		6	1
5	1		5	1
6	1		4	1
7	1		9	1
8	1		7	1
9	9		8	9
10	1		11	1
11	1		12	1
12	9		10	1
13	1		34	1
14	24		31	1
15	13		33	1
16	13		32	1
17	1		35	1
18	20		36	1
19	1		37	1
20	1		38	13
21	1		27	1
22	1		28	1
23	1		29	1
24	9		30	24
25	4		15	13
26	1		16	13
27	1		17	1
28	1		18	13
29	13		23	1
30	1		21	1
31	1		22	1
32	1		20	1
33	1		19	1
34	1		24	1
35	1		25	9
36	1		26	4
37	1		13	1
38	21		14	21
171	Overall Cycle Time		156	Overall Cycle Time

Figure 2. Excel worksheet for user input and program output.

on the minimum CIP time by modifying values in the changeover matrix shown in Table 2.

Extensions

Defining Critical Changeovers

By making minor changes to the FORTRAN code, the problem of obtaining only one optimal production cycle when many may exist can be overcome. First, the problem is solved to determine the minimum CIP time. Then the problem is resolved and every solution whose objective value equals the minimum CIP time is reported prior to it being fathomed from the Branch and Bound tree. Using the minimum CIP time from the initial solution as the upper bound for the resolved problem reduces the computation necessary to obtain all the optimal solutions.

For the example problem being studied, there are 408 production cycles that have the minimum CIP time of 156 time units! While being distinct, many of these optimal solutions differ only in the ordering of products within product families (products with similar active ingredient profiles). Getting any useful information from this many

Table 3. Number of Changeovers Occurring in Optimal Solution Set.

From\To	1	2	3	4	5	6	7	8	9	10	11	12	13	14
1		**408**												
2			120	96	96	96								
3				96	96	120	48	48						
4			96		96	96	56	48	16					
5			96	120		96	48	48						
6			96	96	120		48	48						
7								200	200		8			
8							200			192	16			
9							8	16			384			
10														
11												24	192	192
12										**408**				
13														204
14													204	

Figure 4. Monte Carlo framework.

solutions is a challenge. One approach is to tabulate the number of times each changeover occurred in the optimal solutions (Table 3). For instance, product 8 follows product 7 in 200 of the 408 optimal solutions while product 2 follows product 1 in every optimal solution. Any changeover that occurs at least once in an optimal solution we will refer to as a "critical" changeover. Reducing the changeover time for any critical changeover will reduce the minimum CIP time. Reducing changeover times for non-critical changeovers may reduce the minimum CIP time as well, but it is not ensured.

One way to visualize the critical changeovers is the optimal solution diagram given in Figure 3. A line indicates a critical changeover while a bold line indicates that the changeover is critical in every optimal solution. A box indicates that the ordering of the contained products can be permuted without effecting the minimum CIP time. A line crossing the border of a box limits the precedence relationship of the connected tasks. For instance, product 4 can precede products 7, 8, or 9 while products 3, 5, and 6 can only precede products 7 and 8. This diagram can help the user to see the possible manipulations that can be made to an optimal solution and still maintain the minimum CIP time. For example, switching the order of products 7 and 8 will not effect the cycle time while doing the same for products 36 and 37 will produce a sub-optimal production cycle.

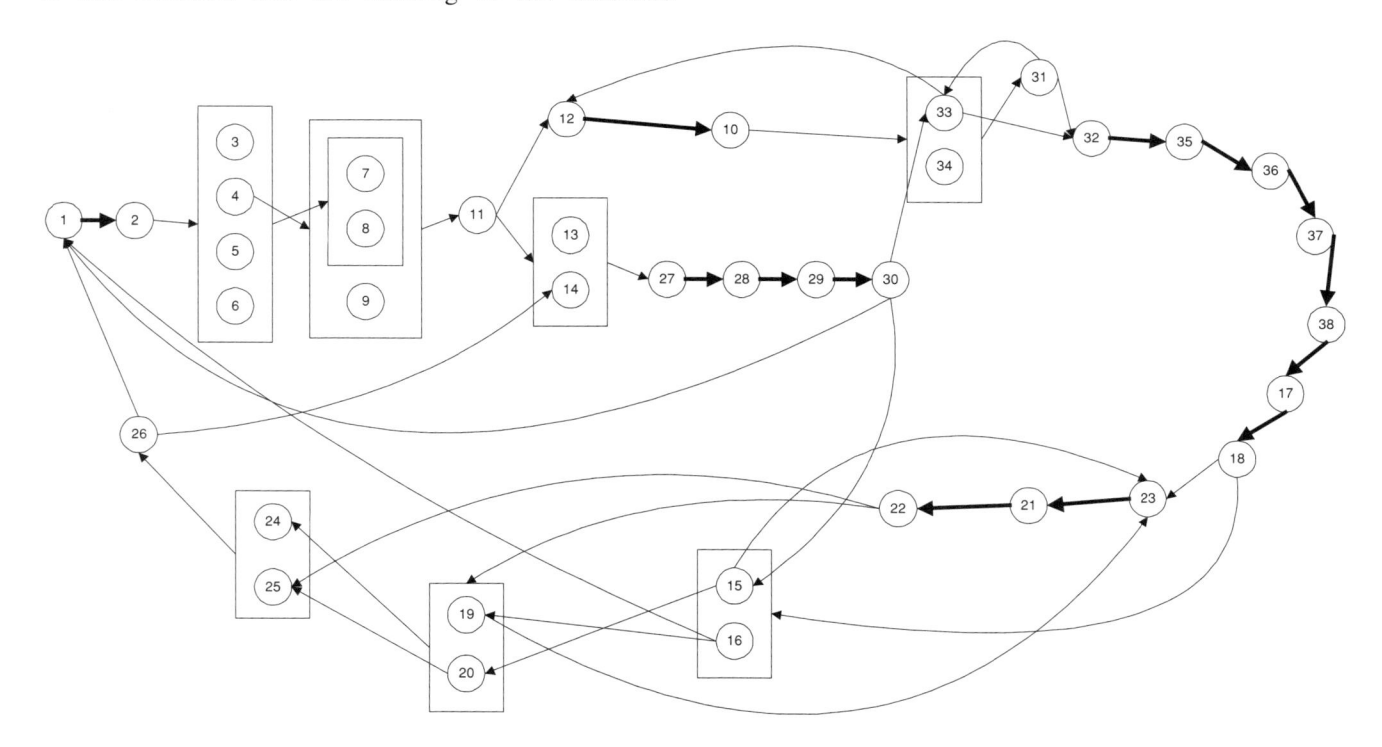

Figure 3. Optimal solution diagram.

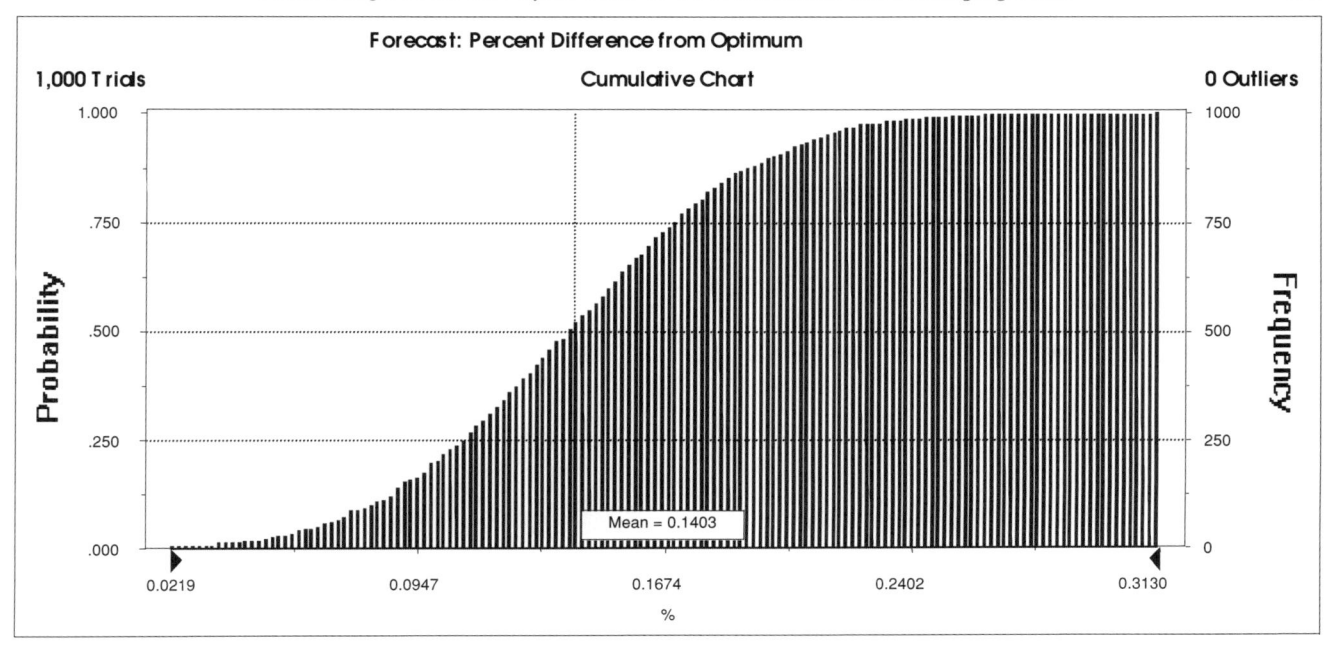

Figure 5. Robustness chart.

Uncertainty Analysis

Using a program such as Crystal Ball®, the user can determine the robustness of a selected (trial) production cycle to variations in the changeover times. Crystal Ball uses Monte Carlo simulation to determine the effect input uncertainties have on outputs. The Monte Carlo simulation proceeds by performing a number of iterations, each consisting of four steps (Figure 4). First, the distributions of all uncertain changeover times are sampled (inputs). Second, the minimum CIP time is calculated by the FORTRAN program using the sampled changeover times. This value is defined as the minimal CIP time since it is minimum only for this specific sampling. Third, the CIP time for the trial production cycle is calculated using the sampled changeover times. Finally, the deviation of the trial production cycle CIP time from the minimal CIP time is determined (output).

Robustness is inversely proportional to the deviation of the trial production cycle CIP time from the minimal CIP time for all sampled changeover times. The smaller the difference between the trial and minimal CIP time solutions, the more robust the trial production cycle. Figure 5. shows the cumulative distribution plot for robustness where the optimal cycle from Fig. 2. is used as the trial production cycle. One thousand iterations were performed with CIP times allowed to vary uniformly ±25%. The choice of this level of uncertainty is based on the variability of the plant data used to calculate the initial changeover times. For this example, there is a 50% probability that the CIP time for the trial solution is within 14% of the minimal CIP time.

In general, Monte Carlo approaches are very computer intensive. This is true for this problem as well. The bottleneck step is the determination of the minimal CIP time within each iteration via the FORTRAN program. Once the solution is found, however, the minimal CIP time can be quickly compared to calculated CIP times for numerous trial production cycles. This approach gives the user another criterion by which to select a production cycle in addition to minimum CIP time. For instance, for the example presented in this paper, there are 408 optimal production cycles based solely on CIP time. Using this robustness measure, these solutions can be further differentiated.

Conclusions

A tool has been presented which determines the optimal production cycle that minimizes CIP time using the Traveling Salesman Problem formulation. With the help of an Excel spreadsheet linked to a FORTRAN program, we have been able to implement a user-friendly interface to obtain a minimum CIP time and production cycle for a user-defined set of products. Whereas the planners previously used their intuition to determine good production cycles (taking several hours per week), using this tool, the planners are now able to determine optimal production cycles in minutes. Additionally, with this tool the planners are able to perform what-if scenarios for changes in both the production cycle and changeover matrix.

For a more in depth examination of the solution, the FORTRAN program has been modified to report all

---

® Trademark of Decisioneering, Inc.

minimum CIP time production cycles. Two approaches for examining the numerous solutions were presented: tabulation and the optimal solution diagram. Finally, the robustness of trial production cycles was calculated using Monte Carlo simulation.

Nomenclature

c_{ij} = changeover time from product i to product j.

x_{ij} = 1 if product j immediately follows product i in the production cycle, 0 otherwise.

n = number of products

References

Carpaneto, G., M. Dell'Amico, and P. Toth (1990). A Branch-and-Bound Algorithm for Large Scale Asymmetric Traveling Salesman Problems. *Technical Report Dipartimento di Economica Politica Facolta' di Economica E Commercio*, Universita' di Modena, Italy.

Nemhauser, G. L. and L. A. Wolsey (1988). In *Integer and Combinatorial Optimization*. John Wiley & Sons, Inc., New York. Part II, Chap. 6.3, pp. 469-495.

Salkin, H. M. and K. Mathur (1989). In *Foundations of Integer Programming*. North-Holland, New York. Chap. 8, pp. 245-297.

INTEGRATION OF PRODUCTION SCHEDULING AND ACTIVITY BASED COSTING MODELS

Vangelis Lionis, Guillermo E. Rotstein and Ebrahim Mohamed
Imperial College of Science, Technology and Medicine
SW7 2BY London, UK

Robert A. Weiss
ICI Paints
SL2 5DS Slough, UK

Abstract

It is now becoming common for manufacturing companies to use Activity Based Costing (ABC) systems for production management and costing. However, standard tools that are currently in use employ a top-down approach, starting with lumped costs that are allocated to activities using over-simplified rules. As a consequence, the results lack accuracy, reliability and relevance to the production floor. The development of novel on-line scheduling systems has opened the potential for significant advances in the attempt to integrate the process and business sides of an organization. This paper exploits this potential, by introducing a costing methodology that employs a bottom-up approach that pushes detailed process data upwards into the organizational structure, thus providing more consistent, reliable and up-to-date information for decision making purposes. The approach generates results that are consistent with established accounting practices and relevant to decision making needs at all business levels. Moreover, it addresses issues of particular significance in a processing environment, such as resource utilization efficiency and the scope for establishing a consistent quantitative approach for assessing the cost saving potential of a given plant.

Keywords

Batch processing, Scheduling, Costing, Computer integrated manufacturing, ABC, Opportunity costs.

Introduction

Recent advances in batch production scheduling are revolutionizing the operation of multi-purpose batch plants. Schedule optimization and on-line production management tools are offering the opportunity to improve dramatically the automation, collection of information and efficiency of batch processes. Optimization tools rely heavily on thorough and accurate cost mapping of the process and enterprises find it increasingly important for such a mapping to be consistent with financial cost data at the management level. On the other hand, management accounting systems are facing the need for more accurate and relevant input from the production environment. Such detailed information can be found in on-line production management tools, however, they have not yet been exploited for the purpose of accurate production costing.

These issues bring up the opportunity for integrating business management systems (accounting oriented) and management execution systems (engineering oriented). The challenge is to provide relevant, consistent and updated information at various business levels, as well as to facilitate this process with a global cost data platform.

This study tackles these issues by proposing a production costing model for batch plants, based on a schedule modeling framework. Here, the modeling standard of BatchManager, an advanced on-line production management tool, is used. The proposed costing framework relies upon the activity based costing

(ABC) accounting rationale. It aims to provide a useful management tool, able to assist production and planning decisions by making financial information accessible to the factory floor supervision *and* promoting the communication of process data upwards to the management levels of the organization.

Production Costing in the Manufacturing Industry

Costing applications currently used in industry are applying the accounting concept of activity based costing (ABC), in an effort to supply the management level of the business with meaningful costing information, suitable for decision making purposes.

ABC formulations have replaced the established conventional aggregated costing methodologies for capturing and distributing costs to the output goods or services. These cost accounting operations employ a "black box" view of the production process, making use of overall process costing data only, disregarding detailed costing representation of various production units, resources, products, recipes, etc. Instead, they use the classic model of cost distribution which was designed around the major factors of production: direct labor, direct materials and overhead. Business and business-like structures have relied upon the historical model of cost accounting for over 100 years. With the recent advent of accounting, it has been discovered that the traditional cost accounting methodology can create a significant difference in output cost because of the manner in which overhead costs are allocated to output rather than traced to output. This difference in distribution can skew the ultimate price of the output and lead to poor management decisions.

What is Activity Based Costing?

ABC is a cost management tool that has emerged from the need to estimate costs in a detailed manner, so that *costs* can be traced to the *activities* that caused them. ABC systems aim to assign costs in a hierarchical structure similar to the structure of the organization and/or the process itself. Underlying ABC is the assumption that activities consume resources and products consume activities. Activities may include establishing vendor relations, purchasing, shipping, machine setup and running, reorganizing the production flow, product redesign and taking a customer order. The execution of these activities triggers the consumption of resources that are recorded as costs in accounts. The activities are performed in response to the need to design, produce, market, and distribute products. (Cooper, 1989)

The basic distinction between conventional costing and ABC is as follows: The former allocate costs to products based on attributes of a single unit, such as the number of direct labor hours required to manufacture a unit, the purchase cost of merchandise resold, or the number of days occupied. Allocations therefore are proportional to the volume of units produced or consumed. In contrast, ABC systems focus on activities required to produce each product, based on their consumption over the entire product volume. The fundamental difference is that conventional costing assumes that products cause costs, whereas ABC assumes that activities cause cost and products create the demand for activities. (Tjiang, 1997) Such an approach is obviously much closer to the one adopted in a typical Management Execution System.

Motivation: ABC and the Industry Needs

Appropriately applied, ABC provides a far more accurate representation of cost than previous accounting methods. In manufacturing companies, in particular, ABC has revolutionized the way analysts interpret process cost information. (Bailey, 1991; Cooper *et al.*, 1992) So far, however, its implementation has been targeted exclusively at business management and has remained largely detached from the production shop floor. This is a considerable limitation, because it affects the quality and validity of the results, the speed and consistency of the data exchange and, ultimately, restricts the flexibility, reliability and scope of the implementation. Indeed, ABC systems implemented so far in multi-purpose batch plants are proven to be inadequate in satisfying the management needs. (Giglio, 1997; Huynh, 1997; Kaplan, 1990)

According to a survey recently completed in a major British process industry employing automated costing models, (Huynh, 1997) a successfully implemented costing system should be able to:

Provide the true product cost for any batch size.
Provide accurate manufacturing costs for new products.
Include auxiliary costs related to production, such as cleaning and testing.
Facilitate regular updates to ensure validity and reliability, without entailing excessive amount of man-hours in maintaining the system.
Assist the users by reporting results in a meaningful format.
Measure the costs and efficiencies of producing different products so that management can prioritize the manufacturing of their products based on their demand and profitability.
Enable model building and 'what if' analyses on a variety of production techniques to determine the impact of the different techniques on the cost and efficiency of production.
Integrate the non manufacturing functions of the business.
Be implemented uniformly and consistently in the various plants of a multi-site company, in order to provide information useful for cost comparisons among the different sites.

Be easily accessible through the company IT facilities.

Allow the users flexibility to make use of the available information in the way they choose and thus contribute to the analysis of the problems identified and to the improvement of the company costing system overall.

A review of previous ABC models published or applied in industry (Lionis, 1997; Mohamed, 1996) has concluded that the reason current ABC implementations have not quite lived up to the expectations is mainly the fact that they employ a *top-down* approach, whereby aggregated high level costs are allocated at the low level, by applying notional allocation criteria, known as cost drivers. E.g. the cost driver for allocating the machine set-up costs would be the number of batches processed by that machine. However, this method fails to describe accurately the way costs and profitability are affected by the realities of the production floor. It therefore fails to provide the management with a good enough insight of the process for meaningful 'what-if' scenarios to be drawn. The impact of modifications in the production floor to the cost picture of the process is hence not correctly modeled and consequently not fully understood and assessed. In the above example, for instance, the machine set-up cost allocation would be misleading, if all batches do not require the same set-up procedure. The use of more sophisticated cost drivers only complicates the system enormously, making it very difficult and costly to implement, run, modify and maintain, user unfriendly and yet never accurate enough.

Integration of ABC and Scheduling Models

The cost model presented in this paper proposes a framework for evaluating production costs and profits, by employing a *bottom-up* approach, starting from detailed low level process information in order to build up the higher level cost information. The detailed process information required is obtained from a typical production scheduling tool—here BatchManager—and a financial database. BatchManager (Macchietto, 1996) is based on data models and scheduling algorithms of the Supervisory Batch Management System (Superbatch) developed at Imperial College, (Cott *et al.*, 1989) incorporating a modeling environment for almost any type of batch plant. A broad parallelism between BatchManager and ABC can be drawn by the way BatchManager splits the plant model information into two distinct areas: plant resources and orders / batches. The resources are consumed in order for the batches to be processed, and can therefore be viewed as sources of cost. The batches require those resources, they can therefore be regarded as recipients of cost. A financial database provides all the data on costs for the resources.

The relationship between the proposed cost Model, BatchManager and the financial database is presented in Fig. 1. The cost model reads the list of scheduled batches and the corresponding recipes, the products and raw materials associated with the procedures, the equipment allocation of the batches and also the services consumed, which are specified in the recipe definitions. Based on those and the schedule profiles of every processing step, the model works out precisely which resources were used, how much, for how long and at what cost. This calculation is performed for every single task that takes place in the plant. All these elemental costs are called *cost elements*, and their aggregation gives the 'low level' production costs, i.e. the costs that can be traced to individual batches.

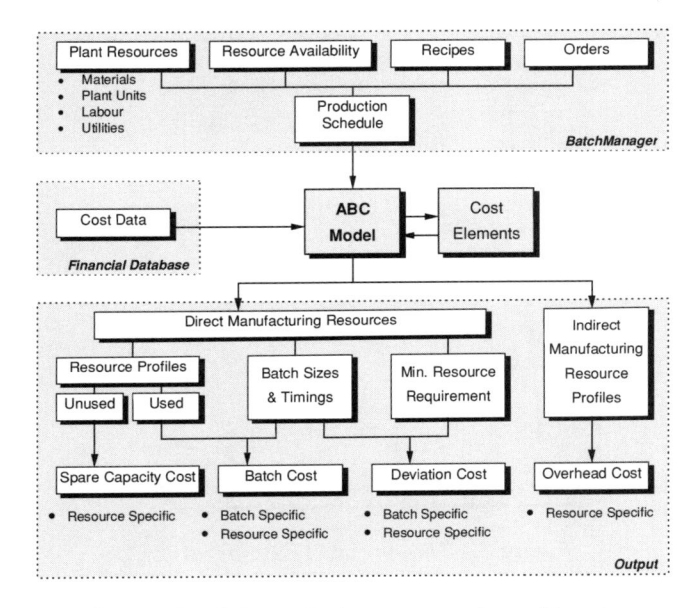

Figure 1. Schematical representation of inputs and outputs.

The cost model also calculates 'high level' costs (i.e. costs budgeted for each plant resource independently of resource utilization). A comparison between these and the low level costs quantifies the level of resources that have remained unused. In addition to working out the costs for each batch, the cost model calculates the minimum level of plant resources that each step requires. Therefore this information can be compared with the actual consumption so that the level of over-utilization—and associated cost— of the resources can be estimated. Profits are also calculated, not only in terms of the real revenue from product sales, but also internal 'virtual' profits allocated to individual batches, to facilitate profitability comparisons among different products. In this context, it must be highlighted that real profits are *not* an aggregated form of the virtual batch profits, as their principle of calculation—and their purpose—is different. (Lionis, 1997)

Cost Hierarchy and Opportunity Costs

By distinguishing low and high level costs, as described above, the proposed cost model is formulated in a two-level hierarchical structure, as shown in Fig. 2. The importance of this hierarchy is exemplified in the way the proposed framework postulates *opportunity costs*. These are process costs reflecting the 'wasted opportunity' of the plant to increase its output (and hence its profit) by increasing the utilization of the resources available. It is therefore a measure of how efficiently the resources are being used. Their computation, however, is far from straightforward and several methods have been proposed. (Giglio, 1997; Heymann & Bloom, 1990; Mohamed, 1996; Tjiang, 1997) These methods tend to be either oversimplified, hence losing accuracy and thoroughness, or complex and case-specific, hence sacrificing flexibility, consistency and generality.

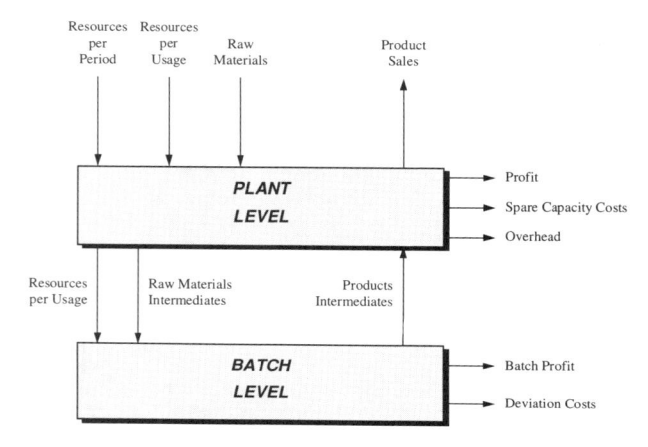

Figure 2. The two-level cost hierarchy.

This study proposes the definition of two distinct types of opportunity costs. Firstly, the ones occurred due to resources which are charged for and remain unused (non-utilized resources) and will be referred to from now on as *spare capacity costs*. These can occur, for example, when a piece of equipment remains idle, for which depreciation is being charged. It should be noted that spare capacity costs are real manufacturing costs, because depreciated equipment which is not being used does cost money to the business, thereby raising production costs.

The second type of opportunity costs will be referred to as *deviation costs*. They occur when more resources than the minimum necessary were consumed for a batch to complete (over-utilized resources). Unlike spare capacity costs, they can be allocated to individual batches. Deviation cost is a part of the batch production cost, it is not additional to it. It reflects the proportion of the total batch production cost that could have been avoided.

The methodology introduced above is based on the concept that opportunity costs arise because the plant resources do not produce the maximum amount of finished product that would be theoretically possible, based on the recipe definitions and batch sizes. In other words, more resources than the minimum required are used to produce a given amount of product. This introduces a direct relation to resource utilization levels. Therefore, any excess consumption of resources constitutes opportunity costs, in the sense that this excess could have been used towards additional production, hence yielding additional profit. This methodology has three important benefits:

1. Opportunity costs are calculated directly according to the extra costs incurred for the additional use of resources, instead of being interpreted according to indirect criteria. (e.g. number of batch reworks, duration of delays, level of inventories, etc.)
2. With the proposed approach, opportunity costs are identified based on a simple and consistent criterion, so it is not necessary to specify and predict their sources beforehand.
3. The detailed breakdown achievable in the proposed cost model allows opportunity costs to be traced in detail, facilitating the cost analysis of the results, in order not only to depict cost saving opportunities but also to quantitatively assess them, so that their relative importance can be prioritized.

A Case Study

The case study is presented in order to demonstrate the way that the cost model displays the various costs involved. Here, the model of a typical yogurt production process is used. The process produces fruit flavored yogurt in a rather simple plant, comprising two major operations: The production stage (entailing pasteurization, incubation, discharging and chilling the yogurt, cleaning the reactor) and the filling stage (packaging and addition of flavoring). A typical flowsheet of this process can be seen in Fig. 3.

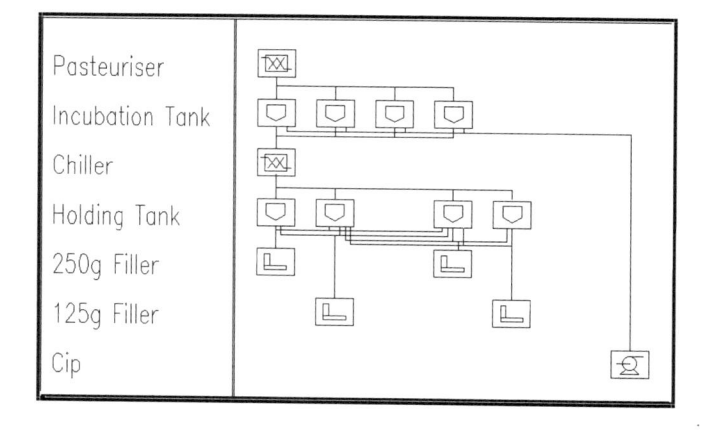

Figure 3. Yogurt process flowsheet.

The modeling and scheduling activities are carried out within BatchManager. The number, size and timings of the batches are determined following the modeling of the plant resources recipes and the required output (normally based on customer orders). This, in turn, yields a production schedule for a given time horizon. All this data constitutes input information to the cost model, which evaluates costs and profits for the same schedule period. The following examples are based on a daily horizon.

The results here are presented in the form of three sample tables (values shown are typical of the dairy industry). Table 1 reports a breakdown of all the costs that are allocated on a batch basis, with deviation costs reported separately. Consequently, revenues and profits are calculated for each batch individually. In Table 2, all the plant level costs are shown together with aggregated batch level costs. The costs are all allocated in terms of *items* (i.e. process / storage units) and *services* (i.e. utilities and labor). The format suggested here treats process and storage items separately. In Table 3, an accounting-style income statement report is shown, summarizing all the manufacturing costs, initial and final stocks and overall profit for the plant. (Note that WIP stands for 'Work in progress' measuring the value of total inventory undergoing processing) It aims to show how the cost model could push upwards to the accounting level detailed costing information. Costs are reported under categories that follow a functional logic. The categories depend on the function performed, thus providing activity-related information. For instance, the depreciation costs for cleaning and storage equipment are grouped together under the headings 'Cleaning' and 'Storage' costs, apart from depreciation costs for processing equipment. It has to be noted, though, that this study is putting forward just one of the possible alternatives; the method for reporting the results can be customized by the user. The formats presented in this paper only aim to give a proposed example that was chosen in order to provide information consistent with recognized accounting practices and also meaningful from a decision-making point of view.

Table 1. Batch Costs.

Batch Name	Batch Revenue (£)	Material Cost (£)	Waste Cost (£)	Deprec. Cost (£)	Labour Cost (£)	Deviation Cost (£)	Batch Cost (£)	Batch Profit (£)	Profit Margin (%)
125 DB_002	3,360	2,696	-	31	27	-	2,754	606	18.0
125 DC_001	4,400	3,751	-	79	69	-	3,899	501	11.4
125 DC_007	4,080	3,478	-	109	65	38	3,652	428	10.5
125 GC_005	4,240	3,615	-	46	40	-	3,700	540	12.7
125 GC_011	4,000	3,410	-	73	64	-	3,546	454	11.3
125 GC_012	4,800	4,092	-	85	74	-	4,251	549	11.4
125 GC_014	1,840	1,569	-	40	35	-	1,644	196	10.7
250 DB_008	6,400	5,136	-	61	53	-	5,250	1,150	18.0
250 DB_013	8,000	6,420	-	73	64	-	6,556	1,444	18.0
250 DB_004	3,920	3,146	-	42	37	-	3,225	695	17.7
250 GB_006	960	770	-	20	17	-	807	153	15.9
250 GB_015	5,600	4,494	-	55	48	-	4,597	1,003	17.9
250 GC_003	2,640	2,251	-	32	28	-	2,311	329	12.4
250DC_009	4,960	4,228	-	50	44	-	4,322	638	12.9
250DC_010	4,160	3,546	-	44	38	-	3,629	531	12.8
DIETB_017	5,500	4,510	83	82	-	2	4,675	825	15.0
DIETB_018	5,500	4,510	83	82	-	2	4,675	825	15.0
DIETB_020	5,500	4,510	83	80	-	-	4,673	827	15.0
GOLDB_016	5,500	4,510	83	82	-	-	4,675	825	15.0
GOLDB_019	5,500	4,510	83	82	-	2	4,675	825	15.0
Totals	~	~	415	1248	703	46	~	~	~

It should be noted that labor is unavailable for the first 8 hours of production, due to the fact that there is no night shift in the factory.

Table 2. Plant Costs.

Specific Item	High (Plant) Level (£)	Low Theor. Costs (£)	Level Deviation Costs (£)	Spare Capacity Costs (£)	Storage Depreciation Costs (£)	Cool Storage Costs (£)	Total Storing Costs (£)
Pasteuriser 1	379	100	-	279	-	-	-
Incubation Tank 1	95	75	2	18	-	-	-
Incubation Tank 2	95	37	2	55	-	-	-
Incubation Tank 3	95	38	2	55	-	-	-
Incubation Tank 4	95	38	2	55	-	-	-
Chiller 1	379	50	-	329	-	-	-
Holding Tank 1	57	-	-	-	57	1,764	1,821
Holding Tank 2	57	-	-	-	57	1,246	1,303
Holding Tank 3	57	-	-	-	57	1,014	1,071
Holding Tank 4	57	-	-	-	57	726	782
250g Filler 1	473	174	-	299	-	-	-
250g Filler 2	473	201	-	273	-	-	-
125g Filler 1	473	237	34	201	-	-	-
125g Filler 2	473	189	-	285	-	-	-
CIP 1	237	62	-	174	-	-	-
Totals	3,492	1,200	42	2,023	227	4,749	4,976
Service Set							
Filler Staff	925	704	4	217			
Totals	925	704	4	217			

The results show that the most dominating cost element of the process is the raw materials (material-intensive process). Spare capacity costs also form a very large proportion. This reveals the fact that the plant facilities are heavily under-utilized. This fact, although readily detectable in the production schedule, would not be apparent outside the production floor. However, the results can show—that cannot be seen from the schedule alone—*which* plant items attract most of these spare capacity costs. This is important, because items that have high depreciation rates, such as the pasteurizer, the chiller and the filler lines incur higher spare capacity costs, for a given level of utilization, than other, less capital-intensive equipment. The scheduling system alone cannot show that, because it only provides utilization level information in terms of time, by looking at the fraction of the schedule period for which a piece of equipment is used. A classical accounting system alone cannot provide such information either, because it does not rely on utilization information and cannot accurately allocate costs at the different hierarchical levels. The proposed cost model not only depicts all the costs related to resources being under-utilized, but also provides a quantitative measure of where these costs are concentrated, so that the user can make use of the information in order to decide the areas of improvement that need to have priority. These can then be allocated to relevant cost accounts.

The same logic is applicable to deviation costs, although these cannot be easily demonstrated by this example, because this case study was carried out using the off-line version of BatchManager, therefore, there is no plant feedback information. By using the on-line version the user can obtain data on any differences of the actual executed plan, as opposed to the theoretical one predicted by BatchManager. However, the small deviation costs depicted at this example reflect very well the capabilities of the cost model: The four delays that occur in four of the incubation batches occupy each incubator for 20 min. in excess of the theoretical time, whereas the 1 hour unavailability of filling staff results a delay of equal time in the 125g Filler. The results reveal that the delay in the filler line incurs a cost nearly 5 times the cost of all the incubation delays altogether, which could not have been depicted by measuring the length of the delay only.

Table 3. Accounting Summaries.

Manufacturing Account	(£)	(£)	(£)
Opening W.I.P.			9,900
Opening Raw Material	0		
Purchases	40,098		
Closing Raw Material	(0)		
Raw Material Used		40,098	
Delivery Cost		208	
Total Material Cost		40,307	
Labour		704	
Depreciation		1,138	
Spare Capacity		2,240	
Batch Deviation		43	
Refrigeration		936	
Cleaning		62	
Storage		227	
Total Expenses			45,657
Total W.I.P. Available			55,557
Material Loss			(413)
Closing W.I.P.			(11,559)
Goods to Trading			43,586

Trading Account	(£)	(£)
Sales Revenue		63,360
Opening Stock	0	
ex. Manufacturing	43,586	
Closing Stock	(0)	
Cost of Goods Sold		43,586
Profit		19,774
Gross Profit Margin		31%

That happens because the filling lines involve higher depreciation costs than the incubators. This example can therefore illustrate how the proposed representation of opportunity costs can be helpful in analyzing and assessing cost trade-offs, as in this case between labor availability and unit utilization.

Conclusions

The cost model presented in this study demonstrates the way that recent advances in the control and scheduling of batch processes can be exploited in order to improve the accuracy, reliability, flexibility and applicability of cost analysis systems at the management level. Not only is its concept simple and consistent, but it can also deliver greater accuracy and huge potential for integrating shop floor and management, promoting the improvement of the communication and understanding between the two functions. The capability of the method to encourage more realistic breakdown of work activities and the proper allocation of costs, makes it possible for potential hidden costs to become more visible. On the other hand, the impact of production floor decisions to the profitability at the high level, can be envisaged at the low decision making level. Moreover, the integration of the proposed costing model with a generic tool, such as BatchManager, ensures that it is widely valid and adaptable.

Still though, we are far from reaching a single universally applicable ABC framework, because of the diversity, complexity and the various costing needs of different companies. The proposed costing methodology is targeted at multi-purpose batch plants, aiming to automatically develop costing equations applicable to all the range of processes that can be modeled using the framework on which BatchManager—or any other similar scheduling tool—is based. Its contribution in the field of production costing is that it elaborates a methodology of deriving a cost model automatically, using the modeling infrastructure *already* available for scheduling, planning and production management purposes.

Further to this work, we are elaborating a prototype software tool incorporating this costing methodology and integrated with BatchManager, with the view to test its implementation in full scale industrial case studies. It is also intended to further develop a mapping between process and accounting data, in order to arrive at optimization methods that would address modern production management issues, where accurate costing information is of critical importance. Typical such decisions include the definition of the company product portfolio, the comparison of relative cost effectiveness among various production sites, or cash flow management.

References

Bailey, J. (1991). Implementation of ABC systems in UK companies. *Management Accounting*, February, 30-32.

Cooper, R. and Kaplan, R. S. (1992). *The Design of Cost Management Systems*, Prentice Hall.

Cooper, R. (1989). What do Activity based cost systems do. *J. Cost Management for the Manuf. Ind.*, Spring, 38-49.

Cott, B.J. and Macchietto, S. (1989). An integrated approach to computer aided operation of batch chemical plants. *Comput. Chem. Engng.* **13**, 1263-1271.

Giglio, R. (1997). Production costing for automated factories. In *Flexible Automation & Intelligent Manufacturing*. Begell House Inc. pp. 820-827.

Heymann, H.G. and Bloom R. (1990). *Opportunity Costs in Finance and Accounting*. Quorum Books, New York.

Huynh, L. (1997). *A user assessment of activity based costing at ICI Paints*. MBA Thesis, Imperial College, London.

Kaplan, R.S. (1990). Limitations of cost accounting in advanced manufacturing environments. In *Measures for manufacturing excellence*. Harvard Business School.

Lionis, V. (1997). *Integration of production scheduling and activity based costing models in multi-purpose batch plants*. MSc Thesis, Imperial College, London.

Macchietto, S. (1996). Batch food processing: the proof is in the eating. *CPCV Proceedings*, January, 7-12.

Mohamed, E. (1996). Financial information for production decision making in a hierarchical manufacturing environment, In *EIASM workshop on production, planning and control*. University of Mons, Belgium.

Tjiang, G., Besant, C.B. and Chang, Y.S. (1997). A practical production planning method in a semiconductor wafer fabrication. In *IFAC/IFIP Conference on Management and Control of Production and Logistics*. Campinas, Brazil.

A TASK-RESOURCE BASED FRAMEWORK FOR PROCESS OPERATIONS MODELING

Gabriela S. Mannarino and Horacio P. Leone
Ingar (CONICET)
GIPSI (Fac. Reg. Santa Fe - Universidad Tecnológica Nacional)
Avellaneda 3657 – 3000 Santa Fe - Argentina

Gabriela P. Henning
Intec – (Universidad Nacional del Litoral - CONICET)
Güemes 3450 – 3000 Santa Fe – Argentina

Abstract

This paper presents a modeling language to describe and analyze process operation activities in a production enterprise. The proposed framework integrates task, domain and dynamic models. *Task Models* represent an organization from a functional point of view. *Dynamic Models* put emphasis on the interaction and evolution of resources when they participate in different tasks. A *Domain Model* includes classes describing the universe of discourse as well as static relationships among the classes.

Keywords

Enterprise modeling, Modeling languages, Enterprise integration, Requirements engineering, Object oriented analysis.

Introduction

The increasing pressure that affects production organizations does impose new constraints on the administration of their resources. Process operation activities, such as Planning, Scheduling, Process Control and Maintenance, need to be properly coordinated. People undertaking these activities require to have not only an overall view of the production enterprise, but also they need to know how the processes they are responsible for are integrated through information and material flows.

Models are basic tools to analyze and evaluate the way activities are performed and to improve and restructure them if needed. Models can be used by organizations as a mechanism to acquire knowledge about the processes carried out at the production floor as well as about the management functions that make possible these production processes. The use of a language shortens the modeling process as it defines the vocabulary to use and the way this vocabulary can be combined to describe an organization.

Several authors have recognized the need for tools to integrate the various process engineering activities both among themselves and with the organization management functions (Lindheim, Totland and Lien, 1996). For example, n-**dim,** proposed by Levy et al. (1993) focuses on the problem of sharing and integrating among different people knowledge pertaining to the design process. Moreover, the need for defining multiple modeling dimensions to describe a complex system is accepted by different authors (Schreiber, Wielinga and Breuker, 1993). This requirement is made explicit by Simensen, Johnson and Årzen (1997), who proposed a framework to support the representation of batch process industries.

This paper presents a modeling language for describing a production organization through its various dimensions. *Task, Domain* and *Dynamic* models represent both the static and dynamic aspects of an industrial environment. Each one gives a partial view, but all of them are required to get a complete outlook of the organization. A *Task Model* describes both the administrative and process operation activities of an organization in terms of a set of resources that participate with specific roles. In this contribution the term resource has a wide interpretation, involving both physical and information resources. *Dynamic Models* (Harel and Gery, 1997) describe how a particular resource evolves from state to state due to its participation in different activities, and also, how it interacts with other resources to achieve complex goals. *Dynamic Models* help to answer questions such as "Which are the effects of a processing unit breakdown"?.

However, an organization cannot be completely described with just the previous two models. Questions such as, "Which is the master recipe of product X?", "To which tanks is connected reactor R-1?", are answered by resorting to *Domain Models*. In a *Domain Model* the characteristics, properties and relationships among the different real and conceptual entities of an organization are described (e.g., equipment items, product recipes, safety specifications, cost and time estimations, etc.).

The Modeling Framework

The Object Oriented Paradigm (OOP) is employed as the basis for specifying the proposed modeling language. Then, every concept of a production enterprise will be expressed by making use of the notion of *object* and by resorting to *static* and *dynamic relationships* among them. In the OOP, every object is defined by a set of *slots* or attributes, which take on some values. Some object attributes may change their values over time as a consequence of receiving different stimuli, in which case it is said that the object changes its state. Thus, the *State of an Object* is defined by a set of characteristic <*SlotDefinition-SlotValue*> pairs that hold under certain conditions. In this contribution, we will concentrate on the states of objects representing enterprise resources. Therefore, a *State* will be informally defined as a condition a *Resource* satisfies during a specific time interval (e.g., an equipment item can be either *"idle"*, *"empty and clean"*, *"empty and dirty"*, etc.

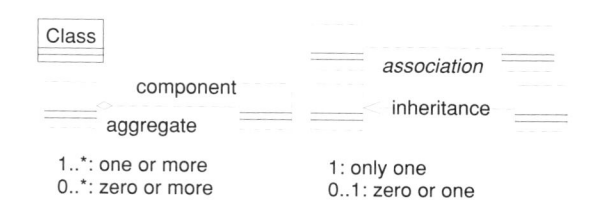

Figure 1. Notation.

To represent the language, the UML graphical notation has been adopted, as shown in Fig. 1. (Rational Rose, 1997). The *Model* class that appears in Fig. 2.

defines the syntax and semantics of the proposed enterprise specification language. All the entities representing the notion of task, resource, goal, and all the concepts pertaining to a production environment appear in the *Model* class. A *View* filters certain characteristics of the overall *Model*. *Task, Dynamic* and *Domain* models represent a *View* of a complete *Model*. Furthermore, a *Dynamic Model* specifies the dynamic characteristics of an organization by resorting to State Transition Diagrams (STD) and State Transition Diagram Nets (STDN).

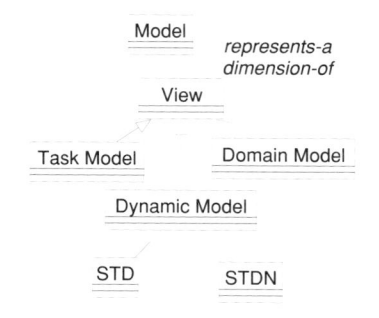

Figure 2. Different types of models are required to represent a production enterprise.

Task Model

A *Task Model* describes an organization from a functional point of view. In a *Task Model* tasks and resources are combined so as to achieve the organization goals. In order, to represent an activity that has a certain characteristic duration, the *Task* construct is employed. The temporal relationships defined by Allen (1983) have been adopted to establish a partial order among the activities (Fig. 3.).

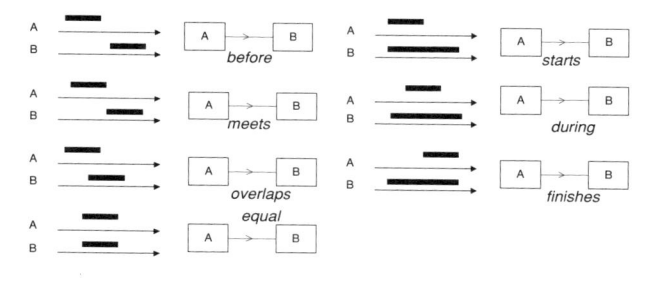

Figure 3. Allen's temporal links (1983).

A *Resource* identifies every entity that participates in the activities an organization performs in order to attain its objectives. Both the physical (e.g., raw materials, products, equipment, personnel, software, energy, etc.) and information resources (e.g., maintenance plans, invoices, etc.) of the organization are abstracted in the class *Resource* (Fig. 4.). A *Resource* may assume different roles; these roles usually depend on the tasks in which the resource participates. Therefore, the *ResourcePerspective* concept is also introduced in the language. It is associated to a *Resource* to encapsulate a particular view of the *Resource* with respect to a *Task* (Fig. 4.).

The effect of a *Task* over a *ResourcePerspective* is expressed by resorting to the links *uses*, *employs*, *creates*, *deletes*, *consumes* and *produces*. A *ResourcePerspective* is *consumed* if it is transformed, according to a conservation law, into one or more *ResourcePerspectives*. Conversely, a *ResourcePerspective* is *produced* if a certain amount of it appears, according to a conservation law, when the *Task* has finished. Thus, the *consumes* link requires the existence of one or more *produces* links. Similarly, a *ResourcePerspective* is *used* if it remains unchanged when the *Task* has finished. The *creates* relationship represents the fact that a new *ResourcePerspective* appears in the domain as a consequence of the *Task* execution and it is applied to those *Resources* that do not satisfy conservation laws. The link *eliminates* is the inverse of the *creates* one. The *task-resource-link* shown in Fig. 4 represents either the *uses*, *employs*, *creates*, *deletes*, *consumes* or *produces* links.

The structure of a *task-resource-link* encapsulates the characteristics of a *ResourcePerspective* that are relevant to a given *Task*. These characteristics are expressed by resorting to the concept of *State*, as previously defined. Fig. 5 shows that a *State* is represented by a particular combination of *SlotDefinition* and *SlotValue*, which is referenced as *StateVar*. A *ResourcePerspective* is associated to its possible states through the *resource-state* link. When a *ResourcePerspective* takes part in a given *Task*, three states need to be identified to express how the *ResourcePerspective* behaves before, during and after participating in the *Task*.

1. The *initial-state* relationship references the state the resource needs to assume in order to participate in the task,
2. The *final-state* relationship references the state the resource assumes when the task has finished,
3. The *intermediate-state* encapsulates the evolution of the resource during the task execution.

TaskMode

Tasks can end up differently, depending on what happens during their execution. In this language, the *TaskMode* concept identifies one possible task ending (Fig. 4). A *Task* is associated with a given *TaskMode* through the *task-mode-link*. The zero or more cardinality at the end of the *TaskMode* construct indicates that each *Task* may have different endings. However, a *TaskMode* is associated to a unique *Task*.

As previously mentioned, the effects of a *Task* are made explicit by a set of one or more *final states* its associated *ResourcePerspectives* may assume. Then, a given *TaskMode* is characterized by the set of ending conditions that the *ResourcePerspectives* associated to such a task assume under a specific mode. As seen in Fig. 4, the *postcondition* link references the final status a *ResourcePerspective* will assume when participating in a task under a specific *TaskMode*. Many *ResourcePerspectives* may participate in a given *Task*. Nonetheless, a *postcondition* link will be established with only a subset of all the resources associated to the task. This subset, that includes the characteristic *ResourcePerspectives*, is the minimal set that allows to distinguish among different *TaskModes*.

Task Version

In general, there is not a unique way to execute a task. Alternative resources can be chosen or alternative solution methodologies can be adopted. The *TaskVersion* construct is introduced in the language to represent a particular way of performing a *Task* under a given *TaskMode* (Fig. 4.). Thus, the *variant* relationship references one of the possible ways a *Task* can be carried out in order to achieve the task final status specified by a certain *TaskMode*.

Since the language has to incorporate facilities to represent tasks at different levels of abstraction, Fig. 4. shows that a *TaskVersion* construct can be decomposed into one or more subtasks by resorting to the *comprises* link. Thus, according to the problem at hand and its required level of detail, a *Task* can be disaggregated into simpler subtasks giving rise to a task structure. The closure of a *Task* structure is represented by the *Body* primitive, which encapsulates a script or a procedure that makes the actual change of state of a set of resources. A *TaskVersion* can be decomposed into either a set of subtasks or a *Body*, but both of them cannot be employed at the same level of detail.

Dynamic models

Dynamic models focus on the behavior of a given resource and on the interactions among a set of resources in a specific scenario. As Fig. 2. shows, two types of models are defined to describe the dynamic behavior of an organization: the State Transition Diagram (STD) which describes the evolution of a unique resource and the State Transition Diagram Net (STDN), showing how different resources interact when they participate in a sequence of tasks. The explicit representation of the states a Resource can assume is very useful for an effective administration of the organization resources. A STD makes possible to know the condition of a resource under a given context, whether it can participate in a specific task or which state it has to reach before taking part in a task.

State Transition Diagram

A *STD* is a digraph in which nodes represent the possible *states* a *ResourcePerspective* can assume and each *Transition* represents the evolution of a resource perspective from a pre-state to a post-state. Figure 5. shows the model of a *STD*. A *STD* is associated to a unique *ResourcePerspective*. In the proposed language, *Tasks* are responsible for changing the state of a *ResourcePerspective*. More specifically, the beginning of a *Task* produces the transition of a *ResourcePerspective* to a new state. Similarly, the end of the task causes its transition to another state. Two types of states can be identified: *ANDStates* and *XORStates*. An *XORState* is a state that represents the behavior of a *Resource Perspective* during the execution of a given *Task*. An *XORState* can be decomposed into different levels of detail according to the dissagregation levels of the *Task* it represents.

It was pointed out before that the structures of a *Task* under a given *TaskMode* are specified by its associated *TaskVersions*. The way a *ResourcePerspective* evolves when it participates in a certain *Task* depends on the particular *TaskVersion* being considered. Thus, a *Resource Perspective* state may abstract many alternative scenarios, each one associated with a particular *TaskVersion*. Therefore, the *XORState* name arises from the fact that this state abstracts alternative decompositions. Each decomposition is a *STD* that represents the *Resource Perspective* evolution under a particular *TaskVersion*. The *Decomp.Alt.* class that appears in Fig. 5. is introduced in the language to represent the decomposition of a state into a given *STD*. Each *Decomp.Alt.* is associated with only one *TaskVersion* and gives rise to a unique *STD*. On the other hand, an *ANDState* cannot be decomposed into *STDs*. It usually represents the state that a *ResourcePerspective* needs to be in, before the task in which the resource is engaged is started. This type of state is also employed to describe the condition a *Resource Perspective* will be in once the task has been finished.

The other component of a *STD* is *Transition*. As a consequence of the *starting* or *ending* of a Task a *ResourcePerspective* changes its state. *BeginTrans* and *EndTrans* reference the transitions that occur when a *Task* is started or finished, respectively. The *Btrigger* links a given *BeginTrans* with the *Task* that causes it. Similarly, the *Etrigger* links an *EndTrans* with a particular *Task*.

State Transition Diagram Net

A *STDN* shows how a set of *ResourcePerspective* interact when they participate in a set of *Tasks*. These interactions are expressed in terms of *States* of the various *ResourcePerspectives* and the *Tasks* that produce specific *Transitions* among them. Thus, *States* and *Transitions* comprise a *STDN*, like they have participated in a *STD* model. Nevertheless, while a STD model concentrates on the states of just one resource perspective, a STDN simultaneously shows all the *ResourcePerspective* states related to the set of *Tasks* that is included in the STDN (Fig. 6.). Since the scope of a *STDN* is given by the set of tasks that it includes, the proposed language incorporates the *bounded-by* relationship to explicitly relate a *STDN* with its associated *Tasks*.

Domain Models

Task and dynamic models are complemented with domain models. A *Domain Model* describes the different entities of the organization, their characteristics and static relationships. Domain models will describe processing plants and business areas, product recipes, enterprise structures, enterprise goals, etc.

Example

To illustrate the proposed language, let us consider a batch process industry that manufactures a set of products in a plant having seasonal demands. A new product grade will be produced for the first time and all the activities associated with such a first time manufacturing need to be modeled to properly coordinate them. Figure 7. depicts a simplified model of the most important activities (gray rectangles) and makes explicit the temporal relationships among them. One of these tasks is exploded in Fig. 8. As seen in this picture, two versions of the task *Prepare Engng. Data* are represented: (i) *Recipe is developed at the corporation* and (ii) *Recipe is licensed*. Due to the lack of space, the model of Fig. 8 is not complete, but it includes some resources (e.g., *Process Engineer, Chemist, Master Recipe*, etc.) and shows their roles in relation to the tasks in which they participate. A detailed model was developed for each of the identified activities. Figures 9 and 10 show partial views of the models associated to the tasks *Develop Process and select ingredients* and *License Recipe*, respectively. The first one is part of the activities comprising the first task version, while the second one comprises the second task version.

Conclusions

The effective administration of a production enterprise requires to properly coordinate all the productive and management activities being executed. Models are invaluable tools to analyze and evaluate the way activities are done. This paper presents a formal modeling language that allows the representation of an organization. To this end, task, domain and dynamic models are proposed. Each one focuses on a specific dimension of the enterprise but all of them are required to get an overall picture of it. The proposed framework is currently being implemented in a CASE environment.

A Task-resource Based Framework for Process Operations Modeling 283

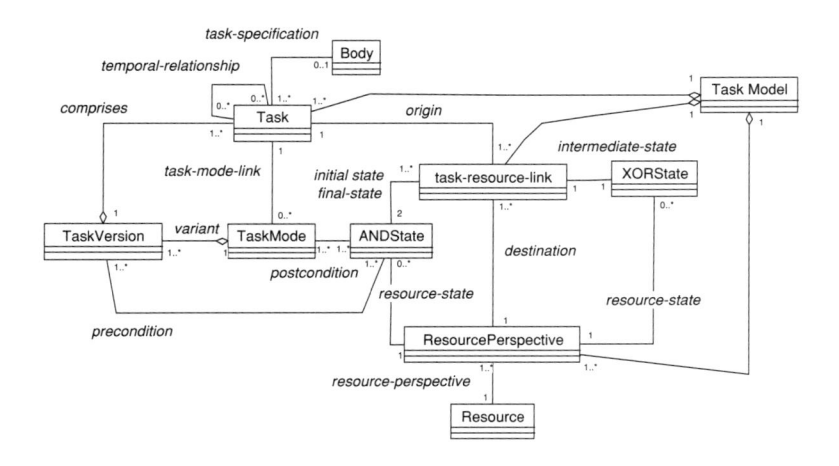

Figure 4. The classes of the Task Model.

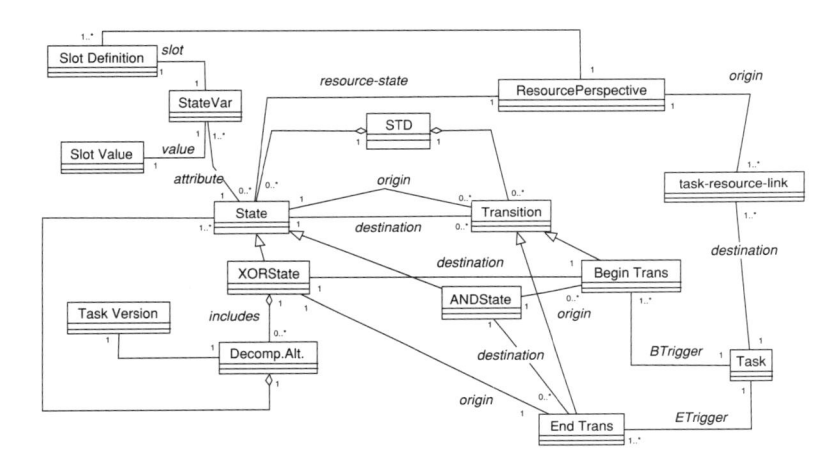

Figure 5. The classes of the STD.

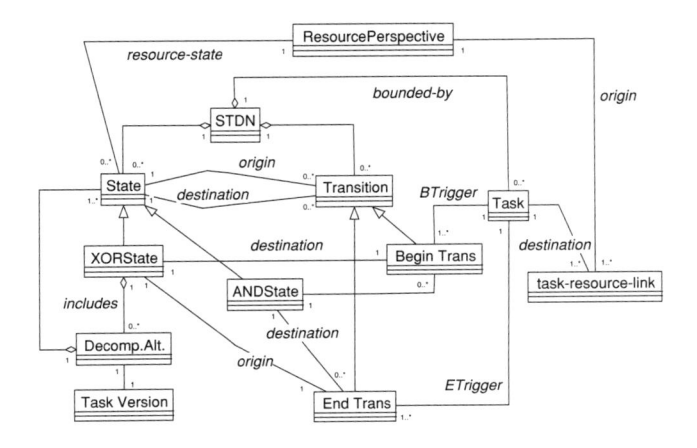

Figure 6. The classes of the STDN.

284　　SCHEDULING

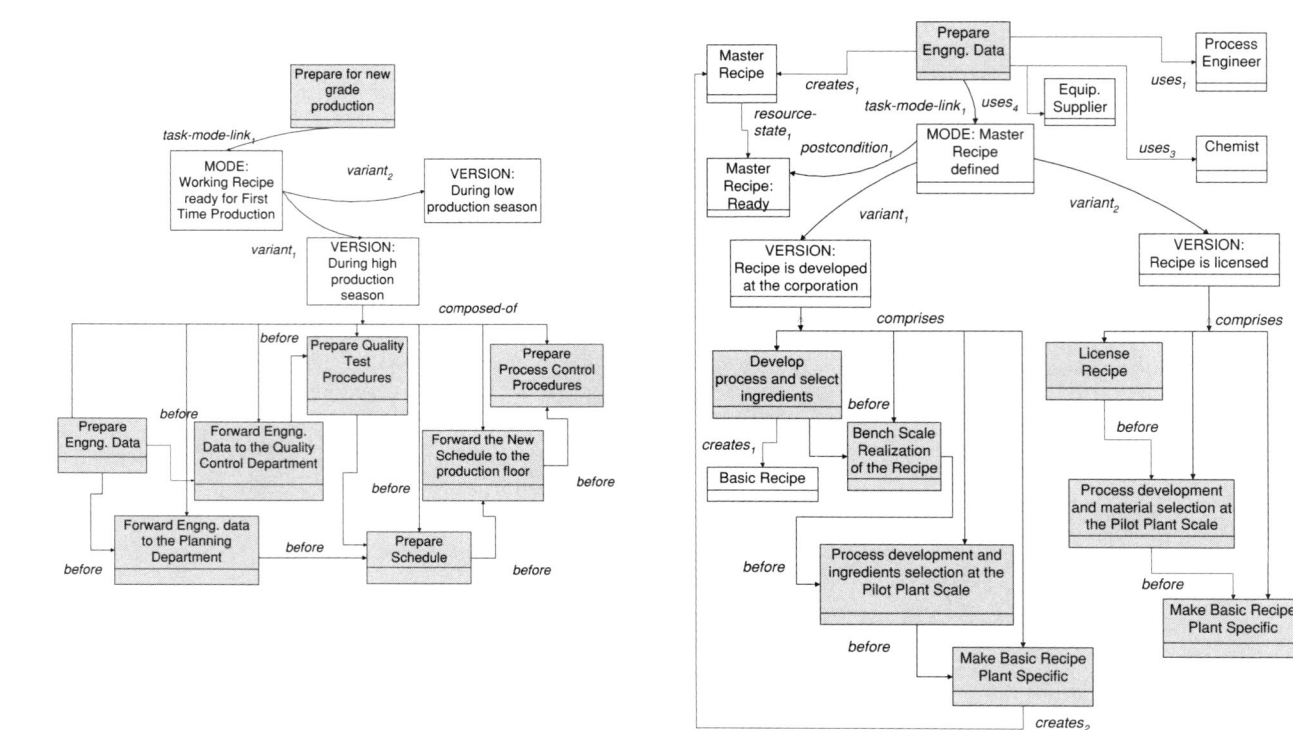

Figure 7. Tasks associated with the manufacture of a new product grade.

Figure 8. Partial view of the task "Prepare Engineering Data".

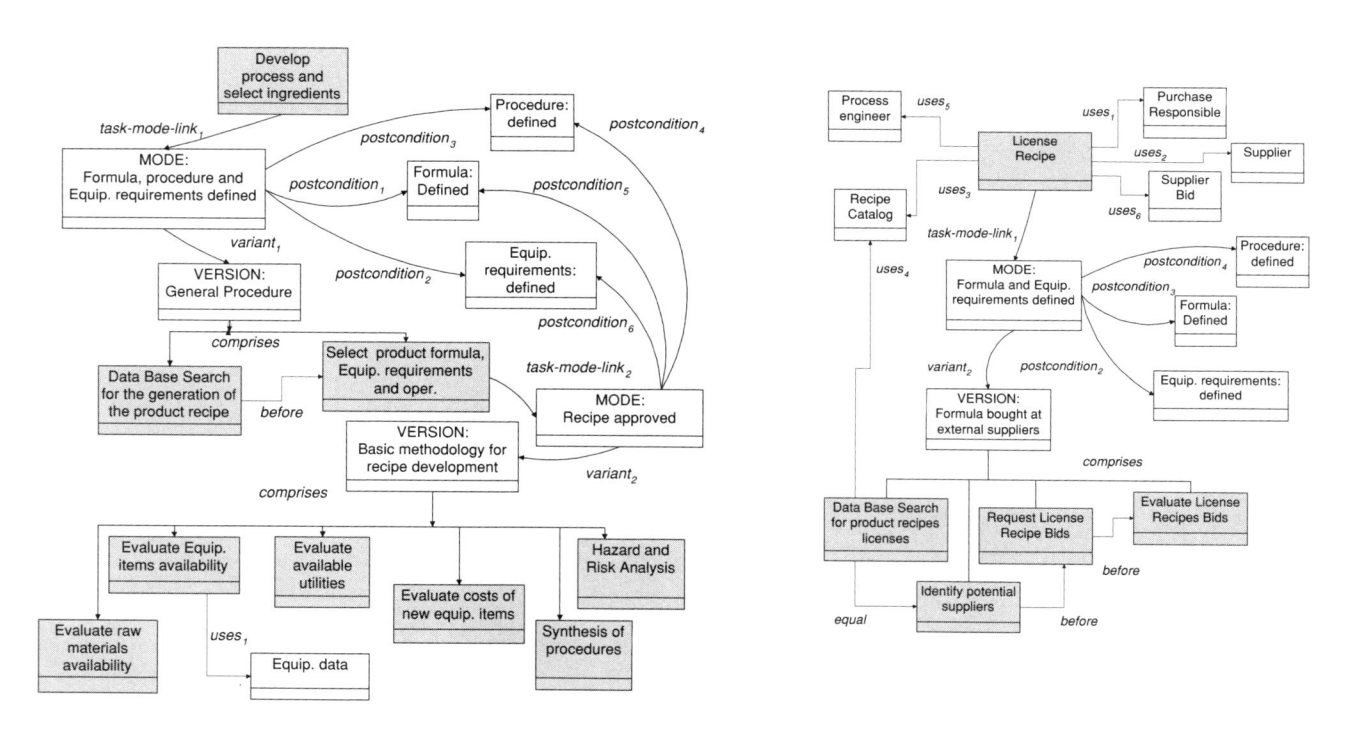

Figure 9. Define the resource requirements for the Basic Recipe specification.

Figure 10. Outsourcing of the Basic Recipe specification.

References

Allen, J. F. (1983). Maintaining knowledge about temporal intervals. *Communications of the ACM*, **26**, 832-843.

Harel, D., and E. Gery (1997). Executable Object Modeling with Statecharts. *IEEE Computer*, **30**, No. 7, 31-42.

Levy, S., E. Subrahmanian, S. Konda, R. Coyne, A. Westerberg, Y. Reich (1993). An overview of the n-dim Environment, EDRC-05-65-93.

Lindheim, C., T. Totland and K. M. Lien (1996). Enterprise Modeling. A new task for Process Systems Engineers? *Computers chem. Engng.*, **20**, S1527-S1532.

Rational Rose (1997). The UML Notation Guide (v. 1.1).

Schreiber, G., B. Wielinga, and J. Breuker (1993). KADS A Principled Approach to Knowledge-Based System Development. Academic Press Inc., Great Britain.

Simensen, J., C. Johnson and K. E. Årzén (1997). A multiple-View Batch Plant Information Model. *Computers chem. Engng.*, **21**, S1209-S1214.

BATCH PRODUCTION SCHEDULING WITH FLEXIBLE RECIPES: THE SINGLE PRODUCT CASE

Moisès Graells, Espen Løberg, Antonio Delgado, Enrique Font and Luis Puigjaner
Chemical Engineering Department
Universitat Politècnica de Catalunya
E.T.S.E.I.B., Diagonal 647, 08028 Barcelona, Spain

Abstract

In this work, batch production scheduling is addressed taking into account the flexibility given by the chance of production recipe adjustment. Operation times and set points are allowed to be manipulated in order to fulfill production objectives. On the other hand, such manipulation is constrained by the inclusion of quality terms in the production objectives, thus originating a new trade-off. Rigorous (NLP) and stochastic (GA) methods have been used in the problem solution. Problem formulation, results achieved and the hypothesis made in both cases are compared and discussed.

Keywords

Batch chemical processes, Scheduling, Flexible recipes, Non-linear programming, Genetic algorithms.

Introduction

Batch production scheduling is a problem commonly addressed by using operation data as a standard problem input. However, optimal resource consumption (time, materials, energy, etc.) for each single operation may not lead to optimum production schedules (delivery times, quality of the products, etc.). Surely, there is certain flexibility associated to the recipe of each product so that a trade-off arises. This flexibility may affect both scheduling and rescheduling so that operation resource consumption at each production step may be adjusted to achieve final production requirements (Puigjaner *et al.*, 1994).

The overall recipe adjustment constitutes a very complex problem. This work addresses the single product production line as a first step towards this end. A scenario is defined in which the relevant properties of products and by-products are summarized by a quality factor. Therefore, starting from the quality of the raw materials, the quality of the intermediate products obtained after stage is given as a function of the quality of the feed and the operation conditions. These conditions are assumed to be controlled by the processing time and an extensive property as energy.

The former may be considered as a theoretical expression for the quality index from which operation set points may be calculated. Furthermore, such a scheme allows the recipe to be modified through control variables t and E and adjusted to meet production objectives taking advantage of the opportunities offered by flexible recipes.

Production objectives are given by different demands to be satisfied (due-dates and product quality). Production costs are given in terms of the energy requirements and stock costs; delivery delay and quality are also taken into account to evaluate the deviation between production achievements and demand requirements. Considering all these features an objective function is proposed to be minimized when determining the optimum production schedule. Since the single product case studied has no assignment or sequence variables, timing adjustment is made depending on the energy supplied within the acceptable product quality limits.

Mathematical Model

The mathematical model formulated to describe this problem is next described. Two basic points cause the model to result in a NLP. On one hand, this first approach is dedicated to the single product case: no binary variables are required since there are no assignment needs or sequencing decisions. Although the single product line is not the general case, it is a common situation in industrial practice. On the other hand, the dependence of the quality factor obtained on the operation time (*TOP*) or the energy (*E*) must clearly be non linear. In this case (Fig.1) normal type functions have been chosen.

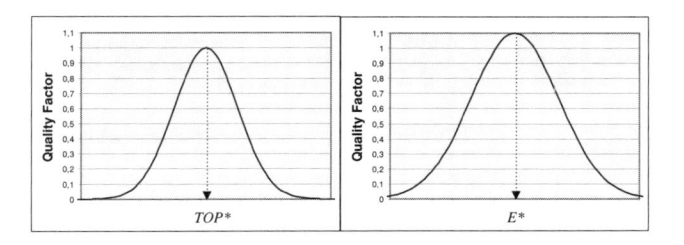

Figure 1. Quality non-linear functions.

The problem is stated as follows. The production line runs *J* production stages for each of the *N* batches to be produced. At each stage $j = 1, 2,... J$ in every production run $n = 1, 2,... N$ operation time TOP_{jn} and energy E_{jn} have to be set so that the output product quality QF_{jn} may be determined from the input product quality $QF_{j-1,n}$. This is illustrated in Fig. 2.

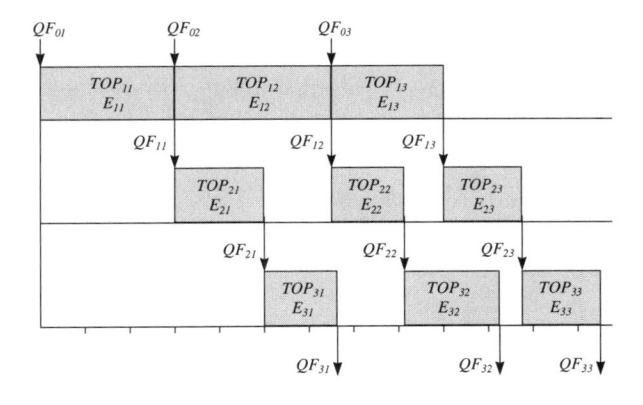

Figure 2. Schedule using flexible recipes.

Thus, given N batches to be produced with due dates DD_n and required quality QD_n the target is to minimize the total resulting costs. The objective function to minimize is given by Eqn. (1) and is made up by the contributions of time deviation cost (Eqn. 2 penalizing both, delay and inventory), quality deviation cost, Eqn. (3) and energy cost, Eqn. (4). Timing constraints are given by Eqn. (5) to Eqn. (8) which limit the values of initial, final, and waiting times (TI_{jn}, TF_{jn} and TW_{jn} respectively) for each operation *jn*. Equations 9 and 10 allow to calculate the contribution to quality factor of both operation time and energy. These contributions, along with the input quality, lead to the quality factor (Eqn. 11) which in this case is assumed to be given by their geometric mean value. Finally, Eqn. (12) and Eqn. (13) give the bounds for the control variables of this problem. Hence:

$$\min Z = C\_TIME + C\_QUALITY + C\_ENERGY \quad (1)$$

$$C\_TIME = \sum_{n=1}^{N} CD_n \left(TF_{Jn} - DD_n\right)^2 \quad (2)$$

$$C\_QUALITY = \sum_{n=1}^{N} CQ_n \left(QF_{Jn} - QD_n\right)^2 \quad (3)$$

$$C\_ENERGY = \sum_{n=1}^{N} \sum_{j=1}^{J} \left(CE_{jn} \, E_{jn}\right) \quad (4)$$

$$TI_{jn} \geq TF_{jn-1} \quad \forall jn \quad (5)$$

$$\begin{cases} TI_{jn} = TF_{j-1,n} & \forall jn \quad \text{for } NIS/FW \\ TI_{jn} \geq TF_{j-1,n} & \forall jn \quad \text{for } UIS \end{cases} \quad (6)$$

$$TF_{jn} = TI_{jn} + TOP_{jn} + TW_{jn} \quad \forall jn \quad (7)$$

$$0 \leq TW_{jn} \leq TW_{jn}^{max} \quad \forall jn \quad (8)$$

$$QT_n = a_j^T + b_j^T \exp\left\{\frac{-\left(TOP_{jn} - TOP_{jn}^*\right)^2}{c_j^T}\right\} \quad \forall jn \quad (9)$$

$$QE_n = a_j^E + b_j^E \exp\left\{\frac{-\left(E_{jn} - E_{jn}^*\right)^2}{c_j^E}\right\} \quad \forall jn \quad (10)$$

$$QF_{jn} = \left(QF_{j-1,n} \cdot QT_{jn} \cdot QE_{jn}\right)^{1/3} \quad \forall jn \quad (11)$$

$$TOP_{jn}^{min} \leq TOP_{jn} \leq TOP_{jn}^{max} \quad \forall jn \quad (12)$$

$$E_{jn}^{min} \leq E_{jn} \leq E_{jn}^{max} \quad \forall jn \quad (13)$$

The NLP problem given by this mathematical model has been solved using two opposite methods: the rigorous approach has been performed by the GAMS/CONOPT package (Brooke *et al.*, 1992) while the stochastic approach has been based on GA (Goldberg, 1989).

Simulation Model and Genetic Algorithms

When the determination of quality parameters as a function of control variables grows in complexity, the difficulty of the mathematical formulation of the problem increases and rigorous solution approach may be given up. Even heuristic approaches may fail if there is no way to find efficient improvement rules. This is another general extreme situation that is usually addressed using general stochastic techniques to guide a systematic trial and error procedures using a simulation model. The optimization of a model applying stochastic techniques needs only a common information structure for the decisions to be made and the system answer (objective function). One of the major advantages of such techniques is their independence of model and objective function structure. Genetic Algorithms (GA) have been chosen for this purpose according to their facility to describe ranges and precision for continuous variables as it will be described.

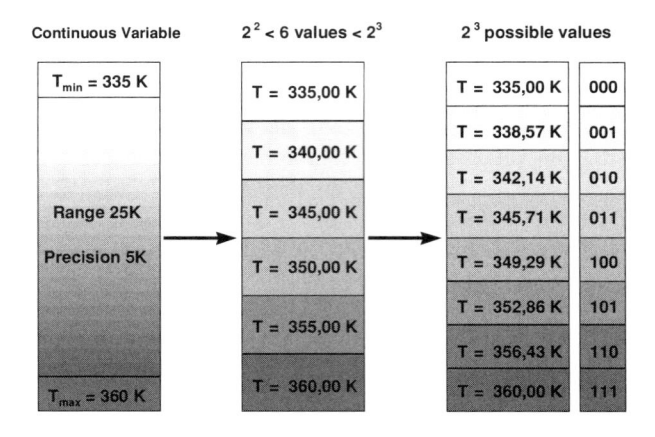

Figure 4. Coding a continuous variable (GA).

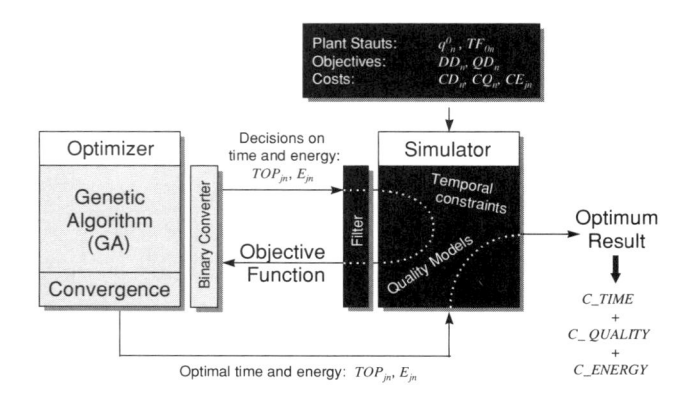

Figure 3. Flexible recipe scheduling using GA optimization of black box model.

The general methodology has been adapted by using the time and energy variables and including a binary filter (Fig. 4). The binary filter is the module for building the chromosomes corresponding to each variable. Later grouping of such chromosomes into a genotype will lead to the information structure describing each system state (individuals) and to their evaluation (phenotype).

Range and Precision

Conversion of continuous variables into chromosomes is the process determining the quality of the solution attained as well as the time cost of the search procedure. The search procedure has to be bounded by using a valid *range* for each variable (maximum and minimum values). Within this range each continuous variable is described by a finite number of binary states. Thus, a *precision* is set according to the quality of the result and the affordable calculation time. Figure 5 shows the binary translation of a continuous variable such as temperature. Precision is automatically adjusted in order to take the maximum advantage of the binary digits with the same calculation effort.

Case Study and Results

A case study is defined by three batches to be produced each one following the same three stage recipe (3x3). This recipe is shown in Table 1. The problem is set once production requirements are defined by due-dates, due-qualities and the quality of the available raw materials (Table 2). Associated unfulfilment costs (CD_n and CQ_n) and energy costs ($CE_1 = CE_2 = CE_3$) defining different scenarios are shown in Table 3 along with the schedules obtained for each one. All schedules are summarized by total amount of time and energy spent (ΣTOP_j and ΣE_j) and the resulting delivery time and quality (TF_{Jn} and QF_{Jn}). Table 3 illustrates the trade off between the different terms involved in the problem (time, energy, quality and associated costs).

Table 1. Parameters for the Three Stage Recipe.

a_j^T	b_j^T	c_j^T	TOP^*_j	a_j^E	b_j^E	c_j^E	E^*_j
0	1	5	10	0.1	1	25	30
0	1	5	20	0.1	1	25	20
0	1	5	30	0.1	1	25	10

Table 2. Raw Material Degree and Production Objectives (Due dates and due qualities).

QF_{0n}	1.0	1.0	1.0
QD_n	0.8	0.8	0.8
DD_n	60	80	100

Table 3. Different Cost Scenarios and Schedules (Lower bounds are underlined).

CE_j	CQ_n	CD_n	ΣTOP_j	ΣE_j	TF_{Jn}	QF_{Jn}	Z
0.01	27	0.01	53.869	54.05	53.87	0.777	
0.01	27	0.01	58.027	53.60	81.90	0.778	3.047
0.01	27	0.01	58.158	51.08	110.05	0.781	
0.01	270	0.01	54.015	54.02	54.02	0.797	
0.01	270	0.01	58.099	54.17	82.11	0.798	3.082
0.01	270	0.01	58.162	52.80	110.28	0.798	
0.01	27	100	57.500	<u>45.00</u>	57.50	0.800	
0.01	27	100	52.500	54.33	80.00	0.024	1284.1
0.01	27	100	52.500	54.33	102.50	0.024	
100	27	0.01	55.147	<u>45.00</u>	55.15	0.723	
100	27	0.01	58.765	<u>45.00</u>	83.91	0.750	13502
100	27	0.01	58.847	<u>45.00</u>	112.76	0.760	

FW policy (TW^{max} = 10) was considered in the scenarios in Table 3. In Table 4, ZW has been considered to compare rigorous mathematical solution procedures with GA. The mathematical programming showed to be very sensible to initial conditions and may be difficult to converge. On the other hand, GA always lead to a solution although (as other stochastic methods such as SA) their performance strongly depends on parameter tuning (number of individuals, mutations, etc.).

Table 4. A ZW Scenario Solved by GAMS and GA.

CE_j	CQ_n	CD_n	ΣTOP_j	ΣE_j	TF_{Jn}	QF_{Jn}	Z
Rigorous NLP (GAMS)							
0.01	27	0.01	53.872	54.02	53.87	0.777	
0.01	27	0.01	58.028	53.58	81.90	0.778	3.047
0.01	27	0.01	58.120	51.89	110.02	0.781	
Genetic Algorithm (GA)							
0.01	27	0.01	54.0	52.8	54.0	0.77	
0.01	27	0.01	56.7	51.8	82.1	0.78	3.100
0.01	27	0.01	57.1	58.8	110.6	0.79	

Conclusions

This work introduces a new sight into the problem of scheduling batch chemical processes. The problem is stated regarding one more level of flexibility such as the recipe. Although deterministic, recipes are considered to be adaptable to production requirements at the expense of products quality. Hence, assuming standard quality models, the problem has been rigorously formulated showing a high degree of flexibility and complexity. However, cases are likely to present more complex models contrasting with a few flexible stages.

Certainly, further work on this direction is needed. There are to basic points to proceed. On one hand, model improvement is required to include more decision levels and to upgrade to the multiproduct and multipurpose cases considering assignment and sequencing opportunities and the associated binary variables.

On the other hand, rescheduling aspects may be included in order to aid process control from an upper decision level. Once finished each stage and the attained quality is determined, the problem may be newly solved so that the redefinition of operation conditions may be used to adjust production runs to fulfill their requirements.

Acknowledgments

Economic support from the CIRIT (Comissió Interdepartamental per la Recerca i la Tecnologia, QFN-4297) is fully appreciated.

Nomenclature

a_j^E	Constant parameter for the energy dependent part of the quality model
a_j^T	Constant parameter for the time dependent part of the quality model
b_j^E	Pre-exponential parameter for the energy dependent part of the quality model
b_j^T	Pre-exponential parameter for the time dependent part of the quality model
CD_n	Penalty cost for due-date deviation (job n)
c_j^E	Constant parameter for the energy dependent part of the quality model
CE_j	Unitary energy cost for stage j
CQ_n	Penalty cost for quality deviation (job n)
c_j^T	Constant parameter for the energy dependent part of the quality model
DD_n	Due-date for job n
E_{jn}	Energy dedicated to stage j in job n
E_{jn}^*	Standard energy in the energy dependent part of the quality model
E_{jn}^{max}	Upper bound for energy (stage j, job n)
E_{jn}^{min}	Lower bound for energy (stage j, job n)
j	Stage number (j = 1,...J)
n	Job number (n = 1,...N)
QD_n	Required quality for job n
QE_{jn}	Energy contribution function to quality index (stage j, job n)
QF_{jn}	Quality index obtained after stage j in job n
QT_{jn}	Processing time contribution function to quality index (stage j, job n)
TF_{jn}	Final time for stage j in job n
TI_{jn}	Initial time for stage j in job n
TOP_{jn}^*	Standard processing time in the time dependent part of the quality model
TOP_{jn}	Operation time for stage j in job n
TOP_{jn}^{max}	Upper bound for operation time (stage j, job n)
TOP_{jn}^{min}	Lower bound for operation time (stage j, job n)
TW_{jn}	Waiting time for stage j in job n
TW_{jn}^{max}	Upper bound for waiting time (stage j, job n)
Z	Objective function: Total cost.

References

Brooke, A., D. Kendrik and A. Meeraus (1988). *GAMS: A User's Guide*. Scientific Press, Palo Alto.

Goldberg, D. E. (1989). *Genetic Algorithm in Search, Optimization and Machine Learning*. Addison-Wesley, New York.

Puigjaner, L., A. Huercio, and A. Espuña (1994). Batch production control in a computer integrated manufacturing environment. *J. Proc. Cont.*, 4, 281-290.

SOLVING TRIM-LOSS PROBLEMS WITH VARIABLE RAW-PAPER AND TRIM-LOSS WIDTHS

Iiro Harjunkoski and Tapio Westerlund
Department of Chemical Engineering
Åbo Akademi University
Biskopsgatan 8, FIN-20500 Turku, Finland

Abstract

One great problem in the paper-converting industry is the trim-loss problem. The problem appears when a customer demand is to be satisfied in a paper-converting mill by cutting a set of product paper reels from raw-paper reels. In its natural form the problem contains integer variables and therefore results in a difficult combinatorial problem. The paper-converting mill needs to adapt itself between the widths specified by customers and the raw-paper widths that are delivered from a raw-paper mill which makes it difficult to avoid material losses during the process. In the present paper, a formulation that solves the trim-loss problem using a variable raw-paper width is presented. Some efficient strategies to find the best width interval will also be discussed.

Keywords

Optimization, Mixed integer linear programming, Trim-loss problems, Scheduling problems, Paper converting.

Introduction

The most common process operations in a paper-converting mill are the coating, printing and cutting of raw paper. To make the production as efficient as possible, it is of interest to do the coating and printing first and the cutting as the last operation. This is owing to the fact that the rolling speeds of machines are often fixed and therefore a wider reel contributes to an increase of the production capacity of the machinery. There are, however, certain restrictions that should be considered. One restriction is for example the widths of the available machines. Another restriction is the number of product reels that can be cut out at one time. The most suitable widths for a converting mill are widths as wide as possible that fit into the machines and the chosen cutting pattern widths.

The cutting procedure is the critical part of the process, for most of the material losses appear there. The main objective is to cut the available raw paper such that customer demands are fulfilled and as little waste, trim loss, as possible, is produced. In real production there also are other important objectives, such as making production fast and economical. The customers often require the order to be delivered in paper reels with specified diameters. While raw-paper reels are often attached together to form an endless reel, the length is less interesting than the width of the reels. Consequently, the most essential matter is how to do the cutting or determine the knife positions. Each alternative of placing the knives is called a cutting pattern. Figure 1, where 5 product reels are cut out from the raw-paper reel, illustrates the simplified process.

In practice there appears always an inevitable trim loss that could be recirculated into the production when appearing in the raw-paper mill. Normally the paper is coated or printed before cutting in a paper-converting mill and cannot therefore be directly taken back into preceding production steps. Thus, the waste has to be either burned or processed in some waste treatment plant which may sometimes be rather expensive.

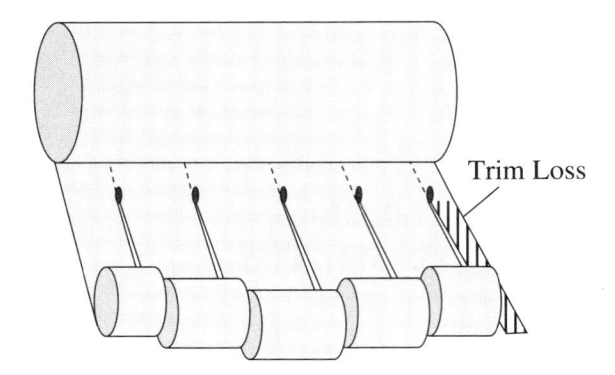

Figure 5. The cutting process.

The appearance of the trim-loss often arises from the fact that raw-paper widths are fixed. One way to improve the waste problem is to introduce more flexible raw-paper widths. This could be done for instance by solving the raw-paper width as a variable or by utilizing some other strategies that make it possible to specify more freely the final width for the raw paper. This kind of strategy would of course demand more flexibility by the raw-paper producer. Therefore, a key issue is the motivation of the paper mill to co-operate in moving part of the inevitable loss from the converting mill to its own production. Here some pricing politics where the paper-converting mill somehow compensates the loss that has been moved upwards in the production could be applied. An even better alternative would be to integrate the converting process with the cutting at the paper machine and try to minimize the loss also at the paper machine end.

An integration requires either some sequential procedure that solves alternating problems or an enlarged problem. This may further complicate already a difficult combinatorial problem since parts of the paper-mill process have to be considered. Owing to the fact that a paper mill often has several customers, the schedule may not allow it to produce just for one customer. In order to simplify the problem we assume in this paper that this is not the case. It should, however, be noted that a kind of "campaign" production where the whole paper-mill production is directed to a certain customer is normally not possible in practice because of the restricted storage capacity in a converting mill and the big differences in the production volumes between a paper mill and a converting mill.

Methodology

The general trim-loss problem can be formulated as an integer nonlinear programming (INLP) problem as follows.

$$\min_{m_j, n_{ij}, y_j} \left\{ \sum_{j=1}^{J} c_j \cdot m_j + C_j \cdot y_j \right\} \quad (1)$$

subject to

$$\sum_{i=1}^{I} b_i \cdot n_{ij} - B_{j,\max} \leq 0 \quad (2)$$

$$-\sum_{i=1}^{I} b_i \cdot n_{ij} + B_{j,\max} - \Delta_{j,\max} \leq 0 \quad (3)$$

$$\sum_{i=1}^{I} n_{ij} - N_{j,\max} \leq 0 \quad (4)$$

$$y_j - m_j \leq 0 \quad (5)$$

$$m_j - M_j \cdot y_j \leq 0 \quad (6)$$

$$j = 1, \ldots, J$$

$$n_{i,order} - \sum_{j=1}^{J} m_j \cdot n_{ij} \leq 0 \quad (7)$$

$$i = 1, \ldots, I$$

$$m_j, n_{ij} \in Z^+ \quad y_j \in \{0,1\}$$

The above objective function minimizes beyond raw-paper usage, also the number of pattern changes. All constraints, except the last one, are linear. The constraint (7) is non-convex owing to its negative bilinear terms. It should be noted that all variables in the bilinear term are integer variables. Furthermore, there are no continuous variables in the basic problem. A pattern is represented by the variables n_{ij} that express how many products i are cut out from the raw paper using the pattern j. The variables m_j give the number of times a pattern j is repeated in the process. The parameter $B_{j,max}$ indicates the raw-paper width, i.e. the maximum width of a cutting pattern and the the parameter $\Delta_{j,max}$ indicates the maximal allowed trim-loss width. Consequently, these parameters are used for defining the width limits of each pattern. The parameter $N_{j,max}$ limits the number of product reels that can be cut out from the raw-paper reel. The limit is machine specific. The variable y_j is a binary variable that is equal to one if the pattern j is used.

The solution of a non-convex INLP problem is very demanding and can only be done by a few methods that are under development (Adjiman et al., 1997; Ryoo and Sahinidis, 1995; Smith and Pantelides, 1997). Nevertheless, the problem can be formulated using a linear strategy similar to Harjunkoski et al. (1996) and solved as a linear mixed integer linear programming (MILP) problem using the Branch & Bound algorithm. Other strategies are, of course, also possible (Harjunkoski et al., 1998) but in this paper the main interest is focused on the linear strategy.

The problem can be written in linear form by first expressing one of the integer variables in bilinear terms with binary variables and then introducing a slack variable that obtains the value of the other integer variable.

Thus, the variable m_j, expressing the number of times a certain cutting pattern is repeated, can be replaced as follows

$$m_j = \sum_{k=1}^{K} 2^{k-1} \cdot \beta_{jk} \qquad (8)$$

Then the n_{ij} variables defining the appearance of each pattern are replaced with slack variables through the following constraints

$$s_{ijk} - n_{ij} \leq 0 \qquad (9)$$

$$-s_{ijk} + n_{ij} - L_{ij} \cdot (1 - \beta_{jk}) \leq 0 \qquad (10)$$

$$s_{ijk} - L_{ij} \cdot \beta_{jk} \leq 0 \qquad (11)$$

$$i = 1, \ldots, I \quad j = 1, \ldots, J \quad k = 1, \ldots, K$$

$$\beta_{jk} \in \{0,1\} \quad n_{ij} \in Z^+ \quad s_{ijk} \in R^+$$

When using this strategy, one main improvement to Harjunkoski et al. (1996) needs to be done. The linear strategy may have a huge integer gap unless the constraints (2-4) are modified. The change is made by first multiplying the constraints by m_j and then replacing the variables m_j and n_{ij} with their substitutes.

By using the new representations the constraints (2-4) can be replaced by

$$\sum_{i=1}^{I} b_i \cdot \sum_{k=1}^{K} 2^{k-1} \cdot s_{ijk} - B_{j,max} \cdot \sum_{k=1}^{K} 2^{k-1} \cdot \beta_{jk} \leq 0 \qquad (12)$$

$$-\sum_{i=1}^{I} b_i \cdot \sum_{k=1}^{K} 2^{k-1} \cdot s_{ijk} + (B_{j,max} - \Delta_j) \cdot \sum_{k=1}^{K} 2^{k-1} \cdot \beta_{jk} \leq 0 \qquad (13)$$

$$\sum_{i=1}^{I} \sum_{k=1}^{K} 2^{k-1} \cdot s_{ijk} - N_{j,max} \cdot \sum_{k=1}^{K} 2^{k-1} \cdot \beta_{jk} \leq 0 \qquad (14)$$

In Westerlund et al. (1998) a parameterization method is used where feasible cutting patterns are generated in advance. The use of this strategy may enforce a considerable increase of variables when the problem is expanded to cover several raw-paper widths. The most critical issue is the size of the defined width region owing to the number of possible combinations. If the raw-paper widths overlap each other i.e. the predefined patterns can be obtained from different raw-paper widths, an extra index r is needed for those pattern variables (m_j, y_j) to denote the raw paper. Such an addition of an extra dimensionality easily multiplies the number of integer variables.

In this paper we use a linear transformation and do not need to add extra indices to the existing integer variables. Instead, we act on the variables considering the widths.

Formulation 1

In the following, different alternatives for generating more flexible formulations are considered. In the first strategy, the width parameter $B_{j,max}$ is simply replaced by a continuous variable, B. Otherwise, the only change that needs to be made is to provide upper and lower bounds for the variable B. Note, however, that the improvements (12-13) cannot be directly applied here while it would result in new bilinearities.

$$B_{j,min} \leq B \leq B_{j,max} \qquad (15)$$

$$B \in R^+$$

The width constraint (2-3) should be modified to allow unused, empty patterns ($n_{ij} = 0$).

$$\sum_{i=1}^{I} b_i \cdot n_{ij} - B \leq 0 \qquad (16)$$

$$-\sum_{i=1}^{I} b_i \cdot n_{ij} + B + B_{j,max} \cdot y_j \leq B_{j,max} + \Delta_{j,max} \qquad (17)$$

Formulation 2

The second alternative changes the allowed trim-loss, $\Delta_{j,max}$ also to a variable Δ_j. An upper bound needs to be specified and it is also recommended to include this variable into the objective, because this deletes a lot of multiple solutions and contributes to the search of the best raw-paper width that causes the least amount of waste. However, the weighting factor of such a supplement should not be too big in order not to dominate over other important issues, such as the number of patterns needed. Let the coefficient for the trim-loss term be γ. The objective, the bounds and the modified constraint (3) are

$$\min_{m_j, n_{ij}, y_j, B, \Delta_j} \left\{ \sum_{j=1}^{J} c_j \cdot m_j + C_j \cdot y_j + \gamma \cdot \Delta_j \right\} \qquad (18)$$

$$0 \leq \Delta_j \leq \Delta_{j,max} \qquad (19)$$

$$-\sum_{i=1}^{I} b_i \cdot n_{ij} + B - \Delta_j + B_{j,max} \cdot y_j \leq B_{j,max} \qquad (20)$$

$$\Delta_j, B \in R^+$$

Formulation 3

In the third alternative a number of fixed raw-paper widths, B_k are used. The formulation is useful for situations where only a few possible widths can be used. This may be the situation when urgent orders have to be satisfied using e.g. material from stock. The width constraints (16-17) are also valid for this case and some extra binary variables are needed for selecting only one of the available widths.

$$B - \sum_{k=1}^{K} B_k \cdot w_k = 0 \qquad (21)$$

$$\sum_{k=1}^{K} w_k \leq 1 \qquad (22)$$

$$B \in R^+ \quad w_k \in \{0,1\}$$

This strategy can also be applied with variable allowed trim-loss.

Full formulation

Now the full formulation for solving the trim-loss problem with a variable raw-paper width and a variable maximum trim-loss width can be written.

$$\min_{m_j, n_{ij}, y_j, B, \Delta_j} \left\{ \sum_{j=1}^{J} c_j \cdot m_j + C_j \cdot y_j + \gamma \cdot \Delta_j \right\} \quad (23)$$

subject to

$$\sum_{i=1}^{I} b_i \cdot n_{ij} - B \leq 0 \quad (24)$$

$$-\sum_{i=1}^{I} b_i \cdot n_{ij} + B + B_{j,\max} \cdot y_j \leq B_{j,\max} + \Delta_{j,\max} \quad (25)$$

$$\sum_{i=1}^{I} n_{ij} - N_{j,\max} \cdot y_j \leq 0 \quad (26)$$

$$y_j - m_j \leq 0 \quad (27)$$

$$m_j - M_j \cdot y_j \leq 0 \quad (28)$$

$$j = 1, \ldots, J$$

$$n_{i,order} - \sum_{j=1}^{J} m_j \cdot n_{ij} \leq 0 \quad (29)$$

$$i = 1, \ldots, I$$

$$B_{j,\min} \leq B \leq B_{j,\max} \quad (30)$$

$$0 \leq \Delta_j \leq \Delta_{j,\max} \quad (31)$$

$$m_j, n_{ij} \in Z^+ \quad y_j \in \{0,1\}$$

$$B, \Delta_j \in R^+$$

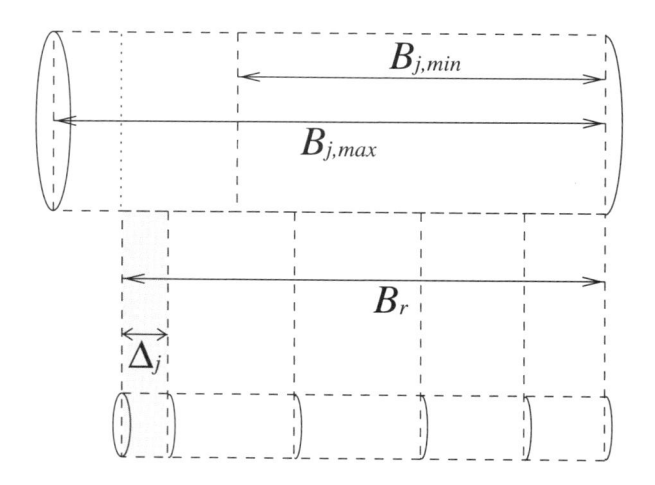

Figure 2. The new variables and limits.

This formulation is general and non-convex. The earlier linear transformation strategy can be used and the constraint (26) may be replaced by Eq. (14). Figure 2 illustrates the new variables.

All of the formulations can also be used with more than one variable raw-paper width. In such a case, the variable B needs to be replaced by a variable B_r. In the width constraints, the variable B is replaced by a slack-variable, s_{jr}, which can be defined through 3 constraints and one binary variable z_{jr}. Assume in the following example that there are R different raw-paper widths, B_r, available. The width constraints (2-3) must be rewritten as

$$\sum_{i=1}^{I} b_i \cdot n_{ij} - \sum_{r=1}^{R} s_{jr} \leq 0 \quad (32)$$

$$-\sum_{i=1}^{I} b_i \cdot n_{ij} + \sum_{r=1}^{R} s_{jr} - \Delta_{j,\max} + B_{j,\max} \cdot y_j \leq B_{j,\max} \quad (33)$$

and the slack-constraints are

$$s_{jr} - B_r \leq 0 \quad (34)$$

$$-s_{jr} + B_r - B_{j,\max} \cdot (1 - z_{jr}) \leq 0 \quad (35)$$

$$s_{jr} - B_{j,\max} \cdot z_{jr} \leq 0 \quad (36)$$

$$j = 1, \ldots, J \quad r = 1, \ldots, R$$

$$\sum_{r=1}^{R} z_{jr} \leq 1 \quad (37)$$

$$j = 1, \ldots, J$$

$$B_r, s_{jr} \in R^+ \quad z_{jr} \in \{0,1\}$$

The change of raw paper is, however, very expensive because it causes a production stop. Therefore it should be avoided. The straightforward way is to form the objective function such that a term related to the number of different raw-paper reels is heavily weighted.

Bounds and Constraints

The problem consists of several integer variables and it is therefore very important to reduce the combinatorial space by specifying tighter bounds and constraints. One technique that often improves the verification times is to diminish the number of multiple solutions. The following constraints contribute to a better optimization performance.

To reduce the number of multiple solutions the patterns can be sorted by their m_j values. It may also be beneficial to sort the y_j variables. This can easily be done by the following constraints

$$m_{j+1} - m_j \leq 0 \quad (38)$$

$$y_{j+1} - y_j \leq 0 \quad (39)$$

$$j = 1, \ldots J - 1$$

The sorting can alternatively be done by pattern widths. By further modifying the coefficients for y_j in the objective function, some improvements can be made. Instead of the coefficient c_j it may be better to use $j \cdot c_j$ which forces the first patterns to be active.

By possibly tightening the individual $N_{j,max}$ values some further constraints may be effective. A lower bound for the sum of patterns m_j can be calculated from the order information.

$$N_{j,\max} = \min\left\{N_{j,\max}, \left\lfloor \frac{B_{j,\max}}{\min_i\{b_i\}} \right\rfloor\right\} \quad (40)$$

$$\sum_{j=1}^{J} m_j \geq \max\left\{\left\lceil \frac{\sum_{i=1}^{I} n_{i,order}}{N_{j,\max}} \right\rceil, \left\lceil \frac{\sum_{i=1}^{I} n_{i,order} \cdot b_i}{B_{j,\max}} \right\rceil\right\} \quad (41)$$

In a similar way, an upper bound for the sum is given, provided that the upper production limits $n_{i,max}$ have been defined.

$$\sum_{j=1}^{J} m_j \leq \left\lfloor \frac{\sum_{i=1}^{I} n_{i,\max} \cdot b_i}{B_{j,\min} - \Delta_{j,\max}} \right\rfloor \quad (42)$$

Example Problem

The problem can finally be formulated using the linear transformation strategy and some additional constraints to improve the optimization performance.

As an example, the following medium-size customer demand is solved.

A main problem in solving trim-loss problems is to choose the raw-paper width. If the bounds for the selected width are too narrow or otherwise unsuitable, even the global optimal solution may not be satisfactory, owing to some poor production parameters.

In the following, we assume the following parameters for all of the formulations: $J = I = 4$, $c_j = 1$, $C_j = 0.1 \cdot j$, $\Delta_{j,max} = 200$ mm and $N_{j,max} = 5$. The $n_{i,max}$ given in the order is only used for generating tighter bounds for the problem.

Table 1. Example Customer Order.

Width (mm)	$n_{i,order}$	$n_{i,max}$
290	15	18
315	28	32
350	21	24
455	30	34

The most difficult problem that appears with the variable raw-paper width is that for a wide possible range $[B_{j,min}, B_{j,max}]$ the tightening constraints may become looser making the problem harder to solve. Therefore, it would be desirable to define not too big a region within which the raw-paper width may vary freely. In the example this is not done. Instead, a very big interval is accepted and the same tightening constraints are used to demonstrate the fact that even if the problem is expanded, only the freeing of the raw-paper width eases up certain combinatorial problems.

For the default case, a raw-paper width of 1800 mm is used. The variable raw-paper should be between 1450 and 2275 mm. These bounds were obtained by multiplying the $N_{j,max}$ with the narrowest and the widest products. The fixed widths that are used in the last case are {2100, 2000, 1900, 1800, 1700} mm. For the cases with a variable maximum trim loss, the coefficient $\gamma = 0.001$ is used.

Case 1 in the following table represents the revised formulation from Harjunkoski et al. (1996). The rest of the cases are ordered according to this paper. Case 2 uses a variable raw-paper width, case 3 considers both the raw-paper and the maximum trim-loss widths as variables. In case 4, a set of fixed raw-paper widths are used and in the last case 5, the fixed raw papers are combined with a variable upper limit for the trim loss.

Table 2. Solution Data for the Example Order.

Case	Obj	Nodes	CPU-s	Loss (mm)
1	20.6	9997	64.08	1515
2	19.6	4429	59.69	1450
3	19.625	6001	94.87	145
4	19.6	7391	96.04	1640
5	19.695	12072	180.96	1475

The loss is expressed as the width of a reel containing all the trim loss. When comparing the results, it should be noted that the objective was not to minimize the total loss. Only in objectives 3 and 5 is the maximal trim-loss width considered. The best raw-paper width was 1830 mm which in case 3 resulted in a trim loss of 0.42%. Note, that the corresponding trim loss for case 1 (1800 mm) was 4.2%. There is also a number of multiple solutions to the problem. For instance, solution 3 is fully valid to case 2 as well.

Significance

The resulting problems were solved with the Branch & Bound algorithm using CPLEX-5.0 with default options on a Pentium Pro 200 MHz running Linux. A lot of effort needed to be put into reducing the combinatorial field and, thus, trying find the optimal solution within a reasonable time. The straightforward way to do this was to introduce new constraints and tighten the former bounds by carefully analyzing the problem. It should be noted that using the strong branching strategy (ILOG Inc., 1997), significantly changed the order of the CPU-times (249.97, 46.30, 57.96, 41.94, 142.85 for cases 1-5).

Also some heuristics were used in defining the bounds and reducing the required number of integer variables, with awareness of the risk of excluding the global optimum from the search space.

The use of MILP methods that provided a global optimal solution to the defined problem made a significant improvement to the amount of trim loss. The formulation with a variable raw-paper width could be expected to compete with other practical MILP-formulations, while already a small degree of freedom may lead to a considerable improvement. This was illustrated in the example case where the chosen raw-paper width, 1800 mm, was replaced by the width, 1830 mm. For solving large-scale problems, further simplification will definitely be needed because of the fact that this methodology involves the combinatorial expansion of an already big problem.

Nomenclature

m_j	the multiple of cutting pattern j used
n_{ij}	number of a product i in pattern j
y_j	binary variable to indicate whether a pattern j is used or not
c_j	cost of a raw-paper reel j
C_j	cost of changing the cutting pattern j
$n_{i,order}$	number of product i ordered
$n_{i,max}$	upper bound of product i ordered
b_i	the width of product i
$B_{j,min}$	the min. width allowed for a pattern
$B_{j,max}$	the max. width allowed for a pattern
$N_{j,max}$	max. number of products that can be cut at a time
$\Delta_{j,max}$	the width tolerance of a cutting pattern
M_j	an upper bound for m_j
B, B_r	the variable raw-paper width
B_k	a fixed raw-paper width
Δ_j	the trim loss on pattern j
γ	the cost of the trim loss
s_{jr}	a slack variable for raw paper
z_{jr}	a binary to denote if raw paper r is used for pattern j

References

Adjiman C.S. and Androulakis I.P. and Floudas C.A. (1997). Global Optimization of MINLP Problems in Process Synthesis and Design. *Computers chem. Engng,* **21**, Suppl. pp. S445-S450.

Harjunkoski I., Westerlund T., Isaksson J. and Skrifvars H. (1996). Different formulations for solving trim loss problems in a paper converting mill with ILP. *Computers chem. Engng,* **20**, Suppl. pp. S121-S126.

Harjunkoski I., Westerlund T., Pörn R. and Skrifvars H. (1998). Different transformations for solving non-convex trim-loss problems by MINLP. *European Journal of Operational Research,* **105**, pp. 594-603

ILOG, Inc. CPLEX Division (1997). Version 5.0 of "Using the CPLEX Callable Library"

Ryoo H.S. and Sahinidis N.V. (1995). Global Optimization Nonconvex NLPs and MINLPs with Applications in Process Design. *Computers chem. Engng,* **19**, pp. 551-566

Smith E.M.B. and Pantelides C.C. (1997). Global Optimization of Nonconvex MINLPs. *Computers chem. Engng,* **21**, Suppl. pp. S791-S796.

Westerlund T., Isaksson J. and Harjunkoski I. (1998). Solving a two-dimensional trim-loss problem with MILP. *European Journal of Operational Research,* **104**, pp. 572-581

CLIFFTENT: DETERMINING FULL FINANCIAL BENEFIT FROM IMPROVED DYNAMIC PERFORMANCE

Pierre R. Latour
SR2
Houston, TX 77079

Abstract

This paper describes the new rigorous nonlinear method to properly quantify financial benefits from improved dynamic performance, smoother operation, reduced variance and improved quality control. CLIFFTENT shows how to quantify missing intangibles, often doubling the quantified benefit. It also connects closed-loop steady-state real-time optimization (CLRTO) to constrained multivariable predictive control (CMVPC) and provides proper values for dependent variable limits, specs and safety margins which usually have much greater influence on profit improvement than model accuracy and controller performance. CLIFFTENT describes new modeling requirements, improves risk management and inspires CMVPC research with the proper objective function: profit. It incorporates statistical uncertainty with rigorous process models, combining SPC to CMVPC. This performance measurement technique allows CIM technology licensing based on value added performance for the first time. Commercial applications to petroleum refining to manufacture clean fuels are given. Benefits exceeding 0.18 $/bbl crude x 200 kbpd were identified from CMVPC of a four draw distillation unit.

Keywords

Dynamic performance measures, Limit setting, Financial benefits, Profit function, CIMFUELS.

Introduction

While there is no doubt people prefer low variability in many things like Wall Street stock values, the major impediment to profitability of the process control business is lack of a rigorous method for quantifying the financial benefit from improved dynamic performance. A corollary is lack of a rigorous method for accounting for uncertainty and spec violation penalties to properly set dependent controlled variable targets or setpoints. A second corollary is lack of a rigorous method for distinguishing the significance among all the controlled dependent variables of multivariable controllers. A third corollary is difficulty in identifying and selecting dependent controlled variables in the first place; recognizing one when we see it. The quality culture revolution of W. E. Deming and J. Juran in the 1980's lacked a rigorous method to quantify the financial value of improved quality, reduced variance and tighter tolerances.

The recent development of the CLIFFTENT function and method by Latour (1992a, 1996) finally allows these problems to be solved and provides guidance for new research directions in optimization and process control.

Situation

People determine the value of things. People care about some things more than others, for different reasons. People set the financial importance of operating variables in HPI plants. People are more interested in some variables than others. People like some variables to be controlled smoothly, they do not like upsets or deviations. They prefer steady process operating conditions. However they do not have a clear method for determining the financial value of their likes and desires for dynamic process control. They have difficulty quantifying its profitability.

With the correct performance measure we can finally keep score on the value of dynamic process control. Further, we need a formal procedure for properly, optimally setting operating limits and targets, a central issue of Constrained Multivariable Predictive Control (CMVPC), Closed - Loop Real Time Optimization (CLRTO), LP planning and scheduling for competitive operation of HPI plants. The key is CLIFFTENT by Latour (1996) and its singular role in defining the performance problem and its solution.

Spending for advanced process control packages and services in 1993 was reported by HPI (1993) for fuels refining to be at rates of 0.018 $/bbl crude oil in the US and 0.017 $/bbl crude in ROW. For the total HPI (fuels, petrochemicals and gas processing) spending was at rates of 0.047 $/bbl crude in the US and 0.040 $/bbl crude outside the US. The benefit potential for fuels refining was reported by Latour (1992, 1994) to exceed 0.4 $/bbl crude and for total HPI probably exceeds 1.0 $/bbl crude. With the world refining about 75 million bpd, it would appear important to identify, measure, capture, sustain and enhance such value properly.

Control variables (CV) are dependent response variables like temperatures, some resulting flows, levels, pressures, compositions, qualities (TFLPCiQ), coking, corrosion, catalyst activity, machine speed, fouling, approach to distillation flood or weep, compressor surge or stall, pump cavitation, two phase separator carryover, coke drum overfill and flaring. CV's are dynamic process response dependent variables we care about, attempt to measure or deduce and are amenable to feedback or feedforward control by adjustment of other directly manipulatable independent variables (MV) like valves and motors

The profitability of CMVPC is primarily determined by proper setting of CV limits. If they are set too tight, narrow, conservative and constraining the controller is almost inactive and generates little value. If they are set too loose, wide, liberal and open beyond the validity domain of the model where big nonlinearities arise, improper bouncing into dangerous or unstable regions generates losses. CV limits should be set by people "just right".

CV are also dependent variables in LP/SQP models. These models are built to predict CV from MV and independent measured and unmeasured Disturbance Variables described by Latour (1979) and Georgiou (1998). LP solutions lie at an intersection of CV limits and MV constraints in n-space for n-MV. LP is a corner picker in n-space, illustrated in Fig. 1. While nonlinear SQP may find an interior top of a hill optima, it is invariably partially (highly partially) constrained at a partial hill corner in n-space because optimum operation of HPI plants is at a combination of CV and MV constraints and limits of equipment. SQP is a corner - hill picker. The art of LP planning and CLRTO goes beyond getting the process model slopes for capacity, yield and operating costs right. It is critical to set dependent variable CV limits right. The profitability of LP planning and CLRTO is primarily determined by proper setting of CV limits. Set them too tightly and profit improvement is invariably small; set them too widely and the solution exceeds the domain of model validity, the solution is not physically realizable and profit is lost. LP and CLRTO do not set CV limits, people do; they should be set "just right".

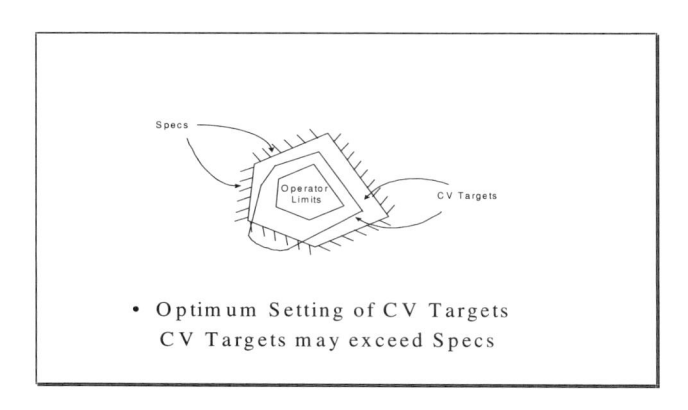

- Optimum Setting of CV Targets
  CV Targets may exceed Specs

Figure 1. Process operating region - dependent controlled variables.

The profitability of an operator, manager or plant is primarily determined by proper setting of operating CV limits. If set too tight with large safety margins and big quality giveaways, yields suffer, operating costs are excessive and capacity is curtailed so the plant is inefficient and uneconomic. If set too loose with inadequate safety margins and excessive quality violations, the plant becomes unreliable, unsafe and uneconomic. A basic job of plant management is regular assessment of the proper tradeoff balance between safe operating margins and technical competitiveness. Setting CV limits properly lies at the heart of decisions between the operating and technical departments in every HPI plant, Latour (1996).

Transients

Process control benefits are traditionally assessed by first estimating the dynamic control performance improvement that reduces CV fluctuations about the same mean, shown in Fig 2. The base case mean is taken as acceptable, optimal at the start (CLIFFTENT shows this is invariably incorrect). This smoother CV operation provides no tangible benefit itself (because we do not know how to quantify it) but is a necessary prerequisite to the second step, move the mean an "appropriate" amount closer to the spec ("appropriate" is arbitrary because we do not know how to properly set the new mean either). Then the CV mean change is multiplied by some flow/CV factor to get improved steady-state yield, capacity or utility consumption. Finally this is multiplied by a $/day per unit flow factor to estimate $/day profit gain. This method is seriously flawed. The missing ingredient shown by Latour (1992a, 1996) is failure to model the penalty for violating the spec limit.

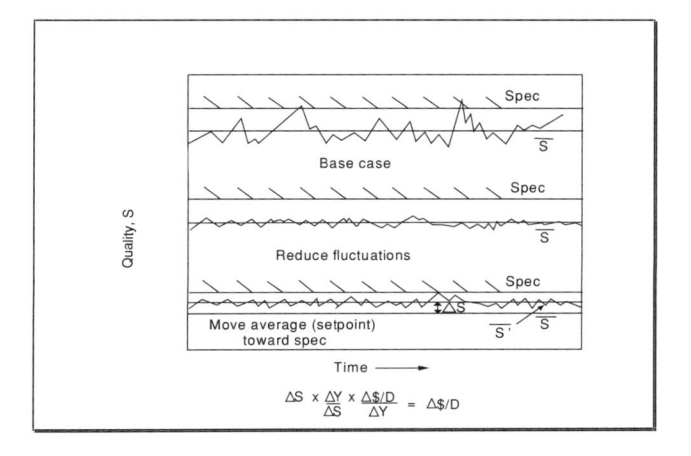

Figure 2. Transients: two steps.

Frequency Distribution Function

The frequency distribution function, f(x), provides the number of units of material at each value, x, of a CV of interest as a function of x. It is the statistical distribution of CV data with mean, μ, and standard deviation, sd, over a time period, T. This function must be provided or assumed; it is a required input, taken directly from the transient data, Fig. 3. It may be Gaussian or arbitrary, provided it is bounded, integrable and its integral is also bounded. It is most useful, of course, if it is stationary for some period into the future until a new f(x) can be determined.

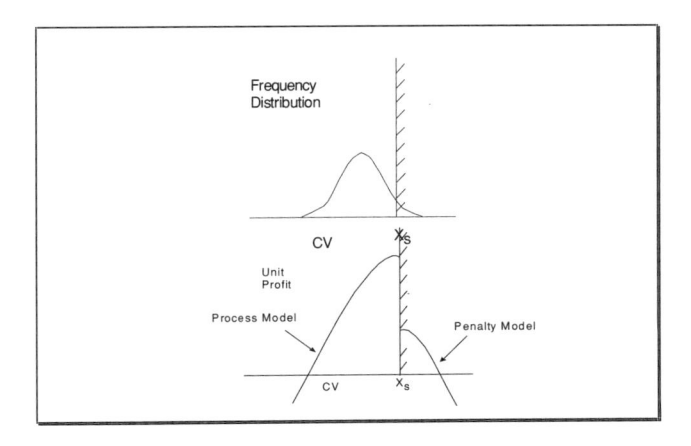

Figure 3. Two input functions.

Unit Profit Function - The Clifftent

The Unit Profit Function, UP(x), provides the profit per unit (bbl, lb., ton, cargo) of material (usually feed or a product) at each value, x, of a CV of interest as a function of x, Latour (1992a, 1996). This function must be provided or assumed; it is an input requirement, Fig. 3. It must become negative in both directions, for small and large x. It may have discontinuities and be arbitrary provided it is bounded, integrable and its integral is also bounded. It is most useful, of course, if it is stationary for some period into the future until a new UP(x) can be determined. The process side slope can come from the CLRTO Jacobian matrix; the curvature from the Hessian matrix.

Now we can answer a common heretofore unanswered question of process control and LP modeling: how does one identify CV? How do we select among the enormous number of possible candidate CV? Are all TFLPCiQ at every point in all processes to become CV? Why not? Our opening definition of CV suggests the answer. Each CV has a UP(x), a CLIFFTENT of consequence. If a variable has or can be assigned a CLIFFTENT it is a CV; if not it is not. A CV is not what we can model or measure; that comes later. A CV is first a variable that matters, one we care about, value, can assign it a CLIFFTENT. Then if UP(x) is interesting, large, significant or crucial we go about measuring and modeling it; then controlling and optimizing it. CLIFFTENT quantifies the purpose of process control; the essence of plant operation.

Time Profit

In view of Fig. 3 is the mean of the CV distribution properly aligned with the Unit Profit function, UP(x)? Is μ optimal or should it be higher or lower, and by how much? It should be raised (shifted right) closer to spec, but not all the way to xs. How much money would be generated by setting it right, optimally, in the first place?

Integration of f(x) and UP(x) provides the Time Profit, TP(μ), average for the period, T.

$$TP(\mu, sd) = \text{INTEGRAL} \; [UP(x) * f\{(x-\mu)/sd\}] * dx$$

We find the entire function TP(μ) in Fig. 4. It is a hill; it has a maximum.

This CLIFFTENT integral is the rigorous method for setting the target mean of a CV near its spec. It accounts for statistical uncertainty and dynamic performance, process performance model, spec violation penalty and profit. The optimum CV target may even exceed the spec when the process incentive slope is steep and spec violation penalty is shallow. This is shown in the lower part of Fig. 1. (This often occurs when highway speeders perceive they can afford a ticket or accident).

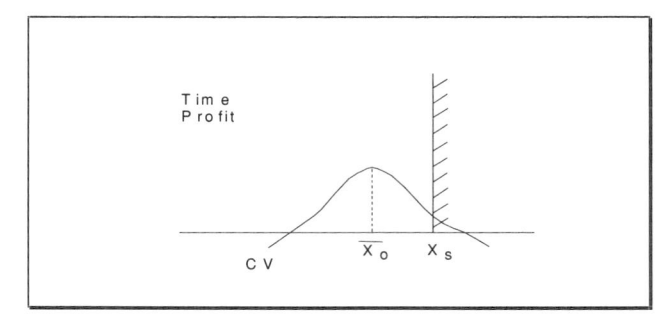

Figure 4. Solution function.

CLIFFTENT provides the rigorous modeling method for taking calculated risks on the RHS and quantifying the intangibles. It shows clearly where the money comes from; less low profit product at the far LHS and far RHS.

The true time profit function near the optimum mean target is a smoother hill, even if UP(x) has a discontinuity. The true profit function with uncertainty at CV limits and LP constraints in Fig. 1 is really not so sharp as a line or cliff, it is a rounder "donut". The slope of the LP profit function at CV constraints is really zero, not the "shadow price". This is why LP "shadow prices" are so unrealistic and useless in practice. The CLIFFTENT modeling concept transcends LP and SQP optimization technology

Multivariable controller CV weighting factors to distinguish the relative ranking importance among all its CV, with dissimilar engineering units and scaling, are easily calculated from TP, representing the amount of each CV limit violation that represents equivalent penalty, say 100 $/d loss.

Comparison

CLIFFTENT provides the rigorous method to compare profits if any of the input parameters change. It quantifies the new proper CV limit and corresponding profit change. These can come from changes in process performance (LHS), customer market or contract (cliff penalty), corporate efficiency (UP(xs)) or dynamic performance, sd.

If UP(x) is linear over the range of the distribution, f(x), reducing sd provides no profit increase; credits from one side equal debits from the other side; they average zero. That is why tighter level control makes no money unless it reduces the frequency of overflow or drainage mishaps. Curvature and discontinuity nonlinearity in the UP(x) function provides the incentive for improving dynamic performance. The greater the nonlinearity, the greater the incentive. The CLIFFTENT concept is illustrated by a commercial example for low sulfur fuel oil in Figs. 5 and 6 and Table 1.

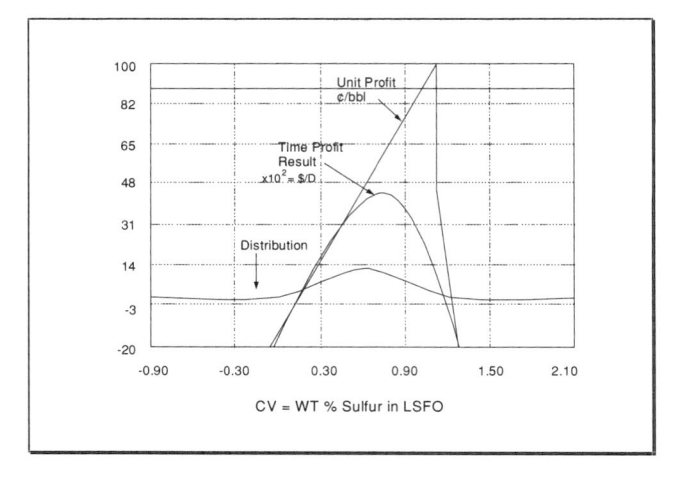

Figure 5. CLIFFTENT: 1.0% LSFO, 10 kbpd.

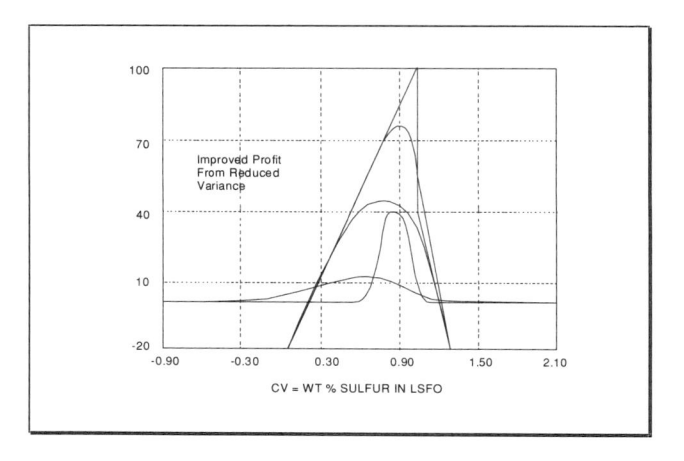

Figure 6. CLIFFTENT solution: 1.0% LSFO.

Table 1. CLIFFTENT LSFO Example.

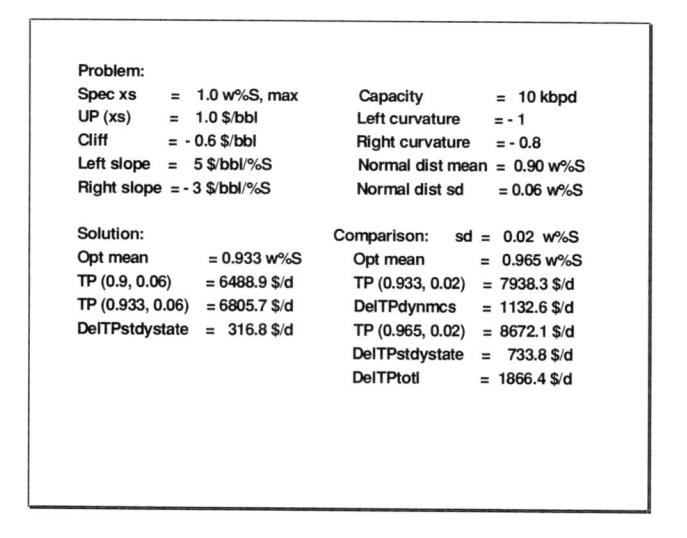

Problem:
Spec xs	= 1.0 w%S, max	Capacity	= 10 kbpd
UP (xs)	= 1.0 $/bbl	Left curvature	= -1
Cliff	= -0.6 $/bbl	Right curvature	= -0.8
Left slope	= 5 $/bbl/%S	Normal dist mean	= 0.90 w%S
Right slope	= -3 $/bbl/%S	Normal dist sd	= 0.06 w%S

Solution:
Opt mean	= 0.933 w%S	Comparison: sd	= 0.02 w%S
TP (0.9, 0.06)	= 6488.9 $/d	Opt mean	= 0.965 w%S
TP (0.933, 0.06)	= 6805.7 $/d	TP (0.933, 0.02)	= 7938.3 $/d
DelTPstdystate	= 316.8 $/d	DelTPdynmcs	= 1132.6 $/d
		TP (0.965, 0.02)	= 8672.1 $/d
		DelTPstdystate	= 733.8 $/d
		DelTPtotl	= 1866.4 $/d

Connection to "RE"

CLIFFTENT provides the rigorous modeling method to connect process models for conservation balances of mass, energy and momentum for yield, utilities and capacity performance to the penalties for breaking rules or violating specs of the surrounding world of RE (Table 2): customers, safety, maintenance, environment, human values, the money balance. RE stands for so many of these activities like recheck, refund, replace, recycle, reprocess listed in Table 2. It provides the financial connecting link between operating condition target setting and risk management from process surroundings like HAZOPS. It provides the connection between statistical quality control and dynamic process control. It provides the connection between CLRTO and CMVPC. It connects quality to value.

Table 2. Spec Violation Modeling - "RE".

Breech	Hazop	Reprocess	1. Entrainment, foaming
Cancellation	Injury	Reprimand	2. Pump
Carryover (1)	Insurance	Reputation	3. Tube metal temperature
Cavitation (2)	Inventory	Rerun	4. Distillation
Citation	Late penalty	Resample	5. Valves
Coking	Litigation	Reship	6. FCC catalyst circulation
Complaints	Losses	Resubmit	7. Compressor
Corrosion	Overpressure	Return	8. Permit, regulation, law
Damages	Plugging	Reversal (6)	
Deactivation	Reblend	Revise	
Default	Recall	Rework	
Demurrage	Recheck	Risk management	
Discount	Recycle	Safety	
Discharge	Redo	Scrap	
Downgrade	Reflux	Settlement	
Explosion	Refund	Spill	
Fatigue (3)	Reject	Stream factor	
Fine	Relief (5)	Surge (7)	
Fire	Repair	Turndown	
Flaring	Repeal	Turnover	
Flooding (4)	Repeat	Venting	
Fouling	Replace	Violation (8)	

CLIFFTENT connects chemical industry process modeling methods to the modeling methods of an equally large industry, insurance, because modeling methods in the land of RE often require techniques of statistics, expected values, life expectancy, cost of occurrence and risk management central to the insurance industry. In light of CLIFFTENT many would agree the chemical process control business may have overemphasized modeling the chemistry and physics of processes and neglected modeling the surrounding world of RE: economic and business issues. CLIFFTENT unifies these two dissimilar modeling approaches to determine profit improvement.

Examples

Several examples of tradeoffs that have been modeled, solved and optimized in oil refining by Latour (1996) illustrate the breadth of generality and usefulness of the CLIFFTENT method. Multivariable control of 200 kbpd atmospheric crude distillation units typically generates conventional analysis benefits of 0.08 $/bbl. A rigorous CLIFFTENT analysis showed the true benefit is more like 0.18 $/bbl, more than double.

High process loads cause distillation flooding, compressor stall and vibration, separator liquid entrainment carryover, packed bed channeling, pump cavitation, reactors to loose conversion or depart from equilibrium, furnace tube rupture, heater huffing and pipe vibration. Low loads cause distillation weeping, compressor surge, and heater flame outs. Each piece of equipment has a maximum capacity indicator and minimum turndown capacity indicator. These phenomena in the world of RE should be modeled financially to create the CLIFFTENT to set CV targets optimally.

Conclusions

A dependent controlled variable is a point phenomenon that people care about, are interested in, value and assign a CLIFFTENT profit function to. Quantifying all the options in Table 3 provides a comprehensive approach to assessing business profitability improvements.

Table 3. CLIFFTENT Conclusions: to Increase Profit.

1.	Optimize miu, mean target -	Know your business and product Make best use of what you have Set targets right
2.	Decrease sigma, variance -	Maintain smooth control Lower variance alone makes money
3.	Increase UPm -	Raise sell price Cut fixed costs
4.	Decrease m1, positive slope -	Raise process efficiency Lower variable costs
5.	Increase m2, negative slope -	Improve customer good will Lower complaints
6.	Decrease CLIFF -	Obey the law Plan for emergencies
7.	Increase C, capacity -	Expand markets Raise production rate
8.	Increase max xs or - decrease min xs	Negotiate for looser spec Reassess basis for spec
9.	If TPmax (optmean) < 0, - no matter what	Sell out fast Liquidate

Table 4. CLIFFTENT Results – Base.

Base - Situation Analysis
1. Time Profit, Function of CV Mean
2. Max Time Profit
3. Optimum CV Mean or Target
4. Profit Increase = Max TP - Base TP
5. Controller Variable Ranking, ECE
6. Amount of Unprofitable Product, %
7. % of Perfect Control Profit

CLIFFTENT shows the merit of people quantifying their financial values - CLIFFTENT functions - first. Then taking calculated risks, analyzing and optimizing tradeoffs, setting CV targets, quantifying benefits from reduced variance dynamic performance is easy. If we do not define CLIFFTENT values well at the outset we can never quantify the value of smoother operation; if we do, we can. People should quantify their values, cares and interests in UP(x) and integrate it properly with their uncertainty, expressed in the frequency distribution, f(x).

Results from CLIFFTENT analyses, Tables 4 and 5, by Latour (1992a) confirm that previously intangible benefits of dynamic process control are of the same order of magnitude as traditional simplified estimates of tangible benefits, so the true merit of MVC, RTO, SCHED and IT to the processing industry is at least double that normally (conservatively) claimed. Table 6 summarizes conclusions from the CLIFFTENT approach.

Table 5. CLIFFTENT results – comparison.

Comparison - Improve Dynamic Performance
1. New Time Profit, Function of CV Mean
2. Dynamic Profit = $TP(\mathcal{Y}_1^0, \mathcal{K}_2) - TP(\mathcal{Y}_1^0, \mathcal{K}_1)$
3. New Max Time Profit = $TP(\mathcal{Y}_2^0, \mathcal{K}_2)$
4. New Opt CV Target = $\mathcal{Y}_2^0$
5. Steady State Profit = $TP(\mathcal{Y}_2^0, \mathcal{K}_2) - TP(\mathcal{Y}_1^0, \mathcal{K}_2)$
6. Total ΔProfit = DYN + SS = $TP(\mathcal{Y}_2^0, \mathcal{K}_2) - TP(\mathcal{Y}_1^0, \mathcal{K}_1)$
7. New Controller Variable Ranking, ECE
8. New Amount of Unprofitable Product, %
9. New % of Perfect Control Profit

Table 6. Summary of CLIFFTENT.

Given: CV Distribution and CLIFFTENT Functions, can:
1. Specify value of CV target to maximize expected profit
2. Quantify financial benefit from improved dynamic performance, reduced variance, smoother CV operation
3. Connect SQC SPC/APC, MVC/RTO, process model/surroundings impact, risk/profit
4. Convert intangibles/experience/human judgement into analytical knowledge modeling for calculated risk taking, to maximize expected profit from dissimilar phenomena
5. Optimize any tradeoff with uncertainty
6. Connect oil industry and computer industry with insurance industry
7. Minimize unforeseen occurances

There is a lot of money to be made by setting CV targets in the HPI right. One large US oil refiner (1 kkbpd) reported concern about "60 kk$/y in unforeseen occurrences in 1995". Modeling CLIFFTENT penalties and setting CV limits for maximum expected profit would go far to reduce the "unforeseen". Deploying CMVPC and CLRTO would go far to reduce the 60 kk$/y.

The manufacturing industries have generally had difficulty seeing and quantifying the financial benefits from MVC, RTO, SCHED, IT and CIM, because traditional benefits analysis methods are shallow and incomplete. They neglect modeling the penalty for noncompliance, violating the specs. CLIFFTENT provides the theoretically sound method for assessing the value of performance correctly and leads people to focus their modeling efforts on the issues that matter, like the penalty side of the UP(x) function, the frequency distribution function and unit profit at spec, UP(xs).

Many (perhaps most?) MVC, RTO, IT and CIMFUELS projects are launched using the "faith theory": computers are good; technology is needed to survive; everybody is doing it; surely we can cut fuel 3 %, improve yield 1 %, increase capacity 2 % and that is plenty of justification. Before embarking with the "faith theory" alone, the CLIFFTENT approach validates what the Greeks taught in 400 BC: do analysis before synthesis, function comes before form. The prerequisites are 1) know your process, how it works, 2) know your objective, purpose, 3) know the rules and limits, 4) know the consequences and penalties for breaking the rules, violating the limits and 5) deploy CLIFFTENT analysis to bring your knowledge together, to set CV targets properly, to optimize profit. If you first know **what** you are going to do, **why** you plan to do it and **how** you will **measure** success and failure, then it becomes much easier to figure out **how** to do it, how to harness computers to do a good job improving profits.

References

Georgiou, A. (1998). Ethylene optimization system reaps operations and maintenance benefits. *Oil & Gas Journal*, **96**, No. 10, 46-50.

HPI Market Data brochure. (1993). *Hydrocarbon Processing*.

Latour, P. R. (1979). Online computer optimization 1: what it is and where to do it & 2: benefits and implementation. *Hydrocarbon Processing*, **58**, No. 6, 73-102 & No. 7, 219-223.

Latour, P. R. (1992). APC & RIS, *FUEL T&M*, **2**, No. 2, 14-23.

Latour, P. R. (1992a). Quantify quality control's intangible benefits, *Hydrocarbon Processing*, **71**, No. 5, 61-66.

Latour, P. R. (1994). Mission: Plan to use RIS/APC for RFG/LSD, *FUEL T&M*, **4**, No. 4, 20-25.

Latour, P. R. (1996) Process control: CLIFFTENT shows it's more profitable than expected, *Hydrocarbon Processing*, **75**, No. 12, 75-80.

MODEL PREDICTIVE CONTROL AND IDENTIFICATION: A NEW APPROACH TO CLOSED-LOOP IDENTIFICATION AND ADAPTIVE CONTROL

Alexander Schwarm and S. Alper Eker
Chemical Engineering Dept.
Texas A&M University
College Station, TX 77843-3122

Michael Nikolaou[*]
Chemical Engineering Dept.
University of Houston
Houston, TX 77204-4792
nikolaou@uh.edu

Abstract

Model predictive control and identification (MPCI) is a new class of adaptive controllers, that employ the following on-line optimization to perform closed-loop identification and controller adaptation: At each time step maximize information for identification subject to (a) standard MPCI input constraints, (b) chance constraints on the outputs and (c) persistence of excitation constraints. The advantages of the purposed MPCI formulation are (a) It does not require any external dithering signals, (b) Time needed for identification is minimized and quality of data used for identification is maximized, (c) Inputs are kept within bounds, (d) Outputs are robustly kept within bounds throughout the identification process, and (e) It can be easily integrated with existing MPC technology. This control philosophy can be used for processes which need short identification times and can tolerate process output variations within preset bounds.

Keywords

Closed-loop identification, Adaptive control, Persistence of excitation, Chance constraints.

Introduction

Model Predictive Control and Identifcation (MPCI) was introduced by Genceli and Nikolaou (1996) as a new adaptive control paradigm based on model predictive control (MPC). Since MPCI relies on on-line optimization, it provides wide flexibility for formulating various constraints and defining objectives other than the standard minimization of 2-norms. For example, one could completely forego the minimization of the sum of square errors (which attempts to keep outputs at their setpoints) and simply attempt to keep outputs within specification bounds during identification, while maximizing the lower eigenvalue of corresponding information matrices. The control objective may be loosened or tightened by adjusting the output constraint bounds. In addition, the MPCI formulation trivially guarantees, by construction, that data are going to be

---

[*] Author to whom all correspondence should be addressed. Email: nikolaou@uh.edu. Phone: (713) 743 4309. Fax: (713) 743 4323.

informative. This flexibility allows us to develop a new mathematical formulation of realistic engineering goals associated with the identification of a process controlled by constrained MPC.

In this work we are considering an MPCI formulation with the following important features: (a) The objective function to maximize on-line is the sum of the lower bounds on the minimum eigenvalues of the information matrices over a finite horizon. Through this modification, the process output is free to move away from setpoint, as long as it remains within specifications. In that way inputs to the process are allowed to excite the process highly enough to generate as much modeling information as possible. (b) Predictions of future process outputs are probabilistic and constraints on them are chance constraints. The reason for this is the following: Future outputs within an MPCI moving horizon have to be predicted on the basis of a process model. Because the model is not yet good enough for control purposes, output predictions are uncertain and may result in adverse violation of output constraints by the closed-loop system, even though predicted outputs over the moving horizon might have been properly constrained. Consequently, constraints other than deterministic must be placed on process outputs.

As a solution, one can use chance constraints on outputs as developed by Charnes and Cooper (1963). Uncertainty information obtained from the on-line identification in the form of a covariance matrix for the process model coefficients can be used. Then, under the assumption that a linear model with normally distributed coefficients predicts the process output, the chance constraints can be reformulated as convex, deterministic constraints on the input. The resulting on-line optimization problem at each time step requires the solution of a semidefinite program, with nonconvex quadratic matrix inequalities. To obtain a guaranteed local optimum, we first linearize each quadratic matrix inequality to get a linear matrix inequality (LMI) which can then be used in standard semidefinite programming codes, such as the code developed by Alizadeh et. al., to find a suboptimal point. The quadratic matrix inequalities are then linearized again at this suboptimal point and the resulting semidefinite program is solved. This process is repeated until a convergence criterion is met. The above procedure (successive semidefinite programming) is proved to converge in Genceli and Nikolaou (1996) and Shouche et. al. (1996).

Chance-Constrained Optimization

The formulation of MPCI proposed in this work efficiently identifies process models while also constraining future output predictions. Future outputs that are to be constrained must be predicted on the basis of a process model. Uncertainty in the process model makes these predicted outputs uncertain. This is especially true in the case where identification of the process model is needed. Uncertainty in output predictions may result in adverse violation of output bounds for the closed-loop system, even though predicted outputs over the moving horizon might have been properly constrained. This is because the output constraints in the moving horizon on-line optimization problem are constraining an inaccurate prediction of the process output. Uncertainty in the output prediction can be determined using covariance information of the model coefficients. This output prediction uncertainty can then be used in a new form of constraint which robustly maintains process outputs within bounds. Chance constraints incorporate parametric uncertainty such that constraints containing uncertain coefficients are ensured to be satisfied with a specified probability. The form used is as follows:

$$\Pr\{y(k+i \mid k) - \varepsilon \leq y_{\max}\} \geq \alpha \qquad (1)$$

where $\Pr\{A\}$ is the probability of event A occurring, $y(k+i|k)$ is the output predicted for time $k+i$ using information up to and including time k, ε is the constraint softening variable, $y_{\max}$ is the specified maximum output, and α is the specified probability, or confidence level, that the output constraint will be satisfied. Under the assumption that the output is predicted by a linear model with normally distributed coefficients, the chance constraint can be reformulated as a convex, deterministic constraint on the input:

$$y(k+i \mid k) - \varepsilon_i + K_\alpha \sigma_y \leq y_{\max} \qquad (2)$$

where σ_y is the standard deviation of $y(k+i|k)$ which is a function of future inputs and K_α is a constant corresponding to the desired confidence level. This constraint can then be easily added to the on-line optimization problem.

QMI Optimization

For the MPCI formulation proposed in this work, the on-line optimization problem is not to minimize a quadratic objective function involving inputs and states, but to maximize the sum of the lower bounds on the minimum eigenvalues of the information matrices over a finite horizon. In that way, inputs to the process are allowed to excite the process highly enough to generate as much modeling information as possible, while the process goes off-spec as little as possible. The on-line optimization problem is as follows:

$$\max_{u(k|k),\cdots,u(k+M-1|k),\rho_1,\cdots,\rho_M,\mu_1,\cdots,\mu_M} \sum_{i=1}^{M}\rho_i - h\sum_{i=1}^{P}\mu_i \qquad (3)$$

subject to

$$u_{max} \geq u(k+i-1|k) \geq u_{min}, \quad i=1,\cdots,M \qquad (4)$$

$$\Delta u_{max} \geq \Delta u(k+i-1|k) \geq \Delta u_{min}, \quad i=1,\cdots,M \qquad (5)$$

where

$$\Delta u(k+i-1|k) \triangleq u(k+i-1|k) - u(k+i-2|k)$$

$$u(k+M+i|k) = u(k+i|k), \quad i=0,\cdots,M-1 \qquad (6)$$

$$\Pr\{y_{min} - \mu_i \leq y(k+i|k) \leq y_{max} + \mu_i\}, \quad i=1,\cdots,P \qquad (7)$$

$$\mu_i \geq 0$$

$$\sum_{j=0}^{s-1} \lambda^j \phi(k-j+i|k)\phi(k-j+i|k)^T \succeq \rho_i \mathbf{I} > \mathbf{0}, \quad i=1,\cdots,M \qquad (8)$$

$$y(k+i|k) = \phi(k+i|k)^T \overline{\theta}(k|k), \quad i=1,\cdots,P \qquad (9)$$

$$\overline{\theta}(k|k) = \left[\sum_{j=0}^{s-1} \lambda^j \phi(k-j|k)\phi(k-j|k)^T\right]^{-1} [\phi(k|k) \quad \phi(k-1|k) \quad \cdots \quad \phi(k-s+1|k)]\mathbf{Y}(k) \qquad (10)$$

$$\mathbf{Y}(k) \triangleq [y(k) \quad y(k-1) \quad \cdots \quad y(k-s+1)]^T \qquad (11)$$

$$\phi(k-j+i|k) = [u(k-j+i-1|k)\, u(k-j+i-2|k)\ldots u(k-j+i-n|k)\, 1]^T \qquad (12)$$

In this work we select

$$s = M \qquad (13)$$

for this on-line optimization problem.

Remarks

- As seen from eqns. (3) and (8), the above MPCI variant maximizes the lower bounds on the minimum eigenvalues of information matrices. In this way, inputs are forced to satisfy a PE constraint, in order to generate as much modeling information as possible.

- The control objective is expressed by the output constraint in equation (7), the chance constraint formulation discussed above.

- The feasibility of the on-line optimization problem is guaranteed by the existence of the softening variables μ_i in the output constraints, equation (7). To reduce the dimensionality of the on-line optimization problem, one may assume that all μ_i are equal to each other. This will result in minor loss of controller flexibility, because output constraints are important only for the first few steps of the prediction horizon.

- In place of equation (10), one can easily use the recursive least squares (RLS) algorithm, namely:

$$\overline{\theta}(k|k) \triangleq \hat{\theta}(k+1) = \hat{\theta}(k) + \frac{\mathbf{P}(k)\phi(k+1)}{1+\phi(k+1)^T\mathbf{P}(k)\phi(k+1)}\left[y(k+1) - \phi(k+1)^T\hat{\theta}(k)\right]$$

$$\mathbf{P}(k+1) = \left[\mathbf{P}(k) - \frac{\mathbf{P}(k)\phi(k+1)\phi(k+1)^T\mathbf{P}(k)}{1+\phi(k+1)^T\mathbf{P}(k)\phi(k+1)}\right] \qquad (14)$$

where $\hat{\theta}(k)$ is the least squares estimate of θ at time k; $\mathbf{P}(k)$ is the covariance matrix at time k; $\hat{\theta}(0)$ and $\mathbf{P}(0)$ are initial values for the algorithm. $\mathbf{P}(0)$ is the covariance of $\hat{\theta}(0)$, and is usually chosen sufficiently large since at the beginning of identification there is assumed to be no knowledge about parameters. Parameter convergence using the recursive least squares algorithm is guaranteed if the following condition is satisfied:

$$\lim_{k\to\infty} \lambda_{min}\left(\sum_{j=1}^{k-1}\phi(j)\phi(j)^T\right) = \infty \qquad (15)$$

- The PE constraint in equation (8) requires PE in a moving horizon with fixed window length. However PE can be also required over a series of windows of increasing length as well. This will give a new variant of MPCI. All the details of the MPCI variants, and the solution of on-line optimization problem can be found in Eker and Nikolaou (1998).

Case Study

Example 1

Let us assume that the real behavior of a linear process is initially described by the equation:

$$y(k) = u(k-1) + 0.5u(k-2) + 0.2u(k-3) + 0.1u(k-4) + d(k) + w(k) \quad (16)$$

where d is a deterministic disturbance and w is white noise with zero mean and standard deviation equal to 0.01. The process input u must satisfy the constraints

$$-0.4 \leq u(k) \leq 0.1 \quad (17)$$

at all times k. Assume that the linear model

$$y(k+i|k) = 1.1u(k+i-1|k) + 0.55u(k+i-2|k) + 0.22u(k+i-3|k) + 0.11u(k+i-4|k) + d(k+i|k) \quad (18)$$

is available for the above process from previous data. The process experiences a step setpoint change

$$y^{sp} = -0.3 \quad (19)$$

at time $k=0$. After this set point change, we want the output y to be constrained as

$$\Pr\{-2 \leq y(k+1) \leq -2\} \geq 0.99 \quad (20)$$

to meet specifications. The process also experiences a change at time $k=26$ such that the real process has changed as follows

$$y(k) = -0.3u(k-1) - 0.2u(k-2) - 0.1u(k-3) + 0.05u(k-4) + d(k) + w(k) \quad (21)$$

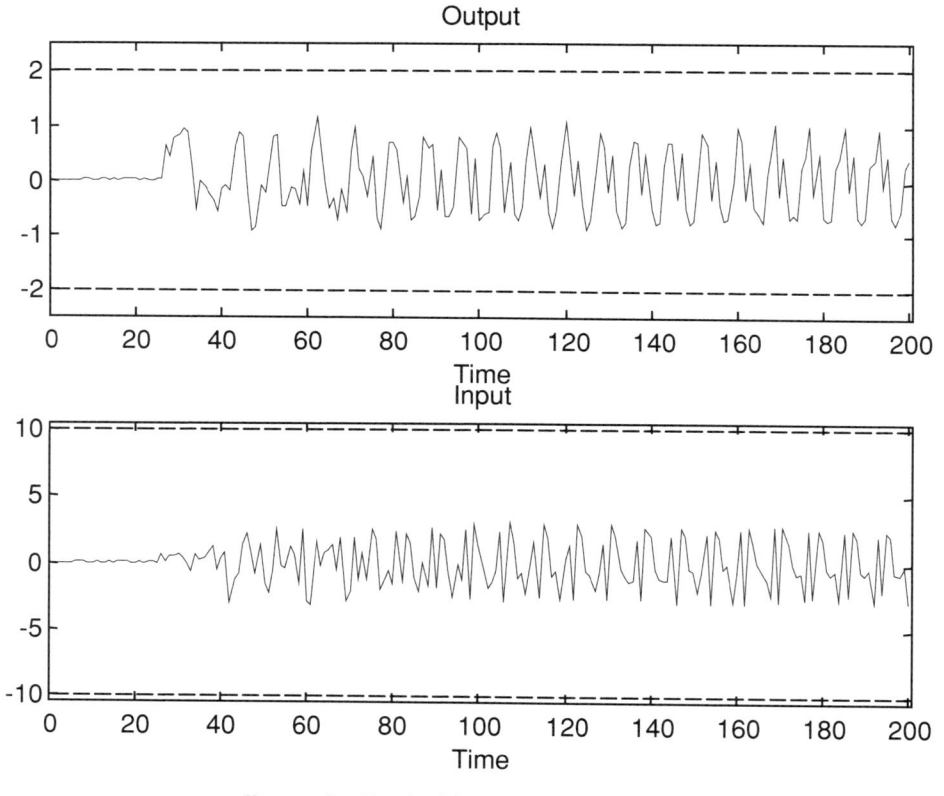

Figure 1. Dashed lines are constraints.

Figure 1 shows that the process output is robustly kept within specification bounds during parameter identification. The input also oscillates to provide maximal information for identication purposes.

Example 2

The same situation as example 1 is examined, but in this case the inequality constraint (17) has been replaced by the following more stringent constraint on u

$$-2 \leq u(k) \leq 10 \qquad (17)$$

at all times k. This bound is chosen to show that the algorithm performs well even while hitting input constraints.

As Fig. 2 shows the input u is maintained above its lower bound of -2, while identification is successfully performed. Notice that the input u behaves assymetrically about the origin.

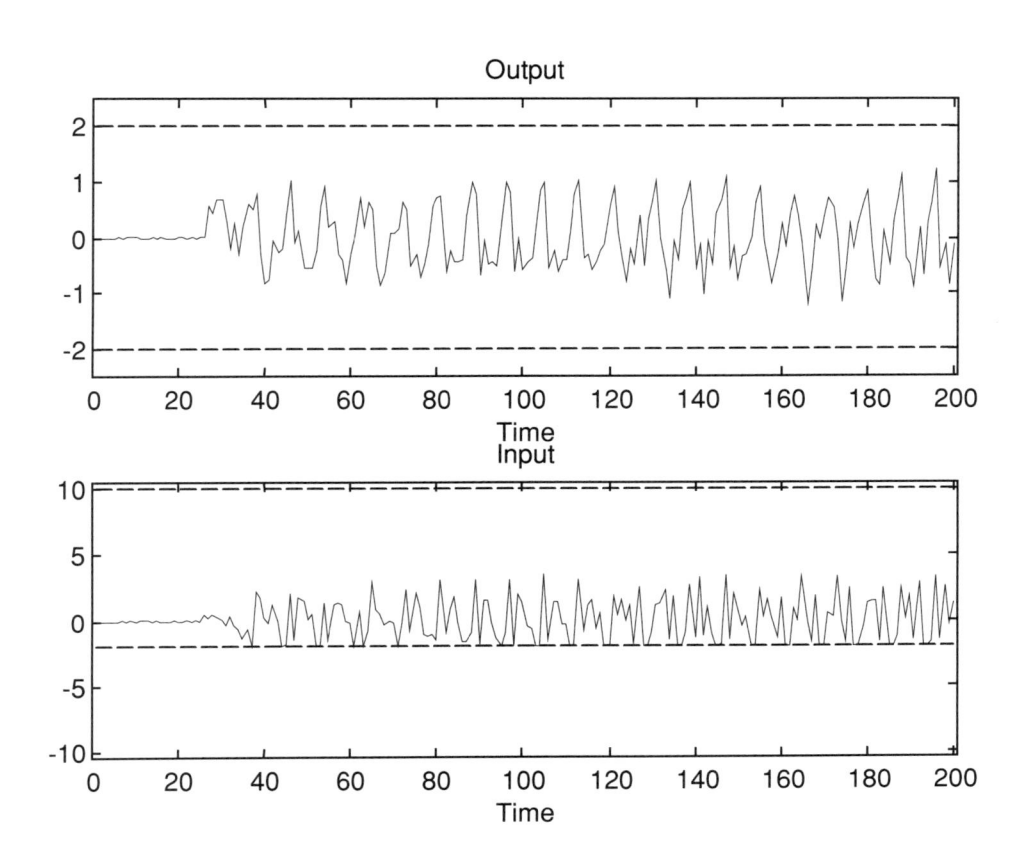

Figure 2. Dashed lines are constraints.

Discussion

MPCI, combined with chance constraints, identifies systems efficiently while satisfying standard MPC constraints and keeping process outputs robustly within preset bounds. For processes with outputs which can deviate from set point, but must stay within present bounds, chance-constrained MPCI maximizes identification information such that identification experiments take as little time as possible.

References

Alizadeh, F., J. A. Haeberly, M. V. Nayakkankuppam, M. L. Overton and S. Schmieta, (1997). SDPPACK user's guide - version 0.9 beta for matlab 5.0, TR1997-737, June 26, 1997.

Charnes, A. and Cooper, W. W., (1963). Deterministic Equivalents for Optimizing and Satisficing Under Chance Constraints, *Operations Research*, **11**, 18-39.

Eker, S. A. and M. Nikolaou, (1998). Adaptive Control through On-line Optimization: The MPCI Paradigm, submitted to *AMCS Int. J.*

Genceli, H. and M. Nikolaou, (1996). A new approach to constrained predictive control with simultaneous model identificatiom, *AIChE J.*, **42**, 10, 2857-2868.

Schwarm, A. T. and M. Nikolaou (1997). Chance constrained model predictive control, AIChE National Meeting, Los Angeles.

Shouche, M., H. Genceli, M. Nikolaou (1996). Simultaneous constrained model predictive control and identification of DARMA processes, submitted to *Automatica*.

OPERATIONAL IMPLICATIONS OF OPTIMALITY

John Bagterp Jørgensen and Sten Bay Jørgensen
Department of Chemical Engineering
Technical University of Denmark
DK-2800 Lyngby, Denmark

Abstract

The main focus in this paper is on the possible occurrence of complex behaviors around optimal operational points as a consequence of the double quest for optimality in process design and in process operation. It will be demonstrated that optimal operation exploits process non-linearities, which may render complex behaviors appear near the optimum. A procedure for investigating these complexities is delineated. Insight into the occurrence of complex nonlinear behaviors is provided through formulation of a necessary condition for optimality and investigation of the occurrence and types of complex behaviors. The implications for control structuring and tuning requirements are discussed. A key conclusion is that in many instances closed loop control is mandatory to enable reliable process operation near the optimum.

Keywords

Optimal operation, Complex behavior, Bifurcation analysis.

Introduction

To maximize the economical potential the process plant should be both designed for optimal return on the investment and operated near the conditions, which optimize productivity at the actual market conditions. Process integration enables efficient utilization of plant capabilities for the benefit of the environment and economy. These capabilities are also being further exploited in process intensification. Nevertheless several chemical processes are designed conservatively to avoid complex operating regimes as pointed out by Seider *et al.* (1990). The main purpose of this paper is to further reveal and investigate some possible operational implications of this double quest for simple optimality in design and in operation. These issues are introduced via examples of the types of behavior, which have been encountered in some processes and plants. Firstly, examples of complex behaviors are given, secondly two hypotheses are formulated for the likelihood of occurrence of complex behaviors, and thirdly a necessary condition for operational optimization is derived and a generic objective function is formulated for operational optimization. Subsequently, the hypothesis concerning occurrence of complex behaviors around operational optima is illustrated via analysis of relatively simple process plant examples. The paper ends with a discussion on the more general question of how to handle the operational issues at the process operations and control design stage(s).

Complex Behaviors

The developments within nonlinear dynamics science of chemical processes have revealed that relatively complicated behaviors are possible in many processes, e.g. separation processes and even in apparently simple isothermal reactors such as CSTR reactors with kinetically limited reactions such as A to B to C. These behaviors may become much more complicated if thermal effects are included. When such processes are combined into even relatively simple plants which for efficient utilization purposes contain recycles of raw material or energy, the possible behaviors may become plentifully more complicated due to the recycle streams. This increase in complexity arises either due to the positive feedback of the process dynamics or due to a dynamic effect of the recycle. The complexities may be revealed both in the static behavior and/or the dynamic behavior. Some of the more common static complexities may be described from an operational perspective in terms of different multiplicities each of which have different dynamic implications and thereby different control implications.

Static Multiplicities

Several types of static multiplicities have been reported. The term multiplicity refers to multiple solutions for given operating conditions. The static multiplicities may be subdivided into three types: Input, output and state multiplicity.

Input multiplicity: Multiple input values exist for a given set of outputs. The dynamic implication may be the occurrence of a right half plane zero, which may give rise to an inverse response. Input multiplicities were discussed by Koppel (1982) who showed that operation after a square wave disturbance may result in an uneconomical manipulated variable value.

Output multiplicity: Multiple output values exist for a given set of inputs. The dynamic implication hereof is a right half plane pole, i.e. instability. Output multiplicities have been shown in a simple exothermal A to B CSTR reactor by Uppal *et al.* (1974), in an A to B to C CSTR investigated by Halbe and Poore (1981), and by Farr and Aris (1986). Further output multiplicities have been demonstrated in a fixed-bed reactor with feed effluent heat exchanger by Bonvin *et al.* (1980) and by Recke and Jorgensen (1997). Output multiplicity has also been shown experimentally in a continuous bioreactor by Jorgensen *et al.* (1992).

State multiplicity: Multiple values of internal states exist for a given set of inputs *and* outputs. State multiplicity have recently been shown in a single infinitely long column at infinitely high reflux rate for heterogeneous azeotropic distillation by Bekiaris *et al.* (1996) and under similar conditions in two different sequences for heterogeneous azeotropic distillation by Esbjerg *et al.* (1998). The latter authors conjecture that state multiplicity in real columns gives rise to high sensitivity rather than state multiplicity.

Nonlinear dynamics

In addition to the dynamic implications of static complexities given above several complex phenomena mainly have dynamic implications. Such phenomena can be the appearance of cyclic behavior, a simple case is loss of stability of a static state which may be combined to the appearance of a limit cycle where the process or plant states oscillate in a periodic mode. Such phenomena are well known for a simple CSTR reactor with an exothermal A to B reaction (Uppal *et al.* 1974). In a more complicated case two periodic modes may coexist which may lead to a torus type solution. Recke and Jorgensen (1997) have demonstrated this case in fixed-bed reactors with recycle of unconverted reactant.

Occurrence of Complex Behaviours

Attempts at integrating process design, control structuring and operations design (scheduling) lead to a rather complex total optimization problem. Due to the small time constants of the market compared to the construction time the above computational complexity issues are often dealt with by using a hierarchical decomposition. In this decomposition the top layer deals with the combined process and control structure design, e.g. Heath *et al.* (1996) and the operations optimization is treated on a separate layer.

Following the above hierarchical structuring two hypotheses are formulated concerning the possible occurrence of complex behaviors. The first hypothesis concerning complex behavior due to operation optimization (CB.O) follows from the fact that a profit optimal operating point exploits nonlinear behavior:

CB.O: *Somewhere around a profit optimal operating point complex behavior will be encountered.*

This statement delineates the occurrence of complex behavior, but not where within the operating window. The second hypothesis concerns occurrence of complex behavior due to process integrated design (CB.D):

CB.D: *Optimizing process-integrated design (e.g. with explicit recycle streams) increases the likelihood of occurrence of complex behavior.*

Especially the first of these hypotheses will be illustrated in this paper by deriving and investigating a general optimality criterion and by using a couple of relatively simple examples.

The Unconstrained Optimal Operations Problem

The SISO steady-state operations problem is for a given set of disturbances, **d**, to select the manipulable variable, u, such that a measure of operational goodness is maximized:

$$\begin{aligned} \max_{u} \quad & J(\mathbf{x},u) \\ \text{s.t.} \quad & \mathbf{f}(\mathbf{x},u) = \mathbf{0} \end{aligned} \quad (1)$$

The steady-state constraint, $\mathbf{f}(\mathbf{x},u) = \mathbf{0}$, can be regarded as a function giving **x** as an implicit function of u, i.e. $\mathbf{x} = \mathbf{x}(u)$. Consequently, the first order optimality conditions can be stated as

$$\frac{dJ}{du} = \frac{\partial J}{\partial u} + \sum_{i=1}^{n} \frac{\partial J}{\partial x_i}\frac{dx_i}{du} = 0 \qquad (2)$$

which in compressed notation corresponds to

$$J_u + J_x \frac{d\mathbf{x}}{du} = 0 \qquad (3)$$

Similarly linearization of the process model and imposition of the steady state constraint gives

$$\delta f = f_x \delta \mathbf{x} + f_u \delta u = 0 \qquad (4)$$

Assuming that f_x is non-singular, the change of states on the steady-state manifold as function of changes in the manipulated variable can be calculated using the implicit function theorem

$$\delta \mathbf{x} = -f_x^{-1} f_u \delta u \qquad (5)$$

which gives the relation

$$\frac{\partial \mathbf{x}}{\partial u} = -f_x^{-1} f_u \qquad (6)$$

Combining Eqn. (3) and Eqn. (6) relates the objective function and the process function at the optimal operating point

$$J_u - J_x f_x^{-1} f_u = 0 \qquad (7)$$

Thus the above condition is a necessary condition for optimal operation.

The regulation problem is to keep the process at the optimal steady state despite disturbances. Accordingly, this problem implies that the equilibrium condition is relaxed. Assuming that one of the states is measured, the linearized state space equation used for linear SISO control is:

$$\delta \dot{\mathbf{x}} = f_x \delta \mathbf{x} + f_u \delta u + f_d \delta d$$
$$\delta y = C \delta \mathbf{x} \qquad (8)$$

Operational Optimization Objective Function

A generic objective function for operational optimization is of the form:

$$J = DC_{prod} - \kappa D C_{raw} = D(Y - \kappa)C_{raw} \qquad (9)$$

Where D is the dilution rate of the plant and C_{prod} and C_{raw} are product and raw material concentrations respectively and $Y = C_{prod}/C_{raw}$ is the production yield. The parameter κ designates the relative cost of raw material to the product prize. The above objective function clearly displays the difference to just optimizing yield Y. The latter case was investigated by Fox et al. (1984). It was shown that a zero is located on the imaginary axis at that optimum, if product concentration is the controlled variable. For the above objective function with $\kappa = 0$ the zero is located at infinity for the optimal dilution rate where C_{prod} vs. D has the slope $-C_{prod}/D$, and approaches zero as κ increases.

In this paper, the optimal operating point is assumed to lie within process constraints. In situations where it does not, the appropriate variables are assumed to be fixed at their constrained values, before further analysis is undertaken.

In the sequel operational optimization of a couple of cases are investigated wrt. the relative locations of the optimal operating point and the point of occurrence of complex behaviors. The latter point(s) are located using bifurcation analysis wrt. a key operational parameter. In this paper the dilution rate represents this key operational parameter. In practice a couple of parameters may be used.

Van de Vusse Reaction

Control of a CSTR with the Van de Vusse reaction has been studied intensively by a.o. Kravaris and Daoutidis (1990) and Sistu and Bequette (1995). The Van de Vusse reaction

$$A \xrightarrow{k_1} B \xrightarrow{k_2} C$$
$$2A \xrightarrow{k_3} D$$

is carried out in an isothermal CSTR with constant volume. This reaction system can therefore be described by the following system of differential equations

$$\frac{dC_A}{dt} = D(C_{A,in} - C_A) - k_1 C_A - 2k_3 C_A^2$$
$$\frac{dC_B}{dt} = -DC_B + k_1 C_A - k_2 C_B \qquad (10)$$

where the parameters in casu are: $k_1 = 50\,h^{-1}$, $k_2 = 100\,h^{-1}$, and $k_3 = 5\,h^{-1}$. The feed concentration of A is $C_{A,in} = 10\,mol/L$.

The steady state of this system is

$$C_{As} = \frac{(k_1 + D_s) - \sqrt{(k_1 + D_s)^2 + 8k_3 C_{A,in,s} D_s}}{-4k_3} \quad (11)$$

$$C_{Bs} = \frac{k_1}{k_2 + D_s} C_{As}$$

Let the dilution rate be the manipulable variable and let the concentration of A in the feed be the disturbance. Further let the concentration of B in the reactor be the controlled variable. The transfer function of the linearized system from the manipulable input and the disturbance to the output is

$$y(s) = G_u(s)u(s) + G_d(s)d(s) \quad (12)$$

where

$$G_u = K_u \frac{s - z}{(s - p_1)(s - p_2)} \quad G_d = K_d \frac{1}{(s - p_1)(s - p_2)}$$

$$K_u = \frac{-1}{C_{Bs}} \quad K_d = k_1 D_s$$

$$z = k_1 \frac{C_{A,in,s} - C_{As}}{C_{Bs}} - (D_s + k_1 + 4k_3 C_{As})$$

$$p_1 = -(k_1 + D_s + 4k_3 C_{As}) \quad p_2 = -(D_s + k_2)$$

The expressions for the poles reveal that they are always negative. Hence the system is stable for all parameter values. The expression for the zero indicates that it may be both positive and negative depending on the operating point and the kinetic parameter values. Hence the system may have a RHP zero limiting the achievable control performance.

Using the parameters and feed concentration of A given previously, Fig. 1 depicts the steady state concentration of A and B as function of the dilution rate. The concentration of A depends monotonously on the dilution rate, while the concentration of B depends non-monotonously on the dilution rate. There is an input multiplicity between the dilution rate and the concentration of B. Further the plot of the zero as function of the dilution rate shows that the maximum concentration of B appears at the same dilution rate as the zero changes from being positive to being negative.

The optimal operating profit for $\kappa = 0.1$ appears at a dilution rate somewhat larger than the dilution rate at which the concentration of B has a maximum and the zero changes sign. In the case considered the optimal profit is very sensitive to the relative raw material price, κ. For the Van de Vusse reaction system it turns out that it is only possible to generate positive optimal profits if the dilution rate is to the right of the maximum of the concentration of B, i.e. corresponding to negative zeros. As the relative raw material prices increases, the optimal profit decreases and the optimal dilution rate moves toward the dilution rate at which the zero changes sign. At an optimal operational profit of exactly zero, the optimal dilution rate is equal to the dilution rate at which the zero changes sign. Accordingly, if it is desirable to operate this process, i.e. the operating profit is positive, the nominal optimal operating point is at a dilution rate at which the zero is negative.

However, including economics or capacity constraints of downstream equipment may force operation at a dilution rate corresponding to a RHP zero. This will limit the possible dynamic performance of the reactor.

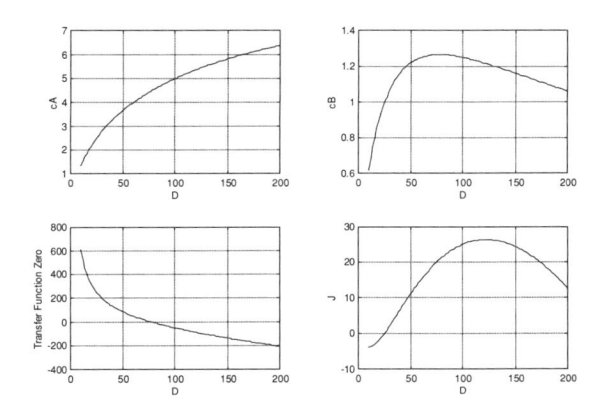

Figure 1. Upper: The steady state of a CSTR with a Van de Vusse reaction as function of the dilution rate. Lower: The zero location and operating profit, i.e. Eqn. (9), with $\kappa = 0.1$ as function of the dilution rate.

Bioreactor with Cell Recycle

A continuous fermentor integrated with a centrifuge concentrating biomass for recycle is depicted in the figure below.

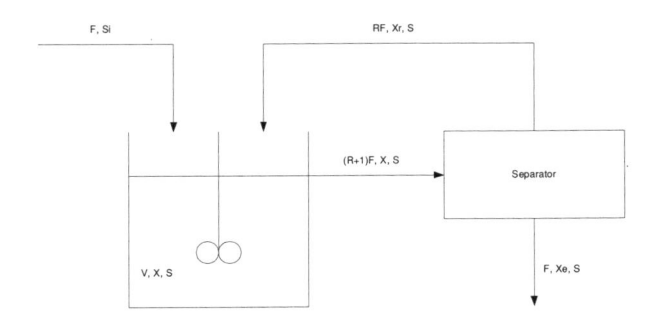

Figure 2. Continuous bioreactor with centrifuge for cell recycle.

By including maintenance in the kinetic model describing the growth of biomass and consumption of substrate, a model of this system is:

$$\frac{dX}{dt} = -\alpha D X + \mu(S) X$$
$$\frac{dS}{dt} = D(S_i - S) - (\gamma_{xs}\mu(S) + m_s) X \quad (13)$$

where

$$\alpha = \frac{X_e}{X} \quad (14)$$

The productive steady state of this continuous bioreactor system is

$$\mu(S_s) = \alpha D_s \qquad X_s = \frac{D_s(S_{is} - S_s)}{\gamma_{xs}\mu(S_s) + m_s} \quad (15)$$

Let the states be the biomass and substrate concentration, X and S, let the manipulable variable be the dilution rate, D, and let the disturbance be the concentration of substrate in the feed, S_i. The linearized model of this system around a productive steady state is

$$\dot{\mathbf{x}} = \mathbf{A}\mathbf{x} + \mathbf{B}u + \mathbf{E}d \quad (16)$$

where

$$\mathbf{A} = \begin{bmatrix} 0 & X_s \dfrac{d\mu}{dS} \\ -(\gamma_{xs}\mu(S_s) + m_s) & -D_s - \gamma_{xs} X_s \dfrac{d\mu}{dS} \end{bmatrix}$$

$$\mathbf{B} = \begin{bmatrix} -\alpha X_s \\ S_{is} - S_s \end{bmatrix} \qquad \mathbf{E} = \begin{bmatrix} 0 \\ D_s \end{bmatrix}$$

The eigenvalues of the Jacobian matrix are

$$\lambda = -\frac{1}{2}\left(D_s + \gamma_{xs} X_s \frac{d\mu}{dS} \right)$$
$$\pm \frac{1}{2}\sqrt{\left(D_s + \gamma_{xs} X_s \frac{d\mu}{dS} \right)^2 - 4(\gamma_{xs}\mu(S_s) + m_s) X_s \frac{d\mu}{dS}}$$

It is immediately apparent by observation of the eigenvalues that a necessary and sufficient condition for stability, i.e. negative real part of the eigenvalues, is

$$\frac{d\mu}{dS} > 0 \quad (17)$$

Furthermore the expression for the eigenvalues show that a Hopf-bifurcation with two purely imaginary eigenvalues will occur only if the following two conditions are satisfied

$$D_s + \gamma_{xs} X_s \frac{d\mu}{dS} = 0 \quad X_s \frac{d\mu}{dS}(\gamma_{xs}\mu(S_s) + m_s) > 0 \quad (18)$$

The first condition can only be satisfied if $d\mu/dS < 0$, which inevitably violates the second condition. Hence, this bioreactor system will never exhibit a Hopf bifurcation.

Turning point bifurcations or simple bifurcations (intersection of two branches of solutions) are possible if just one of the following two conditions are satisfied:

$$\frac{d\mu}{dS} = 0 \quad \mu(S_s) = -\frac{m_s}{\gamma_{xs}} \quad (19)$$

The second condition is obviously never satisfied. Consequently, the only possible bifurcation for this system is a fold bifurcation occuring when $d\mu/dS = 0$.

The necessary first order condition for optimal productivity, $J_1 = DX$, can be rearranged into the following form

$$\frac{d\mu}{dS} = \frac{\gamma_{xs}\mu(S_s^{opt}) + m_s}{\gamma_{xs}\mu(S_s^{opt}) + 2m_s} \frac{\mu(S_s^{opt})}{S_{is} - S_s^{opt}} > 0 \quad (20)$$

Thus the operating point corresponding to optimal productivity is a stable node. This conclusion will also be reached in case of optimization of the produced cells for harvesting, i.e. $J = DX_e$, as $X_e = \alpha X$. Although this condition states that the optimal operating point always will be stable it does not explicitly provide neither the distance to the critical dilution rate nor the effect that uncertain parameters may have on the actual operation of the system.

For cases in which the cost of substrate is not negligible compared to the value of the produced biomass, the operating profit is a realistic normative measure of the operation of this continuous fermentation. With the above expression it can be shown that if the operating profit is positive, i.e. it is desirable to operate the fermentor, then the optimal operating point will be a stable node.

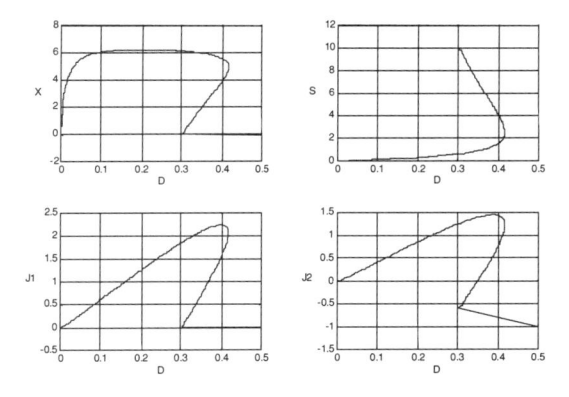

Figure 3. Bifurcation diagram for a continuous fermentor with cell recycle ($\alpha = 0.8$) and substrate inhibition kinetics. $J_1 = DX$ and J_2 is J with $\kappa = 0.2$.

Substrate Inhibition Kinetics

In a medium in which growth is inhibited by the limiting substrate at high concentrations the specific growth rate can be described by (Nielsen and Villadsen, 1994):

$$\mu(S) = \mu_{max} \frac{S}{K_s + S + S^2/K_i} \quad (21)$$

Using the continuation methods, e.g. as described by Allgower and Georg (1990), the steady state biomass, substrate, productivity, and operating profit is computed as function of the dilution rate.

From the graphs in Fig. 3 it is apparent that both the optimal productivity and operating profit is on the stable branch of the equilibrium curve. However, it is also apparent that the optimum lies very close to the critical dilution rate, i.e. the dilution rate at which a vertical slope of biomass wrt. dilution rate occurs.

In practice the system described cannot be operated at the optimum operating point in open loop because disturbances may render the system unstable resulting in washout of the biomass. The consequence being that the system must be shut down and restarted.

Discussion and Control Implications

The above examples illustrate for relatively simple cases how the necessary optimality condition Eqn. (7) may provide insight into the possible occurrence of complex behaviors. If the optimization variable as in this paper is dilution rate then J_u takes a constant value, which has to be matched by the second expression of Eqn. (7). That expression contains three terms of which f_x^{-1} and f_u are most interesting. In the first example above Eqn. (7) was satisfied through variation in f_u. And that variation corresponded to movement of a transfer function zero. In the bioreactor case Eqn. (7) was satisfied through variations in f_x^{-1} which corresponds to movement of poles and where the bifurcation occurs as the pole crosses the imaginary axis. In the former case the zero will deteriorate the achievable performance if it is located in the right half plane. To avoid such deterioration application of multivariate control should be considered. In the case of the single reversible reaction in a CSTR (Fox et al., 1984) the reactor temperature may be used as manipulable variable to reach and maintain the optimal operating point. In the case of possible bifurcations that imply poles moving into the right half plane then closed loop control may be indispensable to avoid disturbances from destabilizing the process operation. In both cases significant advantages may be gained through a closer study of the particular complex behavior in order to apply a suitable nonlinear control design.

The above examples illustrate that complex operational behaviors indeed occur around operational optima. The occurrence of complex behavior is seen to depend upon both the process and upon the cost dependence of the objective function. Even though the complex behavior may be encountered at other operating points than the optimal operating point, the occurrence of disturbances renders it often advisable to perform operation in closed loop in order to ensure profit optimal operation. This paper has demonstrated a procedure for investigating the occurrence of complex behaviors. Additional questions are related to selection of control structures and also how to possibly modify the process design to change the complex behaviors. An extension of this question is how to take advantage of nonlinear behaviors.

References

Allgower, E. L. and K. Georg (1990). *Numerical Continuation Methods. An Introduction.* Springer Verlag

Bekiaris. N., G. A. Meski and M. Morari (1996). Multiple Steady States in Heterogeneous Azeotropic Distillation. *Ind. Eng. Chem. Res.* **35,** 207-237.

Bonvin, D., R. G. Rinker and D. Mellichamp (1980). Dynamic Analysis and control of a Tubular Autothermal Reactor at an Unstable State. *Chem. Eng. Sci.* **35** 603-612.

Esbjerg, K., T. R. Andersen, D. Muller, W. Marquardt and S. B. Jorgensen (1998). Multiple Steady States in Heterogeneous Azeotropic Distillation sequences. Submitted for publication to *Ind. Eng. Chem. Res.*

Farr, W.W. and R. Aris (1986). Yet who would have thought the old man to have so much blood in him? - reflections on the multiplicity of steady states of the stirrred tank reactor. *Chem. Eng. Sci.* **41** 1385-1402.

Fox, J.M., J. W. Schmidt and J. C. Kantor (1984). Comparison and Feedback control for Optimized Chemical

Reactors. In Proceedings of the 1984 American control conference, San Diego, CA; 1621-1627.

Halbe, D.C. and A. B. Poore (1981). Dynamics of the Continuous Stirred Tank Reactor with Reactions A to B to C. *Chemical Engineering Journal* **21**, 241-253.

Heath, J., J. D. Perkins and S. Walsh (1996). Control Structure Selection Based on Linear Dynamic Economics – Multiloop PI Structures for Multiple Disturbances. Tehnical Report, Centre for Process Systems Engineering, Imperial College, London

Jorgensen, S.B.; Moller, H.E; Andersen, M.Y. (1992): Adaptive Control of Continuous Yeast Fermentation. *ACS Conference Proceeding Series: Harnessing Biotechnology for the 21st Century*, 364-369

Kravaris, C. and P. Daoutidis (1990). Nonlinear state feedback control of second order nonminimum-phase nonlinear systems, *Comp. Chem. Engng.*, **14**, 439

Nielsen, J. and J. Villadsen (*1994*). *Bioreaction Engineering Principles*. Plenum Press

Recke, B. and S. B. Jørgensen (1997). Nonlinear Dynamics of a Fixed Bed Reactor with Recycle, *Fractals and Chaos in Chemical Engineering*, Ed. M. Giona and G. Biardi, World Scientific, 652-663

Seider, W. D., D. D. Brengel, A. M. Provost and S. Widagdo (1990). Nonlinear Analysis in Process Design. Why Overdesign to Avoid Complex Nonlinearities? *Ind. Eng. Chem. Res.* **29**, 805-818

Sistu, P. B. and B. W. Bequette (1995). Model Predictive Control of Processes with Input Multiplicities. *Chem. Eng. Sci.*, **50**, 921-936

Uppal, A; Ray, W.H.; Poore, A.B. (1974) On the Dynamic behaviour of continuous Stirred Tank Reactors. Chem.Eng.Sci. **29**, 967

DYNAMIC SIMULATION AND CONTROL STRATEGY EVALUATION FOR MTBE REACTIVE DISTILLATION

Douglas A. Bartlett
Oliver M. Wahnschafft
Aspen Technology, Inc.
Cambridge, MA 02141

Abstract

This paper investigates the control of a reactive distillation column producing methyl tertiary butyl ether (MTBE) from isobutylene and methanol. Extensive literature reports indicate that this process is prone to multiple steady states for the same set of feed and operating conditions, to the point where the conversion of isobutylene to MTBE can vary from nearly 100% to close to zero.

These results can be predicted through the use of rigorous modeling tools. Steady-state sensitivity studies show that the column can produce different results, depending upon the path taken to reach a given set of operating conditions. This phenomenon can be explained through the use of residue curve analysis, by identifying how distillation boundaries affect the overall separation. Finally, dynamic simulation is used to determine how to keep the reactive system operating at high conversion and how to respond should an upset occur. Different control strategies and disturbances are examined, revealing the inherent limitations of traditional feedback control for this process. Using a hybrid feedforward/feedback control strategy, however, it is possible to reliably produce high purity MTBE at maximum conversion of isobutylene and methanol.

This paper demonstrates the insights that can be gained from advanced modeling technologies, including tools for physical properties analysis, steady-state simulation, and dynamic modeling. It also illustrates why a combination of tools is often necessary to arrive at the optimal process and control design.

Keywords

MTBE, Dynamic simulation, Reactive distillation, Control design, Residue curves, ASPEN PLUS®, SPEEDUP™.

Introduction

Methyl tertiary butyl ether (MTBE) is primarily used as a gasoline additive. As an oxygenated fuel, it serves both to increase octane and reduce carbon dioxide emissions from automobiles. The elimination of leaded gasoline in most parts of the world has led to a significant market demand for MTBE, encouraging producers to find ways of maximizing production from existing facilities.

MTBE is produced in a reaction between isobutylene and methanol in the presence of an acid catalyst. The reaction is given by Eqn. (1):

$$CH_3OH + (CH_3)_2C=CH_2 \rightleftharpoons (CH_3)_3COCH_3 \quad (1)$$

Methanol   Isobutylene   MTBE

MTBE was traditionally produced in a liquid-phase fixed-bed reactor system. A typical configuration consists of two fixed-bed reactors in series. Methanol is introduced as a pure feed, while isobutylene is introduced as part of a mixed C_4 stream, typically 35% isobutylene and 65% n-butene. The reaction products are unreacted methanol and isobutylene, MTBE, and the inert n-butene. These are separated by a combination of distillation and extraction columns, with the unreacted methanol and isobutylene recycled back to the reactors.

Isobutylene conversion is limited by thermodynamic equilibrium to between 80 and 85 per cent. As the conversion limit increases at lower temperatures, a combination of effluent recycle and cooling water is required to minimize the exothermic temperature rise. MTBE production is therefore constrained by the tradeoff between conversion and available cooling capacity.

Reactive distillation offers significant improvements over the fixed-bed system. A typical installation includes two nonreactive trays at the top, followed by eight reactive trays and five stripping trays. MTBE is removed as the bottoms product. The primary advantage of this system is that the overall conversion is no longer limited by thermodynamic constraints. By continuously removing the reactants from each stage, conversion rates are increased so that nearly all isobutylene is consumed (methanol is kept in slight excess to avoid the formation of byproducts such as dimethyl ether). In addition, the exothermic heat of reaction (-37.7 kJ/mol) provides much of the heat required for separation, reducing utility costs. Finally, older plants can be upgraded without adding a new column, by replacing the internals of the primary distillation column just downstream of the fixed-bed reactors. Reactive distillation is the single most effective method for increasing capacity of an existing unit, to the extent that virtually all MTBE producers are now using it.

Although heavily favored by process economics, reactive distillation creates a number of operational challenges. The methanol-butene-MTBE mixture is highly nonideal. Methanol forms azeotropes with MTBE, n-butene, and isobutylene, creating distillation constraints at both ends of the reactive distillation column. As shown by the residue curve map in Fig. 1, the methanol/C_4 azeotrope occurs at approximately 9% methanol. This azeotrope lies at one end of a distillation boundary, which traverses to the methanol/MTBE azeotrope and divides the separation space into two regions. In the upper right-hand region, the residue curves start at the methanol/C_4 azeotrope and end at MTBE. Given sufficient boilup and number of trays, pure MTBE can therefore be recovered as the bottoms product, as long as the liquid composition at the top of the stripping section is within this region. If the composition entering the stripping section lies to the left side of the boundary, the relative volatility reverses, leading to methanol, instead of MTBE, as the bottoms product.

The top of the column is similarly constrained. Under normal circumstances, the overhead product is primarily n-butene, with small amounts of unreacted isobutylene and methanol. Although a slight methanol excess is desirable, the methanol/C4 azeotrope serves as an upper limit on the amount of methanol that can be removed through the overhead without causing operational difficulties. This points to the need for tight control of the feed stoichiometry.

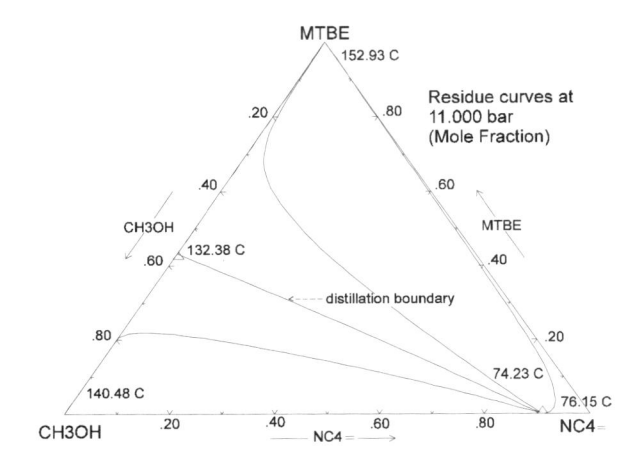

Figure 1. Residue curve map for MTBE, methanol, and n-butene.

The effects of these nonidealities lead to the possibility of multiple steady states. Nijhuis (1993) and Jacobs (1993) used steady-state simulation to show that, for a given set of operational parameters such as feed composition, feed location, and reflux ratio, the column can produce significantly different results, depending upon the starting point used for the simulation. When the methanol feed is introduced near the bottom of the reactive zone, for example, isobutylene conversion can vary from nearly 100% to close to zero for the same set of operating conditions

The purpose of this paper is to build upon these steady-state results by using dynamic simulation. In this instance, steady-state simulation can warn us of a potential problem, but it does not tell us what to do about it. For example, what kind of disturbance would cause the system to shift from high conversion to low conversion? If this happens, how should the operators respond? Finally, can we design a control strategy that will keep the system running at high conversion under most circumstances and return the system to the desired conditions following a large disturbance? These questions can be answered by dynamic simulation.

Problem Description

The reactive distillation column design used for this paper is illustrated by Fig. 2. The column has 17 theoretical stages, including the condenser, reboiler, and 15 column stages. The reactive zone comprises stages 4 through 11 (from the top). The mixed C₄ stream enters on the bottom reactive tray, stage 11, with the methanol one tray above.

Most MTBE plants include both fixed-bed reactors and reactive distillation columns. This is largely a consequence of retrofits – the fixed-bed reactors are there, so they might as well be used – but also because of catalyst costs. The catalytic trays are expensive, making it cost-effective to do the initial reaction using bulk catalyst beds, using the reactive distillation column for final reaction and purification. In this paper we consider reactive distillation only, sending the unreacted feeds to the column directly.

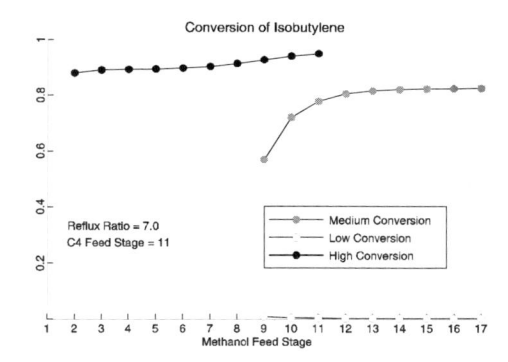

Figure 3. Sensitivity to methanol feed stage.

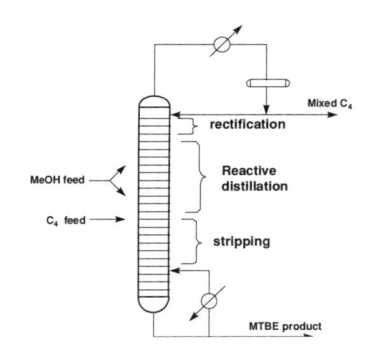

Figure 2. MTBE reactive distillation problem description.

The reactive distillation column model was configured using the rigorous distillation model RADFRAC within ASPEN PLUS®. Kinetic reaction equations were specified for the reactive trays via user subroutines. With this setup we were able to demonstrate the multiple steady-state phenomena noted by the earlier papers. Fig. 3 illustrates a key result. Isobutylene conversion is plotted against the methanol feed stage. Isobutylene is fed to stage 11, while the reflux ratio and bottoms product flow rate are held constant. There are three conversion paths, each generated by starting at a different point and changing the feed location one stage at a time. Multiple steady states exist for all methanol feed locations below stage 8. While it might be safest to operate with the methanol feed above stage 8, the isobutylene conversion increases as the feed stage is lowered within the reactive zone, providing a financial incentive to lower the feed tray.

Similar effects can be seen from sensitivity studies of changes in reflux ratio and mass of catalyst. A composition profile provides additional insight. Fig. 4 shows the composition profile for the high conversion case. While the MTBE concentration stays below 11% throughout the reactive zone, the methanol concentration remains even lower and is well below 1% at the top of the stripping section. This places the composition on the right side of the distillation boundary, allowing for the production of nearly pure MTBE from the bottoms.

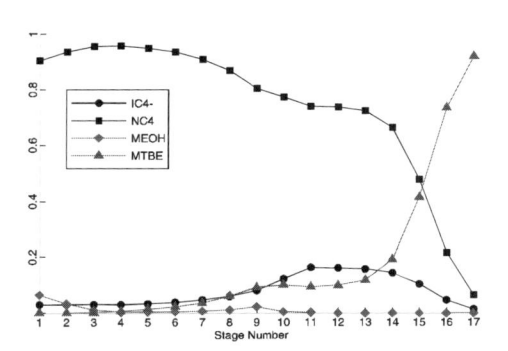

Figure 4. Composition of liquid for high conversion case.

The composition profile is significantly different for the low conversion case, as seen in Fig. 5. Reaction rates are excessive in the top reactive trays, with MTBE peaking at nearly 50% on stage 8. This leads to MTBE decomposition and an increase in methanol on successively lower stages. The composition at the top of the stripping section ends up on the wrong side of the distillation boundary, leading to pure methanol as the bottoms product. MTBE is driven back into the reactive zone where it completely decomposes.

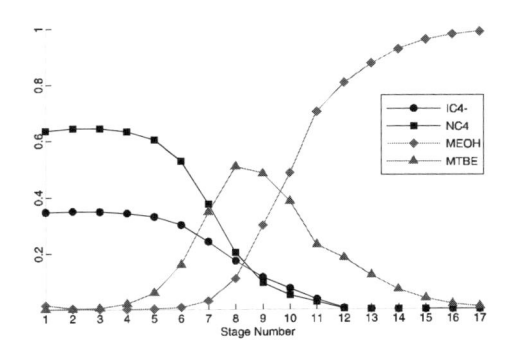

Figure 5. Composition of liquid for low conversion case.

It can therefore be seen that while the external conditions such as feed location and reflux ratio may be the same for the high and low conversion cases, the internal column conditions are much different. Can these internal conditions be measured to provide sufficient information for control? Also, are the changes in these internal variables directionally consistent? For example, does a falling temperature always indicate the need for more heat? If not, automatic control becomes very complex, if not impossible.

Dynamic Simulation Results

To ensure consistency of results, the dynamic model was generated from ASPEN PLUS directly through the use of DynaPLUS™. DynaPLUS produced a dynamic version of the MTBE reactive distillation column that could run in the dynamic simulation program SPEEDUP™. All input parameters, including feed stream definitions, reactions, and physical property routines, were carried over.

The steady-state problem definition provides only a starting point for a dynamic model. The user must also specify equipment sizing and internal configuration details. This includes column diameter, reflux drum dimensions, and condenser and reboiler details. A key specification concerns the internal stage calculations. For the MTBE simulation, we specified the use of structured packing to represent the both the reactive and nonreactive stages. This allows for the dynamic calculation of both pressure drop and holdup on each stage.

The final set of specifications involves the control strategy. Controllers are the dynamic equivalent of a steady-state design specification. The major difference is that a steady-state model will merely provide the final answer, while a dynamic model calculates the path taken between two sets of operating conditions.

After adding these additional specifications, we used the dynamic model to identify how the system responded to disturbances and which variables could be used for control. Starting with the process running at high conversion, producing a nearly pure MTBE bottoms product, we would introduce a disturbance and watch how the process responded. We would then modify the control strategy and retest the same disturbance, or try another disturbance variable. The key disturbance variables were methanol feed flow, butene feed concentration, and reflux flow rate. Individual case studies are presented below.

Dynamic Model Validation

The first test of the dynamic model was to duplicate the steady-state findings by identifying a disturbance that could lead to a second steady state. For this test the column was set up with the same controls that are implicit in the steady-state simulations. The overhead system includes a pressure controller that manipulates the condenser coolant flow and a reflux drum level controller manipulating the reflux flow rate. A reflux ratio controller adjusts the distillate product flow in response to reflux flowrate changes. At the bottom of the column, the MTBE product flow rate is held constant, requiring that the bottoms level controller manipulate the reboiler steam flow. Note that this strategy would not necessarily be recommended for normal operations. It does, however, duplicate the design specifications used in the steady-state studies, which included fixed pressure, reflux ratio, and bottoms flow rates.

Figure 6 is a plot of the process response to a feed disturbance. The methanol feed flow is increased by 15%, held constant for one hour, and is then returned to its starting point. As can be seen in the plot, the bottoms MTBE concentration starts at nearly 100%, begins to fall soon after the pulse is introduced, and never recovers. Thus, even though the feed flow returns to its initial condition, the process ends up in a new steady state in which the bottoms product is less than 65% MTBE.

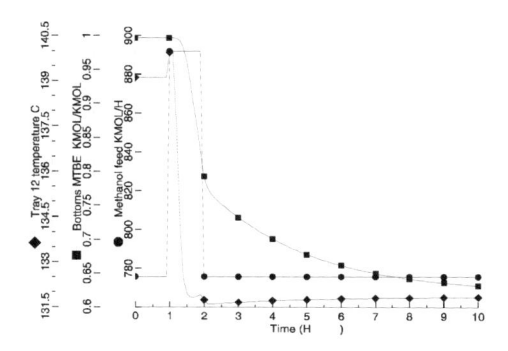

Figure 6. Methanol pulse test with constant bottoms flow.

We have thus shown that multiple steady states are possible, verifying the steady-state results. We can now test alternative control strategies to determine how to keep the process running at high conversion.

Stripping Section Temperature Control

A common distillation control strategy is the use of temperature as an inferential variable for bottoms product purity. A stripping section temperature is typically used, taken at a point where there is a significant temperature gradient across several trays for good sensitivity to compositional changes. As shown in Fig. 6, the temperature on tray 12 appears to mimic the bottoms MTBE concentration, dropping by about 8 degrees C after the disturbance and never returning to its starting point.

As we saw from the residue curve analysis, stoichiometry is also important. A simple mechanism for maintaining the desired stoichiometry is a feed ratio controller that is based upon the mass flowrates of the two feed streams. While this does not account for C_4 feed compositional changes, it is easily implemented and serves as a good starting point for further studies.

The resulting control strategy is shown by Fig. 7. The feeds are kept in mass ratio with a ratio controller, while the temperature on tray 12 is controlled by manipulating the reboiler steam flow. Bottoms level is maintained by manipulating the bottoms product flowrate, allowing the column to adjust to changes in the MTBE production rate.

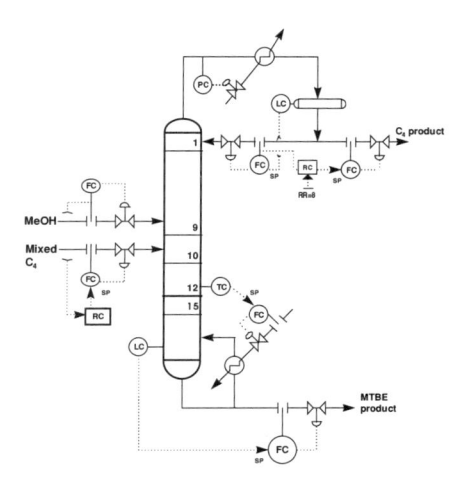

Figure 7. Tray 12 temperature control.

This control is partially successful in that it returns the process to high conversion after the methanol pulse is complete. However, the MTBE purity drops significantly during the pulse itself, to 90%, indicating that the process is not in control.

The shortcomings of this control strategy were even more apparent from the results of a second test, in which the isobutylene feed concentration was changed from 35.7% to 45%. In this test, the MTBE purity fell to below 90% and never recovered. Fig. 8 provides some insight into what happened. The isobutylene concentration was increased one hour into the simulation. The temperatures at tray 12 and in the reboiler both drop by about 10 degrees C. During this time the temperature controller is increasing the heat load to try to raise the temperature, to no avail. In contrast, the temperatures in the reactive zone have increased – by nearly 25 degrees C in the case of tray 8. This indicates that the reaction rates are too high and that the reboiler duty should be decreased, not increased. Increasing the duty only makes the situation worse.

Tray 12 temperature is therefore inappropriate for control. While it correctly returned the process back to high conversion following the methanol pulse test, for the composition change it responded in the wrong direction. A better controlled variable would be the temperature on tray 8.

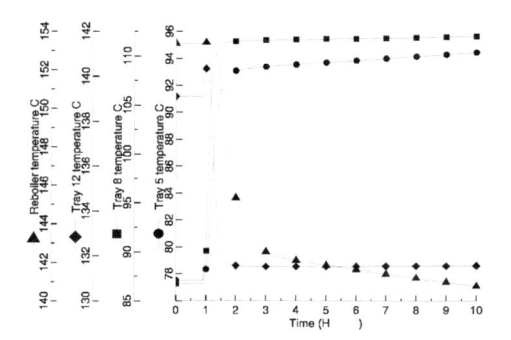

Figure 8. Tray 12 temperature control response.

Reactive Zone Temperature Control

The control strategy was revised by replacing the tray 12 temperature control with tray 8, keeping all other controls the same. The results are shown by Fig. 9. While the tray 8 temperature still rises by 20 degrees C, the controller correctly responds by reducing the reboiler duty, causing the tray 8 temperature to eventually return to its set point. This keeps the reaction rates low, resulting in only a minor variation in the MTBE purity.

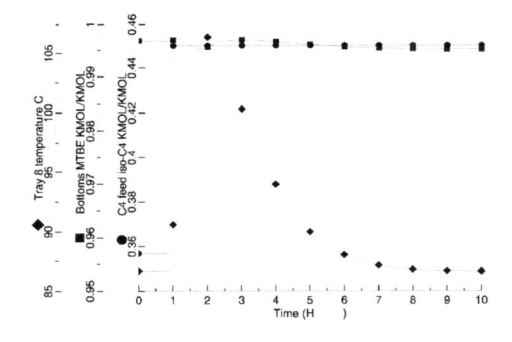

Figure 9. Tray 8 temperature control response.

Further studies tested the performance of the reflux ratio control. These showed that the column responds faster to a feed disturbance if the reflux flowrate is simply kept constant, with the reflux drum level controller manipulating the product flowrate. The reflux ratio controller accentuates the effects of a disturbance by changing the reflux flowrate, which alters the heat balance and interacts with the tray 8 temperature controller. It is much better to keep the reflux flow constant, smoothing out overhead disturbances by sending a constant cooling duty back to the column. This is consistent with the steady-state studies, which showed that changes in reflux ratio had little effect on the overall column performance.

A final set of tests showed that the tray 8 temperature control does have limitations when there is too much methanol in the system. The overhead methanol concentration is limited to approximately 9 mole % by the methanol/C₄ azeotrope. As the methanol concentration approaches this level, it gets driven back into the reactive zone, creating excessive reaction rates, which lead to MTBE decomposition.

The tray 8 temperature controller fails to maintain MTBE purity under these circumstances. Interestingly, the temperature itself can be brought under control, as the increased reaction rates are primarily in the upper stages of the reactive zone. However, the excess methanol must leave the column somehow, and as it cannot be through the top, it leaves through the bottoms. In a step test where the feed isobutylene concentration was reduced from 37.5% to 30%, the bottoms MTBE concentration dropped to 88%, with the balance being methanol.

It is therefore essential to maintain tight control over the feed stoichiometry to avoid excess methanol. However, as a slight excess is required to minimize byproduct formation, the control must operate within a relatively narrow band.

Hybrid Feedforward/Feedback Control

Because the boiling point of the methanol/C4 azeotrope is very close to that of pure n-butene, temperature cannot be used to infer changes in overhead methanol concentration. The stoichiometric balance must therefore be determined from online analyzers or back calculated from downstream product flow measurements.

Two control strategies were tested – overhead methanol concentration control and a feed molar ratio control. Both were configured to adjust the feed mass ratio controller set point. Both responded well to changes in feed isobutylene concentration, correctly adjusting the feed mass ratio to eliminate excess methanol. An advantage of the overhead methanol controller is that it can respond to changes in catalyst activity as well as to stoichiometric imbalance. It could therefore be used as part of a constraint-pushing strategy to maximize throughput. A disadvantage is its response to a reflux disturbance, resulting from operator set point changes or cooling water disturbances that impact the degree of subcooling. Although a change in reflux has little impact on the steady-state results, it does affect the fractionation dynamically, at least until the temperature controller can respond by adjusting the reboiler. This unfortunately sets up an unfavorable interaction between the methanol concentration controller and the tray 8 temperature controller. The former responds by adjusting the feed ratio, which changes the heat of reaction, changing the reactive zone temperatures. This affects the tray 8 temperature controller, which is already responding to the reflux change. The interaction quickly gets out of hand, setting up a massive oscillation that can only be stopped by taking the controller off control.

The molar feed ratio controller avoids this problem completely, as it is a feedforward controller, acting upstream of the column. This feedforward strategy can then be combined with the tray 8 temperature controller in a hybrid control strategy that maintains optimal reaction conditions while responding to external disturbances. As shown in Fig. 10, the tray 8 temperature control can also be supplemented with a bottoms nC₄ concentration controller to fine-tune the MTBE purity. The response to a butene concentration disturbance is shown by Fig. 11.

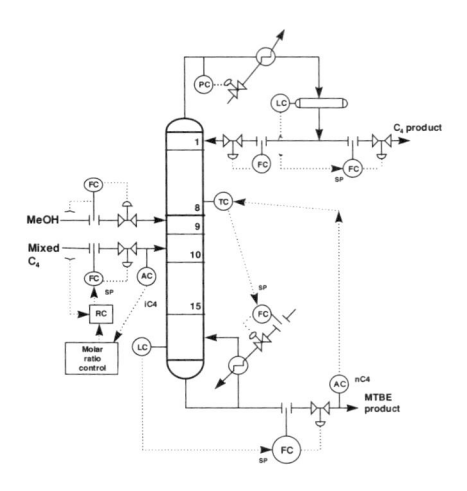

Figure 10. Hybrid feedforward/feedback control.

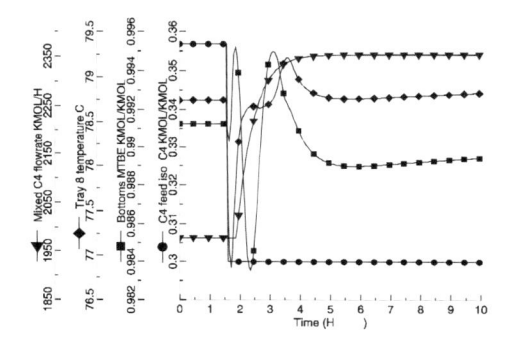

Figure 11. Hybrid control response to feed disturbance.

Conclusions

Dynamic simulation has been shown to be an effective tool for the analysis of a reactive distillation column. Keys to its effectiveness are the insights learned from steady-state modeling and the integration of the two tools to ensure consistent results. Finally, while the study has shown that it is possible to configure an effective control strategy for this process using traditional control techniques, it has also demonstrated the complexity of the MTBE system. The need for tight control of feed stoichiometry, its severe response to process constraints, and the high degree of controller interaction make this process an excellent candidate for multivariable predictive control.

References

Jacobs, R. and R. Krishna (1993). Multiple solutions in reactive distillation of methyl tert-butyl ether synthesis. *Ind. Eng. Chem. Res.*, **32, No. 11**, 1706-1709.

Nijhuis, S.A., F.P.J.M. Kerkof, and A.N.S. Mak (1993). Multiple steady states during reactive distillation of methyl tert-butyl ether. *Ind. Eng. Chem. Res.*, **32, No. 11**, 2767-2774.

REAL-TIME OPTIMIZATION OF A FCC RECOVERY SECTION

John Brydges, Andrew Hrymak and Thomas Marlin
Department of Chemical Engineering
McMaster University
Hamilton, Ontario, Canada L8S 4L7
email: marlint@mcmaster.ca

Keywords

Real-time optimization, Fractionation.

Introduction

Due to improvements in numerical optimization methods and increases in computing speed, Real-Time plant Operations optimization (RTO) of large, integrated plants has become an attractive method for increasing the profit of continuous chemical processes (e.g., Marlin and Hrymak, 1996). A large number of decisions must be made when designing and implementing a RTO system, and much progress has been made in topics such as selecting measurements (Krishnan et al, 1992) and selecting model parameters for updating (Forbes and Marlin, 1996). This paper addresses the ever important issue of model fidelity, specifically the necessary structure of the model so that it represents plant interactions and flexibility resulting from heat integration, material recycles and so forth. In general, one expects a more fundamental and complex model to provide a better representation of the plant but to introduce greater optimization challenges and require longer computing times. The major goal of this study is to evaluate the value of high fidelity models.

The FCC Process

This study investigates the optimization of the fractionation product recovery section of a Fluid Catalytic Cracking (FCC) plant in a petroleum refinery given in Figure 1. The feed to this section is the effluent from a reactor-regenerator, which converts a heavy petroleum fraction to form lower boiling gasoline and fuel oil products, along with gas and heavy oil by-products. This is an excellent example because of the complexity of the equipment, large number of stream components, extensive material and energy recycles, and relatively large number of optimization variables.

A key material integration involves two liquid streams in the naphtha range which are sent forward from the fractionator and recycled back from the debutanizer to the gasoline absorber; all of this liquid is ultimately processed in the distillation section and removed as product in the debutanizer bottoms. In addition, "light cycle" fuel oil is sent forward to the sponge oil absorber and subsequently returned to the fractionator. Heat integration is important for the efficient operation of the process. The depropanizer and debutanizer have feed-effluent heat exchange, and the main fractionator in the process has three "pumparounds", in which liquid is withdrawn from the tower and passed through one or more heat exchangers before being returned to the tower. Thus, pumparounds are mid-tower condensers. Naturally, all condensation could occur at an overhead condenser, which would maximize the fractionation for the given number of trays and vapor flow. However, the recovery of large amounts of energy at high temperatures militates for mid-tower heat recovery. Each of the pumparounds is now described briefly.

- The *top pumparound* provides heat for a distillation tower in another part of the plant. If the distillation tower requires more heat transfer to the reboiler than is possible via the top pumparound, an auxiliary reboiler using steam is available.

- The *fuel oil or "light cycle oil (LCGO)" pumparound* provides heat transfer for the depropanizer reboiler. If the depropanizer reboiler duty is greater than can be supplied by the pumparound, an auxiliary steam reboiler is provided. For situations where the desired cooling of the pumparound is greater than the depropanizer reboiler duty, an air-fin heat exchanger is provided. This process configuration is shown in Figure 2 with bypasses that provide operating flexibility.

- The *"heavy cycle oil" pumparound* reboils the deethanizer and has a process configuration similar to the light cycle oil pumparound.

- The *bottoms pumparound* provides heat at the highest temperature to several consumers in the FCC reactor and fractionator sections and for generating, medium pressure steam as shown in Figure 3. Note that the heat transfer in three of the heat exchangers is determined by control systems for the stripper, distillation tower and reactor feed control. However, flexibility exists to adjust the steam production as well as to distribute the pumparound return flow between the flash zone (feed tray) and the tower bottoms.

Clearly, the material and heat integration results in strong interactions among the distillation towers in this plant section, which complicates the optimization. Some of the key factors found to be important in modeling this process for RTO include 1) material recycle used in absorbers to recover light ends, 2) heat integration including by-passes, 3) flexible responses to disturbances using by-passes, waste heat exchangers (to air) and spare steam heat exchangers, and 4) limitations to equipment performance.

RTO Model and Solvers

A typical design-quality simulation of this process would involve fundamental distillation models with detailed physical property data. However, because equipment sizing would be performed subsequent to completion of the flowsheet, the flowsheet would likely involve short-cut models for heat exchange and ignore equipment limitations. However, these factors are crucial for determining the best operation of an existing process.

The FCC fractionation section in this study separates 47 components/pseudo-components into 6 product streams using 93 ideal fractionator stages, in 7 distillation units, 18 heat exchangers, 11 mixers/splitters, and 1 two-stage compressor. Fundamental models of these units were used, with appropriate efficiencies used to relate the work performed by the compressor to its energy consumption (Brydges, 1995). The standard models were built via the Aspen Plus model manager, and the tailored models were added via the PLTCON equation editor. Accurate physical properties were estimated using the Grayson-Streed method of corresponding states (Aspen, 1993).

The base case solution was evaluated using the Aspen Plus™ sequential modular flowsheeting system. The optimization studies were performed using the RT-OPT™ open-form modeling system with appropriate variable scaling solved with a feasible-path sequential quadratic program using analytical first derivatives (Aspen, 1995). Each optimization case involved over 14,000 equations and variables.

RTO Design

The optimization of the process is determined by a complete definition of the operating objectives; the operating goals and economic values (in US $) for this study are taken from public-domain sources (Anon, 1993). First, the products must be within specification limits for key measured properties, which are given in Table 1. Second, the profit depends on the product values and the energy costs for steam (generated and consumed) and for power to the compressor. The feed rate and composition to the fractionation section is not influenced by the optimization variables, and therefore, the constant contribution due to the feed value is not needed to optimize the operating variables. Finally, the optimization variables are selected, and the allowable ranges for adjustments are defined in Table 2. The RTO model involved from 21 to 23 optimization variables that were allowed values between their upper and lower bounds. Constraints were placed on product qualities and equipment performance, which along with variable bounds defined the operating window for the process. Again, the profit represents the instantaneous difference between the costs and revenues for the steady-state operation.

RTO Results and Interpretation

An important first step in optimization evaluation is the selection of the base case operation, which, if unrealistically poor, will result in the benefits for optimization being overstated. The base case was selected to represent good operation. Where astute judgment would identify that best operation occurred at a constraint, that variable was set to its bound; for other variables, good practice was used in selecting values that satisfied product quality and equipment constraints (Lieberman, 1985; Montgomery, 1993; Watkins, 1979). Naturally, the base case profit of $23836.1 per hour is quite high because constant feed costs, fixed costs and capital recovery are not considered in the calculation.

Optimization Case A reported in Table 1 involves determining the best values for 21 optimization variables. Since the feed was constant, only improving product recovery and/or reducing energy consumption can increase profit. The major improvements were 1) an increased recovery of material in the propane product mostly from the fuel gas product (through increased adsorption liquid flow rates) and 2) reduced energy usage (through increased heat recovery from the fractionator pumparounds). The profit was increased about $102 per hour or about $900,000 per year. Interestingly, the fractionator bottoms temperature was minimized, while literature recommendations indicate that the temperature should be maximized to thermal cracking limits (Lieberman, 1985). A sensitivity case of the full non-linear model indicated a very small effect of this variable on profit (~ $0.10/h per °C), which over a small range indicates that the variable has negligible effect on profit.

Side cases around Case A at moderately different feed rates indicated no change to the key operating variables. These results were expected because the effects of flow rate on efficiencies were not included in the model. Further side cases with different feed compositions, which were based on changes in the FCC reactor conversion, required only one variable, the light cycle oil flow to the sponge oil absorber, to be adjusted significantly in response to the change in composition.

Case B added one optimization variable, the absorber pressure, to Case A. The major effects were an increase in propane recovery and an increase in compressor work. However, deethanizer reboiler duty was increased due to the decreased relative volatility at the higher pressure; reduced fuel oil to the sponge absorber was required; and the fractionation in the depropanizer was deduced. The profit was increased by about $140,000 per year over Case A.

Case C added a different optimization variable to Case A, the debutanizer bottoms naphtha recycle to the gasoline absorber. The recycle was increased by a factor of nearly three from the base case, which resulted in increased propane recovery and a profit increase of about $300,000 per year over Case A. Since Cases B and C obtain benefits from the same general effect of increasing light end recovery, the sum of the profit increases from these cases could not be achieved, were both variables available for optimization.

Case D involved a reduction in the debutanizer condenser duty of 20% from Case A. If this were a short-term situation, the plant operations personnel might want to maintain the operations of the reactor and fractionator nearly unchanged to be able to quickly return to normal operation after the full heat exchanger capacity has been restored. Therefore, this case is defined with the fractionator feed unchanged. Remarkably, the optimization was able to find a feasible operating region, with all product qualities on-specification and equipment within limitations. The optimum operation for the scenario with fractionator feed unchanged has the following characteristics; 1) to reduce the debutanizer overhead, the separation was less pure and the feed preheat was reduced to zero so that all vapor contacted liquid on all trays, 2) the gasoline recycle was reduced to its minimum, 3) greater losses of valuable light ends to fuel gas were required, and 4) the fractionator top pumparound duty was increased to compensate the changes in other duties. The profit at the optimal operation was reduced by only $150 per hour from Case C for this severely distressed operating situation.

Conclusion

While recognizing that the proper decision for a model depends on the specific RTO problem, the results of this study indicate that key models, i.e., those having significant influence on the profit, product qualities and equipment constraints should employ the best available process knowledge. Because of material and energy integration, a unit that has relatively minor effect on the profit, for example, the debutanizer, may need to be modeled accurately because of the interactions with the fractionator and the absorbers. Given the likely occurrence of many disturbances and equipment performance changes over the life of an RTO implementation, it seems advisable to err on the side of higher fidelity models.

Acknowledgment

The authors would like to acknowledge Aspen Technology who provided software and consulting on modeling and the McMaster Advanced Control Consortium who provided financial support.

References

Anon., "Industry Statistics - Table of Refined Product Prices, Oil and Gas Journal", August 16, pg. 92-93, 1993.

Aspen Technology Inc., "Aspen Plus - User's Guide for Release 8", Cambridge, Massachusetts, 1988.

Aspen Technology Inc., "Real Time Optimization System, RT-OPT - User's Manual for Release 1.6", Cambridge, Massachusetts, 1995.

Brydges, J. Issues in Real-Time Optimization: The Recovery Section of a FCC Unit, M.Eng. Thesis, McMaster University, 1995.

Decroocq, D., "Catalytic Cracking of Heavy Petroleum Fractions", Gulf Publishing Company, Houston, Texas, 1984.

Krishnan, S., G. Barton, and J. Perkins, Robust Parameter Estimation in On-Line Optimization-Part I. Methodology and Simulated Case Study, Comp. Chem. Eng., 16, 545-562 (1992).

Forbes, F. and T. Marlin, Design Cost: A Systematic Approach to Technology Selection for Model-Based Real-Time Optimization Systems, Comp. Chem. Eng., 20, 717-734 (1996).

Lieberman, N., Troubleshooting Process Operations, 2nd Ed., PennWell Publishing Company, Tulsa, 1985.

Marlin, T. and A. Hrymak, Real-Time Operations Optimization of Continuous Processes, in Kantor et al, Fifth International Conference on Chemical Process Control, AIChE Symp Ser. 316, 93, 156-164 (1997).

Montgomery, J.A., "Guide to Fluid Catalytic Cracking", Part One, W.R. Grace & Co.-Conn., Baltimore, Maryland, 1993.

Watkins, R.N., "Petroleum Refinery Distillation", 2nd Ed., Gulf Publishing Company, Houston, Texas, 1979.

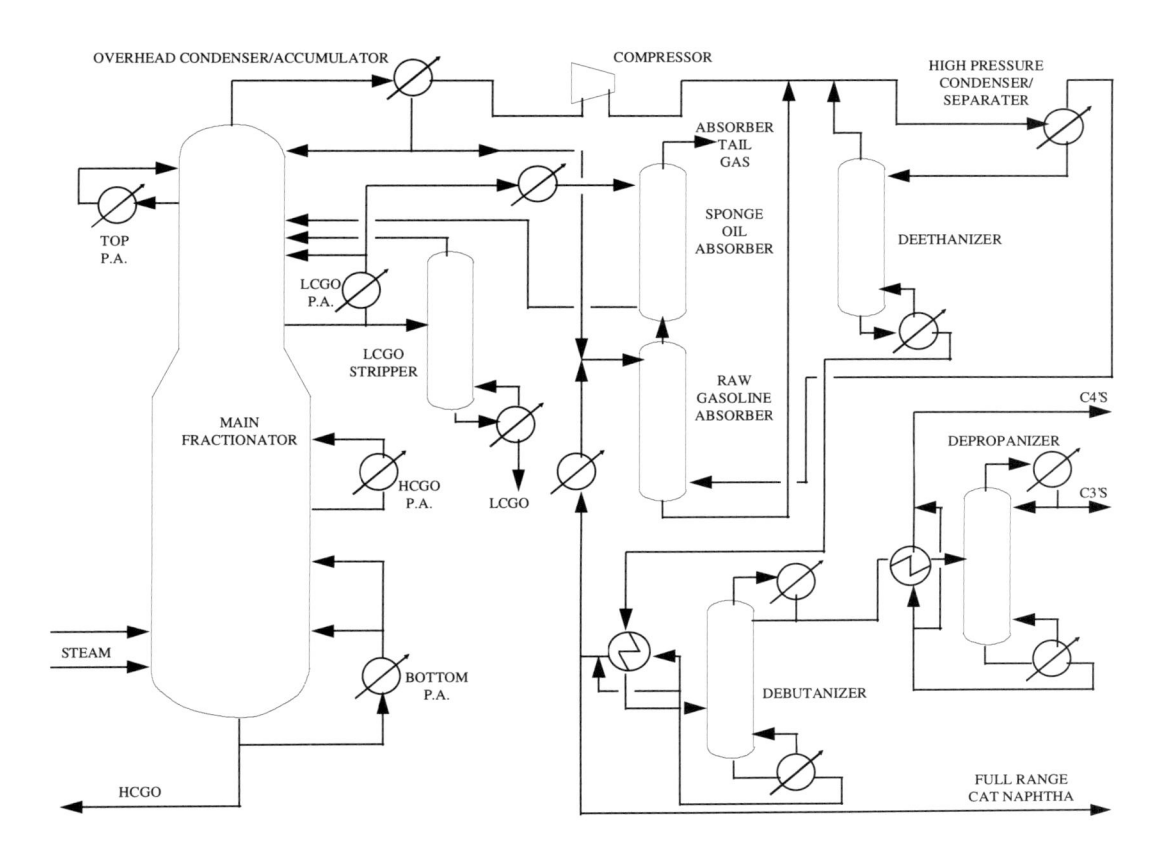

Figure 1. Fluid Catalytic Cracking product recovery fractionation section.

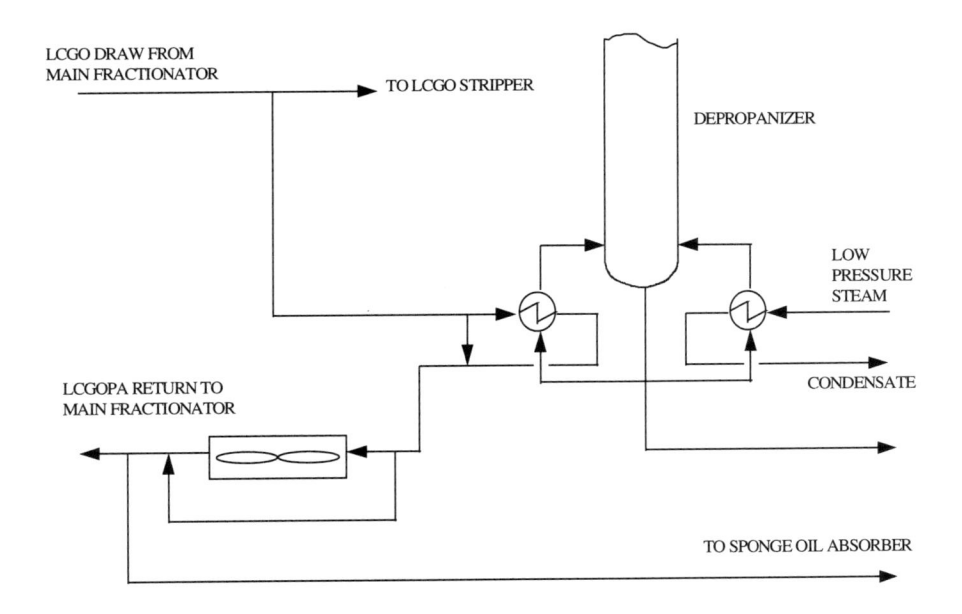

Figure 2. LCGO pumparound.

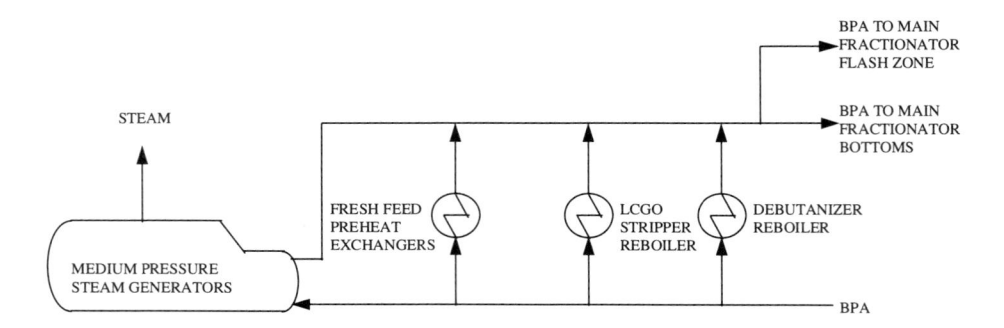

Figure 3. Bottoms pumparound.

Table 1. Product Quality Constraints.

Product	Quality Constraints
C_3	• Mol. % of C_2^- ≤ 4.767×10^{-1} • Mol. % of C_4^+ ≤ 3.222×10^{-1}
C_4	• Vapor Press ≤ 5.8396×10^5 N/m^2 • Vol. % of C_5^+ ≤ 2.0
Naphtha	• Reid Vapor Press. ≤ 4.8263×10^4 N/m^2 • ASTM D 86 End Boil.Pt ≤ 498 K
LCGO	• ASTM D 86 5% Boil Pt ≥ 498 K • ASTM D 86 End Boil Pt ≤ 611 K
HCGO	• ASTM D 86 10% Boil Pt ≥ 498 K • API Gravity ≥ -2.0

Table 2. Optimization Results.

Variable	Units	Case A	Case B	Case C	Case D	Lower bound	Upper bound
Profit	$/h	23938.2	23953.8	23972.4	23824.8		
BPA PA flowrate	kmol/h	881.4	835.2	770.8	889.4	360	54000
LCGO PA flowrate	kmol/h	363.3	315.1	294.4	181.0	90	1800
Fractionator top reflux flowrate	kmol/h	15.8	15.8	15.8	15.8	15.8	100
Debutanizer feed to exchanger	frac.	0.01	0.01	0.01	0.99	0.01	0.99
Depropanizer feed to exchanger	frac.	0.013	0.01	0.01	0.039	0.01	0.99
LCGO to sponge absorber	kmol/h	345.3	297.1	276.4	163.0	36	1800
BPA to fractionator bottoms	kmol/h	301.2	291.0	275.2	304.0	0.36	3600
HCGO PA Duty	GJ/h	-29.47	-32.54	-37.53	-25.17	-360	0
LCGO PA Duty	GJ/h	-16.25	-14.85	-13.93	-9.25	-360	0
Top PA Duty	GJ/h	-47.1	-48.45	-47.91	-60.7	-360	0
Debutanizer reboiler duty	GJ/h	19.91	19.62	22.93	17.4	0	360
Debutanizer reflux ratio		2.32	2.35	2.68	1.74	1	10
Depropanizer reboiler duty	GJ/h	16.25	14.85	13.93	9.35	0	360
Depropanizer reflux ratio		5.48	4.83	4.34	3.02	1	10
LCGO stripper reboiler duty	GJ/h	13.23	12.44	12.01	12.59	0	360
Deethanizer reboiler duty	GJ/h	29.47	32.54	37.53	25.17	0	360
HCGO PA flowrate	kmol/h	260.2	291.7	335.6	218.4	18	1800
LCGO withdrawl flowrate	kmol/h	594.8	537.8	511.1	414.9	18	1800
Top PA flowrate	kmol/h	3584.8	3580.2	3587.1	3604.0	1800	54000
LCGOPA reboiler duty on deprop.	GJ/h	-16.25	-14.85	-13.93	-9.25	-360	0
HCGOPA reboiler duty on deeth.	GJ/h	-29.47	-32.54	-37.53	-25.17	-360	0
Absorber Pressure	$10^6 N/m^2$	2.07(F)	2.28	2.07(F)	2.07(F)	1.86	2.28
Naphtha recycle flowrate	kmol/h	116.5(F)	116.5(F)	337.5	0.36	0.36	1164

(F) = variable fixed at constant value

ONE STEP COLLECTIVE GROSS ERROR IDENTIFICATION AND COMPENSATION IN LINEAR DYNAMIC AND STEADY-STATE DATA RECONCILIATION

Miguel Bagajewicz and Qiyou Jiang
University of Oklahoma
100 E. Boyd St, T-335
Norman, OK 73019, USA

Abstract

A new method of simultaneous gross error identification and size estimation is presented in this paper. A spanning tree of an undirected graph corresponding to a system is first used to capture all gross errors. According to the equivalency theory presented in a previous paper (Bagajewicz and Jiang, 1998), equivalent gross error sets with different size may exist. A mixed integer linear programming model is then constructed and a minimum number of gross errors are obtained as a solution to this model. This method does not require serial compensation, but rather performs all the identification and compensation in one single step.

Keywords

Gross error detection, Data reconciliation.

Introduction

The presence of gross errors invalidates the statistical basis of data reconciliation techniques, either for techniques that assume steady state or perform data reconciliation using dynamic data. Therefore, gross errors must be detected, identified and then either eliminated or corrected to have a valid data reconciliation. Gross error detection has been the object of intense research with various degree of success. As methods based on steady state data reconciliation have not proven to have high power in all situations, attention has shifted lately to the use of dynamic data reconciliation based techniques.

In this paper an integral approach for simultaneously identifying, estimating gross errors and reconciling linear dynamic and steady state plant data is described as an extension of earlier work (Bagajewicz and Jiang, 1997). The basis of the reconciliation procedure is to formally integrate the ODE representing the system and to propose a polynomial representation for all variables. The reconciliation problem reduces then to calculate the coefficients of these polynomials. This paper presents:

a) A generalized model for gross error estimation that includes flowrate biases, volume biases and tank leaks, with extensions and tests to first order biases and leaks.

b) A new Dynamic Measurement Test contains the steady state measurement test as a particular case.

c) An equivalency theory for gross errors. (Bagajewicz and Jiang, 1998)

d) A serial procedure to detect and estimate gross errors.

e) A one step identification and estimation procedure.

Equivalent Set Theory

To motivate the generalized theory we present the following simple examples: (1) For gross error detection purposes, a change in tank volume (or equipment hold up) and a tank or equipment leak can be made equivalent to streams entering/leaving the vessels. The following equivalence holds on the gross error behavior: An order k bias in a flowrate of a stream that feeds/leaves a tank is equivalent to a volume bias of order $k+1$. In turn these two are equal to a tank leak of order k. (2) A constant volume bias is undetectable. (3) A non-constant volume bias cannot be distinguished from a leak of the same vessel. Highlights of the gross error equivalent theory are presented next.

We first define the concepts of Gross Error Cardinality and Basic Subset of a set:

Gross Error Cardinality Definition: A set of variables has Gross Error Cardinality $\Gamma = t$ when every possible set of gross errors in the system can always be equivalent to at least one set of t or less nonzero gross errors.

Basic Subset: A set of variables constitutes a Basic Subset of a system, when every set of gross errors is equivalent to a set of gross errors in the basic set.

As a corollary of this definition of basic subsets, we must note that a) for every system, more than one Basic Subset may exist, and b) all Basic Subsets are equivalent to each other, provided the right size of gross errors is used.

Illustration

Consider the process of Figure 1 and assume that all streams are measured. Measurements for S_1, S_3 and S_6 are 12, 10 and 2. The set $\Lambda = \{S_3, S_6\}$ has Gross Error Cardinality $\Gamma(\Lambda)=1$ as a gross error in one of them can be alternatively placed in the other without change in the result of the reconciliation. We now illustrate the fact that the set $\Lambda = \{S_2\ S_4\ S_5\}$ has Gross Error Cardinality $\Gamma(\Lambda)=2$. As shown in Table 1, a bias of (+1) in S_2, a bias of (-1) in S_4 and a bias of (+2) in S_5 can be represented by three alternative sets of two gross errors (Cases 2,3 and 4).

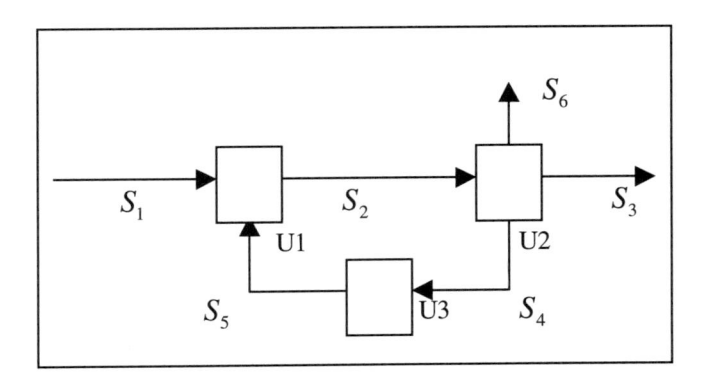

Figure 1. Example process.

Table 1. Illustration of Gross Error Cardinality in $\{S_2, S_4, S_5\}$.

		2	4	5
Measurement		18	4	7
Case 1 (Bias in S_2, S_4, S_5)	Reconciled data	17	5	5
	Estimated biases	1	-1	2
Case 2 (Bias in $S_4,, S_5$)	Reconciled data		8	
	Estimated biases			2
Case 3 (Bias in S_2, S_4)	Reconciled data		9	
	Estimated biases	1	3	
Case 4 (Bias in S_2, S_5)	Reconciled data		6	
	Estimated biases			

The equivalency is also rigorously demonstrated by the following theorems (proofs are not provided for reasons of space):

Gross Error Cardinality Theorem: Let m columns $[d_1\ d_2\ \cdots\ d_m]$ of the system matrix **D** correspond to a set of flowrate variables Λ. The set has Gross Error Cardinality $\Gamma = t$ if rank $[d_1\ d_2\ \cdots\ d_m] = t$.

Corollary 1: A set with Gross Error Cardinality $\Gamma = t$ contains subsets with Gross Error Cardinality $k=1,\ldots,t-1$.

Corollary 2: A l.d. set with Gross Error Cardinality $\Gamma = t$ always contains more than one Basic Subsets with Gross Error Cardinality $\Gamma = t$ and these basic subsets are equivalent to each other.

Corollary 3: When the Data Reconciliation with Gross Error Estimation is applied to a system with gross error cardinality $\Gamma = t$, only the gross errors corresponding to a basic subset of Λ can be introduced simultaneously.

Corollary 4: In a l.d. set of Gross Error Cardinality $\Gamma = t$, a single gross error $\delta_r = \Delta$ in one stream not belonging to a basic subset can be represented by gross errors of size $\xi_{rj}\Delta$ in the basic subset and vice versa.

Some linearly dependent sets of variables correspond to a closed loop in the system. If the environmental node is included, then all l.d. sets of variables correspond to a closed loop. Some of these loops include the environmental node. In addition, the inclusion of the environmental node does not change the cardinality of the set. Thus, by adding the environmental node one can merge both cases: Sets of the same cardinality consisting of streams in a loop and sets consisting of streams in a path from a feed/product stream to any feed/product stream into a single category of sets in a loop. We therefore introduce the notion of augmented graph.

Augmented Graph: The graph consisting of the original graph representing the flowsheet with the addition of the environmental node is called Augmented Graph.

We finally link our notion of cardinality to graph theory.

Loop Theorem: Every loop of *m* streams in the graph of the system is a set with Gross Error Cardinality Γ =m-1.

Augmented Graph Theorem: The Gross Error Cardinality Γ of an augmented graph corresponding to an open system is equal to the number of process units.

Corollary 1: The maximum number of gross errors that can be modeled in an open system is equal to the number of process units.

Corollary 2: The maximum number of gross errors that can be modeled in a closed connected system is equal to the number of process units minus one.

Extension of the Gross Error Cardinality Theorem to Leaks: We also extend the conclusions of the Cardinality Theorem to leaks. As leaks are part of the constraints, the proof of the theorem hold when matrix **H** is augmented to include leak streams. As leaks of order *k* are equivalent to the corresponding tank volume bias of order *k+1*, by including leaks we are actually considering all possibilities. Since zero order bias in volume are undetectable, and in the absence of any other criterion volume biases can be ignored from the analysis if leaks are included.

Gross Error Detection and Collective Compensation

A collective compensation strategy was presented by Bagajewicz and Jiang (1998) and can be summarized as follows: Determine a list of suspect gross errors using the measurement test. Run the compensation model for each suspect and add to the list of confirmed gross errors the one that gives the smallest objective function. Determine a new list of gross errors. Run the compensation method using the confirmed gross errors and adding suspect candidates one at the time. The suspect gross error that produces the smallest objective function is added to the list of confirmed gross errors. Although only biases are detected, the equivalency theory proves that leaks are equivalent to a set of biases, and therefore can be detected once the equivalent set is constructed. We call this method Serial Identification with Collective Compensation (SICC). To compare the effectiveness of this strategy with other methods a new version of OPF (Rollins and Davis, 1992) is used. The OPF is defined as the frequency of successfully identified gross errors. 1992.

We define a new measure (OPFE, Sanchez et. al, 1998) which is defined the same way, but considers an equivalent set of gross errors as a success. The following table shows the comparisons with the UBET strategy (Rollins and Davis, 1992) and the Collective Compensation based on GLR (CGLR, Keller et al, 1994) with the example used by Rollins and Davis,

Table 2. Comparison of Performance for SICC, UBET and CGLR.

Biased Streams	Sizes	OPFE		
		SICC	UBET	CGLR
1-6	0.875, 0.5	0.974	0.923	0.935
1-7	0.875, 0.5	0.974	0.901	0.935
6-7	0.875, 0.5	0.977	0.956	0.798
2-3	2.625, 1.5	0.988	0.980	0.986
2-4	2.625, 0.5	0.876	0.836	0.869
3-4	2.625, 0.5	0.876	0.939	0.870
4-5	0.875, 1.0	0.983	0.949	0.949
4-6	0.875, 0.5	0.981	0.914	0.949
5-6	1.750, 0.5	0.943	0.923	0.926

One-Step Compensation Method

We resort to the connection between graph theory and equivalency theory. A spanning tree is a basic set of maximum cardinality, that is, of cardinality equal to the number of units. Thus, one can run the compensation model using the spanning tree as a set of gross error candidates and capture all gross errors. Those variables in the spanning tree that show a value lower than a threshold can be disregarded using statistical tests. The set of gross errors identified by the spanning tree is given by $\hat{\delta}$.

Assume now that the least number of gross errors is to be identified (this will eliminate degeneracy). Then, we use the following model, which minimizes the number of gross errors:

$$Min \sum_{k=1}^{Ns} Y_k + \sum_{k=1}^{Nu} Z_k$$
s.t.
$$|\delta_k| - UY_k \leq 0$$
$$|\delta_k| - \varepsilon_k Y_k \geq 0$$
$$|l_k| - UZ_k \leq 0 \quad (1)$$
$$|l_k| - \varepsilon_k Z_k \geq 0$$
$$A\hat{\delta} - A\delta - l = 0$$

The binary variable Y_k represents the presence of a gross error of size $\delta_k \geq \varepsilon_k$ in stream *k* (*U* is a large number). Setting $Y_k = 0$, makes 0, whereas $Y_k = 1$ allows a value larger than ε_k. The same type of constraints is used for leaks. The last constraint represents the equivalency between the values of the gross errors in the spanning tree $\hat{\delta}$ and the set of new, hopefully smaller set included in the set (δ, l).

To solve this problem the following is used: $|\delta| = Max(\delta, -\delta)$. By replacing the *Max operator* by a set of mixed integer linear constraints proposed by Bagajewicz and Manousiouthakis (1992), the above problem becomes a MILP problem.

Illustration

The following table shows the results of the application of the technique for the case of Fig. 1. Assume that one spanning tree (S_1, S_2, S_5) is found.

Table 3. Gross Error Identification with the Minimum Cardinality Model.

No.	Gross Error Introduced	Gross Error Identified with Spanning Tree	Minimum Number of Gross Errors
1	S_2 (0.5))	0.0, 0.5, 0.0	S_2 (0.5))
2	S_4 (0.5)	0.0, -0.5, -0.5	S_4 (0.5)
3	U_2 (0.5)	0.5, 0.5, 0.0	U_2 (0.5)
4	S_1 (0.5) S_2 (1.0)	0.5, 1.0, 0.0	S_2 (0.5) S_3 (-0.5)
5	S_5 (0.5) U_2 (1.0)	1.0, 1.0, 0.5	U_2 (0.5) U_3 (0.5)
6	S_1 (0.5) S_2 (1.5) S_4 (0.5)	0.5, 1.0, -0.5	S_2 (0.5) S_3 (-0.5) S_5 (-0.5)
7	S_1 (0.5) S_2 (1.0) S_4 (0.5)	0.5, 0.5, -0.5	S_3 (-0.5) S_5 (-0.5)

Note: S_n means a bias in stream *n;* U_n means a leak in unit n.

In Run 1, 2 and 3, exact errors are identified. In Run 4, 5 and 6, one obtains equivalent sets. Run 7 is a degenerate case.

Conclusions

The theory of equivalencies allows the effective identification of gross errors, including leaks, in steady state data reconciliation. All the theory developed proves that exact identification of the location of gross errors, be it biases or leaks, is not always theoretically possible. Instead, a set of measurements and possible leaks is identified, out of which many alternative sets containing the number of gross errors identified can be selected. For this reason, successful identification is classified as exact when the real gross errors are detected, or correct when an equivalent set is identified. Thus, at the end of any successful identification, a number of equivalent alternatives exist. Our method is capable of identifying such alternatives for both steady state and dynamic integral data reconciliation. This paper unveils for the first time the theoretical intricacies of multiple gross error detection, presents a coherent methodology to identify leaks and biases simultaneously in steady state models and compares these techniques with the advantages offered by dynamic models.

A new method of simultaneous gross error identification and sizing is presented based on the spanning tree of the undirected graph corresponding to the flowsheet. This method does not require serial compensation, but rather performs all the identification and compensation in one single step.

References

Bagajewicz M. and Q. Jiang (1997)."An Integral Approach to Plant Linear Dynamic data Reconciliation." *AIChE J.*, **43**, 2546.

Bagajewicz M. and Q. Jiang (1998). "Gross Error Modeling and Detection in Plant Linear Dynamic Reconciliation. *To be published in Computers and Chemical Engineering.*

Bagajewicz M. and V. Manousiouthakis (1992). "Mass/Heat-Exchanger Network Representation of Distillation Networks." *AIChE J.*, **38**, 1769.

Keller J., M. Dorouach and G. Karzala (1994). "Fault Detection of Multiple Biases or Process Leaks in Linear Steady State Systems." *Comp. and Chem. Engng*, **18**,1001.

Madron F. (1992) *Process Plant Performance. Measurement and Data Processing for Optimization and Retrofits."* Ellis Horwood Series in Chemical Engineering, Chichester, England .

Narasimhan S. and R. S. H. Mah (1987). "Generalized Likelihood Ratio Method for Gross Error Detection", *AIChE J.* **33**,1514.

Rollins D. and J. Davis (1992). "Unbiased Estimation of Gross Errors in Process Measurements." *AIChE J.*, **38**, 563.

Sanchez M., J. Romagnoli, Q. Jiang and M. Bagajewicz (1998)."Simultaneous Estimation of Biases and Leaks in Process Plants." *Submitted to Computers and Chemical Engineering.*

IMPROVED STATISTICAL PROCESS CONTROL USING WAVELETS

Sermin Top and Bhavik R. Bakshi
Department of Chemical Engineering
The Ohio State University
Columbus, OH 43210

Abstract

There are a number of existing methods for statistical process control (SPC) with Shewhart, CUSUM and EWMA being the most popular for univariate SPC. A Shewhart chart is able to detect large shifts quickly but is not good for detecting small shifts. Cumulative sum (CUSUM) and exponentially weighted moving average (EWMA) charts are designed for fast detection of small shifts, but need well-tuned design parameters and are slow in detecting large shifts. Furthermore, CUSUM and EWMA charts are not easy to interpret since they filter the data. Consequently, in practice, it is common to use various control charts together. Another important shortcoming of these control-charting methods is that their performance deteriorates if the measurements are autocorrelated since they give too many false alarms. This paper describes a new approach for SPC based on representing the measurements at multiple scales using wavelets. This approach, called multiscale SPC (MSSPC), provides a common framework for the commonly used methods of Shewhart, CUSUM and EWMA since these methods differ only in the scale at which they represent the measurements. MSSPC exploits the ability of wavelets to separate deterministic and stochastic components and approximately decorrelate autocorrelated. The MSSPC method is explained with examples and average run length (ARL) performance is compared to that of traditional techniques by extensive Monte Carlo simulations. MSSPC does not require much parameter tuning, and is able to retrieve underlying deterministic features in data, enabling the detection of shifts of different types.

Keywords

Wavelets, Statistical process control, Autocorrelated measurements, Average run length, Multiscale SPC.

Introduction

Every process has a natural variability due to unassignable causes, and a variability due to causes that is usually larger than the inherent variability. Statistical Process Control (SPC) aims to achieve improved process operation through the reduction of variability by monitoring the process with the help of control charts for various measured variables. Systematic use of control charts in process monitoring allows quick detection of process shifts and assignable causes so that further investigation of the root cause and corrective action can be taken on time.

Among the large variety of control charts available, Shewhart control charts, developed by Dr. Walter A. Shewhart in 1920s, are the most basic and widely used in a number of existing SPC applications. A Shewhart chart simply plots the measurements themselves along with a pair of control limits at the scale of the sampling interval, it is the easiest and the most widely used method. Shewhart charts are best for detecting large shifts in the process mean. Special control charts such as CUSUM and EWMA are designed for fast detection of small shifts in the mean. They perform well if the shift is smaller than

1.5-2 times the normal standard deviation because they represent the measurements at coarser scales by filtering or accumulating all the information in a new variable. However, they are not the best ones to use for large shifts and changes other than a mean shift. In practice, to detect different types of changes in the signals, different types of control charts need to be used together.

All of the existing methods assume observations to be independently and identically distributed (i.i.d.) random variables. Unfortunately, in practice, the assumption of i.i.d. observations is often violated due to system dynamics, recycle loops, and frequent sampling. Several studies have shown that the performance of traditional SPC techniques deteriorates considerably due to increased number of false alarms even if there is slight autocorrelation in the data (Alwan, 1992; Harris and Ross; 1991; Runger and Willemain, 1995; Montgomery and Mastrangelo 1991,1995). A common approach to deal with autocorrelated observations is to decorrelate the data by fitting a time series model and developing a traditional control chart for the residuals. If the selected time series model fits the data well, residuals should be i.i.d. random variables. The disadvantage of such an approach is that even if an appropriate model exists, it is not very practical to do time series modeling for each variable. Montgomery and Mastrangelo (1991, 1995) approximate the underlying time series model by EWMA and monitor the residuals. This approach considers residuals at only one scale and lacks the benefits of the proposed multiscale approach for SPC.

This paper introduces a new technique called Multiscale statistical process control (MSSPC) based on representing observations at multiple scales, and automatically selecting the scale and detection limits that are best for detecting the abnormal operation. Since existing methods such as Shewhart, CUSUM and EWMA differ only in the scale at which the observations are represented, the multiscale nature of MSSPC unifies these control charts, and can detect any type of change quickly. Furthermore, MSSPC is able to detect changes in autocorrelated observations without too many false alarms. In addition to detecting abnormal operation quickly, MSSPC also extracts the signal feature that is relevant to the abnormality. Such integration of feature extraction with monitoring is expected to permit easier detection of the root cause. This paper focuses on univariate SPC, but may also be extended to multivariate SPC (Bakshi, 1998). In the rest of this paper, existing control charting techniques are reviewed in greater detail and a common framework for SPC charts is presented. The MSSPC methodology is developed and its performance is compared with that of existing methods by Monte Carlo simulations. Finally, conclusions and directions for future work are discussed.

General Multiscale Framework for SPC Charts

Shewhart Charts

A Shewhart chart simply plots the measurements in the time domain. Thus the interpretation of the chart is quite easy. Upper and lower control limits are used to decide if the process is out of control. If any point plots outside these limits, the process is said to be out of control and assignable causes are searched for. These limits could be process specifications or probability limits. If the underlying distribution is normal, 3 sigma limits would embrace ~99.7 % of data when the process is in control. The main disadvantage of this chart is that it only uses information given by the most recent point to decide on the state of the process. Any information given by the entire sequence of points is ignored making the Shewhart chart less sensitive to small shifts. These charts could be sensitized for small shifts by adding pattern recognizing rules, however this results in a more complex chart with increased number of false alarms, reducing in control run length of the control chart dramatically. The Shewhart chart is very effective if the magnitude of the shift is 1.5-2 sigma or larger (Montgomery, 1996).

CUSUM Charts

CUSUM chart plots cumulative sums of deviations from target value. Thus it incorporates information from the whole sequence of data. The two-sided algorithmic CUSUM is based on

$$Sh_t = \max\left[0, (y_t - \tau)/\sigma_y - k + Sh_{t-1}\right] \quad (1)$$

$$Sl_t = \max\left[0, -(y_t - \tau)/\sigma_y - k + Sl_{t-1}\right] \quad (2)$$

where starting values are generally taken as $Sh_0 = Sl_0 = 0$. k is the reference or slack value. If either Sh_t or Sl_t exceed the decision interval h, process is said to be out of control; assignable causes should be searched and corrected before CUSUM is reinitialized at zero. CUSUM also gives a hint to when the shift has started. If one counts backwards from the first time shift was detected backwards, it is possible to find exactly when CUSUM value lifted above zero. Design parameters k and h are chosen such that a good ARL performance can be achieved. Sensitizing rules used for Shewhart charts can not be used since Sh_t or Sl_t are not independent. CUSUM charts have been studied by many authors and there are many analytical studies of CUSUM ARL performance in literature (van Dobben de Bruyn, 1968).

EWMA Charts

The EWMA chart plots the exponentially weighted average of entire set of observations up to current point. It is defined as

$$z_t = \lambda y_t + (1-\lambda) z_{t-1} \qquad (3)$$

where $0 < \lambda \leq 1$ is a constant and $z_0 = \mu_0$. If the observations y_t are i.i.d. random variables with variance σ_y^2, then the variance of z_t is

$$\sigma_{z_t}^2 = \sigma_y^2 \left(\frac{\lambda}{1-\lambda}\right)\left[1-(1-\lambda)^{2t}\right] \qquad (4)$$

Therefore, the EWMA control chart could be constructed by plotting z_t with limits $\pm L \sigma_{z_t}$. L is a parameter that defines the width of the limits. Notice that the second term in brackets approaches unity as t increases. Design parameters L and λ could be chosen such that an ARL performance similar to that of CUSUM could be obtained.

CUSUM and EWMA perform well for small shifts but are late in reacting to large shifts because of the way they are designed. However, the EWMA is often superior to the CUSUM for large shifts, particularly if $\lambda > 0.10$ (Montgomery, 1996). A good way to further improve the sensitivity of control procedure to large shifts while retaining the ability to detect small shifts is to combine a Shewhart chart with any of these methods.

General Multiscale Framework

The filters used for representing the data by various SPC methods are shown in Figure 1. A Shewhart chart represents data at the finest possible scale, as shown in Figure 1a. In contrast, CUSUM charts represent the data at the coarsest scale by incorporating the entire sequence of observations as shown in Figure 1c. The scale of the EWMA filter, shown in Figure 1b, is determined by the value of the filter parameter, λ. A general framework for Shewhart, CUSUM and EWMA charts may be provided by a method that can represent the data at multiple scales and select the best scale adaptively. MSSPC is such a general method since the filters used by MSSPC using Haar wavelets subsume a Shewhart chart, moving average charts at dyadic scales, and a CUSUM chart, as shown in Fig. 1d. At the finest scale, the MSSPC filter reduces to that for a Shewhart chart, whereas for coarser scales, MSSPC specializes to moving average charts of different window lengths. If smoother wavelets are used, they are similar to the EWMA filter (Nounou and Bakshi, 1998), making MSSPC with smoother wavelets analogous to a multiscale EWMA chart. Thus, as discussed in the next section, MSSPC is able to automatically select the best scale and

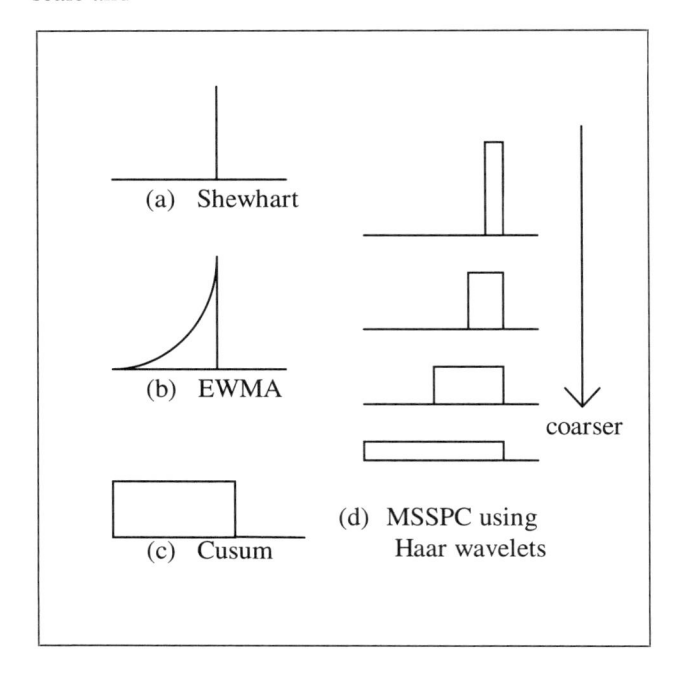

Figure 1. Nature of filters for different SPC charts.

detection limits for detecting a deterministic change in a set of stochastic measurements.

Multiscale SPC

Introduction to Wavelets

Wavelets localize data in both time and frequency by representing them in terms of a set of basis functions $\psi_{s,u}(t)$. These wavelets are generated from a "mother wavelet" $\psi(t)$ as

$$\psi_{s,u}(t) = \frac{1}{\sqrt{|s|}} \psi\left(\frac{t-u}{s}\right) \qquad (5)$$

where s and u are scaling and translation parameters respectively. Scaling spans frequencies by compressing or stretching the mother wavelet, whereas translation spans the time domain by shifting the mother wavelet in time. The discrete wavelet transform of a function f in the space of square integrable functions, $L^2(R)$ is defined as:

$$W_d f(m,n) = \langle f, \psi_{m,n} \rangle \\ = |a_0|^{-m/2} \int_{-\infty}^{\infty} f(t) \psi(a_0^{-m} t - n b_0) dt \qquad (6)$$

where $s = a_0^m$ and $u = nb_0 a_0^m$ are discretized dilation and translation parameters, respectively. Typically, $a_0 = 2$ and $b_0 = 1$ are assumed for practicality. The function f is represented in terms of its projection on wavelets between scales $m = 1$ to $m = L$ (detail signals), and the projection on scaling functions at scale $m = L$ (last scaled signal). The functions, $\phi_{m,n}(t)$ are generated by dilating and translating the scaling function. At each scale, information not captured by detail signal is captured by scaled signal. The function, f can be represented as,

$$f(t) = \sum_{m \in [1,L]} \langle f, \psi_{m,n} \rangle \psi_{m,n}(t) + \langle f, \phi_{L,n} \rangle \phi_{L,n}(t) \quad (7)$$

If orthonormal wavelets are used, the decomposition and reconstruction can be performed in O(N) time for a signal of length N by using Mallat's (1989) fast algorithm. In this paper, only Haar wavelets are used for MSSPC, but the method may be extended easily to using smoother wavelets.

MSSPC Methodology

The MSSPC approach relies on the ability of wavelets to separate deterministic components in a few relatively large wavelet coefficients. The stochastic component is distributed in all the wavelet coefficients according to its power spectrum. Thus, the wavelet transform of Gaussian uncorrelated measurements with zero mean and unit variance is white noise with zero mean and unit variance at all scales. The wavelet transform of autocorrelated measurements, on the other hand, is approximately white, with the variance of the coefficients at each scale being proportional to the power spectrum. These properties have been extensively utilized for noise removal and data compression by eliminating coefficients smaller than a threshold (Donoho et al., 1995).

The methodology of MSSPC consists of decomposing the measurements on a selected wavelet, and constructing a separate control chart for the coefficients at each scale. Separate detection limits are defined at each scale from historical data representing normal operation. The variance of the measurements is decomposed into the variance at different scales as,

$$\text{var}(f) = \sum_{m \in S} \sigma_m^2 \quad (8)$$

For autocorrelated measurements, the detection limits at each scale change according to the power spectrum in the corresponding frequency band. A deterministic change may be detected as coefficients violating the detection limit at one or more scales. Information about the current state of the process is contained in the most recent wavelet and last scaled signal coefficients. If any of the current coefficients violate the detection limit, it may indicate abnormal process operation. The actual state of the process is verified by reconstructing the signal based on only those coefficients that violate the detection limits. This reconstructed signal contains only those features that represent abnormal operation. The detection limit for the reconstructed signal is computed by combining the variance of the normal measurements at the scales at which the current coefficient violated the detection limits. If the current measurement in the reconstructed signal violates the detection limit, then the process is confirmed to be outside its range of normal operation. Without this confirmation of the process state from the reconstructed signal, the MSSPC method suffer from a delay in detecting the return of a process to normal operation (Bakshi, 1998). This delay in analogous to that in CUSUM and EWMA charts for detecting large changes. Verification of the process state from the reconstructed signal also permits MSSPC to adjust the nature of the filter and detection limits to the nature of the measurements. Thus, a large mean shift will first be detected at the finest detail signal, then at the next coarser detail signal, and so on. This is due to fact that coefficients at coarser scales are calculated using more and more information from past. Since previous data come from a process that is in control, recent observations representing abnormal operation will be suppressed at coarser scales. Similarly, when the process returns to normal operation, coarser scales will have a delay in showing that the process is no longer in an abnormal state. This delay is overcome by reconstructing the signal based on only those coefficients that violate the detection limits, and checking whether the most recent point in the reconstructed signal violates its limits.

The MSSPC methodology may be used for on-line monitoring by using the on-line wavelet decomposition approach of Nounou and Bakshi (1998). This approach decomposes the signal in the longest possible moving window of dyadic length so that the most recent measurement is always included in the window. This approach is equivalent to using a non-dyadic discretization of the dilation parameter, and results in N coefficients at each scale for a signal of length N. Since the overall ARL of MSSPC depends on the performance of individual charts at each scale, the control limits at each scale should be chosen such that the overall probability of an out of control measurement will be set to a desired value. Examples of ARL calculation for combined charts can be found in Crowder (1987), Lucas (1982), St. John and Bragg (1991). As the window for on-line decomposition increases in length, the signal can be decomposed to more scales, and a maximum depth needs to be selected. Thus, the tuning parameters for MSSPC are the overall detection limit, maximum depth of decomposition and mother wavelet. Insight into selection of these parameter is provided by the examples in the next section, and in future publications.

Examples

Harris and Ross (1991) tabulated ARL performances of traditional CUSUM and EWMA charts with autocorrelated data. To make a comparison between these methods, data has been simulated using the following model (Harris and Ross, 1991):

$$y_t = \eta_t + \alpha_t \quad (9)$$

$$\eta_t - \tau = \phi(\eta_{t-1} - \tau) + e_t + \delta_{t_0}[\Delta] \quad (10)$$

where, $-1 \leq \phi \leq 1$ and $[\alpha_t, e_t]$ are i.i.d. random variables with mean zero and variances $[\sigma_\alpha^2, \sigma_e^2]$ respectively. α_t is the inherent variability in the process, e_t is the "driving force for the disturbances and η_t is the process mean at time t. ϕ is the autoregressive parameter. This model allows the process mean to wander or drift away from the target value. A single assignable cause need not be associated with this behavior. σ_α^2 was fixed at 0.9. The autoregressive parameter was assigned the values of 0.25 and 0.75. The total variability of $y_t - \tau$ in the absence of shifts was fixed at one by noting that

$$\mathrm{var}(y_t - \tau) = \sigma_e^2 / (1 - \Phi^2) + \sigma_\alpha^2 \quad (11)$$

Detection limits for each scale are estimated using in control process data. A set of training data is used to determine variances and means of detail signals and last scaled signal. Average Run Lengths are calculated by means of Monte Carlo Simulations with 100 replications. Since there are more than one chart employed at a time, depending on the number of scales (thus number of charts to monitor), ARLs would be much less compared to those of a single chart. To correct for this, limit at each scale is increased to make sure that overall probability of getting an out of limit signal would be that of a single chart. Also, limits are adapted for changing window size. Window size starts from the minimum possible of one, and it is allowed to increase to as large as the length of the entire set of data. Wavelet decomposition at each scale is monitored for a shift in mean, if any of the charts indicates an abnormality, the time domain signal is reconstructed using only those wavelet coefficients that violate the limits. Control limits for reconstructed data are calculated by combining the limits of only those scales at which the shift is detected. This avoids the delay found in CUSUM and EWMA because of filtering and enables fast detection of the end of a shift.

ARL performances of Shewhart, CUSUM, EWMA and MSSPC methods are provided in Table 1 through

Table 1. Shewhart* ARL Performance.

$\tilde{\Delta}$	$ARL_{\phi=0}$	$ARL_{\phi=0.25}$	$ARL_{\phi=0.75}$
0	485.2	466.9	446.8
0.5	173.2	179.8	181.5
1	49.4	47.8	54.4
2	7.2	7.3	12.8

* ± 3.1 σ limits

Table 3. Parameters for each chart have been adjusted to obtain comparable In Control Run Lengths (ARL₀). Performance of the EWMA and CUSUM charts were very similar to each other, so the ARL table for CUSUM chart has been omitted because of space restrictions. Shifts of magnitude up to 2 σ have been investigated. Up to this value, Shewhart does not perform well compared to all other techniques, as indicated by the larger ARL values in Table 1. Shewhart chart is not affected by correlation except from a slight decrease in the In Control Run Length. EWMA performs well for smaller shifts, however there is a drastic drop in ARL₀ value with increasing correlation. This type of behavior leads to an increased number of false alarms when the process is in control.

Table 2. EWMA* ARL Performance.

$\tilde{\Delta}$	$ARL_{\phi=0}$	ARL	$ARL_{\phi=0.75}$
0	465.5	385.1	186.0
0.5	41.0	39.1	40.4
1	10.1	10.2	15.1
2	3.5	3.9	6.5

* λ=0.18, L=2.9

In contrast, MSSPC is very consistent and is not adversely affected by ex correlation as shown by the in-control lengths in Table 3. The ability of MSSPC to detect shifts is better than that of Shewhart charts and only slightly worse than that of CUSUM for highly correlated data. The slight delay in detecting the shift is a small price to pay for the absence of false alarms. The increase in in-control run length with increasing autocorrelation in Table 3 may be due to the greater relationship between the wavelet coefficients at different scales.

Table 3. MSSPC* ARL Performance.

$\tilde{\Delta}$	$ARL_{\phi=0}$	$ARL_{\phi=0.25}$	$ARL_{\phi=0.75}$
0	462.6	482.6	539.9
0.5	42.3	44.7	64.6
1	12.0	12.9	23.5
2	3.2	3.8	8.1

* ± 2.85 σ adaptive limits, max. no. of scales=7

The ability of MSSPC to extract the feature representing abnormal operation is depicted in Fig. 2 for a mean shift of magnitude 1 sigma imposed on simulated autocorrelated measurements with ϕ=0.75. The mean shift lasts between 200th-600th observations. The control limits are indicated with the dashed lines.

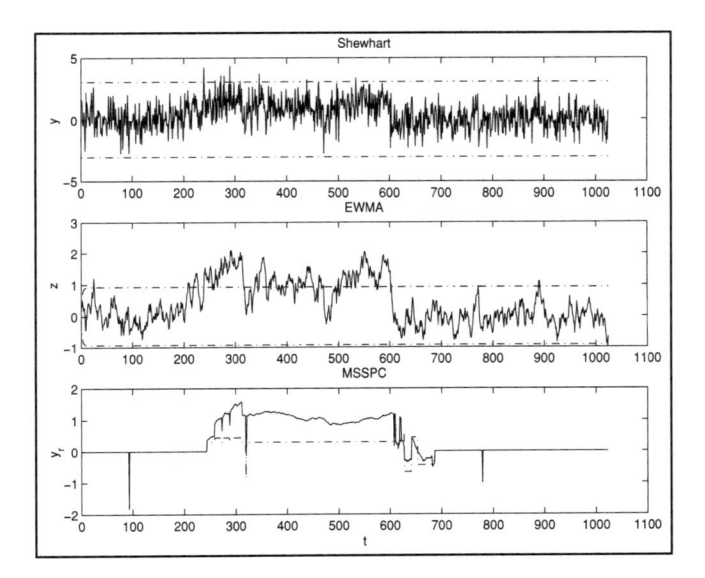

Figure 2. Performance of various control charts for a mean shift in autocorrelated measurements.

Figure 2 also demonstrates the superior ability of MSSPC to extract the feature representing abnormal operation. This property is expected to permit improved fault diagnosis.

Conclusions

A new multiscale approach for SPC is developed in this paper which can specialize to Shewhart, CUSUM or EWMA charts. ARL computation shows that MSSPC can detect both large and small shifts from uncorrelated or autocorrelated measurements. MSSPC may not work as well as a method tailor-made for detecting a certain type of process change, but it is a general method that is effective for a broad variety of data and process changes. Thus, MSSPC is expected to perform better than existing methods for monitoring of processes where the nature of the data and features representing abnormal operation are not known in advance and may be of several different types. Performance of MSSPC for shifts other than a step change as well as for other types of simulated and industrial data, and with wavelets other than Haar will be investigated and reported in future work.

References

Alwan, L. C. (1992). Effects of autocorrelation on control chart performance. *Commun. Statist. -Theory Meth.,* **21,** No. **4,** 1025-1049.

Bakshi, B. R. (1998). Multiscale principal component analysis with application to multivariate statistical process monitoring. *AICHE Journal,* July, 1998.

Crowder, S. V. (1987). Computation of ARL for combined individual measurement and moving range charts. *J. of Quality Technology.,* **19,** No. **2,** 98-102.

Van Dobben de Bruyn, C. S., 1968. *Cumulative sum tests: theory and practice.* London: Charles Griffin & Company Ltd.

Donoho, D. L., I. M. Johnstone, G. Kerkyacharian, and D. Picard (1995). Wavelet Shrinkage: Asymptopia?. *J. of Royal Statistical Society B.,* **57,** No. **2,** 301-369.

Harris, T. J., and W. H. Ross (1991). Statistical process control procedures for correlated observations. *The Canadian J. of Chem. Eng.,* **69,** 48-57.

Hunter, J. S. (1986). The exponentially weighted moving average. *J. of Quality Technology,* **18,** No. **4,** 203-210.

St. John, R. C., and D. J. Bragg (1991). Joint x-bar & R charts under shift in mu or sigma. *ASQC Quality Congress Transactions-Milwaukee,* 547-550.

Lucas, J. M. (1982). Combined shewhart-cusum quality control schemes. *J. of Quality Technology.,* **14,** No. **2,** 51-59.

Mallat, S. G. (1989). A theory for multiresolution signal decomposition: the wavelet representation. *IEEE Transactions on Pattern Analysis and Machine Intelligence,* **11,** No. **7,** 674-693.

Mastrangelo, C. M., and D. C. Montgomery (1995). SPC with correlated observations for the chemical and process industries. *Quality and Reliability Engineering International,* 79-89.

Montgomery, D. C., and C. M. Mastrangelo (1991). Some statistical control methods for autocorrelated data. *J. of Quality Technology,* **23,** No. **3,** 179-193.

Montgomery, D. C., 1996. *Introduction to statistical quality control.* 3rd ed. New York: John Wiley & Sons, Inc.

Nounou, M. N., and B. R. Bakshi (1998). On-line decomposition and rectification without process models. *Technical Report,* Dept. of Chem. Eng., Ohio State Univ., 1998.

Runger, G. C., and T. R. Willemain (1995). Model-based and model-free control of autocorrelated processes. *J. of Quality Technology,* **27,** No. **4,** 283-292.

PROCESS OPERATION IMPROVEMENT BASED ON MULTIVARIATE STATISTICAL ANALYSIS

Dae-Hee Hwang and Chonghun Han
POSTECH
Pohang, Kyungbuk 790-784, Korea

Tae-Jin Ahn
POSCO
Pohang, Kyungbuk 790-784, Korea

Abstract

As distributed control systems (DCS) and plant information systems are widely being installed in chemical process industries, a large amount of process operation data are now available. These large amount of data have led to the need for the efficient and reliable data analysis methods to extract useful process information from operation data. Based on the extracted information, many process improvements such as optimal operation conditions or troubleshooting can be accomplished.

This paper proposes a process operation improvement methodology based on multivariate statistical data analysis. The methodology is based on a combination of various techniques such as clustering, factor analysis, and principal component analysis. The proposed methodology is illustrated by its application to the improvement of the industrial blast furnace operation. From the application, several suggestions for the process improvement have been generated and validated.

Keywords

Process improvement, Multilevel operation, Clustering, Factor analysis, Factor score plot analysis.

Introduction

As the international competition among process industries increases, the accomplishment of a better operation condition to increase the productivity or reduce production cost has become a critical task to survive. Quality control has played an important role on that purpose. The main topic in the field of quality control is to reduce the variability in product quality and further to identify the better operation condition. A variety of methods for the former, usually based on TQM (Total Quality Management) and SPC (Statistical Process Control) using traditional chart analysis, have been introduced and they have been successful in may industrial applications. However, their application has had a limitation that they can be used for those processes that process variables are kept at an operation mode.

In practice, most of chemical processes usually go through the change of operation condition due to set-point changes, leading to many operation modes. Saraiva and Stephanopoulos (1992) proposed a method to divide the operation range to identify the different operation modes using inductive and analogical learning, instead of using traditional charting techniques, and generated suggestions to attain an sub-optimal operation mode. However, its application to the complex chemical plant has a limitation due to the large number of measurements and the strong correlations among them.

Multivariate statistical methods have been developed to resolve the correlation problem in the large number of measured variables with no relation to process improvement, except for a few studies of statistical controller design based on principal component analysis (PCA) to reduce the variability of process variables.

This paper proposes a new approach for process improvement based on Saraiva's procedure with a combination of multivariate statistical analysis that can eliminate the difficulties from the correlations. To divide the operation range, clustering based on self-organizing feature maps (SOM) has been explored. Each cluster that corresponds to a different operation mode has been analyzed using one of multivariate statistical analyses, factor analysis. Then, the differences between the clusters have been investigated and the suggestions for process improvement have been generated. The proposed methodology is illustrated by giving an application to a blast furnace for the purpose of the determination of the optimal operation conditions, when we want to increase the amount of pulverized coal injection (PCI) without sacrificing the product quality and yield.

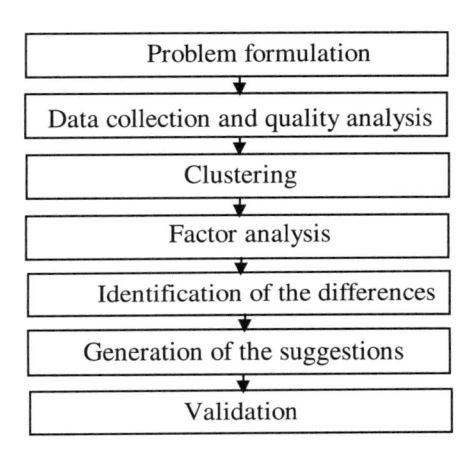

Figure 1. An overview of the proposed methodology.

Process Operation Improvement Methodology

Problem Formulation

The target for process operation improvement is identified as a performance index. It is commonly productivity, production cost, or a quality variable. The relevant variables that influence the performance index are identified at the same time.

Data Collection and Analysis

The good data collection and analysis of its quality are essential prerequisites for the useful analysis. The relevant variables, which describe the process fully enough, should be checked whether all of them are measured using data reconciliation technique (Crowe, Campos and Hrymak, 1983). Otherwise, the additional instrumentation of measurement for the relevant variables not measured will be required. After the collection of operation data (Ljung, 1987), the contaminated data need to be eliminated using the proposed selection algorithm (Hwang, Ahn and Han, 1998). The quality of selected normal operation data is investigated using exploratory data analysis such as checking signal to noise ratio (Santen, Koot and Zullo, 1997). Checking the power spectral density of the key inputs allows us to estimate the feasible operating region in frequency domain where the generated suggestions will be valid.

Clustering

After applying PCA to the selected operation data, the transient operation data are approximately eliminated to increase the resolution of the following statistical analysis using clustering analysis. With only static data, it is likely to know how many operation modes exist in the first PC and second PC plane. With the information on the number of cluster (n), the selected data set is divided into n subset; SOM (Demuth and Beal, 1994) is trained with n output nodes. Since the transient data have been eliminated, each cluster can be considered an operation mode. SOM estimates a sample mean and a sample covariance matrix for every cluster.

The criteria associated with the performance index are defined to label each cluster with 'good' or 'bad'. For instance, if the objective for process improvement were to reduce production cost, fuel cost and process stability may be good candidates for the criteria. A cluster where small amount of fuel is used and where the process is in stable condition will be labeled as 'good'. Otherwise, it will be labeled as 'bad'. With n clusters, each cluster can be labeled sequentially based on its own value of the criteria. Fig. 2 shows clustering and the n-th graded labeling.

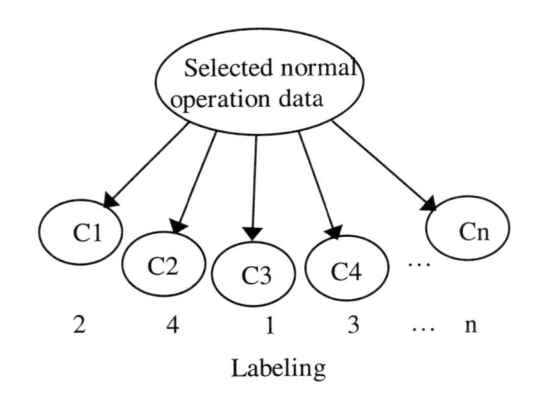

Figure 2. Clustering and labeling.

Statistical Analysis (Orthogonal Factor Modeling)

The essential purpose of factor analysis is to describe the covariance relationships among many variables in terms of a few underlying, but unobservable, random quantities called *factors*. Factor analysis can be considered as an extension of principal component analysis. An orthogonal factor modeling (Johnson and

Wichern, 1992) is applied to each cluster as shown Eqn (1).

$$Z_i = F_{i_m} L_{i_m}^T + \varepsilon_{i_m}$$ (1)

where Z_i is a standardized data matrix for i-th cluster. The F_i, L_i, and ε_i are m common factors, factor loadings and specific factor for i-th cluster, respectively.

The initial factors are rotated to easily interpret the factors under a *varimax* criterion or *promax* criterion. While with a varimax criterion an orthogonal rotation of m common factors is performed, with a promax criterion an orthogonal rotation is performed first and then oblique rotation is performed sequentially. Eqn. (2) shows an orthogonal factor loading decomposition of a correlation matrix. Eqn. (3) shows the rotation under a varimax criterion with $T^T T = I$ or under a promax criterion without the orthogonal constraint of T.

$$R_i = L_{i_m} L_{i_m}^T + \psi_{i_m}$$ (2)

where R_i is a correlation matrix for i-th cluster. The L_i and ψ_i are the first m common factor loadings and specific variance matrix of i-th cluster.

$$\tilde{L}_{i_m} = L_{i_m} T_{i_m}$$

$$\underset{T_{i_m}}{Max}\ V = \frac{1}{p} \sum_{j=1}^{m} \left[\sum_{i=1}^{p} l_{ij}^4 - \frac{\left(\sum_{i=1}^{p} l_{ij}^2\right)^2}{p} \right]$$ (3)

where T_i and V are a rotation matrix and the varimax criterion. The varimax criterion chooses the rotation matrix to make as large as possible the variability of the elements of factor loading in j-th factor. The p is the number of variables and l_{ij} is a i-th element of j-th factor. When the important factors to represent the process are expected to be dependent to one another, a non-rigid rotation with promax criterion should be used.

Generation of Suggestions for Process Improvement

For each cluster, the meanings of the rotated factors are interpreted and labeled with nickname. To interpret a factor is to identify something that brings about the correlation shown in its loading. It is usually a physical phenomenon that can characterize the process.

The differences between the best cluster and the worst cluster are identified in terms of the interpreted meanings for the factors, considering the kind and order of factors and the distinguishing features in those factor loadings. The order is determined by the eigenvalues, that is, the proportions of variance explained by all factors. Then, the suggestions for process improvement based on the differences can be generated. With the differences between the cluster in the middle of the n-th graded labeling and the worst cluster, the differences and suggestions identified between the best and the worst cluster have to be refined or verified. Due to orthogonality in factor loadings under the varimax criterion, the suggestions are independent to each other so that the operation direction including all suggestions can be developed in a single statement with the priority in the magnitude of eigenvalues of the associated factors. However, it is still possible to draw a direction, even when factors share some correlation to one another under promax criterion. The suggestions and direction can be used as an operation guideline. For instance, after the current operation data is classified using a clustering model and the state of operation have been identified, they are applied to move it to that of the best operation.

Validation

Before the generated suggestions and the direction are handed over to decision-makers (process engineers), they need to be validated and refined by the past operation data. Validation can be accomplished by a hypothetical test using the information in the factor score plot. Factor score is the coordinate value in the coordinate system with the axes as the eigenvectors, that is, factor loadings. Factor score values are calculated by projecting the observations into the factor loadings as shown in Eqn. (4).

$$\hat{F}_{i_m} = Z_i L_{i_m} (L_{i_m}^T L_{i_m})^{-1}$$ (4)

Validation procedure is:

1. An observation group of interest is chosen. It would be better to have a group of distinguished or separated observations.
2. The cause they are separated in factor score plot is identified in terms of the interpreted meanings of factors.
3. The cause can produce a hypothetical action one can take as to improve the operation associated with the chosen observation group into a better operation condition, according to the suggestions developed in step 5.
4. The hypothetical action is compared with the action really taken by process operators using the past action data. Applying the same procedure to other observation groups in every cluster, the only valid suggestions that give the hypothetical action identical to the real action will be chosen.
5. The chosen suggestions are combined into a direction.
6. It must be checked whether the path toward the better operating condition based on the

direction is feasible using the past data. It can be confirmed by checking if any past operation along such a path had ever taken place.

After collecting the operation data from the future operation based on the developed direction, applying the step 1-6 and updating the past direction in an evolutionary manner can perform the continuous process improvement. Finally, the process will operate in one of sub-optimal operation condition within the process capacity.

Case Study: A Blast Furnace

A blast furnace (BF) is a large reactor where many reactions occur in complicated ways with physical phase transitions, heat transfer and mass transfer at the same time. Due to those complicated phenomena, the process shows highly nonlinear behaviors and has strong correlations among the measured variables. A variety of mathematical models such as operation diagrams, kinetic model and control model (Omori, 1987) have been developed and are available. They have given a great amount of insight on the internal behavior of the BF and operation conditions to some extent. However, they can not give a sufficient answer for the determination of the optimal operation conditions to improve process operation, such as the increase of pulverized coal injection (PCI). Iron ore is charged down from the reactor top and Pulverized Coal (PC) is charged up from the waist level near the bottom along with oxygen. Recently, as it has been reported that the increase of the amount of PCI will save the amount of expensive coke, many approaches have been taken to increase the PCI level. The proposed methodology based on multivariate statistical analysis has been applied to identify the optimal operation condition to increase the amount of PCI.

Problem Formulation

The objective of process improvement is to find a way to increase the amount of PCI to save fuel cost without disturbing the daily operation significantly.

Data Collection, Selection and Analysis

The last six-month operation data from the fourth blast furnace of POSCO, a steel-making company located at Pohang in Korea, have been collected. The measured variables are composed of 329 temperatures through the whole reactor, 31 pressures in shaft region and the upper part of the reactor, and 4 gas concentrations at the top. The relevant variable search has been done using reconciliation technique, i.e., with graph theory, considering the casual relationship in material and energy balance. As a result, the 130 variables have been selected for the following analysis with the operators' agreement. By applying the proposed algorithm to select the normal operation data from the raw operation data, 4193 original operation data have been reduced to 2506 normal operation data. For the selected data, the feasible regions where the generated suggestions work have been identified by calculating PSD of the key input variables (blast volume). Blast volume has the white noise characteristics so that the suggestions will be feasible within all operation ranges as shown in Fig. 3. Signal to noise ratios of the measured variables are checked using the variability defined as the ratio of the mean to the standard deviation. Most of variables have higher signal to noise ratio than 10, except for the pressure measurements in the upper part of BF. The fact that the least signal to noise ratio is 3 and the average 26 implies that signals have much larger magnitude and play much more powerful role than the disturbances existing inside the BF. Thus, it can be concluded the selected data make possible the useful analyses. Fig. 3 shows PSD for blast volume (V_b) and signal to noise ratio.

Clustering

The criteria for the labeling are defined as the amount of PCI and the K value. K value is an index that indicates the state of the gas flow within the BF. It is proportional to pressure difference between the blast in tuyere and the top. The low K value implies that the operation is stable. The amount of PCI should be selected as one of criteria, because the objective of process improvement is to increase the amount of PCI. The selected dataset has been divided into four clusters using SOM and labeled with 1-4, according to the criteria. The forth cluster is the best cluster and the second cluster is the worst one. Table 1 shows the result.

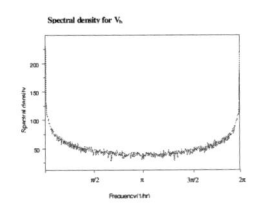

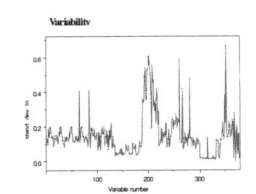

Figure 3. Power spectral density of V_b and the variability of all measured variables.

Table 1. Clustering of the dataset into four subsets.

Class (Labeling)	Observation Number	PCI Level	K Value
C1 (good; 2)	1-760	Middle	Middle
C2 (bad; 4)	761-1545	Low	Low
C3 (moderate; 3)	1546-2149	High	Low
C4 (best; 1)	2150-2506	High	Low

Statistical Analysis (Orthogonal Factor Modeling)

An orthogonal factor modeling for each cluster has been explored with 19 common factors. In case of nonlinear process such as BF, as the factors interact with each other, promax criterion is valid. However, due to some advantages from the orthogonality among the factors, varimax criterion has been used. Fig. 4 shows the rotated factor loadings for all factors.

Generation of Suggestions for Process Improvement

To interpret the first factor in C1, it must be noticed that only the fluidity can influence the temperatures in the hearth and hearth bottom, so the factor can be interpreted as the fluidity in the hearth and bottom. For the second factor in C1, only the permeability can influence the temperatures in the lumpy zone so that it can be interpreted as the permeability in the lumpy zone. The similar interpretation processes have been performed for all factors in every cluster. Table 2 shows the interpretation results of the factors with the nicknames.

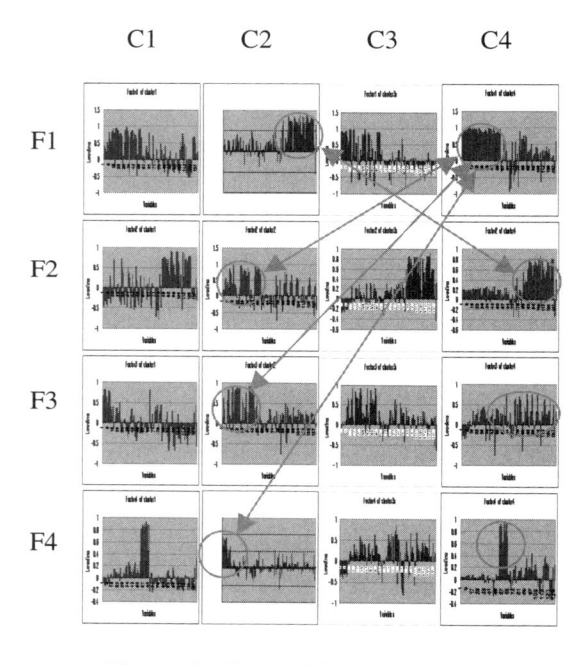

Figure 4. Rotated factor loadings.

Table 2. Interpretation of Factors.

	C1	C2	C3	C4
F1	FL1	P	FL1	FL
F2	P	FL1	P	P
F3	FL2	FL2	FL2	RB
F4	S	FL3	RB	S

where FL, P, S, and RB stand for fluidity, permeability, permeability in shaft, and balance along the radial direction. FL1, FL2 and FL3 represent the split of the fluidity in C1, C2 and C3.

The difference between the best cluster and the worst one has to be identified based on the interpretation. Since the first factor (F1) at each cluster represents the most dominant phenomenon with the largest eigenvalue in the corresponding cluster operation, the difference in the first factor must be the most important. While in the other clusters the fluidity is the first factor that produces the most dominant correlation, in C2 the permeability is the first factor. Another difference is the fact that the fluidity is split into three factors in C2, while all of them become to combine into one factor in C4. It implies that there are three flow-patterns in the hearth and bottom in C2 and, however, only one flow-pattern dominates the whole area through the hearth and the bottom. The one flow-pattern facilitates the movement of the melt materials stream and stabilizes the process operation. It must be noticed that the split of the correlation representing the fluidity has disappeared, rather the correlation becomes much stronger, and a new factor that can be interpreted as radial balance appears, as it goes to C4 through C3. These sources cause the second cluster to be the worst cluster.

The suggestions based on the differences are generated. For instance, for the first factors of all clusters, a good fluidity whose correlation is not split but rather strong must be ensured to go to the best cluster. It is the first suggestion. Since the split and weaker correlation in the factors associated with the fluidity in C1, the first suggestion is verified using the information from the cluster in the middle of the best and worst clusters. The same processes for all factors at the same level have been explored. Table 3 shows the generated suggestions. Specified in a single statement, the operation direction that reflects all suggestions is to ensure a good fluidity as a top priority and permeability in a second priority and then keep radial balance good and facilitate a flow of gas in the shaft.

Table 3. Generation of Suggestions.

Factor	Difference C2	Difference C4	Suggestion
F1	P	FL	Ensure a good fluidity
F2	FL1	P	Ensure a good permeability
F3	FL2	CB	Ensure radial balance
F4	FL3	S	Ensure permeability in shaft

Validation

To validate the suggestions and the operation direction, the hypothetical tests for the groups of separated observations have been performed as shown in Fig. 5. For the first test set circled and labeled in Fig. 5, the amount of PCI can be increased according to the direction, because it has the large score value for its first factor labeled with fluidity. The real action taken by operators during the operation has been checked. It has been confirmed that operation engineer really increased the amount of PCI at that time. The same validation procedure has gone through for six test sets.

As the validation proceeds, it can gives us the information on how process engineers had operated the process historically. For instance, for the first testset of C1, the operators increase the amount of PCI, because score value for the fluidity is large. For the second test set, they maintain the increased PCI level because score value for fluidity is still good. For the third test set, as score value for the fluidity decreases and permeability becomes good, they reduce the amount of PCI. For the fourth test set both of the fluidity and permeability become worse, so they has to reduce the PCI level to the lower level. The score plot analysis based on the factors such as the fluidity and permeability provides us with the information on when the amount of PCI can be increased or be decreased.

For fourteen test sets, the same processes have been applied and resulted in 12 successes and 2 failures. However, for one of two failures the operation objective was not to increase the amount of PCI, but to increase the production rate. For the other failure case, the engineers have admitted that they could not raise blast temperature hot enough due to an equipment malfunction. Fig. 5 shows the absolute valued score plot for C1 and C2 with the test sets marked by the circle. Table 4 shows the results of the hypothetical test.

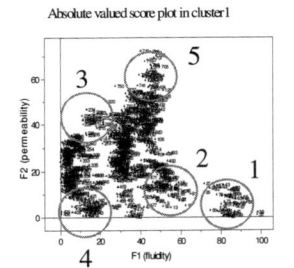

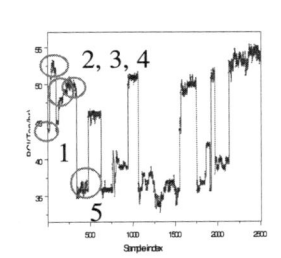

Figure 5. Validation and factor score plot analysis with PCI pattern (right).

Conclusions

The success of developing process improvement strategy based on statistical data analysis depends on the availability of long-term operation data with a good quality and a full description of the process behaviors. Furthermore, it depends on how much knowledge on the process a multivariate statistical model can extract from the collected data. The improvement based on a PCA modeling with no rotation may lead to wrong results, due to the possible mistakes in the interpretation of factors. This paper introduces a factor analysis model instead of the PCA model to enhance the interpretation power, resulting in a better improvement. Also, clustering has been employed to increase the efficiency and stability during the division process of the operation range. The methodology has been applied to the blast furnace operation data to find the better operation conditions to increase the amount of PCI. With no failure in hypothetical test, the suggested operation direction has a promising expectation with a good reliability. The possible increase in the amount of PCI based on the direction is estimated to be 25 ton/hr, resulting in the maximum cost reduction of 5.9×10^6/yr.

Remaining research issue is to extend the application of factor analysis. That is, it can be employed to analyze the process dynamics and reinforce fault diagnosis system using the interpretation of factors. Another issue is to extend the proposed methodology to plant wide scale. A promising idea is to apply a hierarchical clustering to plant-widely collected operation data so that each cluster in the first level contains only the operation data associated with one operation unit. Then, for each cluster the proposed methodology can be applied in the same way as explained in the paper.

Table 4. Hypothetical test for validation.

	Hypothetical action	Real action
Test 1	Increase PCI	Increase PCI
Test 2	Increase PCI	Increase PCI
Test 3	Increase PCI	Increase PCI
Test 4	Decrease PCI	Decrease PCI
Test 5	Increase PCI	Increase PCI
Test 6	Increase PCI	Increase PCI
Test 7	Decrease PCI	Decrease PCI
Test 8	Increase PCI	Increase PCI
Test 9	Decrease PCI	Decrease PCI
Test 10	Increase PCI	Increase PCI
Test 11	Increase PCI	Increase PCI
Test 12	Increase PCI	Increase PCI

Acknowledgment

The authors deeply appreciate the financial support of the Korea Science and Engineering Foundation through the Automation Research Center at POSTECH.

References

Ljung, L. (1987) *System Identification theory for the users*. Prentice Hall, New Jersey. Chap. 14, pp. 358-390.

Johnson, R.A. and D.W. Wichern (1992). *Applied Multivariate Statistical Analysis*. Prentice hall, New Jersey. Chap. 8 and 9, pp. 356-451.

Santen, A., G. L. M. Koot and L. C. Zullo (1997). Statistical data analysis of a chemical plant. *Com. Chem. Eng. Suppl.*, **21**, s1123-1128.

Omori, Y. (1987). *Blast Furnace Phenomena and Modeling*. Elsevier Applied Science, London and New York. Chap. 1-3, pp. 3-280.

Crowe, C. M., Y. A. G. Campos and A. Hrymak (1983). Reconciliation of process flow rates by matrix projection. *AIChE J.*, **29**, 881-888.

Demuth, H. and M. Beale (1994). *Neural Network Toolbox; User's Guide for use of MATLAB*. Math Works Inc., U.S. Chap. 8, pp. 1-21.

Hwang, D. H., T. J. Ahn and C. Han (1998). Process operation improvement based on statistical data analysis. *AIChE J.* (to be submitted).

Saraiva, P. M. and G. Stephanopoulos (1992). Continuous Process Improvement Through Inductive and Analogical Learning. *AIChE J.*, **38**, 161-183.

OPTIMAL ALARM LOGIC DESIGN FOR PROCESS NETWORKS

Chuei-Tin Chang and Chii-Shang Tsai
National Cheng Kung University
Tainan, Taiwan 70101

Abstract

An optimal alarm system design method, which takes full advantage of the inherent data redundancies in the sensor networks of chemical plants, has been proposed in this study. In particular, systematic procedures have been developed to identify independent measurement methods for the flow rate of any stream in the process and to synthesize corresponding alarm generation logic. This system is superior to any of the existing alarm techniques in the sense that it can be appropriately tailored to minimize the expected loss due to misjudgements in generating alarms.

Keywords

Alarm logic, Mass-flow network, Hardware redundancy, Spatial redundancy, Temporal redundancy.

Introduction

Alarm generation is in fact a basic function of the protective system in any chemical process plant. The current practice in the industry is simply to compare measurement data of the variable of interest with a pre-determined threshold value. The decision concerning whether or not to set off an alarm is then made accordingly. Since all measurements are subject to errors, two types of mistakes may be committed in this decision making process. First of all, spurious alarms may be produced due to measurement noises when the variations of the process variables are actually within acceptable limits (type I mistakes). Secondly, the system may fail to detect the existence of hazardous operating conditions and thus no alarms are generated (type II mistakes).

A common industrial practice to reduce the chance of misjudgement is to introduce *hardware redundancy* in the alarm system (Lees, 1980). Specifically, several independent sensors are installed to monitor the same process variable. Any inconsistency identified in the measurement data obtained from different sensors is usually resolved on the basis of operation experience or an arbitrarily chosen alarm logic. The implied objective of such a practice is to achieve a compromise between the conflicting emphases on decreasing type I and also type II mistakes. Although the traditional way of using hardware redundancy is effective on a qualitative basis in this respect, there are still several deficiencies. In particular, the conventional alarm strategy utilizes only the informations obtained from the redundant sensors for measuring the process variable of interest. Thus, other useful informations embedded in the process system are neglected completely. Also, the alarm generating logic are developed on an *ad hoc* basis and thus may not be cost optimal. This drawback can be significant in cases when the finacial loss of misjudgement is large.

One possible way to improve the current operation is to make use of not only the hardware redundancy but also the *spatial* and/or *temporal* redundancy embedded in the sensor network of any process plant (Ali and Narasimhan, 1993, 1995; Stanley and Mah, 1977). Thus, the adjusted data, rather than the raw measurement data, are adopted in this study for alarm generation purpose. However, all such estimates are also subject to errors and thus type I and II mistakes are still possible in generating alarms. To minimize the expected loss of misjudgements due to these two types of mistakes, a systematic method for synthesizing the optimal alarm logics has been developed in this research. Basically, the design techniques for trip systems (Inoue et al., 1982; Kohda et al., 1983) have been extended for this purpose. In addition, in order to implement the proposed logic synthesis method, the probabilities of false alarms and undetected failures must be estimated in advance. An on-line estimation procedure for these parameters has also been developed in this work. In particular, the estimated probabilities were obtained on the basis of reconciled data. Since in essence both hardware and spatial redunduncies are utilized, this approach is considered to be very reliable. Finally, it should be noted that the scope of present article is limited to mass-flow networks for the sake of brevity. The problems concerning both mass and energy balances will be addressed in the future.

The Error Models

The interdependence of flow and inventory data in a process is most naturally expressed in terms of material balances. In order to explore the network characteristics of the process, it is convenient to represent the mass-flow networks with process graphs (Mah et al., 1976). Without loss of generality, let us assume that all arcs in the process graph are measured. This is reasonable due to that fact that, if some arcs are unmeasured, one can always merge the nodes connected by these arcs to produce a graph in which only the measured ones exist. Let us further assume that there are n streams in the process and their respective mass flow rates are represented as x_j^t and $j = 1, 2, \ldots, n$. Due to unknown disturbances, the true flow rates can be viewed as

$$x_j^t = x_j^d + \delta_j \tag{1}$$

where, x_j^d denotes the design value of the flow rate of j th stream and δ_j is the corresponding deviation of true flow rate from its design value. In this study, δ_j is a normally-distributed random variable with *time-variant* mean. Specifically, its expected value is zero when the system is operated at normal steady state and otherwise when faults occur.

The measurement errors are related to the true flow rates according to the following equations:

$$x_j^{(i)} = x_j^t + e_j^{(i)} \tag{2}$$

where, $x_j^{(i)}$ represents the measurement value of the j th stream using sensor i and $e_j^{(i)}$ denotes the corresponding error. Let us assume that the total number of independent sensors used to measure the flow rate of stream j is m_j. Since the measurement errors are in general much smaller than the deviation δ_j caused by abnormal disturbances, it is thus assumed in this work that the means of measurement errors are negligible. Specifically, they are treated as normally-distributed random variables with zero means in this study. Also, if sensor biases develop during operation, we assume that they can always be identified with available detection schemes, e.g. Mah (1990), and the corresponding data can be removed before the generation of alarms. Finally, it is reasonable to believe that the variances of the measurement errors can be acquired from the vendor or an analysis of the historical data.

The Performance Function and Indicator Function

Among the streams in the process, there may be ℓ ($\ell \leq n$) streams on which alarms are installed. Associated with each of these streams, an operational constraint must be satisfied. The constraint associated with one of the streams, say stream a, can be written in a general form as

$$G(x_a^t) \geq 0 \tag{3}$$

where G is referred to a *performance function*.

Notice that an alarm is set off mainly as an indication of constraint violation. Since the true flow rate can never be determined, one has to rely on the measurement data to evaluate the performance function. In other words, the values of the following indicator function $G^{(s)}$ must be computed:

$$G^{(s)} = G(x_a^{(s)}) \tag{4}$$

where, $x_a^{(s)}$ denotes the measurement value of stream a obtained with the s th independent method. Let us assume that the total number of independent measurement methods is M_a and $M_a \geq m_a$. Due to measurement errors, the values of indicator function evaluated with data obtained from different methods are in general not consistent with one another. Nonetheless, one is still required to make a decision concerning whether or not to set off an alarm with these data. In the industries, the alarm generation logics are usually developed on an *ad hoc* basis using only hardware redundancy (Lees, 1980). For example, an alarm may be generated according to the most reliable on-line sensor or an arbitrarily chosen l-out-of-m_a ($l \leq m_a$) logic, and so on.

Obviously, the m_a independent sensors installed on stream a can be used as means of *directly* measuring the flow rate of stream a. According to Ali and Narasimhan (1995), different ways of *indirectly* determining the same mass flow can be found in the cut sets that contain stream a in which the mass flow of every other stream is measured. However, since some of the elements may simultaneously appear in more than two of these cut sets, the corresponding measurement methods are statistically dependent. Consequently, it is necessary to select out cut sets that do not contain common arcs other than a. A simple procedure has been developed to identify a set of such independent indirect measurement methods. For the sake of brevity, this procedure is presented elsewhere (Tsai and Chang, 1997).

Synthesis of Optimal Alarm Generation Logic

As mentioned before, M_a direct and indirect measurement methods can be used together to monitor the same variable of interest, i.e. the flow rate of stream a. Let us assume that the first m_a measurements are obtained from the sensors installed on stream a and the rest are from indirect methods. The flow rate of stream a can be any of the m_a direct measurement values or determined indirectly with measurements of other streams on the basis of material balance constraints. Consequently, a set of binary indicator variables y_s s can be determined, i.e.

$$y_s = \begin{cases} 1 & \text{if} \quad G^{(s)} < 0 \\ 0 & \text{if} \quad G^{(s)} \geq 0 \end{cases} \quad (5)$$

where, $s = 1, 2, \cdots, M_a$. The system alarm can then be generated on the basis of these indicators. The logic for setting off the alarm can be explicitly expressed with a binary alarm function $f(\mathbf{y})$ and $\mathbf{y}$ is a vector whose elements are the M_a indicator variables. The value of this alarm function is set to be one if the system is generating an alram. Otherwise, its value is zero.

Obviously, the values of the indicator variables y_s s may not be consistent with the true state x_a^t. Let us consider the true value of the performance function, i.e. $G^t = G(x_a^t)$. There are two kinds of mistakes that can be identified accordingly, i.e. the indicator y_s is set to be 1 when $G^t \geq 0$ (type I mistake) or it is set to be 0 when $G^t < 0$ (type II mistake). Similarly, the mistakes committed in generating the *system alarm* can also be classified into type I and II. Since both types of mistakes result in financial losses, there are incentives for developing an optimal alarm generation logic which minimizes the expected loss Ψ. This loss can be written as

$$\Psi = C_a(1 - P_F)P_a + C_b P_F P_b \quad (6)$$

where P_F is the demand probability which is defined as the probability of violating the constraint, i.e. $G^t < 0$, P_a and P_b denote respectively the conditional probability of type I and type II mistakes, and C_a and C_b the corresponding losses. Notice that the conditional probabilities P_a and P_b are affected by the selection of alarm function $f(\mathbf{y})$ and, consequently, there must exist an optimal alarm logic which minimizes Ψ.

It was shown (Tsai and Chang, 1997) that the expected loss is minimized if the alarm function can be chosen such that

$$f(\mathbf{y}) = \begin{cases} 1 & \text{if} \quad h(\mathbf{y}) > 0 \\ 0 & \text{if} \quad h(\mathbf{y}) \leq 0 \end{cases} \quad (7)$$

where,

$$h(\mathbf{y}) = C_b P_F \Pr\{\mathbf{y} \mid G^t < 0\} - C_a(1 - P_F)\Pr\{\mathbf{y} \mid G^t \geq 0\}$$

After obtaining the values of $f(\mathbf{y})$ on the basis of Eqn (7) for all possible $\mathbf{y}$, its functional form can be constructed accordingly. With the functional form given, the logic associated with $f(\mathbf{y})$ can be implemented as a hard-wired circuit or as a computer program.

Application Example

Let us consider the process graph presented in Fig. 1. Under normal operating conditions, it is assumed that the system is at its original steady state and can be described with the parameters presented in Table 1. The means and variances of true flow rates are listed respectively in the second and third column. As mentioned previously, the mean of every deviation δ_j is considered to be zero during normal operation. Thus, the means $E[x_j^t]$ s in Table 1 are also used as the designed values x_j^d s in the present example. In this plant, the flows of streams 1, 3, 4, 5, 6 and 8 are measured and each with one sensor only. The variances of measurement errors are given in the fourth column of Table 1. If data reconciliation is performed, the variances of the corresponding estimation errors d_j s can be computed and the results are given in the fifth column of Table 1.

Table 1. System Parameters Under Normal Operating Conditions.

Stream Numbert (j)	$E[x_j^t]$	$Var[x_j^t]$	$Var[e_j]$	$Var[d_j]$
1	34.33	1.069	0.5347	0.2732
2	34.33	1.069	N.A.	N.A.
3	34.33	1.069	0.5347	0.3182
4	20.61	0.830	0.4150	0.2839
5	13.72	0.397	0.1986	0.1686
6	24.03	0.308	0.1540	0.1654
7	10.30	0.083	N.A.	N.A.
8	10.30	0.750	0.3750	0.3051

In the present example, it is assumed that an alarm system is installed on stream 1 to protect against the detrimental outcomes caused by low flow rate. The lower threshold limit selected in the simulation studies is 32.0 Kg/hr. The first independent measurement method is naturally associated with the sensor for directly measuring stream 1, i.e.

$$x_1^{(s)} = x_1^{(1)} = x_1 \qquad (8)$$

where x_1 denotes the measurement data obtained with the sensor on stream 1. Notice that, since there is only one sensor on each stream in this case, the superscript of x_1 is dropped to simplify notation.

The other independent measurement methods were identified according to the cut sets that contain stream 1. In this case, two cut sets were found, i.e. {1, 6, 8} and {1, 4, 5}. Thus, the indirect measurement methods can be expressed with the following material balance equations:

$$x_1^{(2)} = x_6 + x_8 \qquad (9)$$

$$x_1^{(3)} = x_4 + x_5 \qquad (10)$$

The effectiveness of the proposed alarm generation strategy can be demonstrated with simulation studies. The variation in flow rate x_1^t due to an unknown fault was first simulated. Initially, its mean was kept at its designed value, i.e. 34.33 Kg/hr. The fault occurs at time

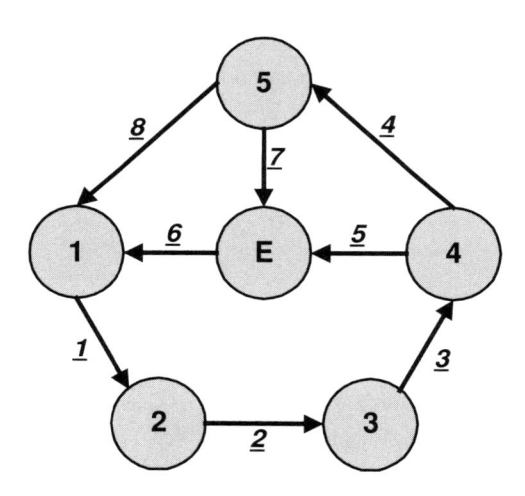

Figure 1. A process graph.

950 Δt and Δt is the sampling interval. As a result, the mean flow rate of stream 1 decreases gradually and reaches a new steady level of 29.96 Kg/hr at time 1050 Δt. The mean values of all other flow rates were generated in such a way that the constraint of material balance is always maintained at every node in the process graph. The random number generator RONNA in IMSL was used for producing the values of $(x_j^t - E[x_j^t])$s. The means of these random variables are zero and the variances are the same as those of the true flow rates (Table 1). Finally, the values of x_j^ts were computed by

adding $\text{E}[x_j^t]$s and ($x_j^t - \text{E}[x_j^t]$)s. A total of 2000 sets of data have been generated in this case. Only half of them, i.e. from sample no. 500 to no. 1500, are shown in Fig. 2. Notice that the solid line in this figure is the location of the mean value of normal flow rate and the dotted line represents its lower limits. Next, the measurement values were simulated. This was done by adding the measurement errors e_js to the corresponding true flow rates. The values of e_js were again created with a random number generator. A sample of the simulation results are presented in Fig. 3.

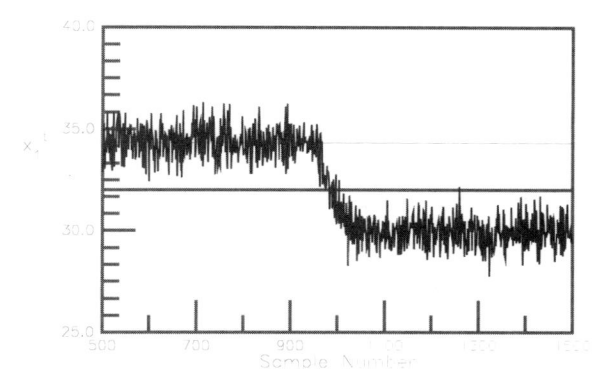

Figure 2. Simulation results of the true flow rate of stream 1.

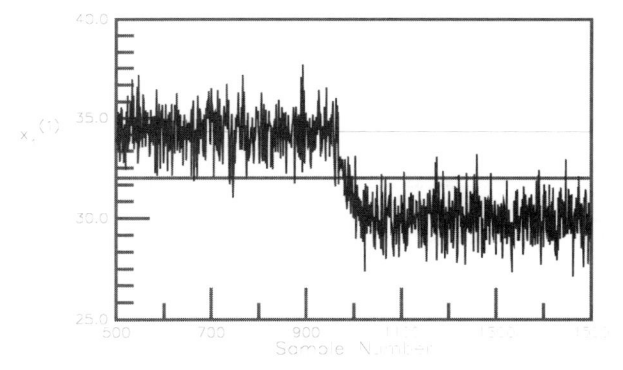

Figure 3. Simulation results of the flow-rate measurements of a normal sensor on stream 1.

Using these measurement data, one can then compute the reconciled flow rates or the estimated flow rates with an Kamman filter by assuming quasi-steady state (Stanley and Mah, 1977). In this study, the traditional approach, i.e. using the raw measurement data of stream 1, was taken first to set off alarm. The proportions of type I and II mistakes in this case were determined to be 0.03242 and 0.02172 respectively. On the other hand, by adopting the reconciled data as the basis for alarm generation (approach A), it was found that, by using the same set of measurement data, the chances of making these mistakes can be lowered significantly to 1.925 % (type I) and 1.481 % (type II). Finally, if the Kalmam-filtered results are used for alarm generation purpose (approach B), the results are even better (0.811 % type I mistakes and 0.395 % type II mistakes).

Each time a new set of measurement data and the corresponding reconciled flow rates are obtained, an optimal alarm logic can be constructed on-line with the proposed synthesis procedure. The computation time needed to build one of these logics is about 2 seconds on a Pentium PC. To implement the implied alarm generation strategy, values of $x_1^{(s)}$ (s = 1, 2, 3) must also be determined with the three independent methods given in Eqns. (8) - (10). The results of adopting the proposed alarm policy (approach C) with different C_b/C_a ratios are summarized in Table 2. In particular, the proportions of type I and II mistakes are presented in this table. For comparison purpose, the results of using the traditional approach, approach A and approach B are also included. From these results, it is clear that the the proposed method is superior in the sense that type II mistakes can be reduced to a negligible level. This is usually the first priority in most cases since the purpose for installing an alarm is almost always to protect against certain catastrophic consequences. However, it should also be pointed out that this improvement is often achieved at the expense of committing more type I mistakes. As the ratio C_b/C_a increases, the percentage of type I errors will inevitably increase also. This is nonetheless still acceptable as long as C_b/C_a is consistent with the primary design objectives.

Table 2. Performance of Optimal Alarm-Generating Strategy when All Sensors Function Properly.

Alarm Strategies	C_b/C_a	Proportion of Mistake	
		Type I	Type II
Traditional	----	0.03242	0.02172
Approach A	----	0.01925	0.01481
Approach B	----	0.00811	0.00395
Approach C	1.0	0.03141	0.02073
	3,0	0.04458	0.01185
	6.0	0.06586	0.00494
	9.0	0.08511	0.00000

In this paper, it is assumed that the gross errors can always be removed in advance. It should be emphasized that this assumption is desirable but not necessary. The proposed alarm strategy should still outperform any of the current practices even when gross errors are present. This is due to the fact that more independent measurement methods are adopted for alarm generation purpose. Additional simulation studies have been carried out to verify this assertion. In particular, let us consider a typical scenario, i.e. the direct measurement values of two streams, i.e. stream 1 and 3, remain "normal" when their true flow rates actually exceed threshold limits. The simulated measurements of the flow rate of stream 1 are presented in Fig. 4. Naturally, if the traditional approach is adopted to generate alarms, the probability of type II mistakes must be very high. Simulation studies have been carried out to confirm this prediction and the results can be found in Table 3. On the basis of the erroneous data of streams 1 and 3 and the measurement data of other streams, the reconciled and Kalman-filtered flow rates of stream 1 can also be computed. It was found that these estimates were grossly inaccurate. Consequently, if approach A or B is used for alarm generation purpose, the possibility type II mistakes should still be unacceptably high (see Table 3). On the other hand, it should be noted that the measurement values obtained by the two proposed indirect methods, i.e. Eqns. (9) and (10), are both still correct. As a result, the chance of type II mistakes can be significantly reduced if approach C is adopted for generating alarms (see Table 3). Finally, notice that it is pointless to compare the proportions of type I mistakes. Since sensor malfunctions have occured in this case, a large number of spurious alarms may even be beneficial. In practice, this phenomenon can often prompt the plant personnel to detect and isolate the failed sensors.

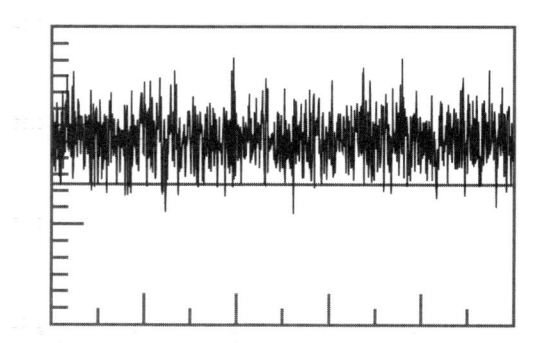

Figure 4. Simulation Results of the Flow-Rate Measurements of a Failed Sensor on Stream 1.

Table 3. Performance of Optimal Alarm-Generating Strategy When Sensors on Stream 1 and 3 Fail.

Alarm Strategies	Proportion of Mistake	
	Type I	Type II
Traditional	0.07700	0.94176
Approach A	0.02128	0.41658
Approach B	0.00507	0.39289
Approach C ($C_b / C_a = 9$)	0.10436	0.15202

Conclusions

From the above discussions, it is clear that the proposed alarm strategy is superior to any of the existing techniques. Such an improvement is brought about mainly by integrating two of the inherent features of mass-flow network, i.e. the hardware and spatial redundancies, into system design. This approach has never been attempted before. Furthermore, from experiences obtained in solving the example problem, one can also conclude that the demand for on-line computation is reasonable especially when the sampling interval is in the range of minutes or longer.

Reference

Ali, Y., and S. Narasimhan (1993). Sensor network design for maximizing reliability of linear processes. *AIChE Journal*, **39**, 820.

Ali, Y., and S. Narasimhan (1995). Redundant sensor network design for linear processes. *AIChE Journal*, **41**, 2237.

Inoue, K., T. Kohda, H. Kumamoto, and I. Takami (Apr. 1982). Optimal structures of sensor systems with two failure modes. *IEEE Trans. Reliab.*, **R-31**, 119.

Kohda, T., H. Kumamoto, and K. Inoue (Apr. 1983). Optimal shut down logic for protective systems. *IEEE Trans. Reliab.*, **R-32**, 26.

Lees, F. P. (1980). *Loss Prevention in the Process Industries*. Vol. 1, Butterworth, London.

Mah, R. S. H., G. M. Stanley, and D. M. Downing (1976). Reconciliation and rectification of process flow and inventory data. *I&EC. Proc. Des. Dev.*, **15**, 175.

Mah, R. S. H. (1990). *Chemical Process Structures and Information Flows*, Butterworth, Boston.

Stanley, G. M., and R. S. H. Mah (1977). Estimation of flows and temperatures in process network. *AIChE Journal*, **23**, 642.

Tsai, C. S., and C. T. Chang (1997). Optimal alarm logic design for mass-flow networks. *AIChE Journal*, **43**, 3021.

ADAPTIVE RANDOM SEARCH AND SHORT-CUT TECHNIQUES FOR PROCESS MODEL IDENTIFICATION AND MONITORING

Gheorghe Maria
Department of Chemical Engineering
University 'Politehnica' Bucharest
P.O. 15 - 253 Bucharest, Romania
E-mail: g_maria@chim.upb.ro

Abstract

Significant benefits in process safety and optimal monitoring can be realized if a sufficiently accurate and reliable process model is available. Mechanistic based models can offer the advantage of good predictions, physical meaning of the parameters and the possibility of storing the results for future process analysis. As their elaboration is less costly, reduced models are preferred even if they require periodical up-dating (e.g. re-estimation of parameters and model structure adjustment). In all the modelling-identification-simulation steps, gradient optimization methods (*GM*) are effective in finding a local problem solution. However, they present frequently convergence difficulties in finding the global solution for multimodal problems, complex constraints, poor-conditioning (over-parametrisation, model - data imprecisions, low variable observability), because the search progress is strongly dependent on the function gradient(Jacobian)/Hessian matrix accurate evaluation. Gradientless random search (*RS*) optimizers can by-pass most of the *GM* difficulties, being very reliable for solving multimodal/non-convex problems while their convergence rate is relatively slow. Recent hybrid multi-level adaptive random searches (*ARS*) improve the convergence rate, and are more robust to poor-conditioning cases. A short survey of *RS* are completed with tests to compare *ARS, RS* and *GMs* performances. Finally, an effective and novel strategy to couple the multilevel *ARS* (e.g. *MMA* of Mihail and Maria, 1986) with a novel short-cut estimator (*MIP* of Maria and Rippin, 1997) able to decompose and to fix the poor-conditioning by using data-bank information is presented. Some on-line model updating problems under isothermal and non-isothermal dynamic operating conditions exemplify this strategy (*Note:* the full paper manuscript of 35 pages is available).

Keywords

Model identification, Adaptive random optimization, Short-cut estimation, Coupled estimation techniques.

Process Model Identification (model update) Problem

Although requiring more experimental and computational effort in identification, the *mechanistic-based models* for process analysis are preferred to the empirical ones because of better quality and confidence in predictions. Moreover, the physical meaning of the parameters offers the possibility of estimate interpretation and storing in data-banks ready for further use. The built model complexity depends on the amount of available information and on the utilization scope. For design and optimization a moderately large model is sufficient, while for real-time-fast-responding process monitoring and control reduced models are preferred. The model *parameter estimation* is based on the estimation statistical theory, e.g. extremization of a suitable objective function which minimize the deviations (residuals) between data and model predictions in terms of output variables. The problem is solved with a nonlinear programming *iterative* numerical method, by considering the parameter and model constraints (usually mass and energy balances). Reliable identification of the global solution depends not only on the data amount/quality but also on the performance of the optimization technique used. The *estimate quality (inference, significance)* based on a local sensitivity analysis, and the *model adequacy* (residual analysis vs. noise level) are statistically defined in direct relationship with the data quality and model structure uncertainty (Bates and Watts, 1988). For instance, if data are very noisy, poorly distributed, intercorrelated, the estimate will become uncertain and meaningless, even if the model adequacy is satisfactory, with repercussions in further model-based predictions. Nonlinear transformations in parameter / model functions can increase the convergence rate of the optimization method (Espie, 1986; Watts, 1994; Bilardello et al., 1993), but the estimate quality is rather influenced by the model structure and data quality (Stewart et al., 1996). An increase in estimate quality is obtained by performing new experiments or by a *model reduction* strategy. Keeping a reasonable adequacy some parameter/observations are rejected or lumped with the corresponding loss in model generality (Tomlin et al., 1995; Vajda et al., 1989). Often model reduction is performed by means of an optimization rule (Maria, 1989; Edwards and Edgar, 1994), and results checked in various variable domains. For on-line process monitoring, possible data - model inadequacy is corrected by up-dating the *reduced model*, the loss of information being compensated by a parameter / structure reactualisation (Filippi-Bossy et al., 1989; Terwiesch, 1995; Fotopoulos, 1996). In mathematical terms all the mentioned modeling steps involve the use of *numerical optimization* to extremise a specific objective function subjected to variable and model constraints:

$$\text{Min } f(\theta), \theta = (x,y); \text{ subject to:}$$
$$x_{min} \leq x \leq x_{max}; y_{min} \leq y \leq y_{max}; g(\theta) \leq 0 \quad (1)$$

(where **x** is an *n*-dimensional vector of continuous parameters, **y** is a vector of integer parameters, *min/max* denotes the minimum / maximum bound of the parameters θ). The used *GM* or *RS* generates a sequence of improved approximations to the optimal solution θ^*, following an iterative scheme:

$$\theta_{j+1} = \theta_j + \lambda_j s_j, \text{ (feasible, in the } j\text{-th iteration)} \quad (2)$$

where s_j is an *n*-dimensional direction vector and the scalar λ_j is the distance moved in that direction. The iteration is terminated when a point θ is 'sufficiently close' to θ^*.

The scope of this paper is I) to present a short survey and classification of the most significant *RS/ARS* by pointing out development trends in this field; a comparison of novel *ARS* performance based on complex numerical tests is also included; ii) to prove the effectiveness (reliability and convergence rate) of a novel identification strategy which couples modern *ARS* (e.g. *MMA*) with novel short-cut estimators (e.g. *MIP*) based on prior data-bank information.

Gradient Method (GM) Class

The *GMs* use the derivatives of *f* and *g* of first order (Gauss-Newton type), second order (Newton type) and hybrid (quasi-Newton type). They are known to be effective in rapidly locating a local optimum, the convergence being dependent on the conditioning number and discretization scheme accuracy in evaluating Jacobian/Hessian matrices. Methods as trajectory sensitivity are effective but computative, while complex constraints or poor-conditioning, and additional difficulties as over-parameterization, model / data imprecision, incomplete variable observability complicate the algorithm usually causing *GM* failure. In spite of a large number of variants (for instance reviews of Lootsma, 1972; Edgar and Himmelblau, 1989; Guay and McLean, 1995) the local convergence remains the main weakness of *GMs*, repeated search starting from various initial solutions being necessary (Forbes et al., 1994). Commercial software as SIMUSOLV (Steiner et al., 1990) includes a generalized reduced *GM*; MATLAB (Grace, 1990) includes two sequential quadratic programming *GMs* (*SQP-HP*), a quasi-Newton (*BFGS*), and a Marquardt's rule for non-linear least-square regression (*NLS*). The global optimization procedures (*GOP*) incorporate *GMs* (Kocis and Grossmann, 1988; Floudas, 1995; Floudas and Visweswaran, 1995).

Optimization Random Search (RS) Methods

RS find the extreme of *f* in (1) by random exploration of the feasible parameter domain, without using function derivatives (McMurtry, 1970). In spite of their low convergence rate, *RS* are simple, *robust* to starting conditions, skillful in handling complicated constrained, multimodal/non-convex optimizations. *RS* generate *P* trial points in the iteration *j*, and change the parameter vector x_{j+1} following random direction s_j with a certain step λ_j:

$$x_{j+1} = x_j + \lambda_j s_j, \text{ (feasible).} \quad (3)$$

The sampling can be fixed or variable, independent, uniform, or following the search progress, on a sphere or hypercube. The step can be *fixed*, *optimum*-size (from one-dimensional minimizations), or *adaptive* (considering search history in the new point generation distribution). Pure *RS* can perform local searches from all/some tried points (*multistart RS*), or parallel search in various *clusters* of the domain. The search stops in the case of no improvement in f, sufficient evidence of a global optimum, or too costly search continuation.

Simulated Annealing (SA)

SA are RS in which step length $\Delta x_j = \lambda_j s_j$ is randomly uniformly generated in a hypersphere $\|s_j\| \leq 1$, or in a hypercube centered on current x_j and of step size $\|s_j\| \leq n$ (Kalivas, 1995). The difference appears for the trial failure, when a probability of conditional acceptance of a detrimental step is computed:

$$p(x_{j+1}) = exp(-\beta \Delta f_j); \rho \in [0,1], \text{uniformly random number} \quad (4)$$

If $p<\rho$ the increment is accepted, otherwise another Δx_j is generated, SA temporarily accepting a poorer solution with a probability expressed by a Boltzmann distribution. Different rules of choosing λ_j, $p(x_{j+1})$, β try to improve the search reliability, but the convergence rate is order of magnitudes lower than of ARS (Banga and Seider, 1995).

Genetic Algorithms (GA)

GA are RS in which the iterative rule presents similarities with a biogenetic mutation process (Rechenberg, 1973; Goldberg, 1989). A j iteration implies: i) generation of L 'individuals' (L = 'population size') uniformly distributed; ii) the obtained set of 'parental individuals' is randomly divided in two subsets: 'mother individuals' and 'father individuals'; iii) two 'individuals' are selected by means of a random (adaptive, proportional, ranking) rule; iv) the selected 'parental' vectors are coded, e.g. rewritten in an (usually) binary representation, having a 'chromosome length'; v) a new 'individual' is created by crossover of the 'parental' individual, then changed by a 'mutation' algorithm (with a certain 'mutation frequency'), e.g. mutation of the point x, of a part of x, or of any position of the x part; vi) the resulted 'individual' is decoded and the corresponding f evaluated. Steps *iii)-vi)* are repeated ('number of generations') until the best iterative point x_{j+1} (minimizing f) is retained. GAs have been recently applied in solving kinetic model identification problems (Moros *et al.*, 1995; Edwards and Edgar, 1994).

RS with Centroid Generation

The Nelder and Mead (NMA, 1965) algorithm, in the j-iteration, randomly generates P+1 points (P= [x]-dim.), detect the min/max values of the f, and calculate a P+1 polyhedron 'centroid'. The 'worst' vertex (corresponding to the largest f) is replaced with a better (feasible) point obtained by successive reflections through the centroid. If the reflection is successful the centroid is expanded, otherwise is contracted and, if necessary, reduced. Although more reliable in solving multimodal problems

Table 1. ARS from LJ Class: Iterative Formula and Procedure Parameters [$x_{j+1} = x_j + M(Z_j)^k r_j/p - e_j$; $r_{j+1} = (1-\varepsilon)r_j$; $r_0 = \lambda (x_{max} - x_{min})$; S = Search Success, F = Search Failure].

LJ, Luus and Jaakola (1973): $k=1$, $M=I$, $p=1$, $\varepsilon=0.05$, $e=0$, $\lambda=1-2$, $P \leq 100$
GG, (see Heuckroth et al 1976): LJ with $k>0$
HGG, Heuckroth et al. (1976): LJ with $k=p>0$
Nair (1976): $k=1$, $M=I$, $p=1$, $\varepsilon=0.05$, $e_j = x_{j-1} - x_j$, $e_0 = 0$, $\lambda = 1-2$
WL, (see Wang and Luus, 1978): $k=p=1$, $M=diag[m_{ii}]$; j-th iteration executes 10 contractions for each variable i, with $m_{ii} = 2$, $m_{kk} = 0.05$; $\varepsilon=0.05$, $e=0$, $\lambda=1-2$, $P \leq 100$
MWL, Wang and Luus (1978): $k=p=1$, $M=diag[m_{ii}]$; WL with $m_{ii} = 1$, $m_{kk} = z_{kk}^4$; $\varepsilon=0.05$, $e=0$, $\lambda=1-2$, $P \leq 100$
ARDS, Martin, Gaddy (1982): θ = uniform random $\in [0,1]$; $k=1$, $M=I$, $p>0$, $\varepsilon=0$, $e=0$, $\lambda=1-2$, $P \leq 100$; N= tried points; $p \in [0.4,1]$; $z_i = (2\theta_i - 1)/(\sum_{k=1}^{N}(2\theta_k - 1)^2)^{0.5}$
Lou (1985): $k=1$, $M=I$, $p=1$, $e=0$, $\lambda=1$, $P \geq 100$; $(1-\varepsilon)=\beta_j > 0$, $\beta_j^i = (1.1 \div 1.2)[(5 \div 7)\|x_{j-1}^i - x_j^i\|/r_{j-1}^i - 1]$
Luus and Brenek (1989): LJ rule with one-dimensional GM search: $x_j^{(s)} = argmin\ f(x_j^{(LJ)} - \lambda \nabla f)$, feasible.
SGA, Salcedo et al. (1990); Salcedo (1992): $k=1$, $M=I$, $p=1$, $e=0$, $\lambda=1-2$, $P=100 \div 300$; $\varepsilon=0.05$ if for two (S) $\Delta x < 5\%$; ε is decreased $\in \{0.1, 0.05, 0.025, 0.02, 0.01\}$; re-start after 20 successive domain compressions; periodic domain expansions. Modification for MINLP (eqn. 1): $y_j = y_{min} + INT\{Z_j r_0 + v\}$; Z= random $\in [0,1]$; $v \in \{0.5\}$
ICRS, Banga et al. (1987): Z_j= Gauss random $\in (0,1)$; $k=1$, $M=I$, $p=1$, $e=0$; increment j only for search success; $r_j = k_1 \lambda_j$ (for S); $r_j = k_2^{\alpha} k_1 \lambda_j$ [for $(\alpha n_e n)$ F]; $k_1 = 1/3$, $\lambda_j = min\{(x_{max} - x_j),(x_j - x_{min})\}$; $k_2 = 1/2$, $n_e = 4$

than GMs, NMA modifications try to reduce the polyhedron size and eliminate cycling near the optimum (review of Maria and Maria, 1997). SIMUSOLV and MATLAB packages include modified NMA.

ARS from Luus-Jaakola (LJ) Class

ARS from this class (Table 1) automatically adapt the steplength and direction based on the success or failure obtained in the last iteration(s). The basic *LJ* generates in each iteration *j+1* at least *P* feasible *n* points *x* (*n*-dim.) by using a diagonal uniform random matrix $Z_j=[z_{km}]_j \in [-1,1]$. Because the convergence is ensured only for small *n* and depend on initial x_o, r_o and ε (Lou, 1985), effective *ARS* variants use different procedure parameters (*k*, *M*, *p*, ε, *e*, λ, *P*). *ARS* are effective in reaching 0,1% f_o, but spent too many iterations in reaching more. Improvements in rapidity and reliability are reported for hybrid *ARS* containing locally convergent exploration coupled with search in the whole domain (Schoen, 1991). *ARS* easily treat equality constraints (Banga and Cesares, 1987), and mixed integer problems (Salcedo, 1992). As the search history used in acceleration of *LJ*, *GG*, *HGG*, *WL* increases the risk of local convergence, new strategies of successive expansions / contractions of the search region overcome these difficulties (*SGA*, *ARDS*, *MMA*).

The Adaptive Step Size Class

ARS from this class (Schrack and Choit, 1976) (Table 2) choose an *adaptive*, random generated, iteration search step λ_j by using search information from the last two iterations, performing periodical expansion / contraction of the search hypersphere and changing search sense (*RS* with reversals). For success, the iteration is accepted and the step length reduced ($\lambda_{j+1}=\alpha\lambda_j$, $\alpha<1$); otherwise, the point is returned back if the immediately preceding step was a reverse step, else the search direction is reversed ($x_{j+2} = x_j - \lambda_j s_j$, feasible). Variants of basic reverse-step *ARS* consider more search history in improving the iterative step, alternative positive and negative trial biasing, parallel search in cells, or hybrid *RS* with region expansions / contractions (*HGG*, *ARDS*, Lou's *RS*, *SGA*). It is to mention the Matyas's *ARS* based on all the search history (Schrack and Borowski in Lootsma, 1972).

Their convergence rate is high in the first iterations, not affected by the noise in *f* but affected by the variable scaling. One of the most effective *ARS* is the *MMA* (Mihail and Maria, 1986, Table 2), which modify Matyas' method to overcome their limitations. *MMA* contains two search levels (N_{uni} and N_{mult} steps, switched with a search efficiency test), in which the multimodal/local search is weighted according to some criteria. The bias vector *b* and region contraction factor α_f $\in [0.8, 0.99]$ are restricted, while rejection r_s, r_f and detection c_s, c_f coefficients are those of the original Matyas *ARS*. The procedure parameters $N_{uni}=N_{mult}$, α_f, *P*, λ depend on the problem complexity / multimodality. Maria (1993) recommends $N_{uni} \neq N_{mult}$, $\lambda=1-2$, and vector (N_{uni}, N_{mult}, α_f, *P*)= (60, 60, 0.90, 3), or (10, 10, 0.99, 400) for low complexity, and (10-15, 50, 0.8-0.9, 150-200) for difficult problems. The *MMA* is very effective compared with other *ARS* and *GMs* both in reliability and convergence rate (see tests of Table 3). For instance, the Meyer and Roth (1972) test is difficult for a *RS* because the optimum is included in a very 'narrow valley': a 0.01% deviation in *x* can produce several orders of magnitude variations in *f*, and a *GM* 'following' the gradient is expected to be more effective. In spite of that, modern *ARS* (*MMA*, *SGA*) are very effective, as for solving *GOP*.

ARS in Off-line Kinetic Model Estimation

Complex chemical kinetics play an important role in accurate process characterization, the physical meaning of the parameters allowing information to be stored in databanks for further studies. A kinetic model is an ordinary differential (*ODE*) equation set describing the species concentration evolution over reaction time. Identification of the model (high correlated) Arrhenius constants from observed species concentrations frequently lead to convergence problems, complicated by some model constraints, data noise and low observability, over-parameterization (Stewart and Sorensen, 1981; Damiano, 1983; Watts, 1994). Being local convergent, the *GMs* present difficulties when poor-conditioning of derivative matrices are present. Because *ARS* are more reliable /robust to such cases, Maria and Rippin (1996) integrated *MMA* in an expert system for complex kinetics identification and coupled with short-cut estimation techniques (*MIP*) in order to increase their effectiveness. *Example 1* (Table 4) proves the rapidity of *MMA-MIP* in estimating six Arrhenius constants: quick and reliable *MIP* solutions from isothermal data are transferred and refined with *MMA*. *Example 2* proves the *MMA* higher reliability compared with *NMA* and *GRG-GM* in estimating ten Arrhenius constants of a complex catalytic kinetics. *Example 3* proves the *MMA-CPEMR* (Maria, 1989) effectiveness in solving a complex polymerization equilibrium problem in which the model dimension (parameter *M*) is also unknown (Blau et al., 1972). *Example 4* (Maria and Rippin, 1996) proves the *MMA* reliability in estimating high intercorrelated (99.74%) Arrhenius constants from non-isothermal data for an irreversible first-order reaction. In opposition to *MMA*, the poorly conditioning of the *NLS*- Jacobian create convergence problems to classical Marquardt-*GM* ('leastsq' of MATLAB) and *NMA* ('search' of SIMUSOLV and 'fmins' of MATLAB) routines, the reliability strongly depending on the initial guess (Fig. 1).

Table 2. Adaptive Step Size ARS Class ($x_{j+1} = x_j + \Delta x_{j+1}$, Feasible; n= dim(x); Z_j= Uniform Random Numbers $\in [-1,1]$; S = Search Success, F = Search Failure).

Bekey et al. (1966) (referred in McMurtry, 1970): $r_j = 0$, (for S); $r_j = r_{j-1} + 1$, (for F); $\Delta x_{j+1} = \Delta x_j$ (for S); $\Delta x_{j+1} = -2\Delta x_j$ (F & $r_j < 2$); $\Delta x_{j+1} = Z_{j+1}$ (F & $r_j \geq 2$)
Schumer and Steiglitz (Lootsma, 1972). Alg.I: (variable λ_j) $\Delta x_{j+1} = \lambda_{j+1} Z_{j+1}$, (S); $\Delta x_{j+1} = -\lambda_j Z_j + \lambda_{j+1} Z_{j+1}$, (F) **Alg.II:** $\Delta x_{j+1} = t_j Z_j$ (for S; $t_{j+1} = \alpha_s f t_j$); $t_o = 1$; $\Delta x_{j+1} = 0$ (F; $t_{j+1} = \alpha_f t_j$); ; $\alpha_s = 1.618$; $\alpha_f = 0.618$; M=3n= successive failures after which t_j is decreased.
Kjellström (in Lootsma, 1972): $x_{j+1} = x_j \oplus t_j M Z_j$ (sum in three simultaneous search directions; $t_{j+1} = \alpha_s t_j$ (S); P= 3n; $t_{j+1} = \alpha_f t_j$ (F); $t_o = 1$; $\alpha_s = 1.115$; $\alpha_f = 0.933$
MA, Matyas (Lootsma, 1972): $\Delta x_{j+1} = M(b_j + t_j Z_j)$ (S); $\Delta x_{j+1} = 0$ (F); $t_j = \alpha_s t_{j-1}$, (S); $t_j = \alpha_f t_{j-1}$, (F); $b_j = r_s b_{j-1} + c_s (b_{j-1} + t_{j-1} Z_j)$, (S); $r_s = r_f = 0.75$; $b_j = r_f b_{j-1} + c_f (b_{j-1} + t_{j-1} Z_j)$, (F); $b_o = 0$; $t_o = 1$; $\alpha_s = 1.1$; $\alpha_f = 0.9$; $c_s = 0.5$; $c_f = -0.25$; M= scaling vector
MMA-ARS of Mihail & Maria (1986): - pseudo one-dimensional search: *initialize b_o, t_o, M_o and perform N_{uni} - Matyas (MA) iterations over n feasible points for each variable x_j direction (b, t of [N_{uni} x n]-dim. are separately incremented for each x_j direction).* - pseudo multi-dimensional search: *initialize b_o, t_o, M_o and perform N_{mult}- Matyas (MA) iterations over P feasible points per step (b,t of [N_{mult}]-dim. are simultaneously incremented for all the variable x directions).* - switch between branches following a search efficiency criterion; $|b_j| \leq 1$, , $M = \lambda(x_{max} - x_{min})$, $\alpha_f \in [0.8, 0.99]$.
Kelly, Wheeling (1962): Z_j = uniform random $\in [0,1]$; $\Delta x_j = \lambda_j [\beta_j h_j / \|h_j\| + (1-\beta_j) Z_j]$; $\lambda_j = \alpha \lambda_{j-1}$; $h_o = 0$; $h_{j+1} = \gamma h_j + (1-\gamma) M_j (x_{j+1} - x_j)$, (history vector); $\alpha > 1$ (S); $\alpha < 1$ (F); M_j = scaling vector; β_j, γ = const.
Jacoby et al. (1972): $\Delta x_j = \beta_1 h_j + \beta_2 Z_j$, feasible; $\gamma_1 \neq \gamma_2$; $h_{j+1} = \gamma_1 h_j + \gamma_2 (x_{j+1} - x_j)$; $h_o = 0$; $\beta_1 = \beta_2 = 0.5$

Table 3. MMA, RS and GM tests (n_f = Avg. Number of Function f Evaluations; f_o, f^*= Initial and Solution Values of f; the GM of Dennis, Dixon, Sargent & Sebastian, and Himmelblau, are Referred in Lootsma, 1972; MMA = Mihail and Maria, 1986).

Rosenbrock's valley problem (Dennis, 1972): $x_o = (-1.2, 1)$; $f_o = 24.2$; $x^* = (-1,1)$; $f^* = 0$; Ref.: MH= *Malik and Hughes*; SA of *Bohachevsky* (in Kalivas, 1995); qN-GM= quasi-Newton GM variant (1) of Dennis (1972), variant (2) of Sargent and Sebastian (1972), variant (3) of Himmelblau (1972), variant (4) of Dixon (1972); DS - RS of Himmelblau (1972)

Method	n_f	Method	n_f
MMA	66-114	qN-GM2	64-159 (failures)
MH	152	qN-GM3	53-2092
ARDS	92	qN-GM4	103-266
SA	8277	DS	156-651
qN-GM1	50-300		

Meyer and Roth problem (1972): $x_o = (0.02, 4000, 250)$; $f_o = 1.7e+9$; $x^* = (0, 6180, 345.2)$; $f^* = 87.9$; Ref.: SGA (*Salcedo, 1990*); Mq1= *Marquardt's GM of STATGRAPHICS* (1990), Mq2 of MATLAB; HP-GM of MATLAB; NMA= *Nelder and Mead* (MATLAB); BFGS-GM (MATLAB, Grace, 1990)

Method	n_f	f	Method	n_f	f
MMA	5800	13000	Mq2-GM	2781	87.9
MMA	18000	97	HP-GM	1124	407.3
SGA	?	88	NMA	1943	87.9
Mq1-GM	9226	87.9	BFGS	fail	-

Kowalik problem (Moore, 1992): $x_o = (0.2, 0.2, 0.2, 0.2)$; $f_o = 1.95e-3$; $x^* = (0.1928334, 0.1908362, 0.1231172, 0.1357659)$; $f^* = 307.48e-6$; Ref.: GOP-GM (*Moore* et al., 1992)

Method	n_f	$10^6 f$	Method	n_f	$10^6 f$
MMA	442	311.5	MMA	112773	307.4
MMA	2400	308.8	GOP	545490	307.4
MMA	62012	307.4			

Spectroscopy problem (Moore et al., 1992): $x_o = (130, 50, 6, 8, 1, 0, 5)$; $f_o = 0.11e+6$; $x^* = (130.9, 52.6, 6.73, 9.34, 1.2, 0.97)$; $f^* = 0.996e-10$; Ref.: GOP-GM (*Moore* et al., 1992)

Method	n_f	f	Method	n_f	f
MMA	554	0.4e-2	MMA	4700	0.4e-8
MMA	806	0.3e-5	GOP	109240	1e-10

Floudas & Visweswaran GOP test (Floudas, 1995): random $x_o \in ([0,2], [0,3])$; $x^* = (0.5, 8)$; $f^* = -8.5$

Method	n_f for 0.1%f_o
MMA	450 (any failure from 100 random starting points)

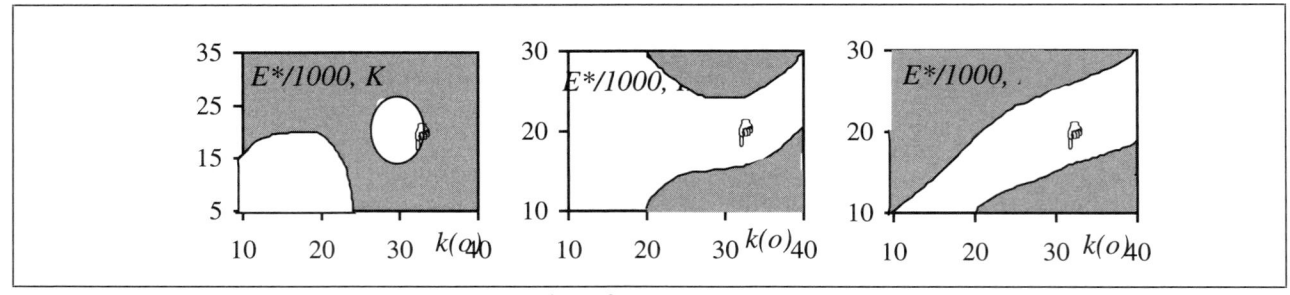

Figure. 1. Ex.4 : convergence regions (white) for NMA-ARS (SIMUSOLV, left; MATLAB, center), and Mq-GM (MATLAB, right).

Table 4. Coupled Short-cut MIP and MMA-ARS for Kinetic Identification (sdSSR =standard deviation of sum of squares of residuals; $k = \exp[A^* - E^*(1/T - 1/T_r)]$).

<u>Ex.1: Kinetics estimation (SIMUSOLV, Steiner et al., 1992):</u>
$A+B \rightarrow C$; $C+B \rightarrow D$; $A \rightarrow E$;
Data= [3 sets x 10 points] of scaled [A,B,C,D,E] at (85°,100°,115°C);
Results: $k = [k_1, k_2, k_3]$;
Procedures: MIP (85°C data):
k = [0.673, 0.16, 0.036]; MIP (100°C data):
k = [1.77, 0.456, 0.11]; MIP (115°C data):
k = [4.11, 1.17, 0.321]; MMA(-MIP):
[$A_1^*, E_1^*, A_2^*, E_2^*, A_3^*, E_3^*$] = [0.6, 8803, -0.838, 9341, -2.247, 9819]; sdSSR = 0.0022; T_r = 373K.

<u>Ex.2: Complex kinetics (Pop et al. 1991; i-butene alkylation):</u>
$A + H^+ \Leftrightarrow CH_3^+ + W$; $A \rightarrow B + W$; $A + B \rightarrow O + W$;
$Bu + CH_3^+ \rightarrow MBu + H^+$; $O + B \rightarrow P$; $MBu + B \rightarrow P$;
Runs(550-650K):[A,Bu,MBu,O,P,W]mol/molA, 0.6-1.4 Bu/A fed, 0-22 g.h/mol;
Results: [$A_{1m}^*, A_{21}^*, A_3^*, A_{45}^*, A_{65}^*, E_{1m}^*,$
$E_{21}^*, E_3^*, E_{45}^*, E_{65}^*$] = [-1.08,-3.53,-3.51,-1.83,0.28,10227, 2544, 1607,10149,15316];
Procedure reliability:MMA=9/9; NMA-ARS (MATLAB)=1/9; GRG-GM (SIMUSOLV)=2/9

<u>Ex.3: Complex equilibrium problem (Blau et al.,1972):</u>
$Min \sum_{k=1}^{P} [\sum_{i=1}^{M} \exp(\sum_{n=1}^{i} \gamma_n + i\mu_k) - y_{ok}]^2$,
subject to
$M \leq P$; $\gamma_1 = 0$; $\sum_{i=1}^{M} i. \exp(\sum_{n=1}^{i} \gamma_n + i\mu_k) = B_{ok}$;
Data: B_o, y_o (P = 24 runs);
Results: [$M; \gamma^T$] = [6;0,5.71,8.95,6.44,5.25,8.68];
Procedures: DFP-GM (Blau, 1972): sdSSR = 4e-4;
MMA-CPEMR (Maria, 1989): sdSSR=1e-4

<u>Ex.4: Single reaction Arrhenius constant estimation:</u>
$dA/d(t/t_{max}) = -\exp(k_o - E^*/T)A$;
Data (n=14): [T; A / A_o];
Results: k_o=30.7; E^*= 20000K;
Procedures reliability: MMA=100%; Nealder & Mead (SIMUSOLV, MATLAB) and Marquardt (MATLAB) less than 100% (Fig. 1)

MIP Short-Cut Estimator for ODE Models

Good initial estimate increases the subsequent NLS regression reliability and overcomes difficulties in reaching a solution with physical meaning. Maria and Rippin (1997) proposed a short-cut estimator (MIP) of parameters $\hat{K}'$ in ODE sets with the following advantages: simplicity, rapidity, reliability for poor-conditioned cases, any convergence problem, any tuning factor required or model linearisations, possibility of using prior information from data-banks. Starting from observed data (C',t') vectors, the MIP principle (Fig. 2) is to transform the ODE set into an algebraic one by performing integral transformations, but taking into account prior information about an analogous process to whom (C,t) and parameters $\hat{K}$ are known. The similarity analysis (e.g. quasi-constancy of the current/historic process rate ratio in a reduced time domain) is coupled with the problem decomposition (by estimating first dominant reaction constants). The poorly conditioning is avoided by use of parameter ratios from the similar process.

Coupled MMA-MIP in on-line kinetic model identification

Classical process monitoring via reduced model up-dating (state-parameter recursive estimators, Extended Kalman Filters, EKF) from on-line data and prior information, use several tuning factors and local quasi-linearizations, being very sensitive to the structural model mismatch, nonlinearity, prior (parameter/variance) knowledge, data noise, variation in species/reaction observability (de Valliere and Bonvin, 1989-1990; Terwiesch et al., 1994). Periodic exact NLS assist EKF in reinitializing the prior knowledge, tuning factors, and even the model structure ('tendency modeling' or 'gray box' model, Fotopoulos, 1996; Jang et al., 1986; Maria et al., 1996a, 1997). Because ARS are very effective in solving complex on-line identification / optimization (Banga et al., 1994; Bojkov et al., 1993; Banga and Seider, 1995), on-line model up-dating can be performed by combining ARS-NLS steps with rapid recursive short-cut MIP and estimate sensitivity analysis (Maria et al. 1996a). The advantage of using MIP is the possibility to

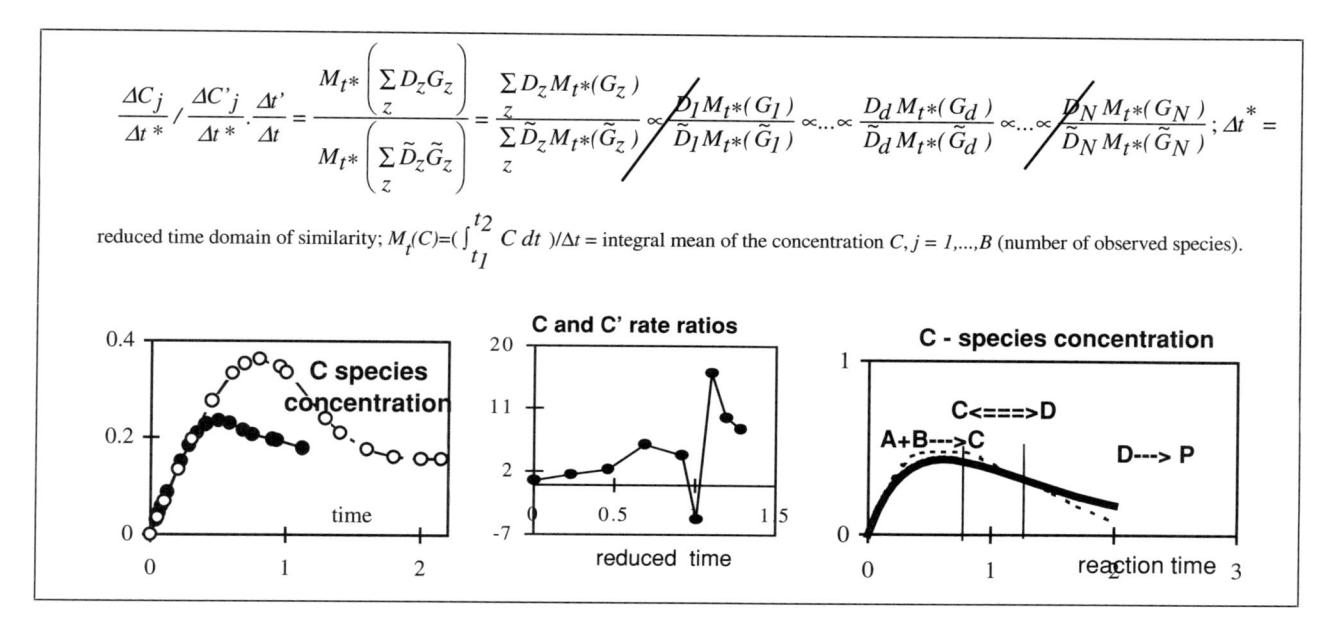

$$\frac{\Delta C_j}{\Delta t^*} \Big/ \frac{\Delta C'_j}{\Delta t^*} \cdot \frac{\Delta t'}{\Delta t} = \frac{M_{t^*}\left(\sum_z D_z G_z\right)}{M_{t^*}\left(\sum_z \tilde{D}_z \tilde{G}_z\right)} = \frac{\sum_z D_z M_{t^*}(G_z)}{\sum_z \tilde{D}_z M_{t^*}(\tilde{G}_z)} \propto \frac{D_1 M_{t^*}(G_1)}{\tilde{D}_1 M_{t^*}(\tilde{G}_1)} \propto \ldots \propto \frac{D_d M_{t^*}(G_d)}{\tilde{D}_d M_{t^*}(\tilde{G}_d)} \propto \ldots \propto \frac{D_N M_{t^*}(G_N)}{\tilde{D}_N M_{t^*}(\tilde{G}_N)} ; \Delta t^* =$$

reduced time domain of similarity; $M_t(C) = (\int_{t_1}^{t_2} C \, dt)/\Delta t$ = integral mean of the concentration C, $j = 1,\ldots,B$ (number of observed species).

Figure 2. MIP short-cut estimator: dominant term identification (up equations and down-right figure); similarity analysis of current data (l) vs. data-bank (m) prior information (down-center and left figures).

store the information in data-bank for use is a subsequent estimation / up-dating rule. Thus, the reliability increases because local solutions with no physical meaning can be avoided and the convergence rate of a *NLS* step is improved. *MIP* is also effective in quick choosing the most suitable model structure from data-bank when big variations in species observability are recorded. In *Example 5* the *MIP-MMA* strategy increases the effectiveness of an on-line dynamic catalytic process identification (Maria et al., 1996b). The *mch* gas-phase dehydrogenation to toluene is a high sensitive-catalytic process. The changes of deactivation mechanism with temperature, model uncertainty, *ODE* set nonlinearity and stiffness make difficult the precise estimation of Arrhenius highly intercorrelated constants. A PC-controlled dynamic experimental policy allows to apply *MIP* for quasi-isothermal domains to estimate four LHHW model constants, followed by *MMA* steps to estimate four constants of the deactivation kinetics. In *Example 6* the *MMA* was applied for on-line model up-dating in a semi-batch process with nonlinear kinetics (Maria *et al.* 1996a). One of the reactants is continuously fed following an optimal policy. On-line measurements allow *MIP* recursive estimation of 3-4 kinetic constants and periodic *MMA* steps for kinetic model updating (after 8 and 19 samples, Fig. 3). The results are better than of classical on-line *EKF* estimators. In *Example 7* a similar coupled *MIP-MMA* estimation procedure was on-line applied to up-date a four parameter extended Monod kinetics for a wastewater treatment process (Ramvall et al., 1994). In *Example 8* a coupled *MIP-MMA* was reported to be very effective in using data-bank information to identify complex kinetic models (up to four reactions) and to estimate their parameters from Differential Scanning Calorimetric data (Maria and Heinzle, 1998). Such a novel approach allows to extend the current in use single-reaction identification methodology when primary information about the process are available in an early stage of process development. The coupled *MIP-MMA* also allows a better calculation of thermal safety assessment indices of the process.

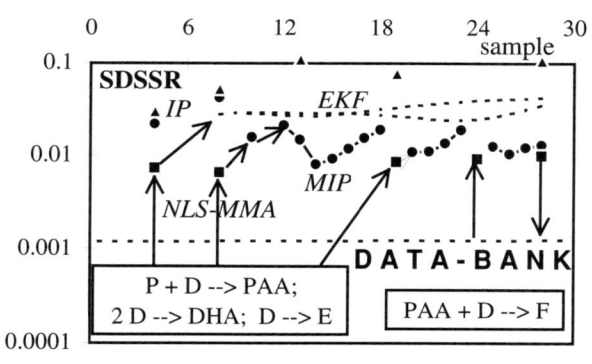

Figure 3. Coupled MMA-MIP effectiveness compared with other recursive kinetic estimators (IP= integral transformation short-cut estimator; EKF= extended Kalman filter estimator; SDSSR= st. dev. of sum of squares of residuals).

Conclusions

On/off-line model developing and up-dating for process monitoring and optimization require effective optimization techniques of increased *reliability* in reaching the global problem solution. This explained the continuous attention paid to more effective *GM*s but also to novel *ARS* optimizers. Recently developed hybrid *ARS*, structured in a multi-level search, can offer considerable advantages: improved speed and reliability, no derivative evaluation, easily treatment of complicated constraints, simple to program even in the mixed integer optimization cases, insensitive to starting solution choice, *robust* to overcome local optima. Some examples prove the good results in applying modern *ARS* (*MMA*) for solving process identification and model-updating problems. Couplation of *MMA* with a novel short-cut estimator (*MIP*) for transferring prior data-bank information increases the solution reliability vs. source of inconsistency / poor- conditioning. Such a novel strategy is effective both in off-/on-line identification of complex kinetic models and process monitoring.

References

Banga, J. and Cesares, J., 1987. *IChemEng. Symp. Ser.*, 100, 183-192.
Banga, J., Alonso, A. and Singh, R., 1994. *AIChE Annual Meeting, San Francisco November 1994*.
Banga, J. and Seider, W., 1995. *SIAM Conference 'State of the Art in Global Optimization', Princeton Univ. April 28.*
Bates, D. and Watts, D., 1988. *Nonlinear Regression Analysis and Its Applications*, New York: Wiley.
Bilardello, P., Joulia, X., LeLann, J.M., Delmas, H. and Koehret, B., 1993. *Comp. Chem. Eng.*, 17, 517-525
Blau, G., Klimpel, R. and Steiner, E., 1972. *Ind. Eng. Chem. Fundam.*, 11, 324-332.
Bojkov, B., Hansel, R. and Luus, R., 1993. *Hung. J. Ind. Chem.*, 21, 177-185.
Damiano, J., 1983. Ph.D. Diss., Carnegie-Mellon, Pittsburgh.
Edgar, T. and Himmelblau, D., 1989. *Optimization of Chemical Processes*, New York: McGraw-Hill.
Edwards, K. and Edgar, T., 1994. *Comp. Chem. Eng.*, 18, 821-828.
Espie, D.M., 1986. Ph.D. Diss., Imperial College, London.
Filippi-Bossy, C., Bordet, J., Villermaux, J., Marchal-Brassely, S. and Georgakis, C., 1989. *Comp. Chem. Eng.*, 13, 35-47
Floudas, C., 1995. *Nonlinear and Mixed-Integer Optimization*, New York: Oxford University Press.
Floudas, C. and Visweswaran, V., 1995. Qaudratic Optimization. *In*: R. Horst and P. Pardalos, eds. *Handbook of Global Optimization*, Dordrecht: Kluwer Academic Publ.
Forbes, J., Marlin, T. and MacGregor, J., 1994. *Comp. Chem. Eng. 18*, 497-510.
Fotopoulos, J., 1996. Ph.D. Diss., Lehigh University (PA).
Goldberg, D., 1989. *Genetics Algorithms in Search, Optimization and Machine Learning*, New York: Wesley.
Grace, A., 1990. *Optimization Toolbox for use with MATLAB*, South Natick: The MathWorks Inc.
Guay, M. and McLean, D., 1995. *Comp. Chem. Eng.*, 19, 1271-1285.
Heuckroth, M., Gaddy, J. and Gaines, L., 1976. *AIChE Jl.*, 22, 744-750.
Jacoby, S., Kowalik, J. and Pizzo, J., 1972. *Iterative Methods for Nonlinear Optimization Problems*, New Jersey: Prentice Hall.
Jang, S., Joseph, B. and Mukai, H., 1986. *Ind. Eng. Chem. Process Des.*, 25, 809-814.
Kalivas, J., 1995. *Adaptation of Simulated Annealing to Chemical Optimization Problems*, Amsterdam: Elsevier.
Kelly, R. and Wheeling, R., 1962. Research Report. Princeton: Mobil Oil Co.
Kocis, G. and Grossmann, I., 1988. *Ind. Eng. Chem. Res.*, 27, 1407-1421.
Lootsma, F., ed., 1972. *Numerical Methods for Nonlinear Optimization*, London: Academic Press.
Lou, Q., 1985. *Int. Chem. Eng.*, 25, 730-737.
Luus, R. and Jaakola, T., 1973. *AIChE Jl.*, 19, 760-766.
Luus, R. and Brenek, P., 1989. *Chem. Eng. Technol.*, 12, 309-318
Maria, G., 1989. *Canad. J. Chem. Eng.*, 67, 825-832.
Maria, G., 1993. Private Comm. *SEG Seminar*, ETH Zürich.
Maria, G., Terwiesch, P. and Rippin, D., 1996a. *Chem. Eng. Comm.*, 143, 133-147.
Maria, G., Marin, A., Wyss, C., Müller, S. and Newson, E., 1996b. *Chem. Eng. Sci.*, 51, 2891-2896.
Maria, G. and Rippin, D., 1996. *Comp. Chem. Eng.*, S20, S587-S592.
Maria, G. and Rippin, D., 1997. *Comp. Chem. Eng.*, 21, 1169-1190.
Maria, G. and Maria, C., 1997. *Science & Technology of Environmental Protection (Bucharest)*, 4, 59-96.
Maria, G. and Heinzle, E., 1998. *J. Loss Prev. Process Ind.*, 11 (in press).
Martin, D. and Gaddy, J., 1982. *AIChE Symp. Ser.*, 78, 99-107.
McMurtry, G., 1970. Adaptive Optimization Procedures. *In*: J. Mendel and K. Fu, eds. *Adaptive, Learning and Pattern Recognition Systems*, New York: Academic Press.
Meyer, R. and Roth, P., 1972. *J. Instr. Math. Applic.*, 9, 218.
Mihail, R. and Maria, G., 1986. *Comp. Chem. Eng.*, 10, 539-544.
Moore, R., Hansen, E. and Leclerc, A., 1992. Rigorous Methods for Global Optimization. *In*: C. Floudas and P. Pardalos, eds., *Recent Advances in Global Optimization* Princeton: Princeton Univ. Press.
Moros, R., Kalies, H., Rex, H. and Schaffarczyk, S., 1995. *Comp. Chem. Eng.*, 19 (submitted for publication).
Nair, G., 1976. *Automatica*, 14, 517-519
Nelder, J. and Mead, R., 1965. *Comp. J.*, 7, 308-313.
Pop, G., Boeru, R., Muntean, O. and Maria, G., 1991. *In*: L. Albright, ed. *Novel Methods of Producing Ethylene, other Olefins and Aromatics*. Boston: M. Dekker.
Ramvall, M., Maria, G. and Rippin, D., 1994. *Proc. CISS Conf. Int. Simulation Society, Zürich 28 August 1994*, 548-552
Rechenberg, J., 1973. *Evolutionsstrategie: Optimierung technischer Systeme nach Prinzipen der biologischen Evolution*, Stuttgart: F. Fromann-Holzboog Verlag.
Salcedo, R., Goncalves, M. and Feyo de Azevedo, S., 1990. *Comp. Chem. Eng.*, 14, 1111-1126.
Salcedo, R., 1992. *Ind. Eng. Chem. Res.*, 31, 262-273.
Schoen, F., 1991. *J. Global Optimization*, 1, 207-228.
Schrack, G. and Choit, M., 1976. *Math. Programming*, 10, 230-244.
STATGRAPHICS 5.2, 1990. *User Manual*. New York: Statistical Graphics System Co.

Steiner, E., Rey, T. and McCroskey, P., 1990. *SIMUSOLV Reference Guide,* Midland: Dow Chemical Co., Suppl. 1992.

Stewart, W. and Sorensen, J., 1981. *Technom.,* 23, 131-141.

Stewart, W., Henson, T. and Box, G., 1996. *AIChE Jl.,* 42, 3055-3062.

Terwiesch, P., Agarwal, M. and Rippin, D., 1994. *J. Proc. Control,* 4, 238-258.

Terwiesch, P., 1995. *AIChE Jl.,* 41, 1337-1340.

Tomlin, A., Turanyi, T. and Pilling, M., 1995, *In*: M. Pilling, ed., *Oxidation Kinetics and Autoignition of Hydrocarbons,* Amsterdam: Elsevier.

Vajda, S., Rabitz, H., Walter, E. and Lecourtier, Y., 1989. *Chem. Eng. Comm.,* 83, 191-219.

de Valliere, P. and Bonvin, D., 1989. *Comp. Chem. Eng.,* 13, 11-20; 1990, 14, 799-808.

Wang, B. and Luus, R., 1978. *AIChE Jl.,* 24, 619-626.

Watts, D., 1994. *Canad. J. Chem. Eng.,* 72, 701-710.

PROCESS MONITORING USING THE CLUSTERING METHOD AND FUNCTIONAL-LINK-ASSOCIATIVE NEURAL NETWORK

Kyung Joo Mo, Dongil Shin, Sooyoung Eo and En Sup Yoon
Division of Chemical Engineering
Seoul National University
Seoul 151-742, Korea

Kun Soo Chang
Department of Chemical Engineering
Pohang University of Science and Technology
Pohang 790-784, Korea

Abstract

This paper presents a new process monitoring methodology that is based on the modified adaptive k-means clustering algorithm and nonlinear principal component analysis(NLPCA). By using the qualitative data interpretation method that is base upon k-means clustering method, qualitative information of the process is extracted. Quantitative information of the process is extracted by NLPCA, that is based on functional-link-associative neural network which has more enhanced learning and generalization capabilities than autoassociative neural network. The usefulness of the proposed methodology is illustrated using its application to the water supply unit of a utility boiler plant.

Keywords

Process monitoring, K-means clustering, Nonlinear PCA, Functional-link-associative neural network.

Introduction

The automated operation-aid system is needed to support decision-making of the operators, especially when the process deviates from the normal state to an abnormal situation. This automated operation-aid system analyzes the real-time data, monitors the process status, and diagnoses the abnormal situation. It provides the operators with the information sufficient to help their decision-making and ultimately assists them in dealing with the problems more efficiently.

Process monitoring means a continuous real time task of determining the conditions of a physical system by recording information, recognizing and indicating anomalies of the behavior. Fault diagnosis implies the determination of the kind, size, location of a fault, and fault detection is defined as the determination of faults present in a system and time of detection (Isermann and Balle, 1996).

Real-time application of an automated fault diagnosis system based on qualitative causal-effect model requires qualitative interpretation (QI) of sensor data, that means conversion of inherently numeric sensor data into useful information(Whiteley and Davis, 1992). Various QI methods such as limit checking, Shewhart control chart, CUSUM control chart and EWMA control chart methods are used in qualitative model based fault diagnosis systems, but they can not easily describe dynamic behavior of the process.

In this paper, qualitative data interpretation method and quantitative fault detection method are proposed. The proposed data interpretation method is based on the adaptive k-means clustering algorithm, so it can handle transient behavior of the process variables and perform the context-dependent qualitative interpretation.

A quantitative fault detection method which is based on nonlinear principal component analysis (NLPCA) is also proposed. In this paper, the functional-link-associative neural network (FLAN) that has more enhanced learning and generalization capabilities than autoassociative network is developed. FLAN is the 5-layer neural network which expands the input space of the autoassociative neural network using the concept of functional-link which is suggested by Pao (1989). It has more knowledge representation space than autoassociative neural network, therefore its learning and generalization capabilities are superior to those of autoassociative neural network.

Qualitative Data Interpretation

Detection of Abrupt Changes

The problem of fault detection for monitoring industrial process involves two types of questions. First, detection of failure or fault should be achieved. But second, and often crucial practical interest, the detection of smaller faults – namely of sudden or gradual (incipient) modifications, which affect the process without going to stop – is also required to prevent the subsequent occurrence of more catastrophic events. Basseville and Nikiforov(1993) showed that both faults and failures can be approached in the abrupt change detection framework. Among the many change detection algorithm, the simplest and most commonly used method is limit checking. The basic form of this method is:

$$Y_{min} < Y(t) < Y_{max} \qquad (1)$$

Where Y(t) = the value of process data at time t.
Y_{min}, Y_{max} = lower and upper limit, respectively.

In this method, if the process value Y(t) is in the range between Y_{min} and Y_{max}, the status of the process is interpreted as normal; otherwise as high or low by the following rule:

IF   Y(t) > Y_{max} THEN 'Y(t) is HIGH'
ELSE IF Y(t) < Y_{min} THEN 'Y(t) is LOW'
                    ELSE 'Y(t) is NORMAL'

Based on the limit checking method, some statistical quality control methods, which are originally used for controlling product quality, are used for qualitative interpretation of process data. Various control chart methods such as Shewart, CUSUM and Moving Average control chart are used for the detection of changes that occur in the process. These control chart methods are focused on the detection of changes from steady state and represents the status of the variable: whether it violates the normal ranges that are represented as upper and lower control limits. However, there are many dynamic behaviors in the chemical process, which are caused by load changes, variation of feedstock or external disturbance, etc. Thus, these methods are not suitable for the dynamic chemical process.

Context-dependent Qualitative Interpretation of Data

Whiteley and Davis (1992) suggested a context-dependent qualitative interpretation of data. It means the simultaneous consideration of more than one variable at the same time. Whiteley and Davis (1992, 1994) used back-propagation neural network and ART(Adaptive Resonance Theory). Mo, et al. (1997) suggest the neural network based on radial basis function and ART for the context-dependent qualitative interpretation of data. The methods can represent dynamic chemical process, and be explained as memorizing the normal trajectories of the process using neural network that has pattern recognition and learning capabilities. Therefore these methods can be applied to the case that the trajectories of the state changes are well-defined such as batch reactor. However, there are so many good trajectories such as boiler plant that frequently changes the load, they cannot be applied because there are so many good trajectories that have to be stored.

K-means Clustering Algorithm

In this paper, modified adaptive k-means clustering algorithm for the data interpretation of the process that has frequent load changes. The standard k-means clustering algorithm is used to find a set of k processing unit centers which represent a local minimum of the total squared Euclidean distance E between the N exemplars (training vectors) x_i and the nearest of the k centers x_α:

$$E = \sum_{\alpha=1}^{k} \sum_{i=1}^{N} M_{\alpha i}(x_\alpha - x_i)^2 \qquad (2)$$

Where $M_{\alpha i}$ is the cluster membership function, which is a k×N matrix of 0's and 1's with only 1 per column which identifies the processing unit to which the given exemplar belongs. The local minimization of E can be formulated as an iterative process on the complete training set or as a real-time, adaptive process (Moody and Darken, 1989).

The adaptive k-means clustering algorithm is actually simpler than the off-line version and yields solutions of similar quality. Furthermore, adaptive clustering methods are more suitable for real-time implementation. The initial centers of clusters are randomly chosen exemplars. At each time step, a random exemplar x_i is chosen and the nearest clustering center x_α is moved by the following amount according to the distance and learning rate η.

$$\Delta x_\alpha = \eta(x_i - x_\alpha) \quad (3)$$

In this paper, a modified version of the aforementioned adaptive k-means clustering algorithm is proposed. In the modified algorithm, the learning rate α and decay ratio β are used and the number of center of processing unit k is set to 1. The initial center value is set by the initial value of the sensor data. When new data are presented, the center of cluster may be updated or newly assigned as follows. The criterion is the Euclidean distance between the center of cluster $c_{i,t-1}$ and the newly acquired sensor value $x_{i,t}$. When applying to the multivariate case, the distances of each sensor value are standardized by their standard deviations σ_i.

$$d = \sqrt{\sum \left((x_{i,t} - c_{c,t-1})/\sigma_i\right)^2} \quad (4)$$

When the distance d is smaller than the detection limit, the center of the cluster is updated following the rule below; otherwise the new center of cluster is assigned as follows:

IF (d ≤ 1) Then
$$C_t = C_{t-1} + \alpha_{t-1}(x_t - C_{t-1})$$
$$\alpha_t = \alpha_{t-1} \times \beta$$
ELSE $C_t = x_{t-1}, \alpha_t = \alpha_0$

Where α is learning rate and β is decay ratio, and both of them have values between 0 and 1.

The detection limits are updated as the center of cluster. As the learning rate α is continuously reduced, the detection limits converges to some value. The updating trajectories of the proposed algorithm for one-dimensional and two-dimensional cases are shown in Fig. 1. Empty circles mean the process data, and filled circles mean the centers of cluster.

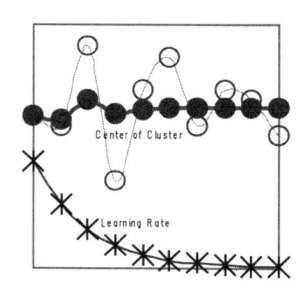

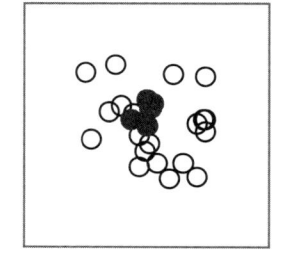

Figure 1. Update trajectory of center of cluster.

Data Interpretation using k-means Clustering Algorithm

The following type of information are extracted from the process data by using the modified adaptive k-means clustering algorithm:

- Status – normal(0), high(+1) and low(-1)
- Tendency– increasing(+1) and decreasing(-1)
- Time landmark – time

Time landmark means the detection time, and the tendency of data is defined as follows:

$$Tend = \begin{cases} \text{IF } C_{new} > C_{old} \text{ Then } Tend = +1 \\ \text{IF } C_{new} < C_{old} \text{ Then } Tend = -1 \end{cases} \quad (5)$$

In addition to those three types of information, the tendency index, TI, is calculated. The non-zero value of TI means that the tendency of the data is changed. Therefore, the trend of the time series data is represented by TI. The meanings of TI are defined in Table 1.

The QI method is used for preprocessing module of the knowledge-based fault diagnosis system. Combined with the two causal-effect digraph model RCED (Reduced Cause Effect Digraph) and PGTT (Pattern Graph Through Time), which are proposed by Mo, et al. (1997), the modified adaptive k-means clustering algorithm is successfully applied to the fault diagnosis of chemical processes. The detailed algorithm of data interpretation is shown in Fig. 2. For the detailed explanation of the RCED, PGTT, and their application to the fault diagnosis refer to Mo, et al. (1997).

Table 1. TI and their Meanings.

Cases	Tend (Old)	Tend (New)	TI	Meaning
1	0	1	1	Increase
2	0	-1	-1	Decrease
3	-1	1	2	Increase(Extreme pt.)
4	1	-1	-2	Decrease(Extreme pt.)

Nonlinear Principal Component Analysis

Functional-link-associative Neural Network

Recently, there have been researches on the multivariate statistical process control of the chemical process. Chemometric methods such as PCA (principal component analysis) and PLS (partial least square) are widely used for the multivariate statistical process control.

While the chemical process has nonlinear characteristics, PCA identifies only linear correlation between variables. Thus many researchers propose nonlinear principal component analysis methods. Kramer (1991) suggests an autoassociative neural network, which performs the identity mapping where the network inputs are reproduced at the output layer. ANN contains an internal bottleneck layer containing fewer nodes than input or output layer, which forces the network to develop a compact representation of the input data, and two additional hidden layers (see Fig. 3). But ANN has some problems in training; Tan and Mavrovouniotis (1995) suggest a three-layer network with input training, and Dong and McAvoy (1996) suggest a method that uses principal curves and two 3-layerd neural networks.

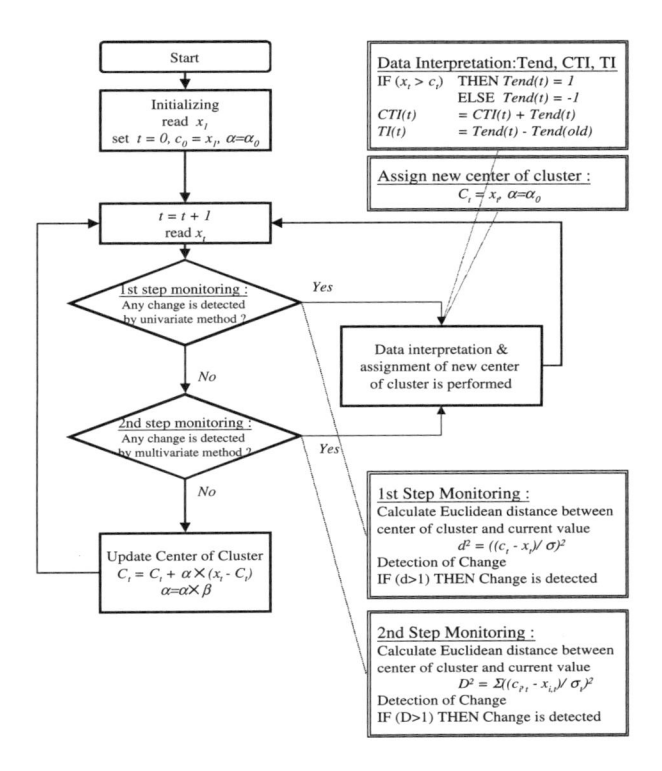

Figure 2. Data interpretation algorithm by using modified adaptive k-means clustering algorithm.

In this paper, a new nonlinear PCA method that is based upon functional-link-associative neural network is proposed. The architecture of the FLAN is similar to that of ANN, but the input layer is expanded by using the concept of functional-link that was suggested by Pao (1989). Using the functional-link of input value x, such as $\sin \pi x$, $\sin 2\pi x$, $\cos \pi x$, $\cos 2\pi x$ or x^2, and so on, the input space of the network is expanded so the knowledge representation space is also expanded. Therefore the learning and generalization capabilities of FLAN are more superior to those of other feed forward neural networks.

Example 1

A simple test of the NLPCA methodology is provided through a data set consisting of two observed variables y_1 and y_2, driven by one underlying parameter, θ.

$$y_1 = 0.8 \sin \theta$$
$$y_2 = 0.8 \cos \theta \qquad (6)$$
$$\theta = U[0, 2\pi]$$

where U[a, b] represents the uniform distribution in the range (a,b). Because only one underlying parameter is required to model these two variables, one nonlinear principal component should be enough to describe the two correlated variables. Sometimes it is not easy to training the ANN to desired error level, but easy to training the FLAN because it has more knowledge representation space. As shown in Fig. 4, FLAN captures the nonlinearity of the example more exactly than ANN does. The learning time of FLAN is also more shorter than that of ANN.

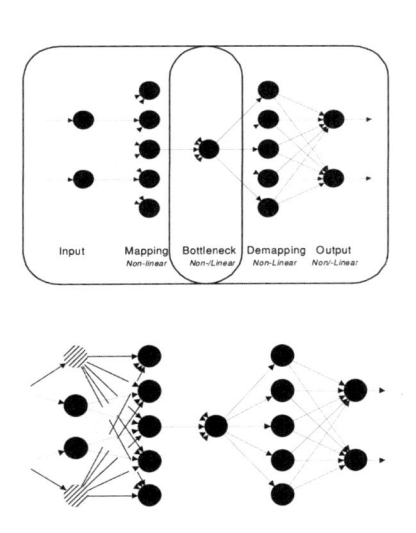

Figure 3. Autoassociative neural network and Functional-link-associative neural network.

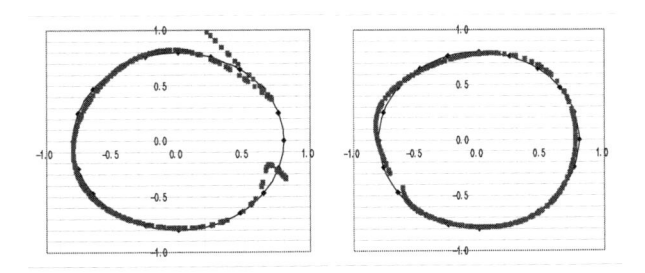

Figure 4. Training result of ANN and FLAN.

Example 2

This example is used by Dong and McAvoy (1995) to illustrate the usefulness of the NLPCA. Consider a system with three variables but only one factor:

$$x_1 = t + e_1$$
$$x_2 = t^2 - 3t + e_2 \qquad (7)$$
$$x_3 = -t^3 + 3t^2 + e_3$$

where e_1, e_2, and e_3 are independent noise N(0, 0.01), $t \in [0.01, 1]$. In the first 100 samples, data are calculated according to these equations, and these data are taken as the normal condition. After the first 100 samples, small changes are made in x_3 and the system becomes

$$\begin{aligned} x_1 &= t + e_1 \\ x_2 &= t^2 - 3t + e_2 \\ x_3 &= -1.1t^3 + 3.2t^2 + e_3 \end{aligned} \quad (8)$$

This condition can be viewed as a fault condition for the system. One hundred samples are calculated for the fault condition. The data are scaled to zero mean and unit variance. The squared prediction error (SPE) is given by

$$SPE = \sum_{i=1}^{m} (x_i - x_i')^2 \quad (9)$$

where x_i is a process variable, and x_i' is the prediction of x_i from PCA model. The resulted SPE charts are given in Fig. 5, where solid lines means 99% control limits. As shown in the figures, FLAN gives the better fault detection result than ANN does.

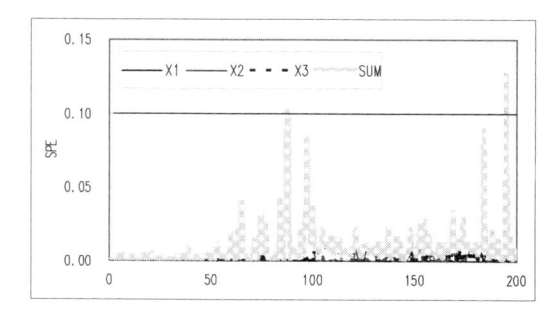

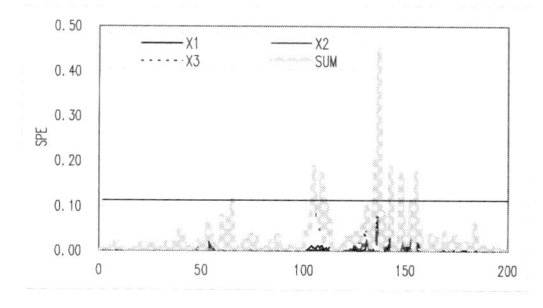

Figure 5. SPE of ANN and FLAN.

Case Studies

Process Descriptions

For the verification of the proposed method, case studies are performed to the water supply unit of a boiler plant. Water supply unit consists of tanks called dearerator, pump, and level and flow controllers.

Case 1 : Load Changes

When the load of the main boiler plant is increased, the pressure PS decreases, and the level of the tank is decreased, which is then increased by the control action. There exist some fluctuation periods; the other variables are also fluctuated. By the control action, level LC is at first decreased then increased to normal range, so conventional QI methods interpret its status as LC(-) and then LC(0). When LC is return to normal range, they lost the symptom related to LC. Thus, the consistency of the causal-effect model is broken, and it fails to explain the behavior of the process. Therefore they lose the real root cause, or the load change.

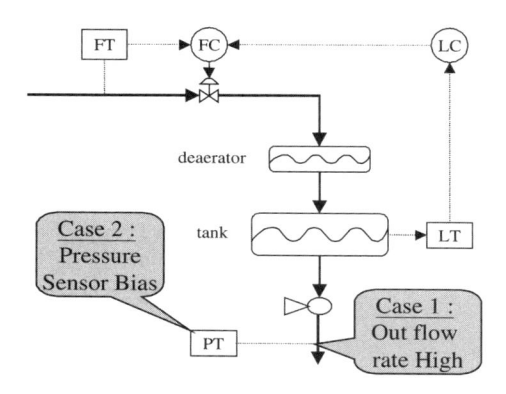

Figure 6. Water supply unit of boiler plant.

Table 2. Process Variables and Meanings of Them.

Variable	Description
FC	Flow Controller PV
FCM	Flow Controller MV
FCS	Flow Controller Set point
LC	Level Controller PV
LCM	Level Controller MV
LCS	Level Controller Set point
PS	Feed Pressure to Upper Drum

In our proposed methodology, after the appropriate detection level of each variable is set, qualitative interpretation using modified adaptive k-means clustering algorithm is applied. At that time, the 3 control outputs, LCM, FCS, and FCM, are treated as one class because they are closely related.

The proposed method handles not the status but the tendency, so it can successfully describe the dynamic behavior using RCED and PGTT as shown in Fig. 7. They don't lose the root cause, the load change, and can fully explain dynamic behavior of all variables.

Case 2 : Sensor bias when load change occurs

In this case, the pressure sensor PS goes out of order at time 70, it indicates a lower value than the real one. As shown in Fig. 8, the qualitative information of the variables is same as that of normal situations, so the sensor fault cannot be detected without using quantitative information. Form the SPE chart of FLAN, the sensor fault can be easily detected at time 82 as shown in Fig. 9.

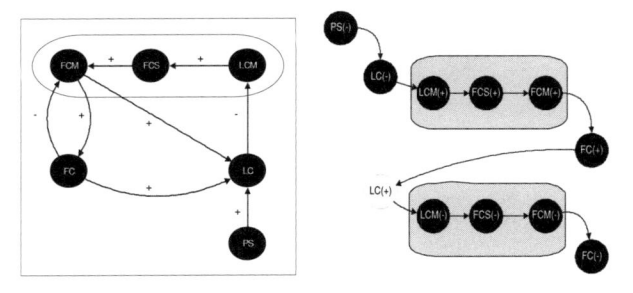

Figure 7. RCED and PGTT of water supply unit when load change occurs.

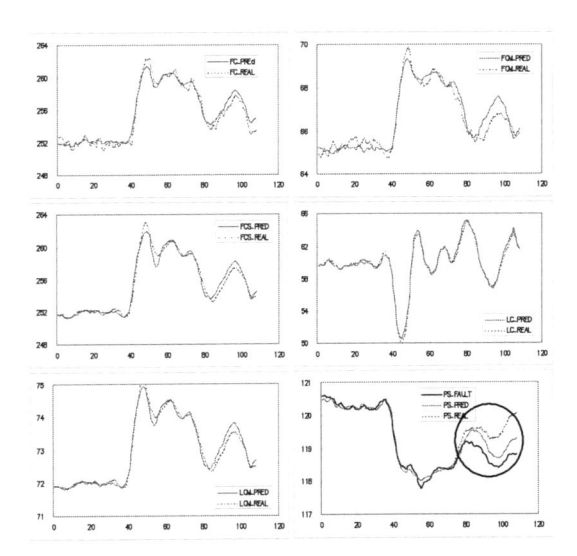

Figure 8. Transient behavior of the process when load change and sensor bias occurs.

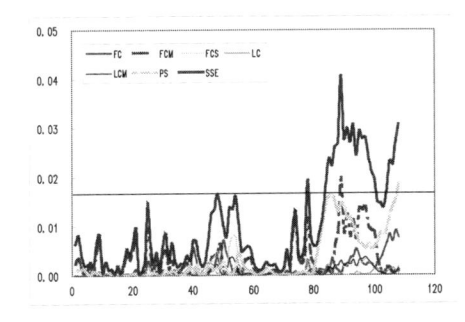

Figure 9. SPE of FLAN model.

Conclusions

In this paper, a new qualitative data interpretation method, which is based on the adaptive k-means clustering algorithm, is proposed. It can handle dynamic behavior of the process and perform the context-dependent qualitative interpretation. Combined with the causal-effect digraph model, it successfully describes the dynamic behavior of the process that other conventional methods cannot. Quantitative information of the process can be added by nonlinear principal component analysis. In this paper, a new neural network model, functional-link-associative neural network that has enhanced learning and generalization capability, is suggested. The usefulness of the proposed method is illustrated using its application to the real data from the water supply unit of a boiler plant.

Acknowledgement

This paper was supported (in part) by Korea Science and Engineering Foundation (KOSEF) through the Automation Research Center at POSTECH.

References

Basseville, M. and I. V. Nikiforov (1993). *Detection of Abrupt Changes*. Prentice-Hall.

Dong, D. and T. J. McAvoy (1996). Nonlinear Principal Component Analysis - Based on Principal Curves and Neural Networks. *Computers chem. Engng.*, **20**, 1, 65-78.

Isermann, R. and Peter Balle (1996). Trends in the Application of Model Based Fault Detection and Diagnosis of Technical Processes. *Proc. of IFAC 13th Triennial World Congress*, pp.7F-01.1-11.

Kramer, M. A. (1991). Nonlinear Principal Component Analysis Using Autoassociative Neural Networks. *AIChE Journal*, **37**, 2, 233-243.

Mo. K. J., Y. S. Oh, C. W. Jeong and E. S. Yoon (1997) Development of Operation Aided System for Chemical Process. *Expert Systems with Applications*, **12**, 4, 455-464.

Moody, J. and C. J. Darken (1989). Fast Learning in Networks of Locally-Tuned Processing Unit. *Neural Computation*, **1**, 281-294.

Pao, Y. H. (1989). *Adaptive Pattern Recognition and Neural Networks*. Addison-Wesley.

Tan, S. and M. L. Mavrovouniotis (1995). Reducing Data Dimensionality through Optimizing Neural Network Inputs. *AIChE Journal*, **41**, 6, 1471-1480.

Whiteley, J. R. and J. F. Davis (1992). Knowledge-Based Interpretation of Sensor Patterns. *Computers chem. Engng.*, **16**, 4, 329-346.

Whiteley, J. R. and J. F. Davis (1994). A Similarity-Based Approach to Interpretation of Sensor Data using Adaptive Resonance Theory. *Computers chem. Engng.*, **18**, 7, 637-661.

APPLYING A PROCEDURAL AND REACTIVE APPROACH TO ABNORMAL SITUATIONS IN REFINERY CONTROL

David J. Musliner and Kurt D. Krebsbach
Honeywell Technology Center
Minneapolis, MN 55418
{musliner, krebsbac}@htc.honeywell.com

Abstract

Oil refineries provide the lifeblood for global economic health, and disruptions to their operations have major worldwide impact. We are developing a large-scale intelligent refinery control system to assist human operators in controlling refineries during abnormal situations. Based primarily on reactive and procedural approaches to intelligent behavior, the Abnormal Event Guidance and Information System (AEGIS) will interact with multiple users and thousands of refinery components to diagnose and compensate for unanticipated plant disruptions. Through intelligent autonomous behavior and improved human situation awareness, the AEGIS project is expected to have a multibillion dollar annual impact on refinery productivity. This paper discusses lessons learned during the initial prototyping efforts of the goal-setting, planning, and plan execution components of AEGIS.

Keywords

Intelligent control, Abnormal situation management, Planning, Plan execution.

Introduction

One of the largest industrial disasters in U.S. history was a $1.6 billion explosion at a petrochemical plant in 1989. This accident represents an extreme case within the spectrum of major process disruptions, collectively referred to as abnormal situations. While most abnormal situations do not result in explosions, they can be extremely costly, resulting in poor product quality, schedule delays, equipment damage, reduced occupational safety, and environmental hazards. The inability of automated control systems and plant operations personnel to control abnormal situations has an economic impact of at least $20 billion annually in the petrochemical industry alone.

At the Honeywell Technology Center, we are building an intelligent, mixed-initiative refinery control system designed to dramatically reduce the frequency, severity, duration, and cost of abnormal situations. The Abnormal Event Guidance and Information System (AEGIS) is a large-scale distributed intelligent system specifically designed both to assist operations personnel (e.g., by displaying the most useful information) and to take diagnostic and compensatory actions autonomously.

This paper describes a portion of the goal-setting, planning, and execution (GPE) components of AEGIS. Although a detailed description of the entire AEGIS system is beyond the scope of this paper, we consider the requirements and constraints that guided our approach, and evaluate the current prototype with respect to them. In particular, we report on the benefits and the challenges raised in our attempt to satisfy the often conflicting requirements inherent in the enormously complex domain of oil refining.

In the next section, we briefly describe the current state of refinery control and the associated problems. We then overview the AEGIS architecture, focus on the goal-setting, planning, and execution components, and discuss the lessons learned in prototyping those functions.

Background: Refineries and Control

Petrochemical refining is one of the largest industrial enterprises worldwide. The functional heart of a refinery is the Fluidized Catalytic Cracking Unit (FCCU). As illustrated in Fig. 1, the FCCU is primarily responsible for converting crude oil (feed) into more useful products such as gasoline, kerosene, and butane. The FCCU cracks the crude's long hydrocarbon molecular chains into shorter chains by combining the feed with a catalyst at carefully controlled temperatures and pressures in the riser and reactor vessels. The resulting shorter chains are then sent downstream for separation into products in the fractionator (not shown). The catalyst is sent through the stripper and regenerator to burn off excess coke, and is used over again.

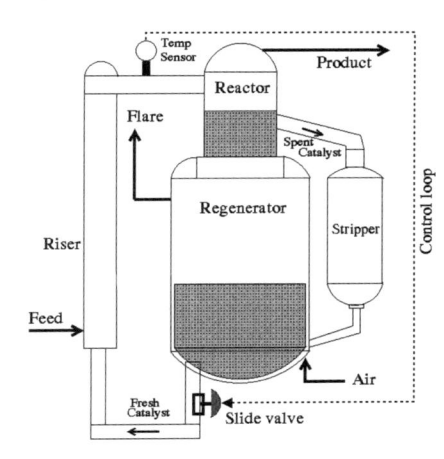

Figure 2. A fluidized catalytic cracking unit.

Figure 2 illustrates how a typical state-of-the-art refinery is controlled. The Distributed Control System (DCS) is a large-scale programmable controller tied to plant sensors (e.g., flow sensors, temperature sensors), plant actuators (e.g., valves), and a graphical user interface. The DCS implements thousands of simple control loops (e.g., PID loops) to make control moves based on discrepancies between setpoints (SPs) and present values (PVs). For example, as depicted in Fig. 1, the dotted line connecting the temperature sensor and the riser slide valve denotes that the position of the slide valve is dependent on the temperature being sensed in the riser. As the temperature drops, the slide valve will be opened to increase the flow of hot catalyst. A typical FCCU will have on the order of 1000 readable "points," and a few hundred writable "points." In addition to PID control loops, the DCS can be programmed with numerous "alarms" that alert the human operator when certain constraints are violated (e.g., min/max values, rate limits). "Advanced control" is the industry term for powerful mathematical control techniques (e.g., multivariate linear models) used to optimize control parameters during normal operations.

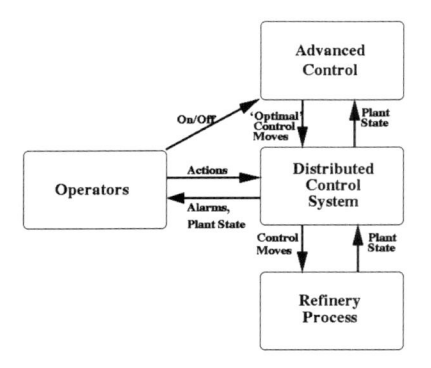

Figure 2. Refinery control without AEGIS.

The human operators supervise the operation of the highly-automated plant. This supervisory activity includes monitoring plant status, adjusting control parameters, executing pre-planned operations activities (e.g., shutting down a compressor for maintenance), and detecting, diagnosing, compensating for, and correcting abnormal situations. The operator has a view of the values of all control points, plus any alarms that have been generated. The actions the operator is allowed to take include changing SPs, manually asserting output values for control points, and turning on or off advanced control modules.

Abnormal Situations

Abnormal situations may cause dozens of alarms to trigger, requiring the operator to perform anywhere from a single action to dozens of compensatory actions over an extended period of time. Major incidents may precipitate an alarm flood, in which hundreds of alarms trigger in a few seconds, leading to scrolling lists of alarm messages, panels full of red lights, and insistent klaxons. In these situations, the operator is faced with severe information overload, which often leads to incorrect diagnoses, inappropriate actions, and major disruptions to plant operations. If left uncontrolled, abnormal situations can be extremely costly, resulting in poor product quality, schedule delays, equipment damage, reduced occupational safety, and environmental hazards.

Because abnormal situations are so serious, many regulatory and administrative structures are already in place to help manage them. Primarily, operators are trained to respond to abnormal situations based on extensive Standard Operating Procedures (SOP) that are written down, checked, and updated regularly. The procedures can be quite long (dozens of pages), with lots of logical control structure and contingencies, since the exact state of the plant is almost never known with certainty. Many procedures involve sampling data, confirming readings, performing diagnostic tests, conferring with other plant personnel, manually manipulating equipment, and adjusting DCS control parameters. Some procedures apply to extremely general contexts (e.g., we're losing air pressure from somewhere), while some are more specific (air compressor AC-3 has shut down).

AEGIS

The Abnormal Event Guidance and Information System (AEGIS) is a large-scale distributed intelligent system designed primarily to improve responses to abnormal situations, both by automating some activities currently performed by operations personnel and by improving human situation awareness. Illustrated in Fig. 3, AEGIS is a distributed software architecture based on blackboard-style communications and several distinguished application roles. Multiple application programs, with varying levels of intelligence and abilities, may fill roles including:

State Estimator—Determines the state of the plant, at varying levels of abstraction, by fusing diverse sensor data and other available information (e.g., prior control moves, known malfunctions, human observations).

Goal Setter—Decides which of the currently-threatened operational goals should be addressed.

Planner—Develops plans to address threatened goals selected by Goal Setter.

Executor—Executes plans, monitors actions, and updates other AEGIS components in progress towards goals.

Communicator—Communicates efficiently and effectively with multiple plant personnel including DCS operators and field personnel located outside the control room.

Monitor—Observes the performance of the AEGIS components and may adjust or adapt the system's behavior in response to observed performance.

These functions interact by exchanging information on shared blackboard data structures. The Plant Reference Model blackboard captures descriptions of the refinery at varying levels of abstraction and from various perspectives, including the plant's physical layout, the logical processing unit layout, the operational goals of each component, and the current state and suspected malfunctions. Figure 4 shows how AEGIS interacts with the existing system.

Advantages of AEGIS

AEGIS emphasizes two main design concepts that form the basis of many of its advantages over the state-of-the-art:

Goal-Centric (not Alarm-Centric) Information—Raw data interpretation and alarm flood management are enormous tasks currently left to the board operator. A hallmark of the AEGIS approach is an abstraction of data and alarms into more useful information such as threatened operational goals, likely malfunctions and their confidence values, relevant symptoms, grouped process data, and trends.

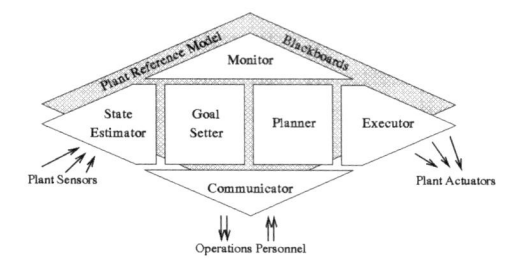

Figure 3. The AEGIS architecture.

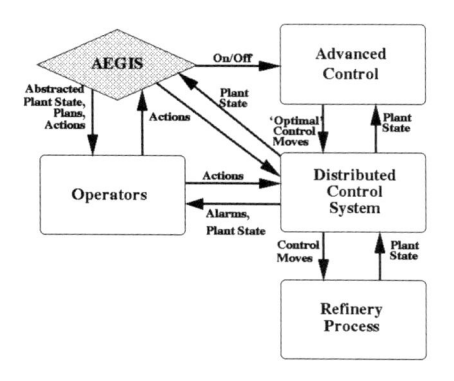

Figure 4. Refinery control with AEGIS.

Mixed-Initiative Plan Execution—Currently, besides being responsible for evaluating the plant state, board operators must choose appropriate courses of action, perform each task or delegate tasks to others, and monitor the progress of these tasks, while simultaneously reevaluating the next context. Many of these tasks are easier for AEGIS to perform. For instance, it is simple to make AEGIS procedures self-monitoring, with little or no loss of attention to concurrent activities.

GPE Requirements

In this paper we focus specifically on the goal setting, planning, and execution components of the larger AEGIS system. We refer to this aggregate functionality as GPE. The major requirements placed on the GPE functions include:

Semi-Autonomy—GPE is semi-autonomous and mixed-initiative: many of the actions it is designed to take can be performed either by AEGIS or by the human operator.

Procedural Orientation—As discussed above, responses to abnormal situations are dictated by formal procedures, many of which are already recorded in plant documentation.

Reactivity—While not hard real-time, the refinery domain requires rapid responses (no more than a few seconds) to rapidly changing environmental conditions; GPE must be able to quickly change its focus of attention and its plans at any time.

Lack of Models—While some partial analytical and simulation models exist for elements of refineries, these models are not tremendously useful in abnormal situation planning for several reasons, including:
- Abnormal situations, the focus of AEGIS, are precisely the times when the plant is behaving outside of its normal, modeled modes.
- Existing models are not sufficiently detailed for first-principles generation of actions spanning large upsets.

GPE: A Procedural Approach

We prototyped the core reasoning engine of AEGIS in C-PRS, the C-based version of the Procedural Reasoning System (Ingrand 1994, Ingrand 1992). As shown in Fig. 5, knowledge in PRS is represented as a declarative set of facts about the world, together with a library of user-defined knowledge areas (KAs) that represent procedural knowledge about how to accomplish goals in various situations. Goals represent persistent desires that trigger KAs until they are satisfied or removed. The intention structure represents currently-selected KAs that are in the process of executing or awaiting execution. The PRS *interpreter* chooses KAs appropriate for current goals, selects one or more to put onto the intention structure, and executes one step from the current intention.

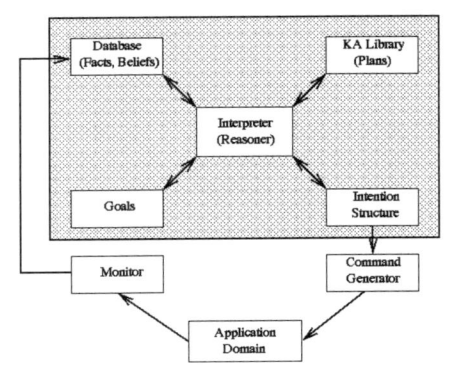

Figure 5. The Procedural Reasoning System architecture.

We chose to use an integrated approach to goal setting, planning, and execution based on the AI community's past experiences with autonomous systems applied to real-world domains (e.g., robotics). That experience has shown that choosing a goal to pursue, planning a course of action, and executing the steps of the plan are inevitably intertwined by the unpredictable and dynamic nature of real-world domains. Execution failures, changing goals, difficult planning problems, and environmental changes all disrupt the ideal of
simple forward information flow. If the GPE functions were separated into distinct programs, the amount of information constantly passing back and forth due to the changing domain, plans, and goals would be overwhelming. In our integrated GPE approach, in contrast, those changes are kept largely local to GPE, so the C-PRS interpreter can be efficient about managing that information.

PRS and AEGIS

The GPE world model consists of a database of facts and beliefs. The database is populated with fairly static information about the plant's physical layout and logical connections between plant components, as well as dynamic data regarding attributes and values of DCS points. The GPE can subscribe to this data on an as-needed basis, but subscribes to some types of information, such as the status of operational goals and malfunction confidence values, on a permanent basis.

As this data changes at run-time, procedures from the plan library are triggered, and new procedural goals are established. As procedures are selected to achieve procedural goals, they are represented on PRS' intention structure. A user-viewable representation is also generated, and is available to the user through an interface called GPEVIEW. From GPEVIEW, an operator can view skeletal plans, authorize or cancel those plans prior to execution, assume responsibility for pieces of them, and so on. These plan modifications are then reflected in the PRS database, and are incorporated into the procedure's runtime behavior.

Many actions on the intention structure can be directly executed by GPE, given authorization from the user. These actions include actual DCS control moves, communication messages with field personnel, and requests for more data.

Benefits

Our PRS-based approach naturally provides several benefits:

Standard Operating Procedures—An obvious benefit to our approach is that much of the knowledge we wish to encode is already available in refineries as paper SOPs. While translating SOPs into PRS procedures is not trivial, they have provided us with a great deal of insight into the role of the operator, and the current state of the art.

Parallel Goal Achievement—During an abnormal situation, a human operator must be focused to properly respond, despite an avalanche of data, cascading effects, and a plethora of pending tasks. This difficult situation can tax even the most experienced operator's time, memory, and communication constraints. The result: a wide variety of errors and inefficiencies in procedure execution.

Because operators represent a scarce resource themselves, SOPs are almost always expressed sequentially, to aid the operator in focusing on one thing at a time. GPE effectively has no such constraint, and can react to multiple goals in parallel, while allowing the operator to focus on the highest priority tasks requiring his attention.

Context-Sensitive Behavior

PRS provides numerous ways to specify context-sensitive triggering of procedures. This is much more flexible than plant SOPs, in which one procedure is often recommended to achieve a goal regardless of the many other factors comprising the current context. For example, one can specify multiple procedures to accomplish the same goal of replacing lost combustion air: one if the secondary pump is available, one if the air loss is below an important threshold, several if the root cause of the malfunction is not yet known, and so on. While several or all of these procedures might be relevant to the goal, the context we describe can distinguish those that truly apply. Further, using (natively available) priorities, or user-defined metalevel reasoning, the interpreter can intelligently select the most preferred among the resulting set of applicable procedures.

Action Effect Monitoring

Many of the hardest tasks for humans to perform reliably involve monitoring the effects of earlier actions. Currently, operators must simply remember to check process data at an appropriate later time to confirm that earlier actions are having their desired effects.

Fortunately, because PRS is not memory, time, or communication-constrained to any significant degree, GPE procedures can quite easily be set up to monitor their actions and the operator's actions, as long as methods exist for confirming goal achievement. In our domain, these methods involve querying the DCS to confirm temperature trends, pressure differentials, and the like, all of which are trivially available. Other more complex confirmatory information can be obtained directly from the operator or other AEGIS components (e.g., state estimators). Early feedback from plant personnel indicates that this automatic monitoring functionality is among the most immediately and widely useful aspects of the PRS approach.

In the following sections we discuss specific challenges with our GPE approach, and some preliminary solutions.

Mixed-Initiative Procedure Execution

During an upset, it is important for GPE to be constantly sensitive to the rapidly-changing plant state, and to respond quickly. On the other hand, a key AEGIS design goal is maintaining user awareness. Unfettered, the lightning fast responsiveness of AEGIS computers could leave users bewildered about what actions the system intends to perform or has already performed. We have spent significant effort addressing this challenge of effectively supporting mixed-initiative, reactive procedure execution. In the following subsections we elaborate on the difficulties in using PRS for a mixed-initiative system, and describe our current solution.

Lack of Projection

Because PRS is reactive, it does not look ahead to determine which procedure it will select to achieve a given goal until that goal has been reached in the procedure. We believe this is "correct" from an engineering perspective, because the precise method of achieving a goal should not be determined until the full environmental context is available for evaluating the alternatives. However, this is insufficient from the operator's perspective, because it provides little insight into what the system is planning globally.

There are several aspects to this problem, within the context of executing a single PRS procedure:

- **Future goal-achieving procedures are not yet selected**. PRS procedures are, in the simplest serial case, executed like a normal computer program. When PRS selects a procedure, it instantiates it, and sets the "program counter" at the first goal. Applicable procedures are determined to achieve that goal, and one is chosen. While this newly-chosen procedure is being executed, however, selection of procedures for goals beyond the program counter is deferred.

- **Future goals and actions are not necessarily meaningful to the user**. Even if future goals and actions could be accessed by the interpreter, some are at the wrong level to be relevant to the user (e.g., binding a local variable), while others are not in a form useful to an operator (code). In general, it is not reasonable to expect the PRS procedure author to use names and constructs that correspond to an operator's understanding, and vice versa.

Pseudo-Projection

To work around this problem, we have developed a "pseudo-projection" method that allows GPE to appear partially projective without making any changes to the reactive PRS interpreter. Pseudo-projection allows the operator to see as far into the future, and with as much detail, as is possible given the reactive procedural paradigm.

We implement pseudo-projection using a procedure annotation syntax that allows the author to annotate each procedure with a series of comments that the AEGIS user will see at runtime when the procedure is chosen by PRS. These annotations, called metacomments, allow PRS to appear partially projective to the user. As soon as a procedure is selected, the user can see the entire structure and status of the procedure.

This metacomment technique is a temporary approach to the problem of user awareness in a reactive system. It unfortunately adds complexity to the process of writing procedures, although the metacomment syntax itself is quite simple and intuitive. Essentially, this added complexity is unavoidable if we wish to maintain two representations of the procedure: one that is machine executable and another that is readily understandable to refinery personnel.

Other Forms of Projection

While pseudo-projection techniques provide a form of look-ahead for the user, other limited forms of model-based projection can be exploited which allow more intelligent control by the reactive system itself. Consider the following simplified procedure segment for responding to a loss of combustion air:

Procedure Novice-Air-Loss-Response

1. Cut riser temperature to 930 degrees F.
2. Eliminate all residual feed.
3. Eliminate all slurry pumparound feed.
4. Cut main feed to 20,000 barrels/day.
5. Add pure oxygen up to 30%.

This procedure fragment is a typical SOP example, intentionally characterized by simple, specific instructions, readily understood by even the least-experienced operator. They are straightforward, safe, static, and suboptimal. In this example, for instance, all residual and slurry feeds are eliminated to allow the operator to concentrate on cutting and monitoring only the main feed. While these procedures provide a starting point for encoding executable procedures, they do not accurately reflect the complexity of most operators' response to an abnormal situation. As operators gain experience, their knowledge of the underlying plant process and DCS response grows, and their response becomes more model-based. For instance, the operators we interviewed noted that they would generally leave in some residual feed to keep the coke component higher, keeping the riser temperature higher. This is an optimization step that, while still safe, maintains a higher level of production, and thus reduces the cost of the disruption.

Mini-Models

In general, the more experienced the operator, the more context-sensitive is his response to an abnormal situation. We view our GPE procedures as evolving in the same way, incorporating more of what we have called *mini-models* directly within PRS procedures. As the authors of the procedures gain a better understanding of the process and control system, we expect the procedures to rely less on static responses, and more on computing over a simplified model to generate a context-sensitive response. For instance, the following is a more model-based version of the same procedure, emulating the expert-operator approach:

Procedure Expert-Air-Loss-Response

1. Compute amount of O_2 in lost air.
2. Add pure O_2 to replace lost O_2, up to 30%.
3. Compute O_2 left to replace.
4. Compute amt. of carbon this corresponds to.
5. For each feed source:
5a. Cut source according to carbon factor.
6. Set riser temperature setpoint based on remaining carbon.

This procedure concentrates on balancing carbon content of the current feed sources with the amount of oxygen available, while staying within safety limits of 30% enrichment. It is based on a simplified mini-model involving a handful of important factors in the process, and is thus much more tailored to the actual abnormal process situation at the time of its invocation. In this example, GPE can greatly assist the operator by easily and automatically computing parameters to the situation response (e.g., correct riser temperature), as well as providing the option for GPE to take the actions autonomously, and monitor the effects of these actions over time.

Existing Predictive Models

In addition to mini-models directly implemented with PRS procedures, small predictive models exist as black-box applications for very limited pieces of the refinery. While these models are quite small (e.g., ten inputs, four outputs), in certain contexts they can be invoked from within a PRS procedure to provide several valuable types of information. By projecting these models forward in time for each option, GPE can more accurately assess the effectiveness of each alternative and choose the best one. Also, the specific results of the projections can provide valuable information to the operator and to GPE. In cases where the results are close, the operator might prefer one method over another for less tangible reasons than GPE is able to consider. From GPE's perspective, the results form a rank ordering of the options, which can be cached and used in case the first goal-achievement method fails.

Conclusions

This paper reports on the current status of an ambitious project to build an intelligent, mixed-initiative refinery control system. The current GPE prototype includes procedures that are successfully able to handle a variety of failures and disruptions to the air feed system of a simulated FCCU. The simulator is a high-fidelity industrial refinery simulator used to train plant personnel. The level of knowledge in the prototype GPE is not yet equivalent to even a rookie DCS operator, but the approach shows promise and has been successfully demonstrated to enthusiastic industry participants. Current GPE-related efforts are centered around limited field tests of the technology in actual oil refineries, as well as research into user interaction semantics and methods for automating user involvement with the system.

Acknowledgment

This research was in part supported by the NIST Advanced Technology Program Award 70NANB5H1073 to the Abnormal Situation Management Joint Research and Development Consortium.

References

Ingrand, F.; Georgeff, M.; and Rao, A. 1992. An architecture for real-time reasoning and system control. *IEEE Expert* 7:6:34-44.

Ingrand, F.F. 1994. *C-PRS Development Environment (Version 1.4.0)*. Labege Cedex, France: ACS Technologies.

PERFORMANCE ASSESSMENT AND DIAGNOSIS OF REFINERY CONTROL LOOPS

N.F. Thornhill
Department of Electronic and Electrical Engineering
University College London
Torrington Place, London WC1E 7JE

M. Oettinger
BP Refinery (Kwinana) Proprietary Ltd, Australia

P. Fedenczuk
BP Oil, Grangemouth Refinery Ltd, UK

Abstract

This paper discusses the application of control loop performance assessment (Desborough and Harris, 1992) in a refinery setting. In a large process it is not feasible to tailor the parameters of the algorithm to every individual control loop. A procedure is illustrated for selecting default values which make it possible to implement the technology on a refinery-wide scale. For instance, it is shown that the prediction horizon parameter in the CLPA algorithm can be set so that the analysis is sensitive to the persistent signals that cause loss of performance. Default values are suggested for refinery applications.

A frequent cause of loss of performance in a control loop is a persistent oscillation due to a valve nonlinearity or a tuning fault. The paper presents an operational signatures in the form of an estimate of the closed loop impulse response that suggest the causes of such oscillations.

Keywords

Control loop performance, Refinery control loop, Oscillation, Signature, Valve fault, Sampling interval.

Introduction

Studies of the performance of single-input-single-output control loops have shown that reasons for poor performance of basic SISO loops include both poor tuning and equipment problems such as sticking valves (Åström, 1991; Ender, 1993; Hägglund, 1995). Oscillations of the process variable either side of the set-point value gives particular cause for concern. Reducing or removing such oscillations yields commercial benefits (Martin *et al.*, 1991) because any reduction in variability means that set points can be held closer to an optimum constraint without the danger of violating that constraint.

Performance indices have been developed by Harris (1989), Desborough and Harris (1992) and Stanfelj *et al* (1993) which provide figures of merit for the performance of a loop. An advantage of these indices is that they can be derived during normal operations without taking loops off-line for special tests. These methods are becoming widely implemented in the petrochemical and chemical sectors (Stanfelj *et al.*, 1993; Kozub and Garcia, 1993) and also in the pulp and paper industry (Perrier and Roche, 1992; Lynch and Dumont, 1996; Jofriet and Bialkowski, 1996; Owen *et al.*, 1996; Harris *et.al.*, 1996).

This paper uses the control loop performance assessment technique (CLPA) proposed by Desborough and Harris (1992) for refinery control loops. It addresses some of the challenges laid down by Kozub (1996), in particular the need for an automated on-line system and the determination of dynamic responses. The main aspects of the paper are:

- Default settings for the parameters in the CLPA algorithm that can be used for automated CLPA of all refinery control loops
- A demonstration of how one of the default settings is selected
- Methods to aid engineers in the diagnosis of loops found to be performing poorly.

Key parameters (prediction horizon, sampling interval, data ensemble length, length of the model) are selected as reported by Thornhill *et. al.* (1998). Here, we demonstrate how the prediction horizon parameter in the CLPA algorithm can be set so that the analysis is sensitive to the persistent signals that cause loss of performance. In particular, the results show that the same default settings can be used for all loops of a similar type.

Operational signatures can be found within routine operating data and used for the purposes of diagnosis. Pryor (1982) presented the use of the power spectrum in the analysis of process data. Desborough and Harris (1992) used the power spectrum to conclude that control loops had a long-term deviation from set point, and also to highlight an oscillatory loop, while Tyler and Morari (1996) have demonstrated a spectral signature arising from a disturbance.

As reported earlier (Thornhill and Hägglund, 1997) the power spectrum helps to distinguish a tuning problem from a limit-cycle oscillation due to non-linearity such as valve friction. New results presented here show that an estimate of the closed loop impulse response also reveals the presence of a limit cycle.

Several authors have reported success in the analysis of disturbances from routine operating data. Stanfelj *et al.* (1993) provided a decision-making tree which included cross-correlation between a feed forward signal and the controlled variable of the loop under analysis. Likewise, Owen *et al* (1996) showed an application which accounts for upset conditions of the whole mill and interactions between control loops. These cases needed a knowledge of the process flowsheet, in particular about which loops might disturb one another. This paper makes use of knowledge of the process flowsheet for a unit in a case where several nearby loops show identical oscillation signatures. The source of the control problem is pinpointed through the use of engineering insight guided by the nature of the signatures.

Methods

Overview

The key variable for CLPA is the controller error, e, given by (sp-pv). If the loop is performing well it should reject disturbances, and the process variable should track the set point. These requirements imply that the controller error should have no predictable component. There should not, for example, be a steady state offset or any persistent oscillation.

Because of the dynamic nature of the process and of the controller itself it takes a little time for the controller to achieve rejection of a disturbance or to bring the process to its set point. Thus the intent of the performance index is to determine how predictable the controller error is beyond some suitable time horizon. If the control error is predictable over this time horizon then the loop is performing poorly and, by contrast, it is performing well if the error is unpredictable over this time horizon.

Theory

Desborough and Harris (1992) devised an index based upon the residuals between the measured controller error denoted by Y and a forward prediction, $\hat{y}$.

$$r(n) = Y(n) - \hat{y}(n) \qquad (1)$$

In a loop that is performing well the controller error has little predictability and the controller error contains only the random noise represented by the residuals. But in a poorly performing loop, one with a significant predictable component, the random residuals are much smaller than the controller error.

Desborough and Harris proposed the following CLPA index. A poorly performing loop has a value of η close to 1 while η for a good loop is close to 0:

$$\eta = 1 - \frac{\sigma^2_r}{mse(Y_i^2)}$$

σ^2_r: variance of the residuals
$mse(Y_i^2)$: mean square value of controller error

The requirement for the prediction model for $\hat{y}$ is just that it is capable of capturing features in the controller error sequence. Desborough and Harris (1992) show that for typical data from process control loops an autoregressive time series model that makes predictions b steps ahead is suitable:

$$\hat{y}(i+b) = a(0) + a(1)Y(i) + \\ a(2)Y(i-1) + \ldots + a(m)Y(i-m+1) \qquad (2)$$

The above model is fitted to an ensemble of n samples of the controller error using a least squares fit procedure.

Strategy for application to a large plant

In a refinery there are large numbers of basic SISO feedback loops. An automated CLPA technique needs a means of providing suitable models for every loop. The autoregressive model needs certain parameters to be specified. These are:

The prediction horizon, b
The number of terms in the model, m
The sampling interval
The data ensemble length

It is important to realise that the sampling interval is much longer that the sample interval used in the on-line PID control algorithm. The issue is to capture the closed loop transient dynamics of the process within the time interval spanned by the *m* terms of the autoregressive model. The proposed strategy for making these choices for a large scale implementation relies upon a classification of control loops into a few generic types. Examples in a refinery would include liquid flow, steam flow, temperature and pressure loops. It is argued (Thornhill *et. al.*, 1998) that the CLPA parameters optimised for a few representative examples of loops of one type can be applied to *all* loops of that type. Such a strategy covers most cases thus allows automated on-line monitoring. The recommended choices for the parameters are given below.

Loop type	Sampling interval	Prediction horizon, b
Pressure	20s	100s
Liquid flow	6s	30s
Temperature	60-120s	360s - 600s
Steam or gas flow	60s	300s
Level	20s	100s
Ensemble length	500-1500 samples	
Length of AR model, m	30 terms	

Choice of Prediction Horizon

The recommended settings for the prediction horizon, b, were arrived at using the method briefly outlined here and which is illustrated in the results section.

Exploration of the effects of different choices of prediction horizon on a selection of representative loops gives an insight into a suitable horizon. In this approach the prediction horizon is regarded as an *engineering criterion*, representing a demand made by the control engineer on the control loop; the criterion is that predictable components of the controller error should be dealt with within the specified time horizon.

CLPA Signatures

Desborough and Harris (1992) indicated the value of the power spectral density of the controller error.

In our work we have used power spectra of the controller error in order to provide an insight into the nature of a problem. The power spectra are computed by the Welch method (Welch, 1967) from a windowed fast Fourier transform

An additional signature is also presented, that of the cross-correlation of the modelling residuals from [1] and the controller errors, Y. The following comments give an insight into this cross-correlation as an estimate of the closed loop impulse response (Tyler and Morari, 1996)

As mentioned, the controller error sequence is modelled as an autoregressive sequence [2]. However, other time series models also suffice, for example, a model of the following form could be used:

$$Y(n) = c(0)u(n) + c(1)u(n-1)+\ldots \quad (3)$$

where the inputs $u(n)$ form a white noise sequence and the coefficients $c(n)$ form the impulse response of the closed loop transfer from the $u(n)$ to the controller error. It is well known that the cross-correlation function of the $u(n)$ and $Y(n)$ sequences gives the coefficients of the impulse response.

For the purposes of closed loop identification the residuals $r(n)$ are identified with the $u(n)$ sequence for the model in equation [3]. For a practical control loop the noise does not usually enter the loop at the set point; it is more often process noise or due to disturbances. However, if the use of the estimated impulse response is restricted to determination of the natural frequency and damping factor the approximations are acceptable because the damping factor and natural frequency are characteristic of the dominant closed loop poles which can be excited by an input at any point in the loop.

Oscillation Detection

Oscillation diagnosis can guide a process control engineer towards suitable special off-line tests. Hägglund (1995) and Thornhill and Hägglund (1997) presented techniques for the characterisation of oscillation in control loops and gave flow charts for diagnosis of the likely cause of an oscillation. They include:

- Interpretation of the CLPA index, a regularity factor and oscillation-detection threshold.
- Examination of features in the power spectrum of the controller error.
- Dynamic *sp-pv* maps for loops where the set point changes often, such as loops in cascade mode.

The following conclusions can follow from the operational signatures

- That an oscillation is present
- That an oscillation is due to poor tuning
- That an oscillation is due to limit cycling caused by a discontinuous non-linearity
- That there may be a disturbance.

This paper presents a further analysis of an example studied in the previous paper by showing how an estimate of the impulse response gives confirmation of a limit cycle caused by a valve fault. It also presents a new case that illustrates the last item on the above list, that of a disturbance.

The issue of automation has not been addressed for the diagnosis step of CLPA. It is supposed that an automated monitoring system will highlight loops with problems, but that process control engineers will want to look at such loops themselves. Tools such as spectral analysis and estimated impulse responses provide meaningful signatures; they also begin to address the call by Kozub (1996) for determination of dynamic responses.

Refinery Examples

Several loops from refineries in Australia, the UK and USA were used to in the selection of prediction horizons. In addition, some loops are inspected in more detail:

Loops 1 to 3: A liquid flow loop having different PID tuning settings. Loop 1 is when the loop is underdamped, Loop 2 has more damping and Loop 3 is over-damped.

Loops 4 and 5: Liquid flow loops known to have, respectively, valve stick slip and a valve dead-band.

Loops 6 to 8: Steam flow loop (Loop 8) which is the slave in a cascade temperature control loop (Loop 7). Loop 6 is an on-line analyser at the top of the column (Fig. 4).

Results

Examples of Prediction Horizon Calculations

Figure 1 shows a set of prediction horizon plots to illustrate the selection of refinery-wide CLPA settings. All the plots show a similar pattern, the key feature of which is the plateau where the CLPA index is constant over a range of values of the prediction horizon.

The significance of the plateau is that the controller error contains a component that is predictable a considerable time ahead. Such predictable components are the cause of concern. Thus Fig. 1 suggests, for instance, that the CLPA prediction horizon parameter should be set to 30s for all refinery liquid flow loops and 5 minutes for refinery steam flow loops.

These prediction horizon plots give an independent means for selection of the sample interval. The decision has been made to use 30 terms in the autoregression, which implies that the impulse response is to be captured within that time span, say in 20 to 30 samples. The settling time of the impulse response is four time constants, τ. Thus $\tau = 5 - 7.5$ sample intervals. But for a well tuned controller with, say, a damping factor of $\zeta > 0.6$ one would not expect significant coherence in the controller error at times beyond τ (this assertion is related to the bandwidth of the closed loop resonance in the frequency domain).

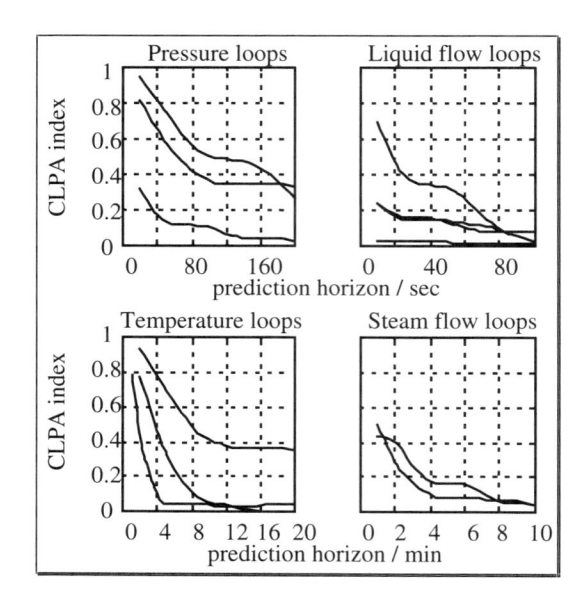

Figure 1. Prediction horizon plots for a selection of typical refinery control loops.

Hence the sampling interval should be chosen so that the prediction horizon represents about 5 to 7.5 sampling intervals. It is concluded, for example, that for liquid flow loops a sampling interval of 6s would be suitable and for steam flow loops, it would be 1min.

Estimated Impulse Responses

Figure 2 shows estimated impulse responses for Loops 1 to 3 made from routine operating data. It was indicated earlier that the estimated impulse responses are able to give good indications of natural frequency and damping factor. The damping factors can be estimated from the first peak of the response (Thaler, 1989) as $\zeta = 0.5$ for Loop 1, $\zeta \approx 0.7$ for Loop 2 and $\zeta > 1.0$ for Loop 3.

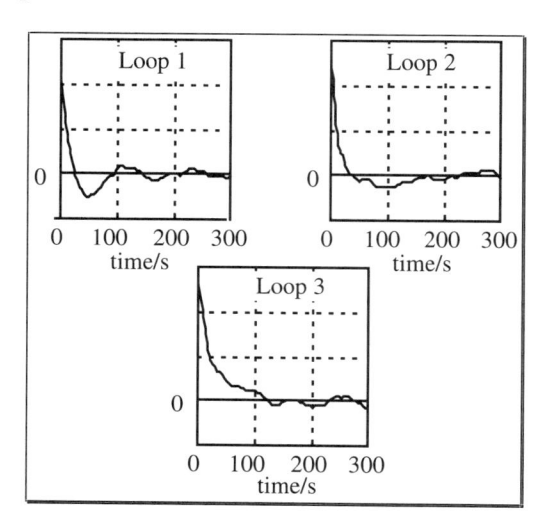

Figure 2. Estimated impulse responses for one liquid flow loop with different tuning settings.

The true impulse response would be expected to die away to zero as the loop settles but the estimated impulse responses do not always have this ideal behaviour. Loops 1 to 3 exhibit a series of small amplitude, random deviations after settling which are within the confidence limits for the estimate. These, therefore are of no importance. For loops 4 and 5 (Fig. 3) however, the long term deviations are regular and large. These loops are known to exhibit limit cycles caused by valve non-linearity. A practical impulse response test on the loop would be expected to initiate the limit cycle, and this is exactly what the impulse response estimation algorithm shows. It is concluded that the presence over a long time scale of a persistent repeating pattern in the estimated impulse response gives an additional diagnostic signature for a nonlinear limit cycle.

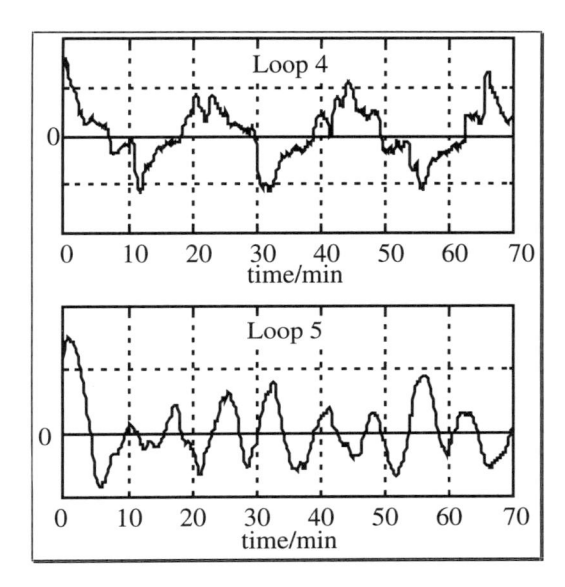

Figure 3. Estimated impulse responses for two liquid flow loops with limit cycles.

Tracking of a Disturbance

Figure 4 shows a schematic of a refinery unit in which the CLPA assessment indicated that several control loops had poor performance.

Figure 5 shows signatures for the spectra and long term estimated impulse responses for three loops. (A good way to present this information in a refinery in order to gain process insight is to fix the process flow sheet to the wall and to pin print-outs of the oscillation signatures at the relevant sensors on the flow sheet).

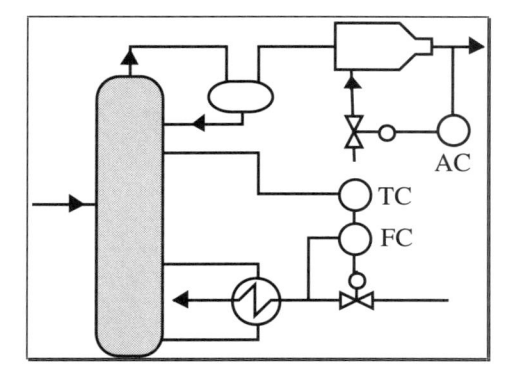

Figure 4. Schematic of a process unit. The three control loops show similar persistent oscillations.

The main features of the spectra are that they all have a sharp spectral peak at exactly the same frequency, $f = 2.5 \times 10^{-3}\ Hz$ (i.e. oscillations having a period of 400s). The estimated impulse responses have the same oscillation period as indicated in the spectra (400s or 6.7 min). The spectral features at very low frequency in Loops 6 and 7 are due to the fact that these process variables exhibit also some long term offsets from the set point.

It is thought that the column-wide oscillatory disturbance is caused by a faulty steam flow sensor in the flow loop. The spectral signature for the steam flow loop shows a harmonic peak which is double the frequency of the fundamental, a clear indication of a non-linearity (Thornhill and Hägglund, 1997). If the flow loop were limit cycling then it might be expected that it will disturb the whole column and that the oscillation will appear at the other sensors. The reason why the second spectral peak is not present at the other sensors is thought to be that the column acts as a mechanical low-pass system and filters out the higher frequency.

This example leads to the conclusion that the use of the spectral and impulse response signatures can help in the analysis of control system disturbances.

Conclusions

The detailed conclusions that can be drawn from the results have been highlighted at the end of each sub-section. The paper has started to address the challenges laid down by Kozub (1996). For instance, it has shown that it is feasible to implement automated CLPA in a refinery setting. It has also illustrated that an estimate of the closed loop impulse response gives information about closed loop dynamics, where the estimate uses only data from routine process operations without the need for special tests. The impulse response signature can be used to determine damping factor and can also indicate when a control loop oscillation is due to a limit cycle.

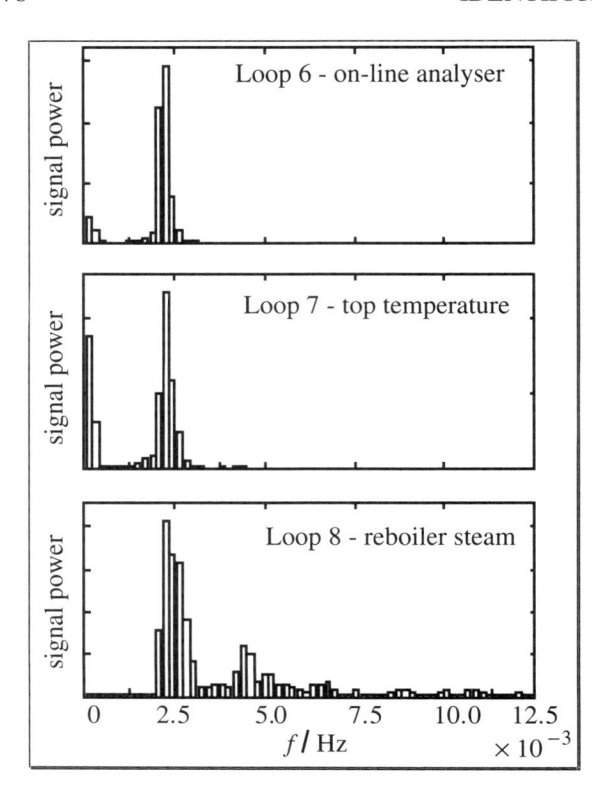

Figure 5. Power spectra for the three control loops of Fig 4 that have related persistent oscillations.

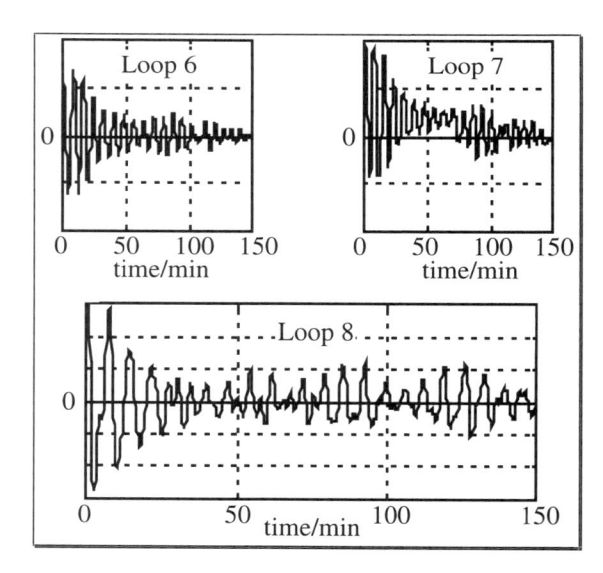

Figure 6. Estimated impulse responses for the three control loops of Fig. 4.

Refinery-wide automated on-line CLPA *monitoring* has been achieved through the use of default settings for the CLPA algorithm of Desborough and Harris (1992). The *diagnosis* of loops found to be performing poorly is not automated, however. The CLPA signatures combined with insight from engineers about their interpretation and the process layout are of value in suggesting testable hypotheses about the causes of poor performance.

Acknowledgements

The authors thank John Davis, Mike Knight, Peter Prichard, Doug Rothenberg and Bruce Varley of BP Oil for directing the work and managing the refinery visits. Thanks also to Roger Lambert and John Warren of BP Oil, Grangemouth Refinery Ltd, as well as to Robert Sadowski and Bradley Scarpness for technical input and support. N.F. Thornhill gratefully acknowledges the financial support of a Royal Society Industry Fellowship and of the Centre for Process Systems Engineering, Imperial College of Science Technology and Medicine.

References

Åström, K.J. (1991). Assessment of achievable performance of simple feedback loops, *Int. J. Adap. Control and Sig. Proc.*, **5**, 3-19.

Desborough, L., and T. Harris (1992). Performance assessment measures for univariate feedback control, *Can. J. Chem. Eng.*, **70**, 1186-1197.

Ender, D.B. (1993). Process control performance: Not as good as you think, *Control Engineering (Sept)*, 180-190.

Harris, T.J. (1989). Assessment of control loop performance, *Can. J. Chem. Eng.*, **67**, 856-861.

Harris, T.J., C.T. Seppala, P.J. Jofriet and B.W. Surgenor, (1996). Plant-wide feedback-control performance assessment using an expert-system framework. *Control Eng. Practice*, **4**, 1297-1303.

Hägglund, T. (1995). A control-loop performance monitor. *Control Eng. Practice*, **3**, 1543-1551.

Jofriet, P.J. and W.L. Bialkowski (1996). Process Knowledge: The key to on-line monitoring of process variability and control loop performance. *Proc. Control. Syst. '96*, Halifax, Novia Scotia, 187-193.

Kozub, D.J. and C.E. Garcia (1993). Monitoring and diagnosis of automated controllers in chemical process industries. *AIChE Meeting*, St Louis.

Kozub, D.J. (1996). Monitoring and diagnosis of chemical processes with automated control, *CPC V meeting*, Lake Tahoe.

Lynch, C.B. and G.A. Dumont (1996). Control loop performance monitoring. *IEEE Trans. Control Systems Technol.*, **4**, 185-192.

Martin, G.D., L.E. Turpin and R.P. Cline (1991). Estimating control function benefits. *Hydrocarbon Processing (June)*, 68-73.

Owen, J.G., D. Read, H. Blekkenhorst and A.A. Roche (1996). A mill prototype for automatic monitoring of control loop performance. *Proc. Control. Syst. '96*, Halifax, Novia Scotia, 171-178.

Perrier, M. and A.A. Roche (1992). Towards mill-wide evaluation of control loop performance. *Proc. Control. Syst. '92*, Whistler, BC.

Pryor, C. (1982). Autocovariance and power spectrum analysis. *Control Engineering (Oct)*, 103-106

Stanfelj, N., T.E. Marlin, and J.F. MacGregor, (1993). Monitoring and diagnosing process control performance: The single loop case, *Ind. Eng. Chem. Res.*, **32**, 301-314.

Thaler, G.J. (1989). Automatic Control Systems, West Publishing Company, St.Paul MN.

Thornhill, N.F., and T. Hägglund, (1997). Detection and diagnosis of oscillation in control loops, *Control Eng. Prac.*, **5**, 1343-1354.

Thornhill, N.F., M. Oettinger, and P. Fedenczuk, (1998). Refinery-wide control loop performance assessment, *submitted to J. Process Control*.

Tyler, M.L. and M. Morari, (1996). Performance monitoring of control-systems using likelihood methods, *Automatica*, **32**, 1145-1162.

Welch, P.D. (1967). The use of fast Fourier transforms for the estimation of power spectra. *IEEE Trans Audio & Electroacoustics*, **AU-15**, 70-73

FAULT ISOLATION IN INDUSTRIAL PROCESSES USING FISHER'S DISCRIMINANT ANALYSIS

Evan L. Russell and Richard D. Braatz
Department of Chemical Engineering
University of Illinois at Urbana-Champaign
Urbana, IL 61801

Abstract

The limited availability of quality data can pose problems for any process monitoring scheme. The problem is especially difficult when the dimensionality of the data is very high, such as the data collected from a paper machine or an entire chemical plant. By projecting the data into a lower dimensional space that more accurately characterizes the state of the process, dimensionality reduction techniques can greatly improve and simplify the process monitoring procedures, fault detection, and isolation. The application of principal component analysis (PCA) as a dimensionality reduction tool for monitoring chemical processes has been studied by several academic and industrial engineers. Although PCA contains certain optimality properties in terms of fault detection, it is not well-suited for fault isolation. Fisher's discriminant analysis (FDA), a dimensionality reduction technique heavily studied in the pattern classification literature, has advantages from a theoretical point of view. In this paper, we develop an information criterion that automatically determines the order of the dimensionality reduction for FDA, and show that FDA is more proficient than PCA for isolating faults, both theoretically and by applying these techniques to paper machine data provided by International Paper and simulated data collected from the Tennessee Eastman problem.

Keywords

Fault isolation, Process monitoring, Pattern classification, Discriminant analysis, Chemometric methods, Fault detection, Large scale systems, Multivariate statistics, Dimensionality reduction.

Introduction

Large amounts of data are collected in many industrial processes. This data can be analyzed to determine whether or not a fault has occurred in the process, where a *fault* is defined as abnormal process behavior whether associated with equipment failure, equipment wear, or extreme process disturbances. This task of determining whether a fault has occurred is called *fault detection,* whereas *fault isolation* is the task of determining which fault has occurred. An important step in fault detection is to quantify abnormal process behavior in terms of the data; similarly, an important step in fault isolation is to quantify the different faults in terms of the data. The accuracy of these quantifications can be improved by using dimensionality reduction techniques, such as principal component analysis (PCA) and Fisher's discriminant analysis (FDA). The lower dimensional representations produced by these techniques can better generalize to new process data than the representations using the entire dimensionality, thereby improving the proficiency of detecting and isolating faults.

Academic and industrial process control engineers have applied PCA for abstracting structure from multidimensional chemical process data (Kresta, Marlin, and MacGregor, 1991; Piovoso, Kosanovich, and Pearson, 1992). PCA determines the most accurate lower dimensional representation of the data, in terms of capturing the data directions that have the most variance. The resulting lower dimensional model has been used for detecting out-of-control status and for diagnosing disturbances leading to the abnormal process operation (Kourti and MacGregor, 1996; Raich and Cinar, 1995). Several applications of PCA to real industrial data were conducted at DuPont over the past six years, with much of the results published in conference proceedings and journal articles. Several academics have performed similar studies based on data collected from computer simulations of processes (Ku, Storer, and Georgakis, 1995; Raich and Cinar, 1995).

FDA provides an optimal lower dimensional representation in terms of discriminating among classes of data, where for fault isolation, each class corresponds to data collected during a specific known fault. Although FDA has been heavily studied in the pattern classification literature and is only slightly more complex than PCA, its use for analyzing process data is not described in the literature. This is interesting, since FDA would seem to have advantages over PCA, when the primary goal is to discriminate among faults. We suspect that part of the reason that FDA has been ignored in the chemical process control literature is that more chemical engineers read the statistics literature (where PCA is dominant) than the pattern classification literature (where FDA is dominant).

The paper is organized as follows. First, the procedure of autoscaling, typically performed before employing the dimensionality reduction techniques, is briefly discussed. Then, PCA and FDA are described, and it is shown how to apply these techniques to monitor processes. An information criterion for FDA is developed for automatically determining the order of dimensionality reduction. Finally, PCA and FDA are compared by theoretical and graphical analysis and by application to paper machine data provided by International Paper and simulated data collected from the Tennessee Eastman problem.

Autoscaling the Data

The data needs to be scaled to avoid particular variables dominating the dimensionality reduction procedure. For instance, when performing an unscaled dimensionality reduction procedure on temperature measurements varying between 300K and 320K and concentration measurements varying between 0.4 and 0.5, the temperature measurements would dominate even though the temperature measurements may be no more important than the concentration measurements in monitoring the process.

Autoscaling the data ensures that each measurement variable is given equal weight before the application of the dimensionality reduction procedure. It consists of two steps. The first step is to subtract each variable by its sample mean because the objective is to capture the variation of the data from the mean. The second step is to divide each variable of the mean-centered data by its standard deviation. This scales each variable to unit variance. For simplicity of presentation only, it is assumed in what follows that the data has been autoscaled.

Principal Component Analysis

PCA is an optimal dimensionality reduction technique in terms of capturing the variance of the data. PCA determines a set of orthogonal vectors, called *loading vectors*, which can be ordered by the amount of variance explained in the loading vector directions. Given n observations of m measurement variables stacked into a training data matrix $\mathbf{X}$, the loading vectors are calculated by solving the singularities of the optimization problem

$$\max_{\mathbf{v} \neq 0} \frac{\mathbf{v}^T \mathbf{X}^T \mathbf{X} \mathbf{v}}{\mathbf{v}^T \mathbf{v}}. \quad (1)$$

The singularities of Eqn. (1) can be computed via the SVD

$$\frac{1}{\sqrt{n-1}} \mathbf{X} = \sum_{i=1}^{m} \sigma_i \mathbf{u}_i \mathbf{v}_i^T \quad (2)$$

where $\mathbf{u}_i$ and $\mathbf{v}_i$ are orthonormal column vectors, and the σ_i are positive real scalars of decreasing magnitude $(\sigma_1 \geq \sigma_2 \geq \cdots \geq \sigma_m)$. The variance of the training data projected along the loading vector $\mathbf{v}_i$ is equal to σ_i^2.

Fault Detection

PCA can characterize normal operations by employing Hotelling's T^2 statistic (Kourti and MacGregor, 1996)

$$T^2 = \mathbf{x}^T \mathbf{P}^T \Sigma^{-2} \mathbf{P} \mathbf{x} \quad (3)$$

where the rows of the matrix $\mathbf{P}$ contain the loading vectors, the diagonal matrix Σ contains the corresponding singular values, and $\mathbf{x}$ is an observation vector of dimension m. The T^2 statistic rotates and scales the variables so that normal operations can be characterized by a scalar threshold. Given a number of loading vectors, a, to include in Eqn. (3), the threshold can be calculated for the T^2 statistic using the probability distribution

$$T_\alpha^2 = \frac{(n^2 - 1)a}{n(n-a)} F_\alpha(a, n-a) \quad (4)$$

where F is the upper $100\alpha\%$ critical point of the F distribution with $(a, n-a)$ degrees of freedom. The

T^2 statistic with Eqn. (4) produces an elliptical confidence region in the measurement space (see Fig. 1). This confidence region defines the normal process behavior, and an observation vector projected outside this region indicates that a fault has occurred.

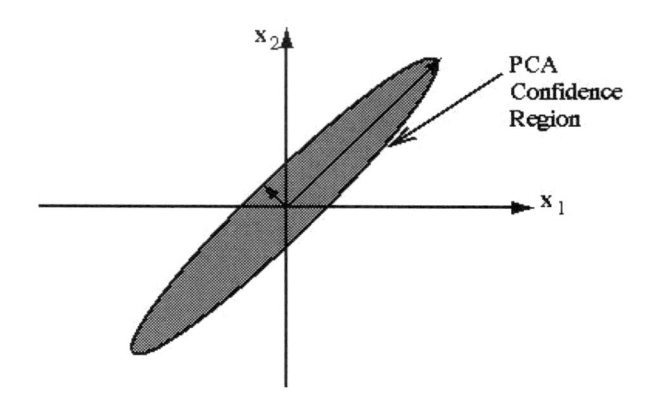

Figure 1. The confidence region for two variables using the T^2 statistic and Eqn. (4).

As the amount of data in the training set decreases, the T^2 statistic tends to be a less accurate representation of the normal process behavior, especially in the loading vector directions corresponding to the lower singular values. In this case, the T^2 statistic is overly sensitive to inaccuracies in the smaller singular values and does not generalize well when applied to data outside the training set. Therefore, only the loading vectors associated with the larger singular values should be maintained in the PCA model.

By including in **P** only the loading vectors associated with the largest a singular values, the portion of the measurement space corresponding to the lowest $m-a$ singular values can be monitored by using the Q statistic developed by Jackson and Mudholkar (1979)

$$Q = \mathbf{x}^T(\mathbf{I} - \mathbf{P}^T\mathbf{P})(\mathbf{I} - \mathbf{P}^T\mathbf{P})\mathbf{x}. \quad (5)$$

Since the Q statistic does not directly measure the variations along each loading vector but measures the total sum of variations in the space corresponding to the lowest $m-a$ singular values, the Q statistic does not suffer from an over-sensitivity to inaccuracies in the lower singular values (Jackson and Mudholkar, 1979). The distribution for the Q statistic has been approximated by Jackson and Mudholkar (1979)

$$Q_\alpha = \theta_1 \left[\frac{c_\alpha(2\theta_2)^{0.5} h_0}{\theta_1} + 1 + \frac{\theta_2 h_0(h_0 - 1)}{\theta_1^2} \right]^{1/h_0} \quad (6)$$

where $\theta_i = \sum_{j=a+1}^{n} \sigma_j^{2i}$, $h_0 = 1 - \frac{2\theta_1\theta_3}{\theta_2}$, and c_α is the normal deviate corresponding to the upper $(1-\alpha)$ percentile. The threshold for the Q statistic can be computed using Eqn. (6).

Reduction Order

A key step in a dimensionality reduction technique is to determine the order of the reduction. There exist several techniques for determining the number of loading vectors, a, to maintain in the PCA model. Several techniques are discussed by Jackson (1991) and Himes, Storer, and Georgakis (1994), but there appears to be no dominant technique. Ku, Storer, and Georgakis (1995) recommend the parallel analysis method, because in their experience, it has performed overall the best. Parallel analysis determines the dimensionality of the PCA model by comparing the singular value profile to that obtained by assuming independent measurement variables. The dimension is determined by the point at which the two profiles cross. This approach is particularly attractive since it is intuitive and easy to automate.

Fault Isolation

Pattern classification theory has been applied to isolate faults (Raich and Cinar, 1995). By collecting data during specific known faults, and categorizing the data into fault classes, a pattern classifier may be used to associate a new observation with the appropriate fault class. A typical pattern classifier consists of a dimensionality reduction step and a discriminant analysis step (Duda and Hart, 1973). Raich and Cinar (1995) investigated the application of PCA with the discriminant functions based on the T^2, Q, and a combination of the two statistics, where an observation is associated with the class exhibiting the minimum discriminant value.

Fisher's Discriminant Analysis

For fault isolation, data is collected from the plant during specific faults and is categorized into classes where each class contains data representing a particular fault. FDA is an optimal dimensionality reduction technique in terms of maximizing the separability of these classes. It determines a set of projection vectors that maximize the scatter between the classes while minimizing the scatter within each class. A mathematical description follows.

Stacking the training data for all classes into the matrix **X** and representing the ith row of **X** with the column vector $\mathbf{x}_i$, the total scatter matrix $\mathbf{S}_t$ is defined as

$$\mathbf{S}_t = \sum_{i=1}^{n}(\mathbf{x}_i - \tilde{\mathbf{x}})(\mathbf{x}_i - \tilde{\mathbf{x}})^T \quad (7)$$

where $\tilde{\mathbf{x}}$ is called the *total mean vector* whose elements correspond to the means of the columns of $\mathbf{X}$. Let $\mathbf{X}_i$ be the rows of $\mathbf{X}$ belonging to class i, then

$$\mathbf{S}_i = \sum_{\mathbf{x}_i \in \mathbf{X}_i} (\mathbf{x}_i - \tilde{\mathbf{x}}_i)(\mathbf{x}_i - \tilde{\mathbf{x}}_i)^T \quad (8)$$

is the *within-scatter matrix* for class i where $\tilde{\mathbf{x}}_i$ is the mean vector for class i. Letting c be the number of classes,

$$\mathbf{S}_w = \sum_{i=1}^{c} \mathbf{S}_i \quad (9)$$

is the *within-class-scatter matrix*, and

$$\mathbf{S}_b = \sum_{i=1}^{c} n_i (\tilde{\mathbf{x}}_i - \tilde{\mathbf{x}})(\tilde{\mathbf{x}}_i - \tilde{\mathbf{x}})^T \quad (10)$$

is the *between-class-scatter matrix* where n_i is the number of observations in class i. The *total-scatter matrix* is the sum of the between-class-scatter matrix and the within-class-scatter matrix,

$$\mathbf{S}_t = \mathbf{S}_b + \mathbf{S}_w. \quad (11)$$

The FDA vectors are calculated by computing the singularities of the optimization problem

$$\max_{\mathbf{v} \neq 0} \frac{\mathbf{v}^T \mathbf{S}_b \mathbf{v}}{\mathbf{v}^T \mathbf{S}_w \mathbf{v}} \quad (12)$$

(assuming invertible $\mathbf{S}_w$). The FDA vectors are equal to the generalized eigenvectors $\mathbf{w}_i$ of the eigenvalue problem

$$\mathbf{S}_b \mathbf{w}_i = \lambda_i \mathbf{S}_w \mathbf{w}_i \quad (13)$$

where the generalized eigenvalues λ_i indicate the degree of overall separation among the classes. Because the direction and not the magnitude of $\mathbf{w}_i$ is important, the norm is chosen to be $\|\mathbf{w}_i\| = 1$.

Fault Detection and Isolation

While Raich and Cinar (1995) applied PCA to each class of faults before evaluating the discriminant functions, the objective of FDA suggests using the dimensionality reduction technique to all the classes together. The lower dimensional representation provided by FDA can be employed with discriminant functions, such as the T^2 statistic, to isolate faults, as discussed in the PCA section. When the data is autocorrelated, time histories can be augmented to the data matrix before performing FDA. This augmentation of the data matrix takes into account the autocorrelations and has been studied using PCA (Ku, Storer, and Georgakis, 1995). FDA can also be used to detect faults by adding a class of data collected during normal process behavior; yet, this method may not be sensitive to faults not contained in the training data.

Reduction Order

When FDA is applied in a pattern classification scheme, the objective of FDA is to determine a lower dimensional space that minimizes the misclassification rate, which is equal to the number of incorrect classifications divided by the total number of classifications. Determining the order to reduce the FDA model plays a key role in minimizing the misclassification rate for new observations outside the training set. If the statistics of the classes are known exactly, from an information point of view the entire measurement space should be maintained for the discriminant analysis step. In reality, inaccuracies in the statistics of the classes exist. Consequently, the amount of information obtained in the measurement space corresponding to the small generalized eigenvalues of Eqn. (13) may not outweigh the inaccuracies in the statistics, and these directions should be eliminated from the FDA model. The pattern classification literature assumes that a testing set is available, and the order of the FDA model is determined by minimizing the misclassification rate of the testing set. This may not always be practical because the training data set may be small, or a testing set may not be attainable.

Akaike's information criteria (AIC) has been utilized to select the model order for system identification (Ljung, 1987). The AIC contains an error term and a penalty term, which penalizes the model complexity, and the two terms rely on information only in the training data. In general, the choice of the penalty term is subjective (Ljung, 1987). The order of the FDA model can be determined by computing the dimensionality, a, which minimizes the information criterion

$$f(a) + \frac{a}{\tilde{n}} \quad (14)$$

where $f(a)$ is the misclassification rate for the training set and $\tilde{n}$ is the number of observations per class. Equation (14), which is similar in form to the AIC, appears to be reasonable, since the penalty term scales relatively well with the error term. This is confirmed later by application.

Comparison of PCA and FDA

The objective functions for PCA and FDA have been stated mathematically in Eqn. (1) and (12), respectively. Equivalently, it can be shown that the loading vectors and the FDA vectors can be calculated by computing the singularities of the optimization problems

$$\max_{\mathbf{v} \neq 0} \frac{\mathbf{v}^T \mathbf{S}_t \mathbf{v}}{\mathbf{v}^T \mathbf{v}} \qquad (15)$$

and

$$\max_{\mathbf{v} \neq 0} \frac{\mathbf{v}^T \mathbf{S}_t \mathbf{v}}{\mathbf{v}^T \mathbf{S}_w \mathbf{v}} \qquad (16)$$

respectively. Equations (15) and (16) indicate that the PCA and FDA vectors are identical for the case when $\mathbf{S}_w = \sigma \mathbf{I}$ where $\sigma > 0$. This situation occurs if the data in each class can be described by a uniformly distributed ball (i.e., circle in 2-D space and sphere in 3-D space), even if the balls are of distinct sizes. Differences between the two techniques occur only if there is elongation in the data used to describe any one of the classes. These elongated shapes occur for highly correlated data sets (Fig. 1), which is typical for data collected from chemical processes. Therefore, when PCA and FDA are applied in the same manner to chemical processes, the PCA and FDA vectors are expected to be significantly different, and the differing objectives, Eqn. (15) and (16), suggest that FDA will be significantly better for discriminating among classes of faults.

The projection of the experimental data taken from (Cheng, Zhuang, and Yang, 1992) onto the first two PCA and FDA vectors are shown in Fig. 2. It is effective at illustrating the difference in the objectives between PCA and FDA. By comparing the limits of the x and y axes and visually inspecting the data, it is clear that the span of the PCA projections is larger than the FDA projections. While PCA is better able to separate the data as a whole, FDA is better able to separate the data among the classes (*, o, x). This is evident in the degree of overlap between '*' and 'o' data regions in the two plots, confirming the optimality of FDA over PCA for fault isolation.

Applications

The Tennessee Eastman test problem was created by the Eastman Chemical Company to provide a realistic industrial process for evaluating process control and monitoring methods (Downs and Vogel, 1993). The test problem is based on an actual chemical process where the components, kinetics, and operating conditions have been modified for proprietary reasons. The simulation code allows 20 preprogrammed disturbances to be selectively added to the process, and these disturbances will serve as the faults of the process. The plant-wide control structure recommended in (Lyman and Georgakis, 1995) was implemented to generate the closed loop simulated process data for each fault.

The training and testing data sets for each fault consisted of $n=100$ observations, sampled starting the first sampling instant after the occurrence of the fault, included all the manipulated and measurement variables except the agitation speed of the reactor's stirrer for a total of $m=52$ variables. The data was sampled every 3 minutes, and the random seed was changed before the computation of the data set for each fault.

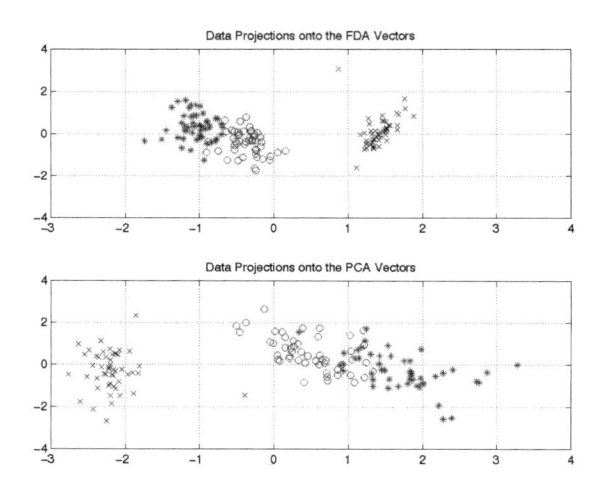

Figure 2. The projection of experimental data onto the first two FDA and PCA vectors.

Using parallel analysis to determine the PCA order, the application of PCA to each fault with the T^2 statistic, Q statistic, and the sum of the $(1-\alpha)$ confidence levels produced by the T^2 and Q statistics (Eqn. (4) and (6)) serving as the discriminant functions, resulted in misclassification rates of 0.90, 0.71, and 0.83, respectively, on the testing set. Classifying the data with the T^2 statistic and no dimensionality reduction resulted in a misclassification rate of 0.47 (Fig. 3). These results indicate that the dimensionality reduction provided by PCA is worse than using no dimensionality reduction and illustrate the fact that PCA has the wrong objective for fault isolation.

The information criterion for the FDA method in Eqn. (12) achieved a minimum at $a=26$ (Fig. 4). The application of FDA with order $a=26$ and the T^2 statistic as the discriminant function produced a misclassification rate equal to 0.45 (see Fig. 3). This misclassification rate compares well with the minimum classification rate equal to 0.41 with $a=38$, indicating that the information criterion is effective for selecting the order for FDA. It is also less than the rates achieved by the PCA measures and the rate achieved with no dimensionality reduction. These results confirm the theoretical analysis, supporting the fact that the objective for FDA is superior to the objective for PCA.

Paper machine data provided by International Paper contained $m=2250$ measurement variables and only $n=50$ to 60 observations for the data sets associated with each operating condition. The application of PCA with the T^2 statistic resulted in a misclassification rate equal to 0.50. The high dimensionality of the data together with the small data sets precluded a direct FDA approach because many of the generalized eigenvalues were singular. By ignoring the denominator of Eqn. (10) and projecting the data onto the vectors connecting the class means, a misclassification rate of 0.19 was achieved. Again, FDA performed much better than PCA.

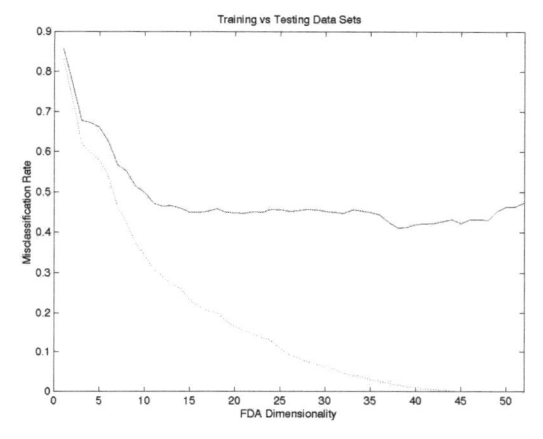

Figure 3. The misclassification rates applying FDA to the Tennessee Eastman problem. The dotted and solid lines represent the training and testing sets, respectively.

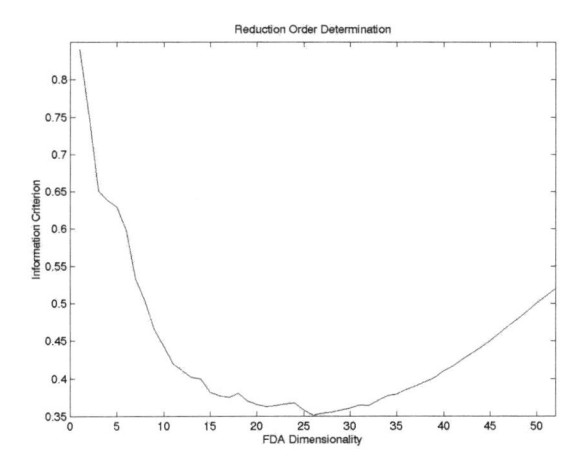

Figure 4. The information criterion from Eqn. (12) applied to the Tennessee Eastman problem.

Conclusions

It has been shown that FDA is a better dimensionality reduction technique than PCA for fault isolation. This has been illustrated from both a theoretical point of view and by applying the two techniques to simulated and real process data.

Acknowledgments

This work was supported by International Paper.

References

Cheng, Y., Zhuang, Y., and Yang, J. (1992). Optimal fisher discriminant analysis using the rank decomposition. *Pattern Recognition*, **25, No. 1**, 101-111.

Downs, J. J. and Vogel, E. F. (1993). A plant-wide industrial-process control problem. *Comp. & Chem. Eng.*, **17, No. 3**, 245-255.

Duda, R. O. and Hart, P. E. (1973). *Pattern Classification and Scene Analysis*. Wiley-Interscience, New York.

Himes, D. M., Storer, R. H., and Georgakis, C. (1994). Determination of the number of principal components for disturbance detection and isolation. *In Proc. of the American Control Conf.* Baltimore, MD. pp. 1279-1283.

Jackson, J. E. and Mudholkar, G. S. (1979). Control procedures for residuals associated with principal component analysis. *Technometrics*, **21**, 341-349.

Jackson, J. E. (1991). *A User's Guide to Principal Components*. Wiley, New York.

Kourti, T. and MacGregor, J. F. (1996). Multivariate SPC methods for process and product monitoring. *J. of Quality Technology*, **28**, 409-428.

Kresta, J. V., Marlin, T. E., and MacGregor, J. F. (1991). Multivariable statistical monitoring of process operating performance. *Can. J. of Chem. Eng.*, **69**, 35-47.

Ku, W., Storer, R. H., and Georgakis, C. (1995). Disturbance detection and isolation by dynamic principal component analysis. *Chemometrics and Intelligent Laboratory Systems*, **30**, 179-196.

Ljung, L. (1987). *System Identification: Theory for the User*. Prentice Hall, Englewood Cliffs, NJ.

Lyman, P. R. and Georgakis, C. (1995). Plant-wide control of the Tennessee Eastman problem. *Comp. & Chem. Eng.*, **19, No. 3**, 321-331.

Piovoso, M. J., Kosanovich, K. A., and Pearson, R. K. (1992). Monitoring process performance in real time. In *Proceedings of the American Control Conf.* IEEE Press, Piscataway, NJ. pp. 2359-2363.

Raich, A. C. and Cinar, A. (1995). Multivariable statistical methods for monitoring continuous processes: assessment of discriminatory power disturbance models and diagnosis of multiple disturbances. *Chemometrics and Intelligent Laboratory Systems*, **30**, 37-48.

ENHANCEMENT OF GROSS ERROR DETECTION WHEN DATA ARE SERIALLY CORRELATED

Ruth Kongsjahju and Derrick Rollins
Departments of Chemical Engineering and Statistics
Iowa State University
Ames, Iowa 50011

Abstract

Chemical process data are typically correlated over time (i.e., serially or autocorrelated) due quite often to recycle loops, large material inventories, sampling lag and dead time, process dynamics created by high order systems, feedback control, and transportation lag. However, many of the approaches that attempt to identify gross errors in measured process variables have not addressed serial correlation, which can lead to large inaccuracies in identifying biased measured variables. Hence, this work extends the unbiased estimation technique (UBET) (Rollins and Davis, 1992) to address serial correlation. The serially correlated gross error detection (GED) study of Kao, *et al*. (1990) is used as a basis for setting up the study and comparison. In their work, the type of autocorrelation was assumed known (*ARMA(1,1)*) and the measurement test (MT) was used for identification of the measurement bias. Kao, *et al*. (1990) attempted to prewhiten the data and used variances of measured variables derived from the knowledge of the time correlation structure. This work presents a different and superior prewhitening method that is shown to truly transform the data to white noise. The UBET and MT are applied to the transformed data and compared in a simulation study.

Keywords

Autocorrelation, Gross error detection, Serial correlation.

Introduction

In chemical processes, we often have recycle loops, large material inventories, sampling lag and dead time, and process dynamics created by high order systems, feedback control, and transportation lag. As a result, process data are often serially correlated. Kao, *et al*. (1990) suggested a method of prewhitening by removing serial correlation from a process data. Prewhitening is the transformation that is needed to improve the modeling by ensuring that residuals are randomly independent (i.e., the residuals exhibit white noise, see Box and Jenkins, 1976). One purpose of this paper is to provide a new perspective on prewhitening in gross error detection (GED).

The first part of this paper replicates results of the measurement test (MT) on serially correlated process data presented by Kao, *et al*. (1990). We also present prewhitened model equation for the unbiased estimation technique (UBET) developed by Rollins and Davis (1992). In the second part of this paper, we discuss some simulation results and compare the UBET and MT. Finally, we conclude our work with some thoughts regarding prewhitening in the context of GED.

Table 1. MT Simulation Results for ARMA(1,1) Comparing Algorithms with $\delta_i = 0.2$, known $\sigma_i^2 = 0.25$, and $\alpha = 0.1$.

	\multicolumn{4}{c}{$\phi_{i1} = 0.4$, $\theta_{i1} = 0.2$}	\multicolumn{4}{c}{$\phi_{i1} = 0.0$, $\theta_{i1} = -0.4$}
i	AVTI	OP	OP*	OPF	AVTI	OP	OP*	OPF
1	0.9657	0.6231	0.61	0.0070	0.9705	0.6331	0.61	0.0056
2	0.9665	0.6420	0.61	0.0067	0.9710	0.6538	0.61	0.0061
3	0.9665	0.6350	0.62	0.0083	0.9716	0.6466	0.62	0.0071
4	0.9685	0.5612	0.48	0.0026	0.9733	0.5755	0.48	0.0025
5	0.9662	0.6097	0.56	0.0046	0.9716	0.6221	0.56	0.0039
6	0.9713	0.5683	0.48	0.0037	0.9752	0.5814	0.48	0.0033
7	0.9663	0.6373	0.61	0.0063	0.9722	0.6497	0.61	0.0053

* OP produced by Kao, et al. (1990), i = biased stream

Measurement Test (MT)

The measurement test is a popular GED test developed by Mah and Tamhane (1982). To compare our results with Kao, et al. (1990), we developed an algorithm for the MT.

The main measure of success in gross error identification that is presented by Kao, et al. is the overall power (OP):

$$OP = \frac{\text{No. of nonzero } \delta s \text{ correctly identified}}{\text{No. of nonzero } \delta s \text{ simulated}} \quad (1)$$

Two indicators of GED performance not included by Kao, et al. are the averaged type I error (AVTI) and overall performance (OPF). These measurements are defined as (see Rollins and Davis, 1992):

$$AVTI = \frac{\text{No. of zero } \delta s \text{ wrongly identified}}{\text{No. of simulation trials (10,000)}} \quad (2)$$

$$OPF = \frac{\text{No. of trials with perfect identification}}{\text{No. of simulation trials (10,000)}} \quad (3)$$

The main goal in identification error in the process is to obtain high OPF (perfect identification) and low AVTI (false identification). Table 1 shows MT results for cases in Kao, et al. using our algorithm for comparison. In general, the agreement between Kao, et al. results and the one generated by our algorithm is good. The variations are most likely due to the randomness of the data sets.

Table 1 reveals very high AVTI and very low OPF. This conclusion agrees with previous studies done on the MT (Rollins, 1990; Rollins and Davis, 1992).

The method of prewhitening presented by Kao, et al. attempts to remove serial correlation from the estimated values of residuals (i.e., prewhitening of the residuals).

The estimated values of the residuals are calculated by Eqn. (4) below:

$$\mathbf{r(t)} = \mathbf{y(t)} - \hat{\mathbf{y}}(t) = \Sigma \mathbf{A}'(\mathbf{A}\Sigma\mathbf{A}')^{-1}\mathbf{Ay(t)} \quad (4)$$

This approach produces biased estimates of $\mathbf{y(t)}$ and therefore produces biased residuals and limited identification accuracy (see Rollins, 1990; Rollins, et al., 1996).

Unbiased Estimation Technique (UBET)

UBET (Rollins and Davis, 1992) attempts to provide unbiased estimates for true values and high OPFs. The original work did not include serially correlated data.

In this paper, we apply the concept of prewhitening to extend the use of UBET to serially correlated process data. Since UBET uses *nodal tests*, the prewhitening is applied directly to the serially correlated measured variable, $y_i(t)$. The measurement model that takes into account measurement biases (δ) and serially correlated measurement errors (ε) is shown in Eqn. (5) with the correlation structure of *ARMA(1,1)*.

$$y_i(t) = \mu_i + \delta_i + e_i(t) \quad (5)$$

where, $i = 1, ..., \#$ of streams

$$e_i(t) = \frac{(1-\hat{\theta}_{i1}B)}{(1-\hat{\phi}_{i1}B)} u_i(t)$$

$$u_i(t) \sim NID(0, \sigma_i^2)$$

The idea of prewhitening $y_i(t)$ is to multiply $y_i(t)$ given by Eqn. (5) with a transfer function that will transform it to the randomly independent $y_i(t)$, $y_i(t)^*$, given by equation below:

$$y_i(t)^* = \mu_i^* + \delta_i^* + u_i(t) \quad (6)$$

where,

$$y_i(t)^* = v(B)y_i(t) = y_i(t) + v_{i1}y_i(t-1) + v_{i2}y_i(t-2)+\ldots$$

$$v(B) = \frac{(1-\hat{\phi}_{i1}B)}{(1-\hat{\theta}_{i1}B)} = 1 + v_{i1}B + v_{i2}B^2 +\ldots$$

$$V_{in} = \hat{\theta}_{i1}^{(n-1)}(\hat{\theta}_{i1} - \hat{\phi}_{i1})$$

$$\mu_i^* = \mu_i(1 + v_{i1} + v_{i2} +\ldots)$$

$$\delta_i^* = \delta_i(1 + v_{i1} + v_{i2} +\ldots)$$

In Eqn. (6), the prewhitened $y_i(t)$, $y_i(t)^*$, is a function of the transformed μ_i, μ_i^*, and transformed δ_i, δ_i^*. Therefore, values of $y_i(t)^*$ are significantly dependent on values of ϕ_{i1} and θ_{i1}. If parameters $\theta_{i1} > \phi_{i1}$, then ($\theta_{i1} - \phi_{i1}$) is positive, and $y_i(t)^* > y_i(t)$. Similarly, if $\theta_{i1} < \phi_{i1}$, then ($\theta_{i1} - \phi_{i1}$) is negative, and $y_i(t)^* < y_i(t)$. Therefore, if $\theta_{i1} = \phi_{i1}$, the difference of these parameters is zero, and Eqn. (6) is reduced to a white noise model (randomly independent process data).

Adequate prewhitening will remove the serial correlation from the process data. One way to verify the removal of serial correlation is to calculate the sample autocorrelation function (ACF). The ACF is considered significantly different from zero if the *t-like* statistics (Eqn. (9) below) is outside specified limits (e.g., ±2 of sample ACF standard error) (the dash lines in Fig. 1, see Meeker, 1996). The closer the ACF is to ±1.0, the more correlated the process data. Conversely, the closer the ACF is to zero, the less correlated the process data. The ACF comparison for the serially correlated process data and the prewhitened results is shown in Fig. 1. Equation (7) is used to calculate the sample ACF for each time lag and Eqn. (8) present the sample ACF standard error.

$$\hat{\rho}_k = \frac{\sum_{t=1}^{n-k}\{y(t)-\bar{y}\}\{y(t+k)-\bar{y}\}}{\sum_{t=1}^{n}\{y(t)-\bar{y}\}} \quad (7)$$

where, $k = 0,1,2,\ldots$

$$S_{\hat{\rho}_k} = \sqrt{\left(\frac{1+2\hat{\rho}_1^2+\ldots+2\hat{\rho}_{k-1}^2}{n}\right)} \quad (8)$$

$$t = \frac{\hat{\rho}_k}{S_{\hat{\rho}_k}} \quad (9)$$

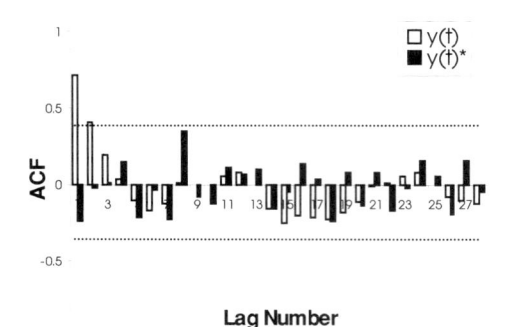

Figure 1. ACF Comparison -- prewhitened process data (y(t)) and serially correlated process data (y(t)).*

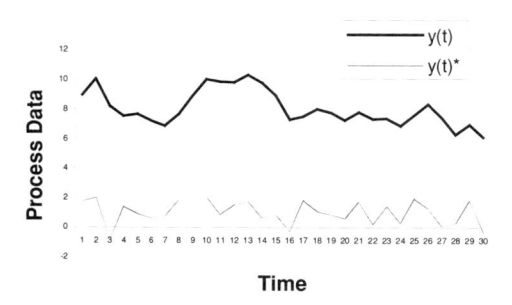

Figure 2. Prewhitening effect on serially correlated process data (y(t) was generated from an ARMA(1,1) with $\phi_{i1} = 0.8$ and $\theta_{i1} = -0.5$.

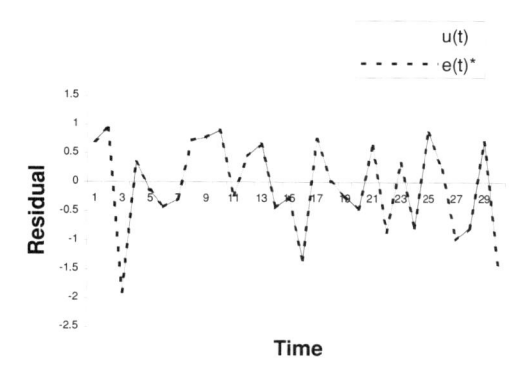

Figure 3. Comparison of prewhitened residuals (e(t)) to the white noise u(t).*

Another way of showing the effect of past values on present or future values is by plotting all the values against time as in Fig. 2. The upper line is the correlated process data and the bottom line is the prewhitened result. We can see the trend line of the correlated process data, which relates current values to its past values.

Table 2. Simulation Results when $\phi_{i1} = 0.2$ and $\theta_{i1} = 0.4$, N=3 for all Streams with $\delta = 5.0$ at ith Stream.

	MT			UBET		
i	AVTI	OP	OPF	AVTI	OP	OPF
1	1.0000	1.0000	0.0000	0.0036	0.9934	0.9899
2	1.0000	1.0000	0.0000	0.0058	0.9978	0.9920
3	1.0000	1.0000	0.0000	0.0042	0.9998	0.9956
4	1.0000	1.0000	0.0000	0.0049	0.9933	0.9885
5	1.0000	1.0000	0.0000	0.0061	0.9996	0.9935
6	1.0000	1.0000	0.0000	0.0050	0.9934	0.9885
7	1.0000	1.0000	0.0000	0.0042	1.0000	0.9958

Table 3. Simulation Results when $\phi_{i1}=0.4$ and $\theta_{i1}=0.2$ for all Streams with $\delta = 5.0$ at ith Stream.

	MT N = 3			UBET N = 3			UBET N = 5		
i	AVTI	OP	OPF	AVTI	OP	OPF	AVTI	OP	OPF
1	1.0000	1.0000	0.0000	0.0031	0.5684	0.5660	0.0002	0.9111	0.9109
2	1.0000	1.0000	0.0000	0.0042	0.5298	0.5268	0.0001	0.9196	0.9195
3	1.0000	1.0000	0.0000	0.0042	0.7483	0.7452	0.0005	0.9894	0.9890
4	1.0000	1.0000	0.0000	0.0025	0.4740	0.4729	0.0001	0.9082	0.9081
5	1.0000	1.0000	0.0000	0.0044	0.6089	0.6065	0.0004	0.9873	0.9870
6	1.0000	1.0000	0.0000	0.0046	0.4873	0.4859	0.0001	0.9069	0.9069
7	1.0000	1.0000	0.0000	0.0037	0.7194	0.7171	0.0004	0.9941	0.9937

In order to check the removal of serial correlation from $y_i(t)$ in Fig. 2, Fig. 3 shows a comparison of the prewhitened process data residuals $(e(t)^*)$ to the white noise $(u(t))$. Since the prewhitened residual is identical to the white noise residual, one concludes the removal of serial correlation and the effectiveness of this prewhitening scheme.

Simulation study

A simulation study was done to show the effect of prewhitening $y_i(t)$ on MT and UBET gross error detection performance. The same flow network as in Kao, et al. (1990) and Rollins and Davis (1992) was used. The true values of (1, 3, 3, 1, 2, 1, 1) were used respectively for each stream flow rate in all situations investigated. For this study, σ_i of 1.0 and the significance level (α) of 0.05 were used, and there was only one biased stream ($\delta_i = 5.0$) in each case. The simulation was repeated 10,000 times and was run by using the FORTRAN® Power Station with IMSL® subroutines in MICROSOFT® Windows®95. The standard normal distributed random numbers were used to generate white noise. The results are shown in Table 2.

As shown by Table 2 and 3, UBET's OP and OPF are high and its AVTI is low. These values present UBET's effectiveness in identifying measurement bias. In contrast, MT has very low OPF and very high AVTI.

Another observation is that OP and OPF are affected by the values of ϕ_{i1} and θ_{i1}. We see that UBET's OP and OPF decrease when the moving average parameter (θ_{i1}) is less than the autoregressive parameter (ϕ_{i1}). Because of that, the quantity of ($\phi_{i1} - \theta_{i1}$) is negative and this causes the v coefficients to have smaller values (see Eqn. (6)). When v's are small, then δ^* becomes smaller as well. This decreases the ability of detecting δ^*, therefore OP and OPF decrease. For this case, UBET becomes less powerful when sample size is 3 (i.e., N = 3). However, this limitation can be reduced by increasing the sample sizes (see Table 3). Note the higher OP and OPF when N = 5.

Concluding Remarks

The idea of prewhitening is not a new concept but the application to the UBET is new. Our approach to prewhiten, is different than Kao, et al. (1990). Here we prewhitened the measured variables directly, while Kao, et al. prewhitened the residuals of the measured variables. In a forthcoming paper we prewhiten nodal balances and compare this approach to the one in this work.

In this study, we have shown that prewhitening the process data suspected of having serial correlation facilitates the use of UBET for effective gross error detection. Furthermore, the performance of UBET was shown superior to MT in handling prewhitened serially correlated data. Although MT seems to give high overall power to detect bias, the large AVTI and small OPF take away its attractiveness. In contrast, UBET can give high OP and OPF, and low AVTI.

Acknowledgement

We are grateful to the National Science Foundation for partial support of this research under grant CTS-9453534.

Nomenclatures

A = constraint matrix
$e_i(t)$ = ith random measurement error at time t
k = lag number
n = the length of the time series
N = sample size
r(t) = measurement test's residual at time t
$S_{\hat{\rho}_k}$ = sample autocorrelation function standard error
t = *t-like* statistic
$u_i(t)$ = ith white noise at time t
$\bar{y}$ = average value of y(t)
$y_i(t)$ = ith measured variable at time t
$y_i(t)*$ = prewhitened ith measured variable at time t
y(t) = measurement test's serially correlated process data at time t
ŷ(t) = measurement test's estimate value of **y(t)**
α = type I error level or the significance level
δ_i = ith measurement bias
δ_i* = transformed ith measurement bias
ϕ_{i1} = the first order of autoregressive parameter at ith stream
μ_i = unknown ith true mean
μ_i* = transformed unknown ith true mean
$v(B)$ = prewhitening transfer function as a function of v_{i1}, v_{i1}, ...
θ_{i1} = the first order of moving average parameter for stream i
$\hat{\rho}_k$ = sample autocorrelation function
σ_i = standard deviation of normal random variable $u_i(t)$
Σ = measurement test's covariance matrix of $e_i(t)$

References

Box, G. E. P., and G. M. Jenkins (1976). In G. M. Jenkins, and Parzen E. (Ed.), *Time Series Analysis, Forecasting and Control.* Rev. ed. Holden Day, San Francisco.

Kao, C. S., A. C. Tamhane, and R. S. H. Mah (1990). Gross error detection in serially correlated process data. *Ind. Eng. Chem. Res.*, **29**, 1004-1012.

Mah, R. S. H., and A. C. Tamhane (1982). Detection of gross errors in process data. *AIChE Journal*, **28, No. 5**, 828.

Meeker, W. Q. (1996). *Applied Time Series.* Class notes for Iowa State University Statistics 451.

Narasimhan, S., and R. S. H. Mah (1987). Generalized likelihood ratio methods for gross error identification. *AIChE Journal*, **33, No. 9**, 1514.

Rollins, D. K. (1990). *Unbiased Estimation of Measured Process Variables when Measurement Biases and Process Leaks are Present* (Ph.D. Dissertation). Ohio State University.

Rollins, D. K., and J. F. Davis (1992). Unbiased estimation of gross errors in process measurements. *AIChE Journal*, **38, No. 4**, 563-572.

Rollins, D. K., Y. Cheng, and S. Devanathan (1996). Intelligent selection of hypothesis tests to enhance gross error identification. *Computers Chem. Eng.*, **20, No. 5**, 517-530.

FASTER DEVELOPMENT OF FERMENTATION PROCESSES - EARLY STAGE PROCESS DIAGNOSIS

João A. Lopes and José C. Menezes
Centre for Biological and Chemical Engineering, IST,
Av. Rovisco Pais P-1096 Lisbon Codex, Portugal
phone: (351-1) 841 7347; Fax: (351-1) 841 9062
email: qmenezes@alfa.ist.utl.pt

Abstract

Multivariate statistical methods provide a wide range of tools to perform database mining and classification, process modeling and process diagnosis. Principal component analysis (PCA) and projection to latent structures (PLS) are two such techniques. In this work the use of PCA and PLS is demonstrated in the areas above for an industrial antibiotic production process at three different scales. With PCA it was possible to detect faulty batches and the fault origin, before they were apparent to plant engineers. Batch clustering produced with PCA was consistent with known differences in scale and in inoculum and raw-material qualities. PCA provided insight as to which process variables had the greater influence in antibiotic production and should therefore be included in a process model.
With linear and non-linear PLS it was possible to predict with less than 2% relative error the final product concentration in the fermentation stage based on inoculum development. In future, combining these techniques in a supervisory system will enable corrective actions to be taken in the early stages of a fault or simply abort transfer of sub-optimal inocula on to the fermentation stage.

Keywords

Database mining, PCA, PLS, Industrial fermentations.

Introduction

Industrial fermentation processes are batch or fed-batch processes that generate large amounts of data. During process operation and in *post-mortem* analysis of historical data univariate techniques are widely used. Comparison of several batches at the same time is thus very difficult and hardly reveals process variable interactions.

Deviations from nominal behavior during process operation are often the result of slight changes in many variables at once. Common problems may include inocula quality and sensor or pump failures. Several disturbances can irreversibly influence the microorganism metabolism and lead to low product concentrations and sub-optimal batches - depending on their intensity and duration. The losses associated with sub-optimal operation increase with process scale, thus early detection of deviations or faults is crucial in large scale industrial bioreactors.

Multivariate statistical methods are very effective at extracting hidden information in problems with multiple correlated variables (Jobson, 1991a,b; Wise and Kowalsky, 1993; Martin and Morris, 1996). Many of these methods perform data and variable compression and can be used in database mining, classification, process modeling and process diagnosis. Principal component analysis (PCA) and projection to latent structures (PLS) are two widely used techniques. Here we report on the use of PCA and PLS in the areas above for an industrial antibiotic production process (inoculum and fermentation steps).

While many multivariate methods are known for many decades, their use on process supervision, diagnosis and as modeling tools is not yet fully exploited (Jobson, 1991a,b; Stephanopoulos et al., 1993; Albert et al., 1996; Martin and Morris, 1996).

Principal Components Analysis

PCA performs data compression by means of a projection of the original variables onto a lower dimensional space in which the projected process variables are orthogonal. In this way variable correlation is eliminated and the new variables (principal components, PCs, or scores) do not contain data redundancy. Clearly, all features in a particular data batch can only be fully accounted for by using the same number of principal components as the number of process variables. However, in many processes a significant amount of the global variance (viz., > 50 %) can be captured by only a few PCs. These PCs can then be used instead of the original variables to describe the process. The number of PCs needed to capture the process main features is obtained from a plot of captured variance per PC as a function of the PC number. The slope of this curve is normally steep for the first PCs and then levels off. In this work PCs capturing less than 5% of the total process variance were considered negligible.

PCA can be either linear or non-linear. In non-linear PCA, process variables are projected on axis transformed with polynomials, splines or artificial neural networks (ANN), while linear PCA variables are projected on straight Euclidean axis. Although non-linear PCA would be best suited to non-linear processes, linear PCA can also be used provided that a reduction on process non-linearities can be achieved (e.g., by scaling the data, mean centering and taking unit variance) as described by Nomikos and MacGregor (1994, 1995).

In the case study presented two phases can be identified in the process: the biomass growth phase and the antibiotic production phase. We postulate that by scaling the data, mean centering and taking unit variance, linear PCA can be performed on each phase.

Confidence bounds in score plots can be calculated through Hotelling's T^2 statistics (Jobson, 1991b). On the score plots throughout this work the ellipses drawn represent 90% and 95% confidence limits. Points lying outside those limits are considered to be outliers.

The coeficcients (loadings) of two principal components can be plotted against each other for each original process variable, thus revealing the correlation between process variables. Variables close to each other in a loadings plot represent strongly correlated variables (i.e., with similar coordinates). This procedure shows which variables can be excluded from subsequent analysis.

Because fed-batch fermentation processes are time-varying processes, observations cannot be considered as single rows on the data matrix because a third dimension - time – must be included. The 3-D process matrix can be unfolded into a 2-D matrix as required by PCA yielding the so called multiway principal component analysis - MPCA (Nomikos and MacGregor,1994,1995). Other details on PCA are beyond the scope of this paper.

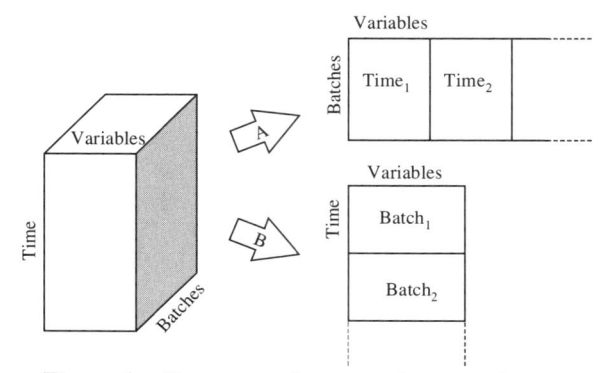

Figure 1. Two ways of process data matrix unfolding [A) suitable for batch classification; B) suitable for process variables correlation].

Projection to Latent Structures (PLS)

PLS are statistical multivariate techniques used to develop regression models between a dependent process variable (e.g., final product concentration) and a set of independent process derived variables (e.g., PCs) (Martin and Morris, 1996). The decomposition of the independent variables can be either linear or non-linear as described earlier for PCA, to yield linear or non-linear PLS models. In this work, in addition to a linear PLS model, three types of non-linear PLS models based on non-linear PCs were also compared (viz., PCs with polynomials, splines and neural networks) [see bellow].

Statistical analysis was performed with the PLS_Toolbox version 1.5 from Eigenvector™ for Matlab™ (Wise and Gallager, 1997).

Case-Study. An Industrial Antibiotic Production Process

In this work the fed-batch fermentation of a β-lactam in an industrial plant is used as case study. Data from three pre-production scales (A, B and C m^3; 1:2:10) was analyzed: A (17 batches), B (63 batches) and C (22 batches).

Monitored variables during fermentation include pH, biomass concentration (PMV), dissolved oxygen, substrate feed rates (Q_i) and substrate and antibiotic concentrations (C_i). Thirteen variables were available for analysis.

Fermentation runs differ in length and were all truncated at the shortest batch duration (viz., the longest batch suffered a 10% reduction in length relative to the shortest one). Because variables were sampled with different frequency, all data batches were interpolated with splines and then resampled at a frequency of 1 hour.

Each fermentation is preceded by inoculum production also in batch tanks. After suitable biomass growth (inoculum stage) the seed is transferred to the main bioreactor (fermentation stage). At the time the database was built monitoring of the inoculum stage was limited to two variables: pH and PMV.

Results and Discussion

Batch Classification

PCA of 22 fermentations at scale C is shown in Fig. 2. The first two PCs accounted for 35 % of the total process variance while the first five PCs captured 57.0 % of the total variance in this 13 variable process. Each of the PCs after the fifth one captured less than 5 % of process variance.

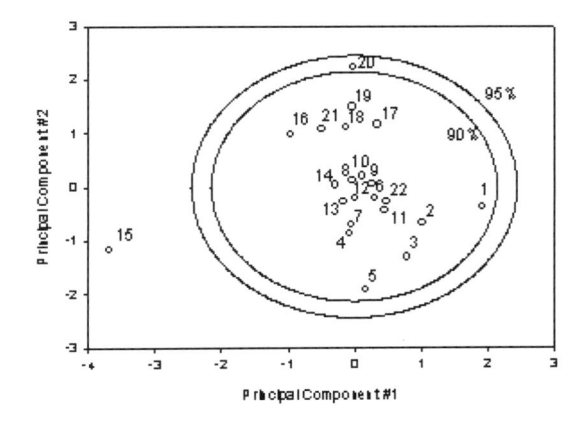

Figure 2. Batch classification through PCA for 22 fermentations at scale C.

Figure 2 shows the presence of at least an outlier (batch 15) and two fermentation clusters. The fault causing the outlier will be identified later on in the sequel.

Batch classification can also help in process modeling. During model development training and validation data sets must be chosen. It is common practice to make this selection randomly without any consideration of the space covered by the training set, thus compromising the model predictive capacity. This becomes even more relevant when developing non-mechanistic models such as statistical or neural networks models due to their limited extrapolation capacity outside the training region.

The same analysis performed on all three scales put together - correcting for different tank geometry and scale dependent feed-rates - reveals a clusterization of the data related to scale (Fig. 3). Probably, important scale dependent variables were left out of the analysis (viz., physical variables such as mixing time or power input, that have a greater impact on the microorganism metabolism and morphology at the higher scale).

Variable Colinearity

The loadings plot for the 22 fermentations at scale C is shown in Fig. 4. The antibiotic concentration is highly correlated with batch age and very little if any at all with operating variables. This is consistent with the known low controllability of bioprocesses as the microorganism itself is in control of most of culture activities at nominal conditions. It also suggests that inoculum development - a batch process with no external control except temperature, aeration and agitation - probably contains enough information about the microorganism state to be correlated with the fermentation stage final outcome.

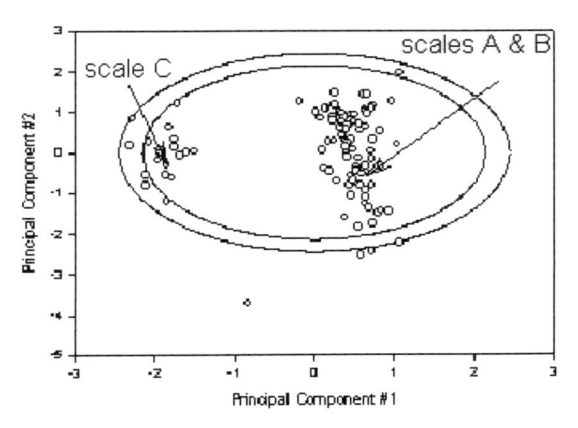

Figure 3. PCA of 102 batches at 3 different scales. Dividing feed-rates by broth height in each of the tanks is not enough to correct for different scales and tank geometry.

Early Fault Detection

The scores plot on process historical data (Fig. 2) showed that batch 15 was an outlier. The same concept can be applied to on-line detect faults in process operation at an early stage.

A PCA was carried out along fermentation time using a moving data window of 3% of the process duration for each of the 22 fermentations at scale C (batch 15 included). From score plots of the first 3 principal components it is clear that batch 15 starts deviating from the nominal region at 10% of final time and that at 20% it is already an outlier with 95% confidence (Fig. 5).

The next step was to find out the fault origin. This was performed by inspecting the contributions of each process variable to the first PCs (Fig. 6). For each PC, contributions of individual variables were scaled to sum 100 %. Plotting the contributions for all batches the fault cause was revealed. It is evident that variable 8 (carbon substrate concentration) has an abnormal contribution to PC-1 of batch 15 relative to the other batches. The same conclusion can be obtained by inspecting the time profile of variable 8 in batch 15 (Fig. 7). It was found that a substrate pump failure was responsible for an abnormal increase in the substrate concentration. It took process engineers many hours to detected this occurrence and many more to return the process to nominal operation (Fig. 7) with an irreversible productivity loss (Fig. 2).

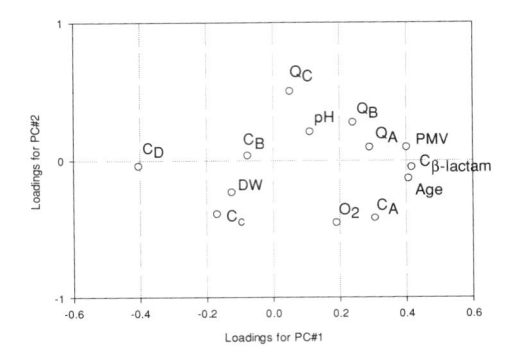

Figure 4. Correlation of process variables through PCA in 22 fermentations at scale C.

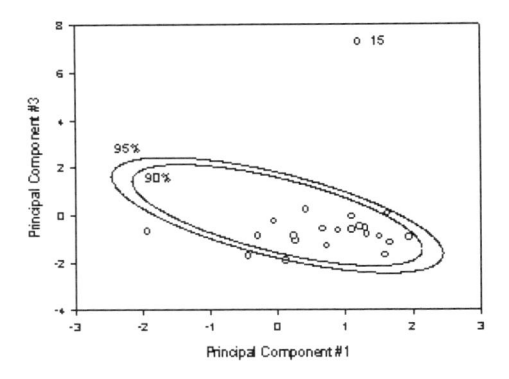

Figure 5. Scores plot for a moving window of process data showing at 10% of fermentation time that a fault is occurring on batch 15 (22 fermentations at scale C).

Figure 6. Original variable contributions for PC-1 of batch 15 and a comparison with the average contributions for all scale C batches.

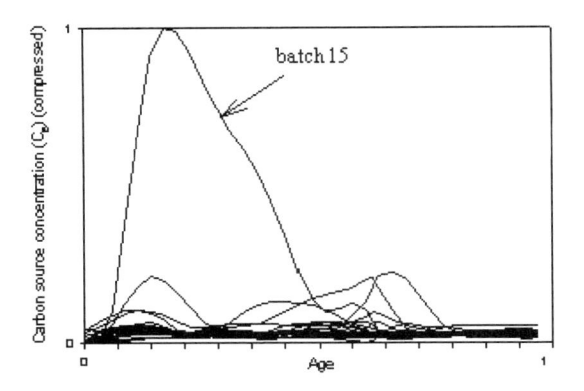

Figure 7. Substrate concentration in 22 fermentations at scale C. The concentration profile for batch 15 is very different from other batches.

Integration on a Real-Time Knowledge Base System

PCA was integrated on a real-time knowledge base system from Gensym (G2®) by means of remote procedure calls to an external software (Matlab) where PCA functions actually run. The G2 ability to monitor and supervise processes was coupled with these statistical tools. G2 is a powerful object oriented platform for development of intelligent modeling and supervision systems. A single G2 knowledge base (KB) running on a PC machine can incorporate real-time process monitoring via TCP/IP. In our G2 application in addition to principal component analysis several process models can be examined (Lopes and Menezes,1998): mechanistic ODE based models, ANN and hybrid models. Current work is being directed to develop a G2 application integrating both multivariate statistical analysis (e.g., PCA and PLS) and the above modeling interface, with on-line process supervision capability, for consistent process control and product quality.

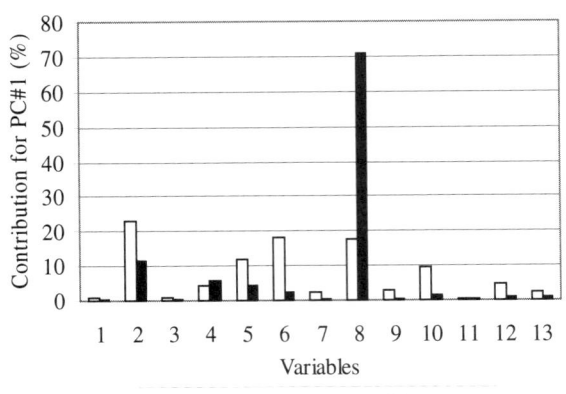

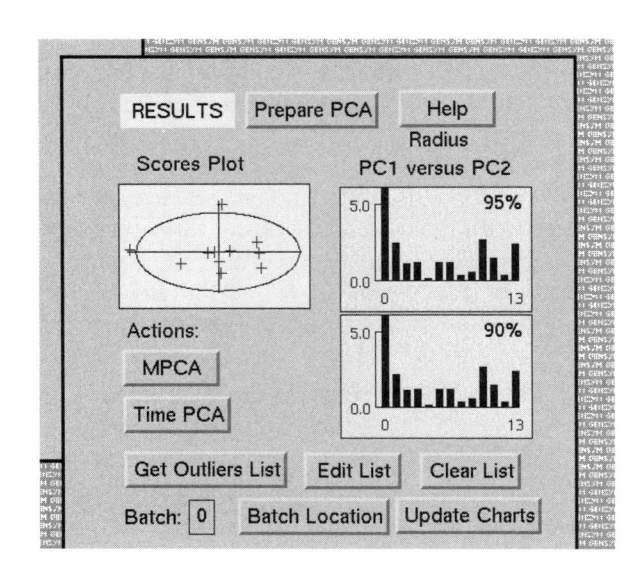

Figure 8. Screenshot of the integrated G2 knowledge base with PCA.

Prediction of Product Concentration from Inoculum Data

As show earlier in loadings plots, the fermentation course is mainly controlled by the microorganism as product formation is highly correlated with fermentation age (Fig. 4). This suggests that product formation can probably be associated with inoculum quality and that a PLS model predicting final product concentration can be established based on inocula development.

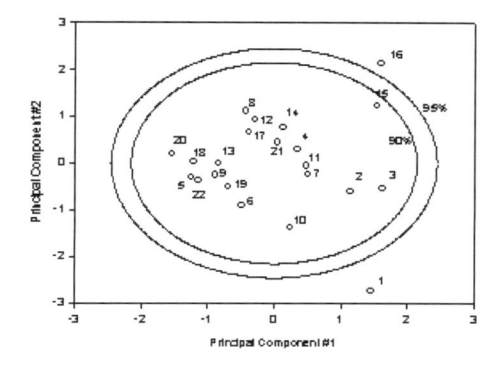

*Figure 9. Inocula classification by PCA
(22 batches scale C)
Batches 1 and 16 are outliers.*

As previous a principal component analysis helps in choosing training and validation inocula sets. In Fig. 9 a plot of the first two principal components shows at least two clusters and two outliers (batches 1 and 16). The first PC accounts for 45 % of the process variance while the second PC for 21 %. Thus, 66 % of the information content of the 22 inocula batches is contained on a single score plot.

Batches 4 and 12 from one of the clusters and batches 9 and 19 from the other cluster, together with the two outliers were chosen as validation set. The remaining 16 batches were used to train PLS models.

Linear and non-linear PLS models were used. The dependent variable in all models was the final β-lactam concentration. The non-linear PLS models included projecting independent variables on axis transformed with polynomials, splines and a simple neural network. Model topologies were optimized by cross validation (training followed by testing to achieve the best fit of training sets without over-fitting their noise) with Matlab PLS toolbox.

Figure 10 shows the predictions of all PLS models for the 6 validation batches. All models gave reasonably accurate predictions (ca. 2 % relative error) for nominal batches (Fig. 10a) although in general non-linear PLS models performed better than the linear model.

It is noteworthy that the best results were obtained with non-linear models with only one principal component. This is consistent with the significant proportion of the total variance captured by the first PC (viz., 45%). Inclusion of more principal components on these models would lead to an improved fitting of the training set but would affect the predictive capacity of the resulting model (i.e., over-fitting the training set).

The prediction errors for outliers (Fig. 10b) were one order of magnitude greater than the errors for batches obtained in nominal operating conditions (Fig. 10a), which is consistent with these batches being outliers (Fig. 9). The marked difference in two the sets of predictions in Fig. 10b is in agreement with the PCA batch classification in Fig. 2 – batch 16 is farther away from the 90% limit than batch 1.

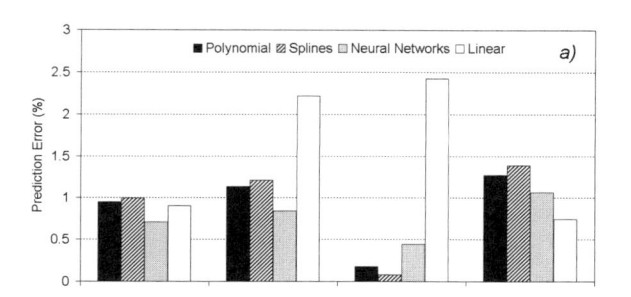

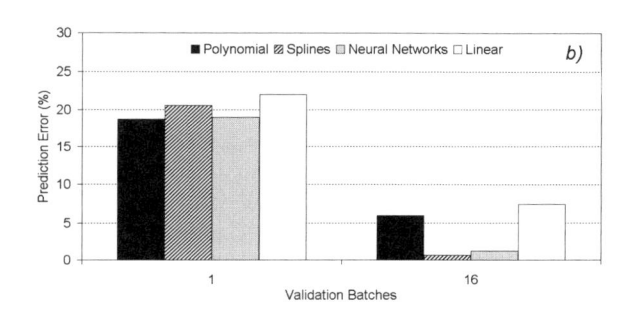

*Figure 10. Comparison of 4 PLS models in 6 validation batche (scale C).
a) 4 nominal batches; b) 2 outlier batches.*

Table 1. Comparison of Four PLS models for Prediction of Final Antibiotic Concentration from Inoculum Data.

PLS results	Linear	Polynomial	Spline	ANN
mean error (%)	1.6	0.9	0.9	0.8
RMS	64	36	39	29
n° of PCs*	1	1	1	1
Order*	--	4	2	--
Knots*	--	--	2	--

*Model topology was optimized by cross validation.

Table 1 summarizes the main quality indices of PLS models investigated for the 4 nominal validation sets (fermentation 4, 9, 12, 19 at scale C). The non-linear models are roughly equivalent to each other and slightly superior to the linear PLS model.

Conclusions

In this work we demonstrated the usefulness of multivariate statistical methods in industrial fermentation process diagnosis and modeling.

Multiway principal component analysis was used in batch classification and as a technique for choosing training and validation data sets for model development. The efficiency of principal component analysis was demonstrated on the detection of process faults even before those faults were apparent to experienced process engineers. The benefits from the application of these methods in on-line process supervision were also demonstrated under a simulation environment to detect a process fault using real industrial data.

Inoculum development data was correlated with the product concentration at the end of the industrial fermentation stage (several days after inoculum production) with linear and non-linear PLS models. The maximum prediction error obtained with all models was 2 % for nominal batches - i.e., bellow the accuracy of the experimental protocol for the antibiotic assay.

A scale dependent separation of various batches was identified which could not be eliminated correcting for tank geometry, scale dependent feed rates or broth volume.

Future work is being directed to exploring the database further (e.g., the reason for scale separation) and to development of a prototype application integrating multivariate statistical techniques with modeling tools to be used in on-line process supervision and diagnosis.

Acknowledgements

The authors gratefully acknowledge Prof. Julian Morris and Dr. Elaine Martin for a staying at the Centre of Process Analysis Chemometrics and Control (CPACC) of the Newcastle Upon-Tyne University (UK) and for several suggestions.

The authors would like to thank Companhia Industrial Produtora de Antibióticos SA (CIPAN) in Portugal for providing the data.

JPL acknowledges a grant from PRAXIS XXI L009-P31B-09/96.

References

Albert, S., E. Martin, G. Montague and A. Morris (1996). Multivariate statistical process control in batch process monitoring. *13th IFAC World Congress*, San Francisco (USA), N, 389-394.

Jobson, J. (1991a). *Applied multivariate data analysis, volume I - Regression and experimental design*, Ed. Springer-Verlag, New York.

Jobson, J. (1991b). *Applied multivariate data analysis, volume II -categorical and multivariate methods*, Ed. Springer-Verlag, New York

Lopes, J.P. and J.C. Menezes (1998), Supervision and diagnosis of pharmaceutical processes with NeurOn-Line[TM]. *Gensym Users Conference*, Rhode Island (USA), May, 1998.

Martin, E. and A. Morris (1996). An overview of multivariate statistical process control in continous and batch process perfomance monitoring. *Trans Inst MC*, **18**, 51-60.

Nomikos, P. and J. MacGregor (1994). Monitoring batch processes using multiway principal component analysis. *AIChE J.*, **40**, 1361-1375.

Nomikos, P. and J. MacGregor (1995). Multivariate S.P.C. charts for monitoring batch processes. *Technometrics*, **37**, 41-59

Stephanopoulos, G., K. Konstantinov, U. Saner, T. Yoshida, (1993), Fermentation Data Analysis for Diagnosis and Control, in *Bioprocessing.*, vol. 3, pg. 355-400, 2nd ed., Ed. G. Stephanopoulos in "Biotechnology", Eds. H.J. Rehm, G. Reed, VCH, Germany.

Wise, B. and B. Kowalsky (1993). Process chemometrics: the case for chemometrics in chemical engineering. *Computing and Systems Technology Division Communications*, **16**, 5-11.

Wise, B. and N. Gallagher (1997). PLS_Toolbox for use with Matlab[TM]. *Eigenvector Research Inc.*, Manson

OPERATIONAL DESIGN AND ITS BENEFITS IN REAL-TIME USE

Yuji Naka, Rafael Batres and Tetsuo Fuchino
Tokyo Institute of Technology
4259 Nagatsuta, Midori-ku, Yokohama 226, Japan
Fax: +81(45)924-5271
E-mail: {ynaka, rafael, fuchino}@pse.res.titech.ac.jp

Abstract

In the operational design methodology, performance and safety are evaluated at different stages of the life-cycle of a plant, such as normal operation, startup, normal shutdown, and emergency shutdown, where the plant structure is modified accordingly in order to improve the future plant operability. This paper presents an approach to the integration between operational aspects of the design and a manufacturing execution system. The proposed approach aims at improving management of change and providing more understandable operations to the operator. Consequently, during plant modifications, operating procedures would be easier to maintain and better integration between operations planning and maintenance would be obtained. The main goal of this research is to integrate the unique technological advances from several closely related projects in the context of a real time situation adaptive integrated production execution system application (SAIPES).

Keywords

Real-time, Procedural control, Operational design, Situation-adaptive integrated production, Manufacturing execution systems, Controlled group unit, CGU, Integration of design and operations.

Introduction

Much effort has been made in the integration of process design and control, as well as integration between operating procedures generation and topology improvement (Aelion and Powers, 1991). Although these efforts have attempted to solve parts of the whole issue separately, the sharing of data to create a common understanding between designers and operators is still not possible. Consequently engineers are faced with construction shortcomings, equipment deficiencies, startup delays, and, more dramatically, losses and incidents during transient operations.

In the operational design methodology (Naka and McGreavy, 1994; Naka et al., 1997), the production and safety requirements of the plant are evaluated at different stages of the life-cycle of the plant, such as normal operation, startup, normal shutdown, emergency shutdown, where operations take the material being processed from one state to another within the plant while the structure is modified accordingly in order to gain better operability. The activities involved in operational design start from a process flow diagram (PFD), and modify the design by updating the flowsheet topology towards a preliminary piping and instrumentation diagram (pre-P&ID). Changes in the structure of the plant are implemented by proposing design alternatives, which are then evaluated against several operational criteria, such as process constraints or safety requirements.

Many kinds of transient operations are common in process plants. The complexity of the operations, and the risk involved during its execution have motivated the development of an adequate integration between a sound design and operations, and the control system that would carry out the transient process.

The underlying thesis in this paper is based on the assumption that the knowledge utilized for the execution of actions during the design can be exported to the plant on stream, and the on-line data used as a vivid representation of the changes on the behaviors that were simulated during the operational design activities.

Motivations

Following the data reported by the High Pressure Gas Safety Institute of Japan, statistics compiled between 1952 and 1990 show that 46% of the accidents that occurred in petrochemical plants took place during transient operations such as startup and shutdown. Furthermore, it has been recognized that improving the control of startups, shutdowns and grade changes presents a potential of world benefits of about 100 billion US$ per year (Benson and Perkins, 1997).

Changes in production requirements or plant redesign require updates to the procedural knowledge that is stored in a DCS. However, it is not unusual to find that problems occur due to inconsistencies in the changes of the procedural knowledge.

On the other hand, in most cases, operations are performed by skilled operators that use their experience to get knowledge about the process that was ambiguous or not specified in the original operating procedures. Determining the tracking and ramping trajectories or even the sequence of operations of some valves are some examples of the use of such knowledge, which may not be incorporated into the procedures.

Operational Design

Operational design is a systematic approach to the detailed process design. In this approach the plant structure is adjusted to improve the future plant operability in terms of safety, reliability, and availability. It starts with a process flow diagram (PFD) that is modified towards a piping and instrumentation diagram (P&ID). This is done by proposing design alternatives, which are evaluated against several operational criteria. The evaluation is carried out by analyzing the operational requirements of the plant at different stages of the plant, such as normal operation, startup, normal shutdown, and emergency shutdown, where the plant design is modified accordingly. As a result, a more *operable* design, and more consistent knowledge for operations are generated. Operational design differs from the traditional approach with the evaluation of the design and operations for startup and shutdown, which is carried out concurrently during the detailed process design. In this way, it becomes possible to identify potential problems while they are still easy to correct. Some of the activities involved in operational design are shown in Table 1. Note that the modeling of the dynamic behavior of the controller is first carried out with perfect control in the definite detailed control design stage and then more rigorously by introducing the parameters of the controller in the final detailed process design stage.

Knowledge for operations

The knowledge for operations that results from the operational design activities is composed of a number of conditions, which are used to produce the exact order of operations. In this sense this knowledge is not defined as a rigid sequence of actions, but as strategies that determine each successive action as a function of information about the current state of the plant structure and process variables. This knowledge constitutes a conditional plan where the actions are conditioned on observations of the *environment* prior to their execution (Peot et al., 1992). These conditional plans are partial plans formed with process specific conditions under which an action is initiated, stopped, paused, or aborted. Human designers provide these process specific conditions from the physical and chemical constraints of the process. If during a simulation of transient operations such as startup or shutdown, the final conditions or goals are not attained, the engineer may modify the entered specific-conditions to guarantee activities occur as expected. Actions contain information about the purpose of the action, the preconditions necessary to perform the action, the effects on the controlled device during the execution of the action (postconditions), the way an action relates to other actions of the same kind (category) and the priority of execution (Batres, et al. 1998).

An activity transforms or transports an entity such as process material, energy or information over a time interval. An action that directly initiates an activity is a *low-level action*. Examples of low-level actions include opening or closing a valve, regulating the position of a control valve, turning a switch on, etc. An action that initiates another action is a *high-level action*. Examples of this kind of action include 'start the commissioning of controllers,' 'start setpoint tracking,' 'ramping,' 'conditional controller execution,' etc. It also could be possible for a high-level action to initiate not one but a series of actions. For example, if a control action modifies the state of a multi-stage compressor from off to on, then a series of other actions will be initiated to modify the state of each stage from off to on.

Control actions might be defined as proposed, actual or assumed. These might be also intended and unintended actions. Intended actions are used for different purposes such as start up and shutdown. Unintended actions might be the result of mistaken decisions such as the result of a precipitated human intervention. An action then changes the state of actionable devices (e.g. valves, pumps) producing, consequently, changes in the behavior that takes place inside the equipment towards a final goal such as driving the plant to a non-operating state as in the case for shutdown. Therefore the appropriate action at the appropriate moment with the suitable supporting facilities produces the desired behavior.

Table 1. Activities Involved in Operational Design.

DESIGN STAGE	OPERATIONS MODE			
	NORMAL OPERATIONS		Corrective Operations for Abnormal Situations	EMERGENCY OPERATIONS
	Steady State	Startup/ Shutdown		
Preliminary detailed process design	• Evaluate performance and controllability • Produce preliminary control logic	• Determine operations policy • Produce preliminary event-driven and sequential control logic • Add any necessary auxiliary equipment	• Determine safety policy management • Produce preliminary event-driven and sequential control logic • Add valves to limit the propagation area	• Determine operations policy • Produce preliminary event-driven and sequential control logic • Add any necessary auxiliary equipment
Definitive detailed Process design	• Evaluate flexibility and operability • Produce final control logic	• Simulate tracking and ramping with perfect control • Produce final sequential control logic	• Carry out process hazard analysis • Control configuration	• Produce final sequential control logic • Produce preliminary safety interlock design
Final detailed process design	• Produce local control tuning	• Produce local control tuning • Simulate tracking and ramping with rigorous control	• Obtain alarm and sensor allocation • Design diagnosis system	• Produce detailed safety interlock design
Control Requirements	Plant-wide regulatory control	Sequential control system for managing different initial/final conditions	Control system for abnormal situations	Sequential control system (interlocks, alarms, emergency shutdown requirements)

Hierarchical Plant Wide Operations

In order to improve the flexibility in the operations and to supply local actions with global knowledge, the plant is divided into smaller units or CGUs (Controlled Group Units) that can be operated relatively independently from the rest of the plant.

A CGU can be identified from the flowsheet topology as an assembly of pieces of equipment surrounded by control valves. Operations are then carried out at multiple levels: plant-wide level, CGU level, equipment level and elementary level (actuator level).

For the execution of the conditional plan, a plant-wide inventory control strategy based on the CGU concept provides a mechanism that is independent for any kind of process. The inventory control is used to coordinate throughput manipulations inside each CGU. Liquid inventory is among the most important controlled process variables, although component inventory plays a key role in overall plant-wide-control (Downs and Siirola, 1997).

Manipulations of flow by means of operating valves and pumps on the boundaries of the CGU and/or on bypass lines inside the CGU allow the generation of stationary states (Fusillo and Powers, 1987) that facilitates bringing the plant to the final state.

At the CGU level the inventory control mechanism operates the surrounding control valves according to the following simplified algorithm:

1. Verify whether the *specific conditions* (downstream restrictions of flow) attached to each control valve are satisfied
2. If conditions are satisfied then grant permission for operating the control valve. Note that this does not necessarily mean to open the valve.
3. If the CGU allows flow to run out of the CGU and permission is granted for operating the control valve then operate the valve (upstream availability of flow).

Specific conditions provide information to coordinate operations among CGUs, while the status of the hold-up of the CGU determines the availability of output. When both the specific conditions and hold-up conditions are satisfied, the CGU fills all the requirements that will put it on-stream. These conditions represent the core of the conditional plan.

If CGU_j is not ready for receiving flow from CGU_i located upstream and next to CGU_j then a *local stationary state* is created inside CGU_i by closing the valves at the supply of CGU_i and by keeping the valves at the outlet closed. If the topology includes recycle lines inside CGU_i these can be used whenever possible to create a kind of stationary state in which the fluid is kept flowing inside CGU_i until the associated conditions at the outlet valves are satisfied. Sometimes, however, it is convenient to create *global stationary-states* to encompass a group of CGUs or even the whole plant. This is done through plant-wide inventory control at the upper level of the hierarchy. This provides more flexibility and efficiency as well as means of isolating a part of the plant (or the whole plant) from the rest of the plant-systems.

As knowledge of operations follows the hierarchical levels, satisfaction of the specific conditions associated to the limiting valves of the CGU grant permission for the action to be executed but it does not necessarily imply the execution of the action. Consequently, the ordering of some plan elements is deferred until more relevant information for making a wise decision is acquired.

A Situation Adaptive Integrated Production Execution System (SAIPES)

An integrated manufacturing execution system (I-MES) bridges the gap between design, high-level management and control systems (MESA, 1997). A SAIPES system extends the I-MES functionality by making control actions to be consistent to changes in the operational objectives through a goal-directed behavior along the whole life cycle of the plant. In other words, a SAIPES system ensures that more specific actions are executed in accordance with its global, longer-term goals for production, maintenance, startup, shutdown and safety. A SAIPES implements the same CGU-based hierarchical levels for operations of the operational design, so as to provide knowledge that guides the initial execution and any required change in the plant-wide goals to be attained (Fig. 1). Therefore a SAIPES can obtain a conditional plan directly from the operational design. Consequently, the procedural knowledge that results from the combination of the knowledge for inventory control and the conditional plan determines not only if an action may execute but when, how and why.

The procedural knowledge that is used by a SAIPES should be able to explain operations (how does a facility interact with other parts of the plant?) and the function of an action (what is the purpose of opening this valve?). A SAIPES then can support management of change: 'what consequences will cause a given change in the plant?', 'are there adequate records of those changes?', 'are those decisions adequately explained?', 'is this decision consistent with the others?', 'how did we do last time?'.

By adequately representing the knowledge and information associated to the plant structure and control we may be able to answer questions such as: 'what caused this valve to open?', 'what caused a change in the operational goal?', 'what caused this change in behavior?', and 'what influences this process variable?'

Explanations of this kind can help human operators analyze a given situation and determine the appropriate operations. Instead of writing a static document containing the operating procedures, the engineer can utilize self-explanatory descriptions to document the design.

When the user asks about why this valve was opened then the associate specific and generic conditions should appear, with the values of the involved process variables. Then it should be possible for the user to get a more complete explanation, by means of an appropriate causal mapping of the valve actions between the hierarchical levels and process variables as well as among variables found in equation models used for the dynamic simulation of the transient process during the operational design. The question of 'which are the conditions that triggered at the execution of the action?' will provide values of the associated process variables contained in specific conditions and/or the state of the inventory control that allowed the valve execution. Analogously, the question of 'how were those process attributes updated?' will tell us about the causal dependencies associated with the simulation models that updated the relevant attributes of the behaviors.

G2 based testbed for modeling conditional plans

A prototype for an engineering design environment, which has been under development, is currently being utilized as an experimental testbed for the operational design theory and methodology research (Batres et al., 1998). The environment is being implemented in the G2 intelligent system shell[7] (Gensym, 1997), which provides GUI development tools, an inference engine, and the necessary elements to build a simulation environment. A multidimensional framework is used to ensure consistency between plant design and operations (Lu et al., 1997). The design environment is being conceived to support the increasing fidelity of the models that dictate the behavior associated to each piece of equipment, combined behaviors, and inheritance of actions. The system is mainly composed of two modules: a control module and a simulator engine.

The control module generates actions that are applied over any actionable structure objects such as valves, pumps or compressors. This module is composed of rules applicable to assemblies of equipment objects, procedures and equations for local controllers and the built-in G2 inference engine. The control module uses knowledge bases that contain plant specific rules for startup and shutdown, which are supplied by the user. There are two sets of rules, one for start up and another for shutdown. Within both sets of rules, there are generic rules employed in any process and process specific rules that are generated automatically from informaion supplied by the user. The specific rules encode the conditions associated to some CGU control valves.

The simulation module contains models for each kind of behavior object and procedures that coordinate the input to the models and the propagation of the properties of the material and energy flows. This module is composed of a simulator, a module for the calculation of physical properties, and a numerical solver. A simulator runs as daemon that can be activated or deactivated as needed. Simulation models are composed of differential equations that simulate the dynamics of the plant during

---

[7] G2 is a registered trademark of Gensym Corporation

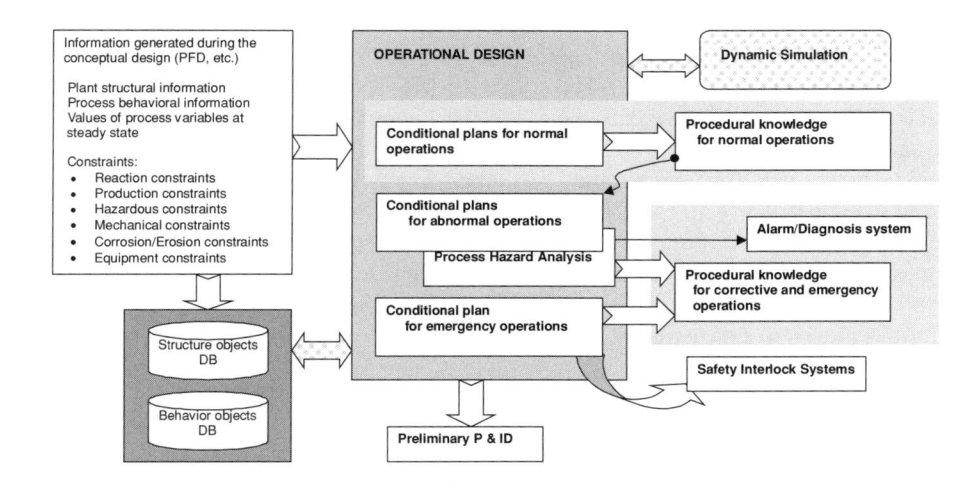

Figure 1. Flow of information and knowledge from operational design to real time operations.

every transition that occurs in the plant (unsteady and steady states). A model manager is composed of procedures used for the propagation of the material and energy flows, and their properties. Any change in the properties of the material flow is propagated through the network of structure objects. The solver used is built into G2.

Industrial case studies

The proposed method has been applied to several practical case studies: the startup of a chemical plant, startup and shutdown of a hydrodesulfurization process and a distillation pilot plant.

Chemical Plant at Yokkaichi

In 1993, applying the PROSEG (Process & Operations Sequence Graph) modeling technique (Naka, 1989) to a chemical plant composed of 4 azeotropic distillation and conventional distillation columns, 3 reactors and 10 heat exchangers, the operating procedures for cold startup were obtained (Yoneda et al, 1997). Then in 1997, based on some concepts derived from operational design, a supervisory system for startup was developed and connected to a DCS. Using the real time process data, this system recognizes the current status of the startup process and performs any relevant decision to send the next action instruction to the DCS. As a result, operations with different initial feed conditions are supported. Additionally, this system can visualize the progress status of the start-up operations and support. Consequently, the system demonstrated a much better maintainability in comparison with the traditional changes in the sequential function charts of the existing DCS.

Startup and Shutdown of a Hydrodesulfurization Plant

The G2 testbed has been successfully tested on an HDS industrial process case study in which initial startup, startup after turnaround, partial, and total shutdown were simulated (Batres et al., 1997a). The following operations can be simulated and observed visually once the user has entered the process conditions, and attributes of equipment and behaviors (Fig. 2): at time t_1 the feed valve BV-01 is opened just after the user has selected to startup the plant from the menu bar. Then the hydrogen feed valve is opened to start the flow of make-up gas through the reactor circuit. While this is taking place, in order, the valves BV-02 on the startup/shutdown bypass line are opened to commence flow towards the high-pressure

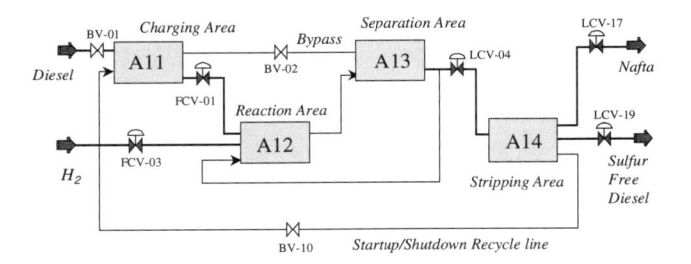

Figure 2. Area view of the HDS process.

separator (separation area). When the diesel reaches normal operating level, the outlet valve of the high-pressure separator is opened by the inventory control mechanism to allow the diesel flow to the low-pressure separator in A13. Analogously, when the oil reaches normal operating level the outlet valve of the low-pressure separator is opened so that the oil flows downstream to the bottoms of the stripper (A14). Then at time t_2 the inventory control around the stripper allows a recycle valve BV-10 to be opened and diesel is driven back to the feed surge drum through the startup/shutdown recycle line. This creates a plant-wide stationary-state that waits for the condition of FCV-01 (the valve between the feed area and the reaction area) to be satisfied. At time t_3 the temperature at the outlet of the reactor makes that the condition of FCV-01 be satisfied and the valve is opened. At time t_4, BV-02 is closed as well as BV-10, terminating the cold-oil stationary state. At this point valve BV-01 is opened to allow diesel oil flow to the surge drum F-01. As the reaction proceeds nafta is obtained at the top of the stripper. Once the inventory control of the stripper has been completed the nafta is driven downstream. See Batres et al. (1997b) for more details.

Distillation pilot plant

A system for executing conditional plans for the startup of a pilot plant has been developed in C++ and implemented in a PC connected to a DCS. The system was intended to support different initial conditions (Pradubsripetch, 1996) resulting in more flexible startup operations. It also implemented a modular approach that allows the user to link building blocks of conditional plans of individual units to produce a combined behavior for the whole process. This modularity was tested when a heat pump was incorporated into the distillation plant. When the operating procedures for the heat pump were added to the operating procedures for the distillation system, the combined unit started up very smoothly.

Future directions

Much work is needed to improve the hierarchical control based on the CGU concept. We believe that the way to go is towards development of an intelligent control system capable of reasoning through a system with an adaptable goal-directed behavior, which implies a combination of reactive and planning strategies. A number of related projects are planned to form the basis of that system. One such project is concerned with the determination of the optimum and safe sequence of operations for mixing control cases that are constrained to avoid certain hazardous states.

Another promising research topic is the development of ontologies for sharing and exchanging knowledge and information between design and operations. Establishing ontologies for plan representation will provide a terminology and structure that facilitates exchange efforts for cooperation and coordination.

We will also require design and operations rationale tools that allow analyzing and streamlining the design and operations decisions as well as improve facilities for the deployment of a distributed and shared environment.

Conclusions

This paper presented the main concepts behind the operational design and its benefits in real time. A formal approach based on hierarchical plant operations, throughput inventory and conditional plan execution was presented. The ideas in this paper have been successfully tested on industrial process case studies from which we have gained important feedback to the projects that aim at supporting real-time operations. Through the implementation of the mechanism for modeling procedural control in a situation adaptive integrated production execution system (SAIPES), we believe it might be possible to gain better consistency between plant design and operations. Consequently a SAIPES would be able to support management of change and provide more understandable operations to the operator.

Acknowledgements

This work was funded by Japan Society for the Promotion of Science, Japan, and by the Ministry of Education of Japan.

References

Aelion, V., G. J. Powers (1991). A Unified Strategy for the Retrofit Synthesis of Flowsheet Structures for Attaining or Improving Operating Procedures. *Comp. Chem. Engng.*, **15**, No. 5, pp 349-360.

Batres, R., Ming Liang Lu and Yuji Naka (1997a). An Agent-Based Environment for Operational Design. *Computers Chem. Engng.*, **21**, Suppl., pp. S71-S76

Batres, R. (1997b). An Investigation of Concurrent Plant Design and Operations. Tokyo Institute of Technology, *Technical Report*, http://www.pse.res.titech.ac.jp/publications/index.html

Batres, R., M. L. Lu, Y. Naka (1998). A Multidimensional Design Framework and its Implementation in an Engineering Design Environment. *Proceedings of the 5th ISPE International Conference on Concurrent Engineering*, Tokyo, Japan, July 15-17, 1998.

Benson, R., J. Perkins (1997). The Future of Process Control—A UK Perspective. In: J. C. Kantor, C. E. Garcia and B. Carnahan, (Eds.), *AIChE Symposium Series 316*, Vol. 93, AIChE, New York.

Downs, J. J., J. J. Siirola (1997). Challenges of Integrated Process and Control Systems Design for Operability. *1997 AIChE Annual Meeting*.

Fussillo, R. H. and G. J. Powers (1987). Synthesis for Chemical Plant Operating Procedures. *Comp. Chem. Engng.*, **11**, No. 4, pp 369-382.

Gensym Co. (1997). *G2 Version 5.0 Reference Manual*.

Lu, M. L., R. Batres, H. S. Li, Y. Naka (1997). A G2 Based MDOOM Testbed for Concurrent Process Engineering. *Computers Chem. Engng.*, **21**, Suppl., pp. S11-S16.

Naka, Y. (1989). PROSEG for Operational Design. Process Systems Engineering Laboratory (Naka-ken). *Tokyo Institute of Technology. Technical Report* (in Japanese).

Naka, Y. and C. McGreavy (1994). Modular Approach for Start-up Operational Procedures of Chemical Plants. In: E.

S. Yoon (Ed.), *proceedings of the PSE '94*, Kyongju, Korea.

Naka, Y., M.L. Lu and H. Takiyama (1997). Operational Design for Start-up of Chemical Processes. *Computers Chem. Engng.* **21**, No. 9, pp. 997-1007.

MESA (1997). Execution-Driven Manufacturing Management for Competitive Advantage. A white paper from the Manufacture Execution System Association. Internet. WWW: http://www.mesa.org/

Peot, M. A., D. E. Smith (1992). Conditional Nonlinear Planning. *Proceedings of the First International Conference on Artificial Intelligence Planning Systems*, College Park, Maryland, USA, 1992.

Predusbsripetch, D. S. Lee, A. Adriani and Y. Naka (1996). Real Time Generation of Operating Procedures to Support Flexible Startup Operation, *Computers Chem. Engng.*, **20**, pp. S1203.

Yoneda, M., Y. Sawai, H. Komatsu and Y. Naka (1998). Development and Validation for a Startup Supporting System Based-on PROSEG (in Japanese), *Kagaku Kogaku Ronbunshu*, **24**, No. 1, pp. 12-17

USE OF DYNAMIC EVENT-DRIVEN SIMULATION FOR THE CHEMICAL INDUSTRY

Heinz Ensen
BASF Corporation
Wyandotte, MI 48192

Joerg Krames, Dirk Wollenweber and Christine Edinger
BASF AG
D-67056 Ludwigshafen, Germany

Abstract

Economically successful operation of multi-product plants or complex and interacting processing chains, depends on the optimal interplay of chemical, technical and logistical parameters. A multitude of tools exist for use in determining the thermodynamics, stream compositions, and equipment sizing/design of new facilities. These tools base their calculations on a steady state condition in the production unit. Additionally tools exist to optimize the operating conditions of existing plants by use of statistically designed experimentation. These tools do not take into account the interactions of materials, equipment, personnel, utilities, and other resources in the operation of a production unit. Computer-aided simulation enables the analysis and evaluation of this material flow dynamics, important for so many processes.

Dynamic simulation of material flow is currently a widely used computer-based tool in part production. In contrast to the simulators more commonly found in the process industries, the dynamic simulation of material flows does not predict the product state and properties of the individual streams in processing units. Its aim is to analyze the dynamics of discrete material flow during waiting and operating time when several production elements or streams are combined.

This article describes the discrete event simulation and compares it to the conventional thermodynamic simulation tools used by the chemical industries. The possible use of this kind of simulation is explained in three examples of the industrial praxis.

Keywords

Debottlenecking, Discrete event simulation, Material flow analysis, Plant design, Simulation.

Introduction

When processes within a plant are optimized, operational logistics are particularly important. Examples of these include optimal utilization of equipment in complex processes, downtime due to maintenance or even high inventory levels in storage tanks. This applies not only to existing processes which are being revamped but also to novel processes under development. The goal of activities in this field is to combine process engineering and logistical methods to identify bottlenecks and weak points in the production sequence and thus demonstrate potential cost savings and/or throughput improvements.

One key tool is a simulator to investigate the interactions between coupled discrete events which is described in the next section (Ensen et al., 1995). To ensure that processes are improved systematically, both material flows and information flows, with their complex and dynamic interactions, must be understood. A difference between discrete event simulation tools and flowsheet simulators is discussed in a subsequent section.

The last section presents a selection of examples. The first example illustrates the problems of linking batch stages to continuous process steps in a process and takes into account utility consumption. Debottlenecking in a multi-product plant is described in the second example. The last example shows the typical use of simulation in accordance with the motto "simulation before investment". It includes hypothetical scenarios ("what would happen if ...").

Discrete Event Driven Dynamic Simulation

Computer simulation is employed to analyze and predict the behavior of chemical production facilities. The objectives of simulation are detailed in Fig. 1. Discrete event driven simulators (DE-Simulators) consider the residence time behavior of discrete "packets" of material and their interaction between several manufacturing or production units.

DE-Simulators can be used to describe and analyze all the relevant process sequences. Stochastic (or random) variables (sampling times, equipment malfunctions etc.) can be modeled by a number of statistical functions. Depending on the problem definition, simulators can supply information such as: the utilization of equipment and personnel, order processing time, buffer stocks and inventories and consumption of resources.

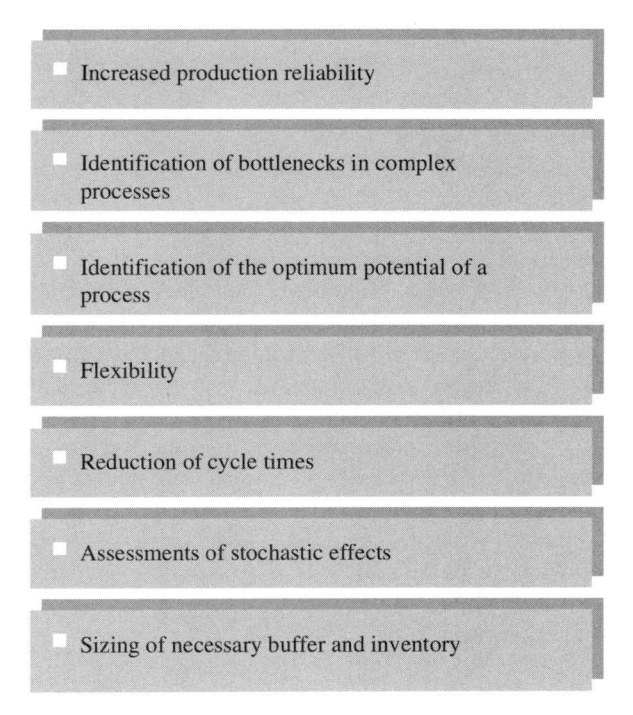

- Increased production reliability
- Identification of bottlenecks in complex processes
- Identification of the optimum potential of a process
- Flexibility
- Reduction of cycle times
- Assessments of stochastic effects
- Sizing of necessary buffer and inventory

Figure 1. Aims of dynamic simulation.

DE-Simulators can be programmed to show the movement of parts, personnel and process states directly on a monitor. With the aid of this animation it is possible to observe closely how, for example, hold-ups and shortages of material can develop on a production line or how equipment states interact to block each other. Based on design input from a group of process experts, simple variations in the operating parameters (product sequence, batch sizes etc.) and changes in the equipment interconnections allow various scenarios to be studied and the effects of deterministic or stochastic variables to be quantified. The result is a better understanding of system behavior, increasingly reliable planning, and ultimately, improved design and operation of industrial processes (Krames, Ensen, 1997).

Application of Dynamic Discrete Event and Flowsheet Simulators

Given the many simulation and flowsheet programs used within chemical industries the question is often posed: What can be achieved through these tools, what are the differences between them, and what are their strengths and weaknesses? Simulation may be defined as follows:

> Simulation is the reproduction of a process in a model capable of experimentation, with the purpose of obtaining insights which can be translated into reality.

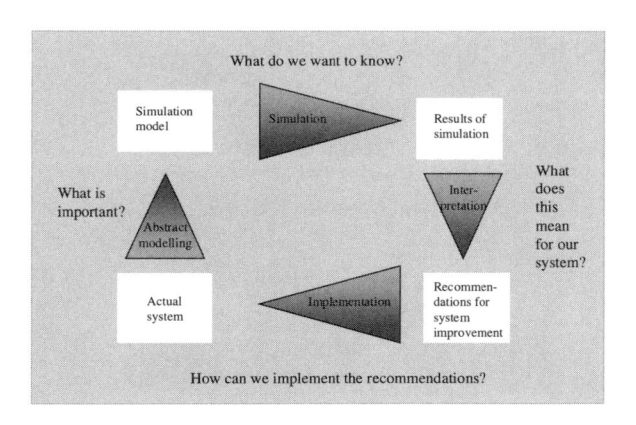

Figure 2. Phases of a simulation project.

Flowsheet Simulators

Flowsheet simulators represent the most commonly used simulation programs within chemical industries and are primarily used for designing processes (AspenPlus®, Hysim®, ChemCad®, etc.). To do this, the thermodynamic states in the process are calculated assuming steady state conditions. To solve the mass and energy balance, a wide range of information is required by the simulation program. To this end, most programs include extensive physical property databases or are able to access external databases.

Discrete Event Simulators

In contrast to flowsheet simulation, discrete event simulation considers the temporal sequence of individual process steps, usually batch-type, and how they affect one another. Pressure, temperature and other thermodynamic variables are immaterial in material flow simulation. The variables entered here include, for example: residence times, transfer times and refitting times, resource availability and their random or seasonal fluctuations.

Discrete event simulators are particularly suitable for visualizing and identifying bottlenecks within a production process and for resolving the type of question "What happens if ..." (e.g. if production strategy, product range or installation configuration are changed). If, for example, several pieces of equipment were available at each stage of a multistage reaction, the discrete event simulator could find the most favorable equipment utilization rate in terms of overall capacity.

The difference between discrete event simulators and flowsheet simulators is shown in Fig. 3.

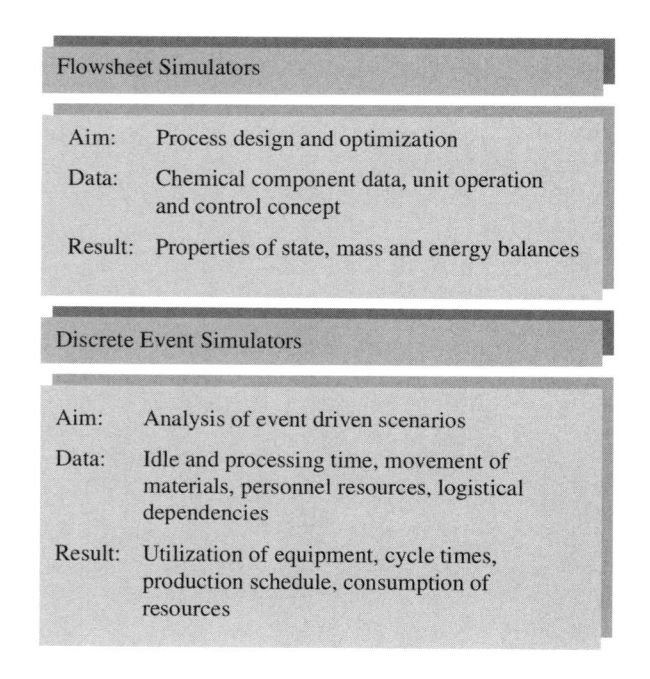

Figure 3. Comparison of simulation instruments.

Examples

The examples shown in this section are typical representatives of the type of study carried out in this area.

Example 1: Linking Discontinuous and Continuous Plant Processes

A multi-product plant is being designed to produce three new products. Simulation is employed to investigate the manufacture of three intermediates A, B and C within the process (Fig. 4). Each intermediate is generated via batch reaction in the two pre-stages R 100 and R 200 and subsequently in the main stage reactor R 300 before storage in tanks T 100, T 200 and T 300. Reaction times are different, both between products and between stages, as are the quantities within the stages. The centrifuges are supplied continuously from the storage tanks. Final product is produced continuously downstream of the centrifuges. Product is required in the approximate ratio of 1:1:3 (A:B:C). The following questions were then asked:

- What is the minimum size of the buffer tanks T 100, T 200 and T 300 to avoid bottlenecks?
- What would be the optimum production strategy?

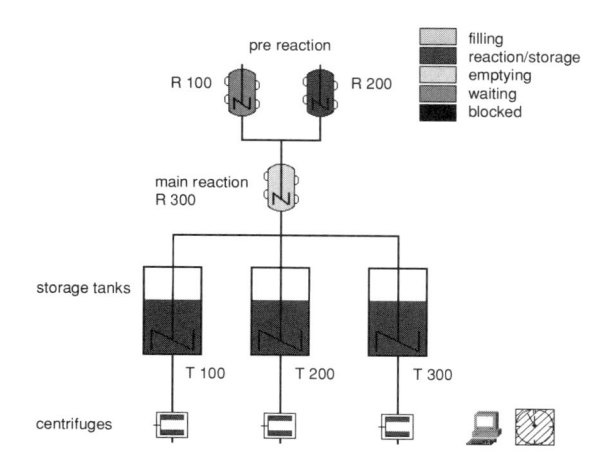

Figure 4. Screen image of the model.

Based on the model shown in Fig. 4 (actual simulator interface) it is possible to study various production strategies and their effects on tank inventories over several days.

Figure 5, the result of a simulation, shows how the inventories in the three tanks vary over time if the production orders are processed in the sequence A, B, C, C, C (roughly corresponding to the production requirement). The "sawtooth curves" are typical for a process having a batch input and continuous output. If the inventory reaches zero, the continuous production section has to be shut down. The simulation covers a period of 20 days, requiring about 20 seconds of desktop computing time.

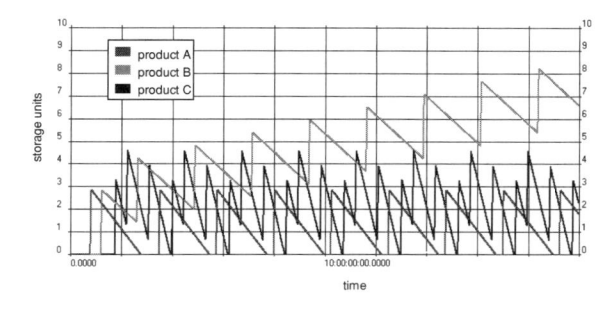

Figure 5. Change in tank inventories over time for the process sequence A, B, C, C, C.

Figure 5 clearly shows that the process sequence A, B, C, C, C has deficiencies. After as little as 3 days availability problems occur for intermediate A, and after 4 days for intermediate C. Intermediate B, on the other hand, is produced in excess of demand. As a result an alternative strategy was developed. The central elements of this strategy are:

- Start-up of the plant using the process sequence (A, A, B, B, C, C, C, C)
- Modifying the basic sequence as a function of the inventory

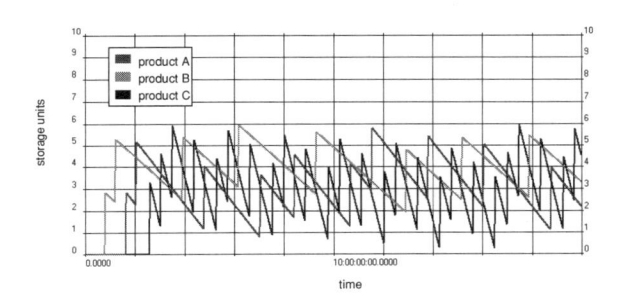

Figure 6. Change in tank inventories over time using a modified strategy.

The results from the simulation are shown in Fig. 6. After 7 days the first corrective intervention took place: instead of intermediate B, intermediate A is manufactured to reduce inventory levels of B and maintain levels of A. After a further 9 days, additional intermediate C is produced to prevent loss in inventory of C. The modified production strategy is expected to require a maximum of only 6 storage units as opposed to 9 storage units predicted by conventional methods, i.e. unaided by simulation.

Example 2: Debottlenecking of a Multi-Product Plant

This plant produces 12-15 different products on five batch-reactor trains, each with varying capacities. The materials (soaps, monomer premixes, catalysts, etc.) required for the different stages of the process are prepared in a number of common and dedicated tanks (see Fig. 7). The raw material composition and time to complete each step of the process (charging, reaction, stripping etc.) is different for each product.

An expansion of the plant capacity had been planned, for which a limited amount of capital was available. A project debottlenecking team was formed to identify and implement the most efficient means to achieve the required capacity increase. This was assisted by the development of a simulation model of the process.

The model dynamically simulates the operating sequence of the solution preparation and charging systems, polymerization (reactors), residual monomer removal (stripping) and product transfer to intermediate storage. The product mix, product cycle times, vessel cleaning schedules etc., together with the available resources (e.g. manpower, maintenance, utilities and supporting processes) were all integrated into the model to give as true a picture of the process as possible. This required the collection of much plant data and determination of how long each individual step of the process took to complete and how much operator time each step involved.

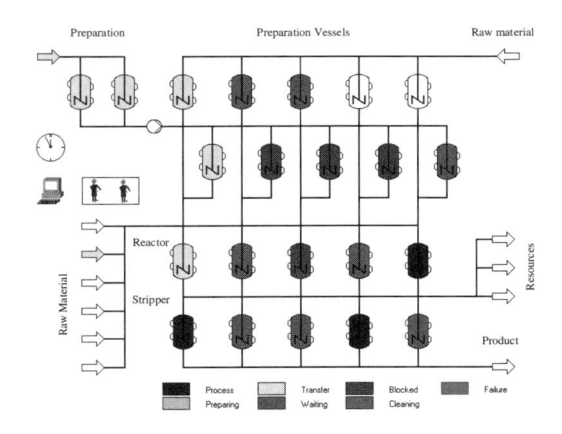

Figure 7. Screen of the simulation model.

During a simulation run, a test is carried out prior to each step to decide whether personnel are necessary and available. If no personnel are available, the step in question has to wait. Priorities are assigned to each part of the process to arbitrate in the event of any conflicts.

Any changes in the number of staff, automation of different parts of the process to free up operator time, product mix, product cycle times, product scheduling etc. can be evaluated by the simulation model in terms of their effect on capacity, either individually or in combination. This allows the bottlenecks to be identified and prioritized, and the determination of where the greatest opportunities for improvement can be found. Table 1 illustrates the effects of some changes in automation on the plant capacity.

Table 1. Effect of Automation.

Simulated change	Effect on capacity
1. Automation of pre mixing system	+ 4 %
2. Automation of reactor monitoring	+ 13 %
3. Automation of soap production	+ 15 %
4. Overall automation (1.-3.)	+ 23 %

In addition to identifying production bottlenecks, the model was also used to determine the steam requirements of the process as production increases. Steam is needed at different flow rates and at different steps of the process. Increasing the number of batches processed per day by eliminating the bottlenecks increases the probability that a number of steam-consuming process steps will proceed in parallel and thus increase the peak steam usage rates. The model gives a computation of how much steam is required at any given time. Figure 8 shows a simulation result as a histogram, describing the percentage of annual steam consumption. This information was required to help size future boiler capacity at the plant, particularly in determining the peak loading requirements.

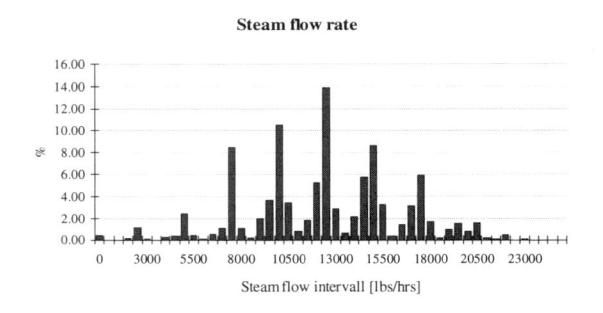

Figure 8. Steam Histogram.

The model has proved to be extremely effective in identifying the "real versus perceived" bottlenecks in the plant. A capital project is being raised which will increase plant capacity by 30 %. It continues to be updated and expanded to identify further bottlenecks upstream and downstream of the main part of the process and has become an important tool in developing other site expansion projects.

Example 3: Bottleneck Analysis in a Single Product Batch Plant

It can also be beneficial to represent and analyze material flows in single product plants as well as multi-product plants with the aid of dynamic simulation, given the often complex interconnections and operating conditions.

A product is produced quasi-continuously in three stages (Fig. 9). These three steps are linked in a highly complex manner, since each stage consists of a number of pieces of equipment operating in parallel (Fig. 10). An increase in capacity is planned for the existing plant.

A large number of ideas were generated. The range of proposed measures extended from simple changes to complex process engineering modifications involving large capital expenditure. The problem with many of these measures was either that the financial risk involved in necessary field trials or equipment capital was too high, or the necessary spare production capacity was not available, given the well-filled order book.

Raw material
↓
Condensation reaction
↓
Crystallization
↓
Centrifugation
↓
Intermediate storage

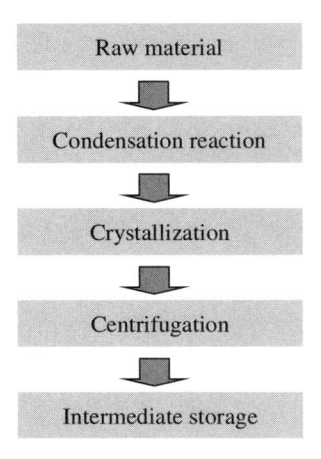

Figure 9. Flow chart.

Agreement was reached to model the process using dynamic simulation. Four phases were defined:

- Accurate representation of current process
- Variation of reaction- and residence times
- Employment of additional equipment
- Variation in the interaction of equipment

On the basis of the simulation results, it was possible to identify those process elements which offer the greatest potential for reducing bottlenecks and increasing throughput. The following table clearly demonstrates that the precipitation units and the mechanical dewatering unit are the bottlenecks. The values in the table represent the percentage changes in plant throughput from the current state.

The simulation study took a total of 5 months, whereas the corresponding field trials would have taken about 5 years, according to plant management.

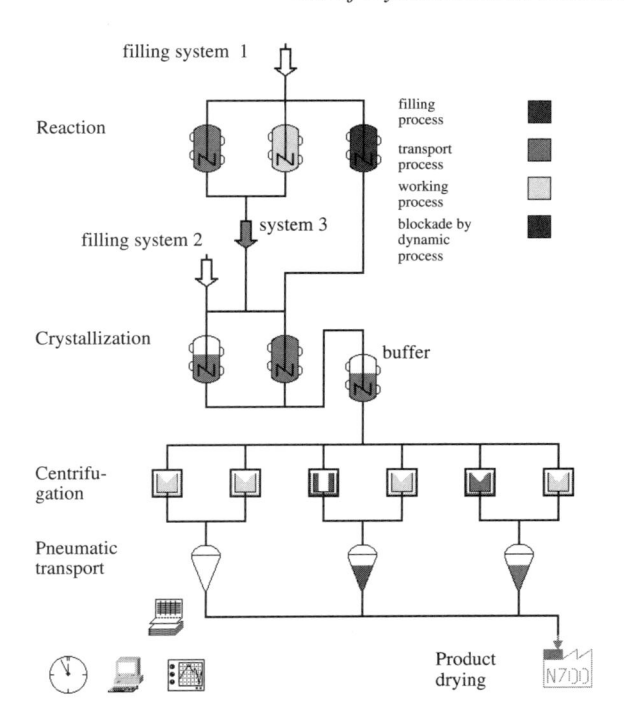

Figure 10. Screen image of the model.

Table 2. Possible Ways to Increase Capacity (simulation experiments).

Additional Centrifuges	0 %
Reduction of filling time of product buffer from 100 to 60 minutes (new pump)	17 %
Modification of filling system 1 and 2 (automatic filling control)	18 %
Modification of filling system (see above) Reduction of crystallization time by 25 % Reduction of centrifugation time by 33 %	54 %

Conclusions

This article explains dynamic discrete event simulation to maximize utilization of equipment, identify bottlenecks and forecast use of resources for chemical processes. There are two additional options for future developments in discrete event simulation.

- The discrete event simulation can be combined with standard thermodynamic simulation tools for a complete description of the process.
- In combination with Gantt-charts, genetic algorithm or other tools, the simulation can be used for scheduling in a plant.

With these additions, dynamic discrete event simulation can have the same value for the chemical industry as thermodynamic simulations.

References

Ensen, H., U. Fuerer, W. Siebenhaar, F. Kurz and G. Koegel (1995). Materialflussanalyse in der Verfahrenstechnik durch dynamische Simulation. *GVC Jahrbuch*, **1995**, 141-149

Krames, J. and H. Ensen (1997). Materialflussanalyse mittels dynamischer Simulation. In J. Rauch (Ed.), *Mehrproduktanlagen*, 1st ed. Wiley-VCH, Weinheim. pp. 185-197.

USING HOMOTOPY CONTINUATION AND GRÖBNER BASES FOR FINDING ALL SOLUTIONS OF STEADY STATE PROCESS DESIGN PROBLEMS

D. A. Harney and N. L. Book
Department of Chemical Engineering
University of Missouri-Rolla
Rolla, MO 65401

Abstract

The problem of finding solutions to process design equations is a classical one in chemical engineering and one in which there is yet considerable room for improvement. Being able to guarantee that all solutions to a particular system of equations have been found is a valuable tool for the design engineer to possess and even more so when global optimization is being attempted.

Two powerful techniques that try to make this guarantee are analyzed here, as individual methods and as one combined technique. The first one is that of homotopy continuation, which has been utilized with considerable success by many authors. Various implementation aspects are discussed, including an algorithm that makes for robust tracking of the homotopy path.

The second technique is newer to the chemical engineering literature and is borrowed from the area of computational algebra. It involves finding a particular basis for the ideal generated by a system of equations, namely a Gröbner Basis. This basis, when calculated under a specific constraint, can create a new system that allows for easy solution (especially when combined with homotopy continuation methods), but still has the same set of solutions as the original system.

A Gröbner Basis for an ill-behaved six variable system is calculated and this basis is then used to find all solution points of the system. Finally, the design equations for two CSTR's in series are solved using the above methods along with several other chemical engineering examples.

Keywords

Homotopy continuation, Gröbner bases, Non-linear equations, Cubic homotopy.

Introduction

Consider the problem of finding all stationary points of the function,

$$g(x) = (x_1^2 + x_2 - 11)^2 + (x_1 + x_2^2 - 7)^2 \quad (1)$$

which requires all solutions of the equation,

$$0.5 g_{x_1} = 2x_1^3 + 2x_1 x_2 - 21 x_1 + x_2^2 - 7 = 0$$
$$0.5 g_{x_2} = x_1^2 + 2 x_1 x_2 - 2 x_2^3 - 13 x_2 - 11 = 0 \quad (2)$$

and has been considered in both Rekalitis et al., (1983) and Kuno and Seader (1988).

A variety of methods have been proposed over the years to solve systems such as this, the earlier ones usually being based on a linearization of the system with Newton-Raphson like methods, which often yielded fast convergence rates for initial points near a solution, but only one solution. Two of the newer methods employed to tackle this problem are investigated here, namely homotopy continuation and Gröbner bases. Both find all nine solutions of this system (shown in Table 1) and the restrictions imposed to make them do so are discussed.

Table 1. Solutions for Equation (2).

(-0.271, -0.923)	(3.0, 2.0)
(-0.128, -1.954)	(-2.805, 3.131)
(3.584, -1.848)	(-3.073, -0.0814)
(3.852, 0.0739)	(-3.779, -3.283)
(0.0867, 2.884)	

Homotopy Continuation Methods

Given the problem of finding solutions to the system of equations:

$$f(x) = 0, \quad f : D \subset \Re^n \to \Re^n \quad (3)$$

a convex linear homotopy is created between this function and a function $g(x)$, where $g(x)$ is a continuous function from Y^n to Y^n with at least one known zero,

$$h(x,t) = tf(x) + (1-t)g(x) = 0 \quad (4)$$
$$h : \Re^n \times [0,1] \to \Re^n$$

By tracing the implicitly defined curve $h^{-1}(0)$ from the $t = 0$ to the $t = 1$ hyperplanes, the homotopy continuously deforms the initial zero of $g(x)$ to a solution of the system $f(x) = 0$. By globalizing the domain of the parameter t from [0,1] to Y, the homotopy can find multiple solutions of the system from the initial starting point. This ability to locate multiple solutions on the homotopy path through the original starting point can be substantially increased by an appropriate choice of the function $g(x)$ and careful path tracking. The specifics of path tracking will not be discussed here, but a good theoretical exposition can be found in Allgower and Georg (1990). What will be considered, are the various ways of applying the homotopy continuation method.

Global Homotopies

The Newton homotopy is defined by Eqn. (4) with the function $g(x)$ defined as $g(x) = f(x) - f(x^0)$, where x^0 is the specified starting point. The Newton homotopy is so called, as it assumes the property of scale invariance possessed by the classical Newton-Raphson methods, and hence scaling of individual equations or variables does not alter the general path behavior. This property is useful when dealing with the ill-conditioned systems of equations that are frequently encountered in chemical engineering, especially when heat and mass balance equations are to be solved simultaneously.

The Newton homotopy finds all solutions of the system given in Eqn. (2) from a starting point of (0.4, 0.4) as shown by Lin, Seader and Wayburn (1987). In fact, moving along the line $x_1^0 = x_2^0$ in the first quadrant, the Newton homotopy finds all nine solutions from each of these starting points. Indeed, the Newton homotopy is often well served by starting points reasonably distant from the solutions, as opposed to the classical iterative techniques, which require a starting point to lie within a 'radius of convergence' of the solution. This improved behavior is due in part, to the fact that the only starting point that will allow solutions of the homotopy function as $t \to \forall \infty$, is one that coincides with a solution, and starting points too near a solution can cause the homotopy path to proceed to very high values of t before a turning point (a point where $dt/ds = 0$) is encountered. However, as the norm of the starting point is increased, the augmented jacobian of the homotopy function becomes increasingly ill-conditioned, and from the starting point (100.0, 100.0), the program HOMES 2.0 (a conversational Homotopy Equation Solver developed at U.M.R.), using the path tracking algorithm of Corvalán and Saita (1991) incorrectly tracks the path and jumps onto another segment of the path, after which, only one solution is found. This is easily remedied by utilizing the determinant monitoring facility, which has been shown to prevent such segment jumping in most cases (Choi et al., 1996; Harney and Book, 1994).

Determinant monitoring also proves useful when searching for solutions of the following system given by Ushida and Chua (1994);

$$f(x) = x_k - \exp[\cos(k \sum_{i=1}^{10} x_i)] = 0, k = 1..10 \quad (5)$$

From the starting point $x_i^0 = 0.0$, $(i = 1..10)$, HOMES 2.0 found 11 solutions, but not before over fifty turning points were encountered and the calculated step size was halved over seventy times due to excessive changes in the determinant of the augmented jacobian from one point to the next. Figure 1 presents a projection of this homotopy path onto the x_1-t plane.

Global Fixed Point Homotopies and Mappings.

The fixed point homotopy is given by Eqn. (4) with $g(x)$ defined as $g(x) = x - x^0$. The main advantage of this homotopy function is that there is a unique solution to $h(x,t) = 0$, on the hyperplane $t = 0$, and consequently, the homotopy function has non-zero topological degree there. This is one of the requirements of the Leray-Schauder theorem, and if in addition, the homotopy path does not asymptotically approach the boundary of the domain of $h(x,t)$, then this theorem guarantees a homotopy path connecting the starting point and a solution of $f(x) = 0$ (Garcia and Zangwill, 1979).

However, the boundary of the domain of definition is frequently crossed when using the fixed point homotopy, and unlike the Newton homotopy, existence of the homotopy path where $t \rightarrow \forall \infty$ is also possible.

If the global fixed point homotopy is used to find the solutions of Eqn. (2) from a starting point of (0.0, 0.0), only one real solution is found, as reported by Kuno and Seader (1988). But if the homotopy path is reflected at the points where $t \rightarrow \forall \infty$ or $*x* \rightarrow \infty$, HOMES 2.0 locates all nine roots after thirteen such reflections. Alternatively, Kuno and Seader (1988) proposed a starting point criterion for the global fixed point homotopy, which minimizes the occurrence of such infinite extent paths, and in this particular case, yields starting points that do indeed have all roots lying on the homotopy path.

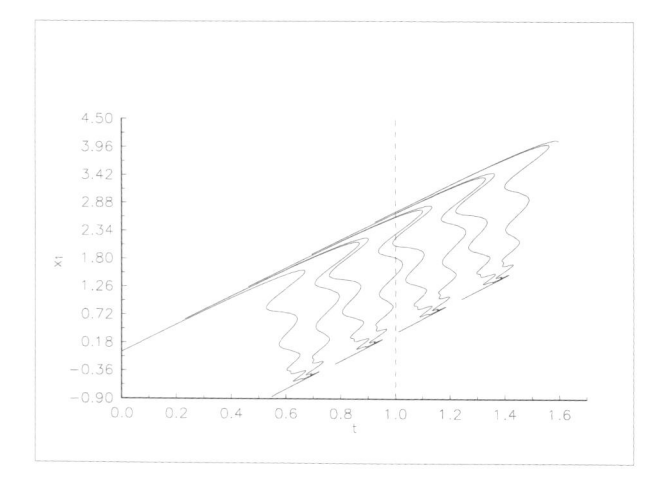

Figure 1. Projection of the homotopy path for Eqn. (5).

The reflections referred to above are not arbitrary manipulations, but rather, are suggested by the fact that the fixed point homotopy can have the same limit when $t \rightarrow \infty$ and when $t \rightarrow -\infty$. This property prompted the mapping functions proposed by Seader et al. (1990), which map the x and the t variables into a compact set, thus guaranteeing, at the minimum, a finite arclength of the homotopy path.

In an effort to preserve the property of non-zero topological degree of the homotopy function on the $t = 0$ hyperplane, but overcome the problem of having only one intersection of the homotopy path with this surface, a cubic homotopy has been proposed, (Harney, 1998):

$$g_k(x) = A(x - \alpha_k x^0)(x - x^0)(x - \beta_k x^0) \qquad (6)$$

where A is a scaling matrix and $0 < \alpha_k < 1 < \beta_k$. It can be shown that the topological degree of this function is a constant ($= \forall 1$). In other words, it possesses the same topological degree as the corresponding fixed point homotopy, but has 3^m intersections with the $t = 0$ hyperplane, where m is the number of the component functions of the vector, g, altered as in Eqn. (6). In contrast with the Newton homotopy, which may also have more than one such intersection, all intersections are known a priori when using the cubic homotopy.

As a first example of an application of the cubic homotopy, consider the fixed point homotopy path for the system in Eqn. (2) from the starting point (3.0, 9.0). The fixed point homotopy finds one real solution and then $t \rightarrow \infty$, and hence, either reflections at infinity or a new homotopy given by one of the mapping functions is required if any of the other eight solutions are to be located. Figure 2 shows the homotopy path for the cubic homotopy with the same starting point and with paramters such that $\alpha = 0.5$ and $\exists = 1.5$ for both equations. (A square indicates a solution point and the circles represent the starting points.) In the positive t direction, four more starting points are located before the homotopy path goes off to infinity, but then tracking the homotopy path from the original starting point, (3.0, 9.0), in the negative t direction, leads to the remaining four starting points being located and then all nine solutions.

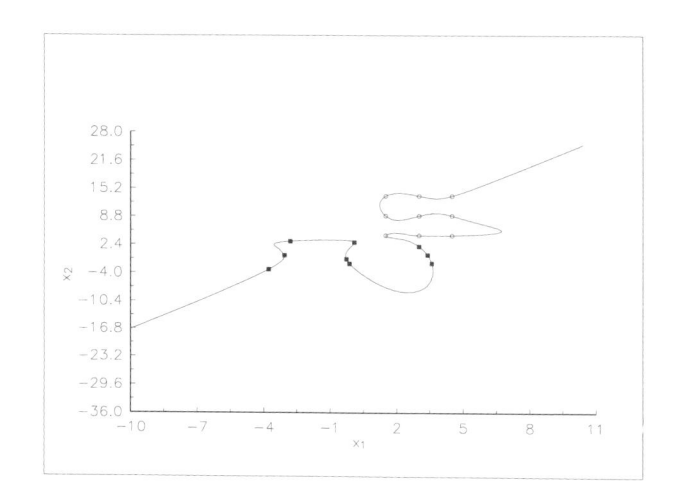

Figure 2. Cubic homotopy path for Eqn.(2) with starting point (3.0, 6.0).

For a second example of an application of the cubic homotopy, consider the system proposed by Choi and Book (1991),

$$\begin{aligned} f_1(x) &= x_1^2 + x_2^2 - 1.5^2 = 0 \\ f_2(x) &= x_2 - x_1^2 + 2 = 0 \end{aligned} \qquad (7)$$

which was originally put forward as a counterexample to the starting point criterion of Kuno and Seader (1988). The Newton homotopy, from a starting point of (3.0, 3.0) locates only two of the four real roots, and neither tracking in the negative t direction nor bifurcating into the complex domain at the located turning points results in the other two solutions being located. Tracking the cubic homotopy path from the same starting point results initially, in five other starting points being located before the original starting point of (3.0, 3.0) is encountered again. Therefore, the point (3.0, 3.0) lies on an isola. When tracking a Newton or fixed point homtopy, a situation such as this would require bifurcating into the complex domain from one of the turning points, in the hope that one of these complex paths will connect the isola to a real path that contains a solutions point. However, as the cubic homotopy has encountered only six starting points, a separate segment of the real homotopy path can be tracked by jumping to one of the three starting points that do not lie on the isola. This results in a path being tracked that contains the three remaining starting points and all four solutions.

A Bounding Function

Rather than concentrating on the fixed point homotopy, Paloschi (1995) proposed a generalized bounding function for the homotopies defined in Eqn. (4), which cause the homotopy path to coincide with its unbounded version when x lies within a defined compact set, Ω, and restricts the homotopy path to values of x that always lie within a compact superset, Ω^*.

$$h_b(x,t) = \pi(x)h(x,t) + v(x) - v(x^0) = 0 \quad (8)$$

$B(x)$ is a $\mathbb{C}^2$ real valued function, that has a value of 1 inside Ω and a value of 0 outside Ω^*. The function $v(x)$ is user-defined, with the restriction that it too, is an element of the $\mathbb{C}^2$ set of functions, and less obviously, that it be a one to one function outside Ω.

As mentioned above, the Newton homotopy locates only two of the four roots of equation (7) from the starting point (3.0, 3.0). The bounding function was applied to the Newton homotopy, with Ω = [-5,-5] x [-5,5] and Ω^* = [-6,6] x [-6,6], resulting in all four roots being located on the homotopy path passing through the starting point (3.0,3.0) as can be seen in Figure 3. In this figure, the squares represent intersections with the $t = 1$ hyperplane, the $t = 0$ hyperplane intersections being represented by circles and the triangles are used to indicate turning points.

Distillation Column

In More (1990), an algorithm was presented that models an n-stage, m-component distillation, with reboiler, condenser and variable feed inlet positions. The algorithm consists of a material balance on each stage and over the column as a whole, vapor-liquid equilibrium and heat balance equations, along with some basic expressions for the temperature dependence of the vapor/liquid equilibrium relationships and the enthalpies. Using standard solution finding methods, the authors were unable to find a solution for one of the given operating conditions. HOMES 2.0 found a steady state for this particular problem with a generic initial guess of 0.5 for all mole fractions and temperature and vapor flowrate initialized to the inlet conditions. In fact, such a starting point was observed to find a solution for all operating conditions that were modeled with this algorithm.

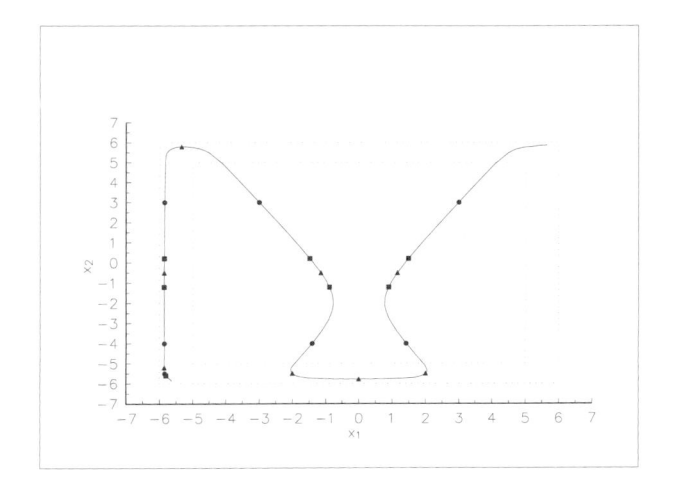

Figure 3. Bounded homotopy path for Eqn.(7).

A Negative Result

A rare counter-example to the excellent functionality of the homotopy continuation methods is given by the following problem suggested by Okonbur (1996),

$$\sum_{i=1}^{3} x_i = 1, \quad \sum_{i=1}^{3} x_i y_i = 1/2$$

$$\sum_{i=1}^{3} x_i y_i^2 = 1/3, \quad \sum_{i=1}^{3} x_i y_i^3 = 1/4 \quad (9)$$

$$\sum_{i=1}^{3} \sum_{j=1}^{i-1} x_i y_j (y_i - y_j) = 1/6$$

$$\sum_{i=1}^{3} \sum_{j=1}^{i-1} x_i y_j y_i (y_i - y_j) = 1/8$$

There are three real solutions to this system of equations, where x_1 has values of approximately -0.0387, 0.5387, 1.3512. From a randomly chosen starting point of (x_1, x_2, x_3, y_1, y_2, y_3) = (.234, .86, 1.445, -.98, .45, 1.1), the Newton homotopy encounters a turning point at $t \approx 0.2$, another starting point before y_3 becomes unbounded as $dt/ds \to 0^-$. Bifurcation into the complex domain at the above turning point results in two of the complex solutions being located, but no further bifurcation onto a real path containing a solution. This behavior is typical of the Newton homotopy initiated from a large number of different starting points. Neither the fixed point, affine or cubic homotopies improve this path behavior. The Newton homotopy, when bounded with a fixed point bounding function in a compact box defined by [6*-5.0] X [6*5.0] finds all three real solutions when the path is tracked in both the positive and negative initial t direction. However, tracking this path requires 19875 LU decompositions and the negotiation of 176 turning points. At present, we can give no satisfactory explanation for the homotopy continuation methods performing so poorly for this particular problem.

In an effort to overcome this unbounded path behavior, the system was reformulated using the theory of Gröbner bases, which are described briefly.

Gröbner Bases

For systems of polynomial equations, $f_i(x) = 0, f_1, \ldots f_s \in k[x_1, \ldots x_n]$, where k is the arbitrary field from which the coefficients of the individual monomials are taken, the affine variety of a set of polynomials, $\mathbf{V}(f_1, \ldots f_s)$, is defined to be the set of solutions in the space k^n, of the polynomial system $f_i(x) = 0$. The affine variety of an ideal I of $k[x_1, \ldots x_n]$, is similarly defined, where an ideal, I, is a subset of $k[x_1, \ldots x_n]$ that satisfies;

(a) $0 \in I$
(b) $f, g \in I \Rightarrow (f + g) \in I$
(c) $f \in I$ and $h \in k[x_1, \ldots x_n] \Rightarrow hf \in I$

Ideals of interest here are sets of the form:

$$<f_1 \ldots f_s> = \{\sum_{i=1}^{s} h_i f_i : h_i \in k[x_1 \ldots x_n]\} \quad (10)$$

It is easily verified that the above set is an ideal. A simple but crucial result with ideals of this sort is that if $<f_1, \ldots f_s> = <g_1, \ldots g_t>$, then $\mathbf{V}(f_1, \ldots f_s) = \mathbf{V}(g_1, \ldots g_t)$. In other words, changing the generating set (the basis) of an ideal preserves the set of solutions of the system of equations.

Hence, given a system of polynomial equations, whose roots are desired, finding a new basis for $<f_1 \ldots f_s>$, that has properties that make for easy solution, could be quite beneficial. Certainly, a desirable property of such a new basis would be finiteness of the generating (or basis) set. This finiteness is guaranteed by the Hilbert existence theorem, as given by Cox et al. (1996);

Theorem. Every ideal $I \subseteq k[x_1, \ldots x_n]$ has a finite generating set. That is, $I = <g_1, \ldots g_t>$ for some $g_1, \ldots g_t \in I$.

In the proof of the theorem presented by Cox et al. (1996), the polynomials are constructed with the property that the leading terms of the g_i's form a basis of the ideal generated by the leading terms of the original ideal, I. This is, in fact, a definition of a Gröbner basis.

Definition. Given a monomial order, a finite subset $G = \{g_1, \ldots g_t\}$ of an ideal I is said to be a Gröbner basis if, $<LT(g_1), \ldots LT(g_t)> = <LT(I)>$

The leading term of g_i, $LT(g_i)$ is the term of the polynomial with the highest order in terms of the given monomial ordering. A monomial ordering is a well ordering (where total ordering is implied), that preserves order under multiplication of the monomials. With the aim of eventual equation solving in mind, the monomial ordering of primary interest is the lexicographic ordering, which is defined as follows;

Lexicographic Ordering. If $\forall = (\forall_1, \ldots \forall_n)$, $\exists = (\exists_1, \ldots \exists_n) \geq 0$. Then $\forall > \exists$, if in the vector difference, $\forall - \exists \geq 0$, the leftmost non-zero order is positive.

(where the vectors $\forall$ and $\exists$ represent the exponents of the two monomials in question.)

Buchberger (1985) formulated an algorithm to calculate a Gröbner basis using a given monomial ordering. A key property of this algorithm is its finiteness, i.e. the fact that ideals in $k[x_1, \ldots x_n]$ form a noetherian ring (which can be proved easily with the Hilbert existence theorem), guarantees eventual termination of the algorithm each time. The Gröbner basis calculated with the Lexicographic ordering has as elements, equations in which the system variables have been eliminated one by one. Therefore, the last element of the basis set will be a function of one variable, namely the last element of the variable set as defined by the particular Lexicographic ordering. This property is formalized in the Elimination theorem, (Cox et al., 1996). Hence, any of the techniques available to find all solutions of a polynomial of one variable (homotopy continuation, for instance) can be used on the last equation, or set of equations, of the Gröbner basis, that are a function of this isolated variable. Each subsequent variable (next on the increasing list of variables in the given lexicographic monomial ordering) can be similarly determined.

For the system given in Eqn. (9), the Gröbner basis was calculated to be a set of six equations in the original variables, with the jacobian matrix being upper diagonal, as guaranteed by the Elimination theorem. The last element of this basis is,

$$6912 y_3^7 - 32832 y_3^6 + 65376 y_3^5 - 70272 y_3^4 + 43644 y_3^3 - 15474 y_3^2 + 2856 y_3^2 + 2856 y_3 - 209 \quad (11)$$

The apparent complexity of the equation is a characteristic of the Lexicographic ordering, but due in part to the set structure of the jacobian matrix, does not result in an ill-conditioned system.

The Newton homotopy, when applied to the six dimensional system produced by the Gröbner basis, found all three real solutions on the one simple path, without encountering the large number of turning points the original system possesses, from every starting point tested.

CSTR Problem

The last example considered here is the design equations for the acid catalyzed esterification reaction carried out continuously in two CSTR's as described in Seader et al. (1990). The phase equilibrium, material balance and reaction kinetics can be written as seven rational expressions, which can be easily converted to polynomial format. These equations were used to form an ideal in $Q[x_1,...x_7]$, for which a Gröbner basis was then calculated using a lexicographic order, which in turn allowed the solution to be evaluated.

It should be noted, however, that the resulting Gröbner basis is very cumbersome, with some of the integer coefficients having in excess of five hundred digits, which caused the evaluation to be extremely slow (over forty eight hours on a HP 9000/735 machine running Maple^{TM}) Herein lies the obvious disadvantage of the Gröbner basis methods. In an attempt to decrease the time taken to calculate this basis, a FORTRAN program was written to do the calculations in floating point arithmetic (thereby changing the field from the rational field to the real number field.) Maple and the other computer algebra systems that perform these calculations all use the rational field. This is probably due to the fact that the division algorithm for multi-variable polynomial equations requires an exact zero in a remainder function for it to terminate. In the FORTRAN program written, the division algorithm was said to have terminated when the norm of the coefficients of the remainder equation is less than * times the norm of the coefficients of the equation being operated on (a value of 10^{-6} was usually used.) For the small number of systems tested, this lead to a calculation of a set of equations that closely approximated the actual Gröbner basis, and in addition, reduced calculation times by up to 50%.

Conclusions

The homotopy continuation method and all its related mappings and functions, constitute, as a whole, a reasonably efficient and extremely robust equation solver. A small amount of experience helps when deciding how to apply the homotopy continuation method to a given system, but when applied properly, it performs very well when compared to other available algorithms. HOMES 2.0 provides the capability to track the homotopy paths with the techniques described here.

For polynomial systems of equations, which are encountered frequently in engineering, the method of Gröbner bases can produce a considerably reduced, yet equivalent system, which when combined with homotopy continuation techniques allow for easy solution of the system. Currently, this method is too slow to be realistically implementable in systems of any substantial size, despite the modest improvements that the real field calculations yielded. However, Amrhein et al. (1997) have recently developed a 'walk' algorithm that allows for significantly faster computation of Gröbner bases, which may allow the technique to become an increasingly useful tool for solving systems of equations.

References

Allgower, E.L. and K. Georg (1990). *Numerical Continuation Methods: An Introduction*, Vol. 13 of *Series in Computational Mathematics*, Springer (Berlin, Heidelberg, New York).

Amrhein, B., A. Gloor and W. Kuchlin (1997). On the Walk. *Preprint submitted to Elsevier Science.* http://www-sr.informatik.uni-tuebingen.de/projects/pareqs/

Buchberger, B. (1985). *Gröbner bases: an algorithmic method in polynomial ideal theory.* In N.K. Bose (Ed.), *Multidimensional Systems Theory*, D. Reidel, Dordrecht, 184-232.

Choi, S.H. and N.L. Book (1991). Unreachable roots for global homotopy continuation methods. AIChE J., **37**, 1093-1095.

Choi, S.H., D.A. Harney and N.L. Book (1996). A Robust Path Tracking Algorithm for Homotopy Continuation. *Computers Chem. Engng.*, **20**, 6/7, 647-655.

Corvalán, C.M. and F.A. Saita (1991). Automatic stepsize control in continuation procedures. *Computers Chem. Engrg.*, **15**, 10, 729-739.

Cox, D.A., J. Little and D. O'Shea (1996). *Ideals, varieties and algorithms: An introduction to computational algebraic geometry and commutative algebra.* Springer-Verlag, New York, 74.

Garcia, C.B. and W.I. Zangwill (1979). An approach to homotopy and degree theory. Mathematics of Operational Research, **4**, 4, 390-405.

Harney, D.A. and N.L. Book (1994). Experiments with Path Tracking Algorithms for Homotopy Continuation Methods. *Proc. Conference on Foundation of Computer-Aided Process Design*, AIChE Symposium Series, **91**, 301, 360-363

Harney, D.A. (1998). Solution Techniques for Process Design Equations. Ph.D. Thesis, University of Missouri-Rolla, *To be published.*

Kuno, M. and J.D. Seader (1988). Computing all real solutions to systems of nonlinear equations with a global fixed point homotopy. *Ind. Eng. Chem. Res.*, **27**, 1320-1329.

Lin, W., J.D. Seader and T.L. Wayburn (1987). Computing multiple solutions to systems of interlinked separation columns. *AIChE J.*, **33**, 6, 886-897.

More, J.J. (1990). A collection of nonlinear model problems. In E.L. Allgower and K. Georg (Ed.), *Computational solution of nonlinear systems of equations*, American Mathematical Society, 723-761.

Okonbur, D. (1996) Personal Communication

Paloschi, J.R. (1995). Bounded homotopies to solve systems of algebraic nonlinear equations. *Computers Chem. Engng.*, **19**, 12, 1243-1254.

Rekalitis, G.V., A. Ravindran and K.M. Ragsdell (1983). *Engineering Optimization*, Wiley, New York.

Seader, J.D., M. Kuno, W.-J. Lin, S. A. Johnson, K. Unsworth and J.W. Wiskin (1990). Mapped continuation methods for computing all solutions to general systems of nonlinear equations. *Computers Chem. Engng.*, **14**, 1, 71-85.

Ushida, A. and L.O. Chua (1984). Tracing solution curves of non-linear equations with sharp turning points. *Circuit Theory and Application*, **12**, 1-21.

DESIGN AND OPTIMIZATION OF CHROMATOGRAPHIC PROCESSES IN PRODUCTION SCALE

Jochen Strube and Henner Schmidt-Traub
University of Dortmund
D-44221 Dortmund, Germany

Michael Schulte and Reinhard Ditz
Merck KGaA
D-64271 Darmstadt, Germany

Abstract

The production of chirality with maximum economy is one of the most challenging tasks of today's pharmaceutical industry. Apart from the use of inherent chirality (starting material from the chiral pool, e.g. amino acid derivatives, carbohydrates) the creation of chiral centers via biocatalysis or asymmetric synthesis are methods commonly used. A third way to obtain pure enantiomers is the separation of racemates via kinetic resolution through preferred crystallization or preparative chromatography on chiral stationary phases (CSP).

This article will emphasize this last method and explain the possibilities of this technique especially in its application form as Simulated Moving Bed (SMB) chromatography and show its benefits and limitations. Therefore, comparisons to classical batch elution chromatographic processes as well as other unit operations (such as crystallization, etc.) have to take cost calculations into account.
The performance of each separation process is quantified by three characteristic objective functions: productivity, dilution and solvent requirement.

Lastly, the specific separation costs or the total costs of separation are calculated as an objective function to lay emphasis on the economy of the separation, including product recovery and solvent recycling.

The comparison of these objective functions, which are determined for batch and SMB processes, leads finally to certain rules of consideration to decide what kind of process (either batch elution or SMB) is preferable as a function of the physical properties of the given binary mixture and the separation task.

Keywords

Separation technology, Process optimization, Simulated moving bed (SMB) chromatography, Life science.

Introduction

Over the last few years chromatographic processes have been gaining increasing interest for preparative and production scale separations in a broad range of pharmaceutical, biotechnological and fine chemical applications. It is expected that life science products are the fastest growing market in chemical production for the next years. For biotechnological productions a sequence of different chromatographic separation steps of diluted and complex mixtures is absolutely necessary. Classical synthesis of complex pharmaceuticals or finechemicals is done over many different reaction and separation steps in which chromatographic separations are an efficient

alternative to other unit operations like cristallisation because the feed losses are minimized. In the last years many adsorbents/stationary phases have been developed to solve chromatographic separation tasks with high selectivity and throughput due to GMP etc. specifications.
In comparison to classical batch separations continuous Simulated Moving Bed (SMB) processes save large amounts of eluent and adsorbent due to the imitated countercurrent movment of solid and fluid flow and the deorbent recycle, see Fig. 1.

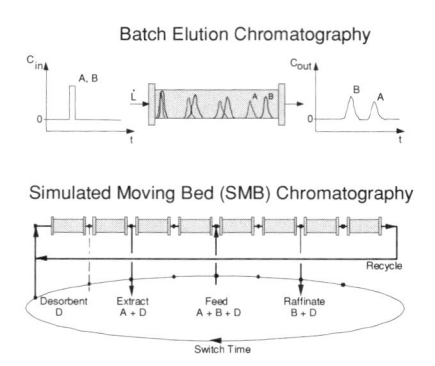

Figure 1. Schemes of batch elution and SMB chromatography.

Furthermore, they make it possible to achieve high resolution and purity also in the separation of species, which exhibit low capacity factors and selectivities - like isomers, proteins, steroids and enantiomers.

The production of chirality with maximum economy is one of the most challenging tasks of today's pharmaceutical industry. Apart from the use of inherent chirality (starting material from the chiral pool, e.g. amino acid derivatives, carbohydrates) the creation of chiral centers via biocatalysis or asymmetric synthesis are methods commonly used. A third way to obtain pure enantiomers is the separation of racemates via kinetic resolution through preferred crystallization or preparative chromatography on chiral stationary phases (CSP).

This article will emphasize this last method and explain the possibilities of this technique especially in its application form as Simulated Moving Bed (SMB) chromatography and show its benefits and limitations. Therefore, comparisons to classical batch elution chromatographic processes as well as other unit operations (such as crystallization, etc.) have to take cost calculations into account.

Design of SMB Chromatography - Process Modeling

The broad applicability of the SMB technology stems from the fact that its application can be based on the scale-up of the analytical method itself, in particular in the important case of enantiomer separations. Empirical scale-up is neither effective nor useful because several phenomena must be accounted for within a sound theoretical frame. These regard competitive adsorption equilibria, as well as axial dispersion and mass transfer resistances, and can be properly described using detailed time-dependent models. In this way not only the scale-up, but also the optimization of the production scale process itself can be performed.

A set of partial differential equations is derived: 1. The fluid mass balance takes into account accumulation in the fluid phase, axial dispersed plug flow and linearized overall mass transfer resistances between the two phases, solid and liquid. In most cases, the inner particle mass transfer resistance is dominating. 2. The particle mass balance consists of accumulation in the particle and mass transfer resistance effects. The large porosity of the particles is taken into account in order to describe their characteristic high loadability and inner diffusion effects. 3. Any equilibrium phase isotherm is needed. 4. The basic assumptions in liquid chromatography are isothermal behavior of competitive adsorption and desorption. 5. The pressure drop over the column length is described by the equations of Ergun, Darcy or Karman. Due to the competitive adsorption behavior and the small amount of axial dispersion at the low Reynolds numbers of liquid chromatography in relation to the convective term it is necessary to solve a set of extrem stiff differential algebraic equations. These are transfered into ordinary differential equations by applying the methods of lines for the axial space domain. Detailed numerical comparisons proved the Leonard difference scheme to be most robust and fastest in relation to orthogonal collocation and other differential schemes. The Gear algorithm proved to be necessary for integration.

Pure empirical design approaches are not effective and in most cases even not possible. An alternative design and optimization approaches based on rigorous dynamic simulation and optimization methods is proposed. Few characteristic experiments based on analytical HPLC methods are necessary to determine the model parameters. The developed simulation system is based on the tools SPEEDUP^{TM} (Aspen Tech., USA), gPROMS/gOPT (Imperial College, GB) and MATLAB^{TM} (The MathWorks Inc., USA).
The comparison of these different design approaches is discussed, judging from their application to different separation case studies.

Process Optimization

On the basis of systematic parameter studies a strategy has been developed to optimize SMB processes with the aid of rigorous dynamic process simulation. At first flow rates are estimated using numerically calculated operating diagrams based on the equilibrium model, see Fig. 2. True countercurrent movement of solid and liquid phase is assumed. The corresponding set of partial differential equations is solved by the method of characteristics.

Operating Diagram: Langmuir Isotherms

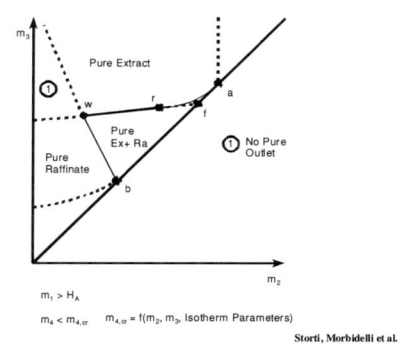

Figure 2. Operating diagram.

Afterwards detailed optimization is carried out in order to achieve maximum feed throughput and minimal dilution of the products at minimal adsorbent and desorbent requirement. The Optimization criteria are shown in Fig. 3.

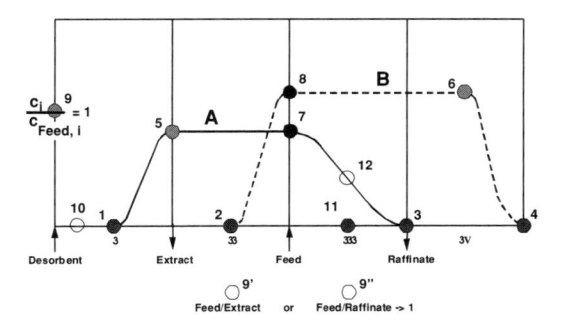

Figure 3. Optimization critera of SMB chromatography.

Today's new applications of SMB processes in the pharmaceutical, agrochemical and biochemical production have to cope with concentration-depending capacity, even separation factors near unity and need to be operated at high resolution, yield and purity. It has been shown that rigorous dynamic process modelling combined with a few characteristic experiments which determine the model parameters describing axial dispersion, multi-component equilibrium with interference of the components and mass transfer resistances are an efficient aid for design and optimization of SMB-processes (Strube, 1996, 1997).

Comparison of optimized batch elution and SMB chromatography

In this paper a comparison of SMB and batch processes is presented for the separation of two different binary mixtures. These examples are chosen to demonstrate the different effects which dominate the applications in large-scale isomer separations and production scale enantiomer separation.

The first example is a fructose/glucose separation with linear isotherms. The model parameters are measured by Nicoud (1992). The second characteristic example is an enantioseparation, Schulte (1997). The corresponding isotherms are of the modified Langmuir type.

The performance of each separation process is quantified by three characteristic objective functions: productivity, dilution and solvent requirement.

Lastly, the specific separation costs or the total costs of separation are calculated as an objective function to lay emphasis on the economy of the separation, including product recovery and solvent recycling.

The comparison of these objective functions, which are determined for batch and SMB processes, leads finally to certain rules of consideration to decide what kind of process (either batch elution or SMB) is preferable as a function of the physical properties of the given binary mixture and the separation task.

Results

A comparison of batch elution and SMB chromatography just by a consideration of the three objective functions does not lead to a consistent answer of the question, "what kind of process should be preferred for a given separation task?"

In the case of *linear isotherms*, the productivities of batch chromatography are much higher than of the corresponding SMB processes, but product dilution and solvent requirements are also much higher.

The case of *nonlinear isotherms* is far more difficult because there is no homogenous result for the process productivity at all. Even product dilution and solvent requirements of batch and SMB may not differ in a significant degree.

Due to that, at least a fourth objective function is calculated unifying productivity and solvent requirement - the separation costs. In this way, the economical relevance of each influencing parameter can be analyzed in a realistic way.

The third objective function, the product dilution, is taken into account. It includes the costs for further upgrading and, therefore, the energy necessary to separate different amounts of solvent from the product. Because batch chromatography generally provides the more diluted fractions, this objective function influences the total separation costs and also the decision about the more economic kind of process.

The total separation costs (TSC) consist of:

1. Costs for stationary phase (SP)
2. Costs for mobile phase (MP)

3. Personal costs (PC)
4. Plant costs
5. Product losses

The separation costs are calculated for typical production quantities. Finally, the decision about the more economic chromatographic process is at first a question of the adsorbent used for separation:

- For resins, zeolithes, and similar low-priced adsorbents, the influence of the stationary phase on the total separation costs can be neglected. Furthermore, the influence of the productivity is not remarkable as demonstrated in the fructose/glucose separation. In this case the SMB process should be preferred - as it is done in the sugar industry since 1960's.
- For expensive adsorbents, the specific separation costs depend on many details which must be considered carefully: Optimum solvent requirement, optimum productivity of each process and, of course, the relation of these objective functions.
- Additionally, the prices of stationary and mobile phase should be taken into account because these prices concisely determine the degree of influence on each objective function of the different processes.
- For enantioseparations in pharmaceutical or fine chemical product development or production, SMB chromatography has major advantages over batch elution like higher productivity, lower product dilution, lower separation costs etc..
- If impurities in the binary separation cut are taken into account, as they occur in reality, the relations between the two chromatographic processes remain at the same magnitudes.

Conclusions

In detail, the more economic process can only be found by detailed process optimization, done by process simulations. Rigorous models are necessary because real effects for peak tailing have to be considered to optimize chromatographic processes.

The feasibility of SMB technology is demonstrated, the process is established in product research and development, and accepted by the involved scientists. Now, chemical engineers are called to develop and prove the profits of SMB chromatography in production and to identify separation task of interest. Therefore, the authors intend to give arguments for the process decision with the presented comparisons of batch elution and SMB chromatography by cost calculations. The design and optimization methodologies are described in detail by Strube (1997, 1998).

Chemical engineers should accept chromatography as an efficient and economic unit operation which has many advantages:

1. Most pharmaceutical and fine chemical products are analyzed by chromatography in the early stage of product development. Therefore stationary, and mobile phases are chosen and experiments are done. Few experiments have to be done at an analytical HPLC column to optimize stationary and mobile phases, and to determine the equilibrium phase isotherms.
2. The design and optimization methodology is developed and can easily be applied to industrial separation tasks as described above.
3. In product development, this methodology is a reliable method to evaluate the separation task by chromatography, in order to benchmark different products. Detailed optimization and consideration of unit operation alternatives could be done if the production decision is made.
4. SMB technology is accepted in laboratory scale for the production of value product. The feasibility is proven (Francotte 1996).
5. The authors have proven Strube (1997) that detailed optimization of SMB chromatography by rigorous simulation studies, in contrast to an empirical trial and error approach, optimizes the operation conditions to about double feed throughput and half desorbent requirements. In production, these efforts for process optimization are of great benefit (Bauer 1997).

References

Bauer, E. R. (1997). A Comprehensive Look at Scaling-up SMB Separations, *ChiraTech Symposium Proceedings* 1997, Philadelphia

Francotte E. R. (1996), Preparative chiral separations by chromatography, *Chiral Europe Symposium Proceedings* 1996

Morbidelli, M., M. Mazzotti, G. Storti (1997). Optimal operation of simulated moving bed units, *J. of Chromatogr. A*, **769**, 3-24

Nicoud, R. M. (1992). SMB chromatography, *LC-GC International*, **5**, 43-47

Schulte, M., R. Ditz, R. M. Devant, J. N. Kinkel, F. Charton (1997), Chiral Separations with SMB, *J. Chromatogr. A*, **769**, 93-100

Strube, J., H. Schmidt-Traub (1996). Modeling of SMB Chromatography, *Comp. chem. Engng.*, **20**, S641-S646

Strube, J, U. Altenhöner, M. Meurer, M. Schulte, H. Schmidt-Traub (1997). Dynamic simulation of SMB chromatography for the optimization of chiral separations, *J. Chromatogr. A*, **769**, 81-92

Strube, J, H. Schmidt-Traub (1998), Modeling and Simulation of SMB Chromatography, *Comp. chem. Engng.*, in press

AN ADVANCED PROCESS ANALYSIS SYSTEM FOR POLLUTION PREVENTION

Xueyu Chen, Kedar Telang and Ralph W. Pike
Louisiana State University
Baton Rouge, LA 70803

Jack R. Hopper, Jamal Saleh and Carl L. Yaws
Lamar University
Beaumont, TX 77710

Thomas A. Hertwig
IMC Agrico Company
Convent, LA 70821

Abstract

A prototype of an advanced process analysis system has been developed to perform comprehensive evaluations on chemical and refinery processes for waste minimization. The system integrates programs for on-line optimization, chemical reactor analysis, flowsheeting, pinch analysis and pollution indices. These programs are used interactively and share plant data through a database. Results from applying the system to a Monsanto/IMC Agrico contact process for sulfuric acid include an increased profit, reduced emissions and process improvements for the reactor and heat exchanger network which demonstrates the applicability of the system for pollution prevention.

Keywords

On-line optimization, Chemical reactor analysis, Pinch analysis, Pollution index.

Introduction

A prototype of an advanced process analysis system has been developed which is to be used by process and plant engineers to develop innovative and economically viable ways to optimize and modify processes significantly beyond their current capabilities. With this system, process engineers interactively and simultaneously use programs for on-line optimization, chemical reactor analysis, flowsheeting, pinch analysis and pollution indices. The results can be process modifications and controls that reduce wastes and energy consumption, in addition to increased profit and improved efficiency of operations. The engineer does not have to learn how to use each of the individual programs and transfer results among the programs.

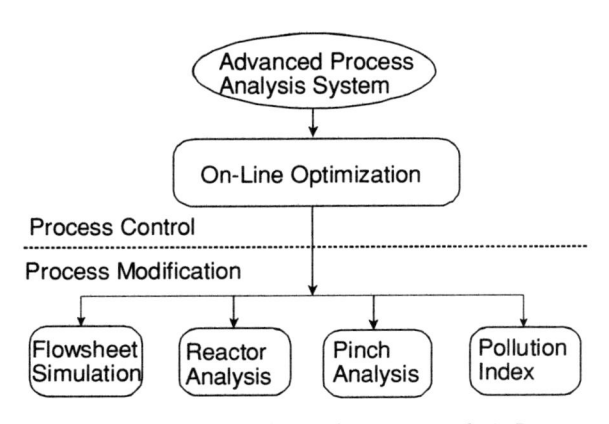

Fig. 1 Overview of Advanced Process Analysis System

An overview diagram of the advanced process analysis system is given in Fig. 1, and referring to this figure, the chemical reactor analysis program (Saleh, Hopper and Walker, 1995) evaluates modifications to have the best chemical reactor type and operating conditions. The flowsheeting program and the EPA pollution index methodology (Hilaly and Sikdar, 1995) identifies pollutants and determines modifications to have the best configuration for separations equipment. The pinch analysis program, THEN, (Knopf, 1993) evaluates modifications to integrate the networks of heat exchangers, boilers, condensers and furnaces for best energy utilization. The on-line optimization program (Chen, 1998) provides accurate plant data to validate the plant descriptions used by the chemical reactor analysis, flowsheeting and pinch analysis programs. Also, it provides the set-points for the distributed control system for the optimal operating conditions for the plant to minimizes costs, energy use and waste generation. The system has an interactive, Windows interface developed using Visual Basic 5.0, and it incorporates a database that maintains process, economic and environmental data which are shared by each program. The database structure is shown in Fig.2.

The program has been developed with industrial and academic collaboration using the Monsanto designed, IMC-Agrico sulfuric acid contact plant. Support has been provided by EPA, and the program will be available through the EPA Technical Assistance Tools program. In the following paragraphs, a brief description will be given of the programs used in the system, and results of applying the program to the IMC Agrico plant will demonstrate the system's capabilities.

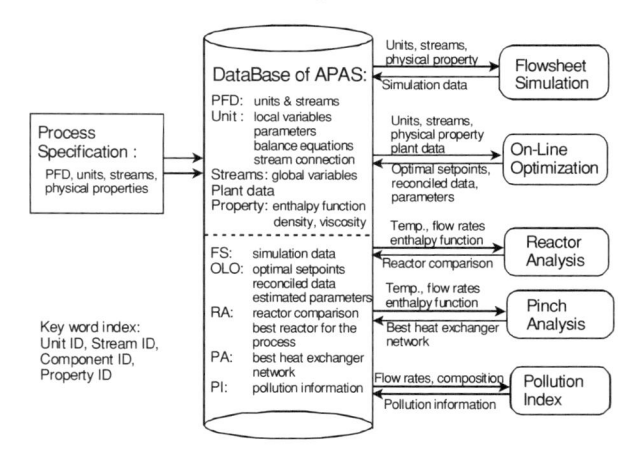

Fig. 2 Database Structure of Advanced Process Analysis System

On-Line Optimization

On-line optimization provides a means for maintaining a plant near its optimum operating conditions by providing set points to the plant distributed control system. This requires the solution of three nonlinear programming problems (NLPs): for combined gross error detection and data reconciliation, for parameter estimation and for process optimization as shown in Fig. 3. The plant model is a set of constraint equations in the NLPs and has to match the current performance of the plant. The parameters in the plant model are updated using data sampled from the distributed control system that has been processed through gross error detection and data reconciliation procedures. The execution frequency for set point updating is based on the settling time of the process, i.e. the time required for the process to move from one set of steady-state operating condition to another, typically four to twelve hours.

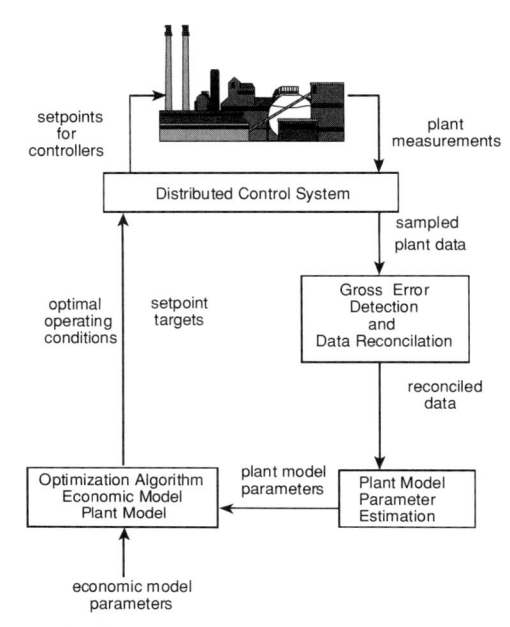

Fig. 3 Simplified Structure of On-Line Optimization

The three nonlinear programs have a similar mathematical statement.

Optimize: Objective function   (1)
Subject to: Constraints from plant model

The objective function is a joint distribution function for data reconciliation and parameter estimation and a profit function (economic model) for plant economic optimization. The constraint equations include material and energy balances, chemical reaction rates, thermodynamic equilibrium relations, among others.

The optimal procedure for on-line optimization is based on the results of Chen (1998). Simultaneous gross error detection and data reconciliation is conducted to detect and rectify gross errors in plant data sampled from distributed control system using the Tjoa-Biegler method (the contaminated Gaussian distribution) for gross errors in the range of 3Φ-30Φ or the robust method (Lorentzian distribution) for larger gross errors. This step generates a set of measurements containing only random errors for parameter estimation. Then, this set of measurements is used for simultaneous parameter estimation and data reconciliation using the least squares method. This step provides the updated parameter values in the plant model for economic optimization. Optimal set points are generated for the distributed control system from the economic optimization using the updated plant and economic models.

The interactive on-line optimization program has an interactive Windows interface for entering the information needed to conduct on-line optimization. Written in Visual Basic 5.0, the program uses this information to write and run the three GAMS optimization programs, and it generates the optimal set points for the distributed control system. Also, summary and detailed reports are prepared. Options include using least squares, the Tjao-Biegler and Lorentzian methods. The process engineer does not need to know the details of the methodology for on-line optimization or the GAMS programming language. Application to the contract process is given in the results section.

Flowsheeting

Process flowsheet development is part of the on-line optimization program. A detailed description of the drawing tools are provided in the users manual with the on-line optimization program (Chen, 1998). As the diagram is prepared, process units and streams connecting the process units are related to the information in the database using an interactive data form. The flowsheet for the Monsanto/IMC Agrico contact process is shown in Fig. 4. Gas and steam can be on separate flowsheets, and the program provides the capability of having multiple flowsheets for a process. Features for developing flowsheets include adding, changing and deleting units and streams and properties of the streams and units. Usual windows features include cut, copy, paste delete, print, zoom, reload, update and grid, among others.

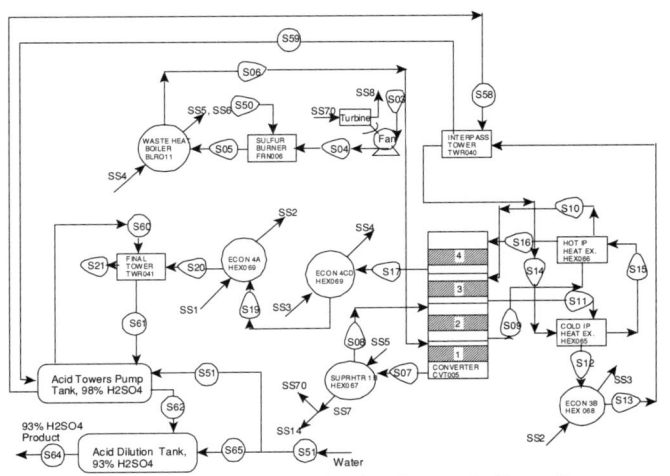

Fig. 4 Flowsheet Diagram for Contact Process

Chemical Reactor Analysis

The chemical reactor analysis program is an interactive windows program that simulates multiple reaction systems with thirty reactions and thirty-six components in the reaction mixture for the types of reactors listed in Table 1.

Table 1. Chemical Reactors in the Simulation Program.

Homogenous
  Single Phase (Gas or Liquid): Plug Flow, CSTR, Batch
Heterogeneous
  Catalytic (Gas or Liquid): Fixed Bed, Fluidized Bed
  Two-Phase (Gas-Liquid): Trickle Bed, Fixed Bubble Bed, CSTR Slurry, Bubble Slurry
  Three-Phase: Fluidized Bed

The reactors can be isothermal, adiabatic or nonisothermal, and Languimir-Hinshelwood and power-law kinetics are included. Also, mass transfer resistance and intraparticle diffusion as catalyst effectiveness factors can be used.

Information required about the stochiometry and kinetics of the reactions taking place in the reactor is entered through interactive windows, and error checking is incorporated to ensure consistent data is provided. Reactor feed rates and compositions are obtained from the advanced process analysis system's database for the process. Results are presented in graphs and tables of conversion, concentration, temperature and pressure. Application to the contract process is given in the results section.

Pinch Analysis

The pinch analysis program, THEN, (Knopf, 1993) is used to evaluate modifications and integrate the networks of heat exchangers, boilers, condensers and furnaces for best energy utilization. An interactive interface has the user extracts the optimal setpoint values for the process variables from the database. This includes stream flow rates, compositions, heat capacities and the enthalpy data. Then the hot and cold streams are selected from the complete list of process streams. Hot streams are streams that need to be cooled, and cold streams are streams that need to be heated. The program then retrieves other necessary data for these streams from the database. If the temperature ranges are small, the enthalpy change can be specified in terms of constant heat capacity values. For larger temperature ranges, the enthalpy coefficient values should be used. Finally, the minimum approach temperature is specified by the user. This is the smallest temperature difference allowed between exchanging streams. Having this information, the pinch analysis algorithm is used to determine the existence of a pinch point, and the minimum hot and cold utilities are calculated. The program draws the grand composite curve showing the heat flows in the system. Also, it uses the pinch design algorithm to synthesize a heat exchanger network that meets the minimum utility requirement. This network is displayed as a grid diagram, showing the placement of the heat exchangers, heaters and coolers.

For an existing process, this network can be compared with the existing one. If the amount of utilities being used is greater than the minimum, excess utilities are being used; and modification to the existing configuration is necessary. If the process is pinched and the existing network has a heat exchanger that operates across the pinch point, this exchanger should be removed; and the resulting imbalance should be adjusted. The configuration should be made as close as possible to the optimum one obtained by the program. Application to the contract process is given in the results section.

Pollution Balances and Indices

The EPA pollution index methodology (Hilaly and Sikdar,1995, Cabezas et al.,1997) provides a quantitative way to identify pollutants and their potential impacts. The Waste Reduction Algorithm (WAR) performs a pollution balance on a plant using a pollution index (mass pollutant/mass product). The pollution index for a plant, I, was defined as:

$$I = wastes/products = -(\Sigma Out + \Sigma Fugitive)/\Sigma P_n \quad (2)$$

and this was used to identify streams and parts of processes to be modified.

This methodology has been extended to the conservation of potential environmental impact in a process. The flow of impact in and out of the process is related to mass and energy flows but is not equivalent to them (Cabezas et al., 1997). A conservation equation can be written as:

$$\frac{dI_{sys}}{dt} = \dot{I}_{in} - \dot{I}_{out} + \dot{I}_{gen} \quad (3)$$

where I_{sys} is the potential environmental impact content inside the process, I_{in} is the input rate of impact, I_{out} is the output rate of impact and I_{gen} is the rate of impact generation inside the process by chemical reactions or other means. Applicaton of this equation to chemical processes requires an expression that relates the conceptual impact quantities to measurable quantities. This is written as:

$$\dot{I}_i = \sum_j \dot{I}_j^{(i)} = \sum_j \dot{M}_j^{(i)} \sum_k x_{kj} \Psi_k \quad (4)$$

where I_i is the total impact flow in the input or the output. The sum over j is taken over all the streams. For each stream, a sum is taken over all of the chemical species. M_j is the mass flow rate of the stream j and the x_{kj} is the mass fraction of chemical species k in that stream. The potential environmental impact of a chemical species is calculated using the following expression.

$$\Psi_k = \sum_l \alpha_l \Psi_{k,l}^s \quad (5)$$

where the sum is taken over the categories of environmental impact. $\forall_l$ is the relative weighting factor for impact of type l independent of chemical k. $\Theta_{k,l}^s$ is the specific potential environmental impact of chemical k for impact of type l.

There are nine different categories of impacts. These can be subdivided into four physical potential impacts (acidification, greenhouse enhancement, ozone depletion and photochemical oxidant formation), three human toxicity effects (air, water and soil) and two ecotoxicity effects (aquatic and terrestrial). The relative weighting factor $\forall_l$ allows the above expression for the impact to be customized to specific or local conditions. The suggested procedure is to initially set all the $\forall_l$ values to one and then allow the user to vary them according to local needs (Cabezas et al., 1997).

To quantitatively describe the pollution impact of a process, the conservation equation is used to define two categories of impact indexes. The first category is based on generation of potential impact within the process. These are useful in addressing the questions related to the internal environmental efficiency of the process plant, i.e., the ability of the process to produce desired products while creating a minimum of environmental impact. The second category measures the potential impact emitted by the process. This is a measure of the external environmental efficiency of the process, i.e., the ability to produce the desired products while inflicting a minimum impact on the environment,.

The pollution index program is called to perform a pollution prevention analysis. It reads all the necessary stream information from the database. This includes the flowrates and the compositions. For each process stream, a stream type has to be selected. Additional data such as specific environmental impact potentials and weighting factors has to be supplied. The program then calculates the various types of pollution indices for the process. These pollution index values can be used to determine either the internal or external efficiency of the process. In addition, it calculates the pollution indexes for individual process streams which are useful in identification of streams with higher pollutant content. These results are presented to the user for evaluation and stored in database for subsequent retrieval. Application to the contact process is given in the results section.

Application to the Contact Process

In this section a description of the results is given for applying the advanced process analysis system to the contact process for sulfuric acid. This is a mature technology and only small increments of improvements are obtained, but the important result is demonstrating the capability of the system on an actual process.

Process Description

The IMC Agrico contact plant in Convent, Louisiana was designed by the Enviro-Chem System Division of Monsanto and began operation in March, 1992. It produces 3200 TPD 93%(wt) sulfuric acid and process steam as a by-product, and it has a Bailey INFI 90 distributed control system. This process incorporates many of the types of process units found in chemical plants such as packed bed catalytic reactors, absorption towers and heat exchanger networks, among others. It represents the state-of-art contact sulfuric acid technology.

In the contact process, molten sulfur is combusted with dry air; and the reaction is exothermic and goes to completion in the sulfur furnace. The gas leaving the burner is composed of sulfur dioxide, nitrogen, and unreacted oxygen at approximately 1400°K. Heat from this gas is recovered in the waste heat boiler as saturated steam at 670 psig. The gas enters the packed bed catalytic reactor that consists of four beds packed with two different types of vanadium pentoxide catalyst. Here sulfur trioxide is produced, and the reaction is exothermic and approaches equilibrium exiting each bed. Heat is removed to shift the equilibrium, and this heat is used to produce steam. Also, the equilibrium conversion is increased the fourth catalyst bed by removing SO_3 in the inter-pass absorption tower. In the final absorption tower, SO_3 is removed from the gas with 98 %(wt) sulfuric acid. Gases exiting the final absorption tower go to the stack with less than 400 ppm SO_2 as required by regulations for emissions, no more than 4.0 lb of sulfur dioxide per ton of sulfuric acid produced.

Process Model

An open form model was developed from the process flow diagram and process design data. The packed bed catalytic reactor was simulated with a kinetic model developed by Richard (1987). The process model has 43 measured variables, 732 unmeasured variables, 11 parameters and 761 linear and nonlinear equality constraints. The model equations were entered in the on-line optimization program, and a comparison of results from the process model with the plant design data was made to assess the validity and accuracy of the simulation. The simulation matched the plant design data within the accuracy of the data. Also, a comparison was made with process data taken from the plant operating five years after start-up, and the simulation with parameters updated with reconciled plant data agreed within the accuracy of the data, e.g. outlet temperatures from the packed bed reactors agreeing within 3EF. Details of these comparisons are given by Chen (1998).

The 43 process measurements obtained from the distributed control system included 25 temperature, 11 flow rate, 2 pressure and 5 composition measurements. The standard deviations were determined based on 61 plant data set from 11 consecutive days. These process variables and their associated standard deviations are given by Chen (1998). Of these 43 measurements, 25 are required to determine the state of the process.

On-Line Optimization

Two sets of plant data from DCS were used to evaluate on-line optimization of the contact process, and the details of these optimal solutions are reported by Chen (1998). Six measurements of the total of 43 were detected as containing gross errors using the contaminated Gaussian function option. These were four temperatures, a flow rate and composition, and they were caused by incorrectly calibrated instruments. These values were replaced by reconciled data, and the simultaneous data reconciliation and parameter estimation program was executed. Then the updated parameters were used in the plant model for economic optimization to obtain the optimal set points. Economic optimization gave an increased profit of 3.0% (or $350,000/ year) and a 10% reduction in sulfur dioxide emissions over current operating condition. This is consistent with other reported applications of on-line optimization and could lead to a typical return on investment of one year.

Chemical Reactor Analysis

The process has four packed bed catalytic reactors that use two different types of vanadium pentoxide catalyst to convert sulfur dioxide to sulfur trioxide. This reaction is exothermic and equilibrium is approached exiting each bed. Heat is removed to shift the equilibrium, and this heat is used to produce steam. Also, the equilibrium conversion is increased the fourth catalyst bed by removing SO_3 in the inter-pass absorption tower. A detailed description of the kinetic model is give by Chen (1998), and it includes an intrinsic reaction rate, pore diffusion temperature gradient between the gas and pellet, and reversible reaction using the equilibrium constant. The kinetic model was entered in the chemical reactor analysis, and a evaluation of the effect of reactor pressure on conversion was made. This showed that the conversion could be increased by 19% in the first reactor and the volume could be decreased by 87% by using a reactor pressure of 10.3 atms. rather than the current operations at 1.3 atms.

Pinch Analysis

The heat exchanger network program was used to apply pinch analysis to the contact process This process is a highly exothermic, and heat released from combustion of sulfur and conversion of sulfur dioxide to sulfur trioxide is used to produce steam which is a valuable product. First, the hot and cold streams were selected from the process database along with the source and target temperatures, flow rates and heat capacities. Then an average heat transfer coefficient was specified to calculate the heat exchange areas, and a minimum approach temperature of 15 K was used. This provided the necessary input, and the program was run obtaining the following results. The process was determined to be below the pinch, and no hot utility was required. The minimum amount of cold utility was 3.703×10^8 KJ /hr, and the grand composite curve for the process is given in Fig. 5. To maximize steam production, a proposed heat exchanger network has thirteen heat exchangers with a total area of 25% less than the current one. The results showed that the existing process is not using any excess utilities, and the energy efficiency can not be improved. However, the network solution provided by the program has less area than the existing network. This shows that the program can be used to check the optimality of existing networks and develop better designs.

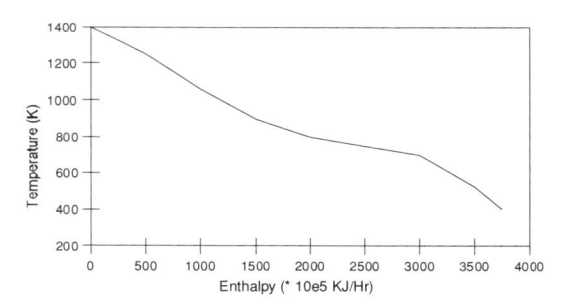

Fig. 5 Grand Composite Curve for the Contact Process

Pollution Indices

The pollution index program was used to demonstrate the use of the pollution prevention analysis with the contact process. In the first step, the process streams were identified that exchange pollutants with the environment, and these streams include the molten sulfur, air, boiler feed water, 93 %(wt) acid, high pressure steam and stack gas. The first three streams are input streams, the second two are product streams and the stack gas is a non-product stream. The composition of these streams was retrieved from the database. Next values were specified for the specific environmental impact potentials for the components in the system and relative weighting factors. Using a value of one for the weighting factors assigned equal importance to all the impact types. With this information, the pollution indices were calculated for the process. The results indicate that the stack gas is the primary pollution impact from the process. The process units with this material in their streams are the sulfur furnace and the converter. They are identified as the candidates for process modification. Thus, the pollution index program can be used to evaluate the environmental efficiency of a plant and assist in making decisions regarding process improvement. As part of the advanced process analysis system, the program can be called at regular intervals to monitor pollution generation with the varying process conditions.

Conclusions

A prototype of an advanced process analysis system has been developed to perform comprehensive evaluations on chemical and refinery processes for waste minimization. With this system, process engineers can use programs interactively and simultaneously for on-line optimization, chemical reactor analysis, flowsheeting, pinch analysis and pollution indices. The engineer does not have to learn how to use each of the individual programs and transfer results among the programs. The results can be process control and modifications that reduce wastes and energy consumption, in addition to increased profit and improved efficiency of operations. Results from applying the system to a Monsanto/IMC Agrico contact process for sulfuric acid showed that on-line optimization could have an increased profit of 3.0% (or $350,000/ year) and a 10% reduction in sulfur dioxide emissions over current operating condition. Process modification using chemical reactor analysis could have the conversion increase and reactor volume decease over current operations, and using pinch analysis could have the total area of the heat exchanger network be 25% less than the current one. The pollution index evaluation identified the stack gas is the primary pollution impact from the process with the sulfur furnace and the converter as the candidates for process modification. These results demonstrate the applicability of the system for pollution prevention.

Acknowledgments

Support from the Gulf Coast Hazardous Substance Research Center and the Environmental Protection Agency are gratefully acknowledged.

References

Cabezas, H., J. C. Bare and S. K. Mallick, (1997), Pollution prevention with chemical process simulators: The generalized waste reduction (WAR) algorithm, *Computers chem. Engng,* **21**, Suppl. s305-s310.

Chen, X., (1998), *The Optimal Implementation of On-Line Optimization for Chemical and Refinery Processes*, Ph.D. dissertation, Louisiana State University, Baton Rouge, LA 70803

Hilaly, A. K., and S. K. Sikdar, (1995), Pollution balance method and demonstration of its application to minimizing waste in a biochemical process," *Ind. Eng. Chem. Res.*, **34**, p. 2051-2059.

Knopf, F. C. (1993), *THEN User's Manual*, Louisiana State University, Baton Rouge, LA 70803.

Richard, M. J. (1987), *The Evaluation of the Applicability of Nonlinear Programming Algorithms to a Typical Process Flowsheeing Simulator,* Ph.D. dissertation, Louisiana State University, Baton Rouge, LA 70803

Saleh, J. M., J. R. Hopper and R. E. Walker, (1995), "Three-phase, catalytic gas-liquid reactors: an interactive simulator," Paper No. 73d, 1995 Spring National Meeting, American Institute of Chemical Engineers, Houston, Texas.

THE COST OF CROSSING REACTION EQUILIBRIUM IN A SYSTEM THAT IS OVERALL ADIABATIC

Willie Nicol, Diane Hildebrandt and David Glasser
School of Process and Materials Engineering
University of the Witwatersrand
Private Bag 3, WITS, 2050, South Africa

Abstract

In a previous paper by the same authors it was shown that for a system with exothermic reversible reaction kinetics it is possible to cross the reaction equilibrium achieved by a single adiabatic reactor in a system that is overall adiabatic. The work in this paper is a direct continuation of the previous work but now we consider the cost effects of such a system. Ideas from the attainable region technique are used to find the optimum process layout and operating conditions for the specified system. The optimum layout consists of complex interconnections between reactors and internal heat exchangers in which reaction is taking place. It is further shown how we can cross the reaction equilibrium with a finite cost comparable to the costs of achieving conversions less than the reaction equilibrium.

Keywords

Adiabatic system, Attainable region, Exothermic reversible, Internal heat exchange.

Introduction

The problem of optimising systems in which exothermic reversible reactions occur is a classical problem in chemical engineering. The choice and control of the temperature profile is a difficult problem as at low conversions high temperatures are desirable to increase the reaction rate while at high conversions a low temperature is preferable to overcome equilibrium constraints. A range of different studies has been done on these reactions due to its interesting qualities and industrial relevance. In a previous paper, Nicol, Hildebrandt and Glasser (1997) looked into optimising conversion for these reactions in a system that is overall adiabatic. It was shown that for an adiabatic system by using different arrangements of heat interchangers one can achieve a conversion higher than that of the adiabatic equilibrium conversion. This implies that the equilibrium conversion achieved in a single adiabatic reactor is not the maximum achievable for an adiabatic system.

The objective in the paper by Nicol et al. (1997) was to maximise the conversion and cost effects were not considered. In this paper we continue the investigation by taking into account the costs involved in crossing the reaction equilibrium. We use concepts from the attainable region (AR) technique (Glasser and Hildebrandt (1987, 1990)) to help us with the optimisation. The problem description of this system does not entirely fit the normal requirements of AR technique so we can only use some of the ideas as guidelines to the optimisation problem.

The System

A simple reaction is considered where pure component A reacts to form component B (A⇔B). First-order reversible kinetics is used. The kinetics are given by:

$$r_x(x,T) = 5\times10^5 \exp(-\frac{4000}{T})x - 5\times10^8 \exp(-\frac{4000}{T})(1-x) \quad (1)$$

By setting $r_x(x,T) = 0$ one can determine the position of the equilibrium line on the conversion-temperature plot.

The feed stream is pure A and is at a temperature of 300K. The conversion is given as the fraction of component A that has reacted. The density as well as the heat capacity of the reaction mixture is assumed to stay constant as the composition of the reaction mixture changes. The energy balance in an adiabatic reactor can now be written as follows:

$$T = T_o + T_{ad}(1-x) \quad (2)$$

where T_{ad} is the adiabatic temperature rise and is equal to the constant $\Delta H/C_p$. T_o is the feed temperature to a reactor for an equivalent feed of pure A. Equation (2) can be graphically represented as a straight line in the conversion-temperature space. Such a line represents the possible set of output temperatures and conversions of an adiabatic reactor and is called an adiabat.

The variables needed for the cost optimisation are conversion (x), temperature (T) and an overall cost value (τ). We define a three dimensional space ($[x, T, \tau]$) of these variables so that the optimisation can be done geometrically. All the heat exchangers in the process layout are referred to as heat interchangers because heat exchange can only be done between internal streams of the process layout (overall adiabatic system). We relate the cost of heat transfer area to the reactor volume with the constant k_c so that the combined reaction and cooling cost is given by $\tau = \tau_r + k_c\omega$ where τ_r is the reactor residence time and ω the area of heat interchange per unit flow of the heat interchanger. For a pure heat interchanger (where heat exchange is the only process occurring in the interchanger) the rate of change of temperature with respect to the cost variable is given by:

$$\frac{dT}{d\tau} = \frac{h_o k_c}{C_p} \cdot \Delta T = K_c \cdot \Delta T \quad (3)$$

A value of $K_c = 150$ was used. All the heat interchangers are countercurrent heat interchangers and the cost (τ) applies for both sides (For one heat interchanger we will only calculate τ once - there is after all just one heat transfer area).

General Optimum Process Layout for Crossing Equilibrium

Before we start defining the general optimum process layout it is necessary to define some terms and definitions we use to describe the heat interchangers. The process layout will consist of one or more heat interchangers. A graphical representation of a heat interchanger in its simplest form is given in Fig. 1. We start from the cold feed at A, heat up to B, react along the adiabat to C and then cool down to D. Point D lies on the adiabat from the feed and the net increase in temperature only resulted from the reaction in section BC. The interchanger for this graph involves heating A to B and cooling C to D. Only countercurrent heat interchangers are considered in this paper. The heat interchanger for Figure 1 has A as a feed stream to the one side and C as a feed stream to the other. The outlet stream B will be in direct contact with stream C (opposite sides at endpoint of heat interchanger) as stream A will be with D. Line AB is referred to as the heating section of the interchanger and line CD as the cooling section. The heat exchanger has the following two requirements to be feasible: Firstly the amount of heating done in AB must be the same as the amount of cooling done in CD, thus the length of line segment AB must be equal to that of CD; secondly the temperature driving force (ΔT) must always be positive (In heat exchanger ABCD ΔT is constant because of the constant heat capacity of the system). Thus a point in the cooling section (take C) must always be at a higher temperature than the corresponding point in the heating section with which interchange is being done (B).

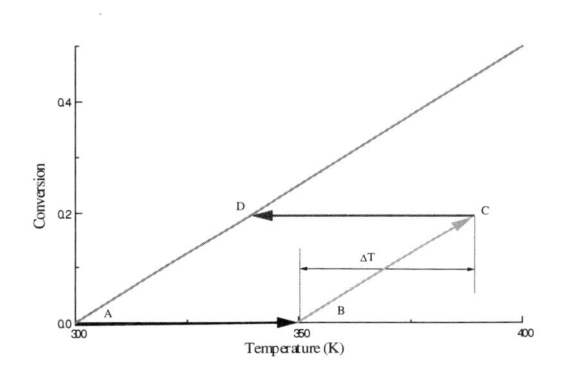

Figure 1. A simple heat interchanger.

All the heat interchangers used in the work are similar to that described in the above paragraph. The only difference is that in some of the heat interchangers the cooling sections of the interchangers will be filled with catalyst. This implies that cooling as well as reaction is taking place simultaneously in those sections. It is for those sections that we are going to use theory from the attainable region (AR) technique to determine the optimum control strategies for the simultaneous operation of reaction and cooling. From the mathematical descriptions we will then determine how the catalyst should be distributed in the cooling section of the interchanger unit.

We need to propose a general layout that will encompass the largest number of possibilities. The proposed layout can be represented graphically in terms of conversion and temperature. This was done keeping all the requirements in mind and the representation of this layout can be seen in Fig. 2. We start from the feed (A) by preheating to some temperature (B), this is followed by reaction along the adiabat in the lowest residence time reactor scheme (BC). From point C we operate along CD where line CD represents the simultaneous occurrence of reaction and cooling in the cooling section of a heat interchanger. (CD does not necessarily represent a single heat interchanger, it rather represents cooling sections of different possible interchangers) From point D we once again react in a single reactor (DE) and we end by heating stream E to F. The first thing that one will notice is that stream F, the output stream, lies on the adiabat from the feed. This must always be the case as the system is overall adiabatic. Thus the difference in temperature between points A and F is only due to the heat of reaction..

Lines AB and EF give the positions of the heating sections of the interchangers (The number of heating sections are not known as the number of heat interchangers are not known, we only know their positions on the x-T plot). Line CD gives the position of the cooling sections of the interchangers. The horizontal length (temperature difference) of CD is smaller than the combined length of AB and EF. This is because the extra reaction heat is slowing down the net cooling process in CD. This does not influence the fact that the amount of heat added to AB and EF are still the same as the amount of heat removed from CD.

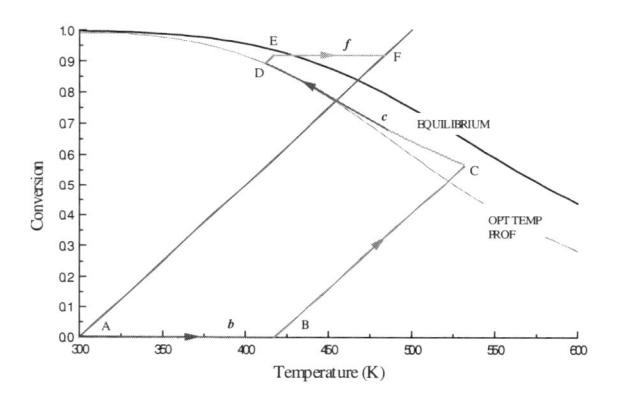

Figure 2. Graphical representation of optimum process layout.

At this stage we are not sure how the interchanger system are going to look like. To find the optimum interchanger layout we are going to divide section CD into evenly sized discrete sections. All the discrete sections will have the same heat flux value where the heat flux refers to the amount of heat transferred from the discrete cooling section. A discrete section on CD will represent a countercurrent heat interchanger which interchanges either with sections of AB or EF. By simulating all the possible interchanger schemes the optimum process layout as well as the optimum process parameters will be obtained.

Conditions for an optimum heat interchanger

We have proposed that the cooling section of the heat interchanger scheme might be simultaneously combined with reaction to reduce overall costs. The heating sections of the interchanger scheme will have no reaction occurring within. There are two reasons for this. The first is that we want to start reaction at a high temperature to increase the initial reaction rate and therefore section AB is done without reaction. The second is that reaction in section EF will decrease the conversion as we have crossed the reaction equilibrium. We begin by representing the processes occurring as vectors in the $c=[x,T,\tau]$ geometric space. The interchange cooling taking place in the cooling section of the heat

$$k_i = \begin{bmatrix} 0 \\ K_c \cdot \Delta T_i \\ 1 \end{bmatrix} \quad (4)$$

interchanger and can be represented by a process vector called k_i. This vector in the c space is given by:
where ΔT_i is the temperature difference between that of the two countercurrent streams directly opposite at some cross section of the heat interchanger. The reaction vector in c space is given by:

$$r = \begin{bmatrix} r_x(x,T) \\ T_{ad} \cdot r_x(x,T) \\ 1 \end{bmatrix} \quad (5)$$

According to AR theory (Godorr et al, 1994) the

$$\varphi = (r \times k_i) \bullet (\frac{\partial r}{\partial c} k_i - \frac{\partial k_i}{\partial c} r) = 0 \quad (6)$$

condition for the optimum simultaneous operation of two vector processes is given by:
This condition can be directly applied to the processes of reaction and interchange cooling to determine the optimum operating conditions for the cooling section of the heat interchanger. This condition will be a function of the temperature driving force within the interchanger and the temperature driving force will in turn be a function of the choice of streams interchanging with one another.

Finding the Optimum Process Layout and Optimum Process Parameters

The general form of the optimum process layout has already been suggested. To find the optimum interchanger layout section CD is divided into discrete sections representing different interchanging units. The question that we must try and answer is which of the discrete sections on CD interchanges with AB and which with EF. Before we can start examining different choices of interchangers we have to examine Fig. 2 again. We notice the letters *b,c,f*. These letters represent the movement along the discrete intervals on the cooling section CD and heating sections AB and EF respectively. The letter *c* will start at C, *b* at B and *f* at F. Let us say the first interchanger is an interaction between a discrete section on CD and AB. On the one side of the interchanger B will interact with C. The points *b* and *c* will represent the interacting points on the other side of the heat interchanger. The position of *b* and *c* will depend on the size of the discrete section. If we have decided to rather interchange the first section of CD with EF, C would have interacted with F in the one side of the interchanger and *c* with *f* on the other side. In such a case *b* would still be at B as section AB has not been utilised for heat interchange.

Each choice of interchanger has to be followed by another until *c* has moved up to D (*b* would have moved up to A and *f* to E as the system is overall adiabatic). We now need to determine what the optimum combination of these discrete options are. There are a lot of options because at each discrete section there are two options for interchange. For each discrete step the number of options therefore doubles. We thus have a tree of possibilities and the size of the tree depends on the number of discrete elements. The system has constraints which can reduce some of the possibilities. The driving force must always be positive and *c* must always be at a higher temperature than *b* and *f*. This constraint cancels some of the options and once an option is cancelled all the options branching from the cancelled option are also cancelled.

Each interchanger unit will have reaction occurring in its cooling section. The optimum operating conditions for each cooling section can be determined from Equations (4), (5) and (6). For interchange with heating section AB the general description of the temperature driving force is:

$$\Delta T_b = (T_c - T_b) + T_{ad} \cdot (x - x_c) \qquad (7)$$

and for interchange with section EF:

$$\Delta T_f = (T_c - T_f) + T_{ad} \cdot (x - x_c) \qquad (8)$$

x is the only variable in this equations. The points *c*, *f* and *b* are fixed for a discrete interval. The result of using Equations (4)-(8) is that each cooling section of an interchanger has a specific conversion-temperature profile (similar to profile CD) that has to be operated along. This profile represents the optimum conditions obtained from the AR theory. If a discrete section for example involves interchange with section AB and the discrete section following that also involves interchange with section AB, the two interchanger units can be combined into one interchanger. The optimum profile along CD for these two units will be smooth as the discrete sections are just parts of one interchanger. This is not the case if we switch from interchanging with AB to interchanging with EF. The optimum profile representing two such interchangers will not be smooth and we will need to operate a small reactor or an interchanger without reaction to join the optimum two profiles.

A program was written which simulated all the possibilities for interchanger schemes. The program firstly tested the constraint of driving forces and cancelled all the impossible options. Thereafter all the other possibilities were simulated to be able to compare their costs. The position of B and E were changed to test the results of the program with different system parameters. The following results were obtained:

> For the simulations of all the system parameters the lowest cost was achieved with only two heat interchangers. The CD profile first interacted with all the discrete increments of EF and then with all the increments of AB. The optimum CD profile was found to consist of only two smooth profiles joined by a small PFR.
>
> It was found that when B and E were at the same temperature, the cost was the lowest for achieving a given conversion. In such a case the two profiles on CD were found to join smoothly.

The only unknown for this optimum system is the amount of preheating that must be done. This was found by simulating the range of possible preheating temperatures. As point B and E must always be at the same temperature to have a smooth CD profile, Point B (amount of preheating) can only differ in the range of possible positions for point E. The range of possible positions for point E is restricted by equilibrium on the right hand side and by the CD profile on the left hand side. The preheating temperature that resulted in the lowest relative cost was found and so all the optimum process parameters were known. The process layout of ABCDEF in Figure 2 can be seen in Table 1 (Curve YZ).

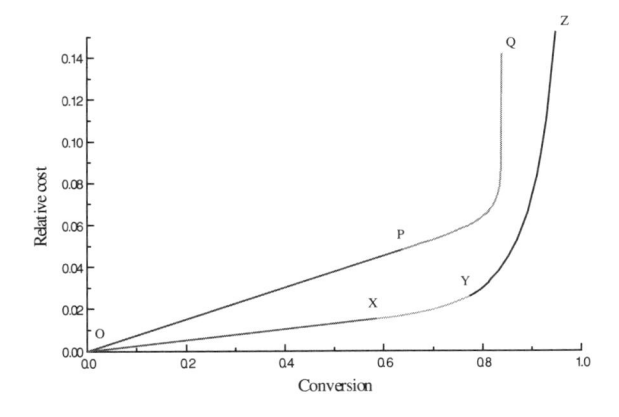

Figure 3. Lowest cost for range of conversions.

The procedure can be repeated for a range of conversions passing the adiabatic reaction equilibrium. The lowest relative cost for these conversions were obtained and are plotted in Fig. 3. The optimum adiabatic reactor configuration (CSTR followed by a PFR) is also plotted in the graph as OPQ. Section YZ on the graph represents the outcomes of the two heat interchanger system. One will see that the equilibrium conversion of 0.84 is crossed in a finite cost. It can also be noted that the two heat interchanger system is also used to obtain conversions less than 0.84. Point Y represents the output from a single heat interchanger. This shows that as one approaches point Y from Z, the interchanger which interacts with EF (Fig. 2) gets smaller and smaller until it vanishes and only the interchanger which interacts with AB (Fig. 2) remains. The process layouts representing the boundary OXYZ can be seen in Table 1.

Conclusions

The heat interchanger system suggested in the previous paper (Nicol et al., 1997) was optimised with respect to cost. The AR theory for the optimum simultaneous operation of two processes was applied to suggest a general optimum process layout. This layout was then simulated with different system parameters and interconnections to obtain the optimum process layout. The optimum layout consisted of two heat interchangers filled with catalyst as well a CSTR and PFR's. The catalyst in the heat interchangers is packed according to the control conditions determined from the AR theory.

It was shown that the reaction equilibrium for an overall adiabatic system could be crossed in a finite cost. The cost of crossing the reaction equilibrium was not much higher than the optimum cost of achieving conversions less than equilibrium.

Nomenclature

c	characteristic vector or state or co-ordinate in AR space
C_p	constant molar heat capacity of reaction mixture (kJ/kmol.K)
h_o	overall heat transfer coefficient (W/m^{2}K)
ΔH	enthalpy of reaction (kJ/kmol)
k_c	constant relation between reactor residence time and heat exchange area
k_i	interchange cooling vector
K_c	$h_o k_c / C_p$
r_x	rate of formation
r	reaction vector
T	temperature of reaction mixture (K)
T_{ad}	adiabatic temperature increase = $\Delta H/C_p$
ΔT	temperature difference over heat interchange area
ΔT_i	temperature driving force over cross section of heat interchanger
x	fraction of initial reactant that has reacted or conversion
τ	overall relative cost
ω	area heat exchange per unit flow of the heat exchanger
τ_r	reactor residence time

References

Glasser, D., Hildebrandt, D. and Crowe, C., 1987, A Geometric Approach to Steady Flow Reactors: The Attainable Region and Optimization in Concentration Space, *Ind. Eng. Chem. Res.*, **26** (9), 1803-1810.

Glasser, D., Hildebrandt, D. and Crowe, C., 1990, Geometry of the Attainable Region Generated by Reaction and Mixing: With and without Constraints, *Ind. Eng. Chem. Res.*, **29**(1), 49-58.

Godorr, S., Glasser, D. and Hildebrandt, D., 1994, The Attainable Region for Systems with Mixing and Multiple-rate Processes: Finding Optimal Reaction Structures, *Chem. Eng. Jour.*, **54**, 175-186.

Nicol, W., Glasser, D. and Hildebrandt, D., 1997, Crossing Reaction Equilibrium in an Adiabatic Reactor System, To appear in *Developments in Chemical Engineering & Mineral Processing*.

Table 1. Optimum Process Layouts.

Curves/ Point	Description and process layout
OX	Bypass from feed to interchanger ending at point X.
XY	Single heat interchanger where reaction is stopped in the middle of the interchanger
Y	Single heat interchanger
YZ	System with two heat interchangers for *crossing reaction equilibrium*

FLEXIBILITY ANALYSIS OF NATURAL GAS PLANTS

Ana M. Eliceche
PLAPIQUI - Chem. Eng. Dept.
Universidad Nacional del Sur CONICET-Camino La Carrindanga, km 7 - 8000
Bahía Blanca, Argentina
Email: eliceche@criba.edu.ar

Laura Fernandez and Maria P. Sanchez
Chem. Eng. Dept.
Universidad Nacional del Comahue
Neuquen, Argentina

Abstract

The main objective of this work is to quantify the variations around the normal feed conditions that a real plant can handle, ensuring a feasible operation. This information is provided by a flexibility analysis of the plant. The active constraints in different directions of the uncertain space are identified. This approach allows the association of bottlenecks with the scenarios of occurrence. The maximum displacements from the nominal point are evaluated solving a nonlinear programming problem. The optimization variables are the main operating conditions of the plant. A rigorous simulation of the natural gas plant has been implemented. Solidification of Carbon Dioxide in the coldest sectors of the plant is not desired. The solubility of Carbon Dioxide in mixtures of light hydrocarbons is predicted. The selection of the operating conditions to avoid carbon dioxide solidification is contemplated adding constraints for each of the top plates of the demethanizer column. The carbon dioxide and methane feed flowrates are assumed as uncertain parameters in the numerical example presented. Once the equipment capacities that limit the flexibility of the plant have been identified, the cost associated with a debottlenecking strategy can be evaluated. Increments in flexibility and cost versus overdesign are reported. The cost is less sensitive than flexibility to equipment over design.

Keywords

Natural gas, Feed perturbations, Flexibility, Bottlenecks.

Introduction

Initial work in the optimization of natural gas plants was done by Eliceche et al (1988). The plants studied do not consider carbon dioxide front removal. The main processes are shown in Figure 3. Ethane a heavier components are separated from natural gas supplied from three different pipelines. The flowrate available and composition for each pipeline change during the year. The plant feed flowrate and composition results from the amount of gas extracted from each pipeline. It is important in that sense to quantify the capacity of the plant to process different feed flowrates and compositions. Thus the main objective of this work is to find the variable feed conditions which the plant can accommodate operating in the feasible region. The optimization of natural gas plants for nominal feed conditions was reported by Fernandez et al (1991).

One of the main concerns is to evaluate the magnitude of the feed perturbations that can be processed avoiding carbon dioxide solidification in the coldest sectors of the plant. A rigorous simulation of the plant was implemented by Fernandez (1993). A procedure was developed to predict the formation of solid carbon dioxide in a mixture of light hydrocarbons. The higher the carbon dioxide composition in the feed the more likely that solidification occurs. Thus carbon dioxide feed flowrate is consider as an uncertain parameter. Methane is the main component of natural gas, thus the methane feed flowrate is also considered as an uncertain parameter.

Flexibility analysis of chemical processes has received increasing attention in the literature. The treatment of parametric uncertainty was introduced by Swaney and Grossmann (1985) where the flexibility index was defined.

The consideration of two dimensions in the uncertain space facilitates the analysis of the behavior of the plant. Different directions of the plane are explored. The maximum deviations from the nominal point quantifies the uncertainty range that the plant can accommodate in a given direction of the uncertain space, detecting simultaneously the equipment which are bottlenecks. The evaluation of the maximum displacement from the nominal points is formulated in the following section.

Problem Formulation

The objective of this work is to analize the behaviour of the plant in the uncertain space of interest. Flexibility represents the ability of the plant to accommodate variations of a set of uncertain parameters. The degree of flexibility is determined by the range of parameter variations that the plant can tolerate remaining in the feasible region. A bounded set of uncertain parameters which vary independently of each other is assumed. The base point is defined by the nominal values of the uncertain parameters θ^N, while expected deviations from the nominal values in the positive ($\Delta\theta^+$) and negative ($\Delta\theta^-$) directions for each parameter are assumed. The uncertain space is represented by a hyper rectangle T centered at the nominal point

$$T = \{\theta / \theta^{Low} \leq \theta \leq \theta^{Upp}\} \quad (1)$$

where the lower and upper bounds are calculated as
$\theta^{Low} = \theta^N - \Delta\theta^-$, $\theta^{Upp} = \theta^N + \Delta\theta^+$
respectively.

The behaviour of a given process can be represented by two sets of equations:

$$\mathbf{h}(\mathbf{d},\mathbf{z},\mathbf{x},\theta) = 0 \quad (2)$$
$$\mathbf{g}(\mathbf{d},\mathbf{z},\mathbf{x},\theta) \leq 0$$

where **h** are the set of equations representing the steady state of the process, and **g** is the set of design and operating constraints which must be fulfilled for a feasible operation. The variables can be classified as:

d: design variables
θ: uncertain parameters
z: control variables
x: state variables.

Once **d** is selected, for any realization of θ, the state variables can be expressed as an implicit function of the variables **z**, solving **h**=0. The set of equations **h** and state variables **x** are solved in a procedure for the natural gas plant simulation. State variables can be eliminated and the process can be described with the set of inequalities:

$$g_j[\mathbf{d},\mathbf{z},\mathbf{x}(\mathbf{d},\mathbf{z},\theta),\theta] = f_j(\mathbf{d},\mathbf{z},\theta) \leq 0 \quad j \in J \quad (3)$$

where j is the index of the inequality constraints. The value of **f** determines the feasibility or unfeasibility of the plant operation. The feasibility for a given θ requires that some control variables **z** exists for which the inequality constraints $\mathbf{f}(\mathbf{d},\mathbf{z},\theta)\leq 0$ are satisfied.

Swaney and Grossmann (1985) presented a scalar index of flexibility F which provides a measure of the feasible operating region. In the uncertain space, the flexibility index F defines the maximum hyper rectangle T that can be expanded around the nominal point in the feasible region. The sides of the hyper rectangle are proportional to the expected deviations. The hyper rectangle can be defined in terms of a scalar variable δ as follows:

$$T(\delta) = \{\theta / \theta^N - \delta\Delta\theta^- \leq \theta \leq \theta^N + \delta\Delta\theta^+\}, \delta \geq 0 \quad (4)$$

The flexibility index F as defined by Swaney and Grossmann (1985) is given by the maximum value of δ in the following semi-infinite programming problem:

$$F = \max \delta \quad \text{s.t.} \quad \max_{\theta \in T} \min_{z} \max_{j \in J} f_j(\mathbf{d},\mathbf{z},\theta) \leq 0 \quad (5)$$
$$T(\delta) = \{\theta / \theta^N - \delta\Delta\theta^- \leq \theta \leq \theta^N + \delta\Delta\theta^+\}, \delta \geq 0$$

The flexibility index corresponds to the maximum deviation of the uncertain parameters from the nominal values for which a feasible operation can be guaranteed by proper manipulation of the control variables. If F is equal to one, the design can accommodate the expected deviations assumed.

Solution of problem (5) is in general quite difficult. Swaney and Grossmann (1985) have shown that for the special case that the constraint functions $\mathbf{f}(\mathbf{d},\mathbf{z},\theta)$ are jointly quasi-convex in z and one dimensional quasi-convex in θ, the solution of problem (5) lies at a vertex of the hyper rectangle T(δ). Then problem (5) can be decomposed in a two-level optimization problem as follows:

$$F = \min_{k \in V} \delta^k \quad (6)$$

with $k \in V = \{k/1 \leq k \leq 2^m\}$, m being the number of uncertain parameters considered. The evaluation of δ^k involves the solution of the following non-linear programming problem:

$$\delta^k = \max_{\delta, z} \delta$$
$$\text{s.t.} \quad f_j(d, z, \theta^k) \leq 0 \quad j \in J \quad (7)$$
$$\theta^k = \theta^N + \delta \Delta\theta^k$$

where δ^k is the maximum displacement in the direction of the vertex k.

The constraint functions **f** are non linear functions for a real plant in operation. Thus the NLP sub problem (7) is solved with a non linear programming code. In this work the successive quadratic programming code OPT of Biegler and Cuthrell (1985) is used.

Numerical Example

The motivating example is a plant in operation processing 16000 molkg/hr of natural gas extracted from three different pipelines with the following nominal molar composition: 91.72 % of methane, 1.25 % of carbon dioxide, 6.76 % of ethane and heavier than ethane components (C2+) and 0.27 % of nitrogen. Expected deviations on feed flow rates are 216 molkg/hr of methane and 40 molkg/hr of CO_2. A minimum ethane recovery of 80 % in the plant is required. The sizing of the equipment for the nominal feed conditions is calculated solving an NLP problem with the objective function of maximizing the profit.

The main operating variables were selected in the previous work by Fernandez et al (1991) and are posed as the optimization variables for problem (7). They are:
- The temperature and pressure of the cold tank that indicates the cooling and compression of the natural gas (Pct, Tct).
- The demethanizer column (DC) pressure (Pdc) that indicates the decompression from the DC to the cold tank.
- The feed splitting factor (Divi), equipment 4 of Figure 3. One fraction of the natural gas is cooled with the top product of the DC column and the remaining fraction is integrated with reboilers of the DC column and the external refrigeration circuit.

Flexibility Analysis

The flexibility index is a scalar number that quantifies the capacity of the plant to accommodate uncertain parameters. For real complex plants it is difficult to prove that the functions $f(d,z,\theta)$ are jointly quasi-convex in z and one dimensional quasi-convex in θ, as it is the case studied in this work. For this reason problem (7) is solved in several directions of the uncertain space.

If a rigorous modelling of the plant is implemented, valuable information is available from the solution of problem (7) in different directions. The values of δ^k and the corresponding active constraints show the operating conditions of the plant in a wide range of scenarios. The analysis of the solution of problem (7) in different directions is called ***Flexibility Analysis***. It allows a systematic approach for detecting common patterns and also bottlenecks and control objectives in different scenarios.

Some of the active constraints that limit the feasible region in the numerical example are:

- Minimum plant ethane recovery.
- Product composition.
- Solidification in some of the six plates of the top sector of the demethanizer column
- Maximum equipment capacities such as heat exchanger areas and compression power.
- Minimum Temperature Difference for Heat Integration.

An active constraint indicates a bottleneck when the maximum or minimum capacity of an equipment is required.

The maximum deviations in different directions are shown in Fig. 1. The values reported correspond to a 10 % over design in the main equipment capacities such as heat exchangers and compression power. Eight directions were explored. Four of them correspond to the vertices directions : k2, k4, k6 and k8. The remaining four correspond to the coordinates :
k1 (positive perturbations of CO_2 feed flowrate), k3 (positive perturbations of C_1 feed flowrate
k5 (negative perturbations of CO_2 feed flowrate),
k7 (negative perturbations of C_1 feed flowrate).

The feasible region in the right half plane, corresponding to an increment in the CO_2 feed flow rate, is smaller than the feasible region in left half plane where the CO_2 feed flowrate decreases. Solidification in the plates of the top section of the demethanizer column is observed when the feed flow of CO_2 increases. This is a common pattern of the right half plane scenarios.

The procedure for the plant simulation has a rigorous modeling of the distillation columns. In the six top plates of the demethanizer column, the vapor-liquid-solid equilibrium is predicted for the CO_2. The CO_2 solubility is evaluated as a function of the plate composition and temperature.

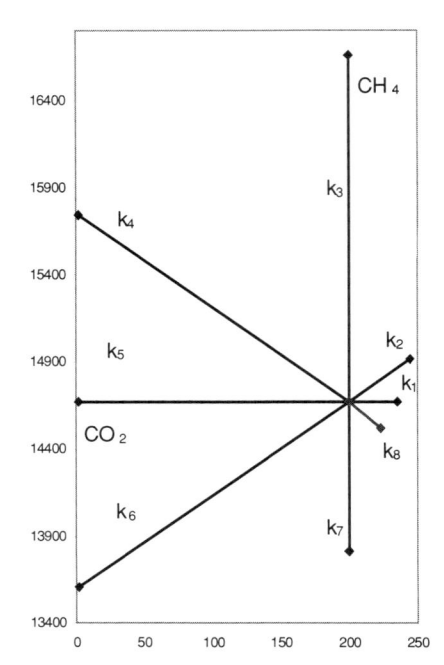

Figure 1. Maximum displacements in different directions of the uncertain space.

The minimum ethane recovery of 80 % is an active constraint in all directions.

The minimum deviation from the nominal point corresponds with the direction k_8 of positive deviation for CO_2 feed flowrate and negative deviation of methane feed flowrate. This is the Worst Case.

Table 1 shows the values for the maximum deviations in directions 1, 2 and 8.

Table 1. Maximum Deviations (δ^k).

Directions	k_1	k_2	k_8
δ^k	0.88	1,11	0.72

The optimal values of pressure of the demethanizer column (Pdc) increases as the CO_2 feed flowrate increases. The temperature profile of the column increases at higher pressures and the solubility of CO_2 in the plates increases. Solidification occurs when the CO_2 composition of a plate is bigger than the CO_2 solubility. Thus higher values of the (Pdc) allows the presence of bigger CO_2 concentration in the cold plates without solidification.

Direction k_8

Only the main results are presented for this direction. Although more information is available. The active constraints for the solution of the NLP problem (7) in the direction k_8 are:

- CO_2 solidification in the second plate from the top of the demethanizer column
- Minimum temperature difference in the gas-gas heat exchanger
- Minimum ethane plant recovery
- Area of the gas-gas (Agg) heat exchanger that allows the integration of the cold methane that leaves the top of the DC with the natural gas to reach the cold tank temperature
- Power of the external refrigeration system.

Due to higher temperature profiles in the demethanizer columns to avoid freezing conditions, the streams leaving the column have less cooling capacities, such as the top stream going to the gas-gas heat exchangers and the reboilers. For this reason more external refrigeration is needed and the power of this circuit is always an active constraint in direction k_8. Due to the higher temperature level of the top stream, less driving force for heat transfer is available and more area (Agg) is required.

The flexibility index or δ^8 is equal to 0.33 for the nominal equipment capacities, calculated solving an optimization problem with the objective function of maximizing the benefit Fernandez (1991,1993). Therefore the plant can accommodate 33 % of the expected perturbations with the nominal capacities. The two bottlenecks are: the area of the gas-gas heat exchanger (Agg) and the power of the external refrigeration (Wer) system. To increase the flexibility of this plant these two sectors should be analyzed.

With 10 % over design, δ^8 is equal to 0.72.
With 20 % over design, δ^8 is equal to 1.09.

The flexibility index increases from 0.33 to 1.09 with a 20% over design of the critical equipment with respect to the nominal values. Thus, the plant can accommodate the perturbations assumed. The other capacities are kept at their nominal values.

The sensitivity of the plant flexibility and the increment in fixed costs are evaluated as a function of the over design of these two sectors. The results are shown in Figure 2. The Flexibility correspond to the Flexibility Index or δ^8 multiplied by 100.

The sensitivity of the flexibility is bigger than the sensitivity of the cost to the over sizing of the heat exchanger area and the power of the refrigeration system. This would indicate that this strategy to overcome the bottlenecks seems viable or at least deserves further studies.

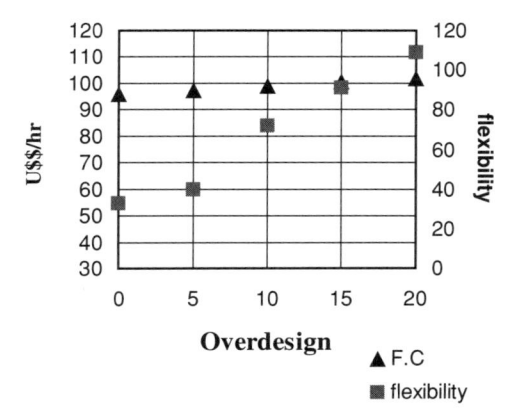

Figure 2. Flexibility index and cost versus over-design.

Direction k_2

The maximum deviation δ^2 or flexibility factor is equal to 1.1, as reported in Table 1, with an over design of 10% in the equipment. The direction k_2 correspond to positive perturbations for CO_2 and methane feed flowrates. The two bottlenecks are the same than in direction k_8: the area of the gas-gas heat exchanger (Agg) and the power of the external refrigeration (Wer) system.

These equipment are active constraints also for 20 % over sizing. If bigger values of than 20 % over design for (Agg) and (Wer) are used, the reboiler area becomes the limiting capacity. This is due to the fact that with a positive perturbation in the methane feed flow rate, the flow from the top of the demethanizer column is bigger and the stream is able to interchange more heat with the natural gas so that no external refrigeration is needed.

For the same direction, the capacities limiting the deviation from the nominal point change. The over design is only applied on the equipment were an increment in the capacity is required.

CO_2 solidification, in the second plate from the top of the demethanizer column, is an active constraint. The plate that is most likely to have solidification problems has been identified.

Minimum temperature difference in the gas-gas heat exchanger and minimum ethane recovery are also active constraints.

Directions k_4, k_5 and k_6

The demethanizer column (DC) operates at lower pressures to increase the decompression ratio from the cold tank to the DC and the recovery of ethane from the natural gas.

The maximum deviations in these directions of the left half plane have a common constraint, which indicates that carbon dioxide composition is zero and cannot be negative. At low CO_2 composition in the natural gas, freezing conditions are not observed.

Conclusions

The flexibility analysis provides valuable information indicating bottlenecks and operating conditions that might disrupt the continuous nature of the processes leading to undesired shut downs of the plant. This would be the case if the demethanizer cannot operate, due to carbon dioxide solidification. Freezing conditions in the second plate is an active constraint in the right half plane. This kind of active constraints would indicate possible control objectives to be monitored for safe operation.

The methodology has been developed for a plant in operation. The equipment that limit the flexibility and the revamping cost have been evaluated. Strategies to overcome the bottlenecks can be implemented.

It will also lead to valuable insight at the design stage of new plants, identifying the equipment and the capacity required in different scenarios. It allows a proper sizing of the equipment to guarantee the desired flexibility.

Feed perturbations have been studied in this work. More uncertain parameters can be incorporated. If a good simulation is available, the flexibility analysis is a valuable tool for the design, operation and control of processes.

References

Biegler L.T. and J. Cuthrell (1985). Improved Unfeasible Path Optimization for Sequential-Modular Simulators II: The Optimization Algorithm, *Comp. and Chem. Eng.* **9**,3.

Eliceche A., Bandoni A., Mabe G. and E. Brignole (1988). Synthesis and Optimization of Ethane Recovery Processes, *The Use of Computers in Chem. Eng. EFCE, Elsevier,* Vol. I, p496.

Fernández L., Eliceche A. and E.Brignole (1991). Optimization of Ethane Extraction Plants from Natural Gas containing Carbon Dioxide, *Gas Separation & Purificaction,* Vol. 5, 229.

Fernández L. (1993). Design of Natural Gas Plants to recover ethane and heavier components, *Ph. D. Tesis, Universidad Nacional del Sur,* Bahía Blanca, Argentina.

Swaney R. and I.E. Grossmann (1985). An Index for operational Flexibility in Chemical Process Design. Part I: Formulation and Theory, *AIChE J.* , **31**, pp 621-630.

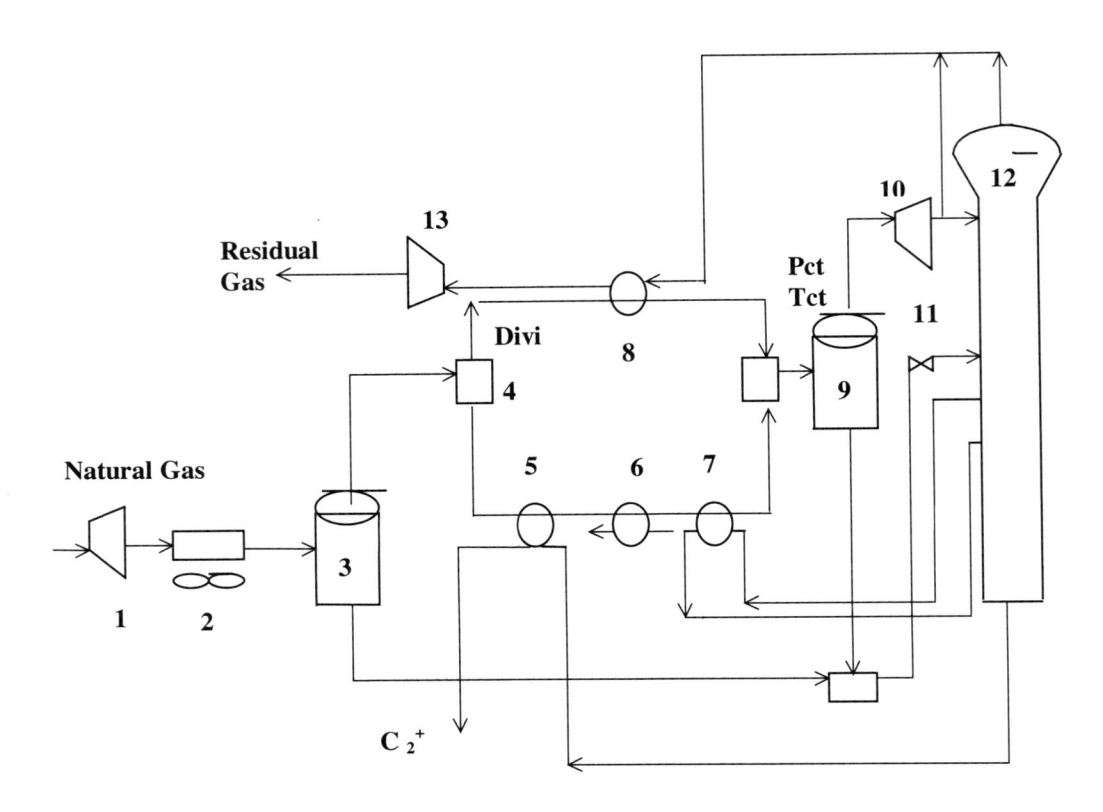

Figure 3. General process flow diagram : 1, inlet compressor; 2, air coolers; 3, flash; 4, flow splitter; 5, reboiler; 6, external refrigeration system; 7, intermediate reboiler; 8, gas-gas heat exchanger; 9, cold tank; 10, turbo expander -booster; 11, JT valve; 12, demethanizer column; 13, compressor.

ON COMBINING OPERATING FEATURES IN THE DESIGN OF A BATCH PROCESS

Girish Joglekar and Steven M. Clark
Batch Process Technologies, Inc.
W. Lafayette, IN 47906

Vikram G. Kalthod
Pharmacia and Upjohn
Kalamazoo, MI 49001

Guy Maineult
Pharmacia and Upjohn
Val de Reuil, France

Abstract

Most of the spent solvent from manufacturing processes is currently not recovered at Pharmacia & Upjohn (P&U), France. Due to the success of the solvent recovery operation at P&U, Kalamazoo, MI, and for environmentally conscious manufacturing, a feasibility study for building a recovery facility in France was initiated. The processing steps in the solvent recovery operation have strong interactions because of internal recycling of solvent streams and shared equipment and resources. Since it is difficult to estimate equipment sizing and operating costs for such complex processes, use of detailed simulation was the only viable approach for quantifying various alternatives. The BATCHES simulator was used for studying the following design alternatives: the number and size of the day tanks, throughput of the batch distillation column, the size of the recoverable solvent storage tanks, amount of water and chemical agent consumed, total amount of material burned and operator needs. Strategies for assigning solvent batches to the distillation column were also studied. Without the simulation study, the design would have been guided by estimates and equipment sizing and operating costs would be difficult to validate. Also, it would not have been possible to determine the effect of operating strategies on equipment sizing and operating costs. A summary of the feasibility study is presented in this paper.

Keywords

Simulation, Batch processes, Design, Operation.

Introduction

The Solvent Recycling and Distribution (SRD) operation at Pharmacia & Upjohn (P&U), Kalamazoo, MI, has been operational for many years. The overall scheme to solvent recovery consists of storage of recoverable solvent in tanks, batch distillation columns for recovering the main components, storage of distilled solvent to be released after assay and storage for pure solvent. The recoverable solvent streams generated by manufacturing processes are sent to specified storage tanks and may vary significantly in composition between manufacturing steps. The approach of aggregating solvent streams differs from the other common approach where only streams with a narrow composition range are mixed. The latter approach requires a larger number of storage tanks but is less complex in operations. The success of the SRD operation is due mainly to the ability to maximize recovery of each distillation batch by determining the fraction end points based on the charge composition (Kalthod, 1995). Once the fraction end points are determined, the columns are operated in intelligent open loop control, obviating the need for analyzing the distillate to switch fractions.

Due to the success of SRD at P&U, Kalamazoo, and for reducing waste generation, a feasibility study for building a similar solvent recycling facility at P&U, France was undertaken. The total projected amount of solvent to be recycled at P&U, France, is smaller than processed at P&U, Kalamazoo. For obtaining a good estimate for the economics of the proposed unit, detailed analysis was undertaken as part of the feasibility study. Thus, the estimates for capital and operating costs were based on sound analysis. Although the recovery of a given solvent consists of few operations, there is strong interaction between various recovered solvent systems due to recycle of streams. Also, there are interactions between various operations due to shared operator resources. Due to the complexity of the underlying operation, it was decided that simulation would provide the most viable approach to studying the problem. Consequently, the BATCHES simulator (BATCHES, 1997) was used for simulating the operation of the proposed facility and evaluating the various design alternatives.

The details of the SRD operation, summaries of the key simulation runs and the final design of the proposed unit will be covered in this paper.

Operation of the Solvent Recycling and Distribution System

The solvent recovery operation consists of the following steps: storage of recoverable solvents, separation of solvents using batch distillation, temporary storage of distilled product in day tanks prior to analysis and release, and storage of pure solvent for delivery to the manufacturing process.

In the proposed facility, five solvents (S1 - S5) are recovered. The spent solvent streams generated in the various manufacturing steps are sent to SRD for recovery. The storage tank to which a recoverable solvent stream is sent is determined by the stream composition. In general, each recoverable solvent stream contains many solvents at low concentration but with one or two dominant solvents. Based on the operating experience, recoverable solvent streams have been divided into several types (RSST_1 - RSST_6 recoverable storage tanks). At P&U, France, six recoverable solvent types were identified, five of which have one dominant solvent, while one has two dominant solvents. The product mix, manufacturing recipe, and schedule of the manufacturing process determine the amount, composition, and the time at which a batch of recoverable solvent is generated. The recovery of solvent(s) from each recoverable solvent type is carried out according to a specific batch distillation recipe. As an example, the BATCHES recipe network for recovery of solvent from the RSST_5 recoverable solvent storage tank is shown in Fig. 1.

The task named PROC models the generation of a batch of recyclable solvent from manufacturing processes to the corresponding recoverable solvent storage tank, RSST_5.

The next step in the recovery of a solvent is the batch distillation operation (DISTILLATION task). The distillation recipe is specific for each recoverable solvent type. Based on the solvent processed in the preceding batch, the column may be cleaned (CLEANING subtask) with water. The spent water from cleaning is sent to the burner. A fixed volume of recoverable solvent is charged to the still pot after it is cleaned. The column is then operated under total reflux for equilibration. Three product fractions are taken after equilibration, each at a pre-determined reflux ratio. For purposes of this work, the amount of these fractions has been calculated based on split ratios for a component in various fractions. In actual operations, a recipe involving temperature trip points and/or volume trip points from detailed simulation is used. The results from the simpler split based analysis have been compared against actual runs and the spilt based approach was found to be a good approximation overall. The distillate from each product fraction satisfies purity specifications. The next fraction (CUT04_RSST subtask) is not as pure but contains significant amount of S2. Therefore, it is recycled to the recoverable solvent storage tank. The material in the column as well as the bottoms is sent to burner.

The outputs from three product fractions are accumulated in the large day tank (DT task). The day tank storage operation represents a strong coupling between the recipes of other solvents. For example, the day tank also accepts (inputs to STORE subtask) solvent S2 product fractions from the distillation of some other recoverable solvent types. After the large day tank becomes at least 60% full, the contents are washed with water and treated with another chemical. A day tank is taken for release if no new material is added to it in more than 72 plant operation hours. Some of the solvent collected in the day tank may be returned to the recoverable solvent storage tank (TORSST_5 subtask). This subtask takes into account the rinsing of a day tank before filling it with the next solvent distilled. The day tank rinsing is done only if the next solvent is a different one to essentially eliminate cross-contamination. The batch is then taken for assay (subtask ASSAY). Based on the analysis, the batch is either released as a successful batch (subtask EMPTY) or rejected. The reject material is sent back for re-processing. A small amount of failed S2 product is burned.

The volume or duration of each fraction in a batch distillation depends on the feed composition and charge volume. Also, the average composition for each fraction depends on the charge composition.

The recoverable solvent storage operations are coupled with its own distillation operation and distillation operations of other recoverable solvent types.

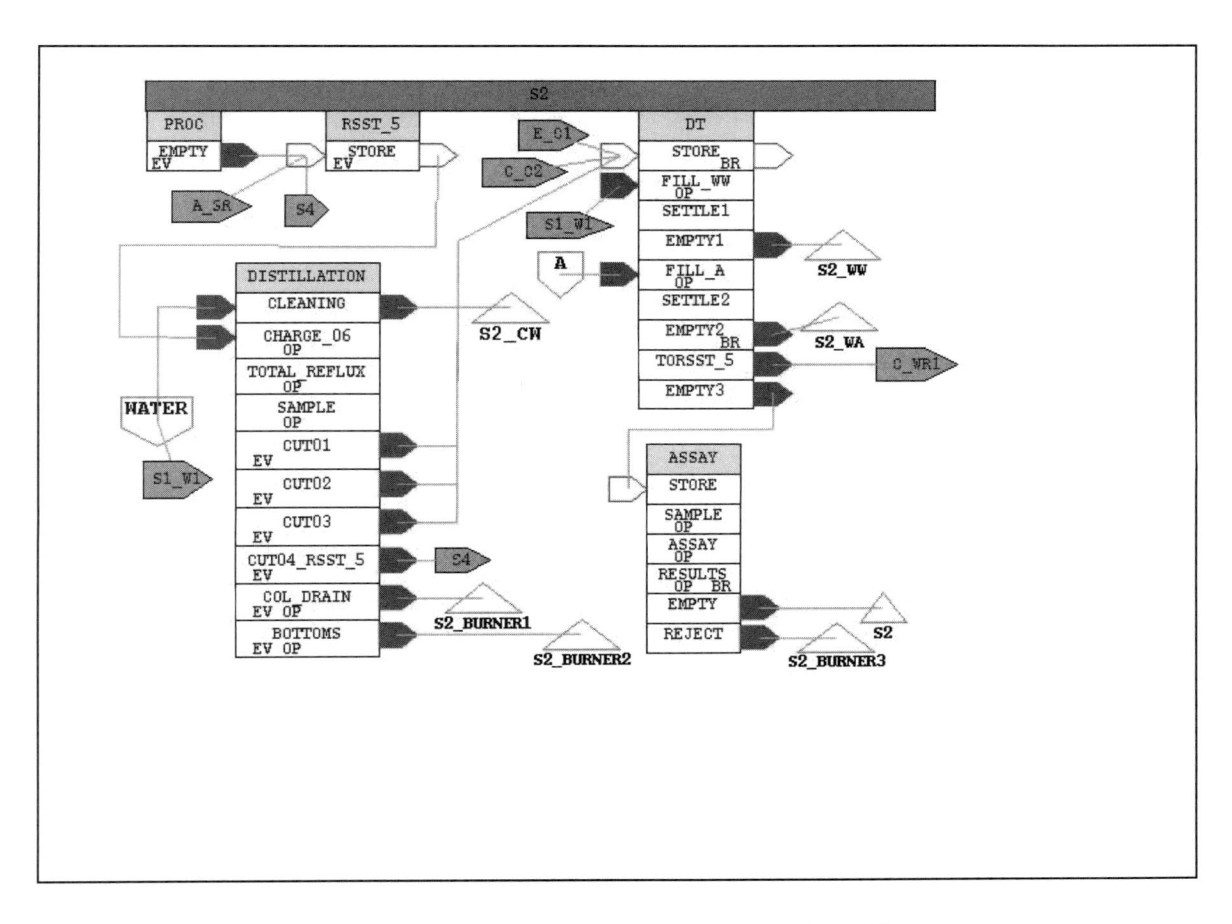

Figure 1. Recipe network for the recovery of solvent S2.

For example, in the recovery of solvent S2 from RSST_5 recoverable storage tank, some distillate is sent back to the recoverable solvent storage tank. Similarly, streams from other distillation operations also are recycled to RSST_5, as represented by the flow connectors to the STORE subtask of RSST_5 task.

The strong coupling between the solvent recipes clearly shows that the recoverable storage tanks must be sized such that they can accommodate the recoverable solvent streams from manufacturing processes as well as internal recycle streams. Also, detailed species mass balances for all operations are necessary so that the yields from distillation can be calculated.

Additional coupling between the various operations exists because of the use of shared resources. For example, the assays performed by analytical testing laboratories require an assay operator and one is available between 7:00 AM and 5:00 AM on a working day (5 working days per week). Also, an SRD operator is required during the following steps: column start-up and shutdown, chemical treatment in the large day tank, and sampling for assay. An SRD operator is available on working days. The plant is shut down on weekends, for a few weeks in summer and December. On the day before a holiday, the operation is halted at 7:00 PM and re-started at 5:00 AM on the day after the holiday. If on a working day before a holiday there is not enough time remaining to complete all fractions of the distillation, then the distillation operation is halted after charging and sampling and operations are resumed on the day after the holiday.

The cost of burning a stream depends on the concentration of chlorinated hydrocarbon in it. Therefore, it is important that the simulation accounts for the composition of the streams, as otherwise, the economic analysis will not be accurate.

Simulation Study

One of the approaches for designing SRD, P&U, France would have been to duplicate the facility in Kalamazoo with some scaling based on production volume. However, since the product mix is different between the two sites, this approach was not used. A detailed simulation based approach was adopted as it allows for validation of the equipment sizing and economics.

Preliminary calculations indicated that one batch column would be adequate to handle the projected

Table 1. Total Amounts for Streams Recycled to the Six Recoverable StorageTanks.

Recoverable Solvent Storage Tank Type	RSST_1	RSST_2	RSST_3	RSST_4	RSST_5	RSST_6
Annual Amount US gallons	160,000	124,370	102,930	132,290	204,600	17,190

recoverable solvent load at P&U, France. Also, due to the nature of the operation, it was imperative that there be at least one storage tank for each recoverable solvent type. The key design decisions to be made were to estimate the charge size for the batch distillation, boil-up rate in the column, volume of the recoverable solvent storage tanks, number and size of day tanks.

The following proposed design was evaluated in most of the simulation runs: 6 recycle storage tanks, one batch distillation column (still pot size 5,000 US gallons charge), boil-up in column of 20 US gallons/minute, five small day tanks (5000 US gallons each) and one large day tank (7,000 US gallons). The large day tank is used for chemical treatment and washing of distillate. The still pot volume includes no vapor space, its volume including vapor space may be between 6,500 to 7,000 US gallons.

The determination of the recoverable solvent type to be assigned for distillation is based on the levels in the recoverable solvent storage tanks. Various assignment policies were tested in the 'what if...' analysis.

The loads considered for the design of SRD were based on the production plan and forecast for the near future. The total amounts for streams recycled to the six recoverable storage tanks are given in Table 1 for the production plan.

In all simulation runs, a time span of one year was used. At the beginning of the simulation, all equipment was assumed to be idle and empty.

Economics of a Simulation Run

In the economic analysis presented here, the positive contributions are made by two items: the saving resulting from recovery of various solvents and the cost associated with disposal of the recovered solvents. The purchase price of solvents recovered ranged from $ 0.85 to $ 8.35 per gallon.

The cost of burning waste solvents is $ 0.7 to $ 1.25/US gallon, depending on the chlorinated hydrocarbon content.

In the economic analysis, negative contributions were made by the following items: the cost of chemicals used for treatment, the cost of burning a waste solvent streams generated from the distillation and direct operating expenses.

The capital cost estimate was based on vendor estimates and estimates done by the engineering contractor. The capital cost included the installation and commissioning costs. Amortization is not included in the analysis provided here.

Evaluation of Simulation Runs

In addition to the economic analysis, the tank sizing obtained as part of a simulation run was also used for evaluation. The maximum level in a recoverable solvent storage tank provided a measure for its size. Similarly, the maximum level in a day tank provided the measure for determining its size. The average utilization of day tanks provided the basis for determining the number of day tanks that would be needed. The average utilization of column was calculated on the available manufacturing time and not the simulation time scale.

A brief summary of the conditions used and a discussion of results for four simulation runs are provided in the following sections. The key results from each simulation run are given in Table 2.

Base Case

The following column assignment rule was used in the base case. The recoverable solvent tank with the highest amount sufficient to fill the stillpot (full charge) was assigned to the column. The base case shows that one of the five day tanks is not used, while the utilization of another is small (2%). The economics show that disposal of waste streams to the burner, most of which is water used for column cleaning, incurs considerable cost.

Scenario 1

In this scenario, a different column assignment rule was implemented in an attempt to reduce the amount of water used for column cleaning. If there was enough quantity, of the same solvent type as previously distilled, to make at least 50% of the charge volume needed and its recoverable solvent storage tank had the highest or second highest level, then the same recoverable solvent batch was assigned as the next run. Note that as the same recoverable solvent type is processed in the column, no cleaning of the column is necessary. Water usage for cleaning is expected to reduce for this reason.

The simulation run showed that there was some lowering in the usage of water for cleaning but the economics were not improved substantially.

Table 2. Summary of Results from Simulation Runs.

	Base Case	Scenario 1	Scenario 2	Near Future
Max. level in RSST_1 (US gal)	9,950	9,940	8,400	8,600
Max. level in RSST_2 (US gal)	14,350	14,200	13,200	9,550
Max. level in RSST_3 (US gal)	10,700	10,800	9,250	11,000
Max. level in RSST_4 (US gal)	10,800	11,200	10,400	7,250
Max. level in RSST_5 (US gal)	9,800	10,600	2,200	9,100
Max. level in RSST_6 (US gal)	7,400	7,400	7,400	6,500
Av. Utilization of Day Tank 1	0.61	0.61	0.57	0.72
Av. Utilization of Day Tank 2	0.39	0.44	0.36	0.50
Av. Utilization of Day Tank 3	0.17	0.16	0.25	0.27
Av. Utilization of Day Tank 4	0.02	0.0	0.05	0.08
Av. Utilization of Day Tank 5	0.0	0.0	0.0	0.01
Av. Utilization of Large Day Tank	0.79	0.77	0.72	0.90
Av. Utilization for Column	0.75	0.76	0.55	0.86
Relative Costs for Economics				
Cost of recovered solvent	+ 2.8	+ 2.9	+ 3.0	+ 3.4
Savings due to not burning solvents	+ 2.4	+ 2.5	+ 2.5	+ 3.1
Cost disposing waste streams	- 1.03	- 1.01	- 1.00	- 1.33
Raw material, assay, operator cost	- 0.3	- 0.3	- 0.3	- 0.4
Total savings	+ 3.9	+ 4.1	+ 4.2	+ 4.8

The maximum levels in the storage tanks were quite similar to the base case.

Scenario 2

In this scenario, a different processing scheme was considered. The P&U, France has a continuous column. This column is suitable for processing one of the recoverable solvent types (RSST_5). Use of the column is expected to reduce the load on the batch distillation column and utilize some of the existing equipment providing some cushion for handling an increased load or upset condition. For Scenario 2, the same column assignment rule as in Scenario 1 was used.

This processing scheme improved the performance some but the overall savings were only slightly higher compared to the two earlier cases. The option did provide a bigger cushion for handling increased loads as noted by the utilization of the batch column (55%). The maximum levels in the recoverable storage tanks are also the lowest for this option.

Near Future

This simulation run was used to determine if the additional load forecasted for the near future can be handled by the proposed design. Projections for the near future were considered. There is an increase of 24% in the amount of recoverable solvent processed for this case. In this simulation, the column assignment rule was the same as the Base Case.

The results show that the proposed design will be able to handle the increases in solvent load forecasted for the near future with some cushion to handle upsets. The economic analysis shows that increase in the load will result in improved savings.

It is noted that operating strategies considered had some impact on the overall economics. One of the options (Scenario 2) considered provided a bigger cushion for handling increased load through integration of an existing continuous distillation column. Also, the maximum level in the recoverable solvent storage tank was the least for this option versus the others based on the production plan.

Conclusions

A BATCHES simulation model provided the necessary framework for integrating the detailed operating features of a multi-product batch process for evaluating the design of a new facility. More importantly, simulation provided a quantitative basis for evaluating the equipment sizing and options and also, operating strategies. A preliminary design was proposed based on the prior experience and some sizing calculations. The simulation study provided refined estimates for equipment requirements. For example, one of the five day tanks in the preliminary design was not needed and the utilization of the fourth tank was low and therefore not essential. Also, the size of the recoverable solvent storage tanks was determined from simulation. The model also accounted for waste stream generation and resource utilization, thus providing good estimates for operator needs and economics.

Equipment sizing based on Scenario 2 has been selected for the final design. Six 12,500 US gallon tanks are considered for storage of recoverable solvents as the period of time for which the level is above 12,500 US gallons is very small for this case. Additionally, the continuous column and associated storage equipment will be integrated into the proposed SRD project.

As more and more accountability is demanded from the teams involved in making capital investment decisions, quantitative justification for the decisions may be needed. Simulation is a convenient approach for a detailed analysis of the inherent complexities of the underlying batch mode of operation.

References

BATCHES User's Manual (1997). Batch Process Technologies, Inc., W. Lafayette, IN.

Kalthod, V.G., S.M. Speaker, S. Kavuri, G.S. Joglekar (1995), Improvement of the recovery of solvents by batch distillation using simulation and optimization at the Upjohn Company. Presented at the AIChE Spring National Meeting, Houston.

NONLINEAR PARAMETER ESTIMATION USING INTERVAL ANALYSIS

Chao-Yang Gau and Mark A. Stadtherr
Department of Chemical Engineering
University of Notre Dame
Notre Dame, IN 46556 USA

Abstract

The reliable solution of nonlinear parameter estimation problems is an important computational problem in chemical process engineering, both in on-line and off-line applications. Conventional solution methods may not be reliable since they do not guarantee convergence to the global optimum sought in the parameter estimation problem. We demonstrate here a technique, based on interval analysis, that can solve the nonlinear parameter estimation problem with complete reliability, providing a mathematical and computational guarantee that the global optimum is found. As an example, we consider the estimation of parameters in vapor-liquid equilibrium (VLE) models. Twelve VLE data sets are fit to the Wilson equation. Results indicate that several sets of published parameter values correspond to local optima only, with new globally optimal parameter values found by using the interval approach.

Keywords

Parameter estimation, Global optimization, Interval analysis, Vapor-liquid equilibrium.

Introduction

Parameter estimation is a common problem in many areas of process modeling, both in on-line applications such as real time optimization and in off-line applications such as the modeling of reaction kinetics and phase equilibrium. The goal is to determine values of model parameters that provide the best fit to measured data, generally based on some type of least squares or maximum likelihood criterion. In the most general case, this requires the solution of a nonlinear and frequently nonconvex optimization problem.

It is not uncommon for the objective function in nonlinear parameter estimation problems to have multiple local optima. However, the standard methods used to solve these problems are local methods that offer no guarantee that the global optimum, and thus the best set of model parameters, has been found. Thus, there is a need for global optimization in nonlinear parameter estimation. One approach that has been suggested is the use of convex underestimating functions in connection with a branch and bound procedure (Esposito and Floudas, 1997). An alternative approach for global optimization is the use of interval analysis (e.g., Hansen, 1992). We demonstrate here the use of interval methods for determining a global optimum in nonlinear parameter estimation problems of interest in process engineering.

As an example, we consider the estimation of parameters in vapor-liquid equilibrium (VLE) models. We demonstrate that even for simple models, such as the Wilson equation, multiple local optima can occur in parameter estimation. It is also shown that for some data sets, published parameter values (Gmehling et al., 1977-1990) correspond to a local but not global optimum. We then demonstrate how a simple global optimization procedure based on interval analysis can be used to reliably determine the globally optimal parameter values. The method used involves the use of an interval Newton technique combined with interval branch and bound. This method provides a mathematical and computational guarantee of global optimality in parameter estimation. The reliability of the method is demonstrated using several VLE data sets, and the globally optimal parameters compared to published values obtained using local methods.

Background

Good introductions to the parameter estimation problem are provided by Bard (1974), Gallant (1987) and Seber (1989). Suppose that observations $y_{\mu i}$ of $i = 1,...,q$ response variables from $\mu = 1,...,p$ experiments are available, and that the responses are to be fit to a model of the form $y_{\mu i} = f_i(\mathbf{x}_\mu, \theta)$, with independent variables $\mathbf{x}_\mu = (x_{\mu 1}, x_{\mu 2},...,x_{\mu m})^T$ and parameters $\theta = (\theta_1, \theta_2,...,\theta_n)^T$. To determine optimal values of the parameters (i.e., the "best" fit), a maximum likelihood criterion is most appropriate in many circumstances. However, with some assumptions this can be simplified to the widely used relative least squares criterion, which requires minimizing the function

$$\phi(\theta) = \sum_{i=1}^{q} \sum_{\mu=1}^{p} [(y_{\mu i} - f_i(\mathbf{x}_\mu, \theta))/y_{\mu i}]^2.$$

This can be treated either as a constrained or, if the experimental observations are substituted directly into the objective function, unconstrained minimization problem. We will consider only the unconstrained formulation of the problem here. For minimizing ϕ, a wide variety of standard minimization techniques are available. However, in general, these are local methods that provide no assurance that a global minimum has been found. What is needed is a technique that can find the global minimum of ϕ, and do so with mathematical and computational certainty. The use of interval analysis provides such a technique.

Interval Analysis

A real *interval* Z is defined as the set of real numbers lying between (and including) given upper and lower bounds; i.e., $Z = [z^L, z^U] = \{z \in \Re \mid z^L \leq z \leq z^U\}$. A real interval vector $\mathbf{Z} = (Z_1, Z_2, ..., Z_n)^T$ has n real interval components and can be interpreted geometrically as an n-dimensional rectangle. Note that in this section lower case quantities are real numbers and upper case quantities are intervals. Several good introductions to interval analysis are available (e.g., Neumaier, 1990; Hansen, 1992; Kearfott, 1996).

Of particular interest here is the interval Newton technique. Given a nonlinear equation system with a finite number of real roots in some initial interval, this technique provides the capability to find (or, more precisely, narrowly enclose) *all* the roots of the system within the given initial interval. For the unconstrained minimization of the relative least squares function $\phi(\theta)$, a common approach is to use the gradient of $\phi(\theta)$ and seek a solution of $\mathbf{g}(\theta) \equiv \nabla \phi(\theta) = \mathbf{0}$. The global minimum will be a root of this nonlinear equation system, but there may be many other roots as well, representing local minima and maxima and saddle points. Thus, for this approach to be reliable, the capability to find *all* the roots of $\mathbf{g}(\theta) = \mathbf{0}$ is needed, and this is provided by the interval Newton technique. In practice, the interval Newton procedure can also be combined with an interval branch and bound technique, so that roots of $\mathbf{g}(\theta) = \mathbf{0}$ that cannot be the global minimum need not be found.

For the system of nonlinear equations $\mathbf{g}(\theta) = \mathbf{0}$ with $\theta \in \Theta^{(0)}$, the basic iteration step in interval Newton methods is, given an interval $\Theta^{(k)}$, to solve the linear interval equation system $G'(\Theta^{(k)}) (\mathbf{N}^{(k)} - \theta^{(k)}) = -\mathbf{g}(\theta^{(k)})$ for a new interval $\mathbf{N}^{(k)}$, where k is an iteration counter, $G'(\Theta^{(k)})$ is an *interval extension* of the Jacobian of $\mathbf{g}(\theta)$, i.e., the Hessian of $\phi(\theta)$, over the current interval $\Theta^{(k)}$, and $\theta^{(k)}$ is a point in the interior of $\Theta^{(k)}$. The interval extension of a real function over an interval is an enclosure of the range of the function over the interval, and can be computed by substituting interval quantities for the corresponding real quantities and using interval arithmetic, or in other ways. It can be shown (Moore, 1966) that any root $\theta^* \in \Theta^{(k)}$ of $\mathbf{g}(\theta) = \mathbf{0}$ is also contained in the *image* $\mathbf{N}^{(k)}$, implying that if there is no intersection between $\Theta^{(k)}$ and $\mathbf{N}^{(k)}$, then no root exists in $\Theta^{(k)}$, and suggesting the iteration scheme $\Theta^{(k+1)} = \Theta^{(k)} \cap \mathbf{N}^{(k)}$. In addition to this iteration step, which can be used to tightly enclose a solution, the following property can be proven (e.g., Neumaier, 1990; Kearfott, 1996): If $\mathbf{N}^{(k)}$ is contained completely within $\Theta^{(k)}$, then there is *one and only one root* contained within $\Theta^{(k)}$. This property is quite powerful, as it provides a mathematical guarantee of the existence and uniqueness of a root when it is satisfied. The foregoing suggests a series of tests to determine whether a stationary point (root of $\mathbf{g}(\theta) = \mathbf{0}$) that might be the global minimum of $\phi(\theta)$ can be contained in $\Theta^{(k)}$:

1. (Function Range Test) Compute an interval extension $\mathbf{G}(\Theta^{(k)})$ containing the range of $\mathbf{g}(\theta)$ over the current interval $\Theta^{(k)}$ and test to see whether it contains zero. Clearly, if $\mathbf{0} \notin \mathbf{G}(\Theta^{(k)}) \supseteq \{\mathbf{g}(\theta) \mid \theta \in \Theta^{(k)}\}$, then there can be no solution of $\mathbf{g}(\theta) = \mathbf{0}$ in $\Theta^{(k)}$ and this interval need not be further tested since it cannot contain a stationary point of $\phi(\theta)$.
2. (Objective Range Test) Compute an interval extension $\Phi(\Theta^{(k)})$ containing the range of $\phi(\theta)$ over the current interval $\Theta^{(k)}$. If the lower bound of $\Phi(\Theta^{(k)})$ is greater than a known upper bound on the global minimum of $\phi(\theta)$, then $\Theta^{(k)}$ cannot contain the global minimum and need not be further tested (see step 3b).

3. (Interval Newton Test) Compute the image $\mathbf{N}^{(k)}$ as described above.

   a. If $\Theta^{(k)} \cap \mathbf{N}^{(k)} = \emptyset$, then there is no root of $\mathbf{g}(\theta) = \mathbf{0}$ in $\Theta^{(k)}$ and it need not be further tested since it cannot contain a stationary point of $\phi(\theta)$.
   b. Evaluate $\phi(\Theta^{(k)})$ and use to determine and update an upper bound on the global minimum for use in step 2.
   c. If $\mathbf{N}^{(k)} \subset \Theta^{(k)}$, then there is *exactly one* root of $\mathbf{g}(\theta)$ in $\Theta^{(k)}$, which may correspond to the global minimum.
   d. If neither of the above is true, then no further conclusion can be drawn.

In the last case, one could then repeat the root inclusion test on the next interval Newton iterate $\Theta^{(k+1)}$, assuming it is sufficiently smaller than $\Theta^{(k)}$, or one could bisect $\Theta^{(k+1)}$ and repeat the root inclusion test on the resulting intervals. This is the basic idea of interval Newton/generalized bisection (IN/GB) methods. A more detailed description of an IN/GB algorithm has been given by Schnepper and Stadtherr (1996). Through the addition of steps 2 and 3b, it has been combined with a simple interval branch and bound scheme. Our current implementation of the IN/GB method is based on appropriately modified routines from the packages INTBIS (Kearfott and Novoa, 1990) and INTLIB (Kearfott *et al.*, 1994). The *worst-case* computational complexity of the IN/GB algorithm is exponential in the number of variables. However, process modeling problems involving over a hundred variables have been successfully solved using this approach (Schnepper and Stadtherr, 1996).

Example

As an example, we consider the estimation from binary vapor-liquid equilibrium (VLE) data of the energy parameters in the Wilson equation for liquid phase activity coefficient. Expressed in terms of the reduced excess Gibbs energy g^E for a binary system and the liquid-phase mole fractions x_1 and x_2, the Wilson equation is $g^E = -x_1 \ln(x_1 + \Lambda_{12} x_2) - x_2 \ln(x_2 + \Lambda_{21} x_1)$, from which expressions for the activity coefficients γ_1 and γ_2 are readily obtained. The binary parameters Λ_{12} and Λ_{21} are given by $\Lambda_{12} = (v_2/v_1) \exp(-\theta_1/RT)$ and $\Lambda_{21} = (v_1/v_2) \exp(-\theta_2/RT)$, where v_1 and v_2 are the pure component liquid molar volumes, T is the system temperature and θ_1 and θ_2 are the energy parameters that must be estimated.

From VLE measurements, experimental values $\gamma_{1,\text{exp}}$ and $\gamma_{2,\text{exp}}$ of the activity coefficients can be obtained. For the parameter estimation problem, the relative least squares objective:

$$\phi(\theta) = \sum_{i=1}^{2} \sum_{\mu=1}^{p} [(\gamma_{\mu i,\text{exp}} - \gamma_{\mu i,\text{calc}}(\theta))/\gamma_{\mu i,\text{exp}}]^2$$

can be used, where $\gamma_{\mu i,\text{calc}}(\theta)$ is determined from the Wilson equation at the same conditions (temperature, pressure and composition) as in the measurement of $\gamma_{\mu i,\text{exp}}$. This parameter estimation problem has been solved for a large number of systems and results presented in the DECHEMA Vapor-Liquid Equilibrium Data Collection (Gmehling *et al.*, 1977-1990), along with the raw VLE data.

For the example here we consider the binary system water(1)–formic acid(2). Twelve VLE data sets, at various pressures, from the DECHEMA Collection were studied. For each data set, the DECHEMA Collection gives the raw VLE data and the results of parameter estimation for θ_1 and θ_2 based on the relative least squares objective. Since Gmehling *et al.*, (1977-1990) use a local method for minimizing $\phi(\theta)$ in this parameter estimation, it is possible that the values of θ_1 and θ_2 obtained do not correspond to a *global* minimum in $\phi(\theta)$. To investigate this, we resolved each parameter estimation problem for the global minimum using our modification of INTBIS.

Results and Discussion

The results for θ_1 and θ_2 and $\phi(\theta)$ from DECHEMA and from the interval approach (INTBIS) suggested here are summarized in Table 1, along with the number of local minima found for each problem (for purposes of determining the number of local minima, the branch and bound steps 2 and 3b were turned off). It can be seen that each problem has multiple local minima, and that in five of the twelve cases (data sets 7–11) the results presented in DECHEMA are not globally optimal. As shown schematically in Fig. 1, the parameter values given in DECHEMA are clustered in two different regions, with results from five data sets falling in one region ($\theta_2 > \theta_1$) and results from seven others falling in the second region ($\theta_1 > \theta_2$). When the global optimum is obtained in the parameter estimation problem the results are much more consistent, with results from ten of the twelve data sets clustered in one region ($\theta_2 > \theta_1$) and only two yielding substantially different results.

We now look more closely at the results for one data set, namely data set 10. For this case, INTBIS (with the branch and bound steps turned off) found five stationary points, three minima and two saddle points, in the initial interval $\Theta_1^{(0)} = \Theta_2^{(0)} = [-10000, 10000]$. These results are summarized in Table 2. The global minimum at $\theta = (-329, 1394)^T$ (root P5) has an objective function value $\phi(\theta) = 0.0819$ that is only about half the magnitude of the local minimum at $\theta = (452, -664)^T$ (root P3) found by Gmehling *et al.* (1977-1990) and reported in the DECHEMA Collection. As is often the case in least squares problems of this sort, all the minima found lie in a relatively flat valley in the parameter space.

The performance of the two different parameter sets, corresponding the local minimum P3 (DECHEMA) and global minimum P5 (INTBIS), in predicting the activity coefficients for water and formic acid with the Wilson equation is shown in Fig. 2. It is clear that when the globally optimal parameter values from INTBIS are used in the Wilson equation, it results in less deviation from the experimental values in comparison to the case in which the locally optimal parameters reported in DECHEMA are used.

It should be emphasized that the sort of difficulties observed in the water–formic acid system, namely the failure of standard local optimization techniques to find the globally optimal parameters, is not restricted to this system and model. This difficulty can be observed in other systems reported in the DECHEMA Collection. This should not be surprising, since with traditional local solution techniques, it is extremely unlikely that the global optimum will always be found.

In determining the global minimum with the interval approach, initial parameter intervals of $\Theta_1^{(0)} = \Theta_2^{(0)} = [-10000, 10000]$ were used for each data set, which should be wide enough to enclose any physically feasible solution. The ability to provide a wide initial interval, as opposed to an initial point guess, means that the method is essentially initialization independent. For each data set, the computation time needed to perform the global optimization was from roughly 10 to 50 seconds on a Sun Ultra 2/1300 workstation. The difference in times is due to the differing number of data points in each data set, and the differing number of stationary points found. It should be emphasized that at this point, no significant efforts have been made to optimize the efficiency of the code. The use of techniques for tightening the evaluation of interval function extensions, as suggested by Tessier (1997) and Hua et al. (1998) can potentially provide an order of magnitude improvement in computational efficiency.

While, in comparison to traditional local methods, additional computation time will typically be required to implement the interval approach, this may be well compensated by the guaranteed reliability of the results. Continuing advances in computing hardware (both in single processor performance and multiprocessing) and software (e.g., compiler support for interval arithmetic) will make this approach even more attractive.

Concluding Remarks

We have described here a new method for reliably solving nonlinear parameter estimation problems. The method is based on interval analysis, in particular an interval Newton/generalized bisection algorithm. The approach provides a mathematical and computational guarantee that the global optimum in the parameter estimation problem is found. We applied the technique here to several data sets in which the Wilson activity coefficient model was used. However, the technique is model independent and can be applied in connection with any thermodynamic model for vapor-liquid equilibrium. The approach presented is general purpose and can also be used in connection with other objective functions, such as maximum likelihood, and other types of VLE measurements. It can also be applied to a wide variety of other nonlinear parameter estimation problems in chemical process engineering.

Acknowledgements

This work has been supported in part by the donors of The Petroleum Research Fund, administered by the ACS, under Grant 30421-AC9, by the National Science Foundation Grants DMI96-96110 and EEC97-00537-CRCD, and by a grant from Sun Microsystems, Inc.

References

Bard, Y. (1974). *Nonlinear Parameter Estimation*. Academic Press, New York

Esposito, W., and C. A. Floudas (1997). Parameter estimation of nonlinear algebraic models via global optimization. Presented *at* AIChE Annual Meeting, Los Angeles, CA, November.

Gallant, A. R. (1987). *Nonlinear Statistical Models*. Wiley, New York.

Gmehling, J., U. Onken and W. Arlt (1977-1990). *Vapor-Liquid Equilibrium Data Collection*. Chemistry Data Series, Vol. I, Parts 1-8, DECHEMA, Frankfurt/Main, Germany.

Hansen, E. (1992). *Global Optimization Using Interval Analysis*. Marcel Dekker, New York.

Hua, J. Z., J. F. Brennecke and M. A. Stadtherr (1998). Enhanced interval analysis for phase stability: Cubic equation of state models. *Ind. Eng. Chem Res.*, in press.

Kearfott, R. B. (1996). *Rigorous Global Search: Continuous Problems*. Kluwer Academic Publishers, Dordrecht, The Netherlands.

Kearfott, R. B., and M. Novoa III (1990). Algorithm 681: INTBIS, a portable interval Newton/bisection package. *ACM Trans. Math. Soft.*, **16**, 152–157.

Kearfott, R. B., M. Dawande, K.-S. Du and C.-Y. Hu (1994). Algorithm 737: INTLIB, a Portable FORTRAN 77 interval standard function library. *ACM Trans. Math. Soft.*, **20**, 447–459.

Moore, R. E. (1966). *Interval Analysis*. Prentice-Hall, Englewood Cliff, NJ.

Neumaier, A. (1990). *Interval Methods for Systems of Equations*. Cambridge University Press, Cambridge, England.

Seber, G. A. F. (1989). *Nonlinear Regression*. Wiley, New York.

Schnepper, C. A., and M. A. Stadtherr (1996). Robust process simulation using interval methods. *Comput. Chem. Eng.*, **20**, 187–199.

Tessier, S. R. (1997). *Enhanced Interval Analysis for Phase Stability: Excess Gibbs Energy Models*. M. S. Thesis, University of Notre Dame, Notre Dame, IN

Table 1. Summary of Parameter Estimation Results for the Water(1) and Formic Acid(2) System, Showing Parameters θ_1 and θ_2 and Objective Function $\phi(\theta)$. (Values shown in bold are globally optimal parameters that differ from those given in DECHEMA).

Data Set	P (mm Hg)	DECHEMA θ_1	θ_2	$\phi(\theta)$	INTBIS θ_1	θ_2	$\phi(\theta)$	No. of Minima
1	760	-195	759	0.0342	-195	759	0.0342	2
2	760	-278	1038	0.0106	-278	1038	0.0106	2
3	760	-310	1181	0.0151	-308	1167	0.0151	2
4	760	-282	985	0.353	-282	984	0.353	2
5	760	-366	1513	0.0257	-365	1509	0.0257	3
6	760	1067	-1122	0.0708	1065	-1120	0.0708	2
7	200	892	-985	0.141	**-331**	**1250**	0.0914	2
8	200	370	-608	0.0459	**-340**	**1404**	0.0342	3
9	100	539	-718	0.165	**-285**	**996**	0.111	2
10	100	450	-663	0.151	**-329**	**1394**	0.0819	3
11	70	558	-762	0.0399	**-330**	**1519**	0.0372	3
12	25	812	-1058	0.0502	807	-1055	0.0502	2

Table 2. Details for Roots (Stationary Points) Found using INTBIS for Data Set 10. (Point P3 is the local minimum presented in DECHEMA, while point P5 is the global minimum).

Root	Position (θ_1, θ_2)	Eigenvalues of Hessian	$\phi(\theta)$	Status
P1	(1958, -1251)	7.55E-5, 2.58E-7	0.164	minimum
P2	(1165, -1083)	6.83E-5, -1.44E-7	0.178	saddle
P3	(452, -664)	6.97E-5, 9.42E-8	0.151	minimum
P4	(-37.8, 38.5)	9.08E-5, -3.54E-7	0.19	saddle
P5	(-329, 1394)	1.23E-4, 1.47E-7	0.0819	global minimum

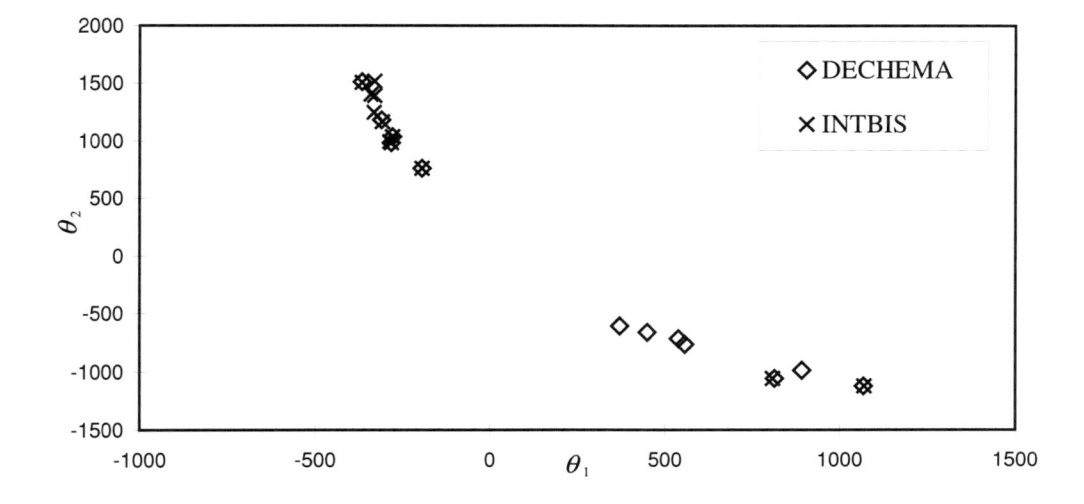

Figure 1. Comparison of estimation results from DECHEMA and INTBIS.

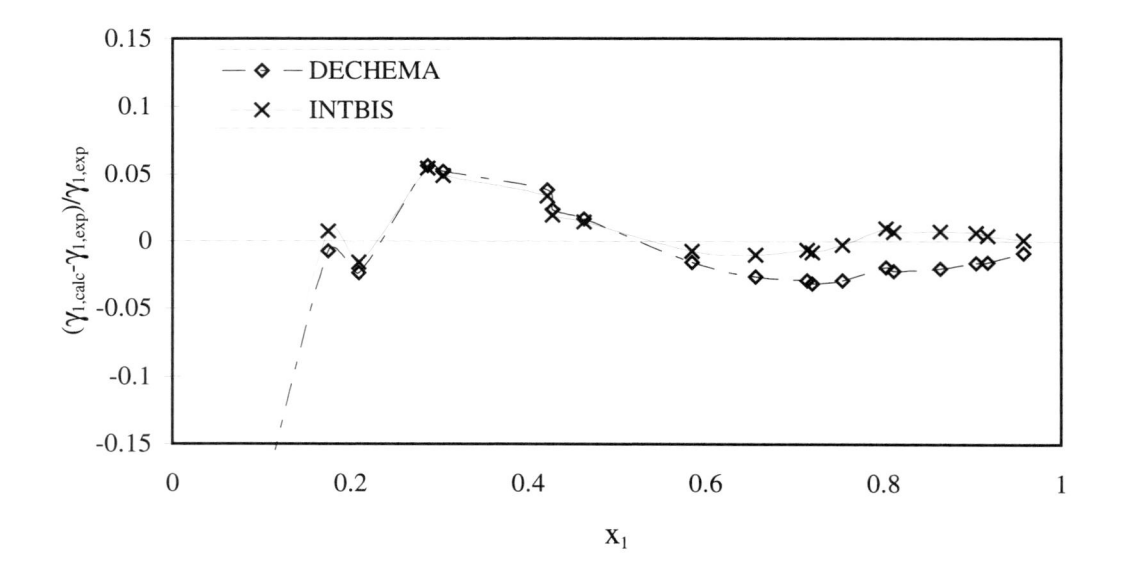

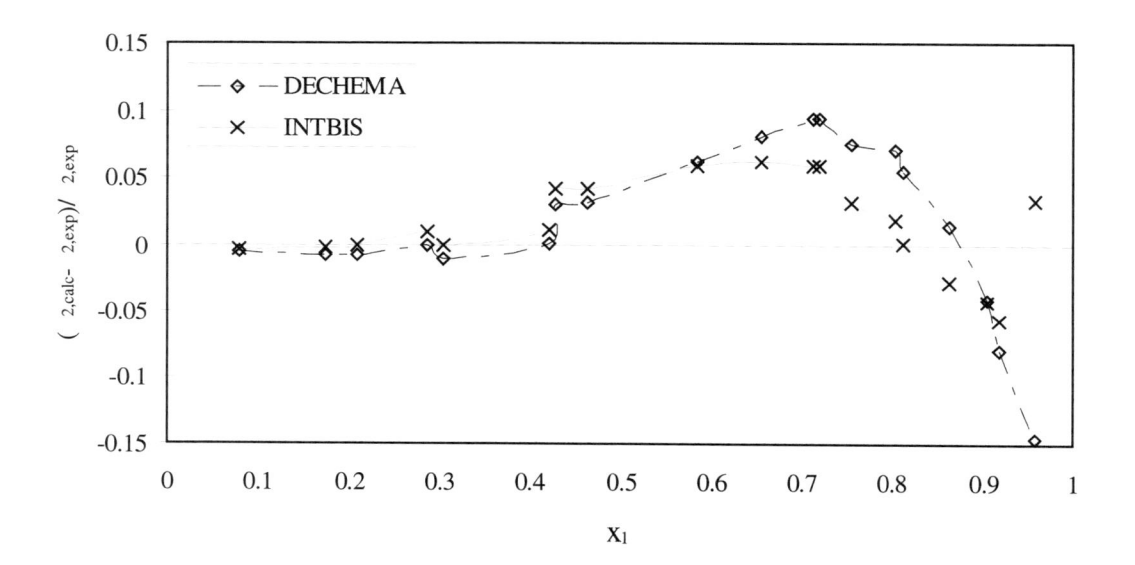

Figure 2. The relative error in data set 10 between calculated and experimental activity coefficients for water (top) and formic acid (bottom) resulting from the locally optimal DECHEMA values and the globally optimal values found using INTBIS. For water (top), the relative error for a data point at $x_1 = 0.0802$ is off scale (roughly at -0.22) for both cases.

A SYSTEMATIC PROCEDURE FOR OPTIMAL OPERATION OF HEAT EXCHANGER NETWORKS

Bjørn Glemmestad*
Norsk Hydro, Research Centre
N-3901 Porsgrunn, Norway

Truls Gundersen
Thermal Energy and Hydro Power
Norwegian University of Science and Technology
N-7034 Trondheim, Norway

Abstract

Heat integration of industrial processes for reduced energy cost introduces interactions that significantly affect the operational aspects of such plants. This paper presents a comprehensive methodology for optimal operation of Heat Exchanger Networks. The extra degrees of freedom normally available in Heat Exchanger Networks should be utilized during operation to minimize energy cost. The first step in the procedure is an analysis of degrees of freedom available for energy optimization during operation. The remaining steps in the procedure span from simple qualitative methods for regulatory control to more involved methods for selection of secondary variables that are used for on-line periodic optimization of setpoints in higher level control to minimize energy cost.

Introduction

Reducing operating cost represents an important challenge for today's process industry. Existing plants may have been designed non-optimal compared to present state of the art or the external conditions for operation may have changed since the plants were built. Redesigning such plants may not be profitable. Even for modern and properly designed plants, operating costs are always traded off against capital cost in order to optimize overall economy. In all cases, one is faced with the problem of operating an existing plant "as good as possible".

In many plants, energy cost constitutes a significant part of the total operating cost. Energy is recovered in Heat Exchanger Networks (HENs) and design (synthesis) of such HENs is a mature research field (see e.g. Gundersen and Naess, 1988). Looking at the large literature on HEN synthesis, the number of publications concerning operation of HENs is quite small in comparison. This paper proposes a systematic procedure for operation of HENs. It is assumed that the HEN structure is given and that heat exchanger areas and targets for the outlet temperatures are fixed. In addition, supply temperatures and heat capacity flowrates including prespecified disturbance ranges are defined.

The purpose of the procedure presented here is to achieve optimal operation for the HEN. Optimal operation is defined according to the following three goals:

1. Satisfy target temperatures at steady state
2. Minimize utility cost at steady state.
3. Satisfactory closed-loop dynamic performance.

During operation of a HEN for maximum energy efficiency it is required that other parts of the plant are not affected, thus inlet temperatures to equipment such as reactors, distillation columns or flash tanks should be unchanged. The primary goal (satisfy targets) ensures that the operating conditions for other parts of the process are not affected by the HEN operation (at steady state). Both

---

* Author to whom correspondence should be addressed. Fax: +47 3556 3686. E-mail: Bjorn.Glemmestad@hre.hydro.com

primary and secondary goal apply at steady state only, indicating that focus is on steady state performance. The third goal concerns dynamics and it is deliberately somewhat more vague than the two first goals. While it is possible to put up quantitative and rigorous requirements for dynamic performance, the procedure in this paper will employ simple rules of thumb to achieve satisfactory closed-loop performance.

The optimal operation problem for HENs basically seeks an answer to the question: "For a deviation in an outlet temperature, which manipulation should be adjusted in order to bring the temperature to its target and at the same time minimize utility cost." Optimal operation of HENs was studied by Marselle *et al.* (1982) and they proposed a method based on graph-theory and on-line experimenting of manipulations to find the optimal operating state. Calandranis and Stephanopoulos (1986) proposed a procedure based on introducing "notional" coolers and heaters in order to find a strategy for optimal operation. They adopted, however, the idea of constraining minimum temperature difference (from the pinch design method in HEN synthesis) into operation where it has no relevance, see Glemmestad (1997, p.23-24). Bojaci *et al.* (1996) suggested a method based on steady state opti-mization, but their objective function, minimizing the 1-norm (sum of absolute values) of temperature deviations, is not directly related to utility consumption. Mathisen (1994, chapter 5) proposed a method for optimal operation based on structural information of the HEN which forms the basis for one of the steps in the procedure presented here.

The rest of the paper is organized as follows: The next section first introduces the overall method before the three main steps are described in more detail in separate subsections. Then an example is presented before the discussion and conclusions.

The Procedure

Before any optimizing control scheme is assessed, it should be verified that there actually are degrees of freedom (DOFs) available for optimization. The first step in the procedure is thus a DOF analysis. If optimization is feasible (DOFs exist), one should proceed developing an optimizing control strategy. The procedure consists of the following four main steps.

1. Analyze the number of DOFs available for utility cost optimization.
2. Develop a control strategy based on structural information only with quantitative verification of flexibility and qualitative evaluation of closed-loop performance.
3. For parts of HENs where step 2 do not yield satisfactory results, develop a control strategy based on steady state optimization and a proper selection of secondary measurements.
4. Simulate the controlled HEN dynamically to verify the suggested control strategy.

While the individual steps of the procedure have been presented in earlier publications, this paper demonstrates how these steps can be combined into one systematic and comprehensive method for optimal operation of HENs.

The two most common manipulations for controlling the outlet temperatures in a HEN are bypass fractions and duty of utility exchangers. Likewise, the outlet temperatures with specified targets may be divided into two corresponding groups. (1) Utility controlled targets are outlet temperatures on process streams with a final utility exchanger. These outlet temperatures are normally controlled by the duty of the final utility exchanger. (2) Outlet temperatures with targets that are located on streams without final utility exchangers are denoted bypass controlled targets, since these are normally controlled using a bypass fraction.

As explained in the introduction, the procedure attempts to identify a control scheme for optimal operation of HENs with given structure, fixed heat exchanger areas, etc. In practice, we have two main applications. For an existing network, the procedure can be used for best possible operation using only the installed bypasses. Alternatively, new bypasses may be identified that if installed would significantly improve operation. The second major application is the grassroot case, where bypasses may be assumed for all heat exchangers, but only those actually utilized in the procedure are installed.

It is emphasized that in operation, each subnetwork (process streams and heat exchangers that form independent networks) should be treated separately. For practical cases, this means that the problem size in operation often is smaller than in synthesis. The following four subsections explains the individual steps in somewhat more depth.

Step 1. Analysis of Degrees of Freedom (DOFs)

In this step the number of DOFs available for optimization of utility cost is determined. It is, however, not always interesting to know the exact number of DOFs, and the main purpose of this step is to decide whether optimization is feasible or not. To begin with, it is important to have a common understanding of what is meant by DOFs. With DOFs in this paper, we mean (1) the extra degrees of freedom after regulatory control is implemented (to satisfy targets), and (2) only the degrees of freedom that can be used for utility optimization. These points are illustrated in the trivial HEN in Fig. 1.

This HEN has four manipulations (two bypass fractions and two utility exchangers with variable duties) and two targets (constraints), thus one might incorrectly expect $N_{DOF}=2$.

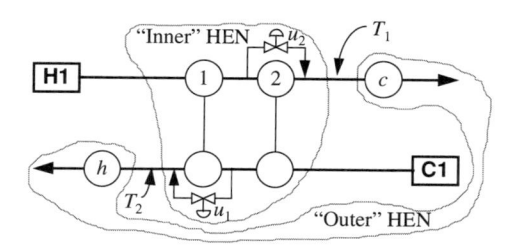

Figure 1. Trivial HEN with $N_{DOF} = 1$.

However, drawing a system boundary around heat exchangers 1 and 2 (the "inner" HEN), it is clear that the two bypasses u_1 and u_2 only span a one-dimensional space in T_1 and T_2 (to satisfy the energy balance). With the two bypasses only contributing with one free variable that affects utility consumption, the two additional variables (utility duties) and the two targets imply that $N_{DOF}=1$ for this HEN. In addition to this, one may shift the duty between heat exchangers 1 and 2. This does not affect the utility consumption, thus it does not increase N_{DOF} (it does, however, explain the "missing DOF"). Such singularities do not only occur when placing two heat exchangers side by side on two streams, they always occur in loops when all heat exchangers have bypasses and all process streams have targets.

The crucial step in order to find the DOFs for a more general HEN is the division into the "inner" HEN and the "outer" HEN as illustrated for the simple network in Fig. 1. The inner HEN consists of all process streams and process-to-process heat exchangers while the outer HEN consists of process streams with targets (after they leave the inner HEN) and the final utility exchangers (if present). Before presenting a general expression for N_{DOF}, a simplified formula will be derived. The utility consumption takes place in the outer HEN where the utility exchangers are located. Each process stream entering the outer HEN may represent a free variable since the temperatures between the inner and outer HEN (T_1 and T_2 in Fig. 1) can be used for controlling a target or for reducing utility consumption. For example, reducing T_1 in Fig. 1 will reduce the cooler duty. However, the energy balance around the inner HEN will restrict the number of free variables from the inner HEN to be at most one less than the number of process streams. Assuming that all process streams have targets, the number of free variables from the inner HEN thus has an upper bound of N_T-1. (N_T is the number of targets). That is, no matter how many manipulations there are in the inner HEN, they can never span a space with dimension larger than N_T-1 in the streams entering the outer HEN. The utility exchangers in the outer HEN serve as additional free variables. However, having more than one utility exchanger on the same utility stream (or type) will form loops that do not contribute to the DOFs. E.g. shifting the duty between two heaters using the same steam quality will not affect utility cost. This implies that the number of free variables introduced in the outer HEN is actually not given by the number of utility exchangers, but by the number of utility streams N_U (different utility types), see Glemmestad (1997, p.62). Since N_T free variables have to be used for control (first goal of optimal operation), an upper bound for the number of DOFs when all process streams have targets, is

$$N_{DOF} \leq N_U - 1 \qquad (1)$$

Outlet temperatures without targets (free outlet temperatures) do not enter the outer HEN. For HENs with free outlet temperatures, the manipulations in the inner HEN may span the whole space of N_T streams entering the outer HEN. The free outlet temperature serves as a sink/source for the energy balance in the inner HEN. For HENs with free outlet temperatures, an upper bound for N_{DOF} is simply

$$N_{DOF} \leq N_U \qquad (2)$$

Note that equation (2) is also valid when there are more than one free outlet temperature. This is due to the fact that free outlet temperatures eliminate the energy balance constraint on the number of free variables entering the outer HEN. Thus, more free outlet temperatures will not increase the DOFs any further.

Assuming that the number of free variables in the inner HEN is only restricted by the number of streams entering the outer HEN and the energy balance constraint, then (1) and (2) turn into equalities that serve as simple expressions for N_{DOF}. If this assumption does not hold, i.e. the number of free variables entering the outer HEN may be limited by the number of manipulations or by singularities in the inner HEN, then N_{DOF} is given by the more general equation (Glemmestad, 1997, p.63):

$$N_{DOF} = R + N_U - N_T \qquad (3)$$

R is the rank of the inner HEN, i.e. the number of free variables entering the outer HEN. This equation is general and accounts for e.g. free outlet temperatures, manipulations on freely selected units and singularities in the inner HEN.

The DOF analysis in this first step may result in one of the three following cases:

1. $N_{DOF} < 0$. The HEN is infeasible (targets cannot be met when disturbances are encountered) and needs redesign in order to meet the first goal of optimal operation.
2. $N_{DOF} = 0$. The HEN is feasible, but optimization is not possible.
3. $N_{DOF} > 0$. Optimization is possible. Proceed with the next steps.

To conclude this subsection, it has been explained that extra manipulations (after regulatory control is implemented) may not always be used for optimization during operation and a quantitative expression for the DOFs available for optimization is presented. Some earlier publications (e.g. Marselle et al., 1982) have presented expressions for N_{DOF} that may fail to find the DOFs that can be used to optimize energy cost. A more thorough treatment of DOFs in operation of HENs is given by Glemmestad (1997, chapter 4). In the remaining steps of the procedure, we assume that there are DOFs available for optimization.

Step 2. Control Strategy from Structural Information

For those subnetworks that have DOFs available for optimization, we first apply a structure based method. That is, the HEN structure is used to derive how the network should be operated in order to minimize energy cost. The method is based on an idea presented by Mathisen (1994, chapter 5) and described more thoroughly in Glemmestad et al. (1996) and Glemmestad (1997, chapter 5). First the HEN is divided into the following two subsystems:

$$y^{\text{BP}} = G^{\text{BP}} u \qquad (4)$$

$$y^{\text{U}} = G^{\text{U}} u \qquad (5)$$

Here, y^{BP} is a vector containing the bypass controlled target temperatures and G^{BP} is the transfer matrix from the input vector u (bypasses) to y^{BP}. In equation (5), G^{U} is the transfer matrix from u to the utility consumption for each of the utility controlled temperatures (assuming that the utility controlled targets are perfectly controlled). In the structure based method, however, the matrices G^{BP} and G^{U} are not normal transfer matrices, they simply contain the signs of such matrices at steady state according to

$$\text{sign}(G)_{ij} = \begin{cases} [+] & \text{if } g_{ij} > 0 \text{ at steady state} \\ [-] & \text{if } g_{ij} < 0 \text{ at steady state} \\ [0] & \text{if } g_{ij} = 0 \text{ at steady state} \\ [\pm] & \text{otherwise (i.e. unknown sign)} \end{cases} \qquad (6)$$

An important feature of the method is that the sign matrices can be easily constructed from the HEN structure only. This is done using simple insight such as the fact that an increased bypass fraction around a heat exchanger decreases its duty. Thus, the corresponding element in G^{BP} from the bypass to the cold exit temperature of that exchanger is [–]. Similarly, if the hot exit temperature has a cooler, the corresponding element in G^{U} is [+], since increasing the bypass fraction will increase the cooler duty. Mathisen (1994, chapter 5) and Glemmestad (1997, chapter 5) give more complete descriptions about how the sign matrices are constructed.

The key issue of this step is to combine the different elements in the sign matrices to find how the different bypasses affect utility consumption when they are used to control the outlet temperatures. The manipulations are divided in three different priorities:

1. Manipulations that decrease utility consumption when used for canceling a control error.
2. Manipulations with zero or mixed effect on utility consumption.
3. Manipulations that increase utility consumption when used for canceling a control error.

To distinguish between manipulations with the same priority, some heuristic rules based on controllability are used. The priorities are used to construct a priority table for the manipulations. This priority table contains information such as: "If outlet temperature 1 is too high, then prefer reducing bypass A, if this bypass saturates then increase bypass B". The control strategy is generated from this priority table, and the control structure may have variable pairing. That is, if the bypass controlling a given output temperature saturates, a new bypass will be assigned (during operation) for control of that output.

The structural method cannot always guarantee that the global optimal operating state is found. It is possible that general rules can be developed to distinguish a priori whether or not the structural method will identify global optimal operation. This is not attempted here, rather it is recommended simply to apply the method and see if the resulting control strategy is unambiguous or if it is unclear.

It is important to note that the structural method only gives recommendations about the priority order for the manipulations. It does not give any information about which manipulations are actually used for different operating conditions. Simple calculations can be performed in order to find which manipulations that will be used for different disturbances. Further, controllability should be evaluated qualitatively. As a rule of thumb, a bypass directly affecting the controlled temperature is good, and a bypass directly affecting the opposite stream of the last heat exchanger (if this is close to counter-current) is good. If other manipulations have to be used (may lead to bad closed-loop performance) or if it is not clear which manipulations should be used in order to minimize utility cost, then step 3 should be performed.

Step 3. Steady-state Optimization and Selection of Secondary Measurements

In this step, a fixed control structure for the controlled target temperatures is assigned. Usually, this can be done simply by assigning the manipulation closest to each target. When this is done, one or more extra manipulations will be present and the key issue is how to use these extra degrees of freedom.

During operation, a steady state model of the HEN is optimized periodically to find the operating state that minimizes utility cost. We do, however, require that the HEN remains feasible (targets can be met) between the optimizations even when unknown disturbances and model errors within prespecified bounds are encountered. If optimum is constrained, it is necessary to back off some-what from the nominal optimum to guarantee feasibility.

The steady state model may compute optimal values for the extra manipulations and these values may be applied directly to the actuator device. However, in many cases it will be better to compute setpoints for some temperatures internally in the HEN and let the extra manipulations control these temperatures through simple feedback loops. Computing the optimal internal temperature setpoints is simpler than computing the bypass fractions since some non-linearities are avoided. More important, however, is the fact that the selection of internal setpoints (or selecting bypass fractions directly) affects how much one has to back off from the nominal optimum and how close to the global optimum it is possible to operate the HEN. Thus, in addition to performing periodic steady state optimization, an important part of this step is the optimal selection of internal setpoints (secondary measurements) for the extra manipulations. The method for selecting internal setpoints is based on a method presented in Skogestad and Postlethwaite (1997, chapter 10), where singular values for a properly scaled system are used. In the procedure described in this paper, a more direct approach is accomplished using tables to select the best secondary measurements.

In short, the method for selecting secondary measurements consists of the following four steps:

a) Select possible candidates for the measurements. The open-loop implementation (applying the control signal directly to the manipulation) should be one of the candidates. Assumptions about unknown disturbances and model errors should also be made.
b) Create a table with one row for each measurement candidate and one column for each unknown disturbance "corner point". For each candidate setpoint, compute the optimal value for "all" unknown disturbances and list these values in the table.
c) From the table in the previous step, select the optimal setpoint (for each candidate) that is feasible for all unknown disturbances. Fix each setpoint (one by one) at this value and compute the objective function (utility cost) for all unknown disturbances.
d) For each candidate, compute the utility cost averaged for all disturbances. Select the candidate that gives the lowest average utility cost.

For the selected measurement(s), the "back-off" constraint (safety margin) on the extra manipulation should also be computed. This secondary measurement selection and back-off calculation is done once prior to operation. During operation, the optimal values for the selected setpoints are computed periodically and applied to the controller. The safety margin for each extra manipulation is included in the steady state model to provide a proper back-off without needing recalculation each time an optimization is performed.

A more thorough description of the method briefly described in this subsection, including the steady state model and a detailed example is given by Glemmestad et al. (1997) and Glemmestad (1997, chapter 7).

An alternative approach to using the "corner points" in steps b and c above is to generate random unknown disturbances (according to suitable probability distributions) and to penalize target deviations instead of considering them infeasible. For some applications, the number of corner points may be large, and for flowrate variations the worst case disturbance may not be located at a corner point. In such cases, this alternative Monte Carlo approach may be recommended.

Step 4. Dynamic Simulation of the Controlled HEN.

When a control strategy is developed in step 2 and/or step 3 above, it is necessary to verify the closed-loop performance through dynamic simulations. While this may certainly be done using a commercial dynamic process simulator, it is also possible to implement it in more general computer tools such as Matlab/Simulink with relatively little effort. Glemmestad (1997, chapter 3) proposed to apply a hybrid model based on the mixing tank principle using the arithmetic mean temperature difference for each mixing tank. This provides a good trade-off between steady state and dynamic properties together with small requirements for computing power.

If the dynamic simulations do not show satisfactory results, one may have to go back and make other selections in step 2 or 3.

Example

The HEN used for the example is shown in figure 2. It is emphasized that the example is included to demonstrate the overall procedure. Each main step is briefly explained, but the example is far from fully developed in terms of showing all the details. For the subsequent discussion, it is important to notice that the heaters use the same steam quality and that cooling water is used in the coolers.

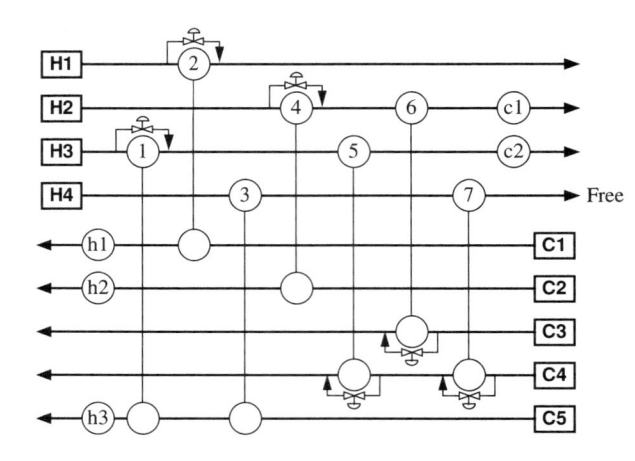

Figure 2. HEN used in demonstration example.

Before applying the procedure, we identify three separate subnetworks: Subnetwork A consists of streams H1 and C1, subnetwork B of streams H2, C2 and C3 and subnetwork C consists of streams H3, H4, C4 and C5. Notice that in design these subnetworks are linked through the utilities, while they act as independent networks in operation.

Step 1. DOF

Subnetwork A has $N_{DOF}=0$ and is excluded from the rest of the procedure since it cannot be optimized. For subnetwork B we have $R=2$, $N_U=2$ and $N_T=3$, thus from equation (3) we have $N_{DOF}=1$. Subnetwork C has $R=3$, $N_U=2$ and $N_T=3$, thus from equation (3) $N_{DOF}=2$.

Concluding step 1, subnetworks B and C can be optimized and control strategies for these will be developed in the next two steps.

Step 2. Structural method

Applying the structural method for subnetwork B results in a unambiguous conclusion for the control strategy that minimizes energy cost. The control strategy switches between u_4 and u_6 and this scheme is implemented simply through split-range control. All details regarding this example is given by Glemmestad (1997, section 5.2 and example 5.1).

For subnetwork C, the structural method does not guarantee global optimum and the suggested control scheme gives rather poor dynamic behavior, see Glemmestad (1997, section 5.8). For this subnetwork, the method in step 3 should be investigated.

Step 3. Steady state optimization

Only subnetwork C remains, and manipulations for the target temperatures (outlet temperatures on all streams except H4) are selected first. For this purpose, final utility exchangers are chosen where present (on streams H3 and C5), and the target temperature on stream C4 is controlled using u_5 (bypass around heat exchanger 5) This leaves the manipulations u_1 and u_7 to be used for optimization of energy cost. Candidates for the secondary (internal) setpoints are hot and cold exit temperatures from heat exchangers 1 and 7 in addition to the bypass fractions (u_1 and u_7) themselves. Due to space limitations, the details for subnetwork C are not shown here.

Step 4. Dynamic simulation

Again, due to space limitations, no results from dynamic simulations are shown here. For subnetwork A, one should not expect any problems related to closed-loop behavior. Glemmestad (1997, p.74) shows dynamic responses for subnetwork B including the split-range control strategy. Dynamic responses for subnetwork C when the control structure suggested by the structural method in step 2 is implemented, are shown in Glemmestad (1997, p.87).

Conclusions

A systematic procedure for energy efficient operation of heat integrated plants is presented. The purpose is to develop a control strategy for operation of a previously designed Heat Exchanger Network (HEN) at minimum energy cost. For HENs that consist of several subnetworks, each subnetwork is treated separately in the procedure.

To check the scope for optimization, an analysis of degrees of freedom is carried out as a first step. It is shown that extra manipulations (after regulatory control is implemented) may not always be used for optimization, and a quantitative expression for the degrees of freedom available for optimization is presented. Provided that optimization is feasible, the next steps are performed to develop the control strategy. First, a structure based method that is easily carried out is applied. The control strategy suggested by this method is evaluated and may be implemented in the final solution. If the result from the structural method is unclear, a more involved method based on periodic steady state on-line optimization is applied. This method also includes optimal selection of some internal setpoints and a back-off calculation to ensure feasibility when unknown disturbances are encountered. The final step includes dynamic simulation of the closed-loop system in order to verify the control strategy.

The main steps in the procedure are presented in earlier publications, and this paper has combined these methods into one systematic overall procedure for optimal operation of HENs.

Nomenclature

- Ci Cold stream number i.
- Hi Hot stream number i.
- R Rank of inner HEN.
- N_{DOF} Number of degrees of freedom.
- N_T Number of targets.
- N_U Number of utility streams (different utility types).
- T Temperature.
- u Manipulated input (often bypass fraction).
- y Output.

References

Bojaci, C., D. Uzturk, A.E.S. Konukman, and U. Akman (1996). Dynamics and Optimal Control of Flexible Heat Exchanger networks. *Comput. Chem. Engng.*, **21** Suppl. B, S775-S780.

Calandranis, J. and G. Stephanopoulos (1986). Structural Operability Analysis of Heat Exchanger Networks. *Chem. Eng. Res. Des.* **64**, 347-364.

Glemmestad, B., K.W. Mathisen and T. Gundersen (1996). Optimal Operation of Heat Exchanger Networks based on Structural Information. *Comput. Chem. Engng.* **20** Suppl. B, S823-S828.

Glemmestad, B. (1997). Optimal Operation of Integrated Processes, Studies on Heat Recovery systems. *Dr.ing. (Ph.D) thesis*, Norwegian University of Science and Technology, Trondheim, Norway.

Glemmestad, B., S. Skogestad and T. Gundersen (1997). On-line Optimization and Choice of Optimization Variables for Control of Heat Exchanger Networks. *Comput. Chem. Engng.* **21** Suppl., S379-S384.

Gundersen, T. and L. Naess (1988). The Synthesis of Cost Optimal Heat Exchanger Networks. An Industrial Review of the State of the Art. *Comput. Chem. Engng.* **12**, 503-530.

Marselle, D.F., M. Morari and D.F. Rudd (1982). Design of Resilient Processing Plants - II. Design and Control of Energy Management Systems, *Chem. Eng. Sci.*, **37**, 259-270.

Mathisen, K. (1994). Integrated Design and Control of Heat Exchanger Networks. *Dr.ing. (Ph.D) thesis*. University of Trondheim - NTH, Norway.

Skogestad, S. and I. Postlethwaite (1997). *Multivariable Feedback Control, Analysis and Design*. John Wiley & Sons, Chichester, UK.

ON THE EFFICIENCY OF MATERIAL FLOW SIMULATION FOR PROCESS AND MULTIPURPOSE PLANT OPTIMIZATION

Philippe Solot, Linus Willmann and Tibor Dudás
AICOS Technologies AG
Efringerstrasse 32
CH-4057 Basel, Switzerland
E-mail: psolot@aicos.com

Abstract

Engineers designing production plants in the process industries traditionally have recourse to the well-known process simulation techniques. However, especially in multipurpose plants, the co-existence of a large number of products offers opportunities to improve productivity by addressing issues that are rather of organizational nature. These in fact belong to the field of production logistics and can be dealt with using a specific type of simulation, the so-called material flow simulation.

The objectives pursued, such as plant debottlenecking and production capacity optimization, can be best achieved if software tools are used that take the numerous particularities of the process industries into account – e.g. batch and semicontinuous manufacturing. Such tools indeed enable engineers to quickly set up accurate models, using a language they are used to, and to easily design and compare alternative production scenarios.

In this paper, the typical data required by these models and the results they provide are first presented. Two practical case studies from the fine chemical industry are then considered: the first one concerns the optimization of a single production process, whereas the second involves a multipurpose plant. For each of them, both the problem description, the recommendations made and their impact on production are discussed. Production capacity improvements of up to 30% could be achieved.

Keywords

Plant (re)design, Debottlenecking, Capacity optimization, Production logistics, Material flow simulation.

Introduction

In all branches of industry, the recent trends to market globalization have been accompanied by the emergence of new competitors that introduce quality products at prices being often much lower than those of the existing offer. The resulting increase in economical pressure has led companies to search ways of improving their business efficiency, as evidenced by the development of "new" ideas and terms such as *business process re-engineering*.

In this context, production undoubtedly constitutes a key process of the whole business chain, where two issues can be distinguished for possible improvements. On the mid- or long term, the *design* of both the products and their production processes can be analyzed in detail so as, e.g., to eliminate time losses by suppressing unnecessarily complex operations. On the short term, *planning* and *scheduling* aspects can be considered, usually with the aim of best using the equipment available in order to guarantee on-time deliveries.

As far as the process – especially the chemical and pharmaceutical – industries are concerned, the design aspects can again be split into two categories. Firstly, at a rather technical level, thermodynamic parameters can be tuned in order to improve production quality and efficiency. Process simulation proves very useful to optimize these parameters and has therefore become a standard technology for engineers designing production processes and plants. However, in multipurpose plants, the scope of such improvements is limited by the product diversity, in the sense that shared equipment can hardly be designed to fit optimally to every product.

Nevertheless, at the logistic level, the high complexity of these plants hides further optimization possibilities. Indeed, material-flow oriented decisions such as:

- an increase of the size of buffer storage used for intermediate products, or
- the introduction of an additional parallel device for performing a given operation,

may result in unexpectedly large production capacity increases, in particular when these changes are designed so as to systematically eliminate conflicts between material flows. Just like at the technical level, simulation methods again provide engineers with the support necessary to efficiently compare different production scenarios; however, the technique used is here different: it is *material flow simulation*.

In this paper, we show how material flow simulation can be used to optimize batch and semicontinuous processes and multipurpose plants. We first present the structure of such a simulation model, using the software tool SIMBAX 4.1 as a reference. We then illustrate these concepts on the basis of two examples derived from practical applications. The first one deals with the optimization of a single process, whereas the second addresses a more complex situation that arose in a multipurpose plant. We finally draw some conclusions about the use of material flow simulation in the process industries.

Material Flow Simulation in the Process Industries

Although quite a lot of software packages have been developed for production optimization in the last fifteen years, most of them can hardly be applied to the batch manufacturing industries due to the numerous particularities of their production processes (e.g. Rippin 1993, Puigjaner and Espuña 1998). Among them, the existence of transfers between devices without any possibility of intermediate storage, the concept of campaigns consisting of several identical batches produced in a row, the wide spectrum of operations going from semicontinuous transfers to container handling, and the long set-up – resp. cleaning – operations required at the beginning – resp. the end – of a campaign can in particular be quoted.

As far as material flow simulation is concerned, these characteristics can only be taken easily into account if a specific software tool such as SIMBAX (Solot and Dudás 1998) is used. A typical model of a batch production facility then decomposes into:

- a model of the plant itself, the so-called layout,
- models of the production processes, i.e., of the recipes, and
- a model of the production orders, i.e., of the production plan.

The layout first represents the equipment available, such as reactors, filters, dryers and tanks, and the connections enabling material transfers between them. Each device is characterized by additional parameters, e.g., its volume capacity. Buffer storage places where containers can be kept temporarily may also be defined. Finally, information about the additional resources required for production (manpower, energy, etc.) belongs to such a model if their availability is expected to be a limiting issue. The plant working schedule may in particular be specified.

The development of such a model can be achieved in a straightforward way on the basis of a usual engineering factory plan. Indeed, only very little additional information – e.g. about manpower – is required.

As for the second type of model component, each recipe is modeled as such and represents a production process with its different stages. Each stage further decomposes into single steps which are carried out sequentially in a same plant device and which may correspond to material transfers, pure processing operations – e.g. heating or filtering –, cleaning processes, synchronization requirements... The definition of a step in general specifies its duration, the energy or manpower required for its execution, and the product and quantity transferred if appropriate: one will for instance *load 200 kg of RawMaterial1, using 1 worker, for 10 minutes*.

Most of the data necessary for setting up the recipe models can usually be derived from documents specifying directions for production. Furthermore, in case of ambiguity or incompleteness of these documents, the missing information can often be obtained over interviews performed on the shop floor, especially with the foremen.

Finally, in multipurpose plants, information about the production orders must be included in the model, since a realistic simulation can only be run if one knows in which quantity, at which moment, and according to which recipe a product is to be manufactured. Such data can easily be obtained by combining historical information issued from the production planning system and sales forecasts made available by the marketing department.

Figure 1 summarizes the input data detailed above and presents the various results that can be derived from the simulation analysis (see also Strohrmann and Krames 1997). The main results consist of the cycle time of each production process, the utilization rate of each device as well as statistics about energy and manpower requirements. They make it possible to determine the various plant bottlenecks and to design logistic modifications leading to their elimination, hence to an increase of the plant production capacity. All the results can of course be translated into costs – e.g. yearly production costs – so as to compare various scenarios on a financial basis.

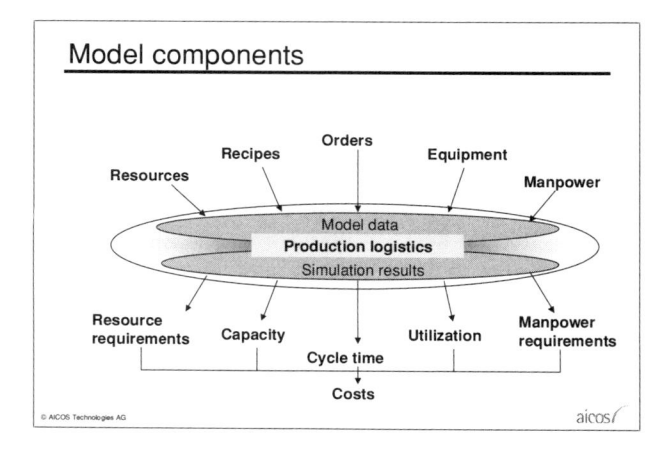

Figure 1. Components of a typical material flow simulation model and results obtained from the analysis.

Logistic Process Optimization: an Example

Problem Description

A chemical synthesis process was considered, with the main objective of reducing the total duration of a typical production campaign. In order to achieve this goal, alternative process scenarios had to be designed and their logistic performance analyzed to determine the most promising one. Additionally, the wish was expressed that the solutions proposed should require little investment and imply no decrease of the utilization of the existing equipment.

The plant, in which other products were also manufactured, consisted of about twenty-five devices, including some fifteen vessels, four separation devices and three dryers. Additional buffer tanks were used to store solvents along a recycling loop. The piping available was that of Fig. 2 which presents the corresponding layout. Note that no production took place during the week-ends (in this case from Saturday 2:00 pm to Monday 6:00 am).

The production process decomposed into four main stages, i.e., Premixing, Reaction, Washing and Distillation, whereas three auxiliary stages dealt with recycled solvents and waste products. The material flow is illustrated by Fig. 3.

Results

Over the SIMBAX Gantt charts that indicate the logistic state of each device at any time (Figure 4) and their statistical evaluation, device occupation bottlenecks could be identified. In general, the devices corresponding to such bottlenecks continuously execute operations defined in the production process, which provokes the occurrence of waiting periods at neighbor pieces of equipment: these have to wait either for a batch delivery from a bottleneck device or as long as such a device is not ready to receive a batch from them. Moreover, volume bottlenecks can be detected by analyzing the device maximum filling rates, a further simulation result. Figure 3 graphically summarizes the bottleneck analysis performed, a gray rectangle being shorter – respectively narrower – than that describing the corresponding process stage if it is an occupation – respectively volume – bottleneck.

Five improvement measures were suggested to reduce the total campaign duration. Three of them aimed at eliminating the most critical occupation bottleneck, the distillation, whereas the two others tackled the capacity limitation due to further bottlenecks, i.e., the reaction and the washing stages. The main changes proposed consisted either in a split of the distillation into two shorter stages or in the use of an additional – new or pre-existent – device for speeding up the process.

All alternative scenarios have been tested by simulation and their feasibility and efficiency discussed with the plant people. Depending on the scenario, the reduction of the total campaign duration was between 2% and 12%. These values in fact hide noticeable improvements of the material flow since 43% of the total campaign duration (in the initial situation) corresponded to final cleaning operations and week-end breaks, and were therefore irreducible. Considering the production time only, the reduction achieved indeed varied between 4% and 20%.

The practical implementation of one of the scenarios designed confirmed the calculations according to which a production time reduction of 12% was feasible with no investment. At this stage, it should furthermore be noted that the whole simulation study required only a few days. More details about this example can be found in a paper by Lüneburg *et al.* (1997).

Capacity optimization in a multipurpose plant

Problem description

In a multipurpose batch plant, a large variety of dyestuffs was produced according to the following material flow:

– synthesis was performed on two similar production lines;

- the dyestuffs were formulated either as liquid products over a single nanofiltration unit or as solid ones over a single dryer.

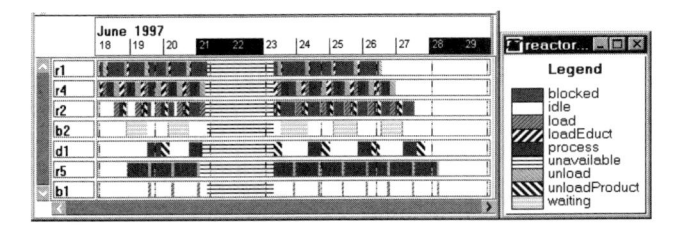

Figure 4. Simulation results for the initial situation.

The plant management suspected the nanofiltration unit to constitute the major occupation bottleneck. This was an important cause for concern since marketing forecasts had predicted an important increase of the demand for liquid forms. A simulation study was carried out with SIMBAX in order to design productivity improvement recommendations for the liquid dyestuffs and to evaluate the feasibility of fully utilizing the capacity of the nanofiltration equipment.

Most of the equipment available is represented in Fig. 5. Besides, since the working schedules were equipment-specific – synthesis was not run on week-ends but formulation was –, these needed to be taken into account in the model in order to get a thorough understanding of the interactions between the different resources and of their impact on the production capacity. Even this issue alone would have justified the use of logistic simulation for the study.

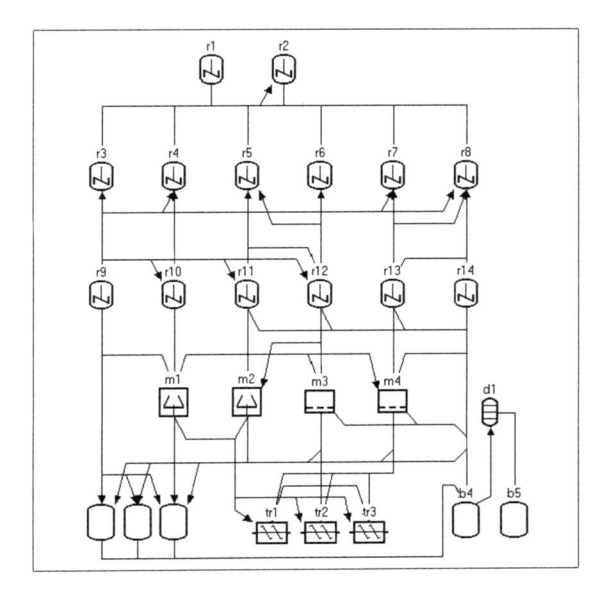

Figure 2. The plant model.

As for the production orders, an analysis of historical data clearly showed that, although the product mix involved very many dyestuffs, only about ten of them accounted for over 90% of the yearly total demand. The scope of the simulation model could therefore be limited to these products. Note finally that the cleaning operations necessary upon switching from a product to another were considered in the recipe models.

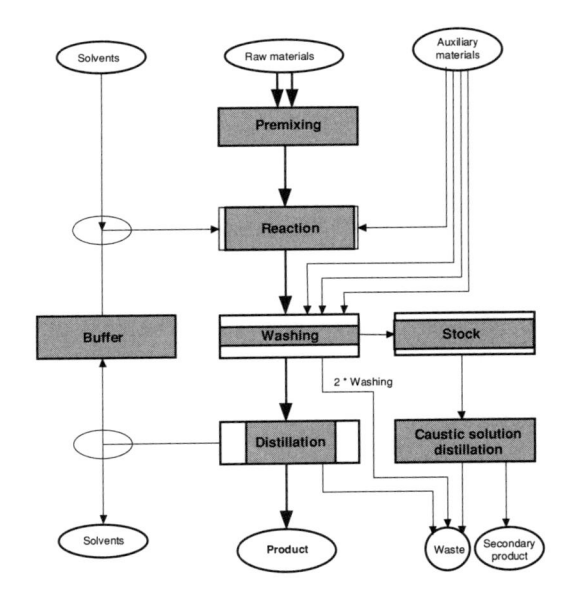

Figure 3. The material flow.

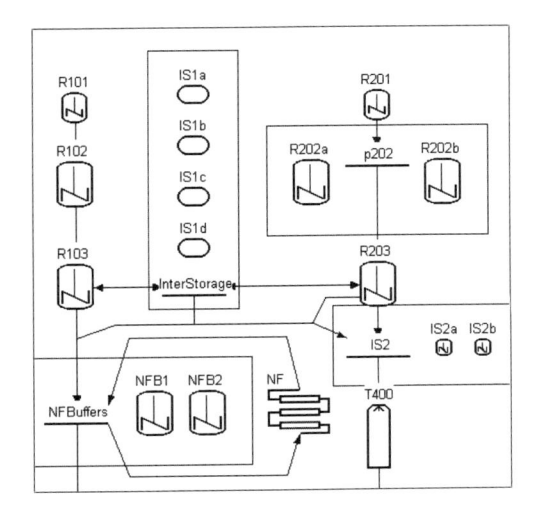

Figure 5. Model of the multipurpose plant.

Results

The simulation results first confirmed that the long nanofiltration times required by liquid products turned the corresponding equipment into the most critical occupation bottleneck of the plant. In other words, if a campaign consisting exclusively of liquid batches was run, important blocking phenomena occurred at synthesis level, since an already synthetized batch could not be further processed before the nanofiltration of the previous batch had been completed. Thus, a campaign-wise assignment of each synthesis line to either liquid or solid forms could only lead to a poor utilization of the synthesis reactors or of the nanofiltration unit – in the unlikely case that the two lines would be simultaneously assigned to solid forms.

Consequently, several strategies for integrating both liquid and solid batches into so-called mixed campaigns have been designed, taking profit of the knowledge gained from the analysis of the single processes. These strategies have been evaluated by simulation and their efficiency compared using the productive utilization rates of the nanofiltration unit and of the synthesis lines as reference criteria. For instance, one of the strategies recommended was inspired from Just-In-Time principles (Groenevelt 1993) and consisted of the following rules:

a. the campaign always begins with a liquid batch;
b. any already synthetized batch having to wait for the nanofiltration unit to become available is assigned to solid formulation, i.e., transferred to the dryer, if the remaining nanofiltration time of the current batch is longer than the synthesis cycle time;
c. the last batch synthetized before a week-end break is always directed towards the nanofiltration, in order to avoid liquid product capacity losses due to the synthesis working schedule;
d. the campaign always ends by a liquid batch, because the last batch does not provoke any blocking at the synthesis level since reactors can already be cleaned up.

Furthermore, complementary simulation runs advocated the use of additional buffers between synthesis and final formulation. Indeed, this enabled a further reduction of the synthesis blocking times, in particular of those not eliminated by rule b., and hence a better utilization of the synthesis lines.

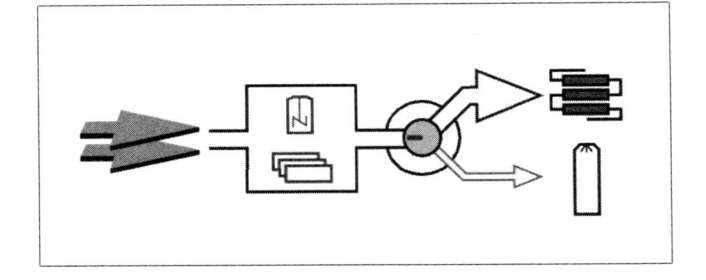

Figure 6. Recommended material flow modifications with intermediate storage and intelligent switching rules.

The logistic improvements proposed, which are illustrated by Fig. 6, resulted in a 30% increase of the production capacity of the whole plant. Moreover, the animated graphical interface of SIMBAX made it possible to convince the production managers of the merits of the recommendations made, in spite of the relatively complex logic inherent to the liquid/solid switching strategies.

Besides, the scheduling look of the strategy presented above leads us to comment on the difference between the approach used and operational scheduling. It is mainly a question of time scale. In the case considered here, the primary objective pursued was to optimize the production of the liquid dyestuffs on the *long term*. Potential improvement measures, which were obviously not known in advance, could therefore concern the processes themselves – as explained in the first case study – and/or the production organization. Both aspects could be analyzed one after the other using material flow simulation.

On the other hand, the purpose of a scheduling software is to help people on the shop floor to take better *short-term* decisions. Such a tool would have been of little use to deal with the process optimization issues that had to be addressed first. Nevertheless, if the production management had judged the improvement strategies proposed too complex to be applied at operational level without a computer system, these could then have been integrated into a scheduling software. In this sense, in the most complex practical situations, material flow simulation and operational scheduling appear to be complementary decision support tools.

Conclusions

Both in the case of single processes and of complex multipurpose plants, material flow simulation constitutes a powerful technique that offers process engineers and production managers the possibility of fully comprehending the logistic aspects of production. The insight gained from the simulation results can then be used to design optimization measures leading to a very efficient use of the equipment and resources available. The examples presented proved that production capacity increases of more than 20% – as much as 100% in some other cases – can be achieved.

From a software point of view, the use of decision support tools like SIMBAX that specifically address the needs of the process industries is essential. Having recourse to general-purpose material flow simulators may indeed lead their user to spend more time on modeling than on problem solving, thus making the work rather inefficient.

References

Groenevelt, H. (1993). The Just-in-Time system. In: S. C. Graves *et al.* (Eds.), *Handbooks in Operations Research and Management Science*, Vol. 4, Elsevier, Amsterdam, 629-670.

Lüneburg, W., D. Reinhold, and Ph. Solot (1997). Rezeptbasierte Simulation zur Optimierung von Mehrzweckanlagen: ein Anwendungsbeispiel. *Automatisierungstechnische Praxis*, **39**, 8, 36-41.

Puigjaner, L. and A. Espuña (1998). Prospects for integrated management and control of total sites in the batch manufacturing industry. *Computers chem. Engng*, **22**, 1-2, 87-107.

Rippin, D. W. T. (1993). Batch process systems engineering: a retrospective and prospective review. *Computers chem. Engng*, **17**, S1-S13.

Solot, Ph. and T. Dudás (1998). SIMBAX reference guide, Version 4.1, AICOS Technologies AG, Basel, Switzerland.

Strohrmann, M. and J. Krames (1997). Synthese und Optimierung verfahrenstechnischer Prozesse unter Einbeziehung dynamischer Materialflußsimulation. *Chemie Ingenieur Technik*, **69**, 9.

PROCESS OPTIMIZATION AND PROCESS ROBUSTNESS USING EXPERIMENTAL DESIGN: STAVEX

Y. L. Grize and W. Seewald
AICOS Technologies AG
CH-4057 Basel, Switzerland
E-mail: ygrize@aicos.com

Abstract

Thanks to user-friendly software and graphics-oriented techniques such as implemented in the expert system STAVEX developed at Ciba-Geigy, the powerful tools of statistical design of experiments (DoE) can now be used easily by non-statisticians.

In an industrial context, the main applications of DoE are those of process optimization and process robustness. In both situations, a quantitative performance measure (response) of the process depends on measurable parameters or factors (process parameters, quality of raw materials...). The problem is to identify the "best" factor combination in order to achieve desirable values for the response. In the optimization case, "best" means that the response is as close as possible to the target value. In the robustness case, "best" means that at these factor settings, the response is as least as possible affected by noise perturbations.

Two practical applications inspired from typical problems at Ciba-Geigy and that illustrate these two cases are discussed. The first one deals with the optimization of a dyestuff process with 11 a-priori important parameters using 29 experiments. The second example deals with the robustification of a pigment production process. Using adjustments of six easily controllable parameters, the impact of three noise factors on the color strength could be substantially reduced with a total of 32 experiments.

Keywords

Expert system, Robust process design, Statistical experimental design, STAVEX, Taguchi method.

Introduction

Statistical design of experiments (DoE) allows scientists and engineers to empirically study a system in an optimal way. With a minimum effort a maximum of reliable, precise and relevant information is quickly obtained from optimal experiments. Although the tools of DoE are not new, it is only with the emergence of user-friendly software and graphics-oriented techniques that these tools can now be used by a large numbers of non-statisticians.

Because of the importance of these methods, Ciba-Geigy, a large Swiss pharmaceutical and chemical company, [which became in 1997 Novartis and Ciba Specialty Chemicals after merging with Sandoz,] decided in the early 90's to develop in-house an expert system, named STAVEX, to encourage chemists and engineers to apply DoE tools without the help of statisticians. The first version of STAVEX was issued in 1993 and thanks to its ease-of-use largely contributed to propagate DoE within the company. Today STAVEX is part of the standard tools used in Novartis and Ciba Specialty Chemicals with a user-base well over 300. After the merger the system has been made commercially available also to other companies.

In an industrial context, the most relevant applications of DoE are those of process optimization and process robustness. In both situations the performance (response) of a system or process is measured quantitatively (quality characteristics, yield...) and depends on partially known measurable parameters or factors (process parameters, quality of raw materials...). The practical problem is to identify the "best" factor combinations in order to achieve desirable values for the responses. In the optimization case, "best" means that the responses are as close as possible to some target value (be it a maximum, minimum, or a number). In the robustness case, "best" means that at these factor settings, the responses are as least as possible affected by noise perturbations. The determination of the "best" parameter combinations can be difficult especially when many responses are simultaneously taken into consideration.

In this paper we first briefly present the basic DoE principles underlying STAVEX. Then we shall describe two practical applications inspired from typical problems at Ciba-Geigy: one dealing with process optimization, the other dealing with process robustness. We do not review the theory of DoE which can be found in numerous good textbooks (e.g. Box 1978, Montgomery 1997). However we will go more into details in the second example where the so-called Taguchi approach is used.

Design of Experiments in STAVEX

The DoE strategy used by STAVEX is essentially based on the following three principles (Seewald 1997):

Principle 1: Sequential Experimentation

In practice many parameters influence the quality of a process. To model their joint influence is in general not feasible because of limited ressources. The goal is therefore to identify and optimize the most important parameters only, i.e. those that have the greatest effects on the process quality characteristics. STAVEX proceeds through a sequence of series of experiments going from parameter screening (phase 1) to response modelling (phase 2) and response optimization (phase 3), thereby always using more refined models and reducing the number of parameters studied. The statistical designs and the methods of analysis for the correponding results are different in each phase.

Principle 2: Local Empirical Models

Simple empirical models such as polynoms of second order have proven to provide a rich enough class to locally model most processes adequately. Shapes such as saddles, ridges, bowls or hills, can all be described using these models. Because of their simplicity, the model building and model estimation can be done reliably and cheaply.

Principle 3: Minimal experimental plans

Once a preliminary analysis of measurement error has been performed, there is no further need for repeated experiments, except in the final stages of optimization where predictions as precise as possible are needed. Therefore, during the screening and modelling phase, the statistical plans used are as saturated as possible. This ensures that the number of experiments is always kept to a minimum.

During each phase, STAVEX asks for appropriate user inputs, proposes a design, performs the analysis, interprets the results and finally suggests how to proceed to the next phase (Fig. 1). The main screen of STAVEX presented in Fig. 2 shows these different steps of a phase as buttons going from top to bottom.

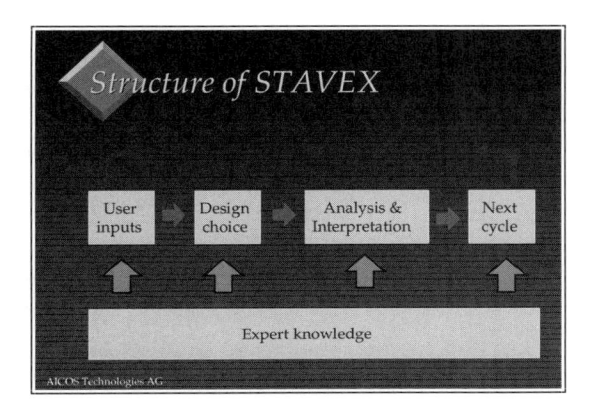

Figure 1. The structure of the expert-system STAVEX.

Figure 2. STAVEX main screen.

A Process Optimization Example

Problem Description

A change in the quality of a dyestuff of an existing process occurred and new parameter settings needed to be found quickly to adjust the process to target. From all the possible factors to be studied, 11 were selected as a-priori important for the two main responses studied: quality deviation (from a standard) and yield. Table 1 lists the 11 initial parameters with their variation range for study chosen initially. Observe that both quantitative and qualitative factors (e.g. producer of salt type or stirrer shape) are considered.

Table 1. Factors and Levels for the Dyestuff Process Example.

	Factor Name	Factor Type	Level
1	Ambient temp.	qualitative	cold/normal
2	Waiting time	quantitative	2-14
3	Salt type	qualitative	WSH/BASF
4	Concentration	quantitative	5-10
5	Ice addition	quantitative	1-5
6	Reaction time	quantitative	10-20
7	Stirrer shape	qualitative	Propeller/Anchor
8	Stirrer speed	quantitative	100-200
9	Filter press	qualitative	1/2
10	Reaction temp.	quantitative	30-40
11	Ice form	qualitative	small/large

The first plan used (screening phase) was a supersaturated design for 16 factors in 12 experiments (Booth and Cox, 1962 and Table 2).

Table 2. Supersaturated Screening Design of Booth and Cox.

	A	B	C	D	E	F	G	H	I	J	K	L	M	N	O	P
1	1	1	1	1	1	1	1	1	1	1	-1	-1	-1	-1	-1	-1
2	1	-1	1	1	1	-1	-1	-1	1	-1	-1	-1	-1	-1	-1	-1
3	-1	1	1	1	-1	-1	-1	1	-1	-1	1	1	-1	1	1	1
4	1	1	1	-1	-1	-1	1	-1	-1	1	-1	1	1	1	1	1
5	1	1	-1	-1	-1	1	-1	-1	1	-1	1	1	-1	1	1	-1
6	1	-1	-1	-1	1	-1	-1	1	-1	1	1	1	1	-1	1	1
7	-1	-1	-1	1	-1	-1	1	-1	1	1	1	1	-1	1	1	1
8	-1	-1	1	-1	-1	1	-1	1	1	1	1	-1	1	-1	1	1
9	-1	1	-1	-1	1	-1	1	1	1	-1	-1	1	1	1	1	-1
10	1	-1	1	-1	1	1	1	1	-1	-1	-1	-1	1	1	-1	-1
11	-1	-1	1	1	1	1	-1	-1	-1	1	-1	-1	-1	1	1	1
12	-1	1	-1	1	1	1	-1	-1	-1	1	-1	-1	-1	-1	-1	1

Since here only 11 factors are studied, 5 additional degrees of freedom are available as dummy factors to help fitting the line of no-effect in the half-normal plot analysis (Fig. 3). In such a plot, a straight line is fitted through the smallest effects and the points that detach themselves below the line indicate the significant effects.

Here the analysis revealed that only 7 factors were really important. Although relevant, the rather impractical factor "stirrer shape" was fixed at its best level and kept constant in the sequel.

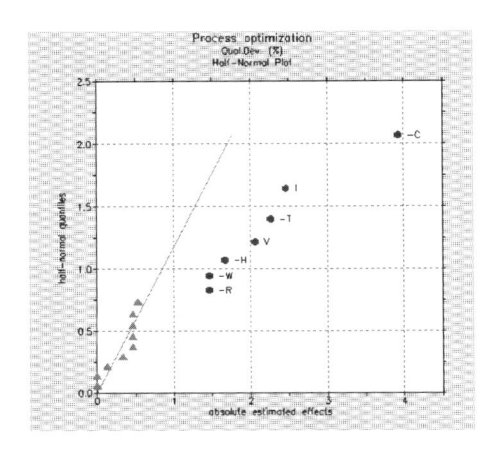

Figure 3. Half-normal plot analysis of the screening design for the dyestuff example.

The next plan used for modeling was an 8-run fractional factorial design (Box, 1978) for 6 factors and one interaction (Concentration*Temperature). A stepwise regression analysis showed that Temperature (T) and Concentration (C) and their interaction were most critical. The final optimization plan was a central composite design (Box, 1978) in 9 experiments for these two factors. The design with the contour lines of the fitted response surface for yield is shown in Fig. 4.

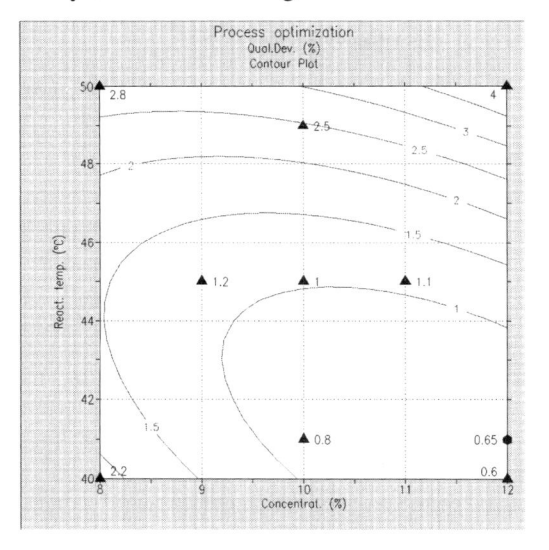

Figure 4. Contour lines analysis of the quality deviation for the dyestuff example.

Conclusion

Three experimental plans with a total of only 29 experiments were used to achieve satisfactory production conditions (Temperature = 41oC, Concentration = 12%). Although the final optimization was done with 2 factors only, 5 other factors have been found important during the course of experimentation and were therefore set at their best level subsequently.

A Process Robustness Example

Problem Description

In an important pigment production chain, the quality of the pigment showed unexplainable fluctuations in color strength (target value 0). Traditional statistical process control techniques (Shewhart control chart, cusum analysis, etc) revealed a high level of batch-to-batch variability which needed further investigation. The most crucial phase of the process was the kneading of the pigment mass. There were six easily adjustable potentially important parameters: kneading temperature, excess of salt affecting the viscosity of the mass, kneading speed, kneading time, order of introduction of the products in the kneader, and charging time. These six parameters are labelled A, B, C, D, E, and F and constitute the design factors. Prior knowledge suggested that one or more of the following factors could have been the source of the process instability: salt quality quantified by the grain size, texture of the resin, and temperature of the cooling water coming from the Rhine river. These three factors could only be controlled for the sake of some experiments but not during routine production. These three factors will be labelled M, N, and O and constitute the noise factors.

Because the noise factors cannot be controlled during production the problem is not an optimization problem like the preceding one but a robustness problem: what is the best design factors combination which simultaneously reduce the influence of the noise factors on strength and at the same time give strength deviation as close as possible to target?

Taguchi Approach to Robustness

Different approaches to solve this problem can be used (Grize, 1995). The one presented here is due to G. Taguchi, a Japanese engineer, who first proposed to use DoE tools to solve robustness problems. His approach consists in studying the effect of the noise factors by varying them in a systematic fashion according to an experimental design (so called outer array). The design factors are themselves varied according to another design, the so-called inner array. The outer array is performed for each run of the inner array, therefore the responses of the inner array consist of the mean and variance of the original response over the outer array. The analysis of these results then show how to set the design factors to have the mean response on target and the variance as small as possible.

Designs and Analysis of Results

Here there are 6 design factors and 3 noise factors. A 2^{6-3} fractional factorial design was chosen for the design factors (inner array D_D) and a 2^{3-1} fractional factorial design for the noise factors (outerarray D_N). Table 3 list the factors and levels chosen.

Table 3. Design and Noise Factors with Levels for the Robustness Example.

		Design factors:	**Levels:**	
1	A	Temperature of mixture	60øC	80øC
2	B	Excess of salt	10%	20%
3	C	Kneading speed	slow	fast
4	D	Kneading time	2 hour	3 hour
5	E	Order of introduction of the materials	A	B
6	F	Transfer time	short	long
		Noise factors	**Levels**	
1	M	Salt grain size	small	medium
2	N	Resin texture	bulk	powder
3	O	Temperature of cooling water	5øC	20øC

Furthermore D_D was chosen so that the estimation of the potentially important two-factor interaction kneading speed * kneading time (C*D) would be possible. D_D and the data obtained (mean and variance averaged over D_N) is shown in Table 4.

Table 4. Inner Array D_D with Mean and Variance Obtained from the Outer Array D_N.

A	B	C	D	E	F		mean	variance
60	10	fast	2	B	longh	-1	2.42	4.22
80	10	slow	2	A	longh	1	5.48	17.44
60	20	slow	2	B	short	1	1.6	5.65
80	20	fast	2	A	short	-1	2.23	6.27
60	10	fast	3	A	short	1	3.23	1.32
80	10	slow	3	B	short	-1	5.35	5.43
60	20	slow	3	A	long	-1	1.8	1.22
80	20	fast	3	B	long	1	3.68	4.04

Because no or few degrees-of-freedom are available to estimate error a graphical analysis using half-normal plots is preferable. Figures 5 and 6 are half-normal plots of the mean response and of the logarithm of the sample variance respectively. It is customary to analyze the logarithm of the variance instead of the variance itself because it is more likely to approximately follow a normal distribution. From the plots one sees that factors A and B affect the mean response and that factors D and A affect the variance. The values of these effects (i.e. how much the response is affected by a factor change from low to high level) are given below:

Effects on the mean: A: +1.9 B: -1.8
Effects on the log(variance): D: -1.1 A: +1.1.

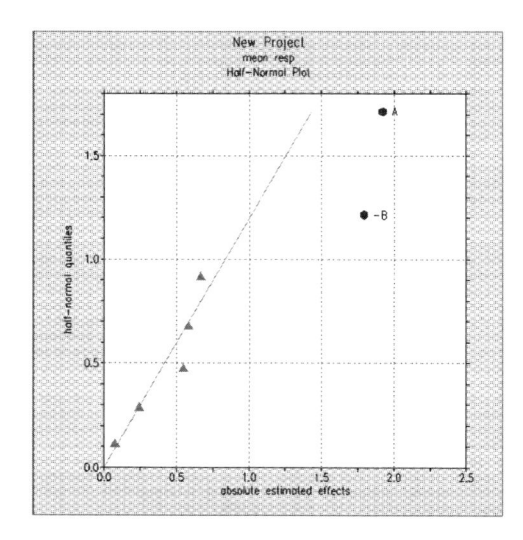

Figure 5. Half-normal plot analysis of the mean response.

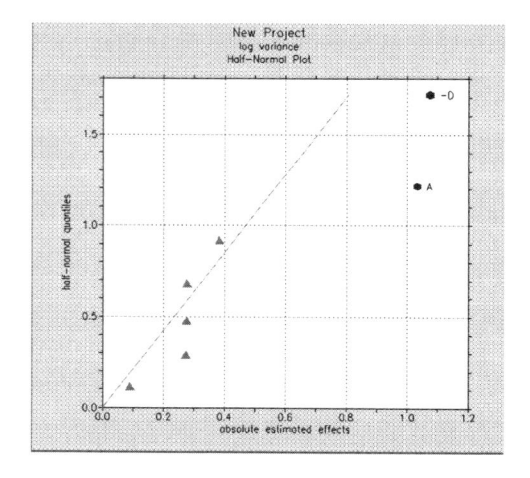

Figure 6. Half-normal plot analysis of the log(variance).

Conclusion

From these results a variance reduction can be achieved by setting D at its high level and A at its low level. This level change of A would also decrease the mean. But since the mean response tends to be well above 0 (see Table 4), a decrease of the mean is desirable. Finally B affects the mean without affecting the variance and so can be used at any time to further bring the mean on target if it is necessary. Here setting B high would decrease the mean further. At this stage a confirmatory experiment with: A low, D high and perhaps also B high, needs to be performed to verify the improvement. A drawback of the approach used here is that one cannot identify which noise factor most affects the variability of Y.

Conclusion

Thanks to the expert system STAVEX sophisticated statistical design of experiments tools can be use by the engineers without statistical knowledge, These tools allows to quickly identify the best process parameters settings, either to optimize a quantity or to optimize the quality of the process by reducing variation.

References

Booth K.H.V.and Cox D. R. (1962), Some systematic Supersaturated designs, Technometrics 4, 489-495.
Box G.E.P., Hunter H. and Hunter S. (1978), *Statistics for experimenters*, N.Y. Wiley.
Grize Y.L. (1995), "A review of robust process design approaches", *J. of Chemometrics* 9, 239-262.
Montgomery D.C. (1997), Design and Analysis of Experiments, N.Y. Wiley.
Seewald W. et al. (1997) Experimental Design and Analysis with STAVEX , Version 4.20, AICOS Technologies AG , Basel, Switzerland.

A NEW CONCEPT TO IMPROVE OPERATION AND PERFORMANCE OF AN INDUSTRIAL SIZE TUBE REACTOR BY MEANS OF CFD-MODELING

Prof. Dr.-Ing. H. J. Warnecke and Dipl.-Chem. M. Schäfer
Universität-GH Paderborn
Warburger Str. 100
D-33098 Paderborn, Germany

Prof. Dr. J. Prüß
Universität Halle-Wittenberg
Thomas Lieser Str.
D-06099 Halle, Germany

Dr. M. Weidenbach
DDI Stade, R&D
Postfach 1120
D-21677 Stade, Germany

Abstract

Process optimization of chemical production plants is becoming increasingly time consuming due to the complex nature of the applied unit operations and constraints to be considered. Laboratory studies are in most cases prohibitive due to the time and money they require, however computer models often appear not reliable enough to fully replace the experimental effort. This paper introduces a concept of combining computational fluid dynamics with the classical area of reactor modeling, which allows to describe a complex chemical reaction in the liquid phase of a gas/liquid tube reactor with physical absorption of the gaseous components into the liquid phase. A three dimensional CFD model is conceptually presented and compared with a simplified model which requires less computational resources.

Keywords

Reactor modeling, Chemical reaction, Mass transfer, Computational fluid dynamics, Two phase flow, Propylene chlorohydrin process, Optimization.

Introduction

The strive to maintain the competitive edge is an ongoing effort for each chemical company which wants to be the lowest cost producer of high quality products, by minimizing unwanted by-products, energy consumption and pollution. Given all the constraints that make this goal so difficult to achieve, a detailed understanding of the markets and a technically sound concept for the plant operation are mandatory. As time is the most precious item nowadays in industry, extensive experimental studies are often prohibitive and the use of computers plays an increasing role in replacing laboratory tests to perform plant optimization. An example process is the chlorohydrin route to propylene oxide, carried out at different productions sites worldwide of The Dow Chemical Company. It is a major commodity for the chemical industry, e. g. a feedstock for polyurethanes or polyols/polyethers. The world's production capacity was 3.9*10^6 t for 1991 and 55 % of the total propylene oxide was produced according to the chlorohydrin chemistry (Kahlich, Lindner, Wiechern, 1993).

This paper concentrates on a fundamental mechanistic model for the tube reactor, employed for the formation of the chlorohydrin pre-cursor to propylene oxide. A new concept for process simulation, combining the fundamentals of computational fluid dynamics (CFD) modeling with the phenomena of fast complex kinetics and with the absorption of two gases will be introduced.

The first step of the propylene chlorohydrin process is characterized by simultaneous absorption of gaseous chlorine and propylene into dispersed water followed by complex exothermic reactions in the liquid phase respectively the liquid side boundary layer. The reaction scheme of the propylene chlorohydrin (PCH) reaction consists of fast equilibrium, parallel and consecutive reactions of different order. The gas volume fraction is strongly reduced during reaction and causes the system to undergo different flow regimes. High absorption rates induce pressure gradients along the vertical section of the employed tube reactor.

For reactor optimization a one dimensional plug flow reactor model and a three dimensional model basing on Computational Fluid Dynamics (CFD) have been developed. The two models which take into account mass, heat and momentum transfer are compared according to their ability to represent the process. Both macroscopic models have been supplemented with a microscopic mass transfer model on the basis of the two film theory.

The major activities in modeling the chlorohydrin process focus on improved PCH-selectivity and increased reactor capacity. Not only should the model represent all qualitatively and quantitatively observed reactor performance data with acceptable preciseness, but also with the lowest possible CPU time requirement. The response value for parameter optimization is influenced by chemical reactions in the dispersed liquid phase on the microscopic scale, by inter phase mass transfer of reactive species and by the hydrodynamics on the macroscopic scale. To develop a mathematical model that can be used to describe reactor performance and to optimize the process by changing operating conditions or reactor geometry, different steps are required: physical and chemical phenomena at different scales need to be determined and translated into mathematical expressions based on the relevant chemical reactions and prevailing mass transfer conditions, as well as the specification of the response values for the observed process. A third step would include the formulation of the objective function, based on economic boundary conditions specified by the business, environmental targets and local production site issues like plant integration, supply and demand etc., which is however not covered in this paper.

Chemical Reaction

The chemical reaction in the dispersed aqueous liquid phase (Buss *et al.*, 1982; Elm, 1977; Shilov 1949) consumes simultaneously absorbed chlorine and propylene to form propylene chlorohydrin and various by-products. A propylene-chloronium ion $C_3H_6Cl^+$ is formed as an intermediate of the reaction between propylene and chlorine in the liquid phase. In the main reaction the chloronium ion reacts with water to give the two propylene chlorohydrin isomers 1-chloro-2-propanol (90%) and 2-chloro-1-propanol (10%) and hydrochloric acid. Because of its high reactivity the chloronium ion participates in several side reactions with nucleophilic compounds. The consecutive reaction with propylenechlorohydrin gives 2,2´-dichlorodiisopropyl ether (DCIPE) and the reaction with chloride ions the main by-product 1,2-dichloropropane (DCP). DCP can also be formed by the direct reaction of propylene and chlorine. Another side reaction is the formation of the two dichlorohydrin (DCH) isomers 1,3-dichloro-2-propanol and 2,3-dichloro-1-pro-panol via allyl chloride. The achieved yields based on propylene in modern plants are (Kahlich, Lindner, Wiechern, 1993):

PCH   88 - 96%
DCP   3 - 10%
DCH   0.3 - 1.2%
DCIPE   0 - 1%

In Fig. 1 the reaction scheme in the liquid phase is presented.

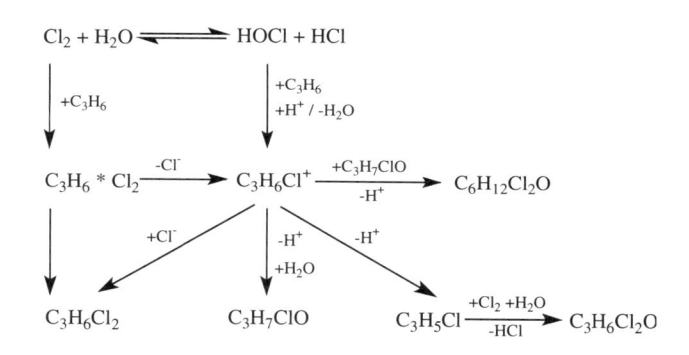

$$Cl_2 + H_2O \rightleftharpoons HOCl + HCl$$

$$\downarrow +C_3H_6 \qquad \downarrow {+C_3H_6 \atop +H^+ / -H_2O}$$

$$C_3H_6 * Cl_2 \xrightarrow{-Cl^-} C_3H_6Cl^+ \xrightarrow[-H^+]{+C_3H_7ClO} C_6H_{12}Cl_2O$$

$$\swarrow{+Cl^-} \quad \downarrow{-H^+ \atop +H_2O} \quad \searrow{-H^+}$$

$$C_3H_6Cl_2 \qquad C_3H_7ClO \qquad C_3H_5Cl \xrightarrow[-HCl]{+Cl_2 +H_2O} C_3H_6Cl_2O$$

Figure 1. Reaction scheme.

The set of differential equations representing reactions in the dispersed liquid phase can be simplified using the Bodenstein approximation for short living intermediates and by relating rate constants of product formation to the rate constant of the main reaction.

The interdependency between chemical reactions in the liquid phase and the design parameter for process optimization is caused by the following factors:

- the total amount of feed water, that affects the molar concentrations of reactive species in the liquid phase,
- the exponential dependency of rate constants from temperature,
- the pressure, that influences the concentration of propylene and chlorine near the G/L-interface.

Mass Transfer

Propylene and chlorine are absorbed simultaneously from the gas phase into the liquid phase. The fast chemical reactions in the liquid phase enhance the absorption rates from both species and consumes propylene completely near the G/L-interface. Due to the complexity of the reaction scheme, analytical solutions for the enhancement factors are not available. For calculation of absorption rates on the basis of the two film theory (Lewis, Whitman, 1924), the stiff set of diffusion-reaction equations of reacting species has to be solved numerically in an inner iteration procedure. The required CPU-time for the inner iteration procedure can strongly be reduced by normalization of the concentrations, concentration gradients and film thickness in a non-linear mesh.

The interdependency of mass transfer parameters and response values for process optimization ranges over the whole microscopic and macroscopic scale. The diffusive mass transport in the laminar film (microscopic scale) leads in dependency of film thickness to accumulation of reactive species and affects reaction rates and selectivities. On the macroscopic scale it is the interfacial area that determines the total absorption rates and influences hydrodynamics, e.g. the local flow regime.

Reactor Models

A three dimensional model on the basis of CFD has been developed in which full compressibility of the continuous gas phase, chemical reactions in the liquid phase, and interface mass transfer of chemical species is included by appended FORTRAN-Routines. As a CFD tool the commercial package CFX Version 4 from AEA Technology is used. Because of the high computational effort when dealing with CFD an one dimensional model of the employed tube reactor has been developed in parallel. Both models take into account chemical reaction, interfacial mass, heat and momentum transfer. The continuous gas phase is assumed to be fully compressible and the dispersed phase is treated as rigid drops. The absorption rates are either calculated by use of constant enhancement factors for both propylene and chlorine or by solving the diffusion-reaction equations in the laminar boundary layer numerically. In addition to the one dimensional model internal friction, heat conduction, diffusive/dispersive mass transport and lateral boundaries can be taken into account with the three dimensional model. The one dimensional model is used to determine the interdependency between model parameters and the response values for optimization. The three dimensional model is used to describe reactor performance of the employed reactor and for geometrical optimization.

Results obtained from CFD simulations are used to improve the one dimensional model with respect to hydrodynamics by correlations, vice versa one dimensional simulations are used to estimate unknown model parameters and to give an initial guess for CFD calculations.

Simulation results obtained with both models for simple reactor geometries coincide well (Fig. 2). It can be shown that in this case the hydrodynamics can be described with constant enhancement factors for calculation of absorption rates instead of solving the diffusion-reaction equations in the film, but not liquid phase concentration profiles. For the description of both hydrodynamics and local product distribution in the liquid phase in the geometry of the employed tube reactor, the inner iteration procedure for calculating the absorption rates has to be used within the CFD model. This is because of the enormous computational effort not practicable nowadays.

The computational grid used for CFD modeling contains about 60.000 cells and the code solves for 40 macroscopic variables. Several days CPU time are required to represent the process in the complex geometry of the technical propylene chlorohydrin reactor with the CFD model on a SGI Indy R5000. In contrast calculation of reactor profiles with the one dimensional model requires some minutes CPU time.

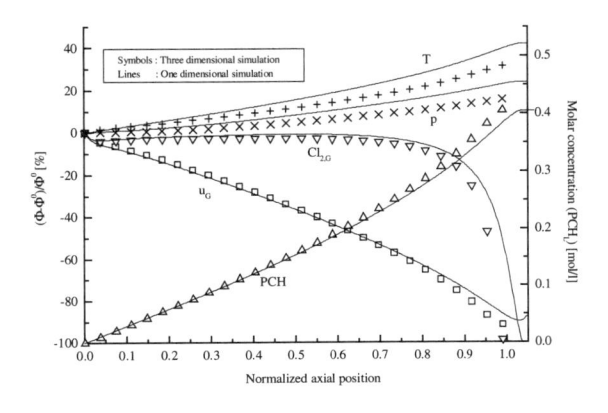

Figure 2. Comparison of simulation results.

Reactor Optimization

Options for optimization of an existing process is the improvement of the reactor geometry and operating conditions. Changes to the reaction system will often require changes in both down-stream and up-stream process sections.

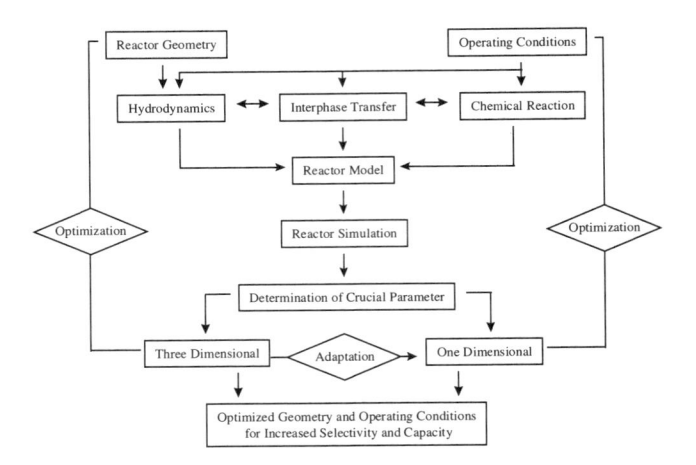

Figure 3. Optimization procedure.

For optimization of complex geometries Computational Fluid Dynamics is advantageous. Up to now the use of CFD in the field of chemical engineering is somehow restricted by the capability of the hard and software. Fig. 3 shows the relationship between the different scales and the optimization procedure.

In reactive absorption processes like the propylene chlorohydrin process an improved reactor capacity can be obtained, when sections of high pressure surge, and sections with radial uneven phase distribution are avoided.

In technical tube reactors narrow sections can be used to establish annular two phase flow. Such sections avoid back-mixing but may be a capacity limiting device due to high pressure head. Fig. 4 shows axial pressure profiles for alternative geometries with narrow sections of different diameters obtained by the CFD model. The narrow section diameter d_S is related to the main diameter d_M of the tube. Δp is the difference between local pressure and the given outlet pressure. The first vertical line represents the transition to d_S, the second one the end of the narrow section and the third vertical line the beginning of the section with diameter d_M.

Figure 5 depicts the radial gas phase distribution in three different sections of the employed reactor. The pictures were obtained from a transient three dimensional simulation of the employed tube reactor under unfavorable operating conditions. The slices through the centerline of the sections show the geometrical optimization potential for the chosen operating conditions.

In the left part of Fig. 5 the liquid phase concentration is lower near the reactor wall than in the middle of the reactor but absorption rates are still high. In the center part of Fig. 5, the liquid phase is flowing down the reactor wall and absorption rates are small because of the reduced interfacial area. In the right part of Fig. 5 the flow regime will cause high rates of back-mixing and hence decrease main product selectivities.

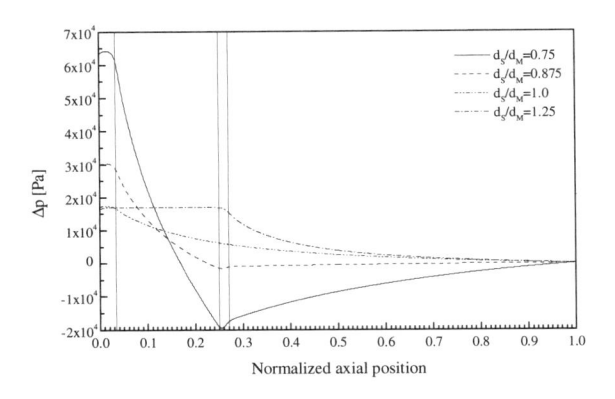

Figure 4. Axial pressure profiles.

Conclusion

The CFD model allows to illustrate the functional behavior of the key parameters, e. g. concentrations, phase velocities, temperature and reactor pressure in an industrial reactor and is validated by experimental data obtained from an industrial plant for different operating conditions. It represents a powerful tool to optimize a large scale production plant with a reasonable computational effort in an acceptable period of time, provided the initial guess from the 1D-model comes close to the final solution. However, the computational effort is still very high and further simplification without sacrificing accuracy is important.

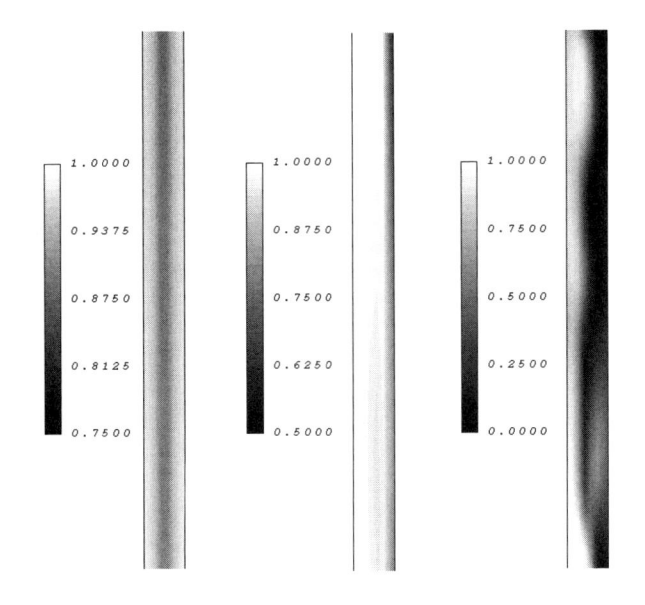

Figure 5. Local gas phase distribution.

Nomenclature

$Cl_{2,G}$	gas phase concentration of chlorine
d_M	main diameter for tube reactor
d_S	section diameter for tube reactor
p	pressure
Δp	pressure difference
T	temperature
u_G	velocity of gas phase
Φ	variable

References

Buss, E., A. Rockstuhl, D. Schnurpfeil (1982). Untersuchungen zum Mechanismus der Chlorhydrinierung von Olefinen. *J. Prakt. Chem.*, **324, No. 2**, 197-208.

Elm, R. (1977). Die Kinetik der Reaktion zwischen gasförmigen Propylen und wässerigen Chlorlösungen. PhD Dissertation, Universität Dortmund.

Kahlich, D., J. Lindner, U. Wiechern (1993). Propylene oxide. In B. Elvers (Ed.), *Ullmann's Encyclopedia of Industrial Chemistry*, Vol. A22, 5th ed. VCH, Weinheim. pp. 239-259.

Lewis, W. K., W. G. Whithman (1924). Priciples of gas absorption. *Ind. Eng. Chem.*, **19**, 1215-1220.

Shilov, E. A. (1949). Synthesis of 2-chloroethanol. VI. Theory of the Gomberg synthesis. *J. App. Chem. U.S.S.R.*, **22, No. 7**, 734-746. (English translation)

INFORMATION MODELS FOR BATCH AND REAL-TIME CHEMICAL PROCESS DATA

Neil L. Book and Arvind Sharma
Department of Chemical Engineering
University of Missouri-Rolla
Rolla, MO 65401

Abstract

The Process Data Exchange Institute (PDXI) has developed information models for process engineering data. The information models were developed to be used in the STandards for the Exchange of Product model data (STEP or ISO 10303). The methods defined in the STEP standards provide a means for the electronic exchange of data captured by an information model. The existing PDXI and STEP models have a product (or snapshot) point of view. For process engineering, the product is the design of the chemical process, which may have several versions. The PDXI information models capture the data for each of the design versions, but do not capture the relationship between the various versions. The PDXI models can be used to capture snapshots that describe batch or real-time operating characteristics of a process at specific points in time. However, there is no way to describe the process that occurs between the snapshots. Also, the snapshots are very inefficient because many data have the same values in many different snapshots.

Information models have been developed to capture information about the process that converts a system from one state to another. A product is a system at a specific state, and so the new information models subsume the existing product models. In fact, large portions of the existing PDXI and STEP models can be used intact. However, the new models allow for the efficient capture of batch and real-time operating data for chemical processes. Various methods exist, including those in the STEP standards, that can be used to provide the means for the electronic exchange of the data captured by the models.

The new models are at a high level of abstraction. The concept of system, state and process can be applied to most any circumstance and is not restricted to chemical processes. As such, the information models have been tested with data that describe games of chess, critical path scheduling problems, recipes for preparing food, steady state chemical processes and batch chemical processes.

Keywords

Information models, System, State, Process, Descriptors.

Introduction

The Process Data eXchange Institute (PDXI) has developed information models (data models) (Book et al. 1992, 1994a, 1994b; Blaha, et al. 1991, 1993; Fielding, et al. 1995; Motard, et al. 1995) for process engineering data that set the standards for defining each data item, its realization and means to access it. These models were designed using the STandards for the Exchange of Product model data (STEP or ISO 10303) methodology. STEP (Doty, 1992; ISO, 1992) information models have a product (snapshot) point of view with means of transmitting and receiving data about the product being the prime concern. PDXI and STEP models do not capture information about the design or manufacturing process that created the product. For example, the PDXI models can be used to describe the state and composition of material in a reactor at the beginning and the end of the batch, however, it cannot describe the process that occurred (eg., isothermal or adiabatic).

Snapshot models describe the product at a given point in time. The disadvantage is that each and every detail has to be stored at every point in time that the product is to be described. The values of many data items do not change during a process. For example, the composition and state of the material in a batch chemical reactor continually change during a batch but the material of construction, the thickness of the reactor walls and the reactor volume are constant throughout.

The existing PDXI and STEP snapshot models can be used to store product information at given points in time. Thus, they are effective if the data for a small number of snapshots are to be stored, e.g., the as-designed, the as-constructed, and the as-operated version of the plant. However, the PDXI models are not efficient for data describing batch or real-time processes where a large number of snapshots are required. Thus, the snapshot model leads to unnecessary storage of replicate data.

Information models have been developed that will describe a process and a product. These models are based on the concepts of system, state, and process. The system remains the same while the state of the system may change during a process. The data describing the system remains static throughout the process and is stored only once. This gives more efficient models for batch or real-time processes as compared to the existing PDXI models. A product is a system at a point in time, therefore large portions of the existing PDXI and STEP models can be used in the new models.

The new models capture similar data to the model proposed by Virrantalo (1994). The new models are much more explicit in the way that the data is captured.

The Information Models

The information models (Figs. 1, 2, 3, and 4) describe the data at the highest levels of abstraction. The lower levels of the models will use large portions of the existing PDXI and STEP models. The object modeling technique (Rumbaugh et al., 1991) (OMT) used in the development of PDXI models is used here. The conventions are adopted from those in PDXI. A complete glossary has been developed for these models.

Figure 1 defines the relationships between system, state and process. Figures 2 and 3 deal with the description of entities. Figure 4 classifies numerical data.

System

Systems are classified as physical systems and abstract systems. A physical system is that portion of the universe set aside for investigation. Thus the physical system is a volume of space. The physical system can be composed of disjoint volumes. The system may or may not be closed and it can be infinitesimally small.

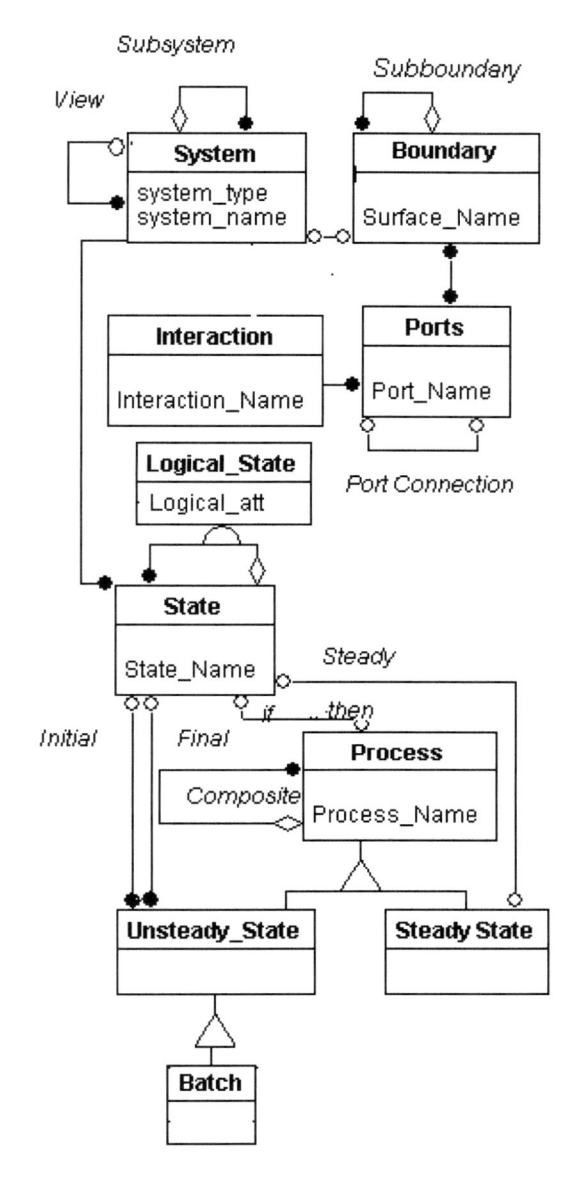

Logical_State.Logical_att=enum{and, or, not, ...}
System. system_type=enum{physical, abstract}

Figure 1. System, state and process.

An abstract system is a representation of a physical system. This includes process flow diagrams, utility line diagrams, process simulations or design versions. The abstract system class represents all the abstract views of real objects.

Many of the existing PDXI and STEP classes such as the process plant site, process train, process simulation, etc. are subclasses of physical and abstract systems. The various relationships between these classes are already established in the PDXI and STEP models.

There are two self-associations for the system class. One is the system_subsystem link and the second one is the system_view link. A process plant site may contain a number of process plants. Similarly each process plant may be an aggregation of a number of process trains. This aggregation relationship is explicitly developed in the existing PDXI models. In the new models, subsystems are separate systems that can be aggregated to form composite systems. Thus, process plants are subsystems of process plant sites. The system_subsystem link establishes the required association.

> Example: A chemical process plant may be divided into a reactor process train, a distillation process train and a purge process train. Similarly the simulation of the plant may be divided into a reactor simulation section, a distillation simulation section and a purge simulation section.

The second link is the system_view link. An abstract system is a view of a physical system. This link provides the association between the abstract view of the system and the physical system.

> Example: There may be a number of unit operations such as a mixing unit operation, a reaction unit operation and a separation unit operation that combine to form a view of a single reactor.

The unit operation and simulation classes of PDXI and the two-dimensional and three-dimensional graphical representation classes of STEP information models are subclasses of abstract systems.

Abstract systems may also represent generic views of physical systems. For example, a catalog description of a piece of equipment is an abstract view of the actual equipment item. There has been a lot of work in STEP developing catalog descriptions of equipment.

PDXI and STEP do not model views well. The PDXI models have simulation and design views. However, for each new view, PDXI has an entire set of classes because the descriptors (attributes) change slightly from one view to another. Adding a new view requires adding a complete set of classes. Some STEP models have a set of unique entities associated with each physical system that indicates each possible view of that physical system. This leads to an inordinate number of entities and each physical system has a restricted number of views. The new models accommodate unlimited views of the physical system.

State

The state of the system is the condition of the system at an instant in time. A system is the same but it can have many states. The dynamic nature of the system is captured by its states.

> Example: Temperature measurements can be taken at various times in a batch process. Each measurement is a state.

Snapshot models, such as PDXI, capture the entire process at each point in time. The new information models only account for the change in the state of the system. Data describing the system remains static, thus reducing data redundancy by eliminating unnecessary data repetition.

There is a self-association on state. The state_substate link allows the state to be defined as a logical aggregate of substates.

> Example: The system state can be an aggregate of spatial and temporal substates as in games of chess where the initial positions of all the pieces is the same for all games but each game occurs at a different time and place.

Interaction

An interaction defines an interchange between two systems. The interchange could be flow, energy, signal, mechanical stress or information.

> Example: An outlet stream symbol on one page and an inlet stream symbol on another page of a P&ID can designate the same stream. This is an example of information interchange between systems.

Boundary

The boundary separates the system from other systems. The system boundaries are free to move with time, thus, the system can change its shape with the passage of time.

The system can have a boundary that is a composite of boundaries. This is captured by the boundary_subboundary link.

> Example: A pipe can have a cylindrical subboundary which forms the real part of the boundary and two imaginary circular subboundaries stretched across each end of the pipe.

> Example: A P&ID may be composed of many pages of diagrams. Each page is a subsystem. Each page has a boundary.

Ports

Ports are entities associated with the boundary of a system through which interactions with other systems can occur. There can be many types of ports, including material, energy, signal, information, or mechanical stress ports.

A port connection is exactly two ports of the same port type through which an interaction occurs.

Example: A material port allows the flow of material and energy contained within the material. Each nozzle of a process plant equipment will be associated with a material port.

Example: An energy port allows the flow of energy in the absence of any material flow. An energy port could be a heat port or a power port.

Example: A mechanical stress port would allow the transfer of stress between systems. A hook supporting a rod would fall under this category.

Process

A process is a period of time during which the state of the system changes and/or interactions with other systems occur. The period of time could be infinitesimal.

An unsteady state process can have an initial state and a final state. A batch process is a special form of an unsteady state process. A steady state process has a single state that is constant during the process. A process can be composed of a number of subprocesses, each of which may be steady or unsteady. This is captured by the process-subprocess link.

Example: A chess game is a batch process composed of a number of moves, each move being a batch process.

A network of sequential and simultaneous processes can be traced through the change in states of the system and the entire history of versions can be captured.

Example: A steady state process may be preceded by an unsteady state startup process and followed by an unsteady state shutdown process. The steady state is the final state for the startup process and the initial state for the shutdown process.

Example: Each move in a chess game starts from an initial state (the position of all the pieces) and ends in a final state. The final state for one move is the initial state for the next move.

Example: Several activities in a critical path network may terminate or originate at the same node. The nodes represent common states (initial or final) for the activities.

There may be processes which are carried out only if a conditional state occurs. The if-then association between processes and states captures these conditions.

Example: An interlock may require that the feed valve be closed and the valve to the cooling water be opened if the temperature in a reactor exceeds a specified value. The simultaneous closing of the reactor feed valve and opening of the cooling water valve is a composite process.

Descriptors:

Descriptors (Fig. 2) are attributes of classes in the model. The descriptor model is designed such that the various descriptor classes can be associated with process, system, boundary, state or interaction entities.

Example: If a process on a system is isothermal, the temperature descriptor can be associated with the system and/or process and set to a value. However, if the system temperature changes during the process, a temperature descriptor can be associated with each state during the process.

In a complete model, the descriptors would be classified into many subclasses, including temporal descriptors, spatial descriptors, material descriptors and process equipment descriptors.

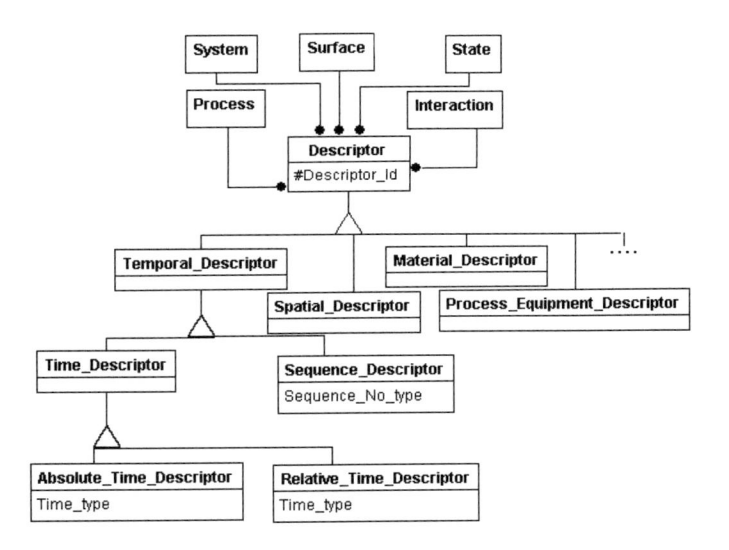

Figure 2. Descriptors.

The temporal descriptors can be used to describe the time taken by a process, the time at which a state occurs or the sequence of a set of states. PDXI does not have the facility to show that a design version is related to some other design version.

The temporal descriptors are classified into time descriptors and sequence descriptors. Time descriptors are further classified as absolute time or relative time descriptors.

Example: In competitive chess matches, each move is clocked. The time descriptor is a relative time from the beginning of the match. The moves are a sequence and are given a sequence number.

Example: The amount of time taken by a process would be a relative time descriptor. The time taken for completing an activity in a critical path problem or for heating a process equipment would also fall in this category.

Example: Version numbers for a piece of software or design of a process are examples of sequence descriptors.

Process equipment descriptors describe characteristics of process equipment other than spatial, temporal, or material characteristics, such as the heat transfer coefficient of a heat exchanger. Process equipment descriptors have a subclass for each type of process equipment. Many of the process equipment descriptors have been identified in the PDXI models.

Shape and/or positions are described by spatial descriptors.

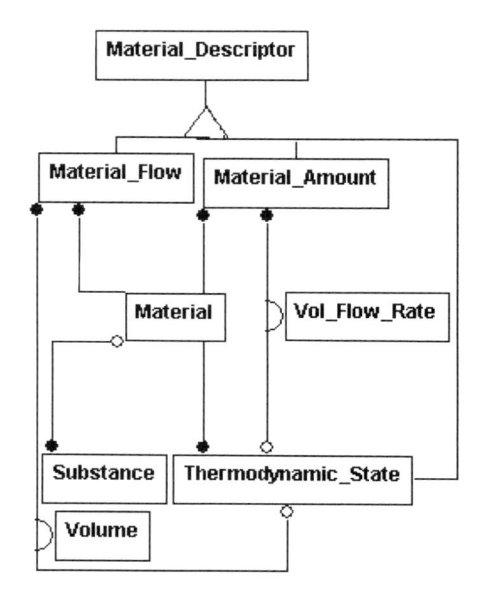

Figure 3. Material descriptors extended to include PDXI Models.

Example: The rank and file positions of a chess piece during a chess game are spatial descriptors as are the thickness, volume and geometric shape of a process vessel.

The product models in STEP focus on describing spatial characteristics. The spatial descriptor entity establishes the link to those models.

Material descriptors (Fig. 3) are classified into material flow, material amount and thermodynamic state descriptors. These are entities from existing PDXI models. The remainder of the model shown in Fig. 3 is taken directly from the Material Model of PDXI and substance is an entity from the PDXI Physical Property Model. Composition and properties of process materials are captured in the PDXI Physical Property Model. Material flow would describe attributes related to any flow process such as the mass flow rate, mole flow rate or the standard volumetric flow rate of the phase. Material amount would describe the total quantity of the material in a system such as its mass, the number of moles or the standard volume. The various interactions are usually in terms of the flow rate or the amount of material being transferred.

The material class is further divided into construction material and process material. There are STEP models for construction materials.

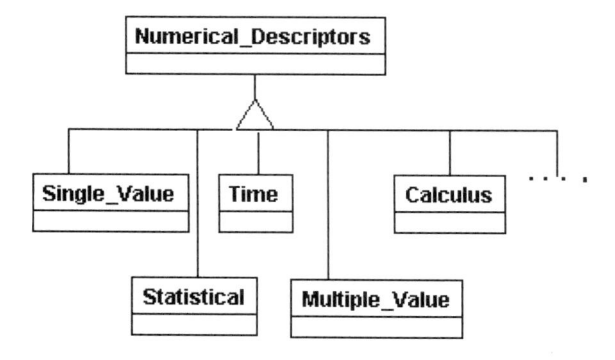

Figure 4. Numerical descriptors.

Numerical Descriptors

Much of the data captured in the descriptor part of the model have numerical values. The numerical descriptors (Fig. 4) can be single, multiple, statistical, time measuring, or calculus values. Single valued descriptors would include values that are equal to, less than, or greater than a particular real number. Subclasses of multiple valued descriptors include arrays, sets, bags, and value ranges. Statistical numbers include average values, number of values, and/or standard deviations. Time measuring values include minutes, hours, days, months, years, am/pm, or military time. Attributes of descriptor classes that have numerical values are designated as being of numerical descriptor type. This effectively creates an association to the numerical descriptor class.

Example: A compound could have an average specific heat in a particular temperature range.

Results

The information models were tested by populating a database with data and making queries against the data. Microsoft Access™, Oracle™ (V 7.0) and Object Store™ (Persistent C++ Object Oriented Database) databases were used. Data to be tested was collected from a batch chemical process, recipes for cooking food, a steady state chemical process, a critical path schedule, and games of chess.

For the batch chemical process, queries were carried out to find the time taken to complete the process, time taken to complete a reaction, the equipment involved, the material composition at each given time, and so forth. Sequential and parellel process were successfuly queried by getting the time stamps for the various subprocess existing for the batch process. For a steady state process, queries for determining the flow rate at a particular port, steady state composition of the reactor, the equipment involved, etc. were carried out. The snapshot models could be successfully accomodated using these new information models. For the chess game queries such as the number of moves, the move number of a particular move , the players involved, the date on which the game was played and the move made by a particular piece were carried out. The concept of the states being an aggregate of substates helped in querying for the various spatial and temporal positions of the chess pieces. The if-then link was successfully tested using the cooking recipes. Change in color of an ingredient when the temperature level exceeded a given limit or steps carried out when the water level fell below a given limit, could be determined using appropriate queries. The concept of the system passing through a series of states, where the final state of a subprocess forms the initial state of its succeeding processes, was tested using the critical path schedule. The queries could describe the simultaneously proceeding processes by getting their time stamps at various nodes.

Conclusions

The information models effectively capture data for steady state, batch and/or real-time operation of a chemical process. These models capture the information about the process that is used to create the product. They can capture the relationships between sequential and/or parallel processes. They are general in the sense that they can be applied to any process, not just chemical processes. The inefficiency of snapshot models is eliminated by using the concept of a state to describe the condition of a system at a point in time. The new models accommodate unlimited views of the physical system. Large portions of existing PDXI and STEP models have links to entities of the new models.

The models were successful in allowing typical queries to be made on databases populated with representative example problem data.

References

Blaha, M. R., R. L. Motard, and J. Mehta (1991). Structure and Methodology in Engineering Information Management, AIChE Annual Meeting, Los Angeles.

Blaha, M. R., N. L. Book and R. L. Motard (1993). Data Modeling Methodology and the PDXI Model, AIChE National Meeting, Houston.

Book, N. L., O. C. Sitton, R. L. Motard, M. R. Blaha, B. L. Goldstein, J. L. Hedrick, and J. J. Fielding (1992). Data Model Based Information Management Systems, AIChE National Meeting, Miami.

Book, N.L, O. C. Sitton, R. L. Motard, M. R. Blaha, B. L. Maia-Goldstein, J. L. Hedrick, and J. J. Fielding (1994a). The Road to a Common Byte. *Chemical Engineering*, pp. 98-110.

Book, N. L., O. C. Sitton, R. L. Motard, M. R. Blaha, B. L. Maia-Goldstein and J. L. Hedrick and J. J. Fielding (1994b). The PDXI Data File Interchange Format Project Deliverables, Vol I-IV, AIChE.

Doty, R. (1992). An Introduction to STEP, Digital Equipment Co., Chelmsford, MA.

Fielding, J. J., N. L. Book, O. C. Sitton, M. R. Blaha, B. L. Maia-Goldstein, J. L. Hedrick and R. L. Motard (1995). Methodology for Data Modeling for the Process Industries, Foundations of Computer-Aided Process Design, AIChE Symposium Series, Vol 91, No. 304, CACHE AIChE.

ISO, (1992). ISO 30303 – Industiral Automation Systems – Product Data Representation and Exchange – Part 1: Overview and Fundamental Principals, Draft International Standard. ISO TC 184/SC4

Motard, R. L., M. R. Blaha, N. L. Book and J. J. Fielding (1995). Process Engineering Databases from the PDXI Perspective, Foundations of Computer-Aided Process Design, AIChE Symposium Series, Vol 91, No. 304, CACHE AIChE.

Rumbaugh J., M. Blaha, W. Premerlani, R. Eddy and W. Lorensen (1991). *Object-Oriented Modeling and Design*, Prentice-Hall.

Virrantalo, A. M. (1994). An Object-Oriented Taxonomy of Declarative Process Knowledge, Computers and Chemical Engineering, Vol. 18, pp. S737-S741.

APPLICATION PROGRAMMING INTERFACES FOR EXPRESS INFORMATION MODELS

Dr. Milind Madhav Khandekar
Hilco Technologies, Inc.
Earth City, MO 63045

Prof. Neil L. Book
University of Missouri at Rolla
Rolla, MO 65401

Abstract

The implementation of advanced manufacturing concepts depends on a company's ability to quickly, efficiently and reliably share information with its vendors, contractors, consultants and other partners. Humanless and paperless electronic data exchange provides the means for integrating software and developing a plant life cycle database upon which advanced manufacturing concepts can be implemented.

The "pdXi Data File Interchange Format Project" involves developing an electronic data exchange system for chemical companies that would save them money and time. The project has three parts: 1) specification of the exchange format, 2) clear context for all the data that is exchanged and 3) a methodology for transferring data to and from the exchange format.

The STandards for the Exchange of Product Model Data (STEP or ISO 10303) is an international standard for exchange of product life cycle data. The STEP physical file specified by Part 21 of the standard was chosen as the exchange format and the context for the data was provided by ten harmonized information models in the object modeling technique and EXPRESS language. Application programming interfaces (APIs) are derived from the information models that transfer data between applications, databases and the STEP physical file. The "pdXi Prototype API," an application developed by the UMR-pdXi research team demonstrates the utility and commercial viability of the research work. The "pdXi Prototype API" application uses vector data structures that combine several entity level C language data structures to get and put data in engineering size blocks.

XpressAPI is a Microsoft® Windows™ application that automates the creation of the source code for an API for any information model written in the EXPRESS modeling language. It is a very stable prototype that produces the API code rapidly and accurately. XpressAPI was used to produce the API for six example information models. Performance comparison on contrasting EXPRESS models and the maintainability of the API are two of the greatest advantages of using XpressAPI. XpressAPI served as the prototype for the commercial pdXi API software that is soon to be released by Simulation Sciences, Inc.

Keywords

Application programming interfaces, EXPRESS information models, STEP physical file, XpressAPI, pdXi, Electronic data exchange, ISO 10303-21, ISO 10303-22.

Electronic Data Exchange

Engineers in the process industry use a number of software packages – some commercial, some developed in-house. According to Winter (1984), at least 50% of a firm's resources are spent on carrying out such routine tasks as moving, copying or converting data from one format to another. Design engineers working with standalone computer programs spend 50 to 80% of their time moving and organizing data between programs (Motard 1989). *Software integration* is needed to replace manual translation of data from one computer program to another.

Due to incompatibilities between different computer systems – hardware and software – data cannot be transferred easily within or across organizations. Any process that involves just data transfer and does not add any value is intrinsically inefficient. *Humanless and paperless exchange of data* saves money.

Regulations put forth by U.S. Environmental Protection Agency (USEPA), Occupational Safety and Health Administration (OSHA) and Clean Air Act Amendments of 1990 increase the amount of data needed to be stored and retrieved by each company. The data must be produced in a timely fashion, be accurate and up-to-date – even after the plant is decommissioned. To meet these requirements, a *plant life cycle database* containing the data from the cradle to the grave is required.

Manual transcription is widely used as a data exchange method. A design engineer manually enters the output of one computer system as the input to another computer system. The manual transcription process is dull and repetitive. It is slow, expensive and requires good knowledge of both the source and the target systems, but worst of all, it is error prone.

Another common way of transferring data between computer systems is to use *direct or point-to-point translators*. Being computer programs, translators are very useful for dull, repetitive and slow tasks. But for N software systems, assuming that the data needs to be exchanged between every possible pair of the systems and in both directions, N*(N-1) translators would be needed. A typical engineer uses 20 to 30 software systems, so this number could go well above 500. Also, if one of the systems were to change, 2*(N-1) translators are required to be updated. Maintenance of the translators is not easy since they have very complex logic. Also, translators are inherently inefficient since they do not create any additional information.

If the entire process industry *standardized on a single software system*, i.e. used the same simulator program, same word processor, same graphics software, etc., standard software programs could be written to exchange the data within the system and be made part of the software system. However ideal, data exchange by standardizing on a single system is impractical. It requires that all the organizations accept a standard system. Evolution of the standard system would be quite disruptive because all the users would have to update simultaneously. The rapidity with which today's computer technology is moving, any standard system will require frequent updates.

Instead of writing point-to-point translators, *translators* are created *to a standard neutral exchange format*. The number of such translators required for N software systems is 2*N which is always less than N*(N-1) for N > 3 (Book 1994a). If one software system changes, only 2 translators are required to be updated. However, if the neutral exchange format changes, all the translators are required to be updated.

If a common data format is accepted as the internal data format by all the software systems, *standardization on a common data format* would occur. Advantages of standardizing on a common data format are the same as those of standardizing on a standard system. The data exchange and storage and access to the data are extremely efficient. This is the ultimate goal, however, there is a great body of existing software with internal data structures that are unique. Using a standard neutral exchange format, these software can exchange data now and, over time, migrate their internal data structures to be the same as the neutral exchange format.

pdXi Data File Interchange Format Project

The Process Data eXchange Institute (pdXi) is a consortium of sponsor companies seeking a way to implement an electronic data exchange system. The project, entitled *pdXi Data File Interchange Format Project,* is aimed at making the envisioned data exchange system an international standard for data exchange in the process industry. The STEP or ISO 10303 project (ISO 10303-1 1994) covers product life cycle data. pdXi chose to adopt the STEP methodology. That way, pdXi could leverage off the work that is already being done towards STEP, and the outcome of the pdXi project could be accepted as an international standard.

The pdXi project consists of at least three parts: 1) specification of a format for the exchange, 2) a clear context for all the data that is exchanged and 3) a methodology for transferring data to and from the exchange format.

Part 21 of STEP (ISO 10303, Part 21), namely, *Clear Text Encoding of the Exchange Structure,* was chosen as the specification for the exchange format (ISO 10303-21 1994). It provides a human-readable, neutral ASCII text file structure for the data exchange. This format is commonly referred to as the STEP physical file (or simply, a neutral file).

The context for the data in a STEP physical file is provided by information models written using the *EXPRESS Modeling Language* (ISO 10303-11 1994).

pdXi produced an information model organized in 10 parts for the process engineering data in the *Object Modeling Technique* (*OMT*) syntax. The OMT models were translated manually to EXPRESS.

To transfer data to and from the exchange format, *Application Programming Interfaces* (API) are provided. pdXi provided Standard Data Access Interface (SDAI) specifications (ISO 10303-22 1994) for its information models written in EXPRESS as a set of pdXi API specifications (Book 1994b). The specifications describe C language data structures that can hold the data in a STEP physical file and a set of C language functions that can retrieve and/or manipulate data in the file.

pdXi API Specifications

The pdXi API specifications document outlines the C language, object level data structures and navigation paths used for the individual *get* and *put* operations for entities in the pdXi EXPRESS model. Figure 1 has an excerpt from the pdXi specifications document.

```
/* EXPRESS Entity Name          C structure Name
SUBSTANCE_NAME                  subst_name
*/

typedef struct subst_name {     //C language data
  char * substance_name;        //structure and unique
} Substance_name;               //attribute declaration

/*Pointer declaration */
typedef Substance_name* Substance_namePnt;

/* Navigation Parameters
substance_name
SUBSTANCE_NAME.substance_name */
```

Figure 1. An excerpt from the pdXi API specifications document.

Each entity in the EXPRESS model has a corresponding C language data structure in the API. The name of the EXPRESS entity in Fig. 1 is SUBSTANCE_NAME and the corresponding C structure name is subst_name. The entity has one attribute that is defined as type *char** for the C structure.

The *get* and *put* function calls for the C structure would be named *get_subst_name* and *put_subst_name*, respectively. The C language data structure and the *get* and *put* functions are collectively called the *object level data structures and functions* and are dependent on the EXPRESS model. XpressAPI, a software system described later in this paper produces these automatically.

The last part of Fig. 1 is a list of arguments which provide a navigation path to an instance of the entity. The navigation path provides the minimum amount of information required to get or put an instance. The mechanism by which the navigation path is traversed and the EXPRESS entity instances are accessed is independent of the EXPRESS model or the data in the STEP physical file. The navigation paths and classes are embedded in the *generic functions*.

The data structures for more complex entities are constructed using the object level data structures. Multi-entity *get* and *put* operations are performed on these complex data structures. The complex data structures and the multi-entity functions are collectively called *convenience functions*.

The *generic functions, object level data structures and functions* and *the convenience functions* together form an API. The *generic functions* need not be recreated unless either the EXPRESS modeling language or the structure of the STEP physical file changes. the *object level data structures and functions* are automatically produced using XpressAPI. The *convenience functions* are produced as needed by the application, using the object level data structures and functions. All application programming interfaces would be based on the pdXi API specification document for consistency.

pdXi API Mechanism

Applications, shown at the very top in Fig. 2 need to exchange data with the STEP compliant repository (such as a STEP physical file, STEP compliant database, etc.). The data exchange to and from the repositories to the applications occurs through several layers.

The EXPRESS model that provides the context for the data in the repository is read into the computer's memory as the *EXPRESS working form*. The EXPRESS working form has initialization functions and class declarations that can create data structures for instances of any EXPRESS entity.

The data from the repository is read into or written from the computer's memory using certain *Read/Write Functions* supplied by NIST software[8]. The read and write functions use the initialization functions in the EXPRESS working form to load the data from the repository into the computer's memory. The in-memory representation of the data is termed as the *STEP Working Form*.

The *generic functions* layer of the API contains functions that do not depend on the EXPRESS model or the data in the STEP physical file. The functions only depend on the syntax of the EXPRESS information modeling language and the structure of the STEP physical file. Therefore, this layer doesn't have to be changed

---

[8] The pdXi API generation process uses several software systems supplied by the National Institute of Standards and Technology (NIST). NIST assists STEP projects within the United States and coordinates activities with the international STEP projects in the process industries.

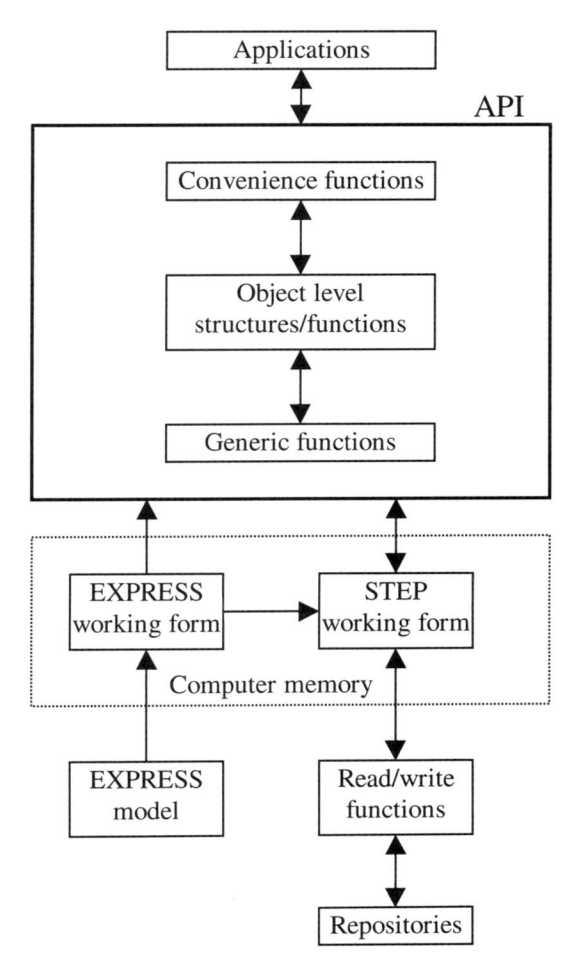

Figure 2. API between applications and STEP compliant repositories.

when either the EXPRESS model or the data in the STEP physical file changes. All the model navigation routines come under the generic functions category. These functions were manually created once. The functions in the *object level data structures and functions* layer can get or put data to the working forms using the functions defined in the generic functions layer.

The *object level structures and functions* layer provides C language data structures and function calls to access the data in the working forms. The data structures hold instances of an EXPRESS entity and the data members hold the attribute values for those instances. Using the functions, the user can perform get, put (or update) operations on the working forms for the instances. Any one object level function can access data for exactly one EXPRESS entity instance and any one object data structure can hold data for exactly one instance of an EXPRESS entity.

The layer above the object level structures and functions layer is formed by a set of application specific *convenience functions*. The role of convenience functions is to group together certain object level function calls and retrieve and/or modify the data which belongs to more than one EXPRESS entity instance. As an example, consider flow and amount specification data for a material stream. Such data belongs to a number of entity instances, however, it is common that all the information about a stream is desired. A convenience function can group calls to all the intermediate single entity functions and provide the composite data as its output. Thus, the object level functions are provided on an individual entity basis whereas convenience functions are a combination of several object level functions grouped together for a specific task.

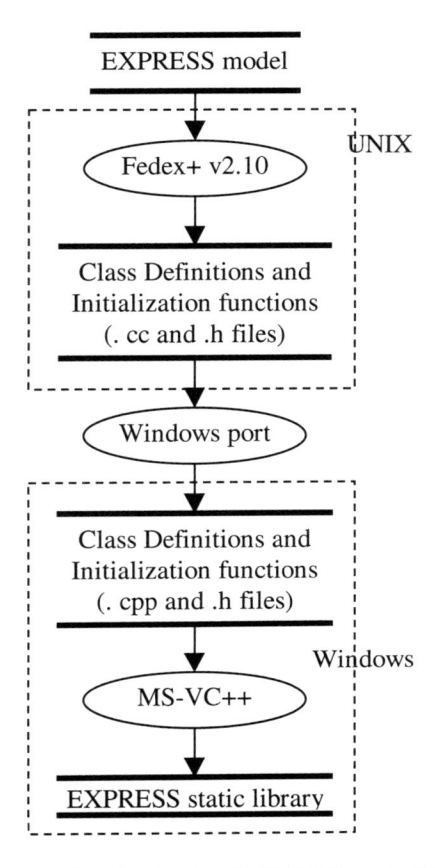

Figure 3. Producing EXPRESS static library.

Producing the API Code

The following section describes the various steps involved in the process of producing the API source code for pdXi EXPRESS models.

Producing the EXPRESS Static Library

The EXPRESS models are compiled using NIST's Fedex+ 2.10, a UNIX based software system that parses and complies EXPRESS models (Fig. 3). The Fedex+ output is a set of *.cc* and *.h* files designed to be compiled with Gnu's G++ compiler on the UNIX platform. The header files (the *.h* files) contain class definitions for the entities in the EXPRESS model. The class definitions provide the templates for the specific instances of the EXPRESS entities from the EXPRESS model. The source files (the *.cc* files) contain initialization functions that have the necessary instructions to create those instances based on the templates from the header files.

The Fedex+ output is ported to the Microsoft® Windows™ platform as a set of *.cpp* and *.h* files. The source files are compiled with Microsoft® Visual C++™ compiler to produce a static library, called the *EXPRESS Static Library* which is saved on the computer disk typically by the name, *schema.lib*.

Producing EXPRESS Working Form

NIST's *Validation and Testing Software (NIST-VTS)* was ported to the Microsoft® Windows™ operating system to compile under Microsoft® Visual C++™. The process of producing the EXPRESS working form uses the functionality provided by the *Registry* class in the NIST-VTS' *STEP Class Library (SCL)*. The object of the *Registry* class loads the static library, *schema.lib* into the computer's memory (see Fig. 4). The in-memory representation of the static library is a set of hash tables and linked lists and is called the *EXPRESS working form*. The object of the *Registry* class maintains a list of all the entities defined in the EXPRESS model.

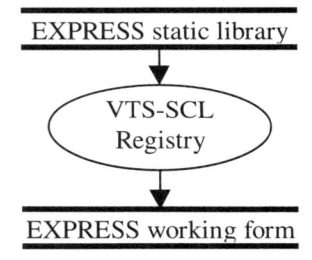

Figure 4. Producing EXPRESS working form.

Producing the Object Level Structures and Functions

Most of the information needed to produce the *Object Level Structures and Functions* is available in the EXPRESS working form and the required functionality is provided by the NIST-VTS-SCL. XpressAPI was developed to produce the source code for the object level structures and functions. Figure 5 depicts the functional model for XpressAPI.

XpressAPI extracts information from the EXPRESS working form and displays it on a graphical user interface. The interface allows the user to supply additional information required to produce the source code. The user-supplied information includes a C language structure name for every EXPRESS entity, a C language attribute name for every attribute of every EXPRESS entity and navigational parameters to uniquely identify an instance of every EXPRESS entity. As described earlier, this information is available in the pdXi

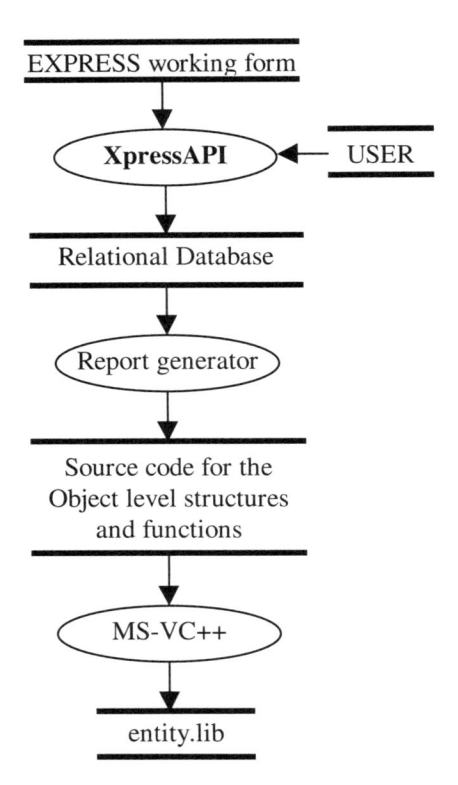

Figure 5. Producing object level data structures and functions.

API specifications documents. The interface provides default names for the C language structure and the attributes and a means to scroll through all the EXPRESS entities defined in the model one by one, providing information for one entity at a time. XpressAPI saves the information extracted from the EXPRESS working form and the additional information that was obtained from the user into a Microsoft® Access™ relational database using Open Database Connectivity (ODBC) drivers. Once the database is populated, a set of XpressAPI supplied reports are run via the report generator functionality provided by Microsoft® Access™. The reports produce a set of *.cpp*, *.c* and *.h* text files. The text files contain declarations and definitions for the object level, C language data structures and *get* and *put* functions. The source files produced by XpressAPI are compiled using Microsoft® Visual C++™ into a static library called *entity.lib*.

Adding Convenience Functions and Producing the API

The *generic functions* are common for each API. The generic functions were compiled into a static library called *generics.lib*. *entity.lib* is obtained using XpressAPI. Multi-entity convenience functions can be written and compiled into the static library, *convence.lib*. All the three libraries, *entity.lib, generics.lib* and *convence.lib* are linked together to produce the API! (Fig. 6).

Applications of XpressAPI

The *pdXi Prototype API* (Fielding,1995) was an API written by hand to demonstrate the viability of electronic data exchange for process industries. A small EXPRESS model was extracted from the pdXi models which had about 50 entities out of more than 600 entities in the pdXi models. The small EXPRESS model allowed the exchange of material stream data. The API was produced by a team of five people (who were already familiar with the NIST software) over a three month period. Much of the code developed was routine, repetitive tasks that suggested automation. A graphical interface was developed for the API that allowed, with a click of the mouse, for stream data from either Aspen Plus or ProII to be written to a STEP physical file. With another click of the mouse, the data could be imported from the STEP

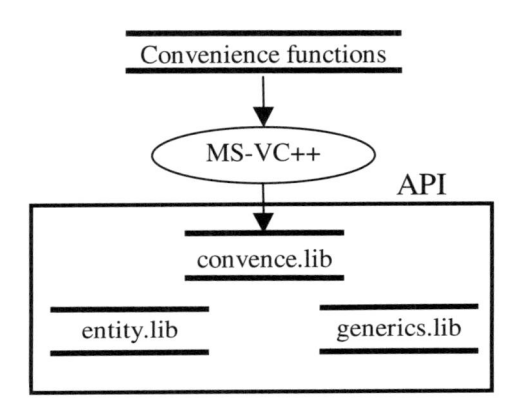

Figure 6. Adding convenience functions and producing the API.

physical file into one of several different applications (HTRI heat exchanger rating programs, a database, a spreadsheet, etc.). The *pdXi Prototype API*, designed to run on a single computer, clearly demonstrated the viability of electronically exchanging process engineering data from one computer to another (perhaps in different organizations) so long as both computers had copies of the API. Based on the success of the prototype, pdXi contracted with Simulation Sciences, Inc., to produce a commercial version of the API (complete with *convenience functions*) for the full pdXi models.

With XpressAPI, the API for the EXPRESS model of the prototype was generated in less than two hours by a single individual. The comparison of the generated code with the hand-written code identified several bugs in the hand-written code.

The existence of XpressAPI has allowed extensive testing to be performed on portions of the pdXi information models. Five EXPRESS models (containing 50 to 100 entities each) have been extracted from the pdXi models and an API created for each. Test data has been written to and extracted from STEP physical files to demonstrate the data exchange capabilities of these portions of the pdXi models.

XpressAPI served as the prototype for the commercial version of the pdXi API soon to be released by Simulation Sciences, Inc. (First Quarter, 1998) Based on experience with XpressAPI and the *pdXi prototype API*, Simulation Sciences chose to develop an EXPRESS parser and C++ class generators rather than use the NIST software. The NIST software generates a static library. A dynamic link library was deemed necessary to obtain acceptable API performance for large EXPRESS models, such as the complete pdXi models.

Conclusions

XpressAPI is an automation tool that provides the means to rapidly produce an API for any EXPRESS information model. The API can be used to demonstrate the electronic data exchange capabilities for data captured by the EXPRESS model. Alternative EXPRESS model that capture the same data can be compared to identify the model with the best performance.

A companion application, eXpressDB was developed (Gidh,1996) that exchanges data between the STEP working form and an ODBC compliant database. These systems create a system where data can be exchanged between applications, databases, STEP physical files.

References

Book, N. L., O. C. Sitton, R. L. Motard, M. R. Blaha, B. L. Maia-Goldstein, J. L. Hedrick, and J. J. Fielding (1994a). The road to a common byte. Chem. Eng. September 1994, pp. 98-100.

Book, N. L., O. C. Sitton, R. L. Motard, M. R. Blaha, B. L. Maia-Goldstein, J. L. Hedrick, and J. J. Fielding (1994b). The pdXi data file interchange format project deliverables. Volumes I-XIV. AICHE.

Fielding, J. (1995) Methodology for data modeling for the process industries: Conception through programming interface. Ph. D. dissertation, Chemical Engineering Dept. University of Missouri, Rolla, MO. 1995.

Gidh, Y. (1996) Development of an application programming interface to STEP databases for chemical process software. Ph. D. dissertation, Chemical Engineering Dept. University of Missouri, Rolla, MO. 1996.

ISO 10303-1 (1994). Industrial automation systems and integration – Product data representation and exchange – Part 1: Overview and fundamental principles. Ed. ISO TC184/SC4, pg. 17, January 1994.

ISO 10303-11 (1994). Industrial automation systems and integration – Product data representation and exchange – Part 11: Description methods: The EXPRESS language reference manual. Ed. ISO TC184/SC4/WG5 P2, pg. 208, January 1994.

ISO 10303-21 (1994). Industrial automation systems and integration – Product data representation and exchange – Part 21: Implementation methods: Clear text encoding of the exchange structure. Ed. ISO TC184/SC4, pg. 57, January 1994.

ISO 10303-22 (1994). Industrial automation systems and integration – Product data representation and exchange – Part 11: Standard data access interface. December 1994.

Motard, R. (1989). Integrated computer aided process engineering. Computers and chemical engineering, 13, pg. 1199.

Winter, P. and C. J. Angus (1984). The database frontier in process design. Proceedings of the second international conference on FOCAPD, CACHE Pub., pp. 75.

A PROCESS PLANT KNOWLEDGE REPOSITORY FOR MULTIPLE APPLICATIONS

Michael J. Elsass, David C. Miller and James F. Davis
Department of Chemical Engineering
The Ohio State University
Columbus, OH 43210

John R. Josephson
Department of Computer and Information Science
The Ohio State University
Columbus, OH 43210

Abstract

There is considerable process plant knowledge and a number of common information processing capabilities shared by diverse decision support applications such as monitoring with diagnosis, hazard analysis, and design criticism. By representing process-plant knowledge in a single repository, it is possible to effectively support multiple decision-support applications, leverage expensive sources of knowledge, speed representation of specific plants, and enhance productivity in building specific applications. Functional Representation (FR), a device-centered representation formalism, consists of readily navigable, hierarchical links among structural, behavioral and functional information which facilitates capturing heterogeneous forms of qualitative and quantitative knowledge. In this paper, we describe the sharable nature of plant knowledge across applications, the commonalties in information processing, and how to use FR to provide a compact, organized, and modular framework for encoding the process-plant information.

Keywords

Functional representation, Knowledge repository, Intelligent systems, Causal process description.

Introduction

Process knowledge useful for a variety of decision-support activities is obtained from a multitude of sources that combine documentation and process expertise. Typical documentation includes process schematics, logbooks, and maintenance records. The usability of these for applications depends heavily on interpretation in the context of the process operation. Any specific decision-support activity (such as abnormal situation management or design criticism) will typically rely on a subset of all available information. There is considerable overlap in the knowledge required for various tasks. In addition to overlap in knowledge content, there is also overlap in knowledge structure. Thus, by organizing process knowledge into a single repository, multiple applications can take advantage of existing information rather than acquiring and structuring this information independently for each application. Such a knowledge repository would promote the use of these decision support activities in plant environments by reducing the investment required for building these systems.

Since many processes contain multiple pieces of the same or similar equipment items, structuring knowledge at an equipment level aids in the reusability within a given process. Such a modular representation allows information to be leveraged across multiple processes since knowledge used for simple devices, such as pumps, valves and tanks is typically similar across many processes. The operations associated with complex devices are heavily dependent upon the process containing the device in order to discern intended behaviors from unintended or abnormal behaviors. However, these devices exhibit a similar set of behaviors regardless of the process. Thus, a representational framework that contains a large number of process-independent behaviors, which are specified by the user as normal or abnormal, provides a powerful tool for representing multiple processes. A library of these device representations makes it possible to model multiple processes for use in diverse applications.

Previous papers (Miller *et al.*, 1997b) have discussed the functional representation and its application. This paper examines the common knowledge requirements for several typical decision-support activities (hazard analysis, fault detection and isolation, and design criticism). Our representational formalism is then reviewed.

Application-Specific Knowledge Requirements

Hazard Analysis – Hazard analyses are performed to identify potential safety hazards within a process plant. Examples of these analyses include HAZOP, What-if analysis, and P&ID review. All incorporate an exhaustive cause and effect study of possible plant malfunctions. Typically this involves propagating qualitatively described behaviors causally forwards and backwards from points of irregularity. Malfunction scenarios are analyzed for potential safety hazards by studying the likelihood, consequences, and importance of the hypothesized irregularities. In addition to causal knowledge of the behaviors associated with device malfunctions, knowledge of probabilities and safeguards is needed. From a knowledge standpoint, a decision-support system needs to be able to consider malfunction causality given a set of equipment and modes of operation. Thus, knowledge of equipment behavior under a variety of circumstances is essential.

By assuming either a process deviation or the abnormal operation of a particular piece of equipment, the effects of a malfunction are propagated through the system. From an initial malfunction such as 'valve-failure' causal transitions propagate the resulting abnormal conditions through the various pieces of equipment which comprise the process. The consequences resulting from the initial fault include process variable deviations as well as other faults and failures. Examples of the consequences could range from 'resulting high pressure at outlet' to 'catalyst damage'. While causal transitions are able to reach these conclusions, functional knowledge is useful in reasoning about the feasibility and probabilities of these consequences. For instance, if a resulting condition is a side effect of the desired mode of operation, there will be a higher probability of this event occurring.

Knowledge of function and mode of operation is also useful in hazard analyses by giving a more descriptive view than can be obtained from simple causal propagation. A wider repertoire of potential faults and failures is produced by explicitly representing operational modes and associated behaviors. For example, an analysis reporting 'reactor high pressure' may designate a rupture disk as a safeguard. While noting that the disk is considered a safeguard, the analysis could continue with representations of the abnormal behaviors associated with the rupture disk such as 'vent toxins to atmosphere'.

Monitoring with diagnosis – Fault detection and diagnosis are important process applications. Diagnostic monitoring may proceed by examining observed process variables to determine if a discrepancy exists between actual process output and process output predicted using a process model assuming normal operations. Representations of behaviors associated with normal operating modes are needed to support such predictions. Once a deviation has been detected, diagnostic reasoning attempts to determine the root cause of the deviation by hypothesis generation and hypothesis discrimination. Hypothesis generation consists of propagating causal events through the process, at varying levels of detail, going backward from observed symptoms to plausible root causes. This causal propagation typically proceeds at the equipment level until a malfunctioning device has been found and then moves to a finer level of detail by investigating causal relationships of device components. Since process variables cannot all be monitored, many such paths may be consistent with observations, resulting in multiple diagnostic hypotheses. To discriminate among these hypotheses, expected behavior can be propagated from each proposed root cause to the anticipated downstream observable consequences, testing a hypothesis by comparing it with process observations.

Design criticism – There are many forms that design criticism may take. All seek to analyze the effectiveness of a given equipment configuration and processing conditions in achieving design goals. These goals can range in scope from entire plants to individual unit operations and range in detail from rough, conceptual design to detailed equipment design. In all of these situations, however, the reasoning process is similar: propose a configuration of equipment to perform the desired functions, critique the design, and modify the design if necessary. In order to perform these tasks, knowledge of equipment must be linked to functions and behaviors.

Miller *et al.* (1997a) reported a specific example of a design decision support system. This system seeks to build and critique a rough design based on a laboratory description of the chemistry. An important initial step in this design criticism is the development of a rough process flowsheet. This flowsheet can then be used to critique both the process and the underlying chemistry. Development of such a rough flowsheet begins by identifying the functions and behaviors required for the process. Equipment is then selected based on its ability to achieve the required functions. These equipment items form the basis of the rough flowsheet. Associated with each equipment item are multitudes of functions that are necessary for its proper operation. These functions are ancillary to the process function which caused the device to be initially selected; however, these additional functions often have major effects on the overall process. Using this rough flowsheet, the process can be analyzed based on cost and environmental considerations that result from both the primary functions and ancillary functions of the equipment.

As the design advances in detail, additional criticism can be performed. Knowledge of potential malfunctions and their probabilities allows the determination of appropriate sensor configurations that can best detect and isolate these potential faults. In addition, a general hazard analysis of the process in a manner similar to HAZOP can be performed so that some potential situations can be corrected by minor modifications to the initial design.

Knowledge Organization and Representation

For the listed applications, there exists a commonality of knowledge types, and of specific knowledge. Descriptions of device behaviors are central to all applications and reasoning about these device behaviors frequently requires knowledge of process-variable state transitions. The relationships among variables involved in state transitions can be expressed qualitatively or numerically.

We believe that process knowledge can be structured most conveniently in a device-centered manner. This facilitates both the acquisition and application of process knowledge. In addition to providing an intuitive method of accessing the information required to build any particular decision-support application, an equipment object template allows for equipment and component topology to be naturally represented.

Our breakdown of process information is based on the following questions: What is it? (equipment), What does it do? (function), When does it do it? (mode of operation), and How does it do it? (behavior). Rather than embedding all of this information into a single, complex data structure for each piece of process equipment, we employ a modularized approach with associative links between the different knowledge types. This allows us to explicitly recognize how similar behavior can be achieved by a variety of different equipment items and how a single equipment item can be employed for a variety of different purposes. In the following paragraphs, we examine what we mean by each type of knowledge and its potential uses.

<u>Equipment</u> – This type of knowledge includes information about a device's structure, such as ports, components, and configurations. It also includes knowledge about the aggregations of different structures, such as aggregations of components to form a device, and of devices to form systems and subsystems.

Modes of Operation – This type of knowledge identifies device behavior as normal or abnormal and further classifies abnormal behavior as fault or failure. Device context is used to determine whether behavioral classifications are generic or based upon the location of a device within a process.

Device operation can be considered as normal or abnormal. This distinction can be based upon the intended function of the device (process specific) or an operation that is clearly abnormal, such as a tank leak. The classification of device operations as normal, fault (readily reversible to a normal mode), or failure (not easily reversed) is dependent upon the device achieving the intended behaviors. The reduction in flow through a centrifugal pump due to high downstream pressure is an example of a fault since it can be easily corrected once the downstream pressure is returned to normal. A component malfunction or catalyst poisoning are indicative of failures since the state cannot be reversed without repairing the component or regenerating the catalyst. HAZOP and design criticism for diagnosability use abnormal operation as an initial hypothesis, and propagate the associated behaviors through the process, whereas design criticism for device selection uses only normal operations. Monitoring uses knowledge of normal device operation to determine if an abnormal situation is taking place and then uses causal propagation among connected devices to locate a malfunctioning device. Since monitoring is concerned with determining if a malfunction is occurring, it requires knowledge about both normal and abnormal behaviors. Representation of operations knowledge provides a means for tagging whether a set of functions is normal, fault, or failure.

Function – This type of knowledge consists of the tasks a device can achieve, and the corresponding side effects related to this. Rather than embedding knowledge of the functions and behaviors of a particular device within the device object itself, functions provides a valuable, device-independent construct which facilitates the creation of a knowledge repository. Functions contain knowledge useful in determining potential behaviors (normal and otherwise) in a given process and provide a means for recording the designer's intent which can be used to classify functions not directly related with desired behaviors.

Behavior – This type of knowledge explicates the causal relationships that occur in order to achieve some function by relating the state transitions of process variables. This knowledge annotates knowledge of state transitions and is used for causal propagation of process states.

Causal Relationships are common to many applications and form a basis for the knowledge structure by providing detailed information of process behaviors.

Functions package behaviors to provide a more comprehensive definition of device behavior. Depending upon the role of a device, the prospective functions may be different. Process operations tag functions (and therefore groups of behaviors) as normal or abnormal. Some of these tags are generic and are true regardless of context as opposed to classifications that are dependent upon the surrounding process.

Function as a Basis for Knowledge Representation

A device-based, knowledge representation can be conveniently organized around multiple modes of operation with each mode describing several functions and each function having associated behaviors (Keuneke, 1991). Although the organization goes from mode to behavior, the functional knowledge provides the representation with the flexibility to be used across multiple processes and applications. For complex devices, function is directly related to the designer's intent. Thus, a device's function may be different for each process in which the device occurs. A knowledge representation suitable for multiple processes must be sufficiently flexible to reflect this context-dependent relationship to function. Several important factors need to be considered when representing function.

Generic vs. Process Specific – Generic functions include behaviors that a device will display regardless of the process. Simple devices such as pumps and valves are examples of devices with generic functions in that they will always exhibit the behaviors 'make flow' and 'restrict flow' respectively. Process specific functions include intended device behaviors that are dependent upon the given process. For example, a heat exchanger may be used for anything from 'heating tube stream' to 'cooling tube stream' to 'condensing vapors'. Heat exchangers are used in all these capacities, but the function depends upon the designer's intent. By incorporating these functions into the heat exchanger representation, the user can designate which function is desired based upon the process.

Abnormal vs. Normal – For simple devices containing many generic functions, there is a clear, intuitive distinction between normal and abnormal behaviors. For a pump, the 'make flow' function is considered normal while 'no flow' would be considered abnormal. However, for a more complex device, such as a jacketed reactor, if the intended behavior of the jacket is to 'cool reactor' then the behavior 'heat reactor' would be considered abnormal and vice versa if reactor heating was intended. As described earlier, functional knowledge makes no distinction between normal and abnormal operation. This is left to knowledge of the mode of operation. Therefore, all possible functions are included in the representation. In the case of complex devices, the user must designate whether a function is normal or abnormal based upon the intended behavior of the device.

Intended vs. Side Effect – Since functions are associated with groups of behaviors, other behaviors and functions are usually associated with the desired one. For example, a designer may choose a catalytic reactor to perform the function, 'effect sulfur reduction.' However, associated with this function is the function, 'emit sulfur dioxide.' Such additional functions are regarded as side effects. Side effect functions can aid a designer in choosing among various possibilities by explicitly identifying undesirable behaviors.

For generic devices in simple processes, normal and abnormal behaviors can be clearly distinguished, and normal behaviors can be clearly identified as intended or side effects. This distinction is less clear for complex, process-specific devices. While there will always be obvious abnormal functions such as 'reactor leak', understanding the device's role within a process may be required in order to designate functions as abnormal and normal. The knowledge of designer's intent is also required to determine whether a normal function is intended or a side effect.

Functional Representation

We have chosen FR as a representational formalism for constructing a multi-process, multi-application knowledge repository. FR is a device-centered representation that organizes process knowledge based upon physical structure, operating mode, function, and behavior. FR can be used to capture knowledge from the component level to the system level. Components are linked to form devices in the same way that devices are linked to form systems. Information is joined among the representations at multiple levels of detail.

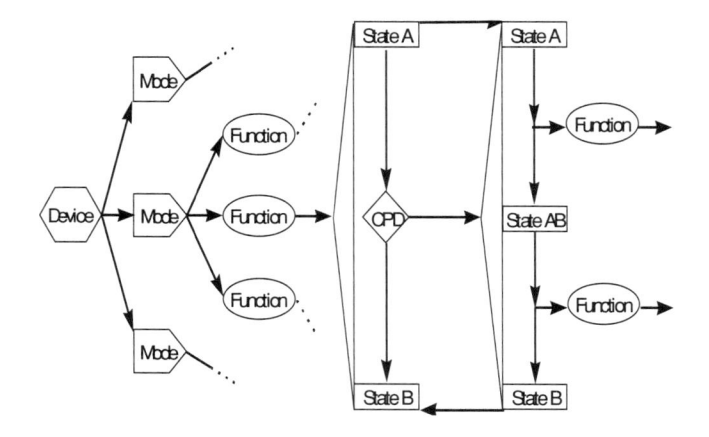

Figure 1. Organization of linked representations in FR.

FR distinguishes four main data types: Structure, Mode, Function, and Causal Process Description (CPD). As shown in Fig. 1, these data structures are linked hierarchically to provide several levels of detail when describing device behavior. The internal behavior of a device or component is represented using the CPD, which shows a digraph type representation of process variable transitions. Associated with these transitions are links to descriptions which assign responsibility for a transition to a physical law, to a function, or to a more detailed set of CPDs.

The Function template is used to organize CPDs. FR contains four different function types: To-Achieve, To-Maintain, To-Prevent, and To-Control. Function templates provide a method for categorizing the behavioral information found in CPDs and use port states to index the associated CPDs. By including the input and output states of the device, Function acts as a coarse description of device behavior that can be propagated to other devices.

Mode templates are used to categorize Functions as abnormal or normal based on the intended use of the device in the process. Abnormal Modes are further broken down into the sub-categories of fault/failure and primary/secondary (Primary malfunctions occur because of a problem with a device, and secondary malfunctions occur due to upstream conditions). When using FR representations to model a process, the user can choose among the many Functions associated with a given Mode.

These templates are organized around the Device template as shown in Fig. 1. The Device template provides a means for storing physical knowledge of a device, such as components, ports, and connections. Applications utilizing FR can access the representation at all levels of the hierarchy.

By not explicitly designating Functions as normal or abnormal, the flexibility of the representation is increased. For simple process-generic devices, most Modes, Functions, and CPD's will be linked in the manner shown in Fig. 1. However, for complex process-specific devices, the Modes and Functions in the FR library are not linked *a priori* since the behavioral classifications (normal, abnormal) resulting from the Mode to Function link are process dependent. Once a process has been chosen, the user can easily choose the relevant Functions and link them with the appropriate Modes.

Depending upon the application, Functions can also be associated with several Devices (linked through Modes) as shown in Fig. 2. This modularity allows the user to tailor the knowledge into a form that best supports the requirements of a specific application. Once the underlying structure has been filled in, the user links the device representations to form a process flowsheet that can be used for several applications. The following examples show how FR is used in different applications.

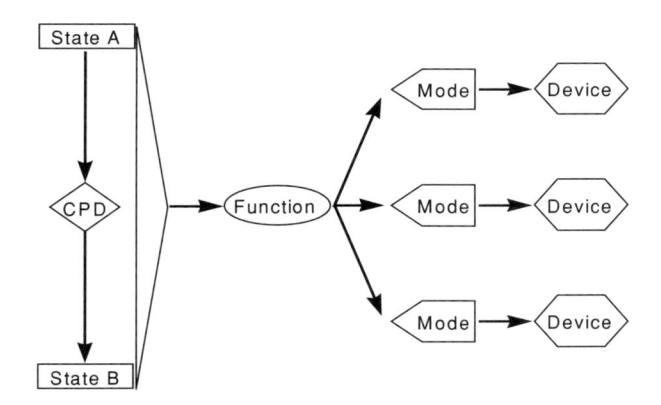

Figure 2. A function achievable by three different devices.

Hazard Analysis – A distillation column serves to illustrate how FR supports hazard analysis. A typical hazard analysis begins by examining a process deviation such as 'high temperature distillation column inlet'. This corresponds to State A of Fig. 1. Since hazard analysis relies primarily on less detailed descriptions of process behavior, the application moves from State A to the Function template, which could designate the column behavior as 'low recovery behavior'. From the function template, the application moves to State B, 'low product in distillate'. State B is then passed directly to State A of a different device. The procedure can also move from State B to State A to search for the cause of a deviation. Hazard analysis can also begin with a device malfunction, corresponding to an abnormal Mode in Fig. 1. The propagation would then proceed to the associated Function and on to the State in the manner described above.

Monitoring with Diagnosis – For the application of fault detection, the observed states of the actual process are modeled with FR states (State A and B of Fig. 1). The procedure then follows the links to the Mode template to determine if an abnormal behavior is occurring. Continuing with the distillation column example, 'low temperature at inlet' is passed up to the 'low recovery behavior' Function. The Function is linked to the Mode that designates 'low recovery' Function as abnormal. Once a fault has been detected, the causal propagation to and from potential root causes is performed at the CPD level in a manner similar to hazard analysis (State A leads to State AB, etc.). When a causal path that matches the actual process is found, the procedure moves up the hierarchy to the Function and associated Mode to obtain behavioral information such as abnormal, intended, etc.

Design Criticism - Rough design begins by querying for specific behaviors and functions. For example, a designer may be looking for a device that will transform a stream form State A to State B such as 'high particulate content' to 'low particulate content'. This behavior is associated with the 'remove particulate' Function found in

Table 1. Comparison of Representational Formalisms.

	Types of Knowledge					
Representation Formalisms	Sensors & Qualitative Interpretations	Causal Relations	Functions & Goals	Procedures	Structure	Process Topology
Tables	X					
Fault Trees Digraphs	X	X				
Device Propagation	X	X				X
Malfunction & Goal Hierarchies	X	X			X	
Procedure Hierarchies	X		X	X		
Functional Representation (FR)	X	X	X	X	X	X

the structure of figure 2. This Function is associated with the wet-scrubber, electro-static precipitator, and cyclone device representations. While this procedure utilizes normal Modes, abnormal modes and side-effects associated with each device can be used to aid in the selection of an appropriate piece of equipment. This results in a table of devices that can be used to achieve the given desire behavior as shown in Fig. 2.

Comparison of Representational Formalisms

While commonalties exist in the knowledge employed by different process applications, traditional applications tend to emphasize different aspects of the knowledge structure. Table 1 compares the representational capabilities of several different formalisms. Causal relations, functions, goals, and structure have all been discussed in this paper. Sensors and qualitative interpretations are a means to describe a process state based on qualitative descriptions of observed variables. Procedures are used in batch processes as a means of describing the actions taken during the process. Process topology provides a description of the process as a whole by showing how devices are connected.

Many formalisms fail to explicitly capture the full range of knowledge required in process support systems. These limited representations were developed specifically for particular applications. For example, digraphs are frequently employed for use in fault detection and diagnosis. Digraph knowledge is structured around the causal relationships between process variables. Methodologies applied to digraphs for diagnosis use process knowledge to directly track process variables. Knowledge not required for this analysis, such as device structure, is not needed and, therefore, not explicitly represented.

For a knowledge representation to be useful for multiple applications, all knowledge types found in process applications must be represented explicitly. In addition, overlapping knowledge must be exploited in order to keep the size of the knowledge repository at manageable levels. Additional types of knowledge required for decision support activities are summarized in Table 1. As can be seen from Table 1, Functional Representation (FR) (Chandrasekaran, 1994) has the ability to explicitly represent the full spectrum of knowledge required for process system applications.

Conclusion

While diverse applications such as hazard analysis, monitoring with diagnosis, and design criticism emphasize different aspects of process knowledge, there exists a significant overlap in the knowledge types used for each. These knowledge types can be categorized as equipment, modes of operation, function, and behavior. Creating a device-centered repository of these types provides a means to represent diverse processes for use in multiple applications. While it would be possible to construct this repository using only behavioral state transitions, the process-specific nature of these descriptions would require that much information be repeated in devices with similar functions and behaviors. Decomposing process knowledge into the types described in this paper enhances the reusability of the representation. By explicitly representing knowledge of device function, and mode of operation, the designation of intended, side-effect, or abnormal behaviors is not required until the process and application context of the device is known. Thus, only a single representation is needed for each device, regardless of its role within a process.

Acknowledgments

We gratefully acknowledge support through the Exxon Research Foundation and through DARPA contract order number D594 administered through the Office of Naval Research, grant number N00014-96-1-0701. In addition, DCM acknowledges support provided under a National Defense Science and Engineering Graduate Fellowship.

References

Chandrasekaran, B. (1994). Functional representation and causal processes. In *Advances in Computers*, Vol. 38, Academic Press, New York. 73-143.

Keuneke, A. (1991). Device Representation: The Significance of Functional Knowledge. *IEEE Expert*, **6**(2), 22-25.

Miller, D. C., J. F. Davis, J. R. Josephson and B. Chandrasekaran. (1997a) A Process Design Decision Support System for Developing Process Chemistry. AIChE National Meeting, Los Angeles.

Miller, D. C., J. R. Josephson, M. J. Elsass, J. F. Davis and B. Chandrasekaran (1997b) Sharable engineering knowledge databases for intelligent system applications. *Computers chem. Engng.* **21**, S77-S82.

AN INITIATIVE FOR INTEGRATED COMPUTER-AIDED PROCESS ENGINEERING

Andreas A. Linninger
Department of Chemical Engineering
University of Illinois at Chicago
Chicago, IL 60607

Helmut Krendl and Helmut Pinger
VAI Industries
A-4020 Linz, Austria

Abstract

Phenomenological process models play an important role in development, planning and design of chemical production processes. The ongoing challenge for systems research is to invent a new breed of computer-based methodologies for assistance and/or partial automation of the creative modeling process. The presentation investigates problems in computer-assisted process engineering from a system theoretical point of view. Complexity of industrially relevant process models call for a novel problem organization structure composed of model declaration, problem instance and mathematical solution. Conceptual elements of formal phenomena-driven modeling language (PML) and a Generic Mathematical Language (GML) are presented. A new class relation paradigm involving the interaction of elementary modeling concepts with meta information promises new avenues for automatic or semi-automatic model generation. Symbolic/numerical methods for the analysis and solution of mathematical models have been demonstrated. The methodology is discussed with reference to TechTool, a computer-aided modeling environment whose final capabilities will constitute a platform for corporate wide information and know-how management.

Keywords:

Computer-aided modeling, Phenomena-driven formal languages, Symbolic/numerical simulation.

Introduction

There are three major approaches in use for industrial process model development: (i) Block-oriented flowsheet simulators (ii) equation-oriented model editors and (iii) direct programming approaches. Flowsheet simulators, e.g. ApenPlus, HySys, ProSim, emanated from the paradigm of associating specific process units with particular physical phenomena. They follow the traditional unit-operations whose purpose was to assist chemical engineers in choosing appropriate standard equipment types and their operations. Today, we see a shift of chemical engineering design from unit operations towards new operations combining elementary physical and chemical operations. From this development it becomes clear that the unit-operations approach is not sufficient for general purpose modeling.

Mathematically oriented modeling CAD tools, e.g. Ascend, Speedup, gPROMS offer a high degree of flexibility but require highly skilled modelers. Direct programming approaches require the collaboration of a domain expert, i.e. an engineer, and a program developer. Phenomenological descriptions of process models need to be converted into a computer code. This approach suffers from major drawbacks:

- Model development is time consuming and expensive

- Models are highly specialized and can hardly be reused
- Model purpose assumptions and simplifications are not included in the final program.

Clearly, simplifying and expediting model development through appropriate automation tools or methodologies would be highly relevant for the industrial practice. Computer-aided process engineering tools have been the focus of several research groups [Siirola and Rudd, 1971; Stephanopolous et al., 1990a; Piela et al., 1991; Barton and Pantelides, 1993; Bogusch and Marquardt, 1995; Banares Alcantara and Lababidi, 1995]. Despite immanent breakthroughs, automation of industrial modeling tasks is still not very advanced and direct implementation is the method of choice in many plant models for process automations, process supervision and capacity planning.

This paper addresses critical system theoretical aspects pertaining to computer-aided model building. The discussion will follow the design of TechTool, a comprehensive computer-aided modeling environment currently under development at the Laboratory of Product and Process Design at UIC and VAI, Austria. The methodology offered by TechTool, is an attempt to put phenomena driven modeling into practice on a corporate wide scale. In this brief presentation we will illustrate the development of three facets computer-aided process engineering: Part I will demonstrate a formal framework for process knowledge declaration. Part II will illustrate on how to evolve a phenomenological model into a consistent mathematical representation. Part III discusses aspects of symbolic and numerical solvability of process models developed through sections I and II.

Part I. Formal Declaration of Process Knowledge

Stephanopoulos (1990b) was the first to present a formal framework for chemical engineering process models. He defines declarative process knowledge as the contextual information about a model such as (i) underlying assumptions, (ii) simplifications (iii) model purpose and (iv) relationships. This view excludes the description of phenomenological aspects by attributing it to the procedural knowledge. Clearly, mathematical abstractions of physical and chemical phenomena exhibit procedural character. However, it is the authors' view that relating physical process entities with their physical phenomena is part of model declaration. This chapter presents a formal framework for phenomenon-driven model declaration alongside the generation of a raw mathematical representation.

A complicated challenge lies in identifying a set of elementary phenomenological building blocks whose combination with physical devices can describe any desired model situation. Hierarchies of physical devices are readily constructed [e.g. Banares-Alcantara, 1995]. It is the compilation of a sufficiently general library of phenomena that makes this task extremely hard.

To overcome the difficulties with concatenation of phenomena, this presentation proposes a new object relation paradigm. This paradigm rests upon a *phenomenological modeling language* as well as *meta information* level. The modeling language offers a vocabulary of fundamental model building blocks. The meta level involves agents that safeguard the consistency of models and their associated phenomena. For model construction under this new view, three concepts are necessary: An association between elementary *physical* and *phenomenological* modeling elements and an *agent*, who safeguards the consistent construction of a raw mathematical representation of the situation described by the two modeling elements.

Phenomenological Modeling Language (PML)

The formal language is composed of two modeling elements: (i) modeling concepts and (ii) semantic relationships.

Modeling Concepts

Modeling concepts possess a specific physical or logical scope and represent the dominant building blocks of process models. Marquardt (1992) distinguishes modeling elements according to substantial and phenomenological complexity. The substantial complexity coordinate describes separate physical entities. Phenomenological complexity stems from the behavior of a substantial element. Equally, modeling concepts can be divided in substantial and phenomenological modeling elements.

Substantial Modeling Elements (SME)

Most SMEs are described by extensive properties or a geometric dimension. These concepts pertain to physical units and their connectivity. Accordingly, there are two major groups of SMEs.

- Devices: A generic balance envelope that encloses a distinct amount of matter or section of plant. Devices extend below the sub-unit scope to material mixtures, their phases, droplets and particles, etc.
- Links: Describe the modeling elements representing the interaction between devices, e.g. material, energy, impulse exchange and signal transaction. In process flowsheets, material streams connecting process units are the most prominent instances of links.

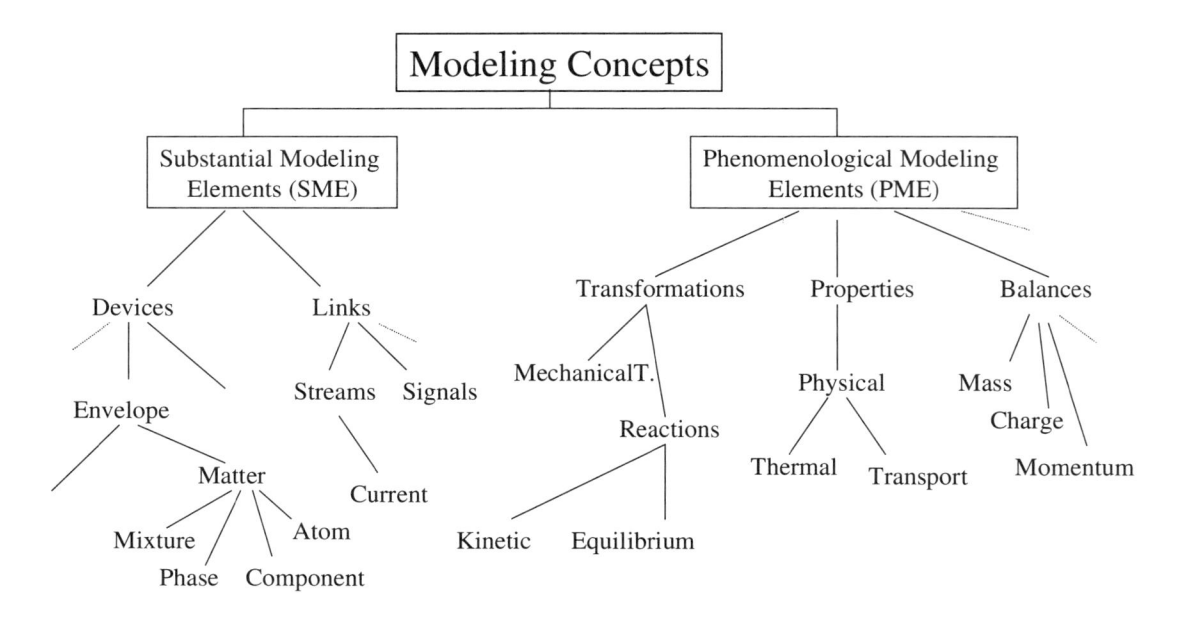

Figure 1. Partial overview of modeling concepts.

Phenomenological Modeling Elements (PME)

PMEs refer to mathematical relations that determine the state or state transition of a substantial modeling element, i.e. devices and links. Typically they involve generic mathematical expressions such as reactions, transport phenomena, balance equations, procedures for physical properties calculations. In principle, PMEs may also include qualitative statements. Quantitative information can be processed through an inference engine similar to numerical solution of quantitative relations through the solver kernel as described in section III.

Scope of this presentation does not allow a more detailed discussion of PML. An excellent discussion on formal languages can be found in Marquardt (1992). Fig. 1. gives a partial view of the modeling concepts for chemical processes and their operation.

Semantic Relationships

Semantic relationships are used to associate modeling concepts with each other. New entities can be formulated as a *concatenation* of ancillary concepts. Existing concept instances can be *refined* breaking the concept into smaller pieces. Through *concatenation* and *refinement* process model evolve hierarchically through a top-down or a bottom-up approach.

The Meta Level

In addition to PML, TechTool uses model generators agents, which perform actions of modeling elements when related to one another. Agents are autonomous programs [Russel and Norvig, 1995] whose task it is to construct a consistent mathematical representation of a phenomenological element in their substantial context. In particular they implement the following context specific instructions:

- Generation of context dependent mathematical representations.
- Consistency maintenance of devices, links and their associated phenomena, e.g. constitutive equations, information on model assumptions and simplifications, model purpose.

Clearly, these model generator agents superintend the underlying modeling language. Therefore model generation agents belong to a meta information level. Model.la [Stephanopolous, 1990a and 1990b] implies the requirement for such a step. The authors note: *A translation task is performed by a procedure that runs after a model characterization step has been concluded.* Jarke and Marquardt (1995) also propose a *meta process model language*.

Agents in TechTool are not dedicated to particular device instances or phenomena, but act on objects with a specific type of interface. This non-traditional class relation between the modeling concepts and the agents resembles the Java *interface-type*.

Interaction of SME and PME

The interrelation of devices and phenomena has been demonstrated for a broad number of applications on a sub-unit model, e.g. dynamic simulation of reaction kinetics in high temperature metallurgical processes [Linninger et al., 1995], state transition of batch pharmaceutical processes [Linninger et al., 1998], and synthesis of waste management policies [Linninger et al., 1996]. The

Material Model, which served in all of the above modeling applications, combines substantial elements, i.e. material mixtures, phases, chemical component and atoms with their phenomenological description, i.e. correlations of mixture properties, equilibrium conditions, etc. The material properties can be found through interpretation of the material state and application of knowledge about property estimation, e.g. Constantinou and Gani (1994). Therefore joining of a PME concept like *Enthalpy* with a SME concept like *Material* leads to an appropriate

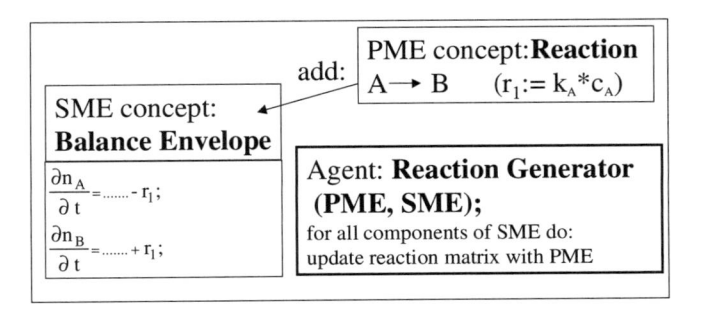

Figure 2. Model Generator Agent joining a reaction concept with a balance envelope

mathematical representation for liquid material mixtures, vapors or a combination of the two. This meta knowledge for the model selection step can conveniently be implemented through a model generator agent. In a similar fashion, more complex tasks such as the choice of a fugacity model for a vapor-liquid equilibrium from state information, i.e. temperature, pressure and composition can be accomplished.

The agent's action may also involve communication with phenomena that have already been associated with the element previously. As an example consider the addition of a reaction concept to a balance envelope. Fig. 2 illustrates this situation: The reaction generator agent implements instructions to update the material balances when a new reaction phenomenon is added. Clearly, neither balance envelope, SME, nor reaction concept, PME, could have accomplished this task alone in a general situation. We expect to significantly enhance the flexibility and reusability of the library of elementary building blocks through the use of meta information.

Completion the Model Declaration

PML allows process developers to declare and refine process model in a hierarchical fashion. Use of PML leads to model declarations whose knowledge is represented as a semantic network in which the nodes establish modeling concepts, and the arcs define relationships between these concepts. A complete model declaration includes the following information:

- Purpose and objective of a process model
- Physical devices and their connectivity
- Physical and chemical phenomena associated with those devices
- Information concerning the evolution of the modeling task (history)

Part II. Declaration of Procedural Process Knowledge

For further evolution of a process model, the declarative model needs to be augmented with procedural information. This task requires transformation of the raw equations into the specific format imposed by solution algorithms. This task may involve conversion of high-level syntactic elements into atomic mathematical expressions, discretization of distributed systems, approximations of partial derivatives, etc. Through the Problem Instance, engineers can define the procedural knowledge of a model through the following specifications:

Problem Type Assignment

- Definition of type and purpose of the problem instance.
- Definition of the objective of a simulation experiment, Optimization, Steady State or Dynamic simulation: ODE, AE, DAE, Optimization Constrained unconstrained – alternative qualitative reasoning
- Assignment of variables, constants, control functions for a simulation experiment.
- Selection of variables, their definition ranges as well as guesses of their initial values

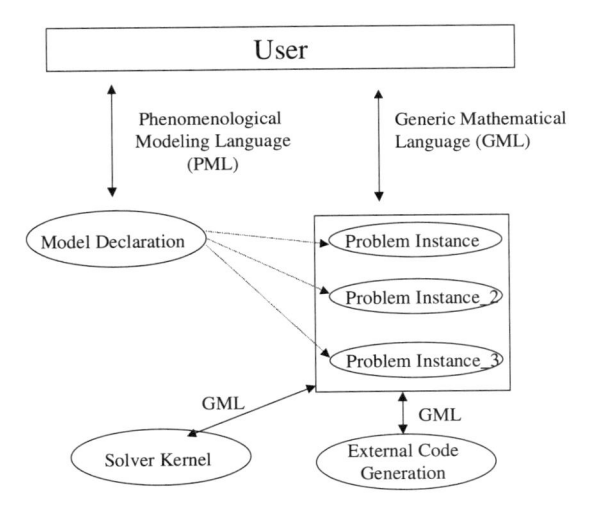

Figure 3. Relation of Model Declaration and Problem Instances.

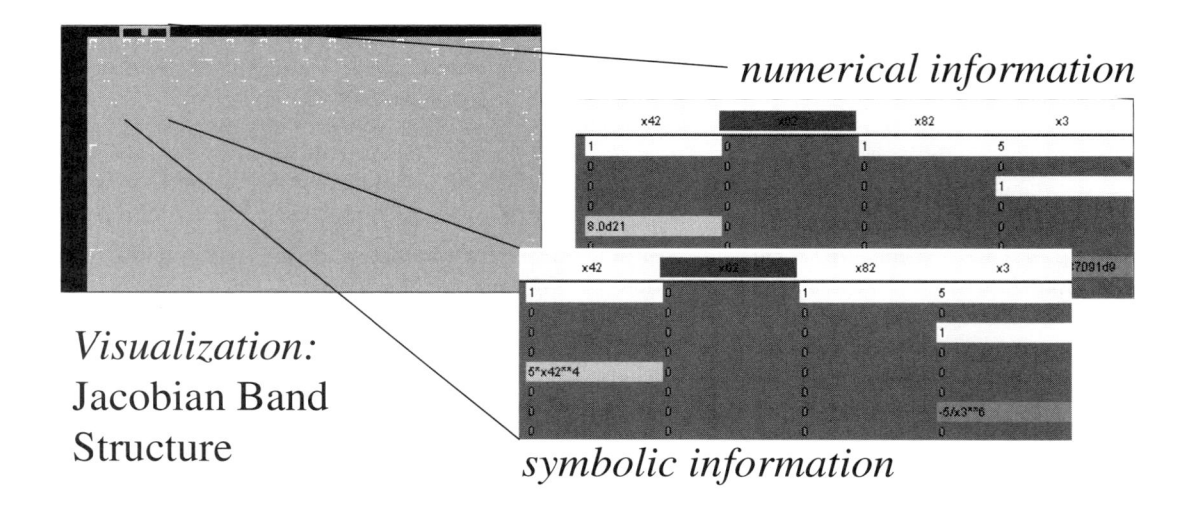

Figure 4. Visual Complexity Analysis in TechTool.

Selection of Solution Strategy

- Automatic or user driven selection of available mathematical solution algorithms
- Symbolic simplifications and automatic generation of ancillary mathematical concepts; Jacobian and Hessian Matrices, linearization of complex expressions

Scheduling and parameterization of a simulation experiment

- Assignment or procedural determination of initial values,
- Values for parameters and schedules for control function such as dynamic inputs. (Charge diagrams, material streams).

For a single model declaration, typically more than one Problem Instance will be formulated (Fig. 3). Distinct versions of problem instances arise form (i) different symbolic manipulations of the raw model or (ii) the choice of different starting values, scaling of variables or (iii) simplification routines for variable and equations reduction, discretization methods. Manipulation of Problem Instances can be done automatically or through a generic mathematical interface.

Generic Mathematical Language (GML)

The mathematical language serves the purpose of manipulating and adjusting problem declarations formulated through PML. The generic mathematical language uses only mathematical abstractions such as vectors, matrices, and their manipulations such as adding, derivation, etc. The object hierarchies defined in the PML are maintained through automatically generated naming conventions. The object structure of the PML is redundant. This level is comparable to equation-oriented modeling tools but suffers from the same shortcomings.

In addition to work on Problem Instances, GML can be used to define new phenomenological concepts. This feature allows for the implementation of new high-level language constructs, e.g. phenomenological modeling element, and aims at supporting dynamic augmentation of the PML vocabulary. The mathematical language also constitutes a logical framework of universal and invariant problem formulation. This generic language interface is beneficial for code generation using external solution algorithms as described the following section.

Part III. Mathematical Problem Solution

A complete model composed of a model declaration as well as a derived problem instance can then evaluation through the Symbolic/Numerical Solver Kernel (Fig. 3.). For the reasons stated, TechTool offers the following features to satisfy the demand of industrial mathematical problems.

- Symbolic Numerical Mathematics
- Discrete Continuous Simulation
- Numerical Stability and Convergence Analysis
- External Code Generation

These features will only presented briefly. For a more detailed discussion of the solver architecture refer to Linninger (1996b).

Symbolic Numerical Mathematics

The complexity of most industrial mathematical problems requires symbolic information in combination with numerical algorithms. Combined symbolic and numerical tools (Maple, Mathcad, Mathematica) are currently finding routine applications in academia as well as industry. Unfortunately, insufficient availability of structural components and lacking property data support prohibit the use of these tools for large modeling tasks.

Many topics in applied mathematics equally involve substantial algebraic manipulations of the underlying mathematical expressions. Currently TechTool supports symbolic differentiation, simplification of algebraic expressions, grid generation for partial differential equations, etc. The library of routines is described in Linninger et. al. (1996b).

Discrete Continuous Simulation

Frequently, discrete actions or events trigger qualitative changes of the underlying process model. This circumstance holds true in particular for startup models whose time trajectories traverse several state transitions. Another situation occurs when modeling reaction systems over a wide range of concentrations. A pseudo-second order kinetic reaction model will rate becomes truly second order when the excess substance falls below a certain threshold concentration. Discrete events delineate a qualitative change of describing equations. Currently, TechTool provides event detection and handling procedures as described by Park and Barton (1996).

Numerical stability and Convergence Analysis

In experience in modeling of industrial systems, construction of a phenomenological model is followed by an intensive testing phase aiming at finding a robust solution strategy. Frequently, there is little help offered from modeling tools for a consistent model with ill-conditioned mathematics. TechTool therefore offers comprehensive support for convergence and stability analysis. Due to the size and complexity of industrial process models, TechTool focuses on qualitative analysis strategies and their visualization. Figure 4 depicts the visualization of a Jacobian matrix as well as symbolic and numerical information. Figure 5 displays an interface for tracing the evolution of residual errors.

External Code Generation

Progress in applied mathematics and rapid changes in software implementations make attempts to compile a comprehensive state-of-the-art solver suite an almost impossible task. The new breed of modeling tools should therefore allow to freely export problem formulations across software and platform boundaries. The significance of this argument is underlined by several national and international activities to develop standardized data exchange formats (e.g. CAPE Open). The mathematical language of TechTool is a contributor to this goal. TechTool also allows for code generation into low-level languages like FORTRAN, C or external programs like Maple. It is important to recognize that only system with symbolic capabilities can do this. Numerical simulators without some sort of GML interface are confined to a black box understanding of library functions.

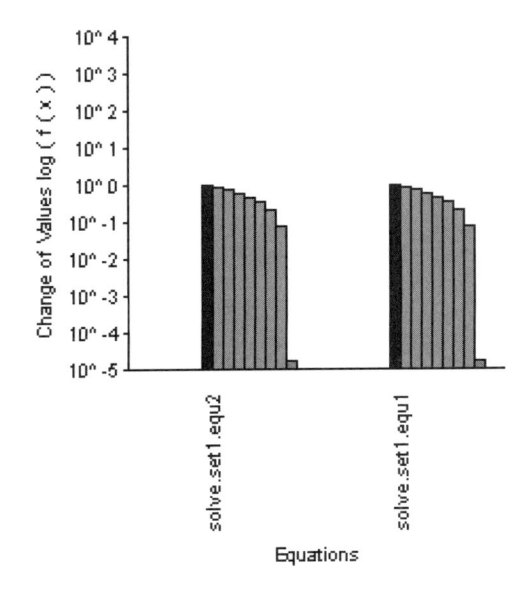

Figure 5. Residual trace of two converging equations.

Conclusions

A phenomenon-driven design methodology was outlined. Generation of equation-oriented mathematical representations was achieved through the association of physical and phenomenological building blocks. Model Generator Agents are agents that safeguard model consistency given by the context, which is determined by substantial and phenomenological concepts. The introduction of meta information enhances the reusability and flexibility of the modeling language. The proposed concepts have been proven effective guidelines for programmatic implementation of process models. For assessing the potential for a computer-aided model generation more experience in practical modeling situations is necessary.

The presented methodology forms the backbone of a corporate wide process modeling environment entitled TechTool. Its ambition is a computer-aided environment, which actively participates in human-decision making in process engineering and optimization.

References

Banares Alcantara, R. and H. M. S. Lababidi (1995). Design support system for process engineering. II. KBDS: an experimental prototype; Comp. Chem Eng., 19, 267-277.

Bogusch, R, and Marquardt, W; (1995) A formal represention of model equations, Proc. ESCAPE-5, Bled.

Marquardt, W. (1992). Rechergestuetzte Erstellung verfahrestechnischer Modelle, Chem. Ing. Tech., 64, pp 25 - 40.

Barton P.I. and Pantelides, C.C. (1993) gPROMS – A combined discrete/continuous modeling environment for chemical processing systems, Sim. Ser. 25, 3, p. 25.

Constantinou, L. and Gani, R. (1994) A New Groups Contribution Method for Estimating Properties of Pure Compounds, Aiche J., 40(10); p 1697.

Jarke, M. and Marquardt, W. (1996). Design and Evaluation of Computer-Aided Process Modeling Tools, Aiche Symposium Series, Vol. 92 (312), pp 97- 109.

Linninger, A. A.; Hofer, M.; Patuzzi, A. A.; (1995) DynEAF - A dynamic modelling tool for integrated electric steelmaking, Iron and Steel Engineer, No 3.

Linninger, A. A. and G. Stephanopoulos, (1996a). Computer-Aided Waste Management of Pharmaceutical Wastes, Paper 23a , *AIChE Meeting*, New Orleans, LA,

Linninger, A.; M. Hofer, H. Krendl, H. Druckenthaner and H. P. Jörgl. (1996b). "M-PROJECT - Organizing problem representation and modeling of steady state and dynamic processes", Comp. Chem. Eng., **20**, p425-430.

Linninger, A. A; E. Salomone, S. A. Ali, E. Stephanopoulos and G. Stephanopoulos. (1998). *"Pollution Prevention for Production Sytems of Energetic Materials"*, Waste Management, Vol. (17) 2/3, pp 165-173.

Park, T. and P. I. Barton, (1996). State Event Location in Differential-Algebraic Models, ACM Trans. on Modeling and Simulation, Vol. 6. No. 2, pp. 137-165.

Piela, P., Epperly, T., Westerberg, K. and Westerberg, A.; (1991). ASCEND. An Object-Oriented Computer Environment for Modeling and Analysis: The Modeling Language. Comp. Chem. Eng., Vol. 15, No. 1, pp 53-72.

Russel, S. and Norvig, P. (1995). Artificial Intelligence – A Modern Approach, Prentice-Hall. New Jersey.

Siirola, J.J. and Rudd, D.F. "Computer-Aided Synthesis of Chemical Process Designs", Ind. Eng. Chem. Fund. 10, 353, 1971.

Stephanopoulos, G.; Henning, G.; Leone, H.; (1990a) MODEL.LA. A Modeling language for process engineering II – Multifaceted Modeling of process systems, Comp. Chem. Eng., Vol. 14, No. 8, pp 847 – 869.

Stephanopoulos, G.; Henning, G.; Leone, H.; (1990b) MODEL.LA. A Modeling language for process engineering I - The formal framework, Comp. Chem. Eng., Vol. 14, No. 8, pp 813 – 846.

A PROCESS ENGINEERING INFORMATION MANAGEMENT SYSTEM USING WORLD WIDE WEB TECHNOLOGY

R. Andrews and J. W. Ponton
Department of Chemical Engineering
University of Edinburgh
Edinburgh, Scotland EH9 3JL

Abstract

The effective management of process engineering information, especially as it is created, manipulated and extended in the context of process design, is a task of increasing complexity, but of critical importance. A wide range of prototype software environments for this activity have been developed in recent years, including Design Kit (MIT), n-Dim (CMU), and KBDS (Edinburgh). A feature of all of these has been the major software effort involved in their creation, and in many cases the use of rather exotic programming languages such as Lisp. We describe an object-based process engineering environment having client server architecture exploiting standard World Wide Web (WWW) servers. Objects can exist, and methods be applied, on a world wide distributed system. Platform independence is achieved through the cross platform availability of web browsers and servers, and through the Java and JavaScript languages. It has proved possible to use standard features of the WWW and related tools to provide many of the facilities which had to be created specially in the earlier environments. The prototype system is in principle accessible to anyone with a WWW browser and an Internet connection. The system has been used to interconnect a range of process engineering tools including AspenPlus, an in house physical property databank and a separately developed process synthesis package, CHiPS. Information may be freely transferred between these tools, and also to and from other standard software e.g. any spreadsheet system. A number of special tools have also been developed for use within the system, including flowsheet graphics and a simple equation based flowsheeting package. The system structure lends itself to a 'componentware', approach that has simplified and speeded the construction of these tools. The paper will describe the approach used in developing platform independent and extensible process engineering objects, their manipulation and management on a client-server system, and the techniques available for tool integration. The advantages of the componentware approach will be highlighted, and future applications and possibilities discussed.

Keywords

Information management, WWW, Integration, Componentware, Object orientation, Open systems.

Introduction

The need for better ways of handling the large amounts of information generated in the whole range of process engineering activities has long been recognised. The requirements may be seen to fall into two categories: the *organisation* of information and its *representation*. Work to date can thus be divided broadly into two areas:

(a) Environments for information management, often in the context of model development, since much of the process engineer's task may be regarded either actually or formally as a modelling activity, e.g. n-dim (Levy at al., 1993). Most environments, e.g. Design Kit (Stephanopoulos et al., 1987), KBDS (Bañares-Alcántara and Lababidi, 1995), have used an underlying object-oriented representation for their information, but have not formalised this in terms of a general data model, nor usually considered its applicability outside of the environment. This may be seen as a weakness of this approach.

(b) The representation of information can be formalised by data modelling. One such formalisation intended to facilitate the exchange of process engineering information is PISTEP, based on STEP (Standard for the Exchange of Product model data) (Kahn, 1995). Experience here (Lu et al., 1995) suggests that a standard data model in itself is not enough. A particular problem is the non-extendable nature of the definition.

The Epée system (Fraga et al., 1995) was an attempt to provide both an environment and an information standard, the latter being extendable. The designer's objectives for this system were the following:

1. The system must be inherently distributed in nature.
2. A mechanism for creating persistent objects is necessary.
3. Efficiency is crucial.
4. The system must be easily extensible.
5. It should support a variety of computer platforms.
6. It must be possible to develop new applications easily and using standard computer languages.
7. An audit trail of the user's activities must be maintained.

The last was not set explicitly in the original design, but was quickly perceived as both necessary and readily implementable. While an interesting prototype, Epée proved too cumbersome in practice, and in particular required major software development and support for its intended multi-platform client-server architecture

In the midst of these special purpose developments, the World-Wide Web (WWW) has come into being. This is an all platform, distributed, client-server based environment. It also has an associated data standard in the HyperText Markup Language (HTML).

In what follows, we will describe a successful attempt to use the WWW and HTML to implement many of the functions of these earlier special purpose systems.

Object Structure

This is an object-oriented system. An object is a structure which encapsulates data in slots, together with procedures, called methods, that act on this data to create new objects. All objects are instances of a class of object, from which they inherit their behaviour. Within the class, objects are grouped into families. Objects are said to be of the same family if they have both descended from a common ancestor.

Objects are divided into three sections: data slots which contain the object data, logistical slots which store information about object creation and methods that can be applied to the object.

Data Slots

Each data slot has four constituents:

- Name of slot
- Value of slot
- Slot description
- Slot status

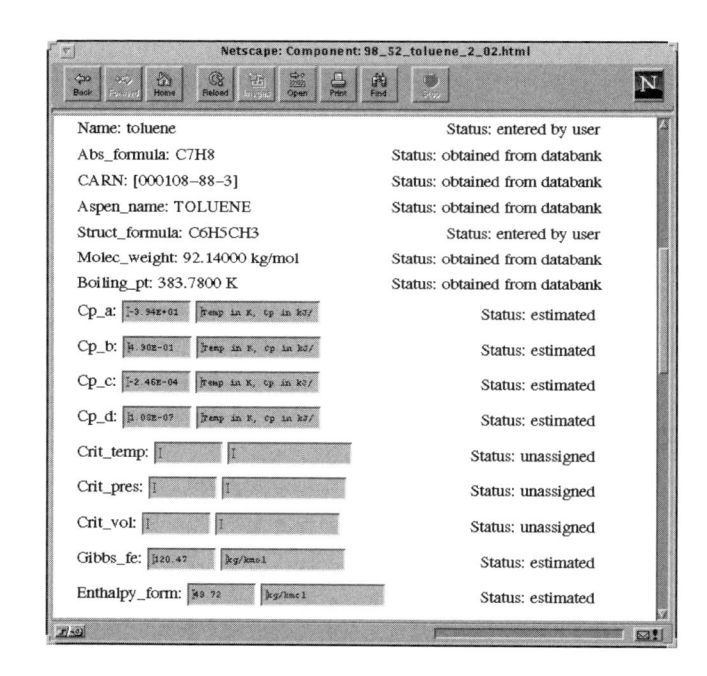

Figure 1. Data slots for a component object.

The value of the slot can either be a number, text or another object. The slot description, which is optional, is normally the units of the slot.

Slot Status

The slot status can be one of a number of values. When created, a slot is empty and thus its status is '**unassigned**'. Data may be typed directly into it. Upon processing its status changes to be '**entered by user**'. The data in the slot is now fixed, and cannot be changed. The same is true if the status is '**obtained from databank**'. However if the status is '**estimated**', the data is still displayed in the slot, but it can be altered. If the data is modified, then the slot status changes to being '**entered by user**'.

There are two more status values, which refer to slots that contain objects. If all of the data slots in the referenced object are complete (i.e. no slots of the referenced object have status '**unassigned**'), then the slot status is '**complete object**', otherwise the status is '**incomplete object**'.

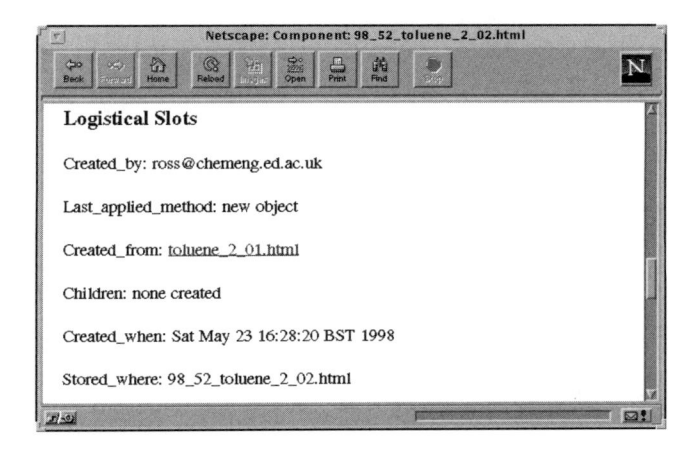

Figure 2. Logistical slots for a component object.

Logistical Slots

This section contains the logistic data for the object, consisting of:

- Who created the object
- The method applied to create the object
- The parent object (if any)
- The children of the object
- Where it is stored
- When it was created

'Children' here refer to objects that were created from the data in the present object.

Methods

This section of the object is divided up into three further sections:

- Methods to apply to the object
- Other selection options
- History options

The following distinction is made between methods, and other selection options.

Methods use the data in the object to create another object. These methods are either specific to each class of object, e.g. 'get property data from databank', for a component object, or generic, e.g. 'add a slot to the object'.

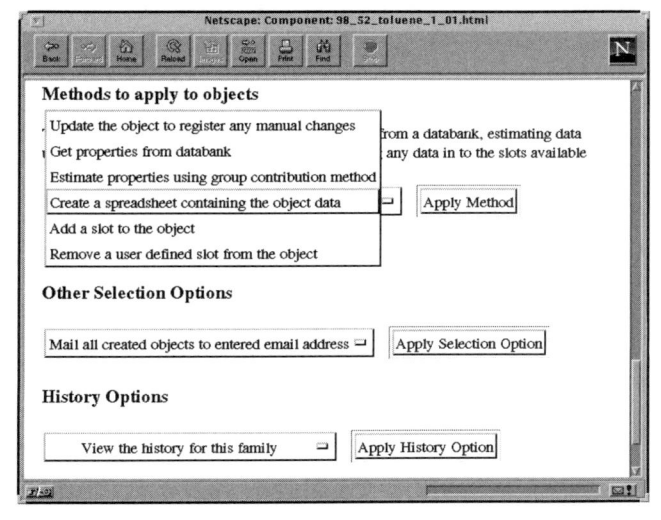

Figure 3. Methods of a component object.

Other selection options are different in that they need to be available at all times and do not use the data contained in the object. Thus they are not strictly methods applied to the current object. These options are common to all objects, and include: 'mail objects', 'create new object', and 'finish the session'.

History options enable the history to be viewed for this family of objects, for all objects of this class, and for all objects created.

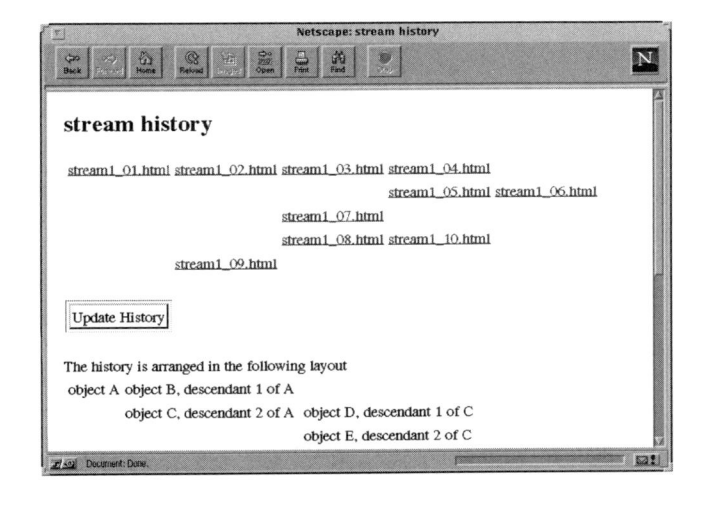

Figure 4. History of a stream family.

Objects and WWW

In the system, each object is an HTML document. HTML is a formatting language interpreted by browsers to display pictures and text across the WWW. There are two

HTML constructs that were used extensively in this project: hyperlinks and forms.

A hyperlink is used to 'link' another file (not necessarily an HTML document) to the current document. Activating this link loads the file to the browser. If it is an HTML document then it is displayed in the browser. If it is not, the browser can be configured to deal with files of different extensions by the use of Multipurpose Internet Mail Extension (MIME) types. Thus for example the browser can be configured to recognise the Lotus .wk1 extension, and to load the file in the spreadsheet of the user's choice.

Hyperlinks are used here to link slots that include objects. They are also used to display parent and child information in the logistical area. Thus the history tree can be traversed at will from parent to child and vice versa.

The other HTML construct of importance here is the form. This allows data to be passed from the browser to a server that can be located anywhere in the world. The data is passed by means of a Common Gateway Interface (CGI) script. This takes the data from the form, processes it, and can return other data back to the browser. It can also run executable programs on the local file system of the server, or indeed on any system accessible to the server. Thus the data that the CGI script returns can be varied indeed.

There are three separate forms in each object. One for the data slots and the methods, one for the other selection options and one for the history options. For the data slots, filled slots are stored as hidden variables of the form. Other slots (to which data may be entered) are displayed as text input areas.

Componentware Approach

CGI scripts can run other programs on the server. These other programs can be written in any language with the only proviso being that they must be able to be executed on the local file system of the server. This encourages the use of small, tightly focussed programs with a defined interface. This is the essence of the componentware approach. To interface an existing program to a CGI script requires the addition of approximately ten to fifteen lines of code to the CGI script.

Methods may be written to run on the server in any language, and with certain restrictions can also be written in Java or JavaScript to run on the client. Typically, new methods are twenty to one hundred lines of code.

Object Management

Objects are created by applying a method. There are generally several parameters necessary to creating an object (e.g. the names of the components for the mixture object), and these can be selected using either a plain CGI form, or Java and JavaScript enhanced forms.

At any stage in the object creation, the user has the option to save the current information. This compresses all of the created objects and sends them to an email address provided by the user. This file can then be uncompressed and have a copy of all objects that have been created. All hyperlinks between objects still work, although any method applied will create objects on the server. It will first check to see whether or not the user's directory on the server still exists. If it does not then the user is again prompted for an email address. The directory will continue to exist on the server until either the user explicitly removes it, or twenty-four hours after the last object has been created.

Once the objects have been emailed to the user, and if the user then puts them on the WWW, they can be used by others in their own workspace. This can be thought of as the equivalent of 'publishing' objects, cf. n-dim (Levy et al., 1993), Epée (Fraga et al., 1995). There is a selection option available that allows the user to upload an object given its Uniform Resource Locator (URL), i.e. its address on the WWW. Thus the user can use objects that have been created by another person and published on the WWW.

System Features

One of the major objectives for this management system was to investigate the possibility of using the WWW to create process models. The topology object, which describes the overall layout of a plant, can be used to model the process.

The topology of the process is entered by means of a 'natural language' description of the unit operations and their connectivity. From the natural language description a three dimensional model is created. After applying the method to create the topology object from this description, the relevant stream and unit objects are also created. The unit objects are created with the relevant slots for the particular unit operation that it will be modelled, e.g. a unit object that will be modelling a reactor has a reaction slot. The data for the units may be entered manually in the slots, or it too may be given using a natural language description. For any information not given but which is considered essential, sensible defaults are chosen. Component and feed stream information has to be created from scratch, i.e. by creating the individual component and stream objects.

Once the model has been created, there are a number of options for the user. An AspenPlus model may be created from this information. Conversely, the user may also take the output from an AspenPlus run and create the corresponding topology object with all associated unit, stream, reaction and component objects. Alternatively there are other models may be created. The models that may be created to date are:

- A spreadsheet model
- A linear equation model

- An AspenPlus model
- A HYSYS model

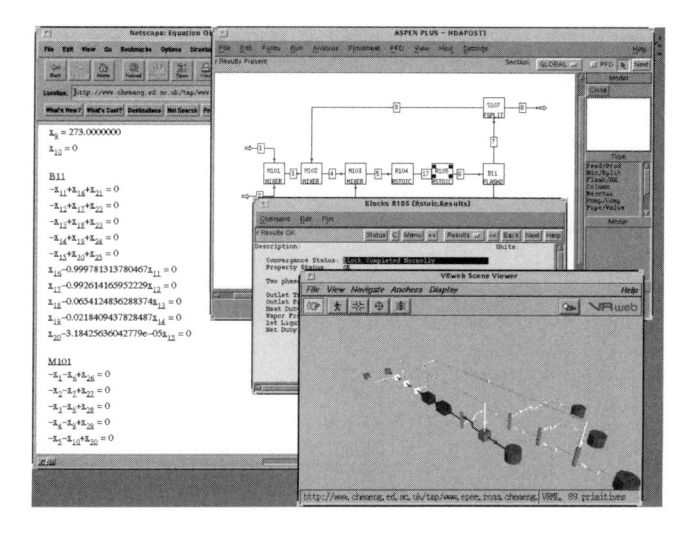

Figure 5. Examples of models that may be created.

These other models are now well integrated. For instance Fig. 5 shows a model initially created by Aspen Model Manager, then transformed into a web object. This included a topology, displayed in the three dimensional viewer, and a linearised model, represented by a set of equations.

Topology objects may also be created by a separately developed process synthesis package, CHiPS (Fraga and McKinnon, 1993).

Application Integration

The system provides a means of creating, and within limits of interchanging between, four different modelling formats: two commercial flowsheeting packages, one standard spreadsheet format and a specially developed, but fairly general, linear equation based system.

The aim has not been to provide complete general, automatic translation of highly detailed models from one format to another. This is not straightforward, and it is not clear how useful it would be in practice, since, for example, the detailed specifications for rigorous distillation in say, AspenPlus, will not necessarily map directly on to those for HYSYS. Instead we have provided the ability to transfer basic information between model formats, allowing the user to complete the model, if necessary, by hand.

Generation of all models starts from the topology object. From this may be created a basic AspenPlus model .inp file, HYSYS script file, or a set of material balance equations. Additional information to create the balance equations, minimally simply the number of components, is drawn from the stream objects. The spreadsheet model is regarded as a special type of equation model in which variables are mapped into spreadsheet cells, and equations are rearranged into a form such that their output set corresponds to the output component flows of each unit.

Additional information to complete the AspenPlus or HYSYS model, and set up the remaining equations for an equation or spreadsheet model, is obtained from the unit object. Minimal information in this case is simply the type of unit: mixer, reactor, separator or splitter. AspenPlus and HYSYS model creation involves scanning the objects for a subset of information which will enable a model to be described. Equation objects and spreadsheets work with linear models and so information on recoveries, extent of reaction and splits is sought. If this is absent variables are left undefined and cells empty for the user to fill in manually.

The resulting model is left under a hyperlink. The user may save it locally or start it directly in an application if this has been given a MIME type.

Translation between model types takes place through the system of objects. These can be created from the browser as previously described, but it is also possible to create a set of objects directly from the AspenPlus input file, or Model Manager backup file. It is further possible to create a linearised equation or spreadsheet model from a solved flowsheet in the form of an AspenPlus summary file. This facility will also be available in the future from HYSYS output.

The physical form of the models is the data input file for the relevant application. In the case of the spreadsheet model, a readable ASCII format is first produced, and then translated into Lotus .wk1 or .wk3 binary form. It is interesting to note that this format translation requires the largest single piece of code in the entire model! The equation model is stored as a file containing matrix and vector elements in a fairly conventional sparse matrix representation. However the user's view of this is a hypertext page (see Fig. 5), with hyperlinks for each unit and variable which bring up the relevant stream or unit object in a separate browser window, providing a useful link between the equation and process level. This facility is not currently available with the spreadsheet model, but as Excel and other spreadsheets now have an HTML export facility this can be added.

Discussion

The system can be evaluated with respect to the criteria set out in the introduction.

1. The requirement to be **inherently distributed** has been met by default by using the WWW.
2. **Persistent object creation** has been demonstrated in the previous sections.
3. The **efficiency** of the system, which will be discussed in the next section, has some limitations.

4. The system is **easily extensible** as the list of slots is not 'hardcoded' into any of the object definitions. Instead they are stored in a plain data file which can easily be altered. New slots may be added interactively to the object by applying a method.
5. There is **wide platform support** for the system as due to the popularity of the WWW, there will almost certainly be a client browser for all but the most obscure hardware-operating system combinations.
6. **New applications may be added in any language** to a particular object by altering the CGI script of the object's form. This application may be in any language, the only requirement being that it can be executed from the server
7. The **audit trail is maintained** both by the logistical slots, and by the creation of history trees.

Limitations

The one criterion that is not satisfied so far is that of efficiency. Nearly all processing is carried out either on the server or on machines local to the server. Thus the time taken to perform the processing will increase with increasing load on the server. Little use is thus made of the potentially large processing power of a typical modern client. One way of exploiting this would be to write methods as Java applets. These applets execute on the user's machine, using its processing power rather than the server from which it was downloaded. This would transfer some of the processing from the server to the client, and should increase the efficiency of the system somewhat. Server scripts would still be necessary as Java applets can only write files on the server from which they were downloaded. There remains some concern about the full portability and security of Java so CGI based methods should remain for those clients either unwilling or unable to use Java.

The other main limitation to the system is paradoxically one of its advantages, namely the WWW. The use and maintenance of objects is completely in the user's hands once they have been emailed to him. They can be changed at will by editing the HTML file, or be published on one URL one week, and a different URL the next. This is clearly a problem if another user links to the object, and then finds that it does not exist anymore. There is not much that can be done about this. When created, all objects are created with read only permissions. These permissions will be maintained after emailing only if the user accesses the objects on a Unix operating system.

Conclusions

We have described an engineering information management system based on the WWW. This approach has several advantages such as global access, and robust standard communication protocols.

Limitations of the system are mainly in respect of efficiency. This will be improved by the use of Java language to transfer processing from the server to the client.

Future work will address this, and client side processing in general. As well as improving the user interface, a programmer's interface will be developed.

Acknowledgements

The authors wish to acknowledge financial support from E.I. DuPont de Nemours, and EPSRC. Also Hyprotech and AspenTech for the use and support of their software.

References

Bañares-Alcántara, R., and H. M. S. Lababidi (1995). Design support systems for chemical engineering II. KBDS: an experimental prototype. *Computers and Chemical Engineering*, **19, No. 3**, pp. 279-301.

Fraga, E.S., and K. I. M. McKinnon (1993). CHiPS: a process synthesis package. *Dept. Chemical Engineering Technical Report TR-1993-06, University of Edinburgh, Edinburgh.*

Fraga, E.S., G. Ballinger, J. Krabbe, D. M. Laing, R. C. McKinnel, J. W. Ponton, N. Skilling, and M. W. Spenceley (1994). The implementation of a portable object oriented distributed process engineering environment. *Dept. Chemical Engineering Technical Report TR-1994-17, University of Edinburgh, Edinburgh.*

Kahn, H. J. (1995). STEP methodology – an overview. *In* J. A. Powell (Ed.), *Object Technology and its Application in Engineering, Glasgow (UK), March 1995.* Rutherford Appleton Laboratory, Chilton (UK). pp. 66-75.

Levy, S., E. Subrahmanian, S. Konda, R. Coyne, A. Westerberg, and Y. Reich (1993). An overview of the n-dim environment. *EDRC Technical Report EDRC-05-65-93, Carnegie-Mellon University, PA*

Lu, M. L., Y. Naka, K. Shibao, X. Z. Wang, and C. McGreavy (1995). A multi-dimensional object-oriented information model for chemical engineering. *In, Concurrent engineering, a global perspective, Virginia Aug 1995*, Concurrent Technology Corporation, USA. pp. 21-29.

Stephanopoulos, G., J. Johnston, T. Kriticos, R. Lakshmanan, M. Mavrovouniotis, and C. Siletti (1987). DESIGN KIT: an object-oriented environment for process engineering. *Computers and Chemical Engineering*, **11, No. 6**, pp. 655-674.

FROM ENGINEERING ANALYSES TO OPERATOR TRAINING

Peter Stanley
Hyprotech Ltd.
Calgary, Alberta, Canada, T2E 2R2

Abstract

Historically, operating companies have had to invest in expensive, single purpose difficult-to-maintain operator training systems. However, recent advances made in the configurability, detail and accuracy of simulation software has enabled rigorous first-principle engineering models to be used as the basis for real-time operations training systems. Rigorous operations training systems can now run under Windows NT systems with links to existing plant or emulated DCS hardware. These "new age" training systems reduce engineering effort, maintenance costs and hardware requirements. This paper will illustrate the wide-range of capabilities and advantages that these operations training systems provide.

Introduction

Since the early 1970's many companies within the hydrocarbon and chemical processing industries have implemented operator training systems. Moreover, a number of these companies have claimed that their facilities experienced improved operating performance and profitability as a direct result of the implementation and use of these systems: reduced startups times, reduced training for new operations personnel, and reduced downtime. Commenting on the successful implementation and use of its operator training system Eastman Chemical stated that, "Not only are operators getting an improved understanding of the operation, they are also learning to troubleshoot the process to prevent mishaps in the first place."[9] In fact there are many industrial accounts of how the implementation and use of operator training systems have provided a wealth of tangible and intangible benefits that positively impact the company's bottom line.

However, while most operating companies would probably agree that better trained employees provide higher-quality products at lower-costs, the number of operator training systems installed worldwide is relatively small. Industry experts indicate that only about 40-60 operator training systems are commissioned per year, with no more than 500 systems currently in operation worldwide. If an operator training system can improve the profitability and operating performance of a given facility then why have so few operating facilities worldwide adopted this technology?

Well historically, operator training systems were expensive, single purpose, difficult-to-develop systems that made their investment unjustifiable to many operating facilities. For example:

- Commercial operator training systems typically cost $500,000 to 2MM US (not including maintenance costs). - Companies wishing to train only four or five operators might find this training investment too expensive.
- In-house operations staff may lack the expertise to maintain their operator training system and have to rely on vendor support, incurring substantial maintenance costs
- Operator training systems based on generalized simulation models may not adequately represent the process.

---

[9] Eastman Chemical Company, "Simulation Success Story", Chemical Processing Technology International 1997/8, Page 91.

- Changing process conditions or control strategies may quickly make an operator training system obsolete.

So, while operator training systems may be an attractive investment for a number of operating companies, many other companies have likely found that an investment in this technology would not deliver an acceptable return to their particular facility. Today however, with better hardware and software technology, the obstacles to developing, maintaining, and updating operator training systems are rapidly disappearing. Many operating companies are discovering that the newer technology is enabling them to develop training systems that meet their particular training requirements, and at the same time provide an attractive and acceptable return.

The "Classical" Operator Training System

In order to gain an appreciation as to why some companies have been hesitant in making substantial investments in operator training systems, it may be helpful to review the components of an operator training system and the technologies of the past.

In its simplest form an operator training system consists of a dynamic simulation model (representing the process plant) interfaced with one or more distributed control system (DCS) operator stations and an instructor station. Actual DCS plant hardware or emulated software is used to control the process model. Operator trainees sit at the DCS operator console and control the process model in a manner identical to the way in which they would control the real plant. Their commands (setpoint changes, controller modes, starting and stopping process equipment etc.) are sent to the dynamic process model, which simulates the process response to these changes. Calculated process variables from the simulation model are returned to the DCS hardware for control response. From the instructor station the instructor manages the training session by starting or stopping the simulation, initiating a number of different operating situations, creating plant disturbances and introducing equipment malfunctions.

Expensive Systems

Historically, the emphasis on operator training systems has been aimed at providing a realistic working environment for trainees and advanced training features for instruction. As such commercial operator training systems promoted the use of actual DCS consoles and MMI graphics to accurately reflect the operator's working environment. And to effect efficient training sessions and record keeping, many commercial operator training systems were equipped with advanced training features, such as the ability to:

- Initialize simulators from preset operating conditions
- Run, freeze and restart operations from instructor's console
- Playback operator actions for reviewing
- Automate numerous operating situations
- Produce Training session reports
- Run the model faster than real-time

While these realistic working environments with their advanced features do provide a superb learning environment the hardware, software and consulting services required to develop, maintain and update these systems does not come cheap. Estimates of the cost of these "classical" operator training systems, not including maintenance and updating services, can range from $500,000 to $2,000,000 US (Giglio, 1997).

Single-Purpose Systems

At the heart of an operator training system is a dynamic simulation model that mimics the actual plant's processes. Because the training system is required to reflect the actual process operation, as a minimum, the dynamic simulator was required to at least run in real-time as the process does. Unfortunately, by their very nature rigorous first-principle dynamic calculations are mathematically intensive. And until recently, hardware and software technology had not evolved to the point where rigorous first-principle dynamic models of entire process flowsheets could run in real-time. To overcome this limitation, early operator training systems used simplified thermodynamic calculations to accelerate the dynamic model's run-time and achieve real-time response. Arguably, because these low-fidelity dynamic models were based on first-principles (the model's response would be generally correct for all process transients) their suitability as an engine in operator training systems was justified.

However, the real limitation of these dynamic models was that they were typically developed for a single-purpose – a training system for a particular process operation. If that process operation changed substantially, the simplifying thermodynamic assumptions made in the original model could become invalid, requiring updating of the process model by either internal company engineering resources or recurring vendor maintenance. Too often the process models were not updated, resulting in a training system that no longer matched the actual plant and rapidly became obsolete. Furthermore, other engineering departments could not leverage the investment made in the development of these dynamic models for purposes other than training. Typically, engineering personnel need to rely on high-fidelity first-principle simulation models for their analyses. Past limitations with hardware and software technology required operating companies, wishing both training and engineering simulation systems, to invest in two simulation technologies; a high-fidelity first-principle simulation platform for engineering analyses and a low-fidelity dynamic simulation platform as the engine for an operator training system.

Difficult-to-Maintain Systems

Until recently, dynamic simulation technology has been a difficult technology to master (Mahoney, 1994). "Due to the unique combination of computer programming, numerical integration, modeling, and sophisticated chemical and thermodynamics skills required, dynamic simulation has remained largely in the hands of "experts" over the years. Additionally, because of the extremely large number of calculations that dynamic simulations require, these simulations have been reserved for very large and powerful mainframe or mini-computers. As such, only the most complex and challenging of process designs seemed to warrant the use of dynamic simulation."[10]

Given these limitations to dynamic simulation modeling, historically most companies that wished to invest in operating training systems typically relied on the expertise of vendors to develop, maintain and update their training systems. As a result, when changes in the process configuration occurred, or a new plant control strategy was designed or a new training scenario was required, the operating companies needed the assistance of their expert vendor to modify their training systems. Any changes to the actual plant or the simulation scenarios either reduced the useful life of the operator training system or increased the need to incur maintenance costs.

In summary, the "classical" operator training system comprised of expensive components, single-purpose functionality and difficult-to-configure/ maintain software has made this technology investment unjustifiable to many operating facilities.

The "New Age" Operations Training Systems

Today however, thanks to substantial advancements in computer hardware and software technology, such as Object-Oriented Programming (Satyro et al., 1996 and Tyréus, 1992), and the development of new ways of packaging dynamic simulation (Mahoney, 1994 and Satyro et al., 1996), this technology is rapidly evolving into a tool for everyday use by engineers and operators. Rigorous first-principle dynamic simulation models developed for engineering analyses can now run in multiples of real-time under Windows NT systems. This advance in dynamic simulation technology is having a tremendous impact on the ability of operating companies to develop, update and maintain their own training systems that suit their specific corporate needs.

Rigorous dynamic simulation models of entire flowsheets developed during engineering analyses can be linked to plant hardware and used in the development of an operator training system. Conversely, rigorous dynamic simulation models developed specifically for training purposes can be taken off-line and used by engineering staff to improve process designs, evaluate alternative control strategies and troubleshoot operating deficiencies. Additionally, easy-to-configure dynamic simulation packages are permitting in-house operations staff to develop, maintain and update their own dynamic models and training systems with reduced need for vendor support and services. Finally, with integrated software technologies, operating companies have the added ability to increase the functionality and detail of their training systems as changing needs arise.

In short today's better hardware and software technology is enabling operating companies to not only develop effective training systems that suit their specific training requirements but to develop training systems that suit their operating budgets as well. With greater flexibility in using dynamic simulation capabilities for both engineering and training purposes, the anachronism OTS is growing to encompass a much broader meaning – that of *Operations* Training Systems.

Operations Training Systems and their Applications

In recent years numerous operating companies have seized this enhanced dynamic simulation technology to create their own unique operations training systems. More and more, companies are breaking away from the "classical" operator training system and developing training systems that meet their specific corporate training requirements. Training systems are being developed with various levels of model detail, a variety of MMI graphic options, and a spectrum of advanced training features (sometimes none). While each company's operations training system is very unique, it is possible to broadly classify these individual training systems as being an Engineering Trainer, a Process Trainer or an Operator Trainer.

Following is a brief description of the different operations training systems, their general purpose, and an industry example illustrating the use and benefits that the system delivered.

Engineering Trainers

As the name implies the concept of an Engineering Trainer is to provide engineers with a system where they can gain a better understanding of their process plant. Whether the objective of the Engineering Trainer is to

---

[10] Don P. Mahoney and Björn D. Tyréus, "Applications of Dynamic Simulation"

- enhance the engineer's knowledge of a process unit and its control strategy
- develop safe and operable control schemes before downloading these new strategies to the plant
- troubleshoot an existing operation

the Engineering Trainer provides a risk-free modeling environment to conduct these analyses. Typically companies interested in developing training systems for this purpose are not necessarily interested in replicating a familiar "working environment" so the need for expensive MMI graphics, instructor stations and advanced training features are usually avoided. However, once developed an Engineering Trainer can be easily expanded to include realistic MMI Graphics, instructor stations, and advanced training features to provide process instruction to operators.

JEC's (Japan Energy Corporation) Distillation Unit

As part of an Advanced Quality Control (AQC) project, JEC was interested in reducing the process variability and increasing the product quality and throughput from their xylene distillation unit. To gain a better understanding of the column's operation JEC constructed a dynamic simulation model of the distillation unit and investigated the operability of the column under different design and operating conditions.

To conduct these studies JEC developed a first-principles dynamic simulation model equipped with standard PID controllers. With their dynamic simulation model they validated its accuracy against plant data and evaluated a number of operating scenarios. The advantage of using dynamic simulation analyses in this project was threefold. First, the project engineer was able to evaluate numerous different operating scenarios without introducing upsets to the actual process operation, which had the potential to reduce operating profits. Secondly, he was able to evaluate operating scenarios that lay outside the column's normal operating region, which could potentially be dangerous to the plant and its personnel. Thirdly, JEC could use the project engineer's dynamic simulation analyses and results as the basis for justifying capital and operating expenses before implementing the recommended process modifications on the actual process.

Subsequent dynamic simulation analyses of several operating scenarios determined that some modifications to the current control strategy would improve the distillate product quality and increase the column throughput. JEC estimated that through improved control performance of the column they could increase operating profits by about $500,000 annually.

In this project, JEC was able to complete its engineering analyses with the use of a standard commercial simulation platform. However, in certain situations commercial simulation packages may be insufficient to properly design, implement and evaluate control strategy alternatives. For example, in situations where the process exhibits excessive deadtime or large interactions between units or significant controller coupling, it may be necessary to implement advanced control strategies to maintain stability in the process. Often the simplest way to implement these advanced control strategies is to link a real-time dynamic model directly to DCS/SCADA hardware. Inside the DCS/SCADA system project engineers can access the comprehensive control capabilities and implement their advanced control techniques. In this arrangement, the dynamic simulation model mimics the actual plant operation and the DCS/SCADA hardware controls the process model.

Solutia's (formerly Monsanto Company) Batch Reactor

In an effort to improve the production yield, payload and batch cycle time for a very important product, Solutia undertook a process improvement project to gain a better understanding of the dynamic behaviour of their batch operation. To do so, Solutia used a commercial simulation platform to develop a dynamic model of their batch operation. Initial dynamic simulation analyses indicated that to better control the batch reactor they would need to employ an advanced control algorithm (a horizon predictive controller algorithm). Unable to easily configure the unconventional control algorithm within their simulation platform Solutia linked their dynamic simulation model directly to their Fisher-Provox DCS system. Inside the DCS system the project engineers configured the advanced control algorithms necessary for the control system evaluation.

Off-line project engineer's used the dynamic simulation model, controlled by the DCS, to evaluate the performance of their batch reactor under changing operating conditions and different control arrangements. In fact, subsequent analyses determined that one of the advanced control strategies substantially improved the controllability and operability of their batch reactor. An added advantage of linking their dynamic model to the actual plant hardware was that Solutia was able to properly tune their controllers before implementing the chosen control strategy in the process. Once implemented, the new control strategy resulted in a 5% increase in product yield and a 5% decrease in batch cycle time, greatly improving the profitability of the batch reactor process.

While direct connections to DCS/SCADA systems are effective means of implementing advanced control strategies, that otherwise could not be implemented in commercial simulation packages, purchasing hardware for this specific purpose can be expensive. As an alternative dynamic simulation models can be connected directly to application programs such as Visual Basic, MatLab and

Excel. Within these packages engineers can develop any number of advanced control strategies, avoiding the need for expensive DCS/SCADA hardware. In these systems the dynamic simulation model mimics the actual plant operation and the application package (Visual Basic, MatLab or Excel) controls the process model in real-time.

Process Trainers

In simplest terms process trainers differ from operator trainers in terms of system detail – process model, operator consoles, instructor stations, advanced features. Generally, the aim of most Process trainers is to capture an appropriate level of process detail and operating scenarios to represent the plant's main operation.

Chevron's Off-shore Turbo-Expander Plant

In 1996 Chevron decided to update two of their offshore platforms. The renovation involved replacing antiquated pneumatic controllers with modern DCS technology - Moore APACS with both PSD and PID control. Early on Chevron recognized that the success of the DCS upgrade project hinged upon their operators' abilities, to use the new DCS technology efficiently and to understand the platform's process interactions - specifically how the platform's compression system and pumps interacted.

Selecting to model only the major unit operations and control strategies, Chevron developed a dynamic simulation model of their offshore platform. To create a realistic training system for the operators, in terms of controller functionality and MMI, the project engineers linked their simulation model directly to the plant's DCS hot spare. Use of the actual DCS hot spare not only provided a true working environment for the operations staff during training exercises, but it reduced the engineering effort required to develop the system. The DCS control logic and MMI graphics, already configured for the platform's actual DCS, were simply downloaded to the hot spare and used in the training system. To complete the Process Trainer, project engineers created a series of normal, abnormal and shutdown simulation cases to illustrate important operating scenarios during the training sessions.

Mimicking the platform's operation, process variables from the simulation model were passed to the DCS. Operating in real-time, control logic performed in the DCS (both PSD and PID) controlled the process model. Instructors, worked one-on-one with the operations staff simulating different operating scenarios and imposing equipment malfunctions to assist the trainees with their process learning and DCS familiarity.

Off-line the on-site engineering staff made use of the dynamic engineering model for their process control and operability studies. Now in active operation, the Process Trainer is used daily by both the engineering and operations staff, to instill process knowledge, develop procedures for abnormal operating scenarios, and troubleshoot the operation.

The main objectives of the Chevron training system were to provide DCS familiarity and process understanding for their operations staff. Thus the use of the actual plant hardware was critical in Chevron's choice of Process Trainer system. However, companies less concerned with the exactness of the actual DCS MMI graphics and its control logic have elected to use alternative third-party platforms to create a Process Trainer.

Comerint's Dynamic Training Simulator (DTS)

Comerint S.p.A., a turnkey training system provider based in Italy, commissioned a Process Trainer for one of its clients. Modeling the process unit operations, process lines and control valves within a commercial simulation package, Comerint was able to accurately imitate the actual plant operation. However, not wishing to invest in expensive DCS hardware, Comerint's client opted to replicate the control functionality and MMI graphics using Intellution's FIX-DMACS system. Operator screens configured for use within Intellution provided a realistic working environment in which trainees were able to develop a deeper understanding of their process and its controllability.

Commenting on the success of the project, Nicola Martinelli Technology Transfer Manager at Comerint stated, "Practicing in a risk-free environment with the DTS system, operator trainees are developing a better understanding of their process operation, so that they can more effectively respond to both normal and abnormal operating conditions. Because DTS is PC-based, it has tremendous flexibility – personnel can be trained in either the classroom or the control room. More importantly, even experienced operators can benefit from the training capabilities provided by DTS. Experienced operators can receive personal refresher training courses as part of their job rotation."

Like Engineering Trainers, Process Trainers are scalable systems that can be enhanced to create sophisticated Operator Trainers. Hyprotech is one process simulation vendor that provides a two-tiered approach to process model development within a common framework. A **HYSYS.Process** modeling package is aimed at the process design activity, providing companies with the option to specify unit operation detail appropriate to capture the major process aspects. A **HYSYS.Plant** modeling package is aimed at the detailed design activity, providing companies with the option to incorporate comprehensive equipment geometry and performance information to develop detailed plant simulations. In addition, integrated simulation platforms provide the means to communicate with virtually any DCS platform add instructor station functionality and incorporate any variety of advanced (software) training features.

Operator Trainers

At the high end of the operations training spectrum in terms of model detail, MMI graphics, instructor stations and advanced training features are Operator Trainers. These systems aim to capture a high degree of process equipment and performance detail to develop highly detailed operator training systems.

Much like the "classical" operator training systems discussed earlier, these systems contain numerous operator consoles, instructor stations and a wealth of advanced training features to provide a sophisticated training environment for trainees. For example instructor stations provide training instructors with the ability to interact with the simulation and examine process variables, introduce process upsets for the trainee to troubleshoot, monitor trainee response actions, initiate interlock trips and view trends and alarms. Additionally, advanced training features can be provided to record students' actions and evaluate their performance. However, unlike the "classical" operator training system described earlier, today's Operator Trainers can be based on rigorous first-principles models that run at real-time or multiples thereof. Moreover, opting to use a rigorous first-principles simulation package as the engine for an Operator Trainer system, the operating company benefits from reduced lifecycle costs. Operating companies have the ability to use the dynamic models off-line for other engineering purposes, and they have the ability to maintain and update the process models in-house as changes are made to the actual facility. This increases the usage of the dynamic process models throughout the organization, and leverages the original modeling efforts and costs across many engineering applications.

Conclusion

Unlike the operator training systems of the past, today's operations training systems are highly versatile. Modern software and hardware technology has improved the rigour and usability of dynamic simulation technology such that it can be used in real-time applications. Already a number of operating companies have adopted this technology and developed their own unique operations training systems. More importantly, operating companies are discovering that they have the in-house capability to develop, maintain and update these systems with reduced vendor support. Additionally, the same dynamic simulation models built specifically for training purposes can be used off-line for engineering analyses. The concept of a plant life-cycle engineering tool is finally here.

References

Broussard, M. R., "Dynamic simulation: User's Perspective", Hyprotech 1996 International Technology Conference, San Antonio Texas, October 13th –16th

Eastman Chemical Company, "Simulation Success Story", Chemical Processing Technology International 1997/8, page 89.

Giglio, R.S. and K.S. Ahluwalia, Foster Wheeler Development Corp., "Use dynamic modeling to improve training programs", Hydrocarbon Processing, October 1997, Page 65

Mahoney, Donald P. and Björn D. Tyréus, "Applications of Dynamic Simulation"

Mahoney, Donald, "HYSYS An integrated system for process engineering and control", Hydrocarbon Asia, November/December 1994.

McMillan, Greg, Kelahan, Robert, Gamel, Mark, Garcia, Alejandra, "The case for dynamic simulation", Hydrocarbon Engineering, March/April, 1997.

Morgan, S.W., Sendelbach, S.P., Stewart, W.B., "Improve process training with dynamic simulation", Hydrocarbon Processing, April 1994, Page 51

Sabharwal, A., Bhat, N.V., Wada, T., "Integrate empirical and physical modeling", Hydrocarbon Processing, October 1997, Page 105.

Sabharwal, Amish, "Dynamic Modeling of a Distillation Column Using HYSYS", Hyprotech 1996 International Technology Conference, San Antonio Texas, October 13th –16th

Satyro, Marco, Wayne D. Sim, William Y. Svrcek, "From Large Computers and Small Solutions to Small Computers and Large Solutions", Hyprotech 1996 International Technology Conference, San Antonio Texas, October 13th –16th

Tyréus, B.D., "Object-Oriented Simulation" In Practical Distillation Control (Editor: Luyben, W.L.), Van Nostrand Reinhold, New York, 1992.

Winter, Peter, Honeywell Hi-Spec Solutions, "Simulator Speeds Operator Training", Chemputers Europe 1 Conference, Milan Italy, November 1994.

OLE AUTOMATION: BRINGING THE POWER OF PROCESS SIMULATION TO PLANT OPERATIONS PERSONNEL

James McGill
Hyprotech Ltd.
Calgary, Alberta, Canada, T2E 2R2
E-mail: James.McGill@hyprotech.com

Introduction

Few groups can apply the power of simulation software more directly and with greater impact than plant operations personnel. Given access to a process simulation, operations personnel could use it to not only to improve the operation of specific plant equipment and areas, but also to increase their own process understanding, which is crucial to optimizing overall plant performance. For example,

- a column simulation could be used to calculate optimal controller setpoints for a feed entered by the operator,
- a recycle loop simulation could be used increase plant production by calculating optimal recycle flow rates, or
- a compressor loop simulation could be used as a "soft sensor" to allow access to data which would otherwise be unavailable due to environment, health or safety reasons.

Given the large potential benefits, why have operations personnel not been given access to simulation software? There are two key reasons:

1. Simulation software can be complex and hard to use for those without the proper background and training, and
2. A process simulation can be inadvertently misused to produce inaccurate results or the process simulation itself damaged or destroyed.

There now exists a way to enable operations personnel to access the power of a process simulation, while shielding them from its complexity and ensuring the integrity of the results and the process simulation itself. Through the use of Microsoft OLE Automation, Windows applications, such as Microsoft Excel™, can be used to access and control OLE Automation capable simulation software, such as *HYSYS*™ from Hyprotech. Using Excel™, a custom front-end can be created that enables operations personnel to access the process simulation in a controlled fashion using a familiar interface. This paper will explore how OLE Automation capable simulation software can be used with Windows application to enhance decision-making, improve process understanding and optimize production.

Description of OLE Automation

Originally OLE stood for Object-Linking-and-Embedding, an example of which is pasting an Excel™ spreadsheet into a Word™ document. However, OLE has now grown far beyond its humble origins and is part of the Microsoft COM (Common Object Model) standard, which includes a host of technologies that enable applications to work more closely with one other and the Windows™ operating system. The OLE Automation portion of this technology (also known as ActiveX Scripting or simply Automation) enables an OLE Automation compliant application to interact with another application by sending commands to it and retrieving information from it. Most major Windows 95/NT applications and many Windows 3.1 applications support OLE Automation, including all products in Microsoft Office, Lotus SmartSuite, and Corel WordPerfect Suite.

OLE Automation Example

In order to use OLE Automation, an object-oriented programming language must be used. Each office suite uses its own built-in object-oriented programming language, but since all are similar to the Visual Basic for

*Table 1. A Slightly Modified Version of a VBA Program, which uses the Excel™ Spreadsheet **Setpoint.xls** to Access a HYSYS™ Methanol Column Simulation and Determine the Setpoints Required for the Column to Meet the Desired Product Specifications.*

```
1. Set hyCase = GetObject("C:\HYSYS\CASES\SETPOINT.HSC", "HYSYS.SimulationCase")
2. Set hyFeed = hyCase.FlowSheet.MaterialStreams.Item("Feed")
3. hyFeed.IdealLiquidVolumeFlow.SetValue 300, "USGPM"
4. Compositions = hyFeed.ComponentVolumeFraction
```

Applications (VBA) language used by Microsoft Office, it will be used in all references throughout this paper. VBA is a subset of the popular and easy-to-use Microsoft Visual Basic language. Once the concept of objects is understood, VBA programs that utilize OLE Automation are easily created and modified. Table 1 is a slightly modified version of a VBA program which uses the Excel™ spreadsheet **Setpoint.xls** to access a *HYSYS*™ methanol column simulation and determine the setpoints required for the column to meet the desired product specifications. **Setpoint.xls** will be discussed in more later in this paper.

1. The variable hyCase is used to refer to the *HYSYS*™ methanol column simulation **Setpoint.hsc**. This line loads the simulation invisibly in the background using the VBA GetObject, sets the object reference using the VBA Set command and informs VBA of the type of *HYSYS*™ object (SimulationCase). All references to hyCase in the rest of the code refer to this *HYSYS*™ simulation case.
2. The variable hyFeed is used to refer to the material stream named Feed on the main *HYSYS*™ Flowsheet using the VBA Set command.
3. Set the value of the ideal liquid volume flow of the stream referred to by hyFeed equal to 300 USGPM.
4. Set the array variable Compositions equal to the component volume fractions of the stream referred to by hyFeed.

The above example also demonstrates the object hierarchy used by *HYSYS*™. For example, the SimulationCase object (referred to by the hyCase variable) contains the Flowsheet object, which in turn contains the MaterialStreams object.

Operations Personnel using Process Simulation via OLE

Though operations personnel can have the most direct impact on plant operations, they are rarely, if ever, given tools that can assist them in improving overall operation and profitability. By giving operations personnel access to a process simulation, which they can combine with their own experience and judgement, they could not only improve the operation of specific piece of equipment or group of equipment, but also to increase their own process understanding, which is crucial to optimizing overall plant performance. The two historical obstacles to giving operations personnel access to simulation software have been

1. Simulation software can be complex and hard to use for those without the proper background and training, and
2. A process simulation can be inadvertently misused to produce inaccurate results or the process simulation itself damaged or destroyed.

Through the use of the OLE Automation technology described in a previous section, there now exists a way to enable operations personnel to fully access the power of a process simulation, while shielding them from its complexity and ensuring the integrity of the results and the process simulation itself. Windows applications, such as Microsoft Excel™, can be used to access and control OLE Automation capable simulation software, such as *HYSYS*™ from Hyprotech. Using Excel™, a custom front-end can be created that communicates with a *HYSYS*™ simulation via OLE and enables operations personnel to access the process simulation in a per-determined and controlled fashion using a familiar interface. The following examples demonstrate this.

Column Setpoint Optimization

The Excel™ spreadsheet **Setpoint.xls** accesses a *HYSYS*™ methanol column simulation (**Setpoint.hsc**) and determines the setpoints required for the column to meet the desired product specifications. Figure 1 shows the custom front-end used.

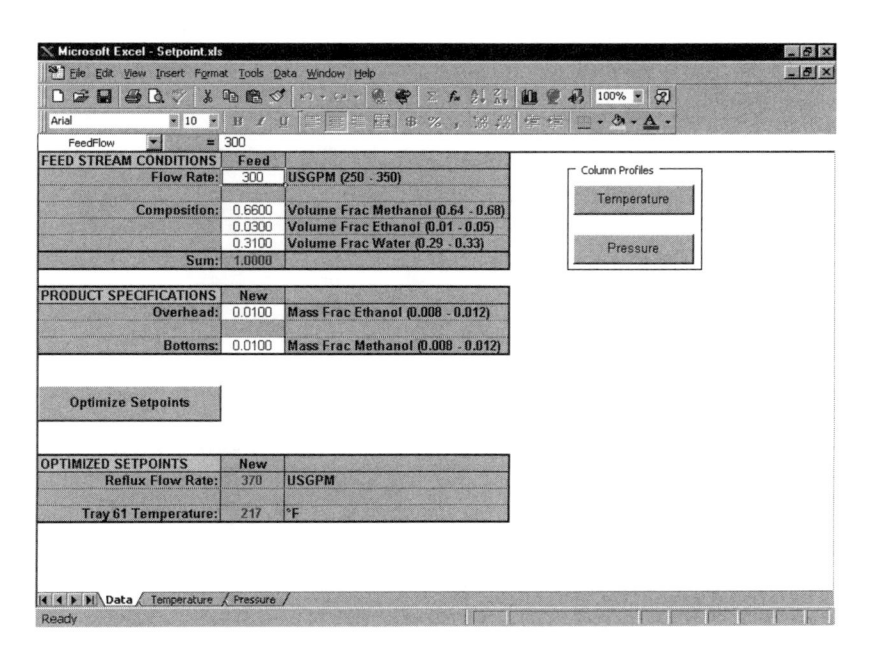

Figure 1. Setpoint.xls custom front-end.

When the user enters a new value for the feed flow rate, feed composition, or the product specifications, the optimized setpoints values are erased and a message is displayed which indicates that the user must press the Optimize Setpoints button to calculate new setpoint values, as shown in Fig. 2.

Figure 2. Press button message.

When the users presses the button, the *HYSYS*™ simulation is loaded invisibly in the background (only the *HYSYS*™ splash screen is displayed), the feed values and product specifications are sent to it, a message is displayed in the status bar informing the user to wait (Fig. 3) and new optimized setpoints are calculated and displayed (Fig. 4).

Figure 3. Please wait message.

Figure 4. Optimized Setpoints results.

Additionally, the temperature and pressure profiles (Fig. 5) for the column are also available for review and are accessible via buttons on the main page or by using the page tabs.

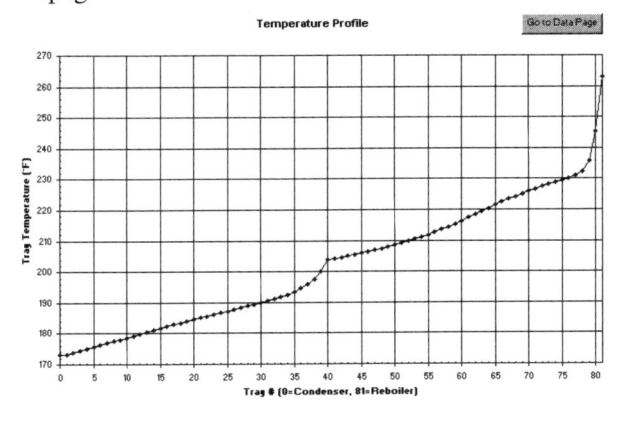

Figure 5a. Temperature profile.

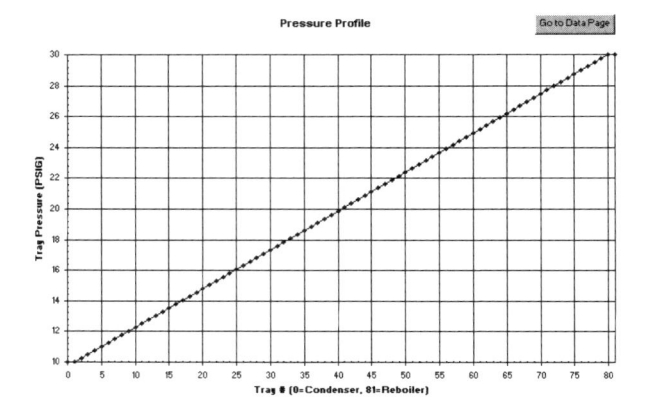

Figure 5b. Pressure profile.

One can easily see how an operator could use this tool, combined with his or her experience, to ensure that the column setpoints are always optimized when a significant change in the feed occurs resulting in a type of real-time optimization. A variation of **Setpoint.xls** called **DataFile.xls** demonstrates how this process can be automated such that the optimal column setpoints are calculated and displayed at set interval.

Recycle Compressor Loop Throughput Optimization / Soft Sensor

The Excel™ spreadsheet **Recycle.xls** accesses a *HYSYS*™ recycle compressor loop simulation (**Recycle.hsc**) and acts as a soft sensor by displaying information about all streams contained in the loop in both a table and a diagram. Additionally, Excel™ calculates the loop ratios and overall methanol production. Figure 6 shows the custom front-end used.

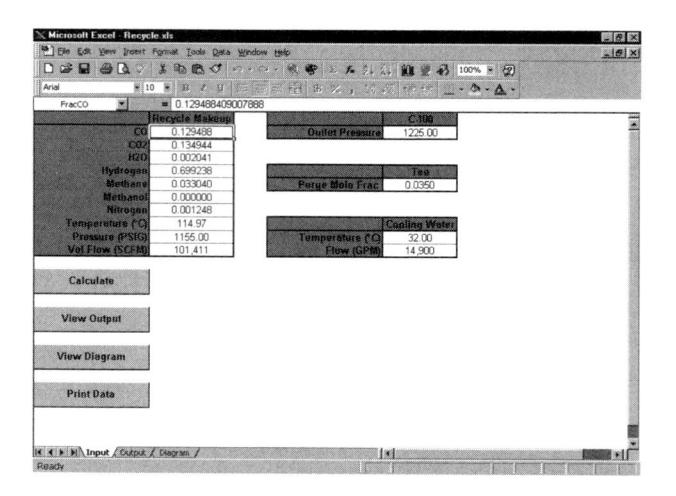

Figure 6. Recycle.xls custom front-end.

Once the user enters values for the recycle makeup values and major pieces of equipment in the loop, he or she presses the Calculate button to load the *HYSYS*™ simulation invisibly in the background and send data to it. A message is displayed in the status bar informing the user to wait (Figure 7) and the tabular results are automatically displayed (Figure 8).

Figure 7. Please wait message.

Figure 8. Recycle compressor loop table.

A diagram of the recycle compressor loop can also be viewed (Fig. 9) by pressing the View Diagram button or by using the page tabs.

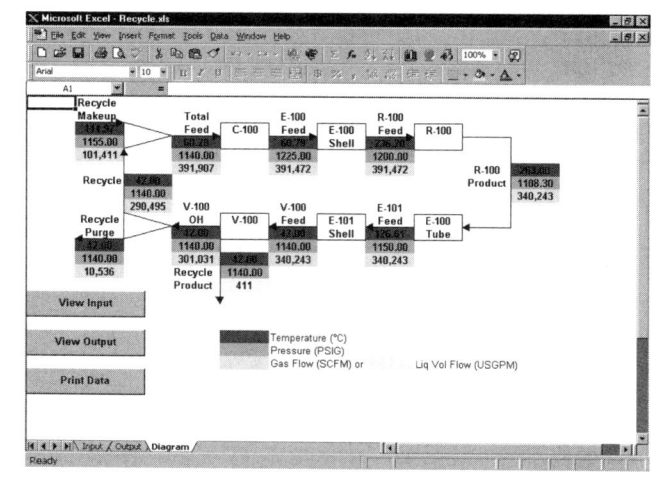

Figure 9. Recycle compressor loop diagram.

With this tool, an operator gains a deeper understanding of the process by accessing data that would otherwise not be available. In addition, the operator can use the recycle compressor loop ratios calculated by Excel™ to "tune" the loop for optimal production.

Conclusion

Giving operations personnel access to a process simulation via OLE is beneficial for the following reasons:
- Operators would have increased process understanding, enabling them to make better day-to-day operations decisions and optimize overall plant performance.
- The operators are shielded from the complexity of the simulation, while being able access to its power.
- The amount of access to the simulation given to the operator via the Excel™ front-end is controlled, ensuring the integrity of both the results and the process simulation itself.

Therefor, using OLE Automation capable simulation software with Windows application will enhance decision-making, improve process understanding and optimize production.

AN OPEN SOFTWARE ARCHITECTURE FOR PROCESS MODELING AND MODEL-BASED APPLICATIONS

Anthony Kakhu, Benjamin Keeping, Yuzhao Lu and Constantinos Pantelides
Centre for Process Systems Engineering
Imperial College of Science, Technology and Medicine
London SW7 2BY, UK

Abstract

This paper considers mechanisms for supporting openness within process modelling tools. It presents an analysis of the requirements for such mechanisms. It also describes a novel concept, that of a Foreign Object, that has recently been developed for achieving openness in the context of *gPROMS*, an equation-orientated process modelling tool. Some of the mathematical and the software issues arising from this development are also considered.

Keywords

Process modeling, Simulation, Software integration, Real-time systems, Component-based software, Standardization.

Introduction

Process modeling tools are nowadays being used extensively at all stages of process design and operation. However, despite the ever increasing sophistication of these tools, there is also a growing realization that no single tool will be able to provide all the functionality that process engineers are likely to require. This has led to a current trend towards the adoption of open software architectures for such tools.

There are many reasons to move towards an open architecture approach for process modelling tools. For example:

1. Users may possess their own proprietary software, developed either for "stand-alone" application or for use in conjunction with in-house modelling tools, which they wish to incorporate into the modelling tool that they are currently using.

2. Third-party software may be more appropriate for the modelling of specific parts of the process. Examples in this category include Computational Fluid Dynamics (CFD) software providing a very detailed simulation of a piece of process equipment within the flowsheet; a spreadsheet computing annualised costs in a design calculation; or a package for physical property calculations.

3. Once the process models are constructed and validated, users may wish to incorporate them within a larger software system, *e.g.* an operator training system or real-time decision support tools.

Of course, most process modelling tools that have been available over the past couple of decades already allow a certain degree of "openness". For instance, most standard sequential or simultaneous modular flowsheeting packages allow users to add their own models of unit operations expressed in a procedural language (*e.g.* FORTRAN or C/C++); and equation-orientated packages, such as SPEEDUP™[11] allow models to incorporate relations among their variables that are also expressed in

---

[11] SPEEDUP™ is a trade-mark of ASPEN Technology Inc., Cambridge, Massachusetts, U.S.A.

terms of external procedural code (Pantelides, 1988); this latter facility is particularly important as a means for interfacing physical property software to these packages.

Albeit undeniably useful, the type of "openness" described above often leaves much to be desired in terms of generality, the ability to check consistency of information transmitted across the interface and to capture any related errors, and other issues such as the precise way in which the interfacing between the two items of software is to be achieved. This paper describes some recent work aimed at addressing these concerns. It has been carried out in the context of *gPROMS*™[12], an equation-orientated process modelling tool.

In the next section, we present some of the background that is necessary to understand the issues discussed in this paper. We then consider the demands posed by openness in modelling physical plant behaviour and present a novel concept that we have developed for achieving this. We also examine some of the software issues that need to be addressed. Finally, we conclude with some remarks on the work presented in the wider context of current efforts towards standardisation of process modelling software.

Background

As described by Pantelides (1996), the process modelling activity can be viewed as encompassing descriptions not only of the physical behaviour of process plant but also of the plant operating procedures, such as those used to start-up the plant, control its normal operation, handle emergencies, shut it down and so on.

gPROMS (Barton and Pantelides, 1994; Oh and Pantelides, 1996) is an equation-orientated process modelling tool that was designed to take account of this dual nature of process modelling. The physical plant behaviour is described by collections of variables and equations, called MODELs; these mathematical descriptions can be quite complex, involving coupled sets of integral, partial and ordinary differential and algebraic equations (IPDAEs). On the other hand, operating procedures are described as sets of actions (called TASKs) executed in sequence, in parallel, conditionally and iteratively.

The dual nature of process modelling described above has implications regarding the form of openness that a modelling tool has to provide. For instance, third-party physical property calculations or a unit described in an external CFD package clearly have to be incorporated within the description of the physical behaviour of the process being modelled, and, therefore, appropriate interfacing mechanisms must be provided. On the other hand, manual actions by human operators or control actions imposed by a real-time control system clearly belong to the operating procedure of the process; since such actions are external to the process modelling tool, some mechanism for interaction with these external agents must also be provided.

The rest of this paper discusses the detailed requirements for openness in descriptions of physical behaviour, and a novel mechanism for achieving this.

Openness in the Description of Physical Behaviour

As has already been mentioned, the description of physical behaviour in equation-orientated (EO) process modelling tools is provided primarily in terms of variables and the equations relating them. One obvious way in which a degree of openness can be achieved is by allowing some of the relations among the variables to be determined externally. Indeed, the approach adopted by some of the earlier EO process modelling tools (see Introduction) is to support *procedures*, *i.e.* relations of the form:

$$x_1 = F(x_2) \qquad (1)$$

where x_1 and x_2 are subsets of the system variables, and the form of the mapping $F(.)$ is provided by external code (*e.g.* FORTRAN subroutines). This type of facility, however, takes too narrow a view of what is really required to support many applications of interest.

Generality

The first issue that needs to be addressed is that of *generality*. For instance, the relation between x_1 and x_2 may be an implicit one of the form:

$$F(x_1, x_2) = 0 \qquad (2)$$

where $F(.,.)$ represents the evaluation of the residuals of a set of equations rather than the calculation of the value of one physically meaningful quantity from that of another. In other cases, $F(.,.)$ could appear as a term in an equation rather than represent an entire equation.

Of course, all of the above complications could be accommodated within the existing framework of procedures of the type described by equation (1) simply by defining an extra set of variables x_3 via the relation:

$$x_3 = F(x_1, x_2) \qquad (3)$$

and setting these to zero (to emulate (2)) or using them in the model equations instead of the term $F(.,.)$. This is achieved at the expense of unnecessarily increasing the problem size.

However, a more fundamental problem with the procedure concept is that it represents relations among variables in a system of equations of a *given* structure; it cannot be used for determining any aspect of the structure

---

[12] *gPROMS*™ is a trade-mark of Process Systems Enterprise Ltd., London, United Kingdom.

of the mathematical problem itself (*e.g.* its size). Consider, for instance, a situation where an external physical property package determines the number of species that are present in a given system in addition to providing facilities for the calculation of values of physical properties of mixtures of these species[13]. Clearly, the number of equations and variables in the mathematical problem will depend on this number of these species.

Another example is provided by the need to incorporate unit operation models described by *external* equation-orientated tools within a "host" process modelling tool. Such models "export" their variables and equations to the host which then proceeds to assemble the information from all the unit operations into a single mathematical system. Again, in many such cases (*e.g.* when multi-stage unit operations are involved), the size of this exported sub-system is often not known *a priori* but is determined at runtime by the external software.

Consistency

Another issue that needs to be addressed in the context of software interfaces is that of ensuring the consistency of the information being passed across the interface. Consistency checks are important in order to prevent errors, some of which may be extremely difficult to diagnose, from occurring during runtime.

Earlier equation-orientated modelling tools, such as SPEEDUP, already enforce certain basic checks on the usage of procedures. For instance, not only the number of arguments of each procedure, but also the type (temperature, flowrate *etc*.) and the size of each argument are checked against a formal declaration of the procedure that is also incorporated within the process model under construction. Whilst this is useful in avoiding certain elementary mistakes in model specification, it does not always prevent subtler errors such as those that arise because of differences in the units of measurement. Moreover, there is always a possibility for discrepancies between the procedure declaration appearing in the model and the actual behaviour of the procedure; such discrepancies are particularly likely to arise as a result of modifications carried out during the model lifetime.

Reliability and Efficiency of Mathematical Solution

Another important issue arising in the use of equations that are (partially) defined externally to the process modelling tool is that the information available on them may be restricted in comparison with that available on equations defined completely within the modelling tool.

This restriction mainly arises from the fact that the symbolic form of the external equations is not normally accessible. Therefore, the process modelling tool cannot analyse them to deduce their structure and to derive exact values for their partial derivatives, and this often has an adverse effect on the reliability and efficiency of the mathematical solution methods. Moreover, the ability of the tool to provide precise diagnostics (*e.g.* to identify terms that lead to illegal arithmetic operations) is also limited.

The Foreign Object Concept

The concept of a *Foreign Object* (FO) represents an attempt to address some of the issues that were identified in the previous section in conjunction with providing openness in the description of physical behaviour.

A FO is a software component that provides computational services to descriptions of physical behaviour in *gPROMS*. In accordance with the well-established concept of software objects, each FO comprises a set of internal data and a set of *methods* which can be used to perform operations on those data. Only the methods are directly accessible to software that is external to the FO. The internal data, which typically determine the detailed behaviour of a FO and/or define its "state" at any given time, are private to it.

Each method of a FO returns a *single* scalar or vector quantity (its "output"). The value of the latter may be real (typically corresponding to a physical quantity, *e.g.* a vector of fugacities), integer (*e.g.* the number of components in a mixture), or logical (*e.g.* a thermodynamic phase stability test).

Each method may also have one or more scalar or vector-valued *inputs*. The method output must be a uniquely defined function of these inputs irrespective of the internal state of the FO.

The set of methods provided, their input-output structure, and their specific software implementation are characteristics of the *class* of a FO. In fact, a *gPROMS* model may make use of services provided by multiple *instances* of the same class. Different instances of a class provide the same external interface but may have different internal data and different sets of methods. For example, a *gPROMS* model may make use of two instances of a FO belonging to a certain class THERMOPACK that provides thermophysical property calculations. The two instances provide the same set of methods, *e.g.* for computing enthalpies, densities, fugacities *etc*. However, they may use different thermodynamic models (*e.g.* an equation of state instead of an activity coefficient model for liquid phase fugacities) or even have different component sets. This may be necessary if the two FOs are used in two different sections of the plant operating in widely different temperature, pressure and/or composition ranges.

---

[13] This is actually a very common situation with electrolytic systems where a special "speciation" calculation has to be performed to determine all ions and other entities likely to exist in the process, as well as the reactions taking place.

The FO methods may appear anywhere in the *gPROMS* model definition where an expression of the appropriate type (real, integer or logical) is expected. Similarly, each input argument of a method can be a general expression of the correct type, perhaps involving invocation of methods provided by the same or different FOs.

An example of the usage of a FO within a model is provided in figure 1 which shows the model of a simple steady-state flash unit. Line[14] 3 declares the fact that this model will make use of services provided by a FO called PPP belonging to class THERMOPACK; this is a PARAMETER as far as this model is concerned. Line 4 declares another parameter, NoC, which represents the number of components in this system; then line 13 equates this parameter to an integer value returned by a method NumberOfComponents provided by the FO. Finally, lines 16-20 make use of other services provided by this FO, this time pertaining to the computation of liquid and vapour phase physical properties.

```
1   MODEL SimpleFlash
2   PARAMETER
3      PPP AS FOREIGN_OBJECT "THERMOPACK"
4      NoC AS INTEGER
5   VARIABLE
6      F,L,V   AS   MolarRate
7      Hf      AS   MolarEnergy
8      Q       AS   EnergyRate
9      T       AS   Temperature
10     P       AS   Pressure
11     x,y,z   AS   Array(NoC) OF MoleFract
12  SET
13     NoC := PPP.NumberOfComponents() ;
14  EQUATION
15     F*z = L*x + V*y;
16     F*Hf = L*PPP.LiquidEnthalpy(T,P,x)
17            + V*PPP.VapourEnthalpy(T,P,y)
18            + Q;
19     x*PPP.LiquidFugacityCoeff(T,P,x)=
20     y*PPP.VapourFugacityCoeff(T,P,y);
21     SIGMA(x) = SIGMA(y) = 1;
22  END
```

Figure 1. Example of Foreign Object Usage.

As a PARAMETER, the FO PPP must eventually be specified if any instance of this model is to be usable. This specification is done by assigning to it a character string "value" that identifies this particular instance of class THERMOPACK. In many cases, this is simply the pathname to a file containing information on this instance (*e.g.* in the case of a physical property package, one that identifies the species involved and specifies the thermodynamic models to be used). However, this is not necessarily always the case as *gPROMS* merely passes this information to the code responsible for the instantiation of the FO (see below).

Handling of Foreign Objects in gPROMS

The foreign object concept satisfies most of the requirements for generality described earlier in this paper. However, as we have mentioned already, another major concern is ensuring consistency between the usage of a FO inside *gPROMS* and the actual behaviour of this external software component.

One possible way of doing this would be to introduce a passive consistency checking mechanism based on declaring the characteristics of each class of FO within the *gPROMS* language, similarly to what was done with procedures in earlier equation-orientated software (see above).

However, in order to address the concerns described earlier, we opted for an *active* mechanism which allows *gPROMS* to establish the consistency of the usage of any particular FO by issuing directly to it requests for information such as:

- The methods that are provided by the FO.
- The length and type of the result returned by each method and, if appropriate, the dimensionality and units of measurement of this result.
- The number of method inputs.
- The length, dimensionality and units of measurement of each of the method inputs.
- The availability of values of the partial derivative(s) of the method result with respect to each of the method inputs.

The dimensionality of each quantity is specified by the FO in terms of the nine fundamental dimensions[15] plus a tenth one denoting money. The addition of the latter was considered to be desirable in order to minimise the possibility of mistakes in the usage of economic information that is computed by several practically important classes of FOs (*e.g.* accounting spreadsheets or equipment costing packages). The units of measurement are specified by the FO in terms of two quantities, an offset and a multiplier such that[16]:

---

[14] Line numbers are not a normal feature of *gPROMS* input language. They are used here only for ease of reference.

[15] Length, mass, time, electric current, temperature, amount of substance, luminous intensity, plane angle and solid angle.

[16] The United States dollar is taken as the "SI Unit" of money.

*SI Value = Offset + Multiplier * FO Value*

The above information supplied by the FO allows *gPROMS* to carry out extensive testing of the correct usage of each FO method within its models, thereby reducing the probability of mistakes remaining undetected and causing problems during execution.

Implementation of Foreign Objects

In order to deliver the functionality described above, each class of FO for use by *gPROMS* is implemented as a software component providing six standard services. The first of these (gFOI) allows the creation of an instance of this FO class using the instance identifier passed to it by *gPROMS*; this is the FO PARAMETER "value" mentioned earlier.

The next two services (gFOCM and gFOCMA) provide information on the methods that the FO makes accessible to the outside world, and their inputs and outputs; this is precisely the information used for consistency checking.

Two further services (gFOM and gFOMD) allow the calculation of the output of a method for given sets of inputs, and also, if available, the partial derivatives of the output with respect to a specified input.

Finally, the last service (gFOT) is used by *gPROMS* to notify the FO that it may terminate.

Most numerical algorithms for dynamic and steady-state simulation and optimisation make use of the values of the partial derivatives of the model equations with respect to the variables occurring in them. In *gPROMS*, these are generated using symbolic differentiation. Although this is not possible for equations involving FO methods, no fundamental difficulty arises provided each such method can supply the exact values of the partial derivatives of its output with respect to all its inputs. When this is *not* the case, we have to resort to numerically generated approximations of the partial derivatives. The obvious way of doing this is by using numerical algorithms that employ appropriate approximation techniques (*e.g.* based on finite differences or least-change secant updates). This is the approach adopted in previous work for dealing with "procedures" (Pantelides, 1988).

An alternative, which is more consistent with the overall concept of component-based software, is to associate the maintenance of partial derivative approximations with the individual method invocations appearing in each model. Consider, for instance, a method $F(x,y,z)$ with three (possibly vector-valued) inputs; the method can compute exact values of partial derivatives with respect to its first input x but not with respect to y or z. Now consider a term appearing in a model equation and involving the following particular invocation of this method:

$$F(f_1(w), f_2(w), f_3(w)) \qquad (4)$$

where w is a (possibly vector-valued) variable and $f_1(.)$, $f_2(.)$ and $f_3(.)$, are given functions specified in symbolic form. Suppose now that we wish to approximate the partial derivative(s) of F with respect to the variable(s) w. These are given by:

$$\frac{\partial F}{\partial w} = \frac{\partial F}{\partial x}\frac{\partial f_1}{\partial w} + \frac{\partial F}{\partial y}\frac{\partial f_2}{\partial w} + \frac{\partial F}{\partial z}\frac{\partial f_3}{\partial w} \qquad (5)$$

The partial derivatives $\partial f_i/\partial w$ may be computed exactly since the symbolic form of the functions $f_i(.)$ is known. Also, $\partial F/\partial x$ will be computed exactly by the method, and this allows the first term in the expression for $\partial F/\partial w$ to be evaluated exactly. However, the last two terms will have to be approximated numerically. If this is to be done using finite differences, there are several options for doing this, such as:

- Estimate the partial derivatives $\partial F/\partial y$ and $\partial F/\partial z$ by perturbing each element of the second and third inputs of the procedure (y and z respectively) independently; then compute the products involved in the last two terms of the above expression for $\partial F/\partial w$ and add them to the first term.
- Perturb each of the w variables, evaluate the corresponding perturbed values of the functions $f_2(.)$ and $f_3(.)$ and then evaluate the method F using the perturbed values of $f_2(.)$ and $f_3(.)$ but the *un*perturbed value of $f_1(.)$. The result of the method then allows us to estimate directly the sum of the last two terms of the expression for $\partial F/\partial w$.

The relative efficiency of the two options will depend on the combined size of the y and z vectors on one hand, and the size of vector w on the other. If $dim(y)+dim(z) < dim(w)$, the former option is preferable; otherwise, the second one may be more efficient; and the optimal choice need not be the same for two different invocations of the same method of the same FO.

Overall, then, the optimal way of approximating the partial derivative information should be determined locally from information pertaining to the individual method invocations. Our implementation, therefore, treats each invocation of a FO method as a separate software object. This also has the advantage that the various numerical solvers (*e.g.* integration algorithms for differential-algebraic equations) can be left to operate at a higher level and need not be concerned with such details.

Software Issues

The implementation of FO software of the type described earlier can be carried out in any procedural programming language such as FORTRAN 77, C or C++. Of course, the use of truly object-orientated languages such as C++ (Stroustrup, 1991) does provide a real advantage in making it easier to handle the case of multiple instances of a FO class occurring in a single process model. By defining a C++ class whose methods are those of the foreign object and whose private data include everything which is specific to a given instance, the FO initialisation routine can consist simply of creating an instance of this C++ class. The private data of this instance will be established in the constructor of this class.

Irrespective of the way in which the FO code is arrived at, it can be compiled and made available to users in a manner which protects the implementors' intellectual property in the software source. This is an important non-technical consideration in promoting the widespread development and use of component-based software.

The final software-related issue that needs to be addressed is that of physically establishing a link between the host modelling tool (in our case, *gPROMS*) and the FO code. In earlier equation-orientated packages which translated the model description into intermediate code written in a procedural computer language (usually FORTRAN), this was simply achieved by compiling and linking the generated code with the externally provided one. However, such static linking is no longer desirable with more modern systems, such as *gPROMS*, which do not create such intermediate code but rely on internal data structures for describing the equations and their partial derivatives.

Instead, we make use of *dynamic loading* mechanisms that allow *gPROMS* to load FO code and make use of its services without the need for a separate linking step. Standardised dynamic loading facilities are nowadays available under both the UNIX and WINDOWS™ operating systems. These provide direct control over the process of loading the FO code into memory during program execution.

In view of the above, each FO class is physically implemented as a UNIX shared object library or a WINDOWS dynamic link library[17]. Once *gPROMS* locates such a file in certain predetermined directories, it uses the facilities provided by the operating system to create and store function pointers to that class's implementation of the six service routines of the FO interface. Thereafter, each instance of that class can be initialised and accessed by making use of these pointers.

A more advanced technology for linking software components is that based on the use of "*middleware*" such as the Object Management Group's CORBA (aimed at all computer platforms, Siegel, 1996) and Microsoft's Active X (thus far mainly WINDOWS-specific, Brockschmidt, 1995). These technologies define standard mechanisms for one piece of software to interact with another based only on an interface definition written in a standard language. Among other things, both provide "local/remote transparency", meaning that the client and server can be running as part of the same process (as assumed so far in this paper), in different processes, or even on separate (possibly quite different) computers. This last is particularly useful for our open architecture where FOs (*e.g.* CFD packages) may have to run on specialised hardware (*e.g.* advanced architecture computers) for both technical and non-technical (*e.g.* licensing) reasons. In any case, the use of middleware is fully compatible with the FO architecture described in this paper.

Conclusions

In this paper, we have discussed the requirements of openness in the description of physical plant behaviour, and introduced the Foreign Object concept as a means for achieving this.

The use of software components as a means for standardisation of process modelling technology has attracted much attention in recent years. For instance, a major collaborative project, CAPE-OPEN, involving 15 industrial and academic partners partially funded by the European Union, is currently under way aiming to establish standards for diverse classes of components, including physical property calculations, unit operations models and various types of numerical codes (CAPE-OPEN, 1997). The FO concept described in this paper can be viewed as providing the necessary infrastructure upon which some of these components and their interfaces can be built.

In view of the dual nature of process modelling, it is natural that mechanisms for achieving openness in the description of plant operating procedures are also necessary. In *gPROMS*, this is achieved via a different mechanism, namely the Foreign Process Interface (FPI). Due to space limitations, this will be described in a future publication.

References

Barton, P.I., and C. C. Pantelides (1994). Modeling of combined discrete/continuous processes. *AIChE J.*, **40**, 966-979.
Brockschmidt, K. (1995). *Inside OLE*. Second Edition. Microsoft Press, Redmond, Washington.
CAPE-OPEN (1997). *Conceptual Design Document for CAPE-OPEN Project*. Available on the World Wide Web at the URL http://www.quantisci.co.uk/cape-open
Oh, M. and C.C. Pantelides (1996). A modelling and simulation language for combined lumped and distributed parameter systems. *Comput. chem. Engng.*, **20**, 611-633.
Pantelides, C.C. (1988). SpeedUp – Recent Advances in Process Simulation. *Comput. chem. Engng.*, **12**, 745-755.
Pantelides, C.C. (1996). Process Modelling Tools in the 1990s. *Proc. 5th World Congress of Chemical Engineering*, San Diego, California.

---

[17] In terms of the example presented earlier, these would typically correspond to files THERMOPACK.so or THERMOPACK.dll respectively.

Pantelides, C.C., and H. I. Britt (1995). Multipurpose Process Modelling Environments. In L.T. Biegler. and M.F. Doherty (Eds.), *Proc. Conf. on Foundations of Computer-Aided Process Design '94*. CACHE Publications, Austin, Texas. pp. 128-141.

Siegel, J. (1996). *CORBA Fundamentals and Programming*. Wiley, New York, New York.

Stroustrup, B. (1991). *The C++ Programming Language*. Second Edition, Addison-Wesley, Reading, Massachusetts.

AUTHOR INDEX

A

Ahn, Tae-Jin 338
Andrews, R. 501
Asakura, Tatsuyuki 185

B

Backx, Ton 5
Badell, M. 217
Bagajewicz, Miguel 328
Bakshi, Bhavik R. 332
Bartlett, Douglas A. 315
Bassett, Matt 267
Basu, Prabir K. 171
Batres, Rafael 397
Bickle, Bruce 261
Biegler, Lorenz T. 6
Blau, Gary E. 127, 197
Book, Neil L. 410, 474, 480
Bosgra, Okko 5
Braatz, Richard D. 380
Brydges, John 322
Bunch, Paul R. 204, 249

C

Chang, Chuei-Tin 345
Chang, Kun Soo 360
Chen, Xueyu 421
Clark, Steven M. 439

D

Davis, James F. 487
De Meyer, Herman 8
Delgado, Antonio 286
Ditz, Reinhard 417
Dua, Vivek 164
Dudás, Tibor 458

E

Edinger, Christine 404
Egan, Erik W. 44
Eker, S. Alper 303
Eliceche, Ana M. 433

E (cont.)

Elsass, Michael J. 487
Engell, Sebastian 224
Ensen, Heinz 404
Eo, Sooyoung 360

F

Fedenczuk, P. 373
Fernandez, Laura 433
Font, Enrique 286
Fuchino, Tetsuo 397
Fujita, Kaoru 185

G

Gau, Chao-Yang 445
Glasser, David 428
Glemmestad, Bjørn 451
Graells, Moisès 286
Grize, Y. L. 464
Grossmann, Ignacio E. 136, 210, 243
Gundersen, Truls 451

H

Han, Chonghun 338
Harjunkoski, Iiro 291
Harney, D. A. 410
Henning, Gabriela P. 279
Hertwig, Thomas A. 421
Hildebrandt, Diane 428
Hopper, Jack R. 421
Houston, R. Brian 171
Hrymak, Andrew 322
Hui, Chi Wai 185
Hwang, Dae-Hee 338

J

Jain, Vipul 243
Jiang, Qiyou 328
Joglekar, Girish 439
Jørgensen, John Bagterp 308
Jørgensen, Sten Bay 308
Jose, Rinaldo A. 152
Josephson, John R. 487

K

Kakhu, Anthony 518
Kalthod, Vikram G. 439
Keck, Owen D. 171
Keeping, Benjamin 518
Khandekar, Milind Madhav 480
Kirschner, Kenneth J. 254
Kohlbrand, Henry T. 112
Kokossis, A. C. 231
Kongsjahju, Ruth 386
Kourti, Theodora 31
Krames, Joerg 404
Krebsbach, Kurt D. 366
Krendl, Helmut 494
Kuenker, Kay E. 127

L

Latour, Pierre R. 297
Lee, P. L. 237
Lee, Yang Gul 146
Leone, Horacio P. 279
Linninger, Andreas A. 494
Lionis, Vangelis 273
Løberg, Espen 286
Lombardo, Stephen P. 60
Lopes, João A. 391
Lu, Yuzhao 518

M

MacGregor, John F. 31
Maineult, Guy 439
Malone, Michael F. 146
Mannarino, Gabriela S. 279
Maria, Gheorghe 351
Marlin, Thomas 322
Marquardt, Wolfgang 5
Masaiwa, Yoshihisa 185
McDonald, Conor M. 62
McGarvey, Bernard 261
McGill, James 513
Menezes, José C. 391
Miller, David C. 487
Mo, Kyung Joo 360
Mockus, Linas 171
Mohamed, Ebrahim 273
Mokashi, S. D. 231
Musliner, David J. 366

N

Naka, Yuji 397
Natori, Yukikazu 1, 185
Nicol, Willie 428
Nikolaou, Michael 303

Noren, Alan R. 171
Nott, H. P. 237

O

Oettinger, M. 373
Oonishi, Haruyoshi 185
Özyurt, Derya B. 190

P

Pantelides, Constantinos 518
Patiño, Hugo 27
Pekny, Joseph F. 91, 204
Perkins, John D. 15
Pike, Ralph W. 421
Pinger, Helmut 494
Pistikopoulos, Efstratios N. 158, 164
Ponton, J. W. 501
Prüß, J. 469
Puigjaner, Luis 217, 286

R

Ramachandran, Bala 142
Ramage, Michael P. 126
Reallf, Matthew J. 190, 254
Reklaitis, G.V. 91
Rollins, Derrick 386
Rosen, Oscar 178
Rotstein, Guillermo E. 273
Rowe, Rex L. 249
Rudolf, Rüdiger 224
Russell, Evan L. 380

S

Saleh, Jamal 421
Sanchez, Maria P. 433
Schäfer, M. 469
Schmidt-Traub, Henner 417
Schnelle, Karl 197
Schulte, Michael 417
Schulz, Christian 224
Schwarm, Alexander 303
Seewald, W. 464
Shah, Nilay 75
Sharma, Arvind 474
Shin, Dongil 360
Solot, Philippe 458
Stadtherr, Mark A. 445
Stanley, Peter 507
Strube, Jochen 417

T

Telang, Kedar 421
Thornhill, N.F. 373

Tjoa, I. Bhieng 1, 185
Top, Sermin 332
Towler, Gavin 42
Tsai, Chii-Shang 345
Turkay, Metin 185

U

Ungar, Lyle H. 152

V

van den Heever, Susara A. 210
Vassiliadis, Constantinos G. 158
Venkatasubramanian, Venkat 136
Vinson, Jonathan M. 171

W

Wahnschafft, Oliver M. 315
Warnecke, H. J. 469
Watson, Doug L. 204
Weidenbach, M. 469
Weiss, Robert A. 273
Westerlund, Tapio 291
Willmann, Linus 458
Wollenweber, Dirk 404

Y

Yaws, Carl L. 421
Yoon, En Sup 360

Z

Zentner, Michael G. 249
Zhu, (Frank) X. X. 42

SUBJECT INDEX

A

ABC, 273
Abnormal situation management, 366
Actives, 197
Adaptive control, 303
Adaptive random optimization, 351
Adiabatic system, 428
AI optimization, 178
Alarm logic, 345
Application programming interfaces, 480
ASPEN PLUS®, 315
Asset productivity, 62
Attainable region, 428
Auctions, 152
Autocorrelated measurements, 332
Autocorrelation, 386
Autonomous order entry system, 217
Average run length, 332

B

Batch chemical processes, 286
Batch manufacturing, 204
Batch process planning, 146
Batch processes, 439
Batch processing, 171, 273
Bifurcation analysis, 308
BOD, 190
Bottlenecks, 433

C

Capacity optimization, 458
Causal process description, 487
CGU, 397
Chance constraints, 303
Chemical kinetics, 44
Chemical process, 44
Chemical reaction, 44, 469
Chemical reactor analysis, 421
Chemometric methods, 380
CIMFUELS, 297
Clean-In-Place, 267
Closed shop, 146
Closed-loop identification, 303

Clustering, 338
Complex behavior, 308
Complexity theory, 91
Component-based software, 518
Componentware, 501
Computational fluid dynamics, 469
Computer integrated manufacturing, 171, 273
Computer-aided engineering, 171
Computer-aided modeling, 494
Consumer goods industry, 142
Control design, 315
Control loop performance, 373
Control structure, 5
Controlled group unit, 397
Costing, 273
Coupled estimation techniques, 351
Cubic homotopy, 410
Cutting-stock problem, 254

D

Data integration, 62, 171
Data reconciliation, 328
Database mining, 391
Debottlenecking, 404, 458
Decentralization, 5
Decision structure, 5
Decision-making, 44
Decision-tree learning, 254
Decomposition, 5
Delivery scheduling, 231
Demand, 197
Descriptors, 474
Design, 210, 439
Dimensionality reduction, 380
Discrete-event simulation, 178, 404
Discriminant analysis, 380
Disjunctive programming, 210
Dispersion, 231
Distributed optimization, 152
Due date, 146
Dynamic modeling, 261
Dynamic performance measures, 297
Dynamic simulation, 8, 315
Dynamic systems, 15
Dynamics in operations, 5

E

Eco-efficiency, 112
Economics, 15
Electronic data exchange, 480
Engineered algorithms, 91
Enterprise integration, 279
Enterprise modeling, 279
Enterprise resource planning, 62, 217
Environment, 112
Environmental modeling, 190
Exothermic reversible reaction, 428
Expert system, 464
EXPRESS information models, 480

F

Factor analysis, 338
Factor score plot analysis, 338
Fault detection, 31, 380
Fault isolation, 380
Feed perturbations, 433
Financial benefits, 297
Financial scheduling, 217
Flexibility, 433
Flexible flowshops, 243
Flexible recipes, 286
Formulated products, 197
Fractionation, 322
Full-cost accounting, 112
Functional representation, 487
Functional-link-associative neural network, 360

G

Generalized disjunctive programming, 185
Genetic algorithms, 286
Geographic information systems, 190
Global optimization, 445
Graph, 231
Gröbner bases, 410
Gross error detection, 328, 386

H

Hardware redundancy, 345
Heuristic methods, 91
Historical data analysis, 31
Homotopy continuation, 410

I

Image analysis, 31
Industrial ecology, 112
Industrial fermentations, 391
Information management, 501
Information models, 474
Information structure, 5

Institutional change, 112
Integration, 501
Integration of design and operations, 397
Intelligent control, 366
Intelligent systems, 487
Intermediates, 197
Internal heat exchange, 428
Interval analysis, 445
Inventory, 146
ISO 10303-21, 480
ISO 10303-22, 480

K

K-means clustering, 360
Knapsack problem, 254
Knowledge repository, 487

L

Lagrangean decomposition, 152
Large scale systems, 380
Life science, 417
Life-cycle analysis, 112
Limit setting, 297
Linear programming, 185
Lot sizing problem, 62

M

Maintenance optimization, 158
Manufacturing execution systems, 397
Mass transfer, 469
Mass-flow network, 345
Material flow analysis, 404
Material flow simulation, 458
Mathematical modeling, 237
Mathematical programming, 1, 91, 178, 204, 249
Maximum dispersion, 231
Microelectronics, 44
Mixed integer linear programming (MILP), 243, 291
Mixed integer nonlinear programs (MINLP), 224
Mixed-integer optimization, 185
Model development, 8
Model identification, 351
Modeling, 1, 44, 204, 249
Modeling languages, 279
MTBE, 315
Multilevel operation, 338
Multiperiod optimization, 210
Multiproduct batch plants, 224
Multiscale SPC, 332
Multivariate projection methods, 31
Multivariate statistics, 380

N

Natural gas, 433
Nearest-neighbor learning, 254
Networked enterprise, 217
New product design, 31
Nonconvex optimization, 224
Non-linear, 210
Non-linear equations, 410
Nonlinear PCA, 360
Non-linear programming, 286

O

Object orientation, 501
Object oriented analysis, 279
On-line optimization, 421
Open shop, 146
Open systems, 501
Operation, 439
Operational design, 397
Operational planning, 190
Opportunity costs, 273
Optimal control, 237
Optimal coordination, 152
Optimal operation, 308
Optimization, 204, 231, 249, 267, 291, 469
Oscillation, 373
Outer-approximation, 164

P

Paper converting, 291
Parameter estimation, 445
Parametric mixed-integer optimization, 164
Pattern classification, 380
PCA, 391
pdXi, 480
Pearson correlation, 27
Penalty-based methods, 152
Persistence of excitation, 303
Phenomena-driven formal languages, 494
Pinch analysis, 421
Plan execution, 366
Planning, 5, 171, 204, 210, 249, 366
Planning and scheduling, 75
Planning under uncertainty, 164
Plant (re)design, 458
Plant design, 404
Plant-wide optimization, 8
PLS, 391
Pollution index, 421
Pollution prevention, 112
Predictive measures, 27
Procedural control, 397
Process, 474
Process control, 27

Process decomposition, 152
Process improvement, 338
Process modeling, 518
Process monitoring, 31, 360, 380
Process operations, 5, 75, 158
Process optimization, 417
Process scheduling, 91
Process uncertainty, 158
Production logistics, 458
Production planning, 62, 185
Production scheduling, 62
Production scheduling and planning, 1
Profit function, 297
Propylene chlorohydrin process, 469

R

Ramp-up, 197
Reactive distillation, 315
Reactor modeling, 469
Real time optimization, 1
Real-time, 397
Real-time optimization, 5, 15, 322
Real-time systems, 518
Refinery control loop, 373
Requirements engineering, 279
Residue curves, 315
Resource prices, 152
Robust operations, 261
Robust process design, 464
RTN, 190

S

Sampling interval, 373
Scheduling, 5, 171, 204, 224, 237, 243, 249, 261, 267, 273, 286
Scheduling problems, 291
Separation technology, 417
Sequencing, 267
Serial correlation, 386
Setup cost, 146
Setup time, 146
Short-cut estimation, 351
Signature, 373
Simulated moving bed (SMB) chromatography, 417
Simulation, 44, 261, 404, 439, 518
Situation-adaptive integrated production, 397
Soft sensors, 31
Software, 44
Software integration, 518
Spatial redundancy, 345
SPEEDUP™, 315
Standardization, 518
State, 474
Statistical experimental design, 464
Statistical process control, 332
STAVEX, 464

Steady-state models, 15
Steel rolling mills, 243
STEP physical file, 480
Supply, 197
Supply chain, 75, 197
Supply chain dynamics, 5
Supply chain management, 62, 142, 249
Supply-chain modeling, 178
Symbolic/numerical simulation, 494
System, 474
System design, 15
System dynamics, 178
Systems dynamics, 142
Systems thinking, 178

T

Taguchi method, 464
Technology management, 44
Temporal redundancy, 345
Theory of constraints, 178
Total site optimization, 185
Traveling salesman problem, 267
Trim-loss problems, 291

Two phase flow, 469

U

Uncertainties, 197
Uncertainty, 44

V

Valve fault, 373
Vapor-liquid equilibrium, 445
Variability reduction, 261
Variable cycle time, 237
Vertical integration, 217

W

Waste elimination, 112
Wavelets, 332
WWW, 501

X

XpressAPI, 480